ANNUAL REVIEW OF ECOLOGY AND SYSTEMATICS

ANNUAL REVIEW OF ECOLOGY AND SYSTEMATICS

VOLUME 27, 1996

DAPHNE GAIL FAUTIN, *Editor*
University of Kansas

DOUGLAS J. FUTUYMA, *Associate Editor*
State University of New York at Stony Brook

FRANCES C. JAMES, *Associate Editor*
Florida State University

http://annurev.org science@annurev.org 415-493-4400

ANNUAL REVIEWS INC. 4139 EL CAMINO WAY P.O. BOX 10139 PALO ALTO, CALIFORNIA 94303-0139

ANNUAL REVIEWS INC.
Palo Alto, California, USA

International Standard Serial Number: 0066-4167
International Standard Book Number: 0-8243-1427-1
Library of Congress Catalog Card Number: 71-135616

∞ The paper used in this publication meets the minimum requirements of American National Standard for Information Sciences—Permanence of Paper for Printed Library Materials, ANSI Z39.48-1984.

TYPESETTING BY TECHBOOKS, FAIRFAX, VA
PRINTED AND BOUND IN THE UNITED STATES OF AMERICA

Annual Review of Ecology and Systematics
Volume 27 (1996)

CONTENTS

(*continued*)

ANNUAL REVIEWS INC. is a nonprofit scientific publisher established to promote the advancement of the sciences. Beginning in 1932 with the *Annual Review of Biochemistry*, the Company has pursued as its principal function the publication of high-quality, reasonably priced *Annual Review* volumes. The volumes are organized by Editors and Editorial Committees who invite qualified authors to contribute critical articles reviewing significant developments within each major discipline. The Editor-in-Chief invites those interested in serving as future Editorial Committee members to communicate directly with him. Annual Reviews Inc. is administered by a Board of Directors, whose members serve without compensation.

Annu. Rev. Ecol. Syst. 1996. 27:1–53

EARLY HISTORY AND PROGRESS OF WOMEN ECOLOGISTS: Emphasis Upon Research Contributions

Jean H. Langenheim
Department of Biology, Sinsheimer Laboratories, University of California, Santa Cruz, California 95064

KEY WORDS: women ecologists' research, women in history of ecology, gender issues

ABSTRACT

Although women are increasingly prominent as ecologists, a report on their progress through the history of ecology in overcoming personal and societal obstacles provides interesting insights regarding their research achievements. Selected, predominantly American, women ecologists are presented within five time frames according to the date of their PhD, an event marking the beginning of their careers. A general view is given for pre-1900 Protoecologists, followed by brief professional sketches for 10 Early Pioneers (1900–1934), 16 Late Pioneers (1935–1960), and 28 members of the First Modern Wave (1961–1975). The relatively large number of women who earned doctorates after 1975 precludes discussion of individuals from this time in this review. The following issues are discussed in the context of their research contributions: 1) motivating factors, 2) graduate education and subfield entered, 3) mentors and role models, 4) employment, 5) marriage and family constraints, and 6) recognition. These issues are compared with data from recent surveys for post-1976 women doctorates. Each selected woman still alive was contacted for her assessment of her research; 156 research citations display the significance and range of subjects studied. A steady, albeit slow, progress since 1900 is evident, although some problems regarding gender equality in professional development of women ecologists persist. These issues, however, are now more clearly recognized and addressed.

INTRODUCTION

In 1988 I hoped that there would be no further need to discuss the contributions of women ecologists because we would be recognized just as ecologists

0066-4162/96/1120-0001$08.00

(81). Although women have become increasingly prominent as ecologists, it still seems timely and useful to consider the progress women ecologists have made in overcoming both personal and societal obstacles, particularly with regard to research contributions. Citations exemplifying their research reveal the significance of women's contributions and the extraordinary range of subjects studied.

Women ecologists clearly, albeit somewhat slowly, have made progress since the beginnings of "self-conscious ecology" (100, 106) near the turn of the century. Initially, they took the opportunities to obtain advanced degrees and used these primarily to teach, although a few succeeded in doing research. The percentage of women obtaining doctoral degrees dropped through the 1930 depression years, the 1940 war years, and the 1950 postwar years. However, some now-influential women ecologists persisted to obtain doctorates through these difficult years. Following the coincidence of the environmental and women's movements in the 1960s and 1970s, the number of women ecologists with doctorates increased as did the positions they obtained in major institutions. Thus these women had increased opportunities to influence ecology through their own research and by having their own doctoral students and postdoctoral fellows at research-oriented universities. Furthermore, the numbers that have received prestigious awards and assumed leadership roles in various professional societies have increased. I discuss the obstacles these successful women have overcome in achieving this status.

Ecology is a notoriously heterogeneous field, and boundaries are even more difficult to define before the recognition of self-conscious ecology in 1894 (100, 106). However, I thought it appropriate to include a few women naturalists from the nineteenth century as "protoecologists" (106). Following general discussion of this pre-1900 period, I emphasize a selected group of women ecologists who obtained PhD degrees before 1976. The large burst of women with doctorates who followed allows in this paper only a general evaluation of their progress, rather than a continued discussion of their individual contributions. Although most of the women discussed are Americans, a few are British and Canadian. To facilitate a sense of progression, the women are grouped in five time frames according to the date they obtained their PhD degrees, an event marking the beginning of a professional career. I communicated with the living ecologists to obtain data regarding them; the significance of their research is derived from their own assessment and suggestions from others working in their subfields.

In the context of their research contributions, the following issues are discussed using different women as examples: 1) motivation to study and to disseminate ecological information (by teaching and/or research), 2) availability of graduate education and the emphasis in subfields, 3) mentors and role models as motivators and/or providers of support, 4) employment, such as availability of jobs at major institutions in which a high level of research is possible, 5) family constraints on careers, and 6) recognition of contributions by election as officers of scientific societies and selection for prestigious awards. Some of these issues are compared with data from a 1987 survey of 200 women (81) and from recent publications that consider various aspects of women ecologists' careers. I conclude by evaluating progress through time.

PRE-1900 PROTOECOLOGISTS AND CONSERVATIONISTS

Women traveled around the world to study natural history during the seventeenth, eighteenth, and nineteenth centuries, e.g. Englishwoman Mary Kingsley (91). Discussion of this period is limited, partially because the exploits of these courageous and adventuresome women should receive separate attention. Bonta (14, 15) points out that natural history studies by American women, even if published, "had been overlooked in standard chronicles of natural history because of women's position in society during the eighteenth, nineteenth, and early twentieth century. Women were viewed as amateurs, even though professional men in the same fields who had achieved great renown often had the same amount or even less professional training than the women" (14, p. xiv).

The seventeenth and eighteenth century exploits of Maria Sibylla Merian are amazing. After divorcing her husband, she obtained a grant from the city of Amsterdam and at age 52 traveled with her daughter to Surinam to illustrate tropical plants and insects; she published her insightful ideas on metamorphosis of insects along with beautifully engraved plates in 1705 and 1719 (110). She had previously published three volumes [1679, 1683, and 1719] on *The Wonderful Transformation of Caterpillars and (Their) Singular Plant Nourishment*. However, because she unfortunately did not integrate her discoveries into the existing body of scientific knowledge she has been considered a "connoisseur rather than an investigator of metamorphoses" (63, p. 22). Her accomplishments are the more incredible because women even in the nineteenth century had difficulty publishing anything other than popularized observations.[1]

Outstanding among women in the nineteenth century who clearly projected her ideas is Ellen Swallow (Richards), whom Robert Clarke identifies as "the

[1]Hutchinson (64) points out that women who studied field natural history in Ireland during the nineteenth century did not publish under their own names. For example, although Mary Bell discovered stridulation in the corixid water bugs—a group of insects that produce sound under water, a process that has complicated functional significance in sexual isolation—her papers were published under her brother's name in 1845 and 1846. Hutchinson (64) suggests that the brother's failure to mention his sister's name was probably related to his feeling that it was immodest for a woman to publish.

woman who founded ecology" (30, title). Swallow, a pioneer woman chemist with a BA from Vassar in 1870, was the first woman admitted to MIT as a special student. She then, with the support of her MIT professor husband, opened the Women's Laboratory there. In 1892 she presented the idea of "oekology"' as an interdisciplinary science concerned with industrial health and water quality, thus anticipating the applied ecological studies of professional ecologists since the 1960s. Her ideas received little attention when in 1894 ecology was formally defined (dropping the initial "o" in the term) as a subdiscipline of plant physiology that emphasized adaptations of organisms to the natural environment (100).

Some early influential ecologists, such as Frederic Clements and Victor Shelford, however, did pose key questions regarding the impact of humans on the natural environment. American women also developed interests and organizations during the late nineteenth century that paved the way for women's prominence in conservation activities during the early twentieth century (109). Literary clubs brought women together, and women's rights movements exposed them to the political process. Leisure time also gave middle- and upper-class women opportunities for botanizing, gardening, bird lore, etc. Some of these women not only recorded their observations on nature, they also wrote articles in a variety of journals encouraging the public to conserve the wonders of nature (15).

Nationally, the General Federation of Women's Clubs, founded in 1890, actively informed women about urgent political, economic and social issues of the day, including environmental concerns. Women were also involved early (starting in 1886) in Audubon Society activities, such as protection of game birds from hunters and protection of plumed birds from extinction due to ladies' fashions. They became particularly active with regard to forestry. For example, recognizing the need for trained men to manage and protect the forests, the women of the California Club in 1903 presented a bill to the state legislature to establish a School of Forestry at the University of California, Berkeley. At that time the only three schools of forestry in the United States were on the East Coast.

"Although the women of the organizations represented at the National Conservation Congresses were public activists in their local communities, they nevertheless accepted the traditional sex roles assigned to them by the late nineteenth century American society as caretakers of the nation's homes, husbands and offspring ..." (109, p. 73). The outstanding achievement of many of these women in bringing to the public the need for preserving the environment was a prelude to the extensive conservation activities of many women who obtained PhDs and managed to secure professional positions in ecology.

1900–1934 DEGREES: EARLY PIONEERS

Near the turn of the nineteenth century women broke through barriers to graduate education in science (154). Simultaneously, universities such as Illinois, Chicago, Nebraska, Minnesota, and Cornell became centers for the new field of ecology. Although some women were actively involved only in volunteer conservation activity, other women in these early days were obtaining a high percentage of the PhD degrees in ecology, e.g. 100% of pre-1920s PhDs, and 50% of those between 1920 and 1929, at the University of Illinois (M Willson, personal communication), and 50% of the pre-1920 doctorates at Cornell (B Peckarsky, personal communication). These are particularly significant figures considering that women did not get suffrage in the United States until 1920.

Women Associated with Research Sponsored by the Carnegie Institution of Washington

Edith Schwartz Clements and Edith Bellamy Shreve are better known for their activities in ecology than are many others who earned degrees just after the turn of the century; they may have become so visible partially because of their famous ecologist husbands. The greater part of their husbands' research was carried out under the auspices of Carnegie Institute of Washington (CIW), a private foundation whose funding has played a significant role in physiological and ecological studies since the turn of the century. These women's research contributions to ecology, however, are quite distinct from one another.

Frederic Clements encouraged Edith Schwartz to do graduate work under his direction (33). With a substantial autecological dissertation on "The Relation of Leaf Structure to Physical Factors," in 1904 she was the first woman to obtain a PhD in botany (as well as ecology) at the University of Nebraska. Edith and Frederic were married in 1899, and after finishing her degree, she became the exemplary helpmate in all aspects of his research. They both preferred being and working together to being separated. In fact, in the 46 years of marriage the Clements were separated only once for more than a few hours (164). Although Edith's dissertation was her only research publication per se (31), because she chose to devote most of her energies to assisting with her husband's work, she utilized her talents in botanical illustration, publishing several books of flower paintings (e.g. 32). She also compiled and edited (with BW Allred) *Dynamics of Vegetation* (2), published after Frederic's death in 1949, and she wrote a book about the Clements' adventures together in the field (33). Frederic apparently recognized Edith's potential as an ecologist in her own right in his comments to the South African ecologist John Philips, "Mrs. Clements would hold that position today ... [like that of Dr. Philips near the top of the world's ecologists] ... had she not devoted herself to furthering my career instead of winning recognition as an ecologist in her own right" (164, p. 242).

Frances Louise Long, another of Frederic Clements' students, took her baccalaureate at Nebraska and her PhD (1917) at the University of Minnesota when Clements moved there. She published her PhD dissertation on the quantitative determination of photosynthesis (94), a topic indicating the importance Clements ascribed to physiological studies in ecology. When Clements moved to CIW in 1918, Long became a research associate there, and they continued their research collaboration, coauthoring several significant volumes on topics such as pollination and adaptation in plants (34, 35). When Frederic received his CIW appointment, Edith remarked that he would also be allowed a (paid) assistant, but a "'mere wife' would work just as hard for nothing" (164, p. 243). Thus was the difference in professional status between Edith Clements and Frances Long.

Edith Bellamy Shreve worked professionally within the framework of her marriage in a very different way and more daringly for her time than did Edith Clements (16). She took her baccalaureate at the University of Chicago in 1902 in chemistry and physics, but never completed her PhD. She then taught physics at Goucher College, where she met Forrest Shreve, who taught botany there. Shortly after their marriage in 1909, they went to the CIW's Desert Research Laboratory in Arizona. Edith accompanied her husband to study distribution of desert vegetation, until the birth of their daughter in 1918. Although it is not known how much she contributed to Forrest's work, sometime in those early years Edith decided to pursue a career of her own. In so doing, she confronted stereotyped ideas of "women's work" and also had to fight the widespread prejudice against married women working independently. Her husband, however, was for that time unusually supportive. Forrest's biographer (17) suggests that his egalitarian attitude probably came from a Quaker upbringing; his mother was also college educated, and he apparently did not feel threatened by a wife with independent achievements.

Edith Shreves' background in the physical sciences led her (after some tutelage from Burton Livingston at Johns Hopkins University and encouragement from DT McDougal, Director of the CIW Desert Laboratory) toward plant physiological studies in what would now be considered physiological ecology. Edith was an experimentalist; most of her studies were conducted under controlled conditions in the laboratory. She was an imaginative, dedicated, and independent-thinking researcher who published papers on the autonomic movements in cactus (cholla joints). She discovered that the plants apparently lost water during the night and took up water during the day, the opposite of patterns expected in 1915 (157). Her continued attempts to understand the mechanisms led her to the verge of discovering crassulacean acid metabolism, a specialized mode of photosynthesis in cactus and other succulents. She also worked on methods for determining leaf temperature and transpiration (e.g. 158, 159), and she produced a significant study of seasonal water relations in desert plants (160).

Early Women in the Chicago-Cowles Genealogy

To trace the origins of several generations of female ecologists that span the time 1900–1975, I describe the major pedagogical genealogy of plant ecologists influenced by HC Cowles at the University of Chicago at the turn of the century. Cowles founded the "School of Physiographic Ecology," which linked long-term landscape changes to vegetational change and emphasized the importance of physiology. This genealogy further shows the role that prominent male ecologists (e.g. Shelford, Cooper, Oosting, Buell, Billings, Marr, and Mooney) played at an ever-increasing number of universities in supporting women who have become prominent contributors to ecology.

Although Emma Lucy Braun took her PhD at the University of Cincinnati in 1914, she was so influenced by Cowlesian views (her dissertation topic was "The Physiographic Ecology of the Cincinnati Region") that she has been included on this pedagogical tree. Braun's interest in ecology was stimulated by her schoolteacher parents, who took their close-knit family to explore forests around Cincinnati. Her older sister Annette was a distinguished entomologist, and the two sisters lived and worked together. This not only provided mutual support but made possible their extensive field studies throughout the eastern deciduous forest. Together they dealt effectively, at a time when women field scientists were uncommon, with such problems as moonshiners and the "backwoods" conditions in the Appalachian Mountains.

Lucy Braun's professional career was also at the University of Cincinnati. She became an Assistant Professor of Botany in 1925 and rose to Professor of Plant Ecology in 1946. She retired in 1948 to have more time for her research and was granted an honorary Doctor of Science degree by the University in 1964. Contrary to the usual situation for women faculty at the time, Braun had 13 MA students and one PhD. Nine of the MA degrees were earned by women; several of their theses were published, but information is not available about the women's careers.

Braun published prolifically, and there was an air of independence in her publications: She was sole author of four books and 180 articles in 20 journals. Her research "coincided with the time when the field of plant ecology was becoming recognized as a scientific discipline" (175, p. 83). In fact, her research was hailed as "instrumental in the development of that discipline;" one of her greatest achievements was her monumental book in 1950, *Deciduous Forests of*

Eastern North America (19), in which many of her significant journal articles are referenced. It represents 25 years of field work, initially in specific areas, followed by traveling 65,000 miles during the last 10–15 years to know the deciduous forest as a whole. Fosberg (55, p.67) wrote "one can only say that it is a definitive work, and that it has reached a level of excellence seldom or never before attained in American ecology or vegetation science, at least in any work of comparable importance." This remarkable woman was constantly a "first woman," e.g. the first woman officer of the Ecological Society of America (ESA)—Vice President (1935) and President (1950)—and the first woman to have an ESA award named in her honor (Braun Award for Excellence in Ecology). She was given the Pope Medal by the Cranbrook Institute of Science as "one of America's major ecologists" with a statement that her book belongs on the same shelf with Kerner's *Des Pflanzenleben der Donaulander*, Schroter's *Das Pflanzenleben der Alpen,* and Tansley's *Vegetation of the British Isles* (51, p. 10).

Although Victor Shelford, the first President of the ESA, was not a student of Cowles, he, too, was sufficiently influenced by him to be considered a part of his pedagogical genealogy. Shelford was apparently supportive of women in graduate education in these early days—he had three women PhDs in ecology between 1918 and 1927, one in 1938, and another in 1941. By contrast, his student Charles Kendeigh, who overlapped and succeeded him at the University of Illinois, had no women among his 51 PhD students. When the women PhDs who were Shelford's students sought jobs for which their degrees seemed to qualify them, however, they generally found their options limited to teaching in high schools or women's colleges, and in some of those colleges only if they remained unmarried. For example, Martha W Shackleford (PhD 1927) became a faculty member at Oklahoma College for Women; there is no evidence that she continued with research.

On the other hand, Shelford's student Minna Jewell (PhD 1918) taught at Kansas State Agricultural College (now Kansas State University) and Thornton Junior College, but in addition made significant experimental contributions to limnology during the Birge-Juday era at the University of Wisconsin when most research focused on surveys (10). Current researchers who specialize in various aspects of limnology have recognized Jewell's work as some of the first and most insightful on topics such as prairie streams (70), pH effects of fishes in acid lakes (74), groundwater-lake interactions (71), and freshwater sponges (72, 73) (T Frost, personal communication). Thus, she was a true pioneer in limnological research despite the barriers for women at this time. She published much of her work in ESA journals, i.e. *Ecology* and *Ecological Monographs,* and these articles are still cited in publications from 1986–1994

(Science Citation Index). Jewell did not have as many opportunities to have graduate students as Braun did, but apparently she "had quite an influence on young people" (10, p. 5). After Jewell retired from Thornton College, she taught in a girls' school in South America and continued to publish papers on freshwater sponges into the 1950s. Despite being a charter member of ESA and publishing significant data, Jewell did not receive recognition by either the Society or the limnological community during her lifetime.

In contrast to Lucy Braun and Minna Jewell, Mildred Faust and Harriet Barclay, two of Cowles's doctoral students, followed more traditional careers as gifted teachers. Mildred Faust obtained her PhD in 1933 and taught in the Botany Department at Syracuse University for her entire career, influencing many graduate students in their research; she was also active in conservation efforts and environmental education. Harriet George Barclay completed her doctorate in 1928 and married another University of Chicago botany graduate. The couple then went to the University of Tulsa where, unlike many married women of her time, Barclay had children but taught in the Botany Department alongside her husband from 1929 until her retirement in 1971. She concentrated much of her effort on carrying out the Cowlesian tradition of inspiring many undergraduate and graduate students through her infectious enthusiasm for field ecological studies of plants, not only at Tulsa University but at various field stations, particularly the Rocky Mountain Biological Laboratory. An alpine enthusiast, Barclay greatly increased the knowledge of Rocky Mountain and Andean plants through her extensive collections. The results of her continuous field research in Oklahoma were published mainly in the *Proceedings of the Oklahoma Academy of Science.* She was also an unstinting leader in conservation, for which she received numerous awards, particularly for her efforts in locating and preserving unique natural areas in Oklahoma, and she was elected to the Oklahoma Hall of Fame.

Women from Nontraditional Ecological Backgrounds

Several women not specifically trained as ecologists made significant research contributions that clearly fit within the framework of ecology. Emmeline Moore was an extraordinary woman who earned a PhD degree at Cornell in 1914. Immediately after obtaining her doctorate, she followed the pattern common among unmarried women of teaching at a women's college (Vassar). However, from 1917 to 1919 she worked on a US government project on primary food relations of fish, which led her into her future research career. She was the first woman given a permanent job in the New York State Conservation Department, where one of her first projects in 1920 was a collaborative survey of the biological, chemical, and physical aspects of Lake George. This study, which was set up by the New York State Legislature to determine how to increase fish productivity, was so successful that the biological survey was expanded to include the entire 60,000 square miles of the New York watershed (21). In 1926, Moore was appointed Director of the Biological Survey in the New York State Conservation Department. The original study and 16 subsequent reports published between 1926 and 1939 (all edited by Moore—112) remain the most comprehensive scientific examination of any state's water resources ever conducted (21); they emphasize her effective collaboration with other scientists. Although she did not teach during these years, Moore created a bond with numerous colleges and universities by hiring students and faculty for her summer field crews. This field training was significant in the development of careers of numerous biologists who would later become well known in their fields (21). Moore's work was recognized by another event unusual for a woman in 1927: election as the President of the American Fisheries Society.

Margaret Morse Nice was another outstanding pioneer who did insightful studies that can be considered ecological. She, like Edith Shreve, did not have a PhD. After receiving her baccalaureate from Mount Holyoke College, she entered Clark University for graduate studies. There she met and married Leonard Blaine Nice; soon after he obtained a doctorate, they moved to the University of Oklahoma where he became head of the Department of Physiology. She returned to Clark in 1915 to obtain her MA in zoology for research done earlier on the food of the Bobwhite. She later received honorary Doctorate of Science degrees from both Mount Holyoke and Elmira Colleges. Her first major publication, *The Birds of Oklahoma* (1925), was coauthored with her husband; her five children enthusiastically supported her field studies as well. In 1927 Nice's husband joined the faculty at Ohio State University. While in Ohio, Nice became a recognized ornithologist; during a very productive period from 1927 to 1936, she published for example, in 1933, a critical paper on territoriality (117). Her ideas regarding territoriality of birds were still heavily cited in 1980, 39 years after a 1941 paper (119) on the subject (R McIntosh, personal communication). It was her monographs on the life history of the Song Sparrow (e.g. 118, 120), however, that established her reputation as an outstanding ornithologist (180). She was elected President of the Wilson Ornithological Club (later Society) in 1938, becoming the first woman to serve as president of any major American ornithological society. In fact, throughout her later career she was highly praised by outstanding authorities, e.g. Tinbergen complimented her for "her cares and sacrifices in the home circle" and for service to science with "remarkable creative power." "Through your works you have become known to ornithologists throughout the entire world as the one who has laid the foundation for population studies now so zealously pursued" (180, p. 438). Ernst Mayer further stated, "I have always felt that she, almost single-handedly, initiated a new era in American ornithology and the only effective countermovement against the list chasing movements.... She was one of the first people in this country (? the first) to analyze a local deme. In other words, she pioneered left and right, as far as the US is concerned" (180, p. 438). Her devotion to research is indicated by the title of her posthumously published autobiography—*Research Is a Passion with Me.* A complete list of Nice's publications apparently does not exist, but the best estimates are that she published more than 250 titles on birds in journal articles, seven of book length, and 3313 reviews of the works of others. She continued her extraordinary activity through her long life (she lived to 90), but as she grew older she increasingly turned her attention toward educating the public about nature and conservation.

Nice wrote about the "tragedy" that women of intellect should have to spend so much time in manual labor. "Our highly educated gifted women have to be cooks, cleaning women and nursemaids ..." (180, p. 433). However, in her case, Trautman (180) indicates that it would be difficult to overemphasize the important role her husband played in Nice's work, in encouraging her and providing the finances for conducting her researches and attendance at meetings. Nonetheless, Nice had to protest constantly that "I am not a housewife, but a trained zoologist," (14, p. 222). Konrad Lorenz wrote that "Margaret Morse Nice was the real founder of ethology" (14, p. 222). She rightfully received much recognition for her research, e.g. being elected to honorary membership in the British Ornithological Union and seven other ornithological or conservation societies. The fact that she never earned a PhD, was never a faculty member of a university, and received few or no grants and little secretarial assistance makes her achievements even more noteworthy.

Ecological Impact of a Woman Natural Science Writer

Even though Rachel Carson never proclaimed herself to be a professional ecologist, the ecological impact of her books makes it mandatory to include her in this historical discussion. She took both BA and MA degrees in zoology (the latter in part studying with Raymond Pearl at Johns Hopkins), but she was always interested in writing. She joined the Bureau of Fisheries (later the Fish and Wildlife Service) and was one of the first two women hired by this agency in other than clerical capacity. Known as a "well trained biologist with a gift for expression..." (20, p. 70), she moved rapidly from Assistant Aquatic Biologist in 1942 to Biologist and Chief Editor from 1949 until her resignation from the Service in 1952 to pursue writing full-time.

Although Carson's books *Under the Sea Wind* (1941), *The Sea Around Us* (1951), and *Edge of the Sea* (1956) were important contributions to natural history, it was *Silent Spring* (24), her last completed book, that had far-reaching ecological impact. Its publication came at a time (1962) when the concept of ecology was just becoming known to the public. *Silent Spring* was recognized within a decade of its publication as one of those rare books that "change the course of history... by altering the direction of man's thinking" (20, p. 227). Its success has been attributed to a combination of her biological background, her boldness in speaking her concerns, and her superb command of the English language. The book took a long time for her to write; ill with arthritis, cancer, ulcers, etc, she persisted in worldwide correspondence with countless experts (e.g. ecologists, ornithologists, physicians) because she said "long and thorough preparation is indispensable to do an effective job" (20, p. 243). She was deeply concerned about simplifying complicated technical data without introducing error. *Silent Spring,* serialized in *The New Yorker* in June, 1962 and published in its entirety in September, instantly created a sensation throughout the country.

The book was bitterly attacked, having initially offended the chemical and related industries as well as the powerful US Department of Agriculture. Continual attempts were made to discredit Carson as a "hysterical woman" (14, p. 271). Throughout the controversy, Carson's chief concern was that *Silent Spring* should have a lasting effect on government policy. Bitter reaction by many scientists to a negative report of the National Academy Committee on Pest Control and Wildlife Relationships led President Kennedy to ask for a study of the whole issue. A Pesticide Committee, set up by the Office of Science and Technology, in 1963 both criticized industry and agencies of the federal government and recognized the service performed by *Silent Spring.* The committee's report endorsed Carson's basic argument that insufficient scientific knowledge was available to assess accurately the risk of these toxic chemicals. By the spring of 1963 the book had become almost as famous in England as it was in America. *Silent Spring* was published during 1963 in France, Germany, Italy, Denmark, Sweden, Norway, Finland, and Holland and shortly afterward in Spain, Brazil, Japan, Iceland, Portugal, and Israel. In 1964, the last year of her life, honors were piled upon Rachel, e.g. Conservationist of the Year (National Wildlife Federation), Audubon Medal (the first to a woman), and American Geographical Society Medal. The most deeply satisfying recognition was her election to membership in the American Academy of Arts and Letters, which at the time had only three women members.

1935–1960 DEGREES: LATE PIONEERS

Much early ecological research was a loose amalgam of concepts, often "heavily influenced by taxonomic, habitat or geographical distribution of the organisms studied" (106). From 1935 to 1960, other avenues of ecological understanding appeared that are evident in women's research at that time.

Although a general downswing occurred in the percentage of women doctorates in ecology from the 1930s through the 1950s, several women who pursued degrees persisted in establishing successful careers, some becoming leaders in the field. It is notable, however, that Barbour (4), in an analysis of conceptual changes in vegetation studies during the 1950s, included only one professional woman ecologist (Langenheim) and two widows of prominent ecologists (Helen Buell and Linda Olswig-Whittaker) among the 34 persons interviewed for their opinions.

A Woman Leader in Aquatic Ecology

Although Ruth Patrick obtained her PhD in 1934 at the University of Virginia, Charlottesville, her prolific research record which spans much of the development of aquatic ecology, seems more appropriately discussed with the women in the late pioneer group. Occupant of the Frances Beyer Research Chair, Patrick Center for Environmental Research, Academy of Natural Sciences of Philadelphia, and Adjunct Professor at the University of Pennsylvania, Patrick has been an active researcher at the Academy since 1933. She apparently found a niche at the Academy and did not suffer the indignities that some women did in academia. She has used her background as a leading authority in the systematics of diatoms to do influential research in aquatic ecology. First, she pioneered the use of diatoms to infer paleoecological conditions (122). Her development of the diatometer permitted her to use a simple device to test important theoretical questions in ecology. For example, using glass slides of different sizes to simulate island size, she tested MacArthur and Wilson's theory of island biogeography. She demonstrated that the size of an area, available species pool and rate of invasion were the important components in defining the establishment and diversity of diatom communities, as had been shown for bird communities (123, 124). She further developed theories regarding the diversity and structure of river ecosystems (124, 126) and how shifts in these systems indicate that pollution is taking place before it becomes acute. Patrick developed the concept that a stream is an integrated community (125). She hypothesized that when pristine stream areas are equated by their structure and physical and chemical characteristics, they support similar numbers of species that remain relatively stable because the number of potential niches is similar. When pollution occurs, however, both numbers and kinds of species change significantly and the relative sizes of the populations become more variable than without pollution (125). Although she did not have a full academic appointment, well-known women researchers today in aquatic ecology and paleoecology mention her influence on the development of their thinking. She rightfully has been widely recognized for her work. For example, Patrick has been President of the Phycological Society of America and American Society of Naturalists, and she is a member of the National Academy of Sciences and of a staggering number of boards and committees. She has received countless awards, including the Tyler Environmental Award, and she was the first woman to receive the Eminent Ecologist Award, the highest award given by ESA to recognize research contributions. She has also received honorary doctorate degrees from 21 colleges and universities. After 62 years there, Patrick is still active in research at the Academy of Natural Sciences of Philadelphia.

A British Woman Pioneer

An Englishwoman, Verona Conway, with her PhD from Cambridge University in 1937, was a pioneer in blending field studies with application of experimental methods in physiological ecology (132). Her dissertation was designed to discover why an important bog sedge (*Cladium*) did not withstand annual cutting. She treated the annual wave of temperature through waterlogged peat quantitatively, predating the general application by ecologists of mathematical analysis to problems of environmental physics (132). She concluded that *Cladium* does not withstand annual cutting because the old leaves provide the pathway for oxygen to diffuse to the rhizome (37). Following appointment to a lectureship at the University of Sheffield, she investigated the ecology and origin of the blanket bogs of the Pennines, from which she critically analyzed dating and significance of recurrent surfaces in relation to current hypotheses of repeated climatic oscillations (39). She also wrote an influential monograph on the bogs of central Minnesota (38). Like many women ecologists, she was involved in conservation activities. In fact, she left her Sheffield lectureship in 1949 to join the newly formed British Nature Conservancy, where she analyzed the effects of burning and draining of peat surfaces on bog hydrology. Conway was recognized for her widely known studies by election to the Council of the British Ecological Society and later as an Honorary Member.

Cowles's Genealogy Revisited

Jane Claire Dirks-Edmunds was another of Victor Shelford's women PhD students (1941) at the University of Illinois. She expected difficulty in being accepted as a woman to do doctoral studies. However, when Victor Shelford was asked about Dirks-Edmunds, he answered, "Send me a good student, I don't care what sex they are" (121, p. 7)—indicating his attitude since the 1920s toward assisting women to obtain PhDs. For her dissertation, Jane compared part of the comprehensive information amassed on the Douglas fir–hemlock forest at Saddleback Mountain, Oregon [carried out by a Linfield College professor and his students (including her) from 1933 to 1938], with an analysis of the oak hickory forests of Illinois (49). She returned to Linfield College after obtaining her doctorate and initially was an assistant in the registrar's office. However, she was listed on the faculty roll as an instructor, partially because the American Association of University Women was beginning to use women faculty members with graduate credentials as one criterion for accrediting colleges (59). In 1946 she began to climb the professorial ladder slowly at Linfield, retiring as a full professor in 1974. She continued the Saddleback Mountain study, following the forest's succession subsequent to selective logging over 20 years, thus setting the framework for understanding future change in these forests. Pacific Northwest foresters have recognized the Saddleback Mountain research as "especially valuable" because it provides comprehensive data over a long period, initiated when little of this kind of ecological research was done on Pacific coast forests (121). Her devoted activities at Linfield College led to the establishment of the Jane Dirks-Edmunds Lecture there.

WS Cooper, one of Cowles' most prominent PhDs, had a number of doctoral students who subsequently sponsored women doctorates. For example, Catherine Keever and Elsie Quarterman completed their PhD degrees in 1949 with Cooper's student HJ Oosting at Duke University. They thought that opportunities opened to them to obtain doctorates because many men were away during World War II. Keever and Quarterman are appropriately presented together as they have done field work and have published together through the years. Keever taught in high school and several small colleges before obtaining her PhD, and afterward she settled as Professor of Biology at Millersville State College, Pennsylvania. Even with heavy teaching loads, she persisted with research, publishing on the causes of old-field succession and the distribution of major forest species in southeastern Pennsylvania (76). Characteristic of her positive professional attitudes, she has stated that, although there was little support for research at Millersville, she did not have the disadvantage of the publish or perish pressure (77). Thus she felt that she had the advantage of doing research because she wanted to, and she has continued to enjoy it since her retirement in 1974. The continuing impact of some of her perceptive successional studies are indicated by part of a title of a 1980 paper by McCormick and Platt "Catherine Keever—you were right!" (105).

Elsie Quarterman, Professor Emerita at Vanderbilt University, also initially taught in high school, but in 1943 she began her long tenure at Vanderbilt. Quarterman focused much of her research on the plant ecology of the Tennessee glade communities. To this end, she and her graduate students investigated the life history and community relationships of some characteristic glade species and endemics. Starting points were seed germination and life cycle studies, as well as interaction of species with physical factors of the environment and interrelationships of the plants with other biotic components, including allelopathic interference (139). Furthermore, with Catherine Keever she published a monographic study of the southern mixed hardwood forests (140), in which they described the pine-hardwood and immediate post pine stages in forest succession of the southeastern coastal plain, in an attempt to clarify existing ideas regarding climax there. The research of both Keever and Quarterman carries on the Cowlesian tradition of understanding the distribution of species in relation to their habitat, in defining successional status of communities and how this relates to possible equilibrium conditions. Quarterman supervised seven doctoral students, including Carol Baskin. Since her retirement, she actively continues conservation work, e.g. with The Nature Conservancy and as President of the Tennessee Environmental Council.

My own career (now as Professor Emerita of Biology, University of California, Santa Cruz) is part of the Cowles' lineage, as the only woman doctorate of WS Cooper (1953). I early followed the Cooperian tradition of doing research at the interface between ecology and geology, working together with my geologist husband in analyzing earthflow succession and vegetational patterns in relation to geology over a wide altitudinal range in western Colorado (e.g. 79). Nepotism regulations (1953–1961) allowed me only an honorary appointment as a Research Associate at the University of California, Berkeley and at the University of Illinois, Urbana. I too, therefore, taught part-time at women's colleges (Mills College and San Francisco College for Women). During this period my position was representative of numerous women scientists, and it underscores the important role of a few male ecologists whose sympathetic support helped circumvent nearly intractable administrative obstacles and thus enabled me to continue research. For example, I collaborated with HL Mason to use language analysis in provocative discussions of such ecological concepts as the environment and natural selection (e.g. 103). However, this kind of an unsupported research position often requires women to be versatile "research opportunists." This can be a deterrent in establishing their own research identity. On the other hand, for me it partially meant broadening my botanical background, which did prove useful later in my own research. Upon the independent pursuit of my own career, as a Research Fellow at Harvard University, I added the interface of chemistry to those of ecology and geology in investigating the paleoecology of amber (fossil resin), in ES Barghoorn's laboratory. The Radcliffe Institute for Independent Study (now Bunting Institute) played an important role in my obtaining the first professorial position for a woman in the sciences at the University of California, Santa Cruz (UCSC) in 1966.

At UCSC a major research interest evolved directly from my novel geochemically oriented paleoecological studies that opened new approaches to the study of amber. Contrary to traditional views, my survey of amber through 300 million years demonstrated the importance of angiosperm resin producers and that the greatest diversity of trees that produce copious amounts of resin occurred in the tropics (80). These results led me to direct my research toward the emerging paradigm regarding plant chemical mediation of some coevolutionary processes, in which I became a pioneer in tropical chemical ecology. My broad botanical background gave me the necessary scope to study in detail model tropical tree genera, which I carried out in collaboration with my graduate students and researchers in institutes in many Latin American and African countries (83). I experimentally analyzed environmental influences on the biosynthesis of terpenoids in both field and laboratory, and then assessed the role of the quantitative variation of these terpenoids in defense of both tropical and Pacific Coast plants against insects, slugs, vertebrates, and fungi. These long-term investigations resulted in an overall analysis of the phytocentric role of terpenoids, in which I challenged some currently held dogma in chemical ecology theory (82).

My years at UCSC further illustrate the trend toward an increasing number of graduate students supervised by women. I have supervised 36 graduate students (12 women) who have completed degrees (24 doctorates, of whom 8 have been women). I also coauthored a textbook, *Plant Biology in Relation to Human Affairs*, in which I emphasize an ecological perspective on the role plants play for humans. The breadth of my research resulted in my serving on numerous international and national committees. It also led to my election as the first woman President of the International Society of Chemical Ecology (1986–1987), as well as the second woman President of the Association for Tropical Biology (1985–1986), the Ecological Society of America (1986–1987), and the Society of Economic Botany (1993–1994).

Women Invade Plant Paleoecology

One of the dominant themes of dynamic ecology has been the long-term historical perspective of change in populations, communities, and environment. In fact, the development of Quaternary paleobotany closely paralleled that of ecology (188). The discovery by European researchers of well-preserved pollen grains in peat and sediments began the detailed and eventually quantitative analysis of post-Pleistocene vegetational and associated climatic change. Paleoecological studies, both Tertiary and Quaternary, initiated during the 1950s have provided important evidence in the revaluation of plant community concepts.

Three women who significantly contributed to the interrelationship between ecology and paleoecology received their degrees in the 1950s. Margaret Bryan Davis, Regents Professor in the Department of Ecology, Evolution and Behavior at the University of Minnesota, has a 1957 doctorate from Harvard University. Davis's ecological research interests have been at interfaces between ecology and geology, and between ecology and paleoecology. She is a long-time leader in the use of palynological data to study the organization of past communities and their dynamics, ecosystem processes, and response to climatic change. She was early influenced by Johannes Iverson, of the Danish Geological Survey, to use pollen for studying plant population and community dynamics in the recent geologic past. Hugh Raup, her Harvard graduate school mentor, perhaps further influenced her thinking with his skepticism regarding the concept of the organismal community and untested assumptions in general. Later, working with Edward Deevey at Yale, Davis demonstrated how accumulation rates in sediments are important in supplementing pollen percentages for inferring past population sizes (43). The first application of this approach (now standard palynological technique) at Roger's Lake, Connecticut not only helped to resolve the character of late-glacial tundra and boreal forest communities in New England, but showed that forest communities are loosely organized populations rather than a product of long evolutionary history units (44). This paper is the only one by a woman in the 1990 Collection of Classic Ecological Papers (147).

While at the University of Michigan, Davis compared pollen in surficial lake deposits with the composition of surrounding forests, which led to studies that she has pursued later in her career of lake circulation and deposition processes affecting the pollen content of the sediment. More recently, she is providing new empirical data regarding the stability of communities—with the fossil record typically demonstrating continuous community change and thereby challenging the concept of community equilibrium (45, 46). Her continued impact in introducing greater scientific rigor into Quaternary palynology via experimental and quantitative approaches, hypothesis testing, and model construction has led to a leadership role for fossil pollen analysis in current discussion of global environmental change (e.g. 47). Throughout her research Davis has guided numerous postdoctoral associates and graduate students. She has served on many national and international committees and advisory groups. Among her honors are election to presidency of the American Quaternary Association (1978), and the National Academy of Sciences (1982), and as a Fellow, American Academy of Arts and Sciences (1991). She was the third woman ESA President (1987–1988) as well as recipient of its Eminent Ecologist award (1993).

Davis points out that the enormous effort she had to expend struggling for opportunities came at the expense of science. As a woman she faced many problems in the process of becoming a professor with salary equivalent to that of men at the University of Michigan and Yale University, before going to the University of Minnesota as head of the department in 1976. However, she became visible in her efforts toward increasing the numbers and recognition of women in science. Her personal statements are poignant. "Now in my sixties, I find I have achieved the goal I was striving for all of my life. I am a professor in a department with a strong graduate program. I have a group of excellent students and my research combines ecology and paleoecology. In this benign environment I spend relatively little time on women's issues, but a decade or so ago I added up all the time I had spent, especially in my early years, maneuvering for laboratory space and for a faculty position, fighting for equal wages, taking on administration in order to improve my bargaining power, serving on committees on equity issues, mentoring women from undergraduates to full professors, and it comes to an appalling 25% of my total investment in science. My experience isn't unusual, either. In many ways I was advantaged by my education—many women have spent more energy than I maintaining their toehold on the academic ladder. Think how many women scientists there are in my generation and younger, all of us expending a quarter or more of our time and energy removing obstacles placed in our paths to slow us down. What a waste for human society that all that time and energy and talent didn't go into science instead" (MB Davis, personal communication).

Estella Leopold, Professor of Botany, University of Washington, was greatly influenced to become an ecologist by her father, the famous wildlife ecologist Aldo Leopold. After graduation from Yale (PhD 1955), Estella worked for the United States Geological Survey until 1976, using palynology to become a specialist in Tertiary history and development of western United States floras. Ruth Patrick was an important role model for her career as a young paleoecologist. Estella demonstrated in various fossil assemblages worldwide the principle that climatic forcing on Cenozoic time scales is a stimulus to evolution and extinction in plants (87). She showed that extinction and evolutionary rates of woody plants during the Late Tertiary were different from those of higher animals, i.e. while climate forced rapid changes in bursts, extinction rates over long periods were relatively slow for woody plants (87, 90). Significant studies of specific vegetation range from grasslands to Arctic tundra. For example, she demonstrated different timing and biogeographic origins for two grassland communities—the Great Plains from Miocene tropical ancestors and the Palouse from pan-Pacific boreal Pliocene ancestors (88). Her palynological studies from the Alaska Range show a warm temperate forest much like the mixed deciduous forest of eastern North America during the Miocene and Pliocene, which suggests surprising youth of tundra plant lineages and that deciduous forests occurred under photoperiod conditions unlike any in their present distributions (89). Leopold's research has been widely recognized; she was elected to the National Academy of Sciences (1974) and as President of the American Quaternary Association (1982). Furthermore, she has carried on her father's legacy of conservation activities, receiving awards such as Conservationist-of-the-Year from the Colorado Wildlife Federation, given jointly to Beatrice Willard (1969).

The third woman in this group is Grace Brush, Professor and Principal Research Scientist in the Department of Geography and Environmental Engineering at Johns Hopkins University, who took her PhD at Harvard University (1956). Her research opportunities varied as she moved to different institutions better suited to her husband's career than to her own. She had part-time appointments at four universities (George Washington, University of Iowa, Rutgers, and Princeton) before obtaining her present position at Johns Hopkins. She has persisted in doing research, however, with an approach alternating between modern and fossil plant distributions, which has served her well in bridging botany and geology. Her recent work has been influenced by the way many of her current engineering associates assess and approach problems. Some of her important contributions to ecology include 1) a vegetation map of Maryland, which shows the close relationship between lithologies that are similar hydrologically and distribution of natural forests (23), and 2) a reconstruction of estuarine history using a stratigraphic record that provides insights on human influence on the environment (22).

Wisconsin Women Plant Ecologists

Six women, supervised by John Curtis, received doctorates from 1953 to 1960 at the University of Wisconsin, Madison, near the time the large pedagogical genealogy of plant ecologists there began (111). They were also part of the re-evaluation of the plant community concept to which Curtis students contributed in the long-term detailed "continuum" analyses of Wisconsin vegetation, supporting the basic individualistic hypothesis of Henry Gleason. Three of these women, Margaret Gilbert (PhD 1953), Bonita Neiland (PhD 1954), and Gwendolyn Struik (PhD 1960), published their dissertations on various aspects of Wisconsin forest-prairie communities (60, 116, 176); then Gilbert became Professor at Florida Southern College, Neiland became Professor and Director of Instruction for the School of Agriculture and Land Management, University of Alaska, and Struik became a private consultant in New Zealand.

Another Wisconsin graduate, Martha Christensen, who was cosponsored by Curtis and mycologist Myron Backus (PhD 1960), not only published important studies on the role of soil microflora in various Wisconsin forests (27, 29) but continued research spanning mycology and ecology while she was a Professor of Botany at the University of Wyoming. She became a leader in the quantitative description of soil microfungal communities. She ordinated microfungal communities and correlated patterns of other biotic and abiotic factors against the patterns dictated by the fungi, thus demonstrating, against commonly held ideas, the existence of habitat specificity for soil microfungi, particularly in forest ecosystems (28). Christensen's numerous achievements in studying the ecology of soil fungi led to her election as the third woman President of the Mycological Society of America (1985–1986).

A Woman Desert Ecologist

Janice C Beatley earned her 1953 PhD at Ohio State University. She taught at several small colleges before becoming a research ecologist at the Laboratory of Nuclear Medicine and Radiation Biology, University of California, Los Angeles (1960–1972). Beatley wrote notable papers on desert community dynamics at the Nevada Test Site with a focus on reproduction of annual plants (8); she also investigated the dependence of rodents on these annuals (9). Her research on the ecological status of introduced brome grasses (7) was one of the first contributions regarding the invasive nature and ecological significance of these annual grasses. Her highly respected research is still used in teaching (B Strain, personal communication). From 1973 until her death in 1987, Beatley was a professor at the University of Cincinnati, where she fulfilled a long-time dream of teaching in the same department where E Lucy Braun had maintained her lifetime affiliation. Here Beatley directed her research toward understanding the wintergreen herbaceous flora of deciduous forests.

Women Animal Ecologists

Margaret Stewart and Frieda Taub are 1950s graduates who have done research in areas of ecology different from those of most contemporaneous women. Margaret Stewart, Professor of Biology, State University of New York, Albany (Cornell PhD 1956), has carried out field studies of behavior and population and community ecology of amphibians in both temperate and tropical ecosystems. She is one of the relatively few women ESA members even today doing active research in animal behavior (181). Her book *Amphibians of Malawi* in 1967 established her reputation in herpetology (172), and she was the first person to spearhead ecological and behavioral studies of the tropical terrestrial frogs (*Eleutherodactylus*), which are major vertebrates in the Antilles. Her work demonstrating that retreat sites, rather than food, are a major factor in population regulation of tropical forest frogs gained wide attention (174); recently she completed a summary of 15 years of fluctuation in a deme of frogs in Puerto Rico rainforests in relation to climate, which has important implications in the assessment of declining amphibian populations (173). Stewart's leadership role in her field has been recognized by her recent election as President of the American Society of Ichthyologists and Herpetologists and by the honorary doctorate she received from the University of Puerto Rico, Mayaguez. She, too, is actively engaged in various conservation projects locally and internationally.

With a zoology doctorate in 1959 from Rutgers, Frieda Taub went immediately to Seattle because her husband took employment there. She stated, "as a result of my inexperience as well as suspicion of women having a PhD in 1959, ... I had to begin at an entry level research position in the College of Fisheries at the University of Washington" (F Taub, personal communication). However, by 1971 she had risen to full professor in the College. She used the concept of microcosms as a tool in analyzing ecological interactions in the regulatory process of releasing new chemicals into the environment (177). She removed a major objection to use of complex communities in testing ecotoxicological responses with a summary of evidence that microcosms allow validation of these tests among laboratories (179). Taub's leadership role in the IBP Coniferous Biome project led to her edited book, *Lakes and Reservoirs* (178), a volume in *Ecosystems of the World,* which takes a comparative approach in emphasizing the similarities of processes underlying the uniqueness of each ecosystem.

1961–1975 DEGREES: FIRST "MODERN" WAVE

Following the coincidence of the environmental and women's movements in the 1960s, there was an increase in number of women ecologists, the institutions from which they obtained doctorates, and the positions they were able to obtain in major universities, where greater opportunity for research existed (Table 1).

A Woman Mathematical Ecologist

The research of EC Pielou represents the emergence of a woman working in an area that continues to be dominated by men. In fact, she has a novel professional history that displays an amazing sense of self-motivation. Pielou was equally interested in natural history and mathematics, and while caring for young children with the help and support of her husband, she did research in mathematical ecology as an amateur. She has written that "as an amateur I was beholden to nobody and could follow my inclinations and make my own decisions without the need to justify them to granting agencies, senior academics, or anybody else..." (EC Pielou, personal communication). In 1962, with no supervisor or committee, she was able to convert several published papers into a PhD from the University of London. She spent four years as a Research Scientist in the Canadian government, then entered academia as a full professor, first at Dalhousie University, Nova Scotia, and then at University of Lethbridge, Alberta. Again in her own words, "Starting at the top has its obvious advantages. I have been my own boss all of my working life..." (EC Pielou, personal communication). Throughout her research that has spanned boreal forests to intertidal marine algae, her aim has been to postulate ecological hypotheses in clear, mathematical form and to design rigorous tests specific for each hypothesis.

Pielou has written six books. Her *Introduction to Mathematical Ecology* (127) and its expanded second edition (131) have had great impact. She also has written *Population and Community Ecology* (1974), *Ecological Diversity* (1975), and *Interpretation of Ecological Diversity* (1984). In journal articles she has developed a mathematical measure of closeness of association among a group of several species, which serves as a measure of the "structure" of many-species communities (128). She extended the application of Leslie matrices to populations of two competing species of sessile organisms in a study of how the interactions between populations affect their spatial patterns (129). She has been equally interested in interrelationships among ecology, biogeography, and their paleo equivalents. Because a mathematical approach to problems in biogeography lagged far behind that in ecology, she analyzed the statistics of biogeographic range maps (130).

Since retiring in 1988, Pielou has written three books for general readers, bridging the gap between research ecologists and interested amateurs who do substantial work on environmental protection. These books put into easily comprehensible form ecology of the world's northern evergreen forests, the return of life to glaciated North America after the Ice Age, and a naturalist's guide to the Arctic. She was given the Lawson Medal of the Canadian Biological Association and was the second woman to receive the Eminent Ecologist Award from ESA (1986).

Cowles' Genealogy Continued: Expansion into Wider Activities

Although the six women within the Cowles lineage in the First Modern Wave (1961–1975) were oriented toward some phase of plant ecological research for their degrees, the careers of four of them tended toward applied research.

Nellie Stark, another Oosting PhD (1962), now Professor Emerita of Forest Ecology, University of Montana, Missoula, was a pioneer, becoming a full professor in a US forestry school in 1979. She had worked for the US Forest Service in the Sierra Nevada Mountains and the Desert Research Institute in Nevada before going to Montana. Stark developed thought-provoking concepts, such as direct nutrient cycling for tropical rain forests on nutrient-depleted soils. This led her to the concept of "biological life of a soil" (170), which she further applied in evaluation of nutrient losses from Rocky Mountain forests in terms of long-term productivity (171). Other studies included the impacts of logging and fire on conifer forests and the resistance and resilience of forest ecosystems to chemical perturbation such as acid rain.

Beatrice A Willard took her PhD in 1963, supervised by Cowles' descendent John Marr, at the University of Colorado. She taught briefly in colleges before answering the call of applied ecology, initiated by the passage of environmental legislation in the 1960s. Willard became Executive Director and then

Table 1 Women discussed 1900–1976, date and university of PhD, area of ecology when degree was taken and later if changed, and primary institution where research was done (M—has been married; C—has children; aC—unmarried, has adopted child; * no complete information regarding family)

Name	Year (PhD)	University	Area of ecology	Current or major institution after degree
Early Pioneers				
Edith S Clements (M)	1904	Nebraska	Plant Ecology	None
Edith B Shreve (M-C)	1902 (BA)	Chicago	Plant physiological ecology	CIW
E. Lucy Braun	1914	Cincinnati	Plant ecology	U. Cincinnati
Emmeline Moore	1914	Cornell	Limnology	New York State Conservation Dept.
Margaret Nice (M-C)	1915 (MA)	Clark	Avian population studies	None
Frances Louise Long	1917	Minnesota	Plant ecology	CIW
Minna Jewell	1918	Illinois-Urbana	Aquatic ecology	Thornton College
Martha Shackleford*	1927	Illinois-Urbana	Community ecology	Okla. College for Women
Harriet G. Barclay (M-C)	1928	Chicago	Plant ecology	U. Tulsa
Mildred Faust	1933	Chicago	Plant ecology	Syracuse U.
Rachel Carson (aC)	1932 (MA)	Johns Hopkins	Natural history, conservation	U.S. Fish & Wildlife Service
Late Pioneers				
Ruth Patrick (M-C)	1934	Virginia	Aquatic ecology	Acad. Nat. Sci. Philadelphia
Verona Conway	1937	Cambridge	Plant physiological ecology	Sheffield U.
Jane C. Dirks-Edmunds	1941	Illinois-Urbana	Community ecology	Linfield College
Elsie Quarterman	1949	Duke	Plant ecology	Vanderbilt U.
Janice Beatley	1953	Ohio State	Plant ecology	UC Los Angeles NM & RB; U. Cincinnati
Catherine Keever	1949	Duke	Plant community ecology	Millersville State College, PA
Jean Langenheim (M)	1953	Minnesota	Plant community & chemical ecology	UC Santa Cruz
Margaret Gilbert	1953	Wisconsin-Madison	Plant community ecology	Florida So. College
Estella Leopold	1955	Yale	Plant paleoecology	U. Washington
Margaret Stewart (M)	1956	Cornell	Animal behavioral ecology	SUNY Albany
Bonita Neiland (M*)	1956	Wisconsin-Madison	Plant community ecology	U. Alaska
Grace Brush (M-C)	1956	Harvard	Plant paleoecology	Johns Hopkins U.
Margaret Davis (M)	1957	Harvard	Plant paleoecology	U. Minnesota
Frieda Taub (M-C)	1959	Rutgers	Microcosm ecology	U. Washington
Gwendolyn Struick (M*)	1960	Wisconsin-Madison	Plant community ecology	Private consultant
Martha Christensen	1960	Wisconsin-Madison	Forest soil fungal ecology	U. Wyoming

Table 1 Continued

Name	Year (PhD)	University	Area of ecology	Current or major institution after degree
First Modern Wave				
E.C. Pielou (M-C)	1962	London	Mathematical ecology	Dalhousie U. & U. of Lethbridge
Nellie Stark (M)	1962	Duke	Plant physiological ecology; forestry	U. Montana
Betty Willard	1963	Colorado	Plant ecology; conservation	Colorado School of Mines
Mary Willson (M)	1964	Washington	Avian and forest ecology	U. Illinois; U.S. Forest Service
Deborah Dexier	1967	North Carolina-Chapel Hill	Marine ecology	San Diego State U.
Carol Baskin (M)	1968	Vanderbilt	Plant ecology	U. Kentucky
Joy Zedler (M-C)	1968	Wisconsin-Madison	Plant & wetland ecology	San Diego State U.
Rebecca Sharitz (M)	1970	North Carolina-Chapel Hill	Plant population & wetland ecology	U. Georgia-Savannah Labs
Frances James (M-C)	1970	Arkansas	Avian ecology	Florida State U.
Judith Myers (M-C)	1970	Indiana	Animal population ecology	British Columbia U.
Katherine Ewel (M)	1970	Florida, Gainesville	Forest-wetland (systems) ecology	U. Florida-Gainesville
Maxine Watson (C)	1970	Yale	Plant population ecology	Indiana U.
Nancy Slack (M-C)	1971	SUNY Albany	Plant & bryophyte ecology	Russell Sage College
Berly Robichaud Collins (M)	1971	Rutgers	Plant ecology; conservation	McGraw-Hill Information Services
Patrice Morrow	1971	Stanford	Plant physiological ecology → plant-insect interaction	U. Minnesota
Sarah Woodin (M-C)	1972	Washington	Marine ecology	U. South Carolina
Pat Werner (M)	1972	Michigan State	Plant population ecology	U. Florida-Gainesville
Barbara Bentley (M-C)	1973	Kansas	Plant-insect interaction	SUNY Stony Brook
Beverly Rathcke (M)	1973	Illinois, Urbana	Insect-plant ecology	U. Michigan
Laurel Fox (M-C)	1973	UC Santa Barbara	Insect-plant interaction; community ecology	UC Santa Cruz
Susan Martin	1973	UC Santa Cruz	Plant (crop) chemical ecology	USDA; Colorado State U.
Karen Porter (M-C)	1973	Yale	Aquatic ecology	U. Georgia
Susan Riechert (M-C)	1973	Wisconsin-Madison	Spider behavioral ecology	U. Tenn.-Knoxville
Frances Chew (M)	1974	Yale	Insect-plant ecology	Tufts U.
Martha Crump (M-C)	1974	Kansas	Animal behavioral ecology	U. Florida-Gainesville
Judy Stamp	1974	UC Berkeley	Animal ecology	UC Davis
Jane Lubchenco (M-C)	1975	Harvard	Marine ecology	Oregon State U.
Deborth Rabinowitz (M)	1975	Chicago	Plant population ecology	Cornell U.

President of Thorne Ecological Institute in Boulder, Colorado (1965–1972), where she built bridges by interpreting ecology and its utility to non-ecologists. She had an important overview role on the effects of the Alaska pipeline while on the Council of Environmental Quality in the Executive Office of the President. Subsequently, she established and became the head of the Department of Environmental Sciences and Engineering Ecology at the Colorado School of Mines. As Director, Industrial Ecology Institute, she formed a bridge between ecology and the mining industry. Despite her administrative duties, she continued to publish significant scientific articles about the impact of human activities on the Rocky Mountain tundra (e.g. 189). Willard has received an impressive number of awards for her environmental work, ranging from engineering institutes (Environmental Conservation Distinguished Service Award—1979) to the US Forest Service (75th Anniversary Award for promoting ecological awareness and understanding—1980) to the United Nations (Outstanding Environmental Leadership Award—1982).

Carol Baskin, a 1968 PhD with Elsie Quarterman, went directly from Vanderbilt to the University of Kentucky as the wife and research partner of Jerry Baskin, who was also a PhD student of Elsie Quarterman. The Baskins are well known for their prolific publication (230 articles as of 1994) on topics such as autecology of endemic limestone glade plants, ecological life histories, and seed germination studies of herbaceous plants, inspired initially when they were students of Quarterman. The Baskins represent an outstanding example of close and continuous family partnership in research. However, despite very high research productivity, Carol had no official appointment at the University of Kentucky until 1984, when she was made an Adjunct Professor in the School of Biological Sciences. Among her many publications, recent autecological syntheses have been of particular interest, e.g. a joint article with her husband (6) in which they first synthesized dormancy cycles of seeds. They point out the way seeds of summer vs winter annuals respond to seasonal temperature, and for the first time they applied the term "continuum" to gradual changes that occur in a seed's physiology as it goes in and out of dormancy. Another synthesis includes many of their individual publications, bringing together information on the germination phenology of 274 herbaceous species in temperate climates and dormancy break and germination for 179 of them. They organized the data by type of life cycle and discussed them with regard to phylogenetic relationships (5). The Baskins' systematic long-term study of a specific aspect of autecology now enables them to reach broad generalizations. Carol has been recognized by election as Secretary of the Botanical Society of America (1980–1984).

Beryl Robichaud Collins has had an unusual career, which displays her persistence in using research to have an impact on conservation. Her undergraduate major was in economics, and she was fully employed in information services (Senior Vice President, McGraw Hill) when she took her PhD with Murray Buell at Rutgers in 1971. Pursuing graduate work on a part-time basis was not the usual practice, but she was highly motivated to obtain a strong scientific background to contribute effectively to public decision-making in environmental affairs. She co-authored *Vegetation of New Jersey: A Study in Landscape Diversity* (152) and recently co-edited the volume *Protecting the New Jersey Pinelands: A New Direction in Land-Use Management* (36). She serves on the Board of Governors of The Nature Conservancy and has received two honorary doctorates for her achievements.

Susan S Martin (PhD 1973) at UCSC also turned toward applied ecology in taking a position with the US Department of Agriculture's Agricultural Research Service (ARS) in Fort Collins, Colorado and as a Faculty Affiliate in the Biology Department, Colorado State University. She applied the chemical ecological approach she had used in her graduate work with Langenheim on resin chemistry of tropical legumes to studies of chemical mediation of plant-pathogen interactions in crop plants such as sugar beet (101). Martin has also been concerned that studies of stress-related compounds in crop plants often have ignored the potential variability that may arise from genetic variation or environmental influence. As a result, she investigated environmental influences and the role of ecotypic differentiation of diverse populations in analysis of phytoalexin accumulation in a common forage legume (102). Also, she is deeply involved with plant conservation in Colorado.

Patrice Morrow, Professor and Head of the Department of Ecology, Evolution and Behavior at the University of Minnesota, took her PhD (1971) in plant physiological ecology with Dwight Billings' student Harold Mooney at Stanford University. In California, Morrow was concerned with effects of drought on plant productivity; however, while a Fulbright postdoctoral fellow in Australia, she became impressed with the effects of insects that consume a large proportion of the photosynthetic surface of *Eucalyptus*. In fact, she became convinced that insect damage was a much greater problem for eucalypts there than adjustment of photosynthesis to seasonal temperature changes. Morrow since has used eucalypts to test aspects of plant defense theory that would be difficult to test where insect effects on plant growth are apparently minor. In her early work on eucalypt tree rings, she demonstrated that insect attack was chronically heavy and suggested that this kind of damage had been rampant in Australia for a long time (114). She pursued this perspective later with comparative estimates of presettlement insect damage in Australian and North American forests, in which higher damage levels were consistently found in Australia (113). This research was part of long collaboration with Laurel

Fox on Australian eucalypts. Although they have had independent research projects, in their joint efforts Fox's perspective in insect population biology complemented that of Morrow's in plant ecophysiology. Another influential paper grew out of their discussions regarding how communities are structured and problems associated with concepts of insect specialization, i.e. whether it was a species property or local phenomenon (58). Morrow has been recognized in ESA by election as a Council member (1986–1988) and as Vice President (1993–1994).

Prominence of Women in Emerging Study of Interactions of Insects and Plants

Morrow's research shift from traditional plant physiological ecology to interactions of insects and plants is representative of the beginnings of an emphasis in this area by a group of women. McIntosh (107, p. 44) refers to this subfield "as a 'growth industry' of ecology ... used extensively in discussion and tests for ecological theory."

The very productive collaboration of Laurel Fox, Professor of Biology, University of California, Santa Cruz, and Morrow is somewhat similar to that between two other women ecologists, Catherine Keever and Elsie Quarterman. Fox earned her PhD at UC Santa Barbara in 1973 with Bill Murdock and Joe Connell, working on predation of generalized stream insects. She spent several years as a postdoctoral and research fellow at the Australian National University where she switched to research on herbivory, not only because of the large, obvious herbivore damage on eucalypts, but because she wanted to test theoretical assumptions regarding specialist insects. Since 1978 she has split a professorial appointment at UCSC in the Biology Department with her ecologist husband; both Fox and Morrow were able to continue research in Australia after they took their US professorships. Fox has challenged some conventional ideas about insect-plant interactions as this area of research has developed. For example, she and Macauley (57) were the first to show that tannins, although abundant in eucalypts, were not an insurmountable defense for at least some insects, and they suggested that gut pH might be part of the reason. They also demonstrated the crucial importance of leaf nitrogen for insects consuming leaves with such low nitrogen levels as eucalypts. This publication became a citation classic and was significant in leading to current advances in thinking about the roles of tannins. She challenged two major ideas about ecology and evolution of insect-plant systems: 1) aspects of apparency and its assumptions about the cost of defense, and 2) pair-wise co-evolution might not be the appropriate model for systems in which large numbers of herbivores feed on the same plants (56). Fox & Morrow (58) argued that researchers used the concept of specialization very loosely—herbivorous insects that might use one or a limited number of plant species in one place might feed on other plants elsewhere. They presented ecological and evolutionary reasons why the diets might be restricted, which led to studies by numerous ecologists that show genetic differentiation among insect populations to food plants (58).

The research of Judith Myers, Professor of Zoology, University of British Columbia (PhD 1970, Indiana University), spans several areas in ecology but has recently emphasized insect-plant interactions. Two themes recur throughout Myers' research: experimental ecology and critiques of overly simplified generalizations (adaptive sex ratios, genetically structured populations, and induced defenses). Her PhD research on voles ultimately led to a review with CJ Krebs, which became a citation classic (78). This work included early hints to later findings that behavior of females should not be overlooked in population studies. Meyers acknowledges the important assistance from a postdoctoral fellow, Kathy Williams, through the hectic years of having two small children. Myers and Williams analyzed the highly popular and oversimplified view that induced chemical defenses control population dynamics of forest caterpillars (190). Recently Myers (115) reviewed her continued significant research on tent caterpillars.

Barbara L Bentley, Professor of Ecology and Evolution at SUNY Stonybrook, is another of the relatively large 1973 crop of PhDs—in her case from the University of Kansas (Table 1). She quickly saw the value of the emerging style of field research that involves manipulation of a natural system, with her dissertation concerning the interhabitat differences of plants bearing extrafloral nectaries and the associated ant community in reduction of herbivore damage (11). Her research has continued to be characterized by field experimentation. In her research on epiphylls, she was the first to bring high technology (gas chromatography) to the Costa Rican rainforest (12), demonstrating how field ecologists could thus greatly expand the range of questions they were asking. Recently, she was among the first to look at effects of elevated CO_2 on plant-insect interactions, particularly including nitrogen fixation and multilevel interactions (13). Part of her goal in these studies has been to develop information pertinent to policy decisions regarding global change. Bentley has been involved with environmental policy in the development of the Decision Makers' Course, when she was Vice President for Education in OTS. She was also elected Vice President of ESA (1989–1990). Bentley's successful career has been accompanied by rearing children; although she admits they have interfered somewhat with her career, she thinks that serving as a good role model in her department has allowed students (and fellow female faculty) to have children in a "much more neutral environment" (B Bentley, personal communication).

A 1973 PhD from the University of Illinois, Beverly Rathcke is Professor of Biology, University of Michigan. Always interested in insects as a child, "her career was settled when she discovered in college that she could study insects and plants for a living" (B Rathcke, personal communication). After obtaining an MS degree at Imperial College, London University, she started her doctorate at Cornell; however, marriage took her to the University of Illinois, where Mary Willson was a female role model. From 1975 to 1978 she had an unsalaried research title at Brown University where her husband taught. Rathcke then accepted an Assistant Professorship at the University of Michigan. Her research on a guild of stem-boring insects contradicted the major predictions from competition theory at a time when it was not popular to do so (144). In this and later studies of flowering phenologies she used random models, which were important because they allowed for testing specific hypotheses and resulted in rejection of then-current dogma generated from competition theory (145, 146). These contradictory results promoted re-evaluation of earlier evidence, the design of more rigorous tests of competition, and the consideration of alternative hypotheses, including facilitation or the positive interactions among species.

Another woman who pioneered research on the interactions of insects and plants with some emphasis on chemical mediation, Frances Chew, Professor of Biology at Tufts University, has a Yale PhD (1974). As a postdoctoral fellow at Stanford, she was intrigued by the concept of coevolution of plants and insects mediated by plant secondary chemistry, then being proposed by Paul Ehrlich and Peter Raven. Chew's research became motivated by the single overarching question of what determines insect-host plant specificity; most of her work has been on *Pieris* butterflies and their host plants (the Cruciferae and allies). With JE Rodman, she was the first to show a "community profile" of plant chemistry (153). She utilized concepts in plant population biology and plant apparency in studies assessing the evolutionary escape from herbivory (25), and she showed that differential host utilization by closely related insect species is mediated by differential sensitivity to plant chemistry (26). Chew's special research contribution has been to establish a strong tie between natural history of a system and critical laboratory work analyzing its parts.

Women Studying Avian Ecology

Studies of birds have always been important in shaping ecological theory (106) and, following the early lead of Margaret Nice, women have added their part.

A 1964 PhD, working with Gordon Orians at the University of Washington, Mary Willson has been a prolific researcher of unusual versatility in ecological and evolutionary studies of plants and animals. Her dissertation, which focused on mating systems in birds, was one of the first studies showing ecological correlates of harem size, and thus it opened the door for the first-generation mating system models with Yellow-headed Blackbirds (191). She was an instructor at Simmons College before going to the University of Illinois where she progressed to full professor in 1976. There she inspired many graduate students and began her studies of plant reproduction, making a controversial suggestion that sexual selection occurs in plants (192); later she compared sexual selection in plants and animals. She also studied the relationship between avian frugivory and seed dispersal (193). During this time she produced two books, one on plant reproductive biology and the other on vertebrate natural history. Although Willson contributed richly to research while at the University of Illinois, in 1989 she opted for the strictly research position of Research Ecologist at the Forest Science Laboratory, Juneau, Alaska. Here she has continued to expand her research horizons to include seed dispersal spectra in comparing plant communities (194), evolution of fruit color in fleshy-fruited plants (195), association of mites and leaf domatia, and endemic birds in fragmented south-temperate rainforests in Chile.

Professor of Biology at Florida State University, Frances James is a 1970 PhD from the University of Arkansas, where she began research on intraspecific size variation in birds and tight negative correlations with environmental variables that are functions of temperature and humidity (66). A simultaneous project on habitat relationships of birds, based on a Gleasonian approach and expressed as multispecies habitat relationships along multivariate axes, introduced the concept "niche gestalt," which won her the Edwards Prize from the Wilson Ornithological Society (67). Subsequent approaches, expanding ideas from her dissertation research, have included cross-fostering experiments with Red-winged Blackbirds (68). Her efforts to standardize sampling procedures and to quantify geographic variation in habitat at the intraspecific level led to an evaluation of applications of multivariate analysis in ecology and systematics (69). James has served in leadership roles on many committees and boards, and was elected a Council member of ESA (1977). She has made her mark in ornithological societies by being the first woman President of the American Ornithologists' Union (1984–1986) and receiving its Elliott Coues Award (1992), given for contributions that had impact on bird research in the western hemisphere.

Women Aquatic and Wetland Researchers

Joy Zedler, Professor of Biology and Director of the Pacific Estuarine Laboratory, San Diego State University, has a professional history reflecting problems that married women have often faced early in their careers, although the difficulties have tended to improve over time. While a graduate student in plant ecology with Orie Loucks at the University of Wisconsin, Madison, she married another ecologist; after her PhD (1968), she and her husband went to the University of Missouri and then to San Diego State University. Zedler took various low-paying jobs until finally she obtained a tenure-track position at San Diego State; her promotion to tenure, however, was delayed because she took time off to have children. Her research has been important in changing ideas about coastal wetlands. Her suggestion that algae under the marsh canopy could be as productive as vascular plants, where high salinity apparently selects for shorter open canopies, was foreign to researchers of the Spartina marshes of the Atlantic and Gulf of Mexico coasts, where most US salt marsh studies have been done (199). Thus, she changed the prevailing dogma that vascular plants are always the base of the estuarine food web and highlighted the importance of epibenthic microalgae. One of the themes in Zedler's career has been to take science into the management arena. Her series of studies on restored and natural wetlands in San Diego Bay have demonstrated the problems with attempts to recreate habitats of endangered species (200). She has worked with the National Research Council on aquatic ecosystem restoration and on its current wetland delineation book; she is also on the governing board of The Nature Conservancy.

Katherine Ewel, Professor of Systems Ecology in the Department of Forestry, University of Florida, Gainesville, also had to adapt to jobs according to where her husband was located. Receiving her degree in vertebrate zoology (PhD 1970, University of Florida), Ewel spent two years as an Instructor at Duke University while her husband finished his PhD at the University of North Carolina. At Duke she became skilled in the use of physical and computer models in biology. Although there were no job prospects for her at the University of Florida where her husband accepted a job, she was fortunate to have colleagues encourage collaboration based on her newly developed talents in computer modeling. After considerable research as a systems ecologist at the University of Florida's Center for Wetlands, she successfully competed for an assistant professorship in the School of Forest Resources and Conservation Engineering Sciences, becoming the first woman to hold a tenure-track position in that unit. Ewel has made two kinds of contributions to ecology. First, she has brought attention to the importance of different kinds of wetlands, demonstrating interrelationships among the different values that society places on these wetlands and compromises a manager must make in choosing among them (52). She has proposed a relationship between productivity and hydrology that establishes a framework for distinguishing quantitatively among different kinds of wetlands within a region (53). Secondly, she synthesized decades of field research together with current understanding of the magnitude and interaction of material flow to propose carbon, water, and nutrient budgets and their interrelationships for a major forest type in the southeastern United States (54).

Nancy Slack, Professor of Biology at Russell Sage College, earned a PhD in ecology following two Cornell degrees, although her degree was somewhat delayed (1971 SUNY Albany) because of marriage and family. Slack has had a career based on highly motivated interest and persistence. She became fascinated with the ecology of bryophytes, a group often ignored, and she adapted ordination methods for use with bryophytes in determining species diversity and community structure (including both vascular plants and bryophytes) along an elevation gradient in the Adirondack Mountains (162). In collaboration with DH Vitt, Slack worked on minerotropically rich fens in Alberta (165). They also discovered that Sphagnum species are remarkable ecological indicators, and they did the first quantitative American bog study including Sphagnum and initial work on niche theory of bryophytes. Slack then developed in depth the concept of niche for bryophytes (163). Thus, despite teaching at a college that does not have graduate students in biology, she has persevered in seeking out collaborators with whom she could expand her research. Best known as a peatland ecologist, she has studied boreal mires from New York to Minnesota, Alberta, British Columbia, and Sweden. She has long been involved in conservation work, serving as a board member and land evaluator for The Nature Conservancy. She also has written two semipopular books on alpine ecology.

Karen Porter, Professor of Zoology in the Institute of Ecology at the University of Georgia, has been fortunate in the opportunities to manage her career and marriage. She and her husband obtained their PhDs from Yale University in 1973 and went immediately to tenure-track positions at the University of Michigan. The good fortune to have two positions enabled them to be professionally independent from the beginning. When they moved to the University of Georgia as Associate Professors, she was a part-time faculty member while she had a daughter, but two years later was returned to full-time. Her research has been focused on aquatic ecology, particularly lake plankton, which has been an expansion of her childhood love of lakes and the sea. She is one of the few women graduate students of Hutchinson. Like Estella Leopold, Porter acknowledges the role of Ruth Patrick during her graduate studies. Although she had early done laboratory studies, e.g. adapting a fluorescent stain for counting free-living bacteria, allergies to laboratory chemicals led her to emphasize field studies, some of which she has done collaboratively with her husband. Some of her most important papers have been syntheses; in fact, one of her first papers was a synthesis of ideas (hers and others) on the potential role of filter-feeding zooplankton in controlling phytoplankton community structure, succession, and co-evolution (133). A second one, evolving from 50 papers done in collaboration with a number of PhD students, is on the microbial-based planktonic food web of Lake Oglethorpe, Georgia, in which she has integrated the microbial loop and classic food chain into a realistic planktonic food web (134). This is one of the first studies to show the effect of microbial production on higher trophic levels based on consumers such as flagellates, rotifers, and larval fish. Since northern lakes are warming, there is considerable interest in a third synthesis, encompassing Porter's 17 years of field data on warm temperate Lake Oglethorpe and the limnology of southwestern monomictic lakes in relation to climate change (135). Porter's first authored research in the 1990s has received 60–190 citations per year, particularly noteworthy in that the study of planktonic food webs does not constitute a large subfield of ecology; her creativity has been recognized by her receiving the University of Georgia Creative Research Medal.

Importance of Women as Plant Population Researchers

Whatever the reasons, studies of plant population dynamics lagged behind human demography and theoretical and experimental studies of animals (106). Terrestrial plant ecologists in the first half of the twentieth century spent much of their effort describing, classifying, and mapping communities. But in the 1970s, plant population biology went through a "revolution"; in fact, Antonovics (3) thought it to be one of the most important events in ecology at the time. It is a paradigm change in which women were early participants, and an aspect of ecology in which they still are prominent (181).

A 1970 PhD from the University of North Carolina, Rebecca Sharitz is one of the pioneer women plant population ecologists. Currently Senior Ecologist at the Savannah River Ecology Laboratory and Professor in the Department of Botany, University of Georgia, her research includes the population dynamics of plants in swamp forest systems and responses of these communities to environmental disturbance. Sharitz early demonstrated experimentally the role of competition in structuring plant communities across a resource gradient, and she presented one of the first examples of using life table techniques in the analysis of plant populations demography as well as relay floristics in primary succession (156). Another important contribution is an eight-year collaborative study that provided rigorous analysis of temporal and spatial patterns of natural seedling recruitment of woody species in bottomland hardwood forests (75). Recently, she suggested ways in which ecological concepts can be brought into management of southern forest resources (155). She was one of the first women elected an ESA Council member (1975); she was Treasurer (1987–1990) and Vice President (1990–1991); and currently she is Secretary General of the International Society of Ecology (INTECOL).

One of the women doctorates supervised by G Evelyn Hutchinson (Yale 1970), Maxine Watson is Associate Professor of Biology at Indiana University.

Her dissertation was designed to test whether plants exhibit niche partitioning in ways similar to those of animals. She showed that closely related moss species differ between genera, and she did the first age-structure analysis for lower plant populations (182). Her innovative research on higher plants examined the interplay between physiology, development, and demography (183). In this work she showed that plants differ from animals due to constraints imposed by their vascular structure; since costs are thus expressed in regard to an integrated physiological unit and may be self-supporting, the limiting resource in plants may often be meristems rather than nutrient or carbon resources. Although unmarried, Watson has an adopted daughter. During a period of serious ill health, she continued to direct many graduate students, and she is now moving actively toward innovative studies of the interplay between developmental phenology and environmental variation in determining plant demographic responses (184). Her goal is to provide foundation papers for the emerging subfield of developmental ecology.

A leader who added novel approaches to the 1970s advances in plant population ecology, Patricia Werner took her PhD in 1972 at Michigan State University, where she remained and rose to Professor of Botany and Zoology. Through analysis of plant characteristics (sessile, "plastic" growth) in field experiments, she modeled population dynamics and calculated population growth rates using mathematical tools developed for animal populations. Her paper demonstrating that the size of the plant is more important than its age in determining its demographic fate led to changed thinking about plant life histories and became a citation classic (185). In research on plant populations in successional environments, she not only captured the ideas of the mid-1970s by incorporating the exciting conceptual advances from recent field experiments, but she put these in the context of the interaction of life history characteristics with a changing environment—thus laying out the applicability of competition theory of the time for plant populations (186). Research on colonization of biennial species formed the basis of her thinking about plant communities and underlies many of the subsequent conceptual and empirical studies developed by her graduate students (who include four women) at Michigan State (187). Werner was elected an ESA Council member (1979) and President of the International Society for Plant Population Biology (1988). Offered the challenge of building a modern research center in the Australasian tropics, in 1985 she became Director of the Tropical Ecosystems Research Center for CSIRO in Darwin, Australia, following which she became Director, Division of Environmental Biology of the National Science Foundation (1990 to 1992). She currently is Professor and Chair of the Department of Wildlife Ecology and Conservation as well as Director of the Center for Biological Conservation at the University of Florida, Gainesville.

An untimely death in 1987 cut short one of the stars among plant population ecologists. Deborah Rabinowitz, a 1975 PhD from the University of Chicago, became the first woman Assistant Professor in the Ecology and Evolutionary Biology Department at the University of Michigan, and subsequently Professor in the Section of Ecology and Systematics at Cornell University. A key characteristic of her research was its "unusual slant on an old problem" (161, p. 86). Her major research fell into three groups: 1) mangrove distribution, in which she analyzed the early growth of mangrove seedlings in Panama with an hypothesis concerning the relationship of dispersal and zonation (141); 2) studies of rarity including empirical studies of rare prairie grasses, aimed at explaining these plants' peculiar ecology (143); and 3) concepts of rarity, in which she clarified its meaning with a classification of rare species based upon range of geographical distribution, degree of habitat specificity, and local population size (142). Rabinowitz was widely recognized as a role model by many younger women ecologists (81). Her obituary said, "she possessed a combination of qualities that are nowhere common. She was among the rarest of the rare" (161, p. 87).

Women Ecologists Studying Animal Behavior

Although the 1987–1988 ESA Survey (185) showed relatively few women studying animal behavior, this may reflect that some women in this field are not members of ESA. Among the ESA membership, Martha Crump, Susan Riechert, and Judy Stamps represent different areas of animal behavior.

Martha Crump, Adjunct Professor of Biology at Northern Arizona University (1974 PhD, University of Kansas) is one of the few women behavioral ecologists whose work has centered on reproductive ecology with tropical amphibians. The first long-term ecological research on a community of tropical amphibians was hers on an Ecuadorian community with the most diverse variety known of reproductive modes in amphibians (41). Her research has been recognized widely as the first to take a community ecology approach to reproductive strategies. Furthermore, she presented a new way of thinking and looking at variability in amphibian egg size (42), a variability that she examined in five species of tree frogs (*Hyla*) as a function of habitat predictability. Crump is married to another tropical field ecologist, and they have two children. Whenever possible, the children accompany and assist their parents in the field (to date in Costa Rica, Ecuador, Colombia, and Argentina). As a Professor at the University of Florida, Gainesville, Crump assured her female graduate students that "field biology and family can be a wonderful, successful union" (M Crump, personal communication). Crump and her ecologist husband decided to leave full professorships at the University of Florida after 16 years to become more involved in field conservation projects and teaching ecology and conservation in Latin America. During the school year they alternate their trips so that one parent stays home with the children. Crump says, "I still see my contribution as a role model for women (in this country and Latin America) who want to combine family and field work. The key is having a supportive husband" (M Crump, personal communication).

A spider population biologist, Susan Riechert, Distinguished Professor of Zoology, University of Tennessee-Knoxville, has a 1973 doctorate from the University of Wisconsin-Madison. She was introduced to natural history at an early age by activities centered on insect collections and trips to local ponds and woods with her grade-school friends. Her career working with spiders was launched from a field zoology course that ultimately led to research on the population biology of a funnel-web-building spider (148), which she has extended to include the evolution of cooperative spider behavior. Since spiders are among the few organisms that lend themselves to testing evolutionary game theory, her series of papers and a review of this theory have gained much attention (151). She also tested the hypothesis of potential limiting effects of gene flow on adaptation with experiments in desert spiders (149), in which she investigated a reason for population deviation from adaptive equilibrium. Riechert has been interested in spider assemblages in managed habitats and recently worked on limiting effects of generalist predators as agents of biological control (150). She discovered manipulations that conserve spider communities in agroecosystems and decrease by 60–80% the plant damage caused by grazing insects. Riechert was elected President of the American Arachnological Society (1983) and a Fellow (1993) and President-Elect of the Animal Behavior Society (1994).

Judy A Stamps, Professor in the Section of Evolution and Ecology, University of California, Davis, received her PhD (1974) at the University of California, Berkeley, where she obtained a strong background in ethology that enabled her to think about problems simultaneously at the proximate and ultimate levels. She rediscovered an idea mentioned by early workers (including Margaret Nice) that newcomers in territorial species prefer to settle next to established territory owners (168). This work alerted ecologists to a behavioral process with important implications for reintroduction programs in conservation biology. She wrote the first paper on parental behavior in birds that suggested avian behavior might be as complicated as that of primates and other mammals (167). Stamps also presented a review (169) using her lengthy series of studies on habitat selection and territoriality in juvenile lizards to illustrate a series of assumptions about territorial animals that are widespread in the literature, but as yet poorly tested. Stamps has been recognized in the Animal Behavior Society as a Fellow (1991) and with their Exemplar Award (1994).

Women Marine Ecologists

Although marine ecology has a long history, American women were slower to enter this area than terrestrial ecology. Deborah Dexter, Professor of Biology, San Diego State University, earned her PhD (1967) at the University of North Carolina. Her area of specialization is marine benthic ecology, especially community structure and population dynamics of dominant species on intertidal sandy beaches. Her research has spanned six continents, and she recently has synthesized 30 years of her own work, along with the last 25 years of literature, about sandy beaches throughout the world (48). She explains why the generalization that diversity increases with tropical habitats is apparently not true for sandy beaches. In a way similar to Catherine Keever, Dexter has continued to carry out research despite her choice to remain at an institution where her major commitment is to teaching.

Sarah Ann Woodin has a 1973 doctorate from the University of Washington. She was an Assistant Professor at the University of Maryland and Johns Hopkins University (1972–1980) prior to marrying and going to the University of South Carolina. She was a Research Professor in the Marine Science Program there before being promoted to Professor of Biology and Marine Science (1987). Much of her research on marine benthic communities has centered around adult-larval interactions. In an early paper she summarized data supporting the importance of interactions between adults and settling larvae or newly settled juveniles in determining the composition of the assembly of infaunal communities (196). Her predictions regarding assemblage characteristics and recruitment success or failure, given the importance of adult-larval interactions, have led to much subsequent research by others. Woodin significantly highlighted the importance of biogenic structures to the composition of the assemblage (197), and she also focused attention, through a series of experiments, on disturbance by organisms and physical forces. More recently she designed an elegant laboratory and field demonstration of the effect of haloaromatic compounds on recruitment (198).

Jane Lubchenco, Wayne and Gladys Valley Professor of Marine Biology, Oregon State University, has a 1975 PhD from Harvard University; she became the first woman Assistant Professor in the Harvard Department of Biology (1975–1977). Lubchenco and her husband went to the Zoology Department at Oregon State University in 1977 with a split position, which worked particularly well while their children were very young (98). They later were elevated to full-time appointments, and now each has been named to an endowed chair in marine biology. She has written that "our philosophy [her husband's and hers] has been that one needn't sacrifice family for career or career for family. Academic couples need to have more CHOICES, more OPTIONS for combining careers and families. For us, combining both in a sane manner meant doing each part-time. Because we were both committed to doing this, it has worked well—beyond our expectation" (J Lubchenco, personal communication).

Lubchenco's research has been directed toward linking ecological patterns and processes occurring at different scales and different levels of organization in rocky intertidal communities. In a paper (95) from her dissertation, which became a citation classic, she experimentally showed that the effect of herbivores on intertidal diversity depended upon the interaction between the food preferences of the herbivores and the competitive abilities of the macroalgae. Jane and her husband often conduct separate but parallel and complementary research projects, with each approach enriching the understanding of the other. One of their collaborative studies (108) draws on comparable experiments conducted in two temperate rocky intertidal communities (Pacific Northwest and New England) and a tropical one (Pacific coast of Panama). The latter is different in lacking a keystone species—characterized instead by a diverse suite of consumers at all trophic levels, each of which can compensate for the removal of the others. Lubchenco presented a new synthesis of plant-herbivore interactions by reviewing marine plant–herbivore ecology (97) and developing a theoretical model for linking ecological patterns and processes from the individual to the population, community, and biogeographic scales. She was the first to propose a mechanism to explain the alternation of morphologies of different phases of seaweeds with complex life histories (96). As is true of many ecologists today, she increasingly is directing her efforts toward resolving the pressing environmental challenges facing humanity. Although the first author on another citation classic (99), known as the Sustainable Biosphere Initiative (SBI), she indicates that the paper resulted from the "collaborative wisdom" of ecologists who served on the ESA Research Agenda Committee. SBI calls ecologists to action in creating the knowledge needed to address urgent ecological problems. Lubchenco serves on numerous national and international boards and panels. Among her many honors are election as the fourth woman President of ESA (1993), as a Fellow in the American Academy of Arts and Sciences (1993), as a John D. and Catherine T. MacArthur Fellow (1993), and as President of the American Association for the Advancement of Science (AAAS) (1996).

CONCLUSIONS REGARDING SUCCESSFUL WOMEN ECOLOGIST DOCTORATES (1900–1975) IN GENERAL COMPARISON WITH POST-1975 DOCTORATES

Personal Characteristics in Common

All of the women discussed were highly motivated. A survey of the living pre-1976 graduates (81) indicated they were motivated almost universally by a childhood interest in nature and doing outdoor activities while growing up, sometimes through scouting. Some also mentioned the influence of teachers, particularly at women's colleges. Of the post-1976 graduates in this survey, 44% indicated the importance of childhood influences, but an equal number pointed to undergraduate courses, 16% to field station experience. The increased motivation derived from ecology courses parallels their increased availability during the 1960s and 1970s (81).

Other prime characteristics clearly necessary for success are willingness to work hard and to persist in doing so. Some sociological studies suggest that positive women role models have a significant impact not only on motivation and career decisions but on the persistence for ultimate success (104). Although most of the pre-1976 PhD women ecologists succeeded without women role models, other support mechanisms were crucial for their persistence to overcome obstacles. For some single women collaboration with relatives or friends provided support (e.g. Braun, Keever, and Quarterman), and in the earlier period this also made some kinds of field work feasible (81). Marriage in some cases has provided this support, i.e. a husband to work with as a team (Baskin and Lubchenco) or to encourage persistence in research (e.g. Shreve, Nice, Crump, Zedler, and Ewel). Assistance with family responsibilities is also important, as managing simultaneously career, marriage, and motherhood demands persistence that is based on adaptability and high levels of energy, enthusiasm, and endurance. However, as discussed in other sections, marriage can lead to problems.

Women frequently have been assisted in their professional struggles by concerned male ecologists who served as mentors. Otherwise, many women would have been unable to progress as far as they have (despite the recent development of a larger group of women ecologists). The lack of role models is still considered by some (104) as an important impediment in the professional advancement of women scientists. However, Brattstrom (18, p, 143) recently suggested "There are role models out there, we just need to talk more about them! . . . And we need to start it now!" Among ecologists, Brattstrom mentions only Nice, but in the context of his remarks, this paper presents a number of possible role models who have been ignored or perhaps undervalued.

Graduate Education

By the turn of the century, women were obtaining PhD degrees in ecology in the major American universities where it was developing as a research field. Along with the expansion of ecology during the 1960s and early 1970s, a corresponding increase occurred in the numbers of women ecology PhDs and of institutions where they were granted. From 1976 to 1987 there was a pronounced increase in women PhDs in ecology throughout American universities, e.g. at Cornell they increased from 11% to 25% (B Peckarsky, personal communication). Increases also occurred in Ecology Programs that generally were established in the 1970s, the period with the greatest number of ecology doctorates (men and women). In fact, the 1980s are a significant period for women doctorates at some universities. For example, women constituted 44% in the University of Minnesota Program (M Davis, personal communication). At Duke University, in Botany, there was a dramatic increase from 12% on average for the previous 38 years to 60% in the 1980s (Billings and Antonovics, personal communication). Indeed, the greatly increased numbers of women doctorates produced in major ecology programs during the late 1970s and into the 1980s precluded my presenting the contributions of women ecologists who obtained their PhD degrees after 1975.

Subfield of Ecology Entered

Women with pre-1960 degrees tended to study plants, either physiologically or in populations and communities (Table 1). Furthermore in a general survey in 1987 of all ESA members, with 3100 responding, 49% of women still indicated plants were their "organism of choice" for study (181), whereas only 37% of males did so. Similarly, 54% of a 1987 survey of 200 established women ecologists with post-1975 degrees were working with plants (81). Studies of animal behavior, marine ecology, and mathematical and systems ecology increased somewhat for successful women with degrees in the 1960s and 1970s (Table 1). In the 1987 ESA survey, some subfields had a similar percentage of males and females (e.g. community ecology, empirical ecosystem studies, and plant physiological ecology), whereas others exhibited dichotomies between the genders. For example, 11% of the females indicated plant population ecology as their best descriptor, whereas only 3% of the males did so. Four percent of the males indicated animal behavior, whereas only a negligible fraction of females did (181).

Permanent Employment After Degree

Many women ecologists with pre-1976 PhD degrees found it difficult to obtain jobs commensurate with their education, especially if they were married. In fact, this has generally been a difficulty for women in science in the past (1). Despite progress, some of these problems continue, as 17% of the total women and 58% of those who were women respondents in the 1992 ESA survey indicated they were not employed as ecologists due to family constraints (85). On the other hand, 29% of the total number of women indicated unavailability of positions in their area of expertise, and another 28% a change in professional interests, as the reasons for being unemployed. However, women progressively have gained positions in major research-oriented universities, and the increasing number of these women with degrees from 1961 to 1975 (the "first modern wave") who married and had children indicates increased possibilities of working out arrangements to combine family and career (Table 1). For example, more attention is being given to the complex problem of two-career marriages (common among ecologists), with various approaches to a solution (98). In a 1990 survey of women ecologists (62), with respondents primarily from large, research-oriented universities, women in dual-career couples (where the woman often takes a soft-money research position), and to a lesser extent those who share positions, have reduced access to resources that facilitate research creativity and productivity. These women further indicate psychological stress from being considered "second-class citizens" because they are "the trailing spouse" (62, p. 150). Some encouraging news comes from women at institutions where policies have become favorable for women to attain full professorships (e.g. Zedler, Ewel, Porter, Woodin and Lubchenco); discouragingly, stubborn problems still persist for other women (e.g. Baskin with an Adjunct Professorship and Fox with a long-shared professorship).

Women, married or single, generally have faced salary inequities that continue today. Salary disparities still occur in some age classes and categories as shown in the ESA 1987–1988 membership survey (181). In the ESA 1992 survey, salary comparisons by age and gender indicated that male respondents earn nearly $6000 more per year than female respondents of the same age (84). Comparisons of income by time since highest degree and gender show that males make $4600 more than females who completed their education the same year. These lower salaries occur at all levels of professional experience (84, 86).

Publication Record: Numbers, Citations, and Recognition as Classics

Recent studies of established scientists have documented that women are less "productive" than men as defined by number of papers published and number of citations. Several studies have focused specifically on ecologists because of the large number of graduate students and comparatively large representation of women (92). One study did not detect significant differences in productivity between men and women (166) , but these results were questioned because of the small sample size and lack of comparisons of academic age and rank (93).

Subsequently, Primack and O'Leary (136) analyzed a large sample of pairs of men and women ecology graduate students who participated in an Organization of Tropical Studies (OTS) course between 1966 and 1979. They used Science Citation Index (1980–1984) to determine the total number of ever-cited publications written by each individual during his/her scientific career. One of their most striking conclusions is that women are unrepresented in the top group of researchers. Whereas women had an average of 47% as many publications that were cited an average of 43% as much as publications by men, the gender difference disappeared when the most productive 9% of men were removed from the analysis. Then no average difference existed in number of cited papers of 91% of the men compared to all of the women. Thus, Primack and O'Leary questioned whether "differences in productivity might be better framed to ask why a small proportion of male scientists are extremely productive" (136, p. 11).

In an extension of the previous OTS studies to include participants from 1966 to 1986, Primack and Stacy (138) confirmed that productivity of women ecologists generally was lower than their male counterparts. However, significantly, most of this difference apparently resulted from a lower percentage of women who continue careers that involve research. Analysis of the group of women who continued by publishing at least one paper indicated that these women are approaching rates of publication and accumulated citations equal to those of the men. Another interesting contrast is that the productivity level of women among the OTS participants trained between 1966 and 1975 (within the same time frame as the "first modern wave" of selected, highly successful women in my discussion) is only about 60% of that of men. On the other hand, women trained from 1976 to 1986 were attaining about 80% to 90% of the productivity level of men. Furthermore, several older women ecologists, in contrast to the men, in the OTS group are "late bloomers" in that they became highly productive after an initial delay. Another indication of increasing productivity of younger women is their ability to achieve their first publication earlier than men.

Although women wrote many significant papers from 1947 to 1979, only two women's names appear among 105 authors or coauthors of the 80 Ecological Citation Classics 53 (ECC) selected by McIntosh (107) during this 30-year period. McIntosh, however, admits his bias in selection of papers published by Current Contents as being ecological, i.e. based on subject matter, recognizable ecologists, and journals known to publish ecological articles. He further suggests that the initial basis of citation frequency does not necessarily express intrinsic merit of a publication and is only one measure of a classic article. Everyone is aware of important advances being published and ignored, only to be recognized much later as classic contributions to science. McIntosh (107, pp. 37–38) quizzically notes, "Most striking is the absence among these [ECC] authors of some very famous names, names that by all criteria belong in the pantheon of ecologists (e.g. GE Hutchinson, RH McArthur, RM May, EP Odum, TW Schoener, RH Whittaker, among others)." He did not know "whether their many articles were actually not cited frequently or whether they chose not to provide the requested biographical statements" that allowed them to be credited as authoring a citation classic. If the giants of ecology listed above have not produced an ECC, perhaps there should be less concern that so few women appear in this study.

On the other hand, it is obvious from this historical review that women with pre-1976 PhD degrees have been publishing papers with the same characteristics as those indicated by ECC authors as a basis for the frequency of citations of their selected articles (107). For example, women have been pioneering ideas and methodologies in subjects of rising interest (e.g. Patrick, Bentley), presenting mathematical approaches (e.g. Pielou), and challenging current dogma in several subfields (e.g. Nice, Willson, Langenheim, Rathcke). In some cases their work became a citation classic (e.g. Fox, Werner, Lubchenco, Myers), and one by Davis was included in a 1990 collection of "classic papers" (147), whereas other women's research has not been recognized for the impact it actually may have.

Recognition

Although opportunities did not come easily to the majority of early women ecologists, they did become members of the Ecological Society of America. Six percent of charter members of ESA in 1915 were women; by 1931, 9% were women, of whom 36% had PhDs (81). Women ecologists claimed full membership, attended banquets, and did not have to present their papers in separate sections—they were not excluded as they were in some other professional societies (e.g. American Chemical Society, Geological Society of America) (154). However, progress was slow from 1915 to 1987 in recognition of women through awards and election as officers, commensurate with their accomplishments and their number in ESA (81). Among the "early pioneers," only Braun was recognized. Among the "late pioneers" are Patrick, Langenheim, and Davis, but there is a perceptible increase among the "first modern wave" with Pielou, Sharitz, Werner, Bentley, Morrow, and Lubchenco. Because of the breadth of ecology, some women ecologists have been recognized by being elected as presidents of related societies (e.g. Moore, Nice, Patrick, Leopold, Davis, Langenheim, Werner, James, Christensen, and Stewart). Prestigious awards, other than those from ESA, have been given to Nice, Carson, Patrick, Davis, Leopold, Willard, and Lubchenco.

To the extent that women receive it, major recognition (e.g. presidencies of societies and the most prestigious awards) would be expected to go to the longer established women in the pre-1976 doctorate category. However, there are encouraging signs for the future in the recognition of women with ESA awards specifically given to younger ecologists. Women have received 61% of ESA's Murray F. Buell Awards for the best paper presented orally and 88% of E. Lucy Braun Awards for the outstanding poster—each given at the Annual

Meeting by an undergraduate or graduate student or a person with a doctorate who has completed defense of thesis within the previous nine months. From 1948 to 1978 there was no woman recipient of the George Mercer Award, given annually to an ecologist under 40 years of age for an outstanding paper published within the past two years. However, from 1979 to 1995, women received almost a third of the awards.

The number of women invited to be members of the editorial board of ESA journals and to participate in symposia at ESA annual meetings has been considered recently. Duffy and Hahn (50) pointed out the relatively low numbers of women on the ESA editorial boards (1980–1993) and questioned whether male dominated boards could contribute to women's lower productivity. Gurevitch (61) demonstrated that women were less likely to be symposium speakers at the 1987 ESA meeting when only men solicited speakers. However, Crowe and King (40) found an increase in the last 10 years in the proportion of women who were first authors in both contributed sessions and symposia in annual ESA meetings. By 1993, the earlier bias (61) of some male organizers in not inviting women speakers apparently had vanished. Crowe and King optimistically suggest that these "results indicate that gender representation can change relatively rapidly and easily" (40, p. 373).

General Comments Regarding Progress

Considerable strides have been made by women ecologists since the turn of the century. Although most overt institutional discrimination policies that older women faced are no longer in place, various "microinequities" may affect the recruitment and particularly the performance of younger women scientists (65). Since 42% of the 21–25 year age group of the 1992 ESA membership is female, recruitment of women into ecology does not appear to be a current problem (84). The progress made in this age group is strong, considering that women over the entire age range of membership make up only 23% of the Society (84). This survey, nonetheless, suggests that it may still be more difficult for women than for men to secure employment in the field, and some of this may relate to problems associated with marriage or family (85). Thus for future increases in the proportion of employed women ecologists, better ways will be needed to combine family and job responsibilities and to find solutions to the difficult career decisions that face dual career couples.

In academia (where most ecologists are employed), male ecologists on average publish more papers, have higher salaries, achieve higher academic positions, and feel generally more successful in their professional lives than do female ecologists (137). Primack and O'Leary concluded that these differences cannot be explained solely by whether the women were married and had children, or by gender differences in attitudes toward careers or time devoted to research; further, women at all stages of their careers face a constant, low-level disadvantage that prevents their competing as successfully in academia as men do. Women ecologists may often face a series of disadvantages early in their careers, leading to lack of recognition that decreases motivation and may eventually stifle high professional attainment and productivity. If a woman succeeds in obtaining a good academic job, she often lacks senior mentoring relationships, which can be crucial in a complex institutional environment. In sum, Primack and O'Leary (137) indicate that these individual disadvantages may seem minor at first because they are temporary and not obviously related to productivity. In fact, the advancement of women ecologists may still be limited due to accumulation of disadvantages and accompanying missed opportunities. Three years later, however, Primack and Stacy (138) add an optimistic note—that various changes in the social climate regarding women ecologists have been crucial in an increasingly positive trend for women to become some of the most productive and honored individuals among the younger generation of ecologists.

Although problems persist for women ecologists, the issues of gender equality in professional development are now being put forward, meaning that they can be more clearly recognized and addressed. This was not the case for the "early women pioneers" nor for the "late women pioneers" until quite recently, or even for the "first modern wave." Thus, these ecologists deserve all the more credit for opening the path with their research contributions, despite many obstacles.

ACKNOWLEDGMENTS

I wish to express appreciation for both personal and professional information provided by each cited living woman ecologist, including her own assessment of her research contributions to ecology. I also acknowledge the enthusiastic assistance in providing various kinds of information of Janis Antonovics, WD Billings, Patricia Brown, Grant Cottam, Margaret Davis, Tom Frost, Dennis Knight, Jane Lubchenco, Barbara Peckarsky, Richard Root, Richard Primack, Boyd Strain, and Mary Willson. I am grateful for the helpful comments on improving the manuscript from Beth Bell, Robert Burgess, Martha Crump, Daphne Fautin, Margaret Davis, Laurel Fox, Pat Lincoln, Susan Martin, Robert McIntosh, Wendy Peer, and Nancy Slack. I also am appreciative of the efforts of Dotty Hollinger in typing the various drafts of the manuscript.

Literature Cited

1. Abir-Am PG, Outram D. 1987. *Uneasy Careers and Intimate Lives: Women in Science 1789–1979.* New Brunswick, NJ: Rutgers Univ. Press
2. Allred BW, Clements ES, eds. 1949. *Dynamics of Vegetation. Selections from the Writings of Frederic E. Clements.* New York: Wilson
3. Antonovics J. 1980. The study of plant populations. *Science* 208:587–89
4. Barbour MG. 1995. Ecological fragmentation in the fifties. In *Uncommon Ground,* ed. W Coronon, pp. 233–54. New York: Norton
5. Baskin CC, Baskin JM. 1988. Studies on the germination ecophysiology of herbaceous plants in a temperate region. *Am. J. Bot.* 75:286–305
6. Baskin JM, Baskin CC. 1985. The annual dormancy cycle in buried weed seeds: a continuum. *Bioscience* 35:492–98
7. Beatley JC. 1966. Ecological status of introduced brome grasses (*Bromus* spp.) in desert vegetation of southern Nevada. *Ecology* 47:548–54
8. Beatley JC. 1967. Survival of winter annuals in the northern Mojave Desert. *Ecology* 48:745–50
9. Beatley JC. 1969. Dependence of desert rodents on winter annuals and precipitation. *Ecology* 50:721–24
10. Beckel A. 1990. Following their own paths: women in UW limnology 1900–1990. *Limnol. News, Univ. Wisc.-Madison,* No. 4:4–9
11. Bentley BL. 1976. Plants bearing extrafloral nectaries and the associated ant community: interhabitat differences in the reduction of herbivore damage. *Ecology* 57:816–20
12. Bentley BL. 1987. Nitrogen fixation by epiphylls in a tropical rainforest. *Ann. Missouri Bot. Gard.* 74:234–41
13. Bentley BL, Johnson ND. 1991. Plants as food for herbivores: the roles of nitrogen fixation and carbon dioxide enrichment. In *Plant-Animal Interactions. Evolutionary Ecology in Tropical and Temperate Regions,* ed. PW Price, TM Lewinsohn, GW Fernandez, WW Benson, pp. 257–72. New York: Wiley
14. Bonta MM. 1991. *Women in the Field. American Pioneering Women Naturalists.* College Station: Texas A& M Univ. Press
15. Bonta MM. 1995. *American Women Afield: Writings by Pioneering Women Naturalists.* College Station: Texas A& M Univ. Press
16. Bowers JE. 1986. A career of her own: Edith Shreve at the desert laboratory. *Desert Plants* 8:23–29
17. Bowers JE. 1988. *A Sense of Place, The Life and Work of Forrest Shreve.* Tucson, AZ: Univ. Ariz. Press
18. Brattstrom BH. 1995. Women in science: Do we ignore women role models? *Bull. Ecol. Soc. Am.* 76:143–45
19. Braun EL. 1950. *Deciduous Forests of Eastern North America.* Philadelphia, PA: Blakiston
20. Brooks P. 1972. *The House of Life. Rachel Carson at Work.* Boston: Houghton Mifflin
21. Brown PS. 1994. Early women ichthyologists. *Environ. Biol. Fishes* 41:9–30
22. Brush GS. 1984. Stratigraphic evidence of eutrophication in an estuary. *Water Resourc. Res.* 20:531–41
23. Brush GS, Link C, Smith J. 1980. The natural forests of Maryland: an explanation of the vegetative map of Maryland (with 1:250,000 map). *Ecol. Monogr.* 50:77–92
24. Carson R. 1962. *Silent Spring.* Boston: Houghton Mifflin
25. Chew FS, Courtney SP. 1991. Plant apparency and evolutionary escape from insect herbivory. *Am. Natur.* 138:729–50
26. Chew FS, Renwick JAA. 1995. Host plant choice in *Pieris* butterflies. In *Chemical Ecology of Insects II,* ed. RT Cardé, WJ Bell, pp. 214–38. New York: Chapman & Hall
27. Christensen M. 1969. Soil microfungi of dry to mesic conifer-hardwood forests in northern Wisconsin. *Ecology* 50:9–27
28. Christensen M. 1981. Species diversity and dominance in fungal communities. In *The Fungal Community: Its Organization and Role in Ecosystems,* ed. DT Wicklow, GC Carroll, pp. 201–32. New York: Marcel Dekker
29. Christensen M, Whittingham WF, Novak RO. 1962. The soil microfungi of wet-mesic forests in southern Wisconsin. *Mycologia* 54:374–88
30. Clarke R. 1973. *Ellen Swallow. The Woman Who Founded Ecology.* Chicago, IL: Follett
31. Clements ES. 1904. The relation of leaf structure to physical factors. *Am. Microscopical Soc.* 102 pp.
32. Clements ES. 1928. *Flowers of Coast and Sierra.* New York: HW Wilson
33. Clements ES. 1960. *Adventures in Ecol-*

ogy, Half a Million Miles . . . From Mud to Macadam. New York: Pageant
34. Clements FE, Long FL. 1923. *Experimental Pollination: An Outline of the Ecology of Flowers and Insects. Carnegie Inst. Wash. Publ. No. 336.* Washington, DC
35. Clements FE, Martin EV, Long FL. 1950. *Adaptation and Origin in the Plant World.* Waltham, MA: Chronica Botanica
36. Collins BR, Russell ER, eds. 1988. *Protecting the New Jersey Pinelands: A New Direction in Land-Use Management.* New Brunswick, NJ: Rutgers Univ. Press
37. Conway VM. 1937. Studies in the autoecology of *Cladium mariscus R. Br.* Pt. III. The aeration of the subterranean parts of the plant. *New Phytol.* 36:64–96
38. Conway VM. 1949. Bogs of central Minnesota. *Ecol. Monog.* 19:173–206
39. Conway VM. 1954. Stratigraphy and pattern analysis of Southern Pennine blanket peats. *J. Ecol.* 42:117–47
40. Crowe M, King B. 1993. Differences in the proportion of women to men invited to give seminars: Is the old boy still kicking five years later? *Bull. Ecol. Soc. Am.* 74:371–74
41. Crump ML. 1974. Reproductive strategies in a tropical anuran community. *Misc. Publ. Mus. Nat. History, Univ. Kansas* 61:1–68
42. Crump ML. 1981. Variation in the propagule size as a function of environmental uncertainty for tree frogs. *Am. Nat.* 117:724–37
43. Davis MB, Deevey ES Jr. 1964. Pollen accumulation rates: estimates from late-glacial sediment of Rogers Lake. *Science* 145:1293–95
44. Davis MB. 1969. Climatic changes in southern Connecticut recorded in pollen deposition at Roger's Lake. *Ecology* 50:409–22
45. Davis MB. 1981. Quaternary history and stability of forest commnunities. In *Forest Succession,* ed. DC West, HH Shugart, DB Botkin, pp. 132–53. New York: Springer-Verlag
46. Davis MB. 1987. Invasion of forest communities during the Holocene: beech and hemlock in the Great Lakes region. In *Colonization, Succession and Stability,* ed. AJ Gran, MJ Crawley, PJ Edwards, pp. 373–93. Oxford, England: Blackwell
47. Davis MB, Sugita S, Calcote RR, Ferrari J, Frelich LE. 1994. Historical development of alternate communities in a hemlock-hardwood forest in northern Michigan, USA. In *Large-Scale Ecology and Conservation Biology,* ed. R May, N Webb, PJ Edwards, pp. 19–39. Oxford, England: Blackwell
48. Dexter DM. 1992. Sandy beach community structure: the role of exposure and latitude. *J. Biogeogr.* 19:59–66
49. Dirks-Edmunds JC. 1947. A comparison of biotic communities of the cedar-hemlock and oak-hickory associations. *Ecol. Monogr.* 17:235–60
50. Duffy DC, Hahn DC. 1993. Letter to editor. *Bull. Ecol. Soc. Am.* 74:229–30
51. ESA Bulletin. 1953. Dr. E. Lucy Braun receives award. *Bull. Ecol. Soc. Am.* 34:9–11
52. Ewel KC. 1990. Multiple demands in wetlands. *Bioscience* 40:660–66
53. Ewel KC. 1990. Swamps. In *Ecosystems of Florida,* ed. RL Myers, JJ Ewel, pp. 281–323. Gainesville: Univ. Presses Florida
54. Ewel KC, Gholz BL. 1991. A simulation model of below ground dynamics in a Florida pine plantation. *For. Sci.* 37:397–438
55. Fosberg FR. 1951. Review of *Eastern Deciduous Forests of North America. Sci. Monthly*, July 66–67
56. Fox LR. 1981. Defense and dynamics in plant herbivore systems. *Am. Zool.* 21:853–64
57. Fox LR, Macauley BJ. 1977. Insect grazing on *Eucalyptus* in response to variation in leaf tannins and nitrogen. *Oecologia* 29:145–62
58. Fox LR, Morrow PA. 198 1. Specialization: species property or local phenomenon? *Science* 211:887–93
59. Gardner F. 1993. Forest acolyte. *Oregonian*, July 1
60. Gilbert HL, Curtis JT. 1953. Relation of the understory to the upland forests in the prairie-forest border region of Wisconsin. *Trans. Wisc. Acad. Sci. Art. Lett.* 42:183–95
61. Gurevitch J. 1988. Differences in the proportion of women to men invited to give seminars: Is the old boy still kicking? *Bull. Ecol. Soc. Am.* 69:155–60
62. Goldberg D, Sakai AK. 1993. Career options for dual-career couples: results of an ESA survey on soft money positions and shared positions. *Bull. Ecol. Soc. Am.* 74:146–52
63. Hutchinson GE. 1977. The influence of the New World on the study of natural history. In *Changing Scenes in the Natural Sciences, 1776–1976,* ed. GE Goulden, pp. 13–34. Philadelphia, PA: Acad. Nat. Sci. (Special Publ. 12)
64. Hutchinson GE. 1982. The harp that once

.... a note on the discovery of stridulation in the corixid water-bugs. *Irish Naturalists J.* 20:457–508
65. Ivey S. 1987. Recruiting more women into science and engineering. *Issues Sci. Technol.* 4:84–86
66. James FC. 1970. Geographic size variation in birds and its relation to climate. *Ecology* 57:356–90
67. James FC. 1971. Ordinations of habitat relationships among breeding birds. *Wilson Bull.* 83:215–36
68. James FC. 1983. The environmental component of geographic variation in the size and shape of birds: transplant experiments with blackbirds. *Science* 221:184–86
69. James FC, McCulloch CE. 1970. Multivariate analysis in ecology and systematics: panacea or Pandora's box? *Annu. Rev. Ecol. Syst.* 21:129–66
70. Jewell ME. 1927. Aquatic biology of the prairie. *Ecology* 8:289–98
71. Jewell ME. 1927. Ground water as a possible factor in lowering dissolved oxygen in the deeper water of lakes. *Ecology* 8:142–43
72. Jewell ME. 1935. An ecological study of the fresh-water sponges of northeastern Wisconsin. *Ecol. Monogr.* 5:461–504
73. Jewell ME. 1939. An ecological study of the fresh-water sponges of Wisconsin. II. The influence of calcium. *Ecology* 20:11–28
74. Jewell ME, Brown H. 1924. The fishes of an acid lake. *Trans. Am. Microscopical Soc.* 43:77–84
75. Jones RH, Sharitz RR, Dixon PM, Segal DS, Schneider RL. 1994. Woody plant regeneration in four floodplain forests. *Ecol. Monogr.* 64:345–67
76. Keever C. 1973. Distribution of the major forest species in southeastern Pennsylvania. *Ecol. Monogr.* 43:303–27
77. Keever C. 1985. *Moving On. A Way of Life.* Stateville, NC: Brady
78. Krebs CJ, Myers JH. 1974. Population cycles in small mammals. *Adv. Ecol. Res.*, pp. 267–399. New York: Academic
79. Langenheim JH. 1962. Vegetation and environmental patterns in the Crested Butte Area, Gunnison County, Colorado. *Ecol. Monogr.* 32:249–85
80. Langenheim JH. 1969. Amber: a botanical inquiry. *Science* 163:1157–67
81. Langenheim JH. 1988. The path and progress of American women ecologists. *Bull. Ecol. Soc. Am.* 69:184–97
82. Langenheim JH. 1994. Higher plant terpenoids: phytocentric overview of their ecological roles. *J. Chem. Ecol.* 20:1223–80
83. Langenheim JH. 1995. Biology of amber-producing trees; focus on case studies of *Hymenaea* and *Agathis.* In *Amber, Resinite and Fossil Resins, Am. Chem. Soc. Symposium Series No. 617,* ed. K Anderson, J Crelling, pp. 1–31. Washington DC: Am. Chem. Soc.
84. Lawrence DM, Holland MM, Morrin DJ. 1993. Profiles of ecologists: results of a survey of the membership of the Ecological Society of America. Part I. A snapshot of survey respondents. *Bull. Ecol. Soc. Am.* 74:21–35
85. Lawrence DM, Holland MM, Morrin DJ. 1993. Profiles of ecologists: results of a survey of the membership of the Ecological Society of America. Part II. Education and employment patterns. *Bull. Ecol. Soc. Am.* 74:153–69
86. Lawrence DM, Holland MM, Morrin DJ. 1993. Profiles of ecologists: results of a survey of the membership of the Ecological Society of America. Part III. Environmental science capabilities and funding. *Bull. Ecol. Soc. Am.* 74:237–47
87. Leopold EB. 1967. Late Cenozoic patterns of plant extinctions. In *Pleistocene Extinctions: The Search for a Cause,* ed. PS Martin, HE Wright, Jr, pp. 223–46. New Haven, CT: Yale Univ. Press
88. Leopold EB, Denton M. 1987. Comparative age of grassland and steppe east and west of the Northern Rocky Mountains. *USA. Ann. Missouri Bot. Gard.* 74:841–67
89. Leopold EB, Lui G. 1994. A long pollen sequence of Neogene age, Alaska Range. *Quaternary Int.* 22/23:103–40
90. Leopold EB, McGinitie HD. 1972. Development and affinities of Tertiary floras in the Rocky Mountains. In *Floristics and Paleofloristics in Asia and Eastern North America,* pp. 147–200. Amsterdam: Elsevier
91. Lloyd C. 1985. *The Traveling Naturalists.* Seattle: Univ. Wash. Press
92. Loehle C. 1987. Why women scientists publish less than men. *Bull. Ecol. Soc. Am.* 68:495–96
93. Loehle C. 1988. Publication rates of men and women: insufficient evidence. *Bull. Ecol. Soc. Am.* 69:143–44
94. Long FL. 1919. The quantitative determination of photosynthetic activity in plants. *Physiol. Res.* 2
95. Lubchenco J. 1978. Plant species diversity in a marine intertidal conununity: importance of herbivore food preference

and algal competitive abilities. *Am. Nat.* 112:23–39

96. Lubchenco J, Cubit J. 1980. Heteromorphic life histories of certain marine algae as adaptations to variation in herbivory. *Ecology* 61:676–81
97. Lubchenco J, Gaines SD. 1981. A unified approach to marine plant-herbivore interactions. I. Populations and communities. *Annu. Rev. Ecol. Syst.* 12:405–37
98. Lubchenco JL, Menge BA. 1993. Split positions can provide a "sane track": a personal account. *Bioscience* 43:243–48
99. Lubchenco J, Olson AM, Brubaker LB, Carpenter SR, Holland NW, et al. 1991. The Sustainable Biosphere Initiative: an ecological research agenda. *Ecology* 72:371–412
100. Madison Botanical Congress. 1894. *Proceedings*. Madison, Wisc., August 23–24, 1893
101. Martin SS. 1977. Accumulation of the flavonoids betagarin and betavulgarin in *Beta vulgaris* infected by the fungus *Cercospora beticola. Physiol. Plant. Pathol.* 11:297–303
102. Martin SS, Townsend CE, Lenssen AW. 1994. Induced isoflavonoids in diverse populations of *Astragalus cicer. Biochem. Syst. Ecol.* 22:657–61
103. Mason BL, Langenheim JH. 1957. Language analysis and the concept of environment. *Ecology* 38:325–39
104. Matyas ML. 1985. Factors affecting female achievement and interest in science and in scientific careers. In *Women in Science. A Report from the Field,* ed. JB Kahle, pp. 27–48. Philadelphia, PA: Falmer
105. McCormick JF, Platt RE. 1980. Recovery of an Appalachian forest following chestnut blight or Catherine Keever—you were right! *Am. Midland Nat.* 104:264–73
106. McIntosh R. 1985. *The Background of Ecology: Concept and Theory*. Cambridge, England: Univ. Press
107. McIntosh R. 1989. Citation classics in ecology. *Q. Rev. Biol.* 64:31–49
108. Menge BA, Lubchenco J, Ashkenas LR, Ramsey F. 1986. Experimental separation of effects of consumers on sessile prey in the low zone of rocky shore in the Bay of Panama: direct and indirect consequences of food web complexity. *J. Exp. Mar. Biol. Ecol.* 100:225–69
109. Merchant C. 1984. Women of the progressive conservation movement: 1900–1916. *Environ. Rev.* 8:57–85
110. Merian MS. 1719. *Dissertatio de Generatione et Metamorphosibus Insectarium Surinamensium.* Amsterdam: Joannen Oosterwyk
111. Mladenoff DJ, Burgess RL. 1993. The pedagogical legacy of John T. Curtis and Wisconsin plant ecology. In *John T. Curtis. Fifty Years of Wisconsin Plant Ecology,* ed. JS Fralish, RP McIntosh, OL Loucks, pp. 145–96. Madison, WI: Wisc. Acad. Sci., Art. Lett.
112. Moore E, ed. 1927–1940. *New York State Conservation Department Biological Survey; No. 1–16.* Suppl. to Ann. Reps. 16–20. Albany, New York
113. Morrow PA, Fox LR. 1989. Estimates of presettlement insect damage in Australian and North American forests. *Ecology* 70:1055–60
114. Morrow PA, La Marche VC. 1978. Tree ring evidence for chronic insect suppression of productivity in subalpine *Eucalyptus. Science* 201:1244–45
115. Myers JH. 1993. Population cycles of tent caterpillars. *Am. Scientist* 81:240–51
116. Neiland BM, Curtis JT. 1956. Differential responses to clipping of six prairie plants in Wisconsin. *Ecology* 37:355–65
117. Nice MB. 1933. The theory of territorialism and its development. In *Fifty Years Progress of American Ornithology 1883–1933,* pp. 89–100. Lancaster, PA: Am. Ornithol. Union
118. Nice MB. 1937. Studies in the life history of the song sparrow. I. A population study of the song sparrow. *Trans. Linnaean Soc. NY* 4:247 pp.
119. Nice MB. 1941. The role of territory in bird life. *Am. Midland Nat.* 26:441–87
120. Nice MB. 1943. Studies in the life history of the song sparrow. II: The behavior of the song sparrow and other passerines. *Trans. Linnaean Soc. NY* 6:328 pp.
121. 1995 LTER scientists connect with Oregon pioneer ecologists. *Pacific Northwest Sci. News* 5(1):3–8
122. Patrick R. 1938. The occurrence of flints and extinct animals in Pluvial deposits near Clovis, New Mexico. V. Diatom evidence from the Mammoth Pit. *Proc. Acad. Nat. Sci.* Philadelphia, 90:15–24
123. Patrick R. 1967. The effect of invasion rate, species pool and size of area on the structure of the diatom community. *Proc. Natl. Acad. Sci. USA* 58:1335–42
124. Patrick R. 1977. The changing scene in aquatic ecology. In *Changing Scenes in Natural Sciences, 1776–1976,* ed. CE Goulden, pp. 205–22. (Spec. Publ. 12) Philadelphia, PA: Acad. Nat. Sci.
125. Patrick R. 1984. Some thoughts concerning the importance of patterns in diverse

riverine systems. *Proc. Am. Phil. Soc.* 128:48–78
126. Patrick R. 1988. Importance of diversity in the functioning and structure of riverine communities. *Limnol. Oceanogr. 33* 6:1304–7. Am. Soc. Limnol. Oceanol.
127. Pielou EC. 1969. *Introduction to Mathematical Ecology*. New York: Wiley Intersci.
128. Pielou EC. 1972. Measurement of structure of animal communities. In *Ecosystem Structure and Function,* ed. JA Weins, pp. 113–35. Corvallis, OR: Proc. 31st Ann. Biol. Colloq.
129. Pielou EC. 1974. Competition on an environmental gradient. In *Mathematical Problems in Biology,* ed. P Van der Driessatie, pp. 184–204. New York: Springer-Verlag
130. Pielou EC. 1977. The statistics of biogeographic range maps: sheaves of one-dimensional ranges. *Bull. Int. Statist. Inst.* XLVII: 111–22
131. Pielou EC. 1977. *Mathematical Ecology*. New York: Wiley Intersci.
132. Pigott D. 1988. Verona Margaret Conway. *J. Ecol.* 76:288–91
133. Porter KG. 1977. The plant-animal interface in aquatic ecosystems. *Am. Sci.* 65:159–70
134. Porter KG. 1995. Integrating the microbial loop and the classic food chain in a realistic planktonic food web. In *Food Webs: Integration of Patterns and Dynamics,* ed. G Polis, K Winemiller, pp. 51–59. New York: Chapman & Hall
135. Porter KG, Saunders PA, Haberhahan KA, Macubbin AE, Jacobsen TR, Hodson RE. 1996. Annual cycle of autotrophic and heterotrophic production in a small, monomictic Piedmont lake (Lake Oglethorpe, GA): analog for the climatic warming on dimictic lakes. In *Regional Assessment of Freshwater Ecosystems and Climatic Change,* ed. D McKnight, P Mulholland, D Brakke. *Limnology and Oceanography*. Special Issue
136. Primack RB, O'Leary V. 1989. Research productivity of men and women ecologists: a longitudinal study of former graduate students. *Bull. Ecol. Soc. Am.* 70:7–12
137. Primack RB, O'Leary V. 1993. Cumulative disadvantages in the careers of women ecologists. *Bioscience* 43:158–65
138. Primack RB, Stacy F. 1996. Women ecologists catching up in scientific productivity, but only when they can join the race. *Bioscience.* In press
139. Quarterman E. 1950. Major plant communities of Tennessee cedar glades. *Ecology* 31:234–54
140. Quarterman E, Keever C. 1962. Southern mixed hardwood forest: climax in the southeastern Coastal Plain. *Ecol. Monogr.* 32:167–85
141. Rabinowitz D. 1978. Early growth of mangrove seedlings in Panama and an hypothesis concerning the relationship of dispersal and zonation. *J. Biogeogr.* 5:113–34
142. Rabinowitz D. 1981. Seven forms of rarity. In *The Biological Aspects of Rare Plant Conservation,* ed. H Synge, pp. 205–217. New York: Wiley
143. Rabinowitz D, Rapp JK. 1984. Competitive abilities of sparse grass species: means of persistence or cause of abundance? *Ecology* 65:1144–54
144. Rathcke BJ. 1976. Competition and coexistence within a guild of herbivorous insects. *Ecology* 57:76–87
145. Rathcke BJ. 1976. Insect-plant patterns and relationships in the stem-boring guild. *Am. Midland Nat.* 96:98–117
146. Rathcke BJ. 1984. Patterns of flowering phenologies: testability and causal inference using a random model. In *Ecological Communities: Conceptual Issues and the Evidence,* ed. DR Strong, D Simberloff, LG Aberle, AB Thistle, pp. 383–93. Princeton, NJ: Princeton Univ. Press
147. Real LA, Brown JH. 1991. *Foundations of Ecology. Classic Papers with Commentaries.* Chicago IL: Univ. Chicago Press
148. Riechert SE. 1974. The pattern of local web distribution in a desert spider: mechanism and seasonal variation. *J. Anim. Ecol.* 43:733–46
149. Riechert SE. 1993. The evolution of behavioral phenotypes: lessons learned from divergent spider populations. *Adv. Stud. Behav.* 22:103–34
150. Riechert SE, Bishop L. 1990. Prey control by an assemblage of generalist predators: spiders in garden test systems. *Ecology* 71:1441–50
151. Riechert SE, Hammerstein P. 1983. Game theory in an ecological context. *Annu. Rev. Ecol. Syst.* 14:377–409
152. Robichaud (Collins) B, Buell MF. 1973. *Vegetation of New Jersey: A Study in Landscape Diversity.* New Brunswick, NJ: Rutgers Univ. Press
153. Rodman JE, Chew FS. 1980. Phytochemical correlates of herbivory in a community of native and naturalized crucifers. *Biochem. Syst. Ecol.* 8:43–50
154. Rossiter MW. 1982. *Women Scientists in America. Struggles and Strategies to*

1940. Baltimore, MD: Johns Hopkins Univ. Press

155. Sharitz RR, Boring LR, Van Lear DH, Pinder JE III. 1992. Integrating ecological concepts with natural resource management. *Ecol. Appl.* 2:226–37
156. Sharitz RR, McCormick JF. 1973. Population dynamics of two competing annual plant species. *Ecology* 54:723–74
157. Shreve EB. 1915. An investigation of the causes of autonomic movements in succulent plants. *Plant World* 18:297–312, 331–43
158. Shreve EB. 1916. An analysis of the causes of variation in the transpiring power of cacti. *Physiol. Res.* 2:73–127
159. Shreve EB. 1919. The role of temperature in the determination of the transpiring power of leaves by hygrometric paper. *Plant World* 22:100–4
160. Shreve EB. 1923. Seasonal changes in the water relations of desert plants. *Ecology* 4:266–92
161. Silvertown J. 1988. Obituary-Deborah Rabinowitz. *Br. Ecol. Soc. Bull.* 19:86–87
162. Slack NG. 1977. Species diversity and community structure in bryophytes: New York State Studies. *NY State Mus. Bull.* 428:1–70
163. Slack NG. 1990. Bryophytes and ecological niche theory. *Bot. J. Linnaean Soc.* 104:187–213
164. Slack NG. 1995. Botanical and ecological couples: a continuum of relationships. In *Creative Couples in the Sciences,* ed. HM Pycior, NG Slack, PG Abir-Am, pp. 235–54. New Brunswick, NJ: Rutgers Univ. Press
165. Slack NG, Vitt DH, Horton DG. 1980. Vegetation gradients of minerotrophically rich fens in western Alberta. *Can. J. Bot.* 58:330–50
166. Sih A, Nishikawa K. 1988. Do men and women really differ in publication rates and contentiousness? An empirical survey. *Bull. Ecol. Soc. Am.* 69:15–16
167. Stamps JA, Clarke A, Arrowood P, Kus B. 1985. Parent-offspring conflict in budgerigars. *Behavior* 94:1–40
168. Stamps JA. 1988. Conspecific attraction and aggregation in territorial species. *Am. Nat.* 131:329–47
169. Stamps JA. 1994. Territorial behavior: testing the assumptions. *Adv. Study Behav.* 23:173–232
170. Stark N. 1978. Man, tropical forests and the biological life of a soil. *Biotropics* 10:1–20
171. Stark N. 1982. Soil fertility after logging in the northern Rocky Mountains. *Can. J. For. Res.* 12:679–86
172. Stewart MM. 1967. *Amphibians of Malawi.* Albany, NY: State Univ. New York Press
173. Stewart MM. 1995. Climate driven populations in rain forest frogs. *J. Herpetol.* 29:437–46
174. Stewart MM, Pough FH. 1983. Population density of tropical forest frogs: relation to retreat sites. *Science* 221:570–72
175. Stuckey RL. 1973. E. Lucy Braun (1889–1971), outstanding botanist and conservationist: a biographical sketch, with bibliography. *Mich. Bot.* 12:83–106
176. Struik GJ, Curtis JT. 1962. Herb distribution in an *Acer saccharum* forest. *Am. Midl. Nat.* 68:285–96
177. Taub FB. 1974. Closed ecological systems. *Annu. Rev. Ecol. Syst.* 5:134–60
178. Taub FB, ed. 1984. Lakes and reservoirs. *Ecosystems of the World,* Vol. 23. Amsterdam: Elsevier
179. Taub FB. 1993. Standardizing an aquatic microcosm test. In *Progress in Standardization of Aquatic Toxicity Tests,* ed. A Soares, P Culow, pp. 159–88. New York: Pergamon
180. Trautman MB. 1977. In memoriam: Margaret Morse Nice. *The Auk* 94:430–41
181. Travis J. 1989. Results of the survey of membership of the Ecological Society of America—1987–1988. *Bull. Ecol. Soc. Am.* 70:78–88
182. Watson MA. 1980. Shifts in patterns of microhabitat occupation along a complex altitudinal gradient. *Oecologia* 47:46–55
183. Watson MA. 1984. Developmental constraints. Effects on population growth and patterns of resource allocation in a clonal plant. *Am. Natur.* 123:411–26
184. Watson MA. 1995. Sexual differences in plant developmental phenology affect plant-herbivore interactions. *TREE* 10:180–82
185. Werner PA. 1975. Predictions of fate from rosette size in teasel (*Dipsacus fullonum L.*). *Oecologica* 20:197–201
186. Werner PA. 1976. Ecology of plant populations in successional environments. *Syst. Bot.* 1:246–68
187. Werner PA. 1977. Colonization of a biennial plant species: experimental field studies in species cohabitation and replacement. *Ecology* 58:1103–11
188. West RG. 1964. Inter-relations of ecology and quaternary paleobiology. *J. Ecol. (suppl.)* 52:47–57
189. Willard BE, Marr JW. 1971. Recovery of alpine tundra ecosystems after damage by

human activities in the Rocky Mountains of Colorado. *Biol. Conserv.* 3:181–90
190. Williams KS, Myers JH. 1984. Previous attack of red alder may improve food quality for fall webworm larvae. *Oecologia* 63:166–70
191. Willson MF. 1966. Breeding ecology of the yellow-headed blackbird. *Ecol. Monogr.* 36:51–77
192. Willson MF. 1979. Sexual selection in plants. *Am. Nat.* 113:777–90
193. Willson MF. 1996. Avian frugivory and seed dispersal in eastern North America. *Curr. Ornithol.* 3:223–79
194. Willson MF. 1992. Ecology of seed dispersal. In *Seeds: The Ecology of Regeneration in Plant Communities,* ed. M Fenner, pp. 61–85. Wallerford, UK: CAB Int.
195. Willson MF, Whelan CJ. 1990. The evolution of fruit color in fleshy fruited plants. *Am. Nat.* 136:790–809
196. Woodin SA. 1976. Adult-larval interactions in dense infaunal assemblages: patterns of abundance. *J. Mar. Res.* 34:25–41
197. Woodin SA. 1978. Refuges, disturbance and community structure: a marine soft-bottom example. *Ecology* 59:274–84
198. Woodin SA, Marinelli RL, Lincoln DE. 1993. Biogenic brominated aromatic compounds and recruitment of infauna. *J. Chem. Ecol.* 19:17–53
199. Zedler JB. 1980. Algal mat productivity: comparisons in a salt marsh. *Estuaries* 3:122–31
200. Zedler JB. 1993. Canopy architecture of natural and planted cordgrass marshes: selecting habitat evaluation criteria. *Ecol. Applications* 3:123–38

Annu. Rev. Ecol. Syst. 1996. 27:55–81

FOREST CANOPIES: Methods, Hypotheses, and Future Directions

Margaret D. Lowman and Philip K. Wittman
Marie Selby Botanical Gardens, 811 South Palm Avenue, Sarasota, Florida 34236-7726

KEY WORDS: forest canopies, canopy access techniques, epiphytes, herbivory, canopy-atmosphere interface, biodiversity, arthropods

ABSTRACT

Forest canopies contain a major portion of the diversity of organisms on Earth and constitute the bulk of photosynthetically active foliage and biomass in forest ecosystems. For these reasons, canopy research has become integral to the management of forest ecosystems, and to our better understanding of global change. Ecological research in forest canopies is relatively recent and has been primarily descriptive in scope. The development of new methods of canopy access has enabled scientists to conduct more quantified research in tree crowns. Studies of sessile organisms, mobile organisms, and canopy interactions and processes have emerged as subdisciplines of canopy biology, each requiring different methods for collecting data. Canopy biology is beginning to shift from a descriptive autecology of individuals to a more complex ecosystem approach, although some types of field work are still limited by access.

Questions currently addressed in canopy research are extremely diverse but emphasize comparisons with respect to spatial and temporal variation. Spatial scales range from leaves (e.g. quantifying the number of mites on individual phylloplanes) to trees (e.g. measuring photosynthesis between sun and shade leaves), to forest stands (e.g. measuring turbulence above the canopy), and entire landscapes (e.g. comparing mammals between different forest types). Temporal variation is of particular significance in tropical forest canopies, where populations of organisms and their resources have diurnal, seasonal, or even annual periodicity. As the methods for canopy access improve, more rigorous hypotheses-driven field studies remain a future priority of this newly coalesced discipline.

There awaits a rich harvest for the naturalist who overcomes the obstacles—gravitation, ants, thorns, rotten trunks—and mounts to the summits of jungle trees. . . (Beebe et al 1917)

0066-4162/96/1120-0055$08.00

INTRODUCTION

Forest canopies have historically remained out of reach of all but the most robust and inquisitive naturalists (6, 83a, 84). This ground-based perception of forests led to some generalizations that were false and, in many cases, resulted in an underestimation of the diversity and abundance of organisms and the complexity of canopy interactions.

The forest canopy is defined as "the top layer of a forest or wooded ecosystem consisting of overlapping leaves and branches of trees, shrubs, or both" (2). Studies of plant canopies typically include four organizational levels of approach: individual organs (leaves, stems, and/or branches), the whole plant, the pure stand, and the plant community (112). Canopy biology is a relatively new discipline of forest science that incorporates the study of mobile and sessile organisms and the processes that link them into an ecological community (74, 76, 96, 97, 122). The definition of canopy biology remains somewhat controversial (71), because the canopy is a habitat or an environment of the forest, not a science in itself (G Parker, personal communication). After lengthy discussions at the First International Conference on Forests Canopies in 1994, most biologists agreed that canopy biology per se is composed of many subsets of science, and canopy scientists are linked by a common physical region of study, namely the crowns of trees (71).

The development of canopy research has been affected by spatial and temporal constraints in this habitat, including: 1) differential use of the geometric space within tree crowns by canopy organisms; 2) heterogeneity of substrates; 3) variability in age classes within the canopy (e.g. leaf cohorts, soil-plant communities accruing unevenly on branches, etc); 4) variability in microclimate of the canopy-atmosphere interface; 5) high diversity of organisms (many as yet unnamed or undiscovered); and 6) lack of protocols to quantify canopy studies (74). Recent advances in the methods of access have greatly accelerated our understanding of forest canopies. More importantly, the use of standardized, replicated methods has expanded the discipline from a descriptive to a more rigorous science in which specific hypotheses can be addressed.

ADVANCES IN METHODS OF CANOPY ACCESS

A Brief History of Methods of Access

The earliest canopy observations were made from ground level, either using binoculars or relying upon material that had fallen from the canopy (46, 84). Despite the range of canopy access techniques now available, ground-based observations remain a preferred method for some studies, such as the behavior of mammals (30) and birds (123), or rapid surveys of trees and epiphytes

(36, 102). Early access methods also included the "reach-and-grab" technique whereby scientists sampled only those branches they could comfortably reach (84). Many scientists erroneously surmised that samples from the understory could be extrapolated to represent the entire forest which may extend several hundred feet overhead.

As biologists became more aware of the importance of reaching the upper canopy, they developed techniques to climb into tree crowns. More daring researchers have attempted the use of chairs suspended on vines (83a, 84), monkeys trained to retrieve samples (18), hot-air dirigibles (11, 41), mobile sleds (75), or even ultra-light planes (89). But the daunting nature of any untested or physically precarious technique can take its toll, both in terms of the safety of the researcher and in the quality of the data collected. Recently, access methods that provide safety and accuracy, and also facilitate collaborative work (rather than solo efforts), have greatly expanded the scope of canopy study (41, 70, 84, 101).

Advantages of Current Methods of Canopy Access

One of the most important advances in canopy methods has been the development of increasingly reliable hardware to facilitate access. The canopy research community has benefited from adaptations of technology developed for arboriculture and mountaineering (28, 70, 98, 103). For their tree work, arborists have adapted harnesses constructed of nylon webbing, light-weight locking carabineers, and jumars that mechanically ascend ropes and offer automatic safety catches (49, 115). In addition, arboricultural materials (e.g. galvanized steel aircraft cable, turnbuckles and clamps, drop-forged aluminum, and pressure-treated wood) for cabling and bracing permanent canopy structures have been developed to maximize structural support while minimizing physical damage to tree crowns (60, 116). Structures are usually affixed to trees with through-bolts and lags, rather than by encirclement which may damage the cambium layer (70). Improved mountaineering technology has also been adapted to canopy research, including polyester ropes with a low degree of stretch, a high tensile strength for a given weight, and resistance to UV degradation (26, 28).

With the advent of safe, reliable hardware for canopy access, researchers can sample effectively and with confidence. They can also work collaboratively and over extended periods. For example, the raft and dirigible apparatus (also called "Radeau des Cimes") enabled 74 scientists to work in tree crowns in Cameroon, Africa, in 1991 (71). Similarly, the recent erection of canopy cranes in old growth forests of the Pacific northwest and in tropical rain forests of both Panama and Venezuela are promoting collaborative research on forest structure, herbivory, photosynthesis, epiphytes, and phenology (85–87, 101, 127; D Shaw, personal communication).

Another methodological advance for canopy study is the portability of many devices. For example, instruments to measure photosynthesis of entire leaves can be transported comfortably up to the canopy via ropes and used in situ by individual researchers. The Biosphere 2 project in Oracle, Arizona, is currently measuring entire tree photosynthesis in its rain forest biome with lightweight, portable, battery-powered Li-Cor gas analyzers (B Marino, personal communication). The biome, experimentally controlled under glass, will be subjected to temperature changes to predict the impacts of global warming on rain forests.

Other advances in methods include the accuracy of certain apparatuses. Traditional equipment such as sweep nets and beating trays to survey arthropod diversity in foliage (4, 32, 73, 79) continually underestimated the real population of the canopy by as much as threefold (73). But the development of canopy fogging, using a nonpersistent insecticide, to harvest insects from the canopies of tall tropical trees has provided more accurate surveys and raised the estimates of the number of species on Earth to at least 30 million (e.g. 31–34).

Long-Term, Collaborative Canopy Field Sites

One of the most important advances in any field of science is the increased sharing of resources and ideas (71, 97). In canopy biology, where scientists often work for long periods of time in remote locations and literally dangle from ropes out of contact with colleagues, opportunities to collaborate are difficult to arrange. The use of common field sites for different projects represents an important means of sharing resources and maximizing the use of their data sets. In the past five years, over 30 field sites have established canopy research (Table 1). A majority of these sites are located in tropical forests, because of global concerns for biodiversity and maintenance of tropical ecosystems (17). But recent interest in quantifying biodiversity has also fueled canopy research in temperate forests, where scientists have easier access to field sites than their tropical-based counterparts (67).

To date, canopy research sites have been established somewhat opportunistically and usually as a consequence of ground-based research. At the First International Canopy Conference (71), the merits of establishing one major long-term experimental canopy research site were discussed. Such a field station, coined "Biotopia" by EO Wilson and Andrew Mitchell, the proponents of this idea, would include different canopy access methods such as dirigible-raft, platforms and walkways, vertical ropes, and perhaps even ultra-light planes and a construction crane (A Mitchell, personal communication). Such a site may be advantageous for the obvious reasons of pooling diverse funding sources, collecting and analyzing detailed measurements of canopy processes, and sharing information.

Sites with a history of canopy research include the British woodlands, where the work of Southwood and others (e.g. 23, 118, 119) over the past two decades has stimulated great interest in insect biodiversity and in the methods to survey arthropods in trees. In his classic work, Southwood (118) used ground-based techniques to measure the arthropod fauna of temperate deciduous trees. He sprayed the crowns with insecticide (called canopy fogging) to provide a snapshot of each tree's insect fauna. Comparisons of insect populations among trees showed a fairly consistent proportion of insect species at various trophic levels: one-quarter phytophagous, one-quarter parasitoids, one-quarter predators, one-sixth scavengers, and the remainder split amongst tourists (visitors), insects in epiphytes, and ants. In total, Southwood logged between 180 and 425 species of insects for *Salix alba* and *Quercus* spp. canopies, respectively, using a composite method of fogging accompanied by observations. Since these initial trials, fogging has become an important method for surveying canopy arthropods, especially in tropical conservation.

Two rain forest sites where long-term studies of canopy foliage have been conducted using single rope techniques (SRT) are Monteverde, Costa Rica and Lamington National Park in Queensland, Australia. Epiphytes and insect-plant relationships, respectively, were studied over two decades, and the methods carefully documented as protocols for other sites. The use of SRT provided a replicated transect into the canopy, and at the Australian site, a prototype canopy walkway was installed that allowed permanent all-weather access to certain crowns.

Nadkarni and her colleagues developed methods for monitoring and measuring epiphyte growth and mortality and their use by birds in the cloud forests at Monteverde, Costa Rica (20, 92, 93). They quantified aspects of nutrient cycling and found that epiphytes served as collectors of dead organic matter, which later fell to the forest floor and comprised 10–15% of the total litterfall. The rates of nitrification were much lower in the canopy than on the forest floor, although microbial biomass (C and N) was similar. By virtue of their diversity and abundance, epiphytes have a greater role in canopy structure and nutrient cycling of tropical forests than in most temperate forests (20).

In Australia, Lowman and her colleagues established methods to measure leaf growth and mortality, insect populations, and herbivory in Australian forests and woodlands. The sites included cool temperate, warm temperate, and subtropical rain forests, and dry woodlands (66, 68, 72). They found that individual trees have different populations of leaves with respect to height and light and age, and that a fairly complex sampling design was required to estimate herbivory accurately. For example, *Ceratopetalum apetalum* (*Cunoniaceae*) averaged as high as 35% annual leaf surface area losses in its young understory

Table 1 Canopy research sites throughout the world

Name of Site	Location	Time Frame	Habitat	Canopy Research Projects
NORTH AMERICA				
Wind River Canopy Crane Research Facility	Washington, USA 45°49′13.76″ N 121°57′06.88″ W	1930's—Exper. forest 1994—canopy studies, canopy crane April 1995	Temperate coniferous forest. Tree heights to 64 m	forest ecology, structure & function, arboreal insects, dwarf mistletoe, avifauna, meteorology, remote sensing, forest health
Hopkins Memorial Forest	Williamstown MA, USA	1990—present	temperate deciduous	herbivory, phenology, mammals, insect diversity
Millbrook School	Millbrook, NY, USA 41°50.75′ N 73°37.26′ W	1995—present	temperate deciduous	meteorology, phenology, insect diversity
Coweeta Hydrologic Laboratory	North Carolina, USA 35° N 83°30′ W	1934—forest hydrology 1968—forest ecology	temperate deciduous	forest hydrology, forest ecology, nutrient fluxes, watershed experiments, vegetation dynamics, net primary production, decomposition, herbivory, streams, biotic regulation
Marie Selby Botanical Gardens	Sarasota, FL, USA 27°19′ N 82°32′ W	1994—present	live oak canopy	herbivory, throughfall, insect diversity, epiphytes
Myakka River State Park	Sarasota, FL, USA 27° N 82° W	1992—present	oak-palm canopy	herbivory, epiphytes, arboreal insects
Penn State Research Forest	near University Park, PA, USA	1988—present	oak forest	plant-insect relationships, leaf chemistry, phenology
Hampshire College	Amherst, MA, USA	1993—present	temperate deciduous	birds, phenology
Findley Lake	King County, Washington, USA	1968—Present	montane forest, west side of Cascade Mountains	production ecology, limnology, physiological ecology, community ecology, canopy processes, soil genesis, remote sensing, biogeochemistry
Smithsonian Environmental Research	Maryland, USA 38°58′ N	1987—present	temperate deciduous	canopy-atmosphere interactions, forest structure, phenology,

Center	67°55′ W			hydrology, radiation, transpiration, canopy interception
Harvard Forest	Petersham, MA, USA		temperate deciduous	canopy-atmosphere interactions, radiation, tree physiology
Vermont Monitoring Cooperative	Mt. Mansfield, VT, USA 44°31′ N 72°52′ W Lye Brook Wild., VT, USA 43°06′ N 73°03′ W	1990—present	temperate northern hardwood & montane conifer forests	forest health, atmospheric chemistry, within-canopy environmental gradients, insect, bird, & amphibian biodiversity
Andrews Experimental Forest	Oregon ~44–45° N	1970— present	old growth Douglas fir forest	forest-atmosphere interactions, canopy-soil interactions, arthropods
Willamette National Forest	Central Oregon Cascades ~44–45° N	1992—present	old growth Douglas fir forest	epiphyte diversity, long term growth of lichens
Duke Forest	Durham, NC, USA	1995—present	temperate deciduous	leaf area dynamics (FACE facility), canopy CO_2 & water flux
Hakalau Forest National Wildlife Refuge	Hawaii, USA	1991–1992	ohia forest	insect diversity
Hawaik Volcanoes NP	Hawaii, USA	1971–1973	ohia forest	insect diversity
Mesita del Buey	Los Alamos, NM, USA 34°30′ N 106°27′ W	1984—present	pinyon-juniper woodland	water balance, soil moisture, solar radiation, tree ecophysiology
BOREAS	Prince Albert N.P., Saskatchewan, Canada ~54° N ~99° W	1994—present	boreal forest	carbon balance, gas exchange & energy flux, canopy architecture & remote sensing, solar radiation
Vancouver Island Forests	Vancouver Island, BC, Canada 48°44′ N 124°37′ W	1990—present	old growth forest	community composition, arthropod diversity
CENTRAL & SOUTH AMERICA				
JASON V Project	Blue Creek, Belize 16°10′ N 89°5′ W	1994—present	tropical wet rain forest	ant gardens, herbivory, arboreal insects

(*Continues*)

Table 1 (*Continued*)

Name of Site	Location	Time Frame	Habitat	Canopy Research Projects
Luquillo Experimental Forest	Puerto Rico 18°19′ N 65°45′ W	~1962	tropical rain forest	predator-prey relationship, lizards, arthropods, herbivory
Coffee Plantations	Central Valley, Costa Rica	~1992—present	coffee agroecosystem	arthropods, agricultural transformation
Finca, La Selva	La Selva, Costa Rica	1974–1983	tropical rain forest	natural history, canopy methods, pollination
Monteverde Cloud Forest Preserve	Monteverde, Puntarenas Province, Costa Rica 10°18′ N 84°48′ W	1980—present	leeward lower montane wet forest	epiphyte ecology, nutrient cycling, bird use of epiphytes
Smithsonian Tropical Research Institute	Barro Colorado Island, Panama 9°10′ N 79°51′ W	1972—present	seasonally dry tropical moist forest	insects, photosynthesis, epiphytes, phenology
Parque Natural Metropolitano	near Panama City, Panama	1990—present	seasonally dry rain forest	photosynthesis, structure, herbivory insects
Operation Drake	near Punta Escoces, Daren, Panama	1979–1980	tropical rain forest	insect diversity, walkway methods, forest structure, pollination biology bats
Radeau de Cimes-Expedition I	French Guiana 5°4′36″ N 53°3′15″ W	1989	tropical rain forest	tree architecture & growth, plant & animal relationships, ecology, canopy-atmosphere interface
French Guiana Research Station	Nouragues, French Guiana 4°5′ N 52°40′ W	1986— present	evergreen tropical forest	growth dynamics, gaps
Smithsonian Fogging Studies	Pakitza, Peru	1988–1992	lowland rain forest	insect diversity
ACEER	Napo River, Peru 3°15′ S 72°54′ W	1991—present	lowland tropical rain forest	epiphytes, herbivory
Surumoni Research Project—European Science Foundation	Orinoco River, Venezuela	1995—present	lowland wet rain forest	structure, epiphytes, ant gardens, birds, phenology

ASIA & SOUTH PACIFIC				
Lamington National Park	Queensland, Australia 28°13′ S 153°07′ E	1979—present	subtropical cool rain forest	insect diversity, herbivory
Dorrigo National Park	NSW, Australia 30°20′ S 153° E	1979–1990	subtropical cool rain forest	herbivory, phenology, insects
New England National Park	NSW, Australia 30°30′ S 152° E	1979–1990	cool temperate forest	herbivory, phenology, insects.
Royal National Park	NSW, Australia 34°10′ S 151°30′ E	1979–1990	warm temperate forest	herbivory, phenology, insects
CSIRO	Atherton, Queensland, Australia	~1980—present	wet tropical rain forest	phenology, herbivory, reproductive biology
Mt. Specd. National Park	Paluma, Queensland Australia	~1980—present	lower montane tropical forest	birds, insect diversity
Cradle Mountain	Central Tasmania 41°35.4′ S 145°55.9′ E	1989—present	cool temperate rain forest	Invertebrate surveys (12 sites)
Operation Drake (Bulolo Forestry College)	Buso, Morobe Province, Papua New Guinea	1979–1980	tropical rain forest	insect diversity, walkway methods, forest structure, pollination biology, bats
Wau Ecology Institute	Wau, Papua New Guinea 7°24′ S 146°44′ E	~1980—present	lower and mid-montane tropical rain forests	herbivory, birds
Operation Drake (Morowali Nature Reserve)	Sulawesi Tengah, Indonesia	1979–1980	tropical rain forest	insect diversity, walkway methods, forest structure, pollination biology, bats
Gomback Watershed	35 km east of Kual Lampur, Malaysia	1960—present	hill, dipterocarp forest	phenology, vertebrates
Aerial Walkway	Bukit Lanjan, W. Malaysia	1968–1976	tropical rain forest	insect vectors, phenology
Canopy Biology Program in Sarawak (CBPS)	Lambir Hills NP, Sarawak, Malaysia 4°20′ N 113°50′ E	1991—present	tropical rain forest	phenology, insect abundance, plant/animal interactions in canopy layers
Nafanua	Savai'i, Western Samoa	1996—	tropical island rain forest	ethnobotany, plant taxonomy

(*Continues*)

Table 1 (*Continued*)

Name of Site	Location	Time Frame	Habitat	Canopy Research Projects
Padang	Padang, Indonesia 0°53′ S 100°21′ E	~1980's—present	tropical rain forest	foliage-canopy structure, height distribution of woody species
Lake Rara National Park	Nepal 29°34′ N 82°5′ E	~1990	alpine tree limit forest	foliage-canopy structure, height distribution of woody species
Yatsugatake	Yatsugatake, Japan 36°5′ N 138°21′ E	~1980's—present	subalpine mixed forest	foliage-canopy structure, height distribution of woody species
Daisen	Daisen, Japan 35°21′ N 133°33′ E	~1980's—present	cool temperate deciduous broad-leaved forest	foliage-canopy structure, height distribution of woody species
Yakushima Island	Yakushima Island, Japan 30°20′ N 130°24′ E	~1980's—present	warm temperate evergreen broad leaved forest	foliage-canopy structure, height distribution of woody species
Hahajima Island	Hahajima Island, Japan 26°39′ N 142°8′ E	~1980's—present	subtropical evergreen broad-leaved forest	foliage-canopy structure, height distribution of woody species
Changbaishan Natural Reserve	Jilin, P.R. China 41°23′ N 126°55′ E	1996—	old-growth temperate spruce-fir forest	canopy processes, structure & modeling
EUROPE				
River Esk Woodlands	Midlothian, Scotland	1990—present	Fagus woodlands	Phyophagous insects, woodland biodiversity, stand structure
Gisbum Forest	Lancashire, England	1955–1992	Uplands mixed deciduous/ conifer plantation	Tree growth, invertebrate populations, soil processes
Imperial College Field Station	Silwood Park, Ascot Berkshire, U.K.	~1977–1982	temperate deciduous	insect diversity and abundance
AFRICA				
Radeau de Cimes-Expedition II	Reserve de Fauna de Campo, Cameroon 2°30′ N 10°0′ E	1991	lowland tropical rain forest	tree architecture & growth, plant & animal relationships ecology, canopy-atmosphere interface
East African Virus Research Institute	Mpanga Research Forest, west of Kampala, Uganda	1961—present	tropical forest	insect vectors, meteorology

leaves, but only 9% in the mature canopy sun leaves (68, 69); and these levels varied among different crowns and sites. Field trials showed that conventional techniques of measuring herbivory—whereby leaves were harvested and measured for holes—underestimated insect damage by as much as three-fold, compared to measurements obtained from long-term observations of leaves in situ (65).

Recently, a series of canopy crane sites were established in various forest types. Cranes offer collaborative opportunities that more than compensate for their relatively high initial costs. The Smithsonian Tropical Research Institute in Panama erected the first construction crane for canopy research in 1990 under the direction of the late Alan Smith (84, 101). This tropical dry Pacific forest was the site of pioneering work on forest canopy structure and photosynthesis (86, 101, 133). Because of the excellent access provided by the crane arm above the canopy, researchers can reach virtually all of the foliage within a tree crown for purposes of whole tree investigations. Similarly, crane research sites have been established in temperate coniferous forest in Washington, USA (Wind River Research Crane Facility) and in lowland tropical rain forest along the Orinoco River in Venezuela (Surumoni Research Project) (85) (Table 1).

Another series of collaborative, long-term canopy sites involves the use of permanent platforms and walkways. Walkways were constructed to investigate arthropod biodiversity in the Carmanah Valley on Vancouver Island, British Columbia, Canada (130). The species diversity and abundance of arthropods in old growth conifers are high and indicate that this structurally complex habitat serves as an important reservoir for temperate biodiversity (130). Using walkways, similar studies of plant-insect interactions are in progress at Coweeta Forest, North Carolina, USA (110), Hopkins Memorial Forest in Williamstown, Massachusetts, USA (67, 77) and Blue Creek Preserve, Belize (70). Neotropical migrating birds in temperate forest canopies are the focus of studies on a walkway at Hampshire College, Massachusetts, USA (121). Walkways offer long-term research opportunities at lower cost than dirigibles and/or cranes (74). The opportunities for long-term comparative studies among walkways and/or cranes provide enormous potential for future studies of canopy dynamics.

HYPOTHESES ADDRESSED IN CURRENT CANOPY RESEARCH

With the advent of safe, reliable canopy access, researchers are proceeding from descriptive studies to the daunting task of studying canopy interactions and testing rigorous hypotheses (27) (Figure 1). Several case studies are included here that represent pioneering research.

Studies of Sessile Organisms

Studies of sessile organisms pose fewer logistic problems than do other aspects of canopy biology. The greatest challenges include: 1) quantifying the distribution and abundance of populations, especially organisms that are cryptic or difficult to identify; and 2) access to the upper canopy region where the growing shoots and reproductive parts of plants are most often found. Methods such as canopy fogging provide access to some groups of sessile arthropods, while other recent access techniques (e.g. canopy crane, raft) facilitate comprehensive observations in the upper crowns. But sampling designs vary for different types of sessile organisms in the canopy. The development of these protocols and the subsequent descriptions of populations of sessile organisms in the canopy continue to dominate research efforts.

Trees represent the largest group of sessile organisms in forest canopies, and they also comprise the major substrate for most other canopy organisms. In their pioneering work on canopy tree architecture, Hallé et al (40) developed 24 models of tree growth and defined tree canopies as architectural units that are iterated or duplicated to comprise a colony (3, 40, 99). Recent observations made with the canopy raft suggest that these architectural units, which collectively comprise entire crowns, may not be synchronous in their growth and reproductive activities (10, 11). In addition, with the increasing harshness of the microclimate in the uppermost canopy, the morphodiversity of tree crowns diminishes (39). Hallé is currently investigating the hypothesis that architectural units in tree crowns also have associated root primordia, which has important consequences for regeneration after tree fall. Models of tree growth and architecture are important for other aspects of canopy research including canopy-atmosphere interactions, photosynthesis, distributions of sessile organisms, and phenological studies such as the availability of fruits (37, 38, 82, 127).

The growth and form of trees is an important basis of comparison both within and between different forest ecosystems (15, 53, 100, 101). In his classic studies of tree geometry in the 1960s, Horn (45) used simple tools, such as "homemade" light meters, to assess foliage density in temperate forests. Using more sophisticated tools, researchers recently surveyed the surface of a tropical forest in Panama with a canopy crane (101). For the first time, surface irregularities and heterogeneity in the upper surfaces of tree crowns were mapped, and most of the variation was correlated to tree species diversity. Surface undulations of the uppermost crowns have important implications for the canopy-atmosphere interface, throughfall patterns and the population dynamics of organisms that

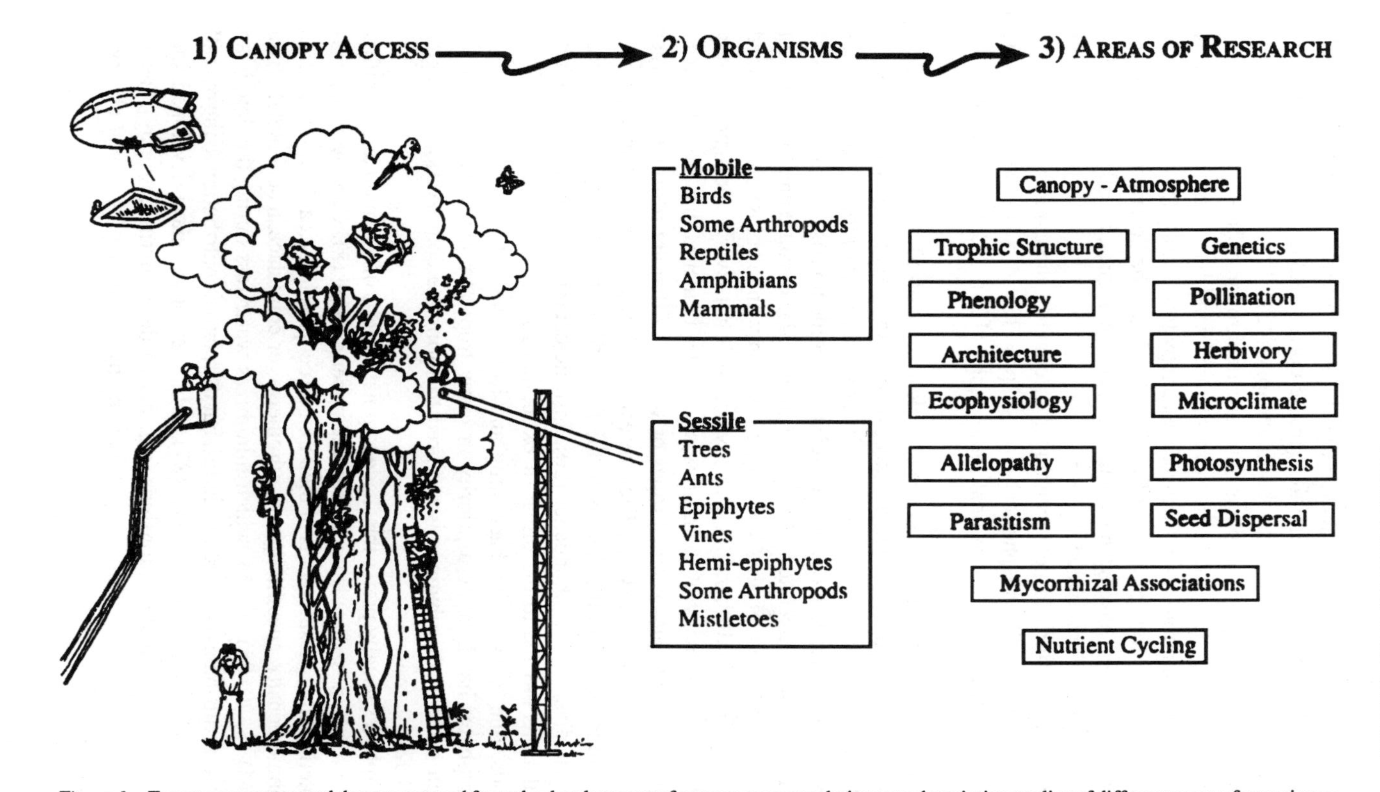

Figure 1 Forest canopy research has progressed from the development of canopy access techniques to descriptive studies of different types of organisms; it is now able to employ a more rigorous, experimental approach to study canopy interactions.

live or roost there. For example, some peaks in the pathway of prevailing winds receive significantly more rainfall than the corresponding "valleys," creating differential rainfall patterns in the understory (43). Physical features of the canopy surface may have impacts that cascade through different trophic levels of the forest ecosystem, such as the case of small wind-blown arthropods whose population fluctuations are influenced by canopy topography (5), which in turn affect the populations of canopy lizards (25). Once the descriptive studies of sessile populations have been measured in forest canopies, broader ecosystem-level studies can be initiated.

Other applications of new techniques to quantify the canopy surface have included the use of hemispherical photographs (13, 14, 111), remote sensing (52, 117, 120), three-dimensional tomography (126), and even fractals (131, 132).

After trees, epiphytes are the most widely recognized sessile organisms in forest canopies. Although canopy access has shown that epiphyte biomass is small relative to total ecosystem pools in most forests, the biomass of epiphytes can actually exceed that of host tree foliage in some forests (e.g. tropical montane forests and temperate wet forests) (20, 91, 92). Because epiphytes are not easy to survey from ground level, their taxonomy and ecology have remained relatively unstudied (8, 9, 78). Canopy access is not always adequate to study epiphytes, since many of them grow on slender branches that are difficult to reach with techniques such as ropes or walkways that are restricted to larger branches for support. In Costa Rica, researchers hired arborists to cut and lower whole branches from the canopy to the forest floor for purposes of surveying epiphyte diversity and abundance (20). Insufficient information about epiphyte populations has led to the threat of extinction for some species, particularly orchids, due to excessive harvesting and loss of habitat (29). Conservation of epiphytes has become a priority in many tropical regions, but further research on their distributions is required to formulate conservation strategies (76, 102).

Epiphytes usually prefer the shaded, moist regions of the mid-canopy, rather than harsh exposed environments in the upper canopy (8). Canopy access has enabled in situ measurements of their unique ecophysiological features, such as trichome density, specific light requirements, or crassulacean acid metabolism (CAM) (8, 20, 83). CAM, which minimizes water loss during the day, is an important attribute for some epiphytes, but its relative contribution to the overall carbon budget remains unknown (87). Many epiphytes have developed specialized plant-animal relationships (e.g. ant gardens, myrmecotrophy). These interactions cannot be studied on fallen epiphytes, but require in situ observations (8, 9).

Ecological studies of epiphytes were pioneered in temperate forests where their role in altering precipitation chemistry was quantified, using a modeling approach (64, 109). With the development of safe canopy access, nutrient cycling in epiphytes of tropical forests ecosystems has also been measured, including the circulation of nutrients and organic matter from the canopy to the forest floor (19, 95). In addition to serving as sinks for nutrients absorbed from precipitation, epiphytes decompose abscised plant litter that is intercepted within the canopy. Dead leaves decayed almost twice as rapidly on a canopy branch surface as on the forest floor in tropical cloud forests (94), representing direct nutrient cycling within the canopy community rather than on the forest floor. Some epiphytes convert atmospheric N to mineral N with above-ground adventitious roots (91). Epiphylly may also be a site of transfer of nitrogen into the host leaf, but the importance of this pathway has yet to be quantified (7). As a result of canopy measurements in situ, epiphytes are now considered keystone organisms that facilitate nutrient cycling within forests at both small (within the epiphyte) and large (within the forest) spatial scales (91). Epiphytes have also been used as potential indicators of air pollution and acid rain, because scientists can now access the canopy to monitor anthropogenic effects (8, 42).

To describe vines as sessile may be somewhat inappropriate due to their mobile nature in traversing the above-ground reaches of tropical forests. Studies of vines represent a major challenge, since they are so abundant in the canopy, yet difficult to measure or count. Most studies of vines have been based on ground observations to date, and this has greatly limited our understanding (105, 106). Vines not only travel throughout the canopy, but their foliage often overtops tree crowns, and their flowers and fruits can dominate food chains that include birds and mammals (30, 61, 62). Stems of *Calamus australis* extended to 33 m in length in Queensland, Australia (106); and one individual of *Entada monostachya* connected 64 trees on Barro Colorado Island, Panama (105).

Studies of Mobile Organisms

Most studies of mobile organisms have been descriptive, including numerous snapshots of diversity and abundance of arthropods (21, 23, 31, 32, 47, 118), mammals (30, 80, 81), and birds (63, 88, 89, 121). All of these baseline studies are important to establish the relative distribution of mobile organisms throughout the canopy, but it remains difficult to survey mobile organisms accurately over time with the methods currently available.

Descriptive studies of lizards in forest canopies were pioneered from canopy towers and walkways (107, 108). Using improved arborist techniques (26), Dial and Roughgarden completed one of the first canopy experiments to test hypotheses about the trophic structure of mobile organisms (24, 25). They manipulated the numbers of lizards in *Dacyrodes excelsa* crowns in Puerto Rico and found subsequent changes in the dynamic balance of the food web. The numbers of insects increased significantly in the absence of their lizard predators, and foliage damage by insect herbivores approximately doubled. Their design was very simple: They removed lizards from seven tree canopies and left seven unmanipulated. Despite this apparent simplicity, the logistics of maneuvering (and in this case, removing lizards) within tree crowns was difficult. Their research showed that experimental study of the ecology of mobile organisms in the forest canopy was not only feasible, but also contributed information on evolutionary ecology in a broader context, by quantifying aspects of forest trophic structure such as 'intra-guild' predation (i.e. spiders competed with lizards for food and in turn were also preyed upon by lizards) (51).

Even less is known about bird populations in the forest canopy. Most studies of canopy birds have involved laborious mist-netting and banding from canopy platforms (e.g. 63, 88, 121). In the eastern Amazon, Lovejoy netted birds to a height of 23 m (63), and Munn netted from the ground to 50 m in the western Amazon and eastern Peru (88, 89). Observations of birds made with binoculars still represent the simplest and most widely used technique for monitoring bird populations, in many cases with minimal disturbance (89, 93). At La Selva Field station in Costa Rica, Loiselle is using binoculars from canopy towers or emergents to address hypotheses linking foraging behavior of small passerines with resources (61, 62). For larger birds such as macaws, innovative access methods such as ultra-light planes (89) or radio-tagging methods (12) prove useful for tracking. As methods improve to study tropical birds discreetly within the canopy, important questions relating to species diversity can be addressed. High species diversity is often attributed to resource availability; and in the case of birds, many of these resources (e.g. flowers, fruits) are concentrated in the canopy.

Invertebrates remain the most controversial aspect of canopy research, as debates continue over the estimates of insect diversity in forests (5, 21, 31–34, 47, 75, 118). Extrapolations of beetle diversity and abundance from tropical trees (with the majority based upon canopy fogging surveys) suggest that insect diversity is extremely high because many invertebrates are specific to certain host trees. This research is still in progress, but the taxonomic work that results from fogging requires many years of laboratory sorting (122, 125). Other invertebrate groups that have been recognized as important residents of the canopy are mites, both herbivorous and predatory species. In Australian rain forests, trees house hundreds of thousands of individual mites inhabiting leaf surfaces, flowers, stems, trunks, epiphytes, hanging humus, and other canopy animals (128, 129). In one of the first in situ controlled experiments of canopy invertebrates, mite abundance was correlated to the presence of domatia on leaf surfaces. The artificial removal of domatia resulted in decreased abundance of predatory mites, lowering the leaf's protection against attack by plant-parasitic mites (128). By grazing on foliar microbes and excreting on the leaf, mites may contribute to nutrient cycling in the canopy, freeing nutrients sequestered by epiphylls, thereby providing rain forest trees with "the equivalent of slow-release fertilization" (129). As the population dynamics of mites and other mobile organisms in canopies become better quantified, their role in complex interactions within the forest will also become better understood.

Canopy Processes and Interactions

One of the biggest contributions emerging from canopy research is to reinforce the notion that forests are linked—through their canopies—to many global cycles, and that these processes can be examined at different scales ranging from leaf to tree to stand and landscape levels. Only with the recent advances in canopy access have scientists been able to examine inter- and intra-crown variation and to quantify forest processes within a global context (76).

Tree canopies exchange carbon dioxide, water, and energy with the atmosphere (35, 44, 104). Interest in the canopy-atmosphere interface has led to the development of sophisticated techniques to measure canopy air movement, particulate material in the atmosphere above the canopy, temperature and air flow and their impact on biological processes (e.g. spore and pollen dispersal, migration of small arthropods), and even emissions of other gases. For example, volatile hydrocarbon fluxes from canopies were recently found to be a major source of reduced photochemically active compounds to the atmosphere (55). Isoprene accounts for 35–55% of the total biogenic flux, and forests are a greater source of emissions than grasslands (56, 57).

Canopy access has increased our understanding of phenology, by enabling direct observations of flowering, fruiting, and leafing activities (62, 68, 82, 88, 93, 127). Leaf longevities in tropical trees range from as short as several months (87, 127) to over 12 years (88). Leaf nitrogen content and photosynthetic rates concomitantly decline with leaf age, while leaf toughness and resistance to herbivory increase with age to a threshold and then decline slightly (16, 66). All of these complex gradients that have been recognized in canopies require periodic sampling over long periods of time, in order to account for temporal and spatial variation (69, 74, 87, 88).

Forests are estimated to account for at least 50% of the global carbon dioxide flux between terrestrial ecosystems and the atmosphere (104). Prior to the development of canopy access, a major problem in estimating the carbon balance of forest ecosystems was the inability to measure photosynthesis in situ in the upper canopy, where the greatest activity was predicted. Factors that affect photosynthesis such as branch architecture (53, 58, 59), light levels (87, 88), and herbivory (69) can now be addressed with real measurements within and between leaves, branches, and entire crowns rather than based on predictions. Photosynthesis, conductance, and water potential were measured and correlated with environmental conditions from the canopy raft at 40 m in lowland tropical rain forests of Cameroon, Africa (48). Leaf water potential was equal to or greater than the gravitational potential at 40 m in the early morning, falling to as low as − 3.0 MPa near midday. Leaf conductance and net photosynthesis commonly declined through midday with occasional recovery late in the day. These patterns were similar to observations in other seasonally dry evergreen forests, suggesting that environmental factors may trigger stomatal closure and limit photosynthesis in tropical rain forest canopies during times of intense sunlight or drought stress. Similar measurements were undertaken from a canopy crane in the seasonally dry tropical moist forests in Panama, where diurnal changes in photosynthesis were hypothesized to be a primary function of incident PPFD (photosynthetic photon flux density) without a midday decline (133).

Herbivory is an important canopy process that directly affects the amount of leaf material available for photosynthesis. Not only is herbivory linked to the forest carbon balance, but herbivory affects other forest dynamics such as tree growth, soil processes, successional status, and nutrient cycling (16, 20, 54, 69). Forest herbivory has typically been measured by collecting samples of leaves growing in the understory and estimating missing leaf area (54, but see 16, 50, 65, 66, 69). With the advent of safe canopy access, estimates of herbivory in forest communities have become more accurate, because they can include the different cohorts of leaves that are stratified vertically throughout the canopy. Whereas original ground-based measurements of herbivory estimated losses of 5% to 7% (50, 54, 69), more comprehensive measurements ranged from 30% annual foliage removal in Australian rain forests, to as high as 300% in dry eucalypt woodlands where beetle outbreaks successively defoliated new flushes of leaves (16, 50, 69, 72). Canopy access has led to improved sampling designs for measurements of herbivory, to account for the vertical stratification of leaves from the understory to the uppermost crown, and spatial scaling from leaf to entire forest (Figure 2).

Studies of reproductive biology and forest genetics have been greatly enhanced by forest canopy access. The surprising importance of a tiny thrip insect in the pollination of some dipterocarps with mast fruiting habits was recognized only from in situ canopy observations (1, 22). The discovery that some strangler figs are formed by the fusion of several individuals initiated in the crown adds a new complexity to forest genetics (124). From canopy platforms,

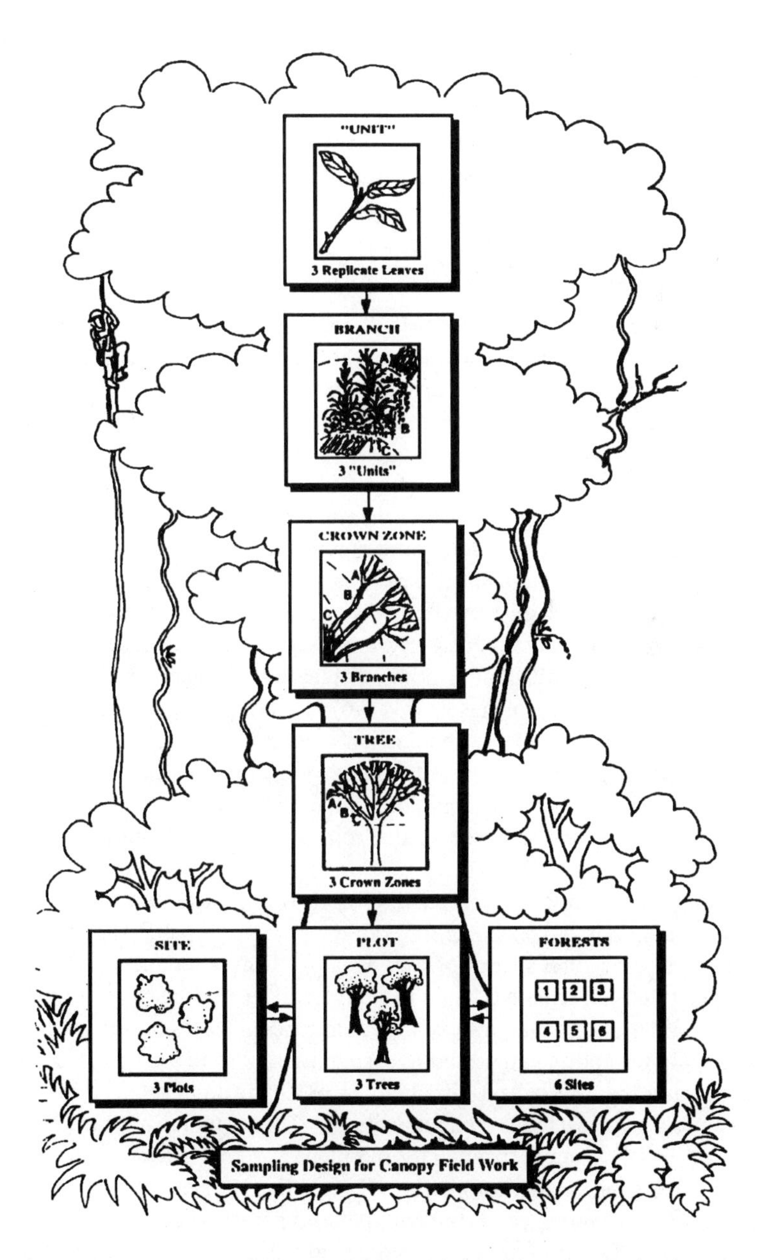

Figure 2 The sampling design for the collection of data in forest canopies has become more comprehensive, as the complexity within tree crowns becomes better quantified. In studies of herbivory, spatial scaling, in order of increasing size, includes leaf to branch to crown zone to individual tree to entire stand and finally, to plots and sites.

biologists can directly observe the reproductive dynamics of trees, especially flowering, fruiting, and pollination activities (82, 90). Since many tropical trees have relatively low population densities, the consequences of rarity on the maintenance of species and genetic diversity remain important questions, especially as many forests are being fragmented or entirely removed (87, 90, 126).

FUTURE DIRECTIONS

Using spectacular cranes, colorful balloons, and OSHA-regulation climbing hardware, canopy scientists have solved the major logistic obstacles to canopy access. Studies of sessile organisms and canopy processes are well under way, although accurate methods to quantify mobile organisms are still being developed. Future directions include the continued measurement of the populations of canopy organisms and the processes that link them, standardization of sampling protocols, more rigorous experimental studies and hypotheses-testing in the field, implementation of long-term studies, and improved technologies to integrate and manage large data bases.

In 1993, an interdisciplinary team was awarded a planning grant from NSF to facilitate the integration of canopy biologists with computer scientists to establish methods for collecting, storing, displaying, analyzing, and interpreting three-dimensional data in tree crowns (96). Future priorities outlined by this group included improved communication among scientists, integration of research projects, examination of potentially applicable information models and software tools already in use, and development of conceptual models and recommendations for the analyses needed to answer questions posed by canopy researchers (76, 97, 122).

Canopies of complex forests such as evergreen tropical rain forests require periodic sampling that accounts for both temporal and spatial variability conducted over long time spans (24, 66, 82, 86). Only after such regimes have been instituted can scientists begin to compare different forests and understand their variation at a global scale. Studies of individual leaves in situ illustrate the small-scale variability in the canopy. Using laser vibrometry, French scientists are studying the role of leaf vibrations in host-parasitoid interactions in tree crowns (J Casas, personal communication). These studies have applications for plant biomechanics, specifically turbulence around a leaf surface, which in turn affects the small arthropods and epiphylly that may inhabit the phylloplane, as well as larger-scale turbulence patterns above the canopy (35, 75, 100). Young leaves of tropical trees seem more vulnerable to insect herbivory than do their tropical counterparts (16, 69). Although leaf-level studies are important to provide information regarding the contribution of specific aspects of forest canopy dynamics, they become more powerful when scaled up to incorporate measurements at the level of whole trees or entire forests. For example, limited data suggest that tropical trees are more vulnerable than temperate trees to cavitation, and that evergreen tropical species are more efficient at supplying water to a given leaf area than are deciduous species (87). Insect herbivores in tropical forests are often nocturnal, while temperate herbivores prefer diurnal feeding patterns (32, 34, 69). Although differential locomotory adapatations confer access to different resources, arboreal mammals have relatively broad diets in tropical forests as compared to temperate forests (81). And as canopy dynamics in natural forest ecosystems become quantified, a new suite of questions arises: How will forest fragmentation affect these patterns?

Ecophysiological studies in tropical forest canopies are challenging our conventional wisdom that respiration and photosynthesis in these ecosystems are in balance (44, 76, 87). Measurements made in situ suggest that old growth tropical forests may be a net carbon sink, which is important in the context of deforestation and rising CO_2 predictions (37a). Future priorities in ecophysiology, all of which require repeated sampling within the canopy with respect to spatial scaling, include: energy capture in leaves, whole-plant patterns of resource allocation and water relations, and refinement of upscaling models to include the functional diversity in tropical canopies (86, 87). As the impacts of human activities continue to alter the environment, responses of tropical tree crowns will undoubtedly remain a research priority (8, 37a, 76, 83, 104). How do tropical leaves process their high energy loads on sunny days? What are the impacts of global warming on tropical ecosystems? How do tropical trees manage their steep resource gradients, such as high solar radiation and heat load in the crown to low light in the understory, or from extremely dry to wet seasons (in cases of seasonally dry forests)? Improved canopy access will also permit experimental manipulation of opportunities for carbon gain (e.g. by defoliation or artificial shading); comparative measurements of plant respiration over time (diurnal, seasonal, annual); and comparisons within and between species and forest types (87).

Future hypotheses-testing will undoubtedly seek to address the relationships between the canopy and the forest floor. Canopy processes such as defoliation may be related to forest floor processes (e.g. decomposition) through inputs of leaf and twig litter, canopy throughfall, and inputs from frass (68, 113). Defoliation by insects in the canopy may also have important consequences for primary productivity and nutrient cycling throughout the forest ecosystem (68, 72, 79). Canopy investigations will also enhance our understanding of other evolutionary relationships in forests: How did the epiphytic habit arise? Do vines create "highways" for herbivores? How do the energetics of strangler figs alter forest community processes? Does canopy topography influence throughfall and, ultimately, seedling recruitment? How will forest fragmentation affect host-specific herbivores? Many questions remain unanswered about mobile organisms in forest canopies, such as home range (12, 30, 77, 81), foraging behavior (24, 62, 123), pollination relationships (1, 22), and their population dynamics (5, 21, 32, 79, 108, 113). The use of satellite radio-tracking, ultralights, and other mobile tracking devices requires refinement to assure the collection of unbiased information (89).

Improved integration with other disciplines of science will greatly expand the scope of forest canopy research. Continued integration of techniques with computer science, with atmospheric chemistry, satellite imagery, and also with medical research on insect vectors will provide a more solid foundation for future research (97, 122). New modeling and mathematical ideas are being applied to canopy data bases. It is possible that fractal dimensions will facilitate the description of ecological and physiological processes in trees, which remain inexplicable for lack of a conceptual framework (131, 132). Again, improved canopy access will facilitate the collection of comprehensive data sets throughout the complex distribution of foliage in a forest. Through programs like National Science Foundation Long Term Ecological Research and permanent field stations maintained by the Smithsonian Institution and Organization for Tropical Studies, integrated canopy research and the shared use of data bases are beginning to produce results (96, 97, 122). As the information collected in forest canopies becomes comprehensive, it also becomes valuable as a tool for the management of biodiversity and sustainable resources throughout world forests (76). One example is the BOREAS project, in which NASA is funding a coordinated effort to understand the canopy of boreal forests (114).

Now that the techniques for safe, periodic canopy access are established, scientists are poised for a decade of comprehensive descriptive and experimental ecology, where the organisms and processes in this above-ground environment can be studied and linked to other components of the forest. Forests are very complex ecosystems. To accurately predict such phenomena as the response of terrestrial ecosystems to rising atmospheric CO_2 concentrations requires examination of the exchange of energy, water and CO_2 between forest canopies and the atmosphere. To understand the impact of forest fragmentation requires knowledge about the life cycles and population dynamics of many organisms, as well as quantifying different components of nutrient cycling and moisture regimes between the canopy and the forest floor. As growing concern for environmental issues accelerates, studies of forest canopies are integral to our understanding of biodiversity, global atmospheric changes, and conservation of forests.

ACKNOWLEDGMENTS

We are grateful to all the respondents of our survey of current canopy research and field sites, conducted on the International Canopy (ICAN) electronic bulletin board. This review benefited from comments and discussion from many canopy researchers, especially Steve Mulkey, Jess Parker, Francis Hallé, Bart Bouricius, Nalini Nadkarni, Jack Schultz, and Guadelupe Williams-Linera. Beryl Black, Ellen Baskerville, and Barbara Harrison assisted with the manuscript. Eddie and James Burgess kindly cooperated in the absence of their Mom while we edited the manuscript.

Literature Cited

1. Appanah S, Chan HT. 1981. Thrips: pollinators of some dipterocarps. *The Malay. For.* 44(2–3):234–52
2. Art HW. ed. 1993. *The Dictionary of Ecology and Environmental Science.* New York: Holt. 632 pp.
3. Barthélémy D, Edelin C, Hallé F. 1991. Canopy architecture. In *Physiology of Trees,* ed. AS Raghavendra, pp. 1–20. New York: Wiley
4. Basset Y. 1988. A composite interception trap for sampling arthropods in tree canopies. *J. Aust. Entomol. Soc.* 27:213–19
5. Basset Y, Aberlenc H-P, Delvare D. 1992. Abundance and stratification of foliage arthropods in a lowland rain forest of Cameroon. *Ecol. Entomol.* 17:310–18
6. Beebe W, Hartley GI, Howes PG. 1917. *Tropical Wildlife in British Guinea.* New York: NY Zool. Soc
7. Bentley, BL Carpenter EJ. 1984. Direct transfer of newly-fixed nitrogen from free-living epiphyllous microorganisms to their host plant. *Oecologia* 63:52–56
8. Benzing D. 1990. *Vascular Epiphytes.* Cambridge, UK: Cambridge Univ. Press
9. Benzing D. 1995. Vascular epiphytes. See Ref. 76, pp. 225–54
10. Blanc P. 1992. Comment poussent les couronnes d'arbres dans la canopée? In *Biologie d'une Canopée de Forêt Equatoriale. II. Rapport de Mission: Radeau des Cimes* Octobre/Novembre 1991. Réservo de Campo, Cameroun, ed. F Hallé, O Pascal, pp. 155–72. Paris: Found. ELF
11. Blanc P, Hallé F. 1990. Timidité et multiplication végétative d'un arbre guyanais: Taralea oppositifolia Aublet. (Légumineuse-Palpilionaceae). In *Biologie d'une Canopée de Forêt Equatoriale.* Rapport de Mission: Radeau *des Cimes* Octobre-Novembre 1989. Guyane Francaise. ed. F. Hallé, P. Blanc pp. 125–35. Paris: Xylochimie
12. Bloom PH, McCrary MD, Gibbons MJ. 1993. Red-shouldered hawk home-range and habitat use in southern California (*Buteo lineatus elegans*). *J. Wildl. Manage.* 57:258–65
13. Chazdon RL, Field CB. 1987. Photographic estimation of photosynthetically active radiation: evaluation of a computerized technique. *Oecologia* 73:525–33
14. Chen JM, Black TA, Adams RS. 1991. Evaluation of hemispherical photography for determining plant area index and geometry of a forest stand. *Agric. For. Meteorol.* 56:129–43
15. Chen JM, Franklin JF, Spies TA. 1995. Growing season microclimatic gradients extending into old-growth Douglas-fir forests from clearcut edges. *Ecol. Appl.* 5(1):74–86
16. Coley PD. 1983. Herbivory and defensive characteristics of tree species in a lowland tropical forest. *Ecol. Monogr.* 53:209–33
17. Connell JH. 1978. Diversity in tropical

rain forests and coral reefs. *Science* 199:1302–10
18. Corner EJH. 1992. *Botanical Monkeys.* Edinburgh: Pentland
19. Coxson DS, McIntyre DD, Vogel H. 1992. Pulse release of sugars and polyols from canopy bryophytes in tropical montane rain forest (Guadeloupe, French West Indies). *Biotropica* 24:121–33
20. Coxson DS, Nadkarni NM. 1995. Ecological roles of epiphytes in nutrient cycles of forest ecosystems. See Ref. 76, pp. 495–543
21. Coy R, Greenslade P, Rounsevell D. 1993. *A Survey of Invertebrates in Tasmanian Rainforest. Rep. No. 9.* Tasmanian Govt. Printer. 104 pp.
22. Dayanandan S, Attygalla DNC, Abeygunasekera AWWL, Gunatilleke IAUN, Gunatilleke CVS. 1990. Phenology and floral morphology in relation to pollination of some Sri Lankan dipterocarps. In *Reproductive Ecology of Tropical Forest Plants*, ed. KS Bawa, M Hadley, pp. 103–34. Carnfort, England: Parthenon
23. Dennis P, Usher GB, Watt AD. 1995. Lowland woodland structure and pattern and the distribution of arboreal, phytophagous arthropods. *Biodiv. Conserv.* 4:728–44
24. Dial R. 1992. *A food web for a tropical rain forest: the canopy view from Anolis.* PhD thesis. Stanford Univ., Stanford, CA
25. Dial R, Roughgarden J. 1995. Experimental removal of insectivores from rain forest canopy: direct and indirect effects. *Ecology* 76:1821–34
26. Dial R, Tobin SC. 1995. Description of arborist methods for forest canopy access and movement. *Selbyana* 15:24–37
27. Diamond J. 1986. Overview: laboratory experiments, field experiments and natural experiments. In *Community Ecology,* ed. J. Diamond, TJ Case. New York: Harper & Row
28. Donahue P, Wood T. 1995. A safe, flexible and non-injurious technique for climbing tall trees. *Selbyana* 16:196–200
29. Dressler RL. 1993. *Phylogeny and Classification of Orchids.* Portland, Oregon: Dioscorides
30. Emmons L. 1995. Mammals of rain forest canopies. See Ref. 76, pp. 199–223
31. Erwin TL. 1982. Tropical forests: their richness in Coleoptera and other arthropod species. *Coleopt. Bull.* 36:74–75
32. Erwin TL. 1989. Canopy arthropod biodiversity: a chronology of sampling techniques and results. *Rev. Peru. Entomol.* 32:71–77
33. Erwin TL. 1991. How many species are there? Revisited. *Conserv. Biol.* 5:330–33
34. Erwin TL. 1995. Measuring arthropod diversity in the tropical forest canopy. See Ref. 76, pp. 109–27
35. Fitzjarrald DR, Moore KE. 1995. Physical mechanisms of heat and mass exchange between forests and the atmosphere. See Ref. 76, pp. 45–72
36. Foster RB, Lowman MD. 1994. Crown status of tropical trees. *Selbyana* 16:A8
37. Gautier-Hion A, Michaloud G. 1989. Are figs always keystone resources for tropical frugivorous vertebrates? A test in Gabon. *Ecology* 70:1826–33
37a. Grace J, Lloyd J, McIntyre J, Miranda AC, Meir P, et al. 1995. Carbon dioxide uptake by an undisturbed tropical rain forest in southwest Amazonia, 1992 to 1993. *Science* 270:778–80
38. Hallé F. 1986. Modular growth in seed plants. In *The Growth and Form of Modular Organisms,* ed. JL Harper, pp. 77–87. London: Philos. Trans. R. Soc
39. Hallé F. 1995. Canopy architecture in tropical trees: a pictorial approach. See Ref. 76, pp. 27–44
40. Hallé F, Oldemann RAA, Tomlinson PB. 1978. *Tropical Trees and Forests: An Architectural Analysis.* Berlin: Springer-Verlag
41. Hallé F, Pascal O. 1991. *Biologie D'Une Canopée de Forêt équatoriale II. Rapport de Mission: Radeau Des Cimes.* Réserve de Campo, Cameroun. Avec Le Parrainage Du Ministére de la Recherche et de la Technologie
42. Henderson A. 1993. Literature on air pollution and lichens. XXXVII. *Lichenologist* 25:191–X202
43. Herwitz SR, Slye RE, Turton SM. 1994. Lateral shading and the differential interception of angular direct-beam radiation among neighboring tropical rain forest canopy trees. *Selbyana* 15:A10
44. Holbrooke NM, Lund CP. 1995. Photosynthesis in forest canopies. See Ref. 76, pp. 411–30
45. Horn H. 1971. *The Adaptive Geometry of Trees.* Princeton, NJ: Princeton Univ. Press. 144 pp.
46. Ingram SW, Lowman MD. 1995. The collection and preservation of plant material from the tropical forest canopy. See Ref. 76, pp. 587–603
47. Kitching RL, Bergelsohn JM, Lowman MD, MacIntyre S, Carruthers G. 1993. The biodiversity of arthropods from Australian rainforest canopies: general in-

troduction, methods, sites and ordinal results. *Austr. J. Ecol.* 18:181–91
48. Koch GW, Amthor JS, Goulden ML. 1994. Diurnal patterns of leaf photosynthesis, conductance and water potential at the top of a lowland rain forest canopy in Cameroon: measurements from the Radeau des Cimes. *Tree Physiology* 14:347–60
49. Laman T. 1995. Safety recommendations for climbing rain forest trees with 'single rope technique'. *Biotropica* 27:406–10
50. Landsburg J, Ohmart CP. 1989. Levels of defoliation in forests: patterns and concepts. *Trends Ecol. Evol. TREE* 4:96–100
51. Lawton JH. 1995. Tabonuco trees, trophic manipulations and Tarzan: lizards in a tropical forest canopy. *TREE* 10:392–93
52. Leckie DG. 1990. Advances in remote sensing technologies for forest surveys and management. *Can. J. For. Res.* 20:464–83
53. Leigh EG Jr. 1990. Tree shape and leaf arrangement: a quantitative comparison of montane forests with emphasis on Malaysia and South India. In *Conservation in Developing Countries: Problems and Prospects,* ed. DC Daniel, JS Serrao, pp. 119–74. Bombay: Oxford Univ. Press
54. Leigh E, Smythe N. 1978. Leaf production, leaf consumption, and the regulation of folivory on Barro Colorado Island. In *The Ecology of Arboreal Folivores,* ed. GG Montgomery, pp. 33–50. Washington, DC: Smithsonian
55. Lerdau M. 1991. Plant function and biogenic terpene emission. In *Trace Gas Emissions by Plants,* ed. T. Sharkey et al, pp. 121–34. San Diego: Academic
56. Lerdau M, Alexander S, Keller M. 1995. Isoprene emission from tropical trees. *Eos* 76:107
57. Lerdau M, Matson P, Fall R, Monson R. 1996. Ecological controls over monoterpene emissions from douglas fir (*Pseudotsuga menziesii*). *Ecology* In press
58. Leverenz JW. 1995. Shade-shoot structure, photosynthetic performance in the field, and photosynthetic capacity of evergreen conifers. *Tree Physiol.* 16: In press
59. Leverenz JW. 1995. Shade shoot structure of conifers and the photosynthetic response to light at two CO_2 pressures. *Funct. Ecol.* 9:413–21
60. Lilly S. 1994. *The Climbers Guide.* Int. Soc. Arboriculture
61. Loiselle BA. 1987. Migrant abundance in a Costa Rican lowland forest canopy. *J. Trop. Ecol.* 3:163–8
62. Loiselle BA. 1988. Bird abundance and seasonality in a Costa Rican lowland forest canopy. *Condor* 90:761–72
63. Lovejoy TE. 1974. Bird diversity and abundance in Amazon forest communities. *Living Bird* 13:127–91
64. Lovett GM, Lindberg SE, Richter DD, Johnson DW. 1985. The effects of acidic deposition on cation leaching from three deciduous forest canopies. *Can. J. For. Res.* 15:1055–60
65. Lowman MD. 1984. An assessment of techniques for measuring herbivory: Is rainforest defoliation more intense than we thought? *Biotropica* 16:264–68
66. Lowman MD. 1985. Temporal and spatial variability in insect grazing of the canopies of five Australian rain forest tree species. *Austr. J. Ecol.* 10:7–24
67. Lowman MD. 1992. Canopy research: an invitation. *Tropinet* 3(2):1
68. Lowman MD. 1992. Leaf growth dynamics and herbivory in five species of Australian rain forest canopy trees. *J. Ecol.* 80:433–47
69. Lowman MD. 1995. Herbivory as a process in rain forest trees. See Ref. 76, pp. 431–55
70. Lowman MD, Bouricius B. 1995. The construction of platforms and bridges for forest canopy access. *Selbyana* 16:179–84
71. Lowman MD, Hallé F, Bouricius B, Coley P, Nadkarni N, et al. 1995. What's up? Perspectives from the First International Forest Canopy Conference at Sarasota, Florida 1994. *Selbyana* 16(1):1–11
72. Lowman MD, Heatwole H. 1992. Spatial and temporal variability in defoliation of Australian eucalypts. *Ecology* 73:129–42
73. Lowman MD, Kitching RL, Carruthers G. 1996. Arthropod sampling in Australian subtropical rain forests—How accurate are some of the more common techniques? *Selbyana* 17: In press
74. Lowman MD, Moffett M. 1993. The ecology of tropical rain forest canopies. *Trends Ecol. Evol.* 8:103–08
75. Lowman MD, Moffet MW, Rinker HB. 1993. A new technique for taxonomic and ecological sampling in rain forest canopies. *Selbyana* 14:75–79
76. Lowman MD, Nadkarni NM, ed. 1995. *Forest Canopies.* San Diego: Academic. 624 pp.
77. Lowman MD, Taylor P, Block N. 1993. Vertical stratification of small mammals and insects in the canopy of a temperate deciduous forest: a reversal of tropical forest distribution? *Selbyana* 14:25
78. Luther HL, Sieff E. 1994. De Rebus

Bromeliacearum I. *Selbyana* 15:9–93
79. Majer JD, Recher HF. 1988. Invertebrate communities in Western Australian eucalypts: a comparison of branch clipping and chemical knockdown procedures. *Aust. J. Ecol.* 13:269–78
80. Malcolm JR. 1991. Comparative abundances of Neotropical small mammals by trap height. *J. Mammal.* 72:188–92
81. Malcolm JR. 1995. Forest structure and the abundance and diversity of neotropical small mammals. See Ref. 76, pp. 179–223
82. McClure H.E. 1963. Flowering, fruiting and animals in the canopy of a tropical rain forest. *Malays. For.* 29:182–203
83. Medina E, ed. 1984. *Physiological Ecology of CAM Plants*. Caracas: IVIC—UNESCO, CIET
83a. Moffett MW. 1993. *The High Frontier.* Cambridge, MA: Harvard Univ. Press. 192 pp.
84. Moffett MW, Lowman MD. 1995. Canopy access techniques. See Ref. 76, pp. 3–26
85. Morawetz W. 1995. *Towards an Understanding of the Structure and Function of a Neotropical Rain Forest Ecosystem with Special Reference to its Canopy.* For: Fonds zur Forderung der wissenschaftluichen Forschung
86. Mulkey SS, Kitajima K., Wright SJ. 1995. Photosynthetic capacity and leaf longevity in the canopy of a dry tropical forest. *Selbyana* 16:169–73
87. Mulkey SS, Kitajima K, Wright SJ. 1996. Plant physiological ecology of tropical forest canopies. *TREE* In press
88. Munn CA. 1991. Tropical canopy netting and shooting lines over tall trees. *J. Field Ornithol.* 62:454–63
89. Munn CA, Loiselle BA. 1995. Canopy access techniques and their importance for the study of tropical forest canopy birds. See Ref. 76, pp. 165–77
90. Murawski DA. 1995. Reproductive biology and genetics of tropical trees from a canopy perspective. See Ref. 76, pp. 457–93
91. Nadkarni NM. 1981. Canopy roots: convergent evolution in rain forest nutrient cycles. *Science* 214:1023–24
92. Nadkarni NM. 1984. Epiphyte biomass and nutrient capital of a Neotropical elfin forest. *Biotropica* 16:249–56
93. Nadkarni NM, Matelson TJ. 1989. Bird use of epiphyte resources in Neotropical trees. *Condor* 69:891–907
94. Nadkarni NM, Matelson TJ. 1991. Litter dynamics within the canopy of a neotropical cloud forest, Monteverde, Costa Rica. *Ecology* 72:849–60
95. Nadkarni NM, Matelson TJ. 1992. Biom ass and nutrient dynamics of epiphytic litterfall in a Neotropical montane forest, Costa Rica. *Biotropica* 24:24–30
96. Nadkarni N, Parker GG. 1994. A profile of forest canopy science and scientists—Who we are, what we want to know and obstacles we face: results of an international survey. *Selbyana* 5:38–50
97. Nadkarni N, Parker, GG, Clement J, McConnell M. 1994. Initiation of interdisciplinary and intercontinental links to enhance forest canopy research, education, and conservation. *Selbyana* 15:A17
98. Oates R. 1994. Climbing to the canopy, a mixture of arborist, climber and caver methods. *Selbyana* 15:A17
99. Oldeman RAA. 1974. *L'architecture de la Forêt Guyanaise, Mém. ORSTOM, N. 73.* Paris: ORSTOM
100. Parker GG. 1995. Structure and microclimate of forest canopies. See Ref. 76, pp. 73–106
101. Parker GG, Smith AP, Hogan KP. 1992. Access to the upper forest canopy with a large tower crane. *Bioscience* 42:664–70
102. Parker TA III, Holst BK, Emmons LH, Meyer JR. 1993. A biological assessment of the Columbia River Forest Reserve, Toledo District, Belize. *Conserv. Int. Rep. #3.81* pp.
103. Perry D. 1978. A method of access into the crowns of emergent and canopy trees. *Biotropica* 13:283–5
104. Potter CS, Randerson JT, Field CB, Matson PA, Vitousek PM, et al. 1993. Terrestrial ecosystem production: a process model based on global satellite and surface data. *Global Biogeochem. Cycles* 7:811–41
105. 1 Putz FE. 1988. Woody vines and tropical forests. *Fairchild Trop. Gardens Bull.* 43:5–13
106. Putz FE. 1990. Growth habits and trellis requirements of climbing palms (*Calamus* spp.) in north-eastern Queensland. *Aust. J. Bot.* 38:603–8
107. Reagan DP. 1991. The response of *Anolis* lizards to hurricane-induced habitat changes in Puerto Rican rain forest. *Biotropica* 23:468–74
108. Reagan DP. 1995. Lizard ecology in an island rain forest canopy. See Ref. 76, pp. 149–64
109. Reiners WA, Olson RK. 1984. Effects of canopy components on throughfall chemistry: an experimental analysis. *Oecologia* 63:320–30

110. Reynolds BC, Crossley, DA Jr. 1995. Use of a canopy walkway for collecting arthropods and assessing leaf area removed. *Selbyana* 16:21–3
111. Rich PM. 1990. Characterizing plant canopies with hemispherical photographs. *Remote Sens. Environ.* 5:13–29
112. Ross J. 1981. *The Radiation Regime and Architecture of Plant Stands.* The Hague: W. Junk
113. Schowalter TD, Sabin TE, Stafford SG, Sexton JM. 1991. Phytophage effects on primary production, nutrient turnover, and litter decomposition of young Douglas fir in western Oregon. *Forest Ecol. Manag.* 42:229–43
114. Sellers P, Hall F, Margolis H, Kelly B, Baldorchi D, et al. 1995. The boreal ecosystem—atmosphere study (BOREAS): an overview and early results form the 1994 field year. *Bull. Am. Meteorol. Soc.* 76:1549–77
115. Shigo AL. 1991. *Modern Arboriculture.* New Hampshire: Shigo & Trees Assoc.
116. Shigo AL, Felix R. 1980. Cabling and bracing. *J. Arboriculture* 6:5–9
117. Smith NJ, Borstad GA, Hill DA, Kerr RC. 1991. Using high-resolution airborne spectral data to estimate forest leaf area and stand structure. *Can. J. For. Res.* 21:1127–32
118. Southwood TRE. 1961. The number of species of insect associated with various trees. *J. Anim. Ecol.* 30:1–8
119. Southwood TRE, Moran VC, Kennedy CEJ. 1982. The assessment of arboreal insect fauna: comparisons of knockdown sampling and faunal lists. *Ecol. Entomol.* 7:331–40
120. Spanner MA, Pierce LL, Peterson DL, Running SW. 1990. Remote sensing of temperate coniferous forest leaf area index—the influence of canopy closure, understory vegetation and background reflectance. *Int. J. Remote Sens.* 11:95–111
121. Stokes AE, Schultz BB. 1995. Mist netting birds from canopy platforms. *Selbyana* 16:144–6
122. Stork N, Best V. 1994. European Science Foundation: results of a survey of European canopy research in the tropics. *Selbyana* 15:51–62
123. Terborgh J, Robinson SK, Parker TA III, Munn CA, Pierpont N. 1990. Structure and organization of an Amazonian bird community. *Ecol. Monogr.* 60:213–38
124. Thomson JD, Herre EA, Hamrick JL, Stone JL. 1991. Genetic mosaics in strangler fig trees: implications for tropical conservation. *Science* 25:1214–16
125. Tobin JE. 1995. Ecology and diversity of tropical forest canopy ants. See Ref. 76, pp. 129–147
126. Vanderbilt VC. 1985. Measuring plant canopy structure. *Remote Sens. Environ.* 18:281–94
127. Van Schaik CP, Terbough JW, Wright SJ. 1993. The phenology of tropical forests: adaptive significance and consequences for primary consumers. *Annu. Rev. Ecol. Syst.* 24:353–78
128. Walter DE, O'Dowd DJ. 1992. Leaves with domatia have more mites. *Ecology* 73:1514–18
129. Walter DE, O'Dowd DJ. 1995. Mites on the phyllopane. See Ref. 76, pp. 325–51
130. Winchester NN. 1996. Centilan extinctions: extirpation of northern temperate old-growth forest arthropod communities. *Selbyana* 17: In press
131. Zeide B. 1993. Primary unit of the tree crown. *Ecology* 74:1598–1602
132. Zeide B, Pfeifer P. 1991. A method for estimating the fractal dimension of tree crowns. *For. Sci.* 37:1253–65
133. Zotz G, Winter K. 1993. Short-term photosynthesis measurements predict leaf carbon balance in tropical rain forest canopy plants. *Planta* 191:40

Annu. Rev. Ecol. Syst. 1996. 27:83–109

EXTINCTION BY HYBRIDIZATION AND INTROGRESSION

Judith M. Rhymer
Department of Wildlife Ecology, University of Maine, Orono, Maine 04469

Daniel Simberloff
Department of Biological Science, Florida State University, Tallahassee, Florida 32306

KEY WORDS: genetic mixing, gene flaw, introduced species, habitat modification, outbreeding depression

ABSTRACT

Nonindigenous species can bring about a form of extinction of native flora and fauna by hybridization and introgression either through purposeful introduction by humans or through habitat modification, bringing previously isolated species into contact. These phenomena can be especially problematic for rare species coming into contact with more abundant ones. Increased use of molecular techniques focuses attention on the extent of this underappreciated problem that is not always apparent from morphological observations alone. Some degree of gene flow is a normal, evolutionarily constructive process, and all constellations of genes and genotypes cannot be preserved. However, hybridization with or without introgression may, nevertheless, threaten a rare species' existence.

INTRODUCTION

Most attention to species extinction has focused on the "evil quartet" (45): overkill, habitat destruction, impact of introduced species, and chains of extinction. Introduced species, in turn, are seen as competing with or preying on native species or destroying their habitat (e.g. 19).

Introduced species (or subspecies), however, can generate another kind of extinction, a genetic extinction by hybridization and introgression with native flora and fauna. Habitat modification can also break down reproductive isolation between native species, with subsequent mixing of gene pools and potential

0066-4162/96/1120-0083$08.00

loss of genotypically distinct populations. These phenomena can be especially problematic for rare species contacting more abundant ones. Conservation geneticists have largely focused not on the loss of distinct gene pools by mixing but on the potential decreasing fitness of individuals in small populations who suffer inbreeding depression and on the possible difficulty of evolution in small populations that have lost alleles to genetic drift (129).

We define "hybridization" as interbreeding of individuals from what are believed to be genetically distinct populations, regardless of the taxonomic status of such populations (cf 127). "Hybridization" most commonly refers to mating by heterospecific individuals but has been applied to mating by individuals of different subspecies and even of populations that, though not taxonomically distinguished, differ genetically. Arnold et al (18) suggest restricting "hybrid" to matings between species and using "intergrade" for matings between subspecies and "cross" or "interbreed" for matings between individuals of geographically distinct populations. Although such distinctions might clarify future discussions, all these terms seem so widely used in the literature for matings at every taxonomic level that they are unlikely to be restricted. Instead one must depend on accurate taxonomic description of the entities between which mating occurs.

Introgression is gene flow between populations whose individuals hybridize, achieved when hybrids backcross to one or both parental populations. Beyond F1 hybrids, the point at which an individual is no longer viewed as a hybrid but rather as a member of one of the parental populations that has undergone introgression is arbitrary (9, 141). A hybrid swarm is a population of individuals in which introgression has occurred to various degrees by varying numbers of generations of backcrossing to one or both parental taxa, in addition to mating among the hybrid individuals themselves. Hybridization need not be accompanied by introgression; for example, offspring of hybrid matings might all be sterile. Introgression can be unidirectional, with backcrossing to one parental population only. But hybridization can pose a threat to small populations even if gene pools do not mix.

Botanists have paid more attention than zoologists to the evolutionary consequences of hybridization and introgression, apparently because these are much more common phenomena in plants than in animals, at least at the interspecific level (cf 72). Further, botanists have more frequently addressed the creative role of hybridization, although they have also pointed to such outcomes of potential conservation consequence as the role of introgressive hybridization in the evolution of aggressive weeds (e.g. 71). Zoologists, on the other hand, have generally been more exercised about the loss of distinct gene pools than about hybridization as a creative force. We show in our examples that the same sorts of conservation problems arise in both plants and animals.

It is often difficult to identify hybrids on morphological grounds alone, particularly after several generations of backcrossing. Morphological observations may suggest that hybridization or introgression has occurred, but introgression is not always reflected morphologically. Often individuals not previously identified as products of introgression can be shown by molecular techniques to be introgressed (115). With the advent of molecular genetic analyses, one can now often document not only the extent of hybridization and introgression between populations but also the gender of hybridizing individuals from each population. In this review, we primarily discuss studies for which some molecular analyses have been done.

With few exceptions, mitochondrial DNA (mtDNA) is maternally inherited (20). Thus a first approach to a study of hybridization in animals is to identify population-specific mtDNA haplotypes of the parental populations and to screen suspected hybrid individuals for those haplotypes. In studies of plant hybridization, chloroplast DNA (cpDNA) is more often used; it is maternally transmitted in most plants (20). These data reveal the direction of hybridization: Does it consist only of males of one population mating with females of the other, or does the reciprocal cross also occur? However, mtDNA or cpDNA alone allows an incomplete picture of the extent of hybridization and introgression (39). In order to identify unequivocally all hybrid individuals, one must also analyze biparentally inherited nuclear markers such as those revealed by studies of allozymes, microsatellite DNA, random amplified polymorphic DNA (RAPDs), and single-copy nuclear DNA. First generation (F1) hybrids will be heterozygous at all loci with population-specific alleles. Backcross individuals from crosses between hybrids and the parental populations will have various combinations of genotypes depending on the cross and the results of Mendelian segregation.

MODIFICATION OF GENE POOLS

Mixing Through Introductions

Mixing of gene pools of formerly distinct taxa by introgression has been called "genetic assimilation" (38, 52, 116), "contamination" (4), "infection" (48), "genetic deterioration" (B Heredia, in 144), "genetic swamping" (74, 80), "genetic pollution" (35), "genetic takeover" (68), and "genetic aggression" (71). The latter seven terms have pejorative connotations. They imply either that hybrids are less fit than the parentals, which need not be the case, or that there is an inherent value in "pure" gene pools. "Genetic assimilation" need have no such connotation, but this term is widely applied to an entirely different phenomenon (143). "Mixing" need not be value-laden, and we use it here to denote mixing of gene pools whether or not associated with a decline in fitness.

A well-known example of genetic mixing through introductions involves mallard ducks (*Anas platyrhynchos*), which have a holarctic breeding range. They have hybridized with several closely related, more narrowly distributed endemic species, with subsequent introgression. Hybridization has been facilitated by introductions of mallards into various localities (e.g. New Zealand, Hawaii, Australia, south Florida, eastern USA). For instance, hybridization with introduced mallards has been implicated in population declines of the New Zealand grey duck *A. superciliosa superciliosa* (115), and mtDNA evidence suggests that loss of this indigenous species through introgressive hybridization is a real possibility (115). Many populations are increasingly mallard-like in appearance (153). Hybridization with introduced mallards has contributed to the decline of the endangered, endemic Hawaiian duck (*A. wyvilliana*) and has hampered attempts to reintroduce this species to Oahu and Hawaii (63). Domesticated nonmigratory mallards that escaped or were released for hunting breed with the endemic Florida mottled duck (*A. fulvigula fulvigula*), and the resultant introgression threatens the existence of the latter subspecies (98). Introgression also occurs between domesticated introduced mallards and the native Australian (Pacific) black duck, *A. superciliosa rogersi* (92).

The North American ruddy duck, *Oxyura jamaicensis,* similarly threatens Europe's rarest duck, the white-headed duck (*O. leucocephala*) (14, 144). Formerly widespread in the Mediterranean region, the white-headed duck in Europe was reduced by habitat destruction and hunting to 22 individuals in southern Spain. Rigorous protection and captive breeding have resulted in growth of this Spanish population to nearly 800 birds. The ruddy duck was first bred in captivity in England in 1949 but quickly escaped; it now numbers some 3500 individuals (109, 144). They invaded the adjacent Continent and recently reached Spain, where at least ten fertile hybrids have hatched (144); the extent of backcrossing has not yet been determined. It is not inconceivable that the ruddy duck could expand its range to contact the only other white-headed duck population, in Kazakhstan.

Among birds, such problems are hardly restricted to ducks. The Seychelles turtle dove (*Streptopelia picturata rostrata*) has so massively hybridized with *S. p. picturata,* introduced from Madagascar, that a hybrid swarm now inhabits the Seychelles, and the average phenotype resembles that of the invader (38). In South Africa, the endemic red-eyed dove, *S. semitorguata australis,* whose populations had waned because of deforestation, has hybridized with introduced *S. s. semitorguata* from Mozambique with introgression to the extent that there is now a hybrid swarm (32). Even the most publicized threatened species of the United States, the northern spotted owl (*Strix occidentalis caurina*), may be at risk from this process. Although destruction of its old growth habitat has been the main cause of its decline (137), the recent invasion of the Pacific Northwest by the barred owl (*S. varia*) is ominous. It is gradually expanding its range and numbers (13), and nest observations already show that it hybridizes and produces fertile offspring with the spotted owl and that there has been productive backcrossing in one direction (67).

There are several well-documented mammalian examples as well. Feral housecats (*Felis catus*) threaten the existence of the wildcat (*F. silvestris*) through hybridization (77). Remote areas of northern and western Scotland were believed to harbor among the purest wildcats in Europe, but 80% of individuals studied by a number of genetic analyses had domestic cat traits. An identical problem is occurring even in remote areas of southern Africa as feral housecats breed with the African wildcat, *F. libyca* (133).

The attempt to reintroduce the red wolf (*Canis rufus*) into the eastern United States (146) may already be doomed. mtDNA analysis of individuals from much of its current and historical range indicate that all of them have either gray wolf (*C. lupus*) or coyote (*C. latrans*) haplotypes, suggesting the red wolf may be a species of hybrid origin (148). Red wolves have also hybridized with coyotes as the latter increased in density and range in the East in this century. The continued increase in Eastern coyote populations bodes ill for the red wolf, as the coyote will ultimately be far more numerous wherever captive-reared red wolves can be released. Hybridization between coyotes and red wolves or gray wolves has probably also affected coyote genotypes in the East (89). Whether the red wolf has a hybrid origin remains controversial (47, 106, 111, 147), but recent microsatellite DNA analysis (122) supports this hypothesis. Either way, it is difficult to be optimistic that genes of individuals released now will not mix extensively with those of coyotes. Such a hybridization between an introduced red wolf and a coyote was recently reported (R Wayne, personal communication).

In Europe, what had been considered pure wolf (*C. lupus*) populations have turned out to be largely hybrids between wolves and domestic and feral dogs (*Canis familiaris*), leading to the formation of a "Wolf Federation" to protect wolves from "genetic pollution" from dogs (35). By contrast, the identical hybridization in the United States has led to recent calls to protect humans from the hybrids (76) and also to delist the wolf because it no longer satisfies the criteria of the Endangered Species Act (ESA) (75).

In Britain, the European polecat (*Mustela putorius*) had declined nearly to extinction because of predator control to enhance gamebird shooting (30). As this pressure is easing, the polecat is recovering, but there is widespread concern (30, 94) for its genetic integrity because of hybridization with domesticated, escaped, and feral ferrets (*M. furo*). The ferret is itself a domesticated form of *M. putorius, M. eversmanni,* or both.

Simien jackals (*Canis simensis*) in Ethiopia were so greatly reduced by habitat destruction and hunting that they are now restricted to habitat fragments (105) and are outnumbered by domestic dogs by about 10:1 (60). Introgression occurs through matings between male dogs and female jackals (60), as demonstrated by mtDNA and microsatellites. As in the red wolf example, the great disparity in the population sizes of the two parental taxa puts the genome of the rare one at risk far more than that of the common species.

In Scotland, sika deer (*Cervus nippon nippon*) introduced from Japan about 80 years ago hybridize with native red deer (*C. elaphus*), and allozyme, microsatellite DNA, and mtDNA analyses show extensive introgression (1). A broad, non-equilibrium hybrid zone is moving, and the genetic integrity of Scottish red deer is threatened (1).

A unique genetic threat has recently been described among amphibians. The hybridogenetic hybrid frog *Rana esculenta* and its two parental species *R. lessonae* and *R. ridibunda* have been introduced to sites in central Spain (15). There is a substantial threat that *R. esculenta* will outcompete the native *R. perezi,* which has a restricted distribution, and there is also concern that *R. ridibunda* will hybridize with *R. perezi.* This cross has occurred in nature in northern Spain and produced a hybridogenetic taxon whose presence has affected the genetic structure of the local *R. perezi* populations and whose individuals may be expected to be more vigorous than those of the latter species (15).

Genetic mixing occurs among fish species because of the prevalence of introductions for sport or commercial fishing, biological control, or through accidental introductions of bait species. Smallmouth bass (*Micropterus dolomieui*) introduced into the Guadaloupe River system in central Texas have hybridized with endemic Guadaloupe bass (*M. treculi*), threatening the continued survival of the latter species (50); allozyme analysis confirms backcrossing (151).

Extensive, multiple introductions of rainbow trout (*Oncorhynchus mykiss*) and cutthroat trout (*O. clarki*) in western US watersheds have resulted in loss of diversity of native species because of massive introgression (2, 46, 90). Rainbow trout hybridize extensively with threatened Apache trout (*O. apache*) and endangered Gila trout (*O. gilae*) in the southwestern United States (46). Of Apache trout populations, 65% contain individuals with rainbow trout alleles, and one native population is now completely composed of rainbow trout. Similar molecular studies in northern Italy show that stocking of domestic forms of brown trout (*Salmo trutta*) has resulted in introgression of domestic strains with unique native forms (*S. fario, S. marmoratus,* and *S. carpio*) (59). In Poland, hybridization and introgression of the introduced *Coregonus peled* with the native *C. lavaretus* is so extensive that only hybrids occur in about 70% of lakes; pure forms are difficult to find (156).

The arctic char (*Salvelinus alpinus*) in Sweden hybridizes with the introduced American lake trout (*S. namaycush*) (28), a cross that occurs with introgression confirmed by allozyme and mtDNA analysis in Canada, where the two species's native ranges overlap (69, 154). Similarly, arctic char have been shown by allozyme electrophoresis to hybridize with brook trout (*S. fontinalis*) where their native ranges overlap in Labrador, raising concern that char gene pools will be modified in numerous parts of its range where brook trout have been introduced (70).

Casual release of bait fish by sport fishermen, even without a stocking program with tens or hundreds of thousands of fishes, sufficed to generate a hybrid swarm through introgression between introduced sheepshead minnows (*Cyprinodon variegatus*), a bait fish, and the endemic Pecos pupfish (*C. pecosensis*) in less than five years. This occurred over a 430-km stretch of the Pecos River in Texas, about one half of the original geographic range of *pecosensis* (49). Massive introgression was demonstrated by allozyme analysis.

Interspecific gene flow is a major problem in plant conservation because so many plant species, occasionally even heterogeneric ones, can hybridize (52, 62). For example, many endangered sunflower (*Helianthus*) species are threatened by hybridization with the weedy *H. annuus* as the latter has spread with human sowing and disturbance (121). A measure of the problem can be gleaned from the fact that more than 90% of all threatened and endangered plants in California occur sympatrically or parapatrically with at least one congener (12). Consider the Catalina Island mountain mahogany (*Cercocarpus traskiae*), now consisting of 11 adult trees and some 75 seedlings. Allozyme and RAPD analyses show five of the adults and some of the seedlings to be of hybrid origin with the more widely distributed and much more common *C. betuloides* var. *blancheae* (117).

The Role of Habitat Change

Several kinds of habitat change can increase the probability and rate of hybridization. First, local habitat modification can lead to the mixing of previously distinct gene pools, and this mixing can occur between two native populations as well as between a native and an introduced one. Wiegand (152) and later authors have noted that local anthropogenic habitat disturbance promotes introgressive hybridization in plants (references in 118). Anderson (7) refers to "hybridization of the habitat" by disturbance as the means by which interfertile taxa, formerly reproductively isolated by different habitat requirements, achieve close enough spatial proximity for hybridization to occur. For example, in the Ozark Mountains, zigzag spiderwort (*Tradescantia subaspera*) typically grows in rich lime soil at the bases of bluffs, while another native spiderwort (*T. canaliculata*) grows in rocky soil on exposed cliff edges. Hybrids in nature are very rare, but when bottomlands are cleared, leaf cover erodes and the sunlight increases; hybridization then becomes common, and a hybrid swarm can replace both parental species (9, 104). Hybridization and subsequent introgression between *Iris fulva* and *I. hexagona* in Louisiana have been similarly facilitated by human disturbance (16).

Local habitat change has similarly broken down reproductive isolation between the Chatham Islands subspecies of the yellow-crowned parakeet (*Cyanoramphus auriceps forbesi*) and the red-fronted parakeet (*C. novaezelandiae chathamensis*), although the exact habitat features that formerly separated these taxa are not well understood (38). Further, the general change in the habitat has been in favor of the red-fronted parakeet, and the growing disparity in numbers must exacerbate the probability that individual yellow-crowned parakeets will hybridize. A hunting effort has been mounted against hybrids and red-fronted birds (38). The yellow-crowned parakeet is similarly threatened on other islands, a situation exacerbated by release of hybrids reared in captivity (138).

Construction of artificial ponds and the disturbed nature of their surroundings have fostered hybridization between the native tree frogs *Hyla cinerea* and *H. gratiosa* in Alabama (88, 125). The latter species calls while floating or partially submerged near pond banks. In undisturbed areas, the former species calls from an elevated position on shrubbery near pond edges or from emergent vegetation. In disturbed areas, such emergent vegetation is absent, and *H. cinerea* calls from the banks or occasionally from trees overhanging the water. In these positions, most hybridization takes place when *cinerea* males intercept *gratiosa* females. Backcrossing occurs, generating introgression, although hybrids and backcrosses may have somewhat reduced fitness. Hybridization is not observed in undisturbed habitat.

A second form of habitat modification that can lead to hybridization is regional habitat change that allows geographic range expansion of one taxon into the range of another. The blue-winged (*Vermivora pinus*) and golden-winged (*V. chrysoptera*) warblers were allopatric in North America before European settlement but have long been sympatric over wide areas because of the abandonment of old fields and reforestation in the northeastern United States (38, 56). Both birds use shrubby stages of old field succession. Hybridization has been frequent wherever *pinus,* which is expanding its range northward, encounters *chrysoptera,* and the latter species has declined greatly (38, 51, 56, 57). The hybrids are fertile, and backcrossing and introgression have occurred (41, 58). The extent to which the *chrysoptera* decline is due to direct competition with the blue-winged warbler, habitat change favoring the latter species, hybridization, or a combination of these factors is unknown (41).

In North America, range extensions of mallards with habitat changes wrought by agriculture have facilitated high levels of hybridization and introgression with Mexican ducks (*A. platyrhynchos diazi*) in the Southwest (78). The Mexican duck in the United States was delisted as an endangered species, and the American Ornithologists' Union declared it to be conspecific with the mallard in 1983, because introgression was so extensive throughout northern Mexico and the southwestern United States (5). Similarly, habitat change associated with agriculture and urbanization has greatly increased the possibility of introgression between mallards and American black ducks (*A. rubripes*) in the Northeast (11, 74, 81). This potential has also increased because of massive releases of game-farm mallards for hunting on the Eastern Shore of Maryland (over one million birds in recent years—J Serie, personal communication).

Several bird subspecies on the Great Plains may disappear as distinct entities because of regional habitat modification (87, 124). The allopatry that resulted from forests spreading westward through grasslands and retracting during warmer interglacials left specifically and subspecifically distinct populations on either side of the prairie. Fire control and planted trees on the Great Plains have formed stepping stones and movement corridors exploited by at least six eastern taxa with subsequent hybridization: eastern and western races of the rufous-sided towhee (*Pipilo erythrophthalmus*), yellow shafted (*Colaptes auratus auratus*) and red-shafted (*C.a. cafer*) flickers, Baltimore (*Icterus galbula galbula*) and Bullock's (*I.g. bullocki*) orioles (cf 75), blue (*C. cristata*) and Steller's (*C. stelleri*) jays, indigo (*Passerina cyanea*) and lazuli (*P. amoena*) buntings, and rose-breasted (*Pheucticus ludovicianus*) and black-headed (*P. melanocephalus palpago*) grosbeaks. For the flickers and orioles, at least, morphological analysis suggests introgression (6, 102). The spread of the barred owl across North America, noted above, might well have been facilitated by the same tree plantings on the Great Plains plus opening and fragmentation of old growth forest in the Pacific Northwest (R Gutierrez, personal communication).

A third kind of habitat change that can engender hybridization is simply the construction of a permanent corridor allowing the continual movement of a taxon into the range of another. For example, several fish species native to the Sumjin River in Korea were introduced to the Dongjin River when two power plants were constructed on the Dongjin, diverting water from the Sumjin (86). One of these species, the loach *Cobitis tainia striata*, hybridized with the endemic subspecies *C.t. lutheri*. High levels of hybridization and introgression have occurred over a 20-km length of the Dongjin River, an area comprising half the historic range of the endemic subspecies.

Outbreeding Depression and the Loss of Locally Adapted Genotypes

Our examples to this point have not discussed whether the new, mixed genotypes are inferior in any way to the original native ones. Sometimes they are not. For example, experimental evidence suggests that hybrid and backcross individuals resulting from hybridization between mallards and North American black ducks and between mallards and New Zealand grey ducks are as fertile and viable as pure parental individuals (65, 110). Also, the formation of a hybrid swarm (e.g. the Pecos pupfish example) suggests that hybrids are not at a substantial disadvantage, particularly when pure parentals disappear and are replaced by hybrids.

Some authors (e.g. 92) would view genetic mixing as a tragedy even if the new genotype were better adapted than the original one to the ambient environment. However, there is every reason to think that hybridization sometimes leads, at least initially, to a population less well-adapted to the local environment. Outbreeding depression is lowered fitness in offspring, or later generations, of crosses between genetically different sources (135). Occasionally the decline in adaptation to the local environment is dramatic, as when the Tatra mountain ibex (*Capra ibex ibex*) population in Czechoslovakia was eliminated through interbreeding with introduced ibexes of different subspecies (135, 136).

Anadromous salmonid fishes have numerous genetically determined adaptations to local environments, including orientation behavior of newly emerged fry and timing of spawning (3). Because of the migratory life cycles of these fishes, these adaptations are crucial to survival and reproduction. Widespread transfers of artificially propagated salmon within the native ranges of these species may cause outbreeding depression through both loss of local adaptation and breakdown of coadapted gene complexes (3).

In another example, widespread introductions of two clam species, *Mercenaria mercenaria* and *M. campechiensis* into one another's ranges, of cultured hybrids into pure single-species populations, and of both species together into previously uncolonized locations have provided many opportunities for hybridization (29). Hybridization between these two species in a Florida coastal lagoon has contributed to a chronic, epizootic incidence of gonadal neoplasia there. Hybrids are more susceptible to the disease, resulting in reduced hybrid fitness through increased mortality and reduced reproduction.

Outbreeding depression is known in several other animals (references in 135) and many plants (references in 52), and observed decreases in fitness are often substantial. However, because the demographic data required to establish a decline in fitness are so scarce, it is likely that other examples, such as some of those cited in the section Mixing through Introductions as instances of simple mixing, also entail a loss of fitness. However, as we discuss in the section Introgressive Hybridization as an Evolutionary Constructive Process, hybrid vigor or heterosis, with exactly the opposite outcome as outbreeding depression, has been far more frequently documented.

Implications of Mixing for Translocation, Reintroduction, and Stock Enhancement Programs

The fact that genetic mixing can occur suggests that certain conservation programs, though promising to solve one problem, may generate others. As is clear from the ibex example, a well-meaning reintroduction project, aiming to reestablish a species now locally extinct or to prevent inbreeding depression in a small isolated population, can produce catastrophic results even if the hybrids are fertile. The salmon example suggests that stock enhancement programs can be similarly disastrous. The threat posed to marine turtles by recently detected hybridization is not known, but it is possible that well-meaning captive-rearing and transplantation activities are the cause (85). In the section Conservation Implications of Hybrid Sterility and Unidirectional Introgression, we give an example of ungulate hybridization, but one in which the hybrids are sterile and nevertheless the hybridization threatens the parental populations. Hybrid sterility in that instance is caused by segregational difficulties during meiosis, in part owing to differences between the parental species in diploid number of chromosomes. Geographic chromosomal variation may not even correlate with variation in phenotypes within some species of mammals and should be taken into account in translocations of animals among populations, lest wasted reproductive effort doom an entire project (119).

Numerous stock enhancement programs for freshwater fishes such as trout or salmon release fry produced in hatcheries; generally the fry are not obtained from broodstock originating from the population to be enhanced (91). This fact leads to concern that fitness in the local environment will be compromised even if the genotypes of added individuals adapt them to some other environment, because coadapted gene complexes may be destroyed (3, 18). Similar questions have been raised (40) about enhancement programs for sea turtles. Likewise, marine fish stocks such as those of cod in the north Atlantic and invertebrates such as shrimp in Italy, Japan, and China are replenished from hatcheries, and the hatchery strains used for these purposes generally differ genetically from the target populations; they are often also somewhat domesticated by culture conditions (91). Frequently, genetic diversity and effective population size of hatchery stock are low because of inbreeding and/or because relatively few males are used or certain pairs are more successful (23). Mixing of hatchery stock with wild populations may also lead to increased variance in reproductive success and a decrease in genetic diversity in the wild.

Domesticated stock—individuals selected for survival and reproduction in culture conditions rather than in the wild—may have characteristics that are maladaptive in the target environment. Captive propagation may select for very different traits than nature would (cf 3, 64), and the net result may be lower average fitness in the target population. Additionally, captive propagation in small populations may increase the frequency of correlated traits that are not under selection in the stock but could be maladaptive in the wild.

CONSERVATION IMPLICATIONS OF HYBRID STERILITY AND UNIDIRECTIONAL INTROGRESSION

The fact that individuals of two different taxa mate (hyridization) does not automatically mean that introgression occurs. The hybrids might all be sterile. One might imagine that, if mixing is not an issue, the consequences for conservation are nil, but wasted reproductive effort can threaten a population. For example, the red hartebeest (*Alcelaphus buselaphus*) and the blesbok (*Damaliscus dorcas*) produce viable hybrid offspring in nature, but the male hybrids are sterile and the female hybrids are probably sterile (120). Though gene pools are not mixed in this example, the reproductive effort expended on hybrids is wasted. This loss is not inconsequential because these species are both declining; they are often found in small, isolated populations. The gestation period for both species is eight months, and only a single calf is normally born (105).

The European mink (*Mustela lutreola*) is threatened and declining almost throughout its range. A chief cause is its interaction with the introduced and more numerous American mink (*Mustela vison*). Where they are sympatric, the larger American mink males mate with European mink females, which then do not permit other males to approach them. But the embryos resorb at an early stage, and the female leaves no offspring for that year, while the American mink females are productively mated by conspecific males (123).

Introduced brook trout (*Salvelinus fontinalis*) are displacing resident bull trout (*S. confluentus*) in areas of northwestern North America (90). According to mtDNA analysis, males of each species mate with females of the other, yet allozyme data indicate that about 97% of detected hybrids are F1 individuals. The apparent meagre amount of introgression may result from near sterility of hybrids, their poor mating success, and/or low survival of their progeny. However, even without substantial mixing, this hybridization can have conservation consequences. The more numerous introduced brook trout may have an advantage because a smaller fraction of the reproductive effort is wasted in the production of hybrids.

Ellstrand (52) describes several examples in plants in which hybrid progeny are either sterile or have such reduced vigor that mixing of either parental gene pool is unlikely. However, if either species is rare, the burden of producing such progeny may threaten its populations with extinction. The fitness costs associated with hybridization of this sort may be severe enough to select for the evolution of secondary isolating mechanisms, but a rare species will often lack the genetic variation necessary for such evolution (52). In any event, several theoretical considerations argue against selection of isolation by this means (37).

Even if introgression occurs, it can be limited in various ways. For example, it can be unidirectional. Either hybrids can fail to mate with individuals of one parental taxon, or such matings can be sterile. Further, hybridization itself can be unidirectional (i.e. males of one species breeding with females of the other species, but not the opposite cross). For example, mtDNA analysis shows that, in the blue-winged/golden-winged warbler example cited above, blue-winged genes are moving into golden-winged populations, but not vice versa (58). Similarly, in the mixing of native Apache trout genomes with those of introduced rainbow trout noted above, rainbow genes have moved into Apache populations, but the reverse introgression has not occurred (46). Even if hybridization is fertile in both directions, it is possible that such hybridization can produce fertile offspring of only one sex. Biased introgression of mtDNA indicates a partial barrier to gene flow, possibly owing to assortative mating or selection against hybrids of one of the crosses (46). In short, the existence of hybridization and even introgression need not mean that there are no barriers to gene flow.

Reduction of the frequency of alleles from introduced species could be attained by elimination of populations or individuals based on morphological or allozyme evidence of introgression. In the absence of such data, culling is presently used to contain hybridization between ruddy and white-headed ducks (144) and between yellow-crowned and red-fronted parakeet hybrids (38). However, culling of hybrids may also result in the loss of locally adapted (nuclear) genetic variation and unique mtDNA haplotypes. If the number of individuals in a population is so low that it is possible that the population could disappear entirely from causes other than introgression, culling could eliminate the very individuals whose genomes might permit the partial "reconstruction" of the species. Extensive genetic data (including mtDNA analysis) are therefore sometimes required to make informed management decisions about the possible eradication of hybrids.

ARE HYBRIDIZATION AND INTROGRESSION MAJOR CAUSES OF EXTINCTION?

What is Important?

If introgression is the perceived threat, there is likely to be disagreement about how concerned we should be. What should we be worried about the extinction of? No one would argue that every individual genome should be protected, although almost all individuals carry a certain amount of genetic variability (genotypic if not allelic) not found elsewhere. A huge gray area—subspecies, races, populations—generates controversy. To adherents of the biological species concept, species are reproductively isolated, although the degree of isolation required for specific status differs among authors. Given this view of species, particularly its most extreme interpretation, introgression would, by definition, be intraspecific. One might then argue that concern over possible outcomes of genetic mixing by introgression is misguided. Why should we worry about loss of such infraspecific entities as subspecies, races, and local populations? On the other hand, often one of the major components of genetic variation in a species is among populations, and so, to conserve genetic variation in a species, one should at least aim to save different populations (20).

There should be no such disagreement about cases in which hybridization without introgression threatens the existence of intersterile taxa (see section on Conservation Implications of Hybrid Sterility and Unidirectional Introgression), since these are clearly not infraspecific units.

Conservation scientists have agreed that entities that do not qualify as species can undergo extinction and that we need to worry about them. The existence of genetically distinct populations, whether or not the differences among them are adaptive, has fostered widespread concern for the conservation of infraspecific entities. Even the ESA allows endangered or threatened status for "subspecies" and "distinct population segments" (75, 140). Such concern leads directly to unease at the prospect of loss of such entities to introgression. The National Research Council Committee on Scientific Issues in the Endangered Species Act (CSIESA) advocates the concept of the "evolutionary unit," "a group of organisms that represents a segment of biological diversity that shares evolutionary lineage and contains the potential for a unique evolutionary future," to replace "distinct population segment" in a future ESA (103). A similar concept, the "evolutionarily significant unit," has been proposed by the National Marine Fisheries Service for managing anadromous salmonids (145). Neither concept forbids gene flow between the units, although the latter emphasizes a substantial (but not absolute) degree of reproductive isolation.

The ESA does not confer protection on hybrids (53), a fact that led to proposals to delist such taxa as the wolf (75) and the Florida panther (53) on the grounds that they have undergone introgession. In fact, the Fish and Wildlife Service (FWS) withdrew its policy that hybrids cannot deserve protection in 1990, but they have not replaced it and are operating on a case-by-case basis (53, 75, 107). The CSIESA proposes that a modified ESA should protect, as evolutionary units, taxa that undergo introgression so long as they remain phenotypically much like the endangered parent taxa, whatever the taxonomic level of the latter (103).

How Common Are These Phenomena?

Just how common is extinction by hybridization and/or introgression? As noted in the introduction, these processes are not usually listed among the major extinction threats. Most general discussions of extinction problems omit any mention of hybridization and introgression, with a few exceptions (e.g. 95, 104). However, there are grounds for thinking these phenomena are more important than is commonly realized. For example, 24 animal species listed under the ESA are now extinct; of these, at least 11 were in fact extinct before the Act was passed (99). Of those 24, introgression was thought to be at least a substantial contributing factor for three taxa, all fishes—the Tecopa pupfish (*Cyprinodon nevadensis calidae*), the Amista gambusia (*Gambusia amistadensis*), and the longjaw cisco (*Coregonus alpenae*) (99).

It is also suggestive that many of the most famous "poster children" among endangered vertebrates are perceived as potentially threatened or irrevocably "contaminated" by hybridization and introgression. In addition to species already cited—the gray and red wolves, the European mink, the Hawaiian, New Zealand, and white-headed ducks, the northern spotted owl, several genera of marine turtles—one can mention the Florida panther (53), Przewalski's horse (112), orangutan subspecies (10), both mountain zebra subspecies (31–33), wisent (54), wood bison (101), black robin (36), and black stilt (113).

We believe that the examples cited are probably the tip of the iceberg. Because of the complexity and extent of the genetic techniques that are the major tools for detecting introgression, for most populations, even those where introgression might be suspected, genetic analysis has been insufficient or non-existent. There is every reason to think that the great majority of introgression has been undetected (8, 118), and the advent of molecular tools is just beginning to redress this situation. Detecting effects of hybridization where introgression does not occur may be equally difficult. For example, assessing the extent to which American mink mate with European ones (123) is no mean feat, and, of course, cannot be aided by post facto genetic analyses.

The only comprehensive massing of literature reports of hybridization and introgression is of fishes (141). Of 42 interspecific hybridizations suspected on morphological grounds, all but two were confirmed by molecular techniques, while backcrossing was demonstrated by molecular techniques in 22 of these cases. The majority of these examples come from North America. Twenty-four new cases of hybridization not previously suspected on morphological grounds were reported as confirmed by molecular techniques, and backcrossing was demonstrated in 19 of these.

At the intraspecific level, examples are legion. For example, among African ungulates of conservation concern, habitat fragmentation and deliberate introduction and translocation have caused hybridization among various taxa of wildebeest, springbok, impala, blesbok, oryx, and zebra (32).

We appealed to Nature Conservancy land stewards in their internal informational bulletin (*Stewardship Newsletter*) for information on suspected threats from hybridization and introgression. This query elicited numerous examples in which introgression was believed to be threatening a vulnerable taxon. Striking was that most of these were unpublished or "gray literature" reports by experts in the field working on threatened taxa, with minimal or no laboratory support.

Just among plants, for example, at the Lanphere-Christensen Dunes Preserve (California), a native lupine (*Lupinus littoralis*) is hybridizing with an aggressive introduced species, yellow bush lupine (*L. arboreus*), as confirmed by morphology and pollination experiments, with backcrossing less firmly determined (149). No molecular tests have been conducted. In the Roy E. Larsen Sandyland Sanctuary (Texas), it is feared that the candidate endangered white firewheel, *Gaillardia aestivalis* (Walt.). Rock var. *winkleri,* is hybridizing with Indian blanket (*G. pulchella* Foug.), widely planted on roadsides by the Texas Department of Transportation (W Ledbetter, personal communication). No molecular evidence is available. At the Kern Lake Preserve (California), the last population of the candidate endangered Bakersfield saltbush (*Atriplex tularensis*) is suspected through morphological evidence of disappearing in an introgressed hybrid swarm with the widespread *A. serenana* (55; T Kan, personal communication). The process is engendered by anthropogenic habitat change. Along the Sacramento River and its tributaries, the California sycamore (*Platanus racemosa*) appears in the process of being lost to introgression with the London plane (*P. acerifolia*), while the California black walnut (*Juglans hindsii*) may have been hybridized with numerous congeners imported from all over the world for commercial purposes (FT Griggs, personal communication). In neither instance is molecular evidence available.

In Oregon, the federally endangered western bog lily (*Lilium occidentale*) is suspected of hybridizing with the Columbia lily (*L. columbianum*); the federally threatened Nelson's sidalcea (*Sidalcea nelsoniana*) is suspected of hybridizing with the rose checker-mallow (*S. virgata*); and the candidate endangered peacock larkspur (*Delphinium pavonaceum*) is suspected of hybridizing with the Columbia larkspur *D. trolliifolium* (L Gooch, personal communication). Evidence on introgression is lacking in all three of these cases, and all three may result from natural or anthropogenic range expansion.

It is a common perception that both hybridization and introgression are more frequent in plants than in animals (72) and among freshwater fishes than among other vertebrates (3). For both these groups, we have cited numerous examples of potential conservation consequences. However, it is striking that we found numerous cases of conservation interest, some noted above, among birds and mammals, and a few among reptiles and amphibians. We know of far fewer examples among invertebrates. However, this lacuna may not reflect a smaller incidence of such problems among invertebrates. It may not be coincidental that the problem is well known among the higher vertebrates and poorly known among invertebrates; this disparity exactly mirrors a difference in attention paid by conservation biologists to the different taxa. In general, much more is known about threatened mammals and birds—about which species of both are threatened, and why they are threatened—than about invertebrates (66).

Finally, the key forces conducing to hybridization—anthropogenic species (and subspecies) introductions and habitat modification—are increasing with burgeoning human populations and mobility (139, 150). Habitat modification, in addition to juxtaposing previously disjunct habitats as noted above, leads to fragmentation and isolation of many populations; this phenomenon has been a main theme of conservation biology for over a decade (100). Individuals in small, isolated populations in contact with other taxa are much more likely to hybridize if only because of the difficulty of finding mates of the same species (subspecies, variety, etc). This situation even obtains when the different population is also small, as in the hybridization of blesbok and red hartebeest discussed above. But it is greatly exacerbated when the other population (whether native or introduced) is much larger, as for the European mink, Simien jackal, red wolf, yellow-crowned parakeet, and bull trout. When introgression occurs, barring specific mechanisms opposing interbreeding, a relatively greater fraction of the small population will hybridize each generation, leaving an ever-smaller fraction that has undergone no genetic mixing. Thus, the very factors that threaten extinction by hybridization and introgression—habitat destruction, fragmentation, and species introductions—are all increasing and often act synergistically. The problem is especially likely on islands, on which there are frequently disproportionate amounts of habitat destruction and relatively more introduced species, many of which are more common than the natives (130). Island plants are often at still greater risk because they tend more than mainland species to be reproductively isolated by habitat rather than by genic or chromosomal barriers (117).

Under what circumstances will introgressive hybridization lead to the genetic extinction of one or both parental taxa as opposed to simply a stable hybrid zone? Although some hybrid zones may be maintained in habitat mosaics that include patches to which hybrids are more suited than parental individuals, broadly speaking, lack of fitness of F1 hybrids, later generations, or backcrosses is evident in many stable hybrid zones (72). This is a form of outbreeding depression. The lack of fitness may result from chromosomal differences, the breakdown of coadapted gene complexes, or both (126). In any event, selection against the hybrids stabilizes the zone and prevents mixing. In this instance, there is an inherent weakness in recombinants of the two parental taxa, independent of the habitat. Hybrid zones also form at boundaries or narrow gradients between distinct habitats, so that each habitat favors one parental taxon, and hybrids are selected against in both (7, 72). There is an equilibrium between increasing dispersal of hybrid individuals away from this "tension zone" and increasing selection against them (24). However, lack of fitness of hybrids is far from a universal phenomenon, and many recent studies of particular taxa show various hybrid classes to have fitness greater than or equal to that of parentals (17). It does not appear possible yet to predict which introgressive hybridizations will lead to stable zones and which to massive introgression and even hybrid swarms; many more empirical studies, particularly those assessing fitness of hybrids of various classes, may lead to such generalizations (17).

What Should Conservationists Do?

The less the genetic distinctness, the less concern is merited, and this is independent of the fact that some conservation "poster children," like the Florida panther and red wolf, that are probably not very distinct genetically may be useful emblems for a larger conservation effort. The amounts of natural gene flow at low taxonomic levels are probably often simply too great to allow optimism that we can maintain segments of the gene pool as distinct entities, whatever ethical considerations might dictate. If natural gene flow is sufficiently great, it is hard to imagine that it should be an ethical concern at all. To some extent, subspecific designations are arbitrary, and most systematists stopped naming subspecies in the 1960s or earlier (18); this tendency was encouraged by the recognition that different traits often show different patterns of geographic variation within species (155). Conservationists are loath to abandon a concept that might be seen as a useful way to save populations, and, as noted above, the ESA allows protection of subspecies. However, modern systematics and molecular techniques cast doubt on the validity of such taxonomy.

Also, it is important for conservationists to choose their fights carefully. One cannot be exercised over every situation in which new genes are flowing into a distinctive population, or economic and emotional resources will be insufficient to win most of these battles. Concern over introgression can lead to absurd situations (79). The dusky seaside sparrow (*Ammodramus maritimus nigrescens*) was listed as an endangered subspecies by the FWS as its population declined because of habitat change. Although the Service spent $5 million to purchase remaining habitat, the decline continued and five of the last six individuals, all males, were captured (82, 84). A proposal to salvage part of the gene pool of this subspecies by breeding these males with females of another subspecies was stymied for two years by the FWS on the grounds that such hybridization would dilute the dusky seaside sparrow gene pool and would, in any event, create birds not covered by the ESA because they would not be duskies (38, 44, 79, 82). In the end, the Service washed its hands of the affair, allowing the project to proceed with private funds at a Disney World facility (44, 83), but the Service slashed funding to protect habitat for reintroducing the fertile hybrids, on the grounds that the hybrids were not duskies (51).

In retrospect, the FWS could have chosen less controversial grounds for withdrawing from the attempt to salvage the dusky seaside sparrow gene pool. An mtDNA analysis showed no basis for distinguishing the last dusky seaside sparrow population as a subspecies distinct from other Atlantic coastal populations of *A. maritimus* (22) and indicated that the main concern, in terms of genetically distinct entities, should be to preserve representatives of Atlantic coastal populations on the one hand and of Gulf Coast populations on the other (21). This tempest in a teapot illustrates an important point. What does it mean to speak of saving the gene pool of a tiny population, whether subspecifically distinct or not? After all, the population is an evolving entity, and its gene pool will change even without human intervention, if not by gene flow then by mutation, selection, drift, etc.

The Florida panther (*Felis concolor coryi*) is a subspecies of the cougar listed as endangered by the FWS. Fewer than 40 individuals are believed to remain in the wild, all in south Florida, leading to enormous concern about the viability of this taxon (128). Of the two largest groups, one, in the Everglades, consists exclusively of hybrids between the Florida panther and individuals of other subspecies recently introduced from South or Central America (108). The other, in the Big Cypress swamp, appears to be primarily composed of "pure" Florida panthers, although several individuals have morphological and/or mtDNA traits suggesting a hybrid origin with South or Central American cats. Occasional migration is believed to occur between the two groups. These findings led some to question as to whether the Florida panther should be delisted (53).

Ironically, before the introgression with Latin American cats was known, the prospect of deliberately using other subspecies in a captive breeding program as part of a recovery plan was denounced because "such miscegenation would contaminate the *F. c. coryi* bloodlines" (42, p. 9), its status under the ESA would be jeopardized (27, 42), and *F. c. coryi* is believed to be the subspecies best adapted for survival in the Florida environment (27). For this reason, feasibility studies for translocation of Florida panthers into northern Florida were carried out with sterilized Texas mountain lions (*F. c. stanlevana*) (26).

Now the general thinking has turned completely around. Arnold et al (18) cautiously asserted that it may be a good idea to introduce individuals from Texas or possibly other subspecies so that hybridization will occur. This stance was largely rationalized by arguments that 1. hybridization has already occurred anyway, and the hybrid population appears cursorily to be less plagued than others with possible indications of inbreeding depression; 2. the numbers of Florida panthers are so low and signs of possible inbreeding depression so ominous that the subspecies will become extinct without the proposed introduction; and 3. the Florida panther is not genetically very different from other subspecies anyway, so that outbreeding depression is unlikely. Thus, eight female Texas cougars were released into south Florida in 1995, prompting a claim that the evidence of an important role for inbreeding depression is inconclusive and that the possibility of outbreeding depression was not sufficiently studied (96).

This case also seems to be a tempest in a teapot; there is so much evidence of ecological threats to the Florida panther (128) and so little evidence of either inbreeding or outbreeding depression (96) that it is difficult to believe that introduction of Texas cougars is crucial one way or the other to the survival of this population.

INTROGRESSIVE HYBRIDIZATION AS AN EVOLUTIONARILY CONSTRUCTIVE PROCESS

If one is at pains to conserve the processes of evolution, as well as its products (e.g. 100), it is also important to recall the constructive roles played by hybridization. Hybridization can allow for rapid evolutionary change by producing novel gene combinations (25, 93) even in small, fragmented populations (84a). Botanists recognize that it may lead to increased genetic variation at both the genic and genotypic levels, increased fitness, and adaptation to new environments in existing taxa (52, 116, 118).

Hybrid vigor, or heterosis, is a well-known phenomenon (e.g. 73) and has been documented more frequently than outbreeding depression has. Perhaps the relative importance or likelihood of outbreeding depression and hybrid vigor is correlated with the degree of genetic differentiation of the parent taxa. In any event, sometimes particularly vigorous hybrids thrive in habitats inimical to both parental taxa, in ways that would normally be construed as benefits to conservation. The London plane tree (*Platanus* x *acerifolia*) is a hybrid of the American sycamore (*P. occidentalis*) and oriental plane (*P. orientalis*), both of which were introduced to England in the seventeenth century, and it is well known as a majestic tree particularly tolerant of coal dust, smoke, compacted soils, and other aspects of the urban environment (132). It is found in many large cities both inside and outside the ranges of both parental species, often where neither parental species can survive. If a tree grows in Brooklyn, it is most likely a London plane (34). That it threatens another taxon in California with introgressive extinction, as noted above, is ironic.

In animals, most of the closely studied examples come from domestic or laboratory species because of the difficulty of measuring fitness differences in the field. However, there is every reason to believe the phenomenon is common in nature. For example, a genetically depauperate sexual topminnow species, *Poeciliopsis monacha,* with consequent low heterozygosity, was consistently outcompeted in a stream for many generations by an asexual congener. Laboratory studies showed reduced developmental stability and resistance to anoxic stress associated with reduced heterozygosity, but the mechanism of competition in the field was not determined. However, addition of a few pregnant *P. monacha* from another, genetically variable population led rapidly to a great increase in genetic diversity and relative frequency of *P. monacha* and a great decrease in relative frequency of the asexual form (142).

In addition, as many as 70% of angiosperm species may have arisen through the formation of polyploid species by hybridization (97), and introgressive hybridization can foster speciation in plants even without the production of polyploids (16). Hybrid speciation may also have played an important role in the evolution of birds (61, 114), fishes (43, 134), mammals (122), and insects (131).

CONCLUSIONS

Hybridization, with or without introgression, frequently threatens populations in a wide variety of plant and animal taxa because of various human activities. Probably cases reported in the literature do not adequately convey the magnitude of the problem. Increased use of molecular techniques reveals examples not manifest from morphological analysis. Further, the increasing pace of three interacting human activities—habitat modification, fragmentation, and the introduction of exotic species—that contribute to this problem suggests it will worsen. It is thus surprising that most conservation texts and reviews barely mention it as a general problem.

If there is no introgression, but reproductive effort lost to fruitless hybridization threatens a species' existence, management actions may frequently be warranted. Introgression may also be a concern, but conservationists need not raise an alarm every time that populations exchange genes. First, some degree of gene flow is a normal and evolutionarily constructive process. Second, as a practical matter it is difficult to detect and often costly or impossible to prevent some gene flow. Third, particularly at the level of local populations, alleles and genotypes will be lost (and others will arise) even without gene flow; it is fruitless to have as a goal the long-term preservation of every constellation of genes and genotypes. However, often introgression between an introduced and native taxon may lead to less fit populations, perhaps even to a threat of extinction (as in the anadromous salmonids). Even where there is no evidence of fitness decline, it is surely worth attempting to prevent deliberate introductions from causing introgression into a morphologically well-defined, evolutionarily isolated taxon, such as the New Zealand grey duck.

ACKNOWLEDGMENTS

We thank Dan Graur, Dave Heckel, Fran James, Carey Krajewski, Margaret Ptacek, and Joe Travis for comments on an early draft of this manuscript; Jim Cox, Frank Gill, Paul Gray, and numerous Nature Conservancy stewards for discussion of certain cases; and John Avise and Johan Hammar for numerous preprints and reprints.

Literature Cited

1. Abernethy K. 1994. The establishment of a hybrid zone between red and sika deer (genus *Cervus*). *Mol. Ecol.* 3:551–62
2. Allendorf FW, Leary RF. 1988. Conservation and distribution of genetic variation in a polytypic species, the cutthroat trout. *Conserv. Biol.* 2:170–84
3. Allendorf FW, Waples RS. 1996. Conservation and genetics of salmonid fishes. In *Conservation Genetics: Case Histories from Nature*, ed. JC Avise, JL Hamrick. New York: Chapman & Hall. In Press.
4. American Ornithologists' Union. 1979. Resolutions. *Auk* 97 (1, suppl.):10AA
5. American Ornithologists' Union. 1983. *Check-list of North American Birds*. Washington, DC: Am. Ornithol. Union. 6th ed.
6. Anderson BW. 1971. Man's influence on hybridization in two avian species in South Dakota. *Condor* 73:342–47
7. Anderson E. 1948. Hybridization of the habitat. *Evolution* 2:1–9
8. Anderson E. 1949. *Introgressive Hybridization*. New York: Wiley
9. Anderson E, Hubricht L. 1938. Hybridization in Tradescantia. III. The evidence for introgressive hybridization. *Am. J. Bot.* 25:396–402
10. Angier N. 1995. Orangutan hybrid, bred

to save species, now seen as pollutant. *New York Times,* Feb. 28, pp. B5, B9
11. Ankney CD, Dennis DG, Bailey RC. 1987. Increasing mallards, decreasing American black ducks: coincidence or cause and effect? *J. Wildl. Manage.* 51:523–29
12. Anonymous. 1989. *1988 Annual Report on the status of California's state listed, threatened, endangered plants and animals.* Sacramento: State of Calif. Dep. Fish & Game
13. Anonymous. 1992. *Recovery Plan for the Northern Spotted owl—Draft.* Washington, DC: USGPO
14. Anonymous. 1993. UK ruddy duck working group. Information. Peterborough: Joint Nature Conserv. Com.
15. Arano B, Llorente G, García-Paris M, Herrero P. 1995. Species translocation menaces Iberian waterfrogs. *Conserv. Biol.* 9:196–98
16. Arnold ML, Bennett BD. 1993. Natural hybridization in Louisiana irises: genetic variation and ecological determinants. See Ref. 72a, pp. 115–39
17. Arnold ML, Hodges SA. 1995. Are natural hybrids fit or unfit relative to their parents? *Trends Ecol. Evol.* 10:67–71
18. Arnold SJ, Avise JC, Ballou J, Eldridge J, Flemming D, et al. 1991. Genetic management considerations for threatened species with a detailed analysis of the Florida panther (*Felis concolor coryi*). Washington, DC: USFWS
19. Atkinson I. 1989. Introduced animals and extinctions. See Ref. 151, pp. 54–75
20. Avise JC. 1994. *Molecular Markers, Natural History and Evolution.* New York: Chapman & Hall
21. Avise JC. 1994. A rose is a rose is a rose. See Ref. 100, pp. 174–75
22. Avise JC, Nelson WS. 1989. Molecular genetic relationships of the extinct dusky seaside sparrow. *Science* 243:646–48
23. Bartley D, Bagley M, Gall G, Bentley B. 1992. Use of linkage disequilibrium data to estimate effective size of hatchery and natural fish populations. *Conserv. Biol.* 6:365–75
24. Barton NH, Hewitt GM. 1985. Analysis of hybrid zones. *Annu. Rev. Ecol. Syst.* 16:113–48
25. Barton NH, Hewitt GM. 1989. Adaptation, speciation and hybrid zones. *Nature* 341:497–503
26. Belden RC, Hagedorn BW. 1992. *Feasibility of Translocating Panthers into Northern Florida.* Tallahassee: Fla. Game & Freshwater Fish Comm.
27. Belden RC, Hines TC, Logan TH. 1986. *Florida Panther Re-establishment: A Discussion of the Issues.* Tallahassee: Fla. Game & Freshwater Fish Comm.
28. Bernes C, ed. 1994. *Biological Diversity in Sweden—A Country Study.* Stockholm: Swedish Environ. Protect. Agency
29. Bert TM, Hesselman DM, Arnold WS, Moore WS, Cruz-Lopez H, Marrelli D. 1993. High frequency of gonadal neoplasia in a hard clam (*Mercenaria*) hybrid zone. *Mar. Biol.* 117:97–104
30. Birks J. 1995. Recovery of the European polecat (*Mustela putorius*) in Britain. *Small Carnivore Conserv.* 12:9
31. Breytenbach G. 1986. Impacts of alien organisms on terrestrial communities with emphasis on communities of the southwestern Cape. See Ref. 95a, pp. 229–38
32. Brooke RK, Lloyd PH, de Villiers AL. 1986. Alien and translocated terrestrial vertebrates in South Africa. See Ref. 95a, pp. 63–74
33. Brown CJ, Gubb AA. 1986. Invasive alien organisms in the Namib Desert, Upper Karoo and the arid and semi-arid savannas of western southern Africa. See Ref. 95a, pp. 93–108
34. Bumiller E. 1995. Go out, go forth and count. *New York Times,* July 29, p. 16
35. Butler D. 1994. Bid to protect wolves from genetic pollution. *Nature* 370:497
36. Butler D, Merton D. 1992. *The Black Robin. Saving the World's Most Endangered Bird.* Auckland: Oxford Univ. Press.
37. Butlin R. 1989. Reinforcement of premating isolation. In *Speciation and Its Consequences,* ed. D Otte, JA Endler, pp. 158–79. Sunderland, MA: Sinauer
38. Cade TJ. 1983. Hybridization and gene exchange among birds in relation to conservation. See Ref. 126, pp. 288–309
39. Compton DE. 1990. Application of biochemical and molecular markers to analysis of hybridization. In *Electrophoretic and Isoelectric Focusing Techniques in Fisheries Management,* ed. DH Whitmore, pp. 241–64. Boca Raton, FL: CRC
40. Carr AF III, Dodd CK Jr. 1983. Sea turtles and the problem of hybridization. See Ref. 125a, pp. 277–87
41. Confer JL. 1992. Golden-winged warbler. In *The Birds of North America,* ed. A Poole, P Stettheim, F Gill, No. 20. Philadelphia: Acad. Nat. Sci. Washington, DC: Am. Ornithologists' Union
42. Cristoffer C, Eisenberg IF. 1985. *On the captive breeding and reintroduction of the Florida panther into suitable habitats.*

Tallahassee, FL: Fla. Game & Freshwater Fish Comm.

43. DeMarais BD, Dowling TE, Douglas ME, Minckley WL, Marsh PC. 1992. Origin of *Gila seminuda* (Teleostei: Cyprinidae) through introgressive hybridization: implications for evolution and conservation. *Proc. Natl. Acad. Sci. USA* 89:2747–51
44. Diamond JM. 1985. Salvaging single-sex populations. *Nature* 316:104
45. Diamond JM. 1989. Overview of recent extinctions. See Ref. 151, pp. 37–41
46. Dowling TE, Childs MR. 1992. Impact of hybridization on a threatened trout of the Southwestern United States. *Conserv. Biol.* 6:355–64
47. Dowling TE, Minckley WL, Douglas ME, Marsh PC, DeMarais BD. 1992. Response to Wayne, Nowak, Phillips and Henry: use of molecular characters in conservation biology. *Conserv. Biol.* 6:600–3
48. DuRietz GE. 1930. The fundamental units of biological taxonomy. *Svensk Bot. Tidskr.* 24:333–428
49. Echelle AA, Connor PJ. 1989. Rapid, geographically extensive genetic introgression after secondary contact between two pupfish species (*Cyprinodon*, Cyprinodontidae). *Evolution* 43:717–27
50. Edwards RJ. 1979. A report of Guadaloupe bass *Micropterus treculi* x smallmouth bass *Micropterus dolomieui* hybrids from 2 localities in the Guadaloupe River, Texas USA. *Tex. J. Sci.* 31:231–38
51. Ehrlich PR, Dobkin DS, Wheye D. 1992. *Birds in Jeopardy*. Stanford, CA: Stanford Univ. Press
52. Ellstrand NC. 1992. Gene flow by pollen: implication for plant conservation genetics. *Oikos* 63:77–86
53. Fergus C. 1991. The Florida panther verges on extinction. *Science* 251:1178–80
54. Fisher J, Simon N, Vincent J. 1969. *Wildlife in Danger*. New York: Viking
55. Freas KE, Murphy DD. 1988. Taxonomy and the conservation of the critically endangered Bakersfield saltbush, *Atriplex tularensis*. *Biol. Conserv.* 46:317–24
56. Gill FB. 1980. Historical aspects of hybridization between blue-winged and golden-winged warblers. *Auk* 97:1–18
57. Gill FB. 1987. Allozymes and genetic similarity of blue-winged and golden-winged warblers. *Auk* 104:444–49
58. Gill FB. 1994. Rapid, asymmetrical introgression of mitochondrial DNA in hybridizing populations of the blue-winged and golden-winged warblers. In *Proc. XXI Int. Ornithol. Congress,* August 20–25, Vienna, Austria
59. Giuffra E, Bernatchez L, Guyomard R. 1994. Mitochondrial control region and protein coding genes sequence variation among phenotypic forms of brown trout *Salmo trutta* from northern Italy. *Mol. Ecol.* 3:161–71
60. Gotelli D, Sillero-Zubiri C, Applebaum GD, Roy MS, Girman DJ, et al. 1994. Molecular genetics of the most endangered canid: the Ethiopian wolf *Canis simensis*. *Mol. Ecol.* 3:301–12
61. Grant PR, Grant BR. 1992. Hybridization of bird species. *Science* 256:193–97
62. Grant V. 1981. *Plant Speciation.* New York: Columbia Univ. Press. 2nd ed.
63. Griffin CR, Shallenberger RJ, Fefer SI. 1989. Hawaii's endangered waterbirds: a resource management challenge. In *Proc. of Freshwater Wetlands and Wildlife Symp.*, ed. RR Sharitz, IW Gibbons, pp. 155–69. Aiken, SC: Savannah River Ecol. Lab.
64. Gross MR. 1991. Salmon breeding behavior and life history evolution in changing environments. *Ecology* 72:1180–86
65. Haddon M. 1984. A re-analysis of hybridization between mallards and grey ducks in New Zealand. *Auk* 101:190–91
66. Hadfield MG. 1993. Introduction to the symposium: the crisis in invertebrate conservation. *Am. Zool.* 33:497–98
67. Hamer TE, Forsman ED, Fuchs AD, Walters ML. 1994. Hybridization between barred and spotted owls. *Auk* 111:487–92
68. Hammar J. 1989. Freshwater ecosystems of polar regions: vulnerable resources. *Ambio* 18:6–22
69. Hammar J, Dempson JB, Sk'old E. 1989. Natural hybridization between Arctic char (*Salvelinus alpinus*) and lake char (*S. namaycush*): evidence from northern Labrador. *Nordic J. Freshw. Res.* 65:54–70
70. Hammar J, Dempson JB, Verspoor E. 1991. Natural hybridization between Arctic char (*Salvelinus alpinus*) and brook trout (*S. fontinalis*): evidence from northern Labrador. *Can. J. Fisheries Aquatic Sci.* 48:1437–45
71. Harlan JR. 1983. Some merging of plant populations. See Ref. 126, pp. 267–76
72. Harrison RG. 1993. Hybrids and hybrid zones: historical perspective. See Ref. 72a, pp. 3–12

72a. Harrison RG, ed. 1993. *Hybrid Zones and the Evolutionary Process,* New York: Oxford Univ. Press

73. Hartl DL, Clark AG. 1989. *Principles*

of Population Genetics. Sunderland, MA: Sinauer. 2nd ed.

74. Heusmann HW. 1974. Mallard-Black Duck relationships in the Northeast. *Wildl. Soc. Bull.* 2:171–77
75. Hill KD. 1993. The Endangered Species Act: What do we mean by species? *Environ. Affairs* 20:239–64
76. Hope J. 1994. A wolf in pet's clothing. *Smithsonian* 25(3):34–45
77. Hubbard AL, McOrist S, Jones TW, Boid R, Scott R, Easterbee N. 1992. Is survival of European wildcats *Fells silvestris* in Britain threatened by interbreeding with domestic cats? *Biol. Conserv.* 61:203–8
78. Hubbard JP. 1977. *The biological and taxonomic status of the Mexican duck. Bull. New Mexico Dep. Game & Fish, No. 1* Albuquerque, New Mexico: New Mexico Dep. Game & Fish
79. James FC. 1980. Miscegenation in the dusky seaside sparrow? (letter). *BioScience* 30:800–1
80. Johnsgard PA. 1961. Evolutionary relationships among North American Mallards. *Auk* 78:3–43
81. Johnsgard PA. 1967. Sympatry changes and hybridization incidence in mallards and black ducks. *Am. Midl. Natur.* 77:51–63
82. Kale HW II. 1981. Dusky seaside sparrow—Gone forever? *Fla. Naturalist* 54(4):3–48
83. Kale HW II. 1983. Duskies transferred to Discovery Island. *Fla. Naturalist* 56(4):3
84. 84. Kale HW II. 1987. The dusky seaside sparrow: Have we learned anything? *Fla. Naturalist* 60(3):2–3

84a. Kaneshiro KY. 1995. Evolution, speciation and the genetic structure of island populations. In *Islands, Biodiversity and Ecosystemn Function,* ed. PM Vitousek, LL Loupe, H Adsersen, pp. 23–33. Berlin: Springer-Verlag

85. Karl SA, Bowen BW, Avise JC. 1995. Hybridization among the ancient mariners: characterization of marine turtle hybrids with molecular genetic assays. *J. Hered.* 86:262–68
86. Kim JH, Yang SY. 1993. Systematic studies of the genus *Cobitis* (Pisces: Cobitidae) in Korea. IV. Introgressive hybridization between two spined loach subspecies of the genus *Cobitis*. *Korean J. Zool.* 36:535–44
87. Knopf FL. 1986. Changing landscapes and the cosmopolitanism of the eastern Colorado avifauna. *Wildl. Soc. Bull.* 14:132–42
88. Lamb T, Avise JC. 1986. Directional introgression of mitochondrial DNA in a hybrid population of tree frogs: the influence of mating behavior. *Proc. Natl. Acad. Sci. USA* 83:2526–30
89. Lariviere S, Crete M. 1993. The size of eastern coyotes (*Canis latrans*): a comment. *J. Mammal.* 74:1072–74
90. Leary RF, Allendorf FW, Forbes SH. 1993. Conservation genetics of bull trout in the Columbia and Klamath River drainages. *Conserv. Biol.* 7:856–65
91. Lester LJ. 1992. Marine species introductions and native species vitality: genetic consequences of marine introductions. In *Introductions and Transfers of Marine Species*, ed. MR DeVoe, pp. 79–89. Charleston, SC: S Carolina Sea Grant Consortium
92. Lever C. 1987. *Naturalized Birds of the World*. Harlow, Essex: Longman Sci. Tech.
93. Lewontin RC, Birch LC. 1966. Hybridization as a source of variation for adaptation to new environments. *Evolution* 20:315–36
94. Lynch JM. 1995. Conservation implications of hybridisation between mustelids and their domesticated counterparts: the example of polecats and feral ferrets in Britain. *Small Carnivore Conserv.* 13:17–18
95. Macdonald IAW, Kruger FJ, Ferrar AA, eds. 1986. *The Ecology and Management of Biological Invasions in Southern Africa*. Cape Town: Oxford Univ. Press

95a. Macdonald IAW, Loope LL, Usher MB, Hamann 0. 1989. Wildlife conservation and the invasion of nature reserves by introduced species: a global perspective. In *Biological Invasions: A Global Perspective,* ed. JA Drake, HA Mooney, F Dicastri, RH Groves, FJ Kruger, et al, pp. 215–55. Chichester UK: Wiley

96. Maehr DS, Caddick GB. 1995. Demographics and genetic introgression in the Florida panther. *Conserv. Biol.* 9:1295–98
97. Masterson J. 1994. Stomatal size in fossil plants: evidence for polyploidy in majority of angiosperms. *Science* 264:421–24
98. Mazourek JC, Gray PN. 1994. The Florida duck or the mallard. *Florida Wildl.* 48(3):29–31
99. McMillan M, Wilcove D. 1994. Gone but not forgotten: Why have species protected by the Endangered Species Act become extinct? *Endangered Species Update* 11(11):5–6
100. Meffe GK, Carroll CR. 1994. *Principles of Conservation Biology.* Sunderland, MA: Sinauer

101. Middleton S, Liittschwager D. 1994. *Witness. Endangered Species in North America.* San Francisco: Chronicle Books
102. Moore WS, Keening WD. 1986. Comparative reproductive success of yellow-shafted, red-shafted, hybrid flickers across a hybrid zone. *Auk* 103:42–51
103. National Research Council, Committee on Scientific Issues in the Endangered Species Act. 1995. *Science and the Endangered Species Act.* Washington: Natl. Acad. Press
104. Nigh TA, Pflieger WL, Redfearn PL Jr, Schroeder WA, Templeton AR, Thompson FR III. 1992. *The Biodiversity of Missouri. Definition, Status, Recommendations for Its Conservation.* Jefferson City, MO: Missouri Dep. Conserv.
105. Nowak RM. 1991. *Walker's Mammals of the World..* Vol. 2. Baltimore: Johns Hopkins Univ. Press. 5th ed.
106. Nowak RM. 1992. The red wolf is not a hybrid. *Conserv. Biol.* 6:593–95
107. O'Brien SJ. 1994. When endangered species hybridize: the U.S. hybrid policy. See Ref. 100, 69–70
108. O'Brien SJ, Roelke ME, Yuhki N, Richards KW, Johnson WE, et al. 1990. Genetic introgression within the Florida panther *Felis concolor coryi. Natl. Geogr. Res.* 6:485–94
109. Owen M, Atkinson-Willes GL, Salmon DG. 1986. *Wildfowl in Great Britain* Cambridge: Cambridge Univ. Press. 2nd ed.
110. Phillips JC. 1915. Experimental studies of hybridization among ducks and pheasants. *J. Exp. Zool.* 18:69–144
111. Phillips MK, Henry VG. 1992. Comments on red wolf taxonomy. *Conserv. Biol.* 6:596–99
112. Possehl S. 1994. Rare Przewalski's horse returns to the harsh Mongolian steppe. *New York Times,* Oct. 4, p. B9
113. Reed CEM, Murray DP, Butler DJ. 1993. Black stilt recovery plan *Himantopus novaezealandiae. Threatened Species Recovery Plan No. 4.* Wellington, NZ: Dep. Conserv.
114. Rhymer JM. 1994. Reticulate evolution in Hawaiian island ducks. In *Proc. XXI Int. Ornithol. Congr.*, August 20–25, Vienna, Austria
115. Rhymer JM, Williams MJ, Braun MJ. 1994. Mitochondrial analysis of gene flow between New Zealand mallards (*Anas platyrhynchos*) and grey ducks (*A. superciliosa*). *Auk* 111:970–78
116. Rieseberg LH. 1991. Hybridization in rare plants: insights from case studies in *Cercocarpus* and *Helianthus.* In *Genetics and Conservation of Rare Plants,* ed. DA Falk, KE Holsinger, pp. 171–81. New York: Oxford Univ. Press
117. Rieseberg LH, Gerber D. 1995. Hybridization in the Catalina Island mountain mahogany (*Cercocarpus traskiae*): RAPD evidence. *Conserv. Biol.* 9:199–203
118. Rieseberg LH, Wendel JF. 1993. Introgression and its consequences in plants. See Ref. 72a, 70–109
119. Robinson TJ, Elder FFB. 1993. Cytogenetics: its role in wildlife management and the genetic conservation of mammals. *Biol. Conserv.* 63:47–51
120. Robinson TJ, Morris DJ. 1991. Interspecific hybridization in the Bovidae: Sterility of *Alcelaphus buselaphus* x *Damaliscus dorcas* F1 progeny. *Biol. Conserv.* 58:345–56
121. Rogers CE, Thompson TE, Seiler GJ. 1982. *Sunflower Species of the United States.* Bismarck, ND: Natl. Sunflower Assoc.
122. Roy MS, Geffen E, Smith D, Ostrander EA, Wayne RK. 1994. Patterns of differentiation and hybridization in North American wolflike canids, revealed by analysis of microsatellite loci. *Mol. Biol. Evol.* 11:553–70
123. Rozhnov VV. 1993. Extinction of the European mink: ecological catastrophe or a natural process? *Lutreola* 1:10–16
124. Samson F, Knopf F. 1994. Prairie conservation in North America. *BioScience* 44:418–21
125. Schlefer EK, Romano MA, Guttman SI, Ruth SB. 1986. Effects of twenty years of hybridization in a disturbed habitat on *Hyla cinerea* and *Hyla gratiosa. J. Herpetol.* 20:210–21
126. Schonewald-Cox CM, Chambers SM, MacBryde B, Thomas L, eds. 1983. *Genetics and Conservation.* Menlo Park, CA: Benjamin/Cummings
127. Searle JB. 1993. Chromosomal hybrid zones in eutherian mammals. See Ref. 72a, pp. 309–53
128. Short LL. 1969. Taxonomic aspects of avian hybridization. *Auk* 86:84–105
129. Shrader-Frechette KS, McCoy ED. 1993. *Method in Ecology. Strategies for Conservation.* Cambridge: Cambridge Univ. Press
130. Simberloff D. 1988. The contribution of population and community biology to conservation science. *Annu. Rev. Ecol. Syst.* 19:473–511
131. Simberloff D. 1995. Why do introduced species appear to devastate islands more

than mainland areas? *Pac. Science* 49:87–97
132. Spence JR, Gooding RH, eds. 1990. Evolutionary significance of hybridization and introgression in insects. *Can. J. Zool.* 68:1699–1805
133. Spongberg SA. 1990. *A Reunion of Trees.* Cambridge, MA: Harvard Univ. Press
134. Stuart C, Stuart T. 1991. The feral cat problem in southern Africa. *Afr. Wildl.* 45:13–15
135. Svärdson G. 1970. Significance of introgression in coregonid evolution. In *Biology of Coregonid Fishes,* ed. CC Lindsey, CS Woods, pp. 33–59. Winnipeg: Univ. Manitoba Press
136. Templeton AR. 1986. Coadaptation and outbreeding depression. In *Conservation Biology: The Science of Scarcity and Diversity,* ed. ME Soulé, pp. 105–16. Sunderland, MA: Sinauer
137. Templeton AR. 1994. Coadaptation, local adaptation, outbreeding depression. See Ref. 100, pp. 152–53
138. Thomas JW, Forsman ED, Lint JB, Meslow EC, Noon BR, Verner J. 1990. *A Conservation Strategy for the Northern Spotted Owl.* Washington, DC: USGPO
139. Towns DR, Daugherty CH, Cromarty PL. 1990. Protocols for translocation of organisms to islands. In *Ecological Restoration of New Zealand Islands,* ed. DR Towns, CH Daugherty, IAE Atkinson, pp. 240–54. Wellington: NZ Dep. Conserv.
140. United States Congress. Office of Technology Assessment. 1993. *Harmful Non-Indigenous Species in the United States.* Washington, DC: USGPO
141. United States Fish and Wildlife Service. 1988. *Endangered Species Act of 1973. As Amended Through the 100th Congress.* Washington, DC: Dep. Interior
142. Verspoor E, Hammar J. 1991. Introgressive hybridization in fishes: the biochemical evidence. *J. Fish Biol.* 39 (Suppl. A):309–34
143. Vrijenhoek RC. 1989. Population genetics and conservation. See Ref. 150, pp. 89–98
144. Waddington CH. 1956. Genetic assimilation of the *bithorax* phenotype. *Evolution* 10:1–13
145. Waite TL. 1993. Relative puts rare European duck at extinction's door. *New York Times,* April 13, p. B-8
146. Waples RS. 1991. Pacific salmon, *Oncorhynchus* spp., the definition of "species" under the Endangered Species Act. *Mar. Fisheries Rev.* 53:11–22
147. Warren RJ. 1994. An emerging management tool: large mammal predator reintroductions. See Ref. 100, pp. 346–47
148. Wayne RK. 1992. On the use of morphologic and molecular genetic characters to investigate species status. *Conserv. Biol.* 6:590–92
149. Wayne RK, SM Jenks. 1991. Mitochondrial DNA analysis implying extensive hybridization of the endangered red wolf *Canis rufus. Nature* 351:565–68
150. Wear KS. 1995. Hybrid lupine (*Lupinus arboreus* x *L. littoralis*) on the Samoa Peninsula, Humboldt County, CA. Arcata, CA: Nature Conservancy
151. Western D, Pearl M. 1989. *Conservation for the Twenty-First Century.* New York: Oxford Univ. Press
152. Whitmore DH. 1983. Introgressive hybridization of smallmouth bass (*Micropterus dolomieui*) and Guadalupe bass (*M. treculi*). *Copeia* 1983:672–79
153. Wiegand KM. 1935. A taxonomist's experience with hybrids in the wild. *Science* 81:161–66
154. Williams MJ, Roderick C. 1973. Breeding performance of grey duck (*Anas superciliosa*), mallard (*Anas platyrhynchos*) and their hybrids in captivity. *Int. Zoo Yearbk.* 13:62–69
155. Wilson CC, Hebert PDN. 1993. Natural hybridization between Arctic char (*Salvelinus alpinus*) and lake trout (*S. namaycush*) in the Canadian Arctic. *Can. J. Fish. Aquatic Sci.* 50:2652–58
156. Wilson EO, Brown WL. 1953. The subspecies concept and its taxonomic application. *Syst. Zool.* 2:97–111
157. Witkowski A. 1989. Fishes introduced to Polish waters and their effect on environment. *Prezglad Zoologiczny* 33:583–98

Annu. Rev. Ecol. Syst. 1996. 27:111–33

EVOLUTIONARY SIGNIFICANCE OF RESOURCE POLYMORPHISMS IN FISHES, AMPHIBIANS, AND BIRDS

Thomas B. Smith
Department of Biology, San Francisco State University, 1600 Holloway Avenue, San Francisco, California 94132

Skúli Skúlason
Hólar Agricultural College, Hólar, Hjaltadalur, Sandár-Kŕokur, Iceland

KEY WORDS: trophic polymorphism, niche use, speciation, phenotypic plasticity

ABSTRACT

Resource polymorphism in vertebrates is generally underappreciated as a diversifying force and is probably more common than is currently recognized. Research across diverse taxa suggest they may play important roles in population divergence and speciation. They may involve various kinds of traits, including morphological and behavioral traits and those related to life history. Many of the evolutionary, ecological, and genetic mechanisms producing and maintaining resource polymorphisms are similar among phylogenetically distinct species. Although further studies are needed, the genetic basis may be simple, in some cases under the control of a single locus, with phenotypic plasticity playing a proximate role in some taxa. Divergent selection including either directional, disruptive, or frequency-dependent selection is important in their evolution. Generally, the invasion of "open" niches or underutilized resources requiring unique trophic characters and decreased interspecific competition have promoted the evolution of resource polymorphisms. Further investigations centered on their role in speciation, especially adaptive radiation, are likely to be fruitful.

INTRODUCTION

Resource-based or trophic polymorphisms are likely more common and of greater evolutionary significance than is currently appreciated. Work on diverse taxa suggests that these polymorphisms, in which discrete intraspecific

0066-4162/96/1120-0111$08.00

morphs show differential resource use and often varying degrees of reproductive isolation, may represent important intermediate stages in speciation (14, 150, 190). Many of the mechanisms and conditions that produce and maintain resource polymorphism are similar, even among highly divergent taxa (150). Yet, until recently, little effort has gone toward examining this phenomenon across different taxonomic groups. For example, important studies on resource polymorphisms involving fishes, amphibians, and birds have been published in recent years, but there have been few comparisons of the processes involved (but see 51, 56, 150, 197). In this review we summarize the diverse nature of the phenotypic differences involved, and through selected examples explore the ecological and evolutionary implications of this phenomenon. We hope to promote an integrative cross-taxon approach to the study of resource polymorphisms and greater efforts toward identifying additional examples. Numerous other examples likely exist, but because discrete phenotypes may be subtle, they are easily overlooked or discounted. We do not review sex-based polymorphisms because they are beyond the scope of the present review. We begin by describing the nature of resource polymorphisms in each taxon, then we discuss circumstances and mechanisms maintaining them, and finally we discuss their evolutionary significance.

TYPES OF ALTERNATIVE ADAPTIVE PHENOTYPES

We define resource polymorphisms as the occurrence of discrete intraspecific morphs showing differential niche use, usually through discrete differences in feeding biology and habitat use. Morphs may differ in morphology, color, behavior, or life history traits, and in many instances they may differ in more than one characteristic (Table 1).

Fishes

Many examples of resource polymorphisms appear across diverse fish taxa, including mostly freshwater and anadromous fishes. Resource-based morphs of fishes may differ in behavior, life history, morphology, and color, and they may coexist within the same freshwater system, i.e. they are intralacustrine (155) and may even be found within small landlocked lakes. Frequently, different forms show varying degrees of reproductive isolation, even to the level of being classified as distinct biological species (56, 133, 141). Since the segregation among some of these forms is clearly correlated with resource use and represents part of a continuum, we include them in our discussion.

Resource polymorphisms in fishes are common in lakes in recently (10,000–15,000 years) glaciated areas of the northern hemisphere (127, 141). The arctic charr (*Salvelinus alpinus*), a circumpolar salmonid, often has from two to four sympatric intralacustrine resource morphs. Morphs differ in adult body size and shape, and in life history characteristics (e.g. 49, 66, 68, 69, 105, 122, 125, 147, 167, 188). In some cases resource segregation is clear and stable, such as between benthic and limnetic habitats (79, 135), while in other cases, habitat and food segregation are less dramatic and often seasonal (57, 125, 168, 188). The degree of phenotypic differences may also differ among lakes (148, 167). For example, in Thingvallavatn, a lake in Iceland where resources are unusually well defined and discrete (67), four morphs display substantial behavioral, life history, and morphological differences (69, 78, 152, 167). The lake has an extensive littoral zone made structurally complex by volcanic substrate. A small benthivorous charr (7–22 cm long) occupies the subbenthic habitat consisting of porous volcanic rubble, while a large benthivorous charr (20–50 cm) occurs in the epibenthic habitat. Both morphs specialize on snails. In addition, there are two limnetic morphs, one smaller (14–22 cm) and planktivorous, the other larger (adults 25–60 cm) and piscivorous (37, 79). The limnetic morphs also have more streamlined bodies, more pointed snouts, and more gillrakers than do the benthic morphs (78, 153, 167). Some spatial and temporal variability in spawning among the morphs can be related to their diet and the availability of spawning grounds (152), and some evidence suggests assortative mating (146). Molecular genetic studies, combined with the distribution of morphs in other lakes, show that the morphs are closely related and arise locally, with benthic forms derived from the more common limnetic form (148, 167; see also 59).

In whitefish (*Coregonus* and *Prosopium* spp.), numerous examples of intralacustrine forms are found in Europe and North America (76, 170). As in arctic charr, the degree of discrete differences among sympatric whitefish forms in resource use and morphology varies among lakes (e.g. 2, 9), but morphs are typically benthic or limnetic. They may differ in adult size and other life history attributes and in morphology such as jaw length and bluntness of snout (76, 83). The number and morphology of gillrakers (associated with type of prey) have most commonly been used to discriminate forms, which can reach five within a single lake (2, 9, 33, 76, 80, 170). An extensive molecular genetic survey in North America suggests that morphs may be genetically divergent within single lakes but do not represent separate invasions from adjacent rivers or lakes (10, 70; but see 5).

The Pacific salmon (*Oncorhynchus nerka*), native to the northern Pacific Ocean, exhibits two morphs, the anadromous sockeye and the nonanadromous kokanee. The former typically spends the first year in a lake before migrating to the ocean, whereas the latter remains in lakes throughout its lifetime (5, 35). Kokanee matures at a smaller body size and often at a younger age than sockeye, and where they coexist, they may display distinct morphological differences such as more gillrakers in the former (35, 99, 198). Kokanee have originated from sockeye independently and repeatedly. For instance, kokanee

Table 1 Resource polymorphisms in selected vertebrate species and the nature of the ecological segregation among morphs

Species	Nature of discrete ecological differences	PD[a]	References
FISHES			
Arctic charr (*Salvelinus alpinus*)	Benthivory, planktivory, piscivory, and migration	m,b,l	(47, 49, 57–59, 66, 69, 79, 105, 106, 122, 125, 149, 153, 167, 168, 188)
Atlantic salmon (*Salmo salar*)	Migration	l	(8, 183)
Brown trout (*Salmo trutta*)	Benthivory, planktivory, piscivory, and migration	m,b,l	(34, 87, 127)
Brook charr (*S. fontinalis*)	Benthivory, planktivory, swimming activity	m,b	(7, 43, 84)
Sockeye salmon (*Oncorhyncus nerka*)	Benthivory, planktivory, and migration	m,b,l	(35, 127, 198)
Coho salmon (*Oncorhyncus kisutch*)	Lake versus stream habitat	m,b	(172)
Lenok (*Brachymystax lenok*)	Benthivory, planktivory, and piscivory	m,1	(71, 110)
Lake whitefish (*Coregonus clupeaformis*)	Benthivory, planktivory, piscivory, and migration	m,b,1	(9, 27, 33, 76, 127, 141, 185, 197)
Least cisco (*Coregonus sardinella*)	Benthivory, planktivory	m,1	(80)
Pygmy whitefish (*Prospium coulteri*)	Benthivory, planktivory,	m,1	(76, 83, 197)
Scandinavian whitefish (*Coregonus* spp.)	Benthivory, planktivory, and piscivory	m,b,1	(27, 127, 141, 197)
Rainbow smelt (*Osmerus mordax*)	Benthivory, planktivory, piscivory, and migration	m,b,1	(175)
Stickleback (*Gasterosteus aculeatus*)	Benthivory and planktivory	m,b	(23, 86, 127, 141, 197)
Bluegill sunfish (*Lepomis macrochirus*)	Benthivory and planktivory	m,b	(28, 29, 127, 197)
Pumpkinseed sunfish (*Lepomis gibbosus*)	Benthivory and planktivory	m	(127, 128, 197)
Tui chub (*Gila bicolor*)	Benthivory and planktivory	m	(39, 127)
Cichlids (*Perissodus* spp.)	Eating scales from left versus right side of live fish	m	(63)
Cichlid (*Cichlasoma minckleyi*)	Feeding on snails and plant material	m,b	(72, 74, 127, 197)
Cichlid (*Cichlasoma citrinellum*)	Feeding on snails and soft-bodied prey	m,b	(90–92, 127, 197)
Cichlid (*Astatoreochromis alluaudi*)	Feeding on snails and soft-bodied prey	m	(46, 94)
Goodeid fish (*Ilyodon* spp.)	Strong indication of differences in food (lake form)	m	(48, 127, 182, 197)

(*Continues*)

Table 1 (*Continued*)

Species	Nature of discrete ecological differences	PD[a]	References
Neotropical fish (*Saccodon* spp.)	Different techniques in eating algae	m	(126, 127)
AMPHIBIANS			
Salamanders and newts			
Notophthalmus v. viridescens	Habitat, metamorphosing	m,1	(55)
N. v. dorsalis	Habitat, metamorphosing	m,1	(53)
Taricha granulosa	Habitat, metamorphosing	m,1	(81)
Ambystoma tigrinum	Habitat/diet, cannibalism	m,1	(19, 22)
A. talpoideum	Habitat, metamorphosing	m,1	(54, 112, 143, 144)
A. lermaensis	Habitat, metamorphosing	m,1	(145)
A. amblycephalum	Habitat, metamorphosing	m,1	(145)
A. rosaceum	Habitat, metamorphosing	m,1	(17, 145)
A. ordinarium	Habitat, metamorphosing	m,1	(3, 145)
A. gracile	Habitat, metamorphosing	m,1	(145)
Frogs and toads			
Spadefoot toad (*Scaphiopus multiplicatus*)	Omnivory, carnivory, and cannibalism	m,1	(113, 115, 119)
Pacific treefrog (*Pseudacris regilla*)	Habitat selection by color morphs	m,b	(97)
BIRDS			
Pacific reef heron (*Egretta sacra*)	Differences in hunting techniques associated with color	m,b	(130)
Little blue heron (*Egretta caerulea*)	Foraging success and vulnerability to predators	m,b	(16)
Buteo hawks	Proposed differential hunting success of color morphs	m	(129, 131)
Hook-billed kite (*Chondrohierax unicinnatus*)	Feeding on different size tree snails	m	(164)
Oystercatchers (*Haematopus ostralegus*)	Different feeding techniques on mussels	m,b	(40, 64, 107, 169)
Woodcock (*Scolopax rusticola*)	Ecological correlates of different bill types unknown	m	(12)
Blackcap warbler (*Sylvia atricapilla*)	Differences in migratory behavior	b	(6)
Robin (*Erithacus rubecula*)	Differences in migratory behavior	b	(6)
Seedcracker (*Pyrenestes ostrinus*)	Feeding on soft- and hard-seeded sedges	m	(156, 160, 162)
Cocos finch (*Pinaroloxias inornata*)	Feeding behavior, food type	b	(189)
Darwin's Finch (*Geospiza conirostris*)	Diet, ephemeral in population	m,b	(44)

[a]PD, Phenotypic difference; m, morphological; b, behavioral; and l, life history.

have appeared after sockeye were introduced to lakes previously lacking the species (35). The timing and locality of spawning may or may not differ (73), but molecular genetic studies suggest that distinct genetic differences exist between sympatric morphs (35) and that intralacustrine morphs arise locally (35, 176). In the volcanic lake Kronotskiy in Kamchatca, benthic and limnetic morphs of kokanee differ in the number of gillrakers and show spatial and temporal segregation in spawning (73).

Rainbow smelt (*Osmerus mordax*) exhibit extensive life history diversity throughout northeastern North America. There are both sea-run and lake-resident populations, with the latter often diversified in single lakes into dwarf and normal-sized forms. The dwarf smelt, which is limnetic, has more gillrakers, larger eyes, and a shorter upper jaw than the normal benthic piscivorous form, which is similar in morphology to anadromous smelt (175). In two of the study lakes, molecular genetic analysis showed that forms are reproductively isolated and that the segregation had occurred independently within each lake (175).

The pumpkinseed (*Lepomis gibbosus*) and the bluegill sunfish (*L. macrochiru*) co-occur and occupy distinct niches in many North American lakes. The adult bluegill is an open-water planktivorous generalist, while the pumpkinseed specializes on snails and occurs in shallow water (128). In a lake where the pumpkinseed is rare, a shallow water morph of bluegill is found coexisting with the typical open-water form. The shallow-water morph has a deeper body and longer fins than the open-water form. Clear differences appear between morphs in the flexibility of their feeding behavior, which correlates with differential foraging success in their respective habitats (28, 29). In lakes where only pumpkinseeds occur, they may segregate into two morphs, differing in the structure of gillrakers and body shape; the typical form feeds on snails, and an open-water form seemingly occupies the bluegill niche (128). Morphs tend to breed in somewhat different habitats (128). The phenotypic differences of the sunfish morphs are relatively subtle and went unnoticed in numerous ecological studies (29, 128).

The threespine stickleback (*Gasterosteus aculeatus*) is widely distributed in coastal regions throughout the northern hemisphere, occurring in marine, brackish, and freshwater and expressing a variety of ecological forms. In six small landlocked lakes in British Columbia, Canada, pairs of limnetic and benthic forms coexist, showing a high degree of ecological segregation. The limnetic form is slim-bodied with many long gillrakers and a narrow mouth, while the benthic is larger, deep-bodied, with a few short gillrakers and a wide mouth (86, 88, 140). The pairs show positive assortative mating (86, 124), but there is a persistent low level of hybridization. The forms are thus recognized as good biological species that evolved after the last glaciation (86). Research shows that they may represent two separate invasions that have subsequently diverged (86), sympatric divergence, or both (177). Benthic and limnetic stickleback morphs were recently identified in a lake in Alaska; they are believed to have arisen sympatrically (23). Morphological differences between them are much less pronounced than those between the species pairs in British Columbia, and the percentage of intermediate forms is much higher.

Cichlid fishes are celebrated for their high species richness in lakes of the African rift and in Central America (27, 38). *Cichlasoma minckleyi* of Cuatro Ciénagas, Mexico, exhibits two morphs, one vegetarian and the other feeding on snails. The former has a narrow head, long intestine, and small papilliform pharyngeal jaw dentition. The latter has short intestines, a wider head, stouter jaw, and larger molariform pharyngeal teeth, used for crushing snails (72, 74, 134). That the morphs interbreed is clear from molecular genetic and spawning ground studies (72 , 134). Behavioral trials show that feeding segregation of morphs is most pronounced when resources are limited (74). Another cichlid (*Cichlasoma citrinellum*) has similar morphs in Nicaraguan lakes, as does an African cichlid (*Astatoreochromis alluaudi*) found in East Africa (46, 90–92). In general, variability in jaw morphology is frequently noted in African cichlids (197), and it has been suggested that some of the numerous species that have been described in the African lakes represent resource morphs (72, 94). A clear case of resource polymorphism with little or no genetic isolation is seen in the scale-eating cichlid *Perissodus microlepis* in Lake Tanganyika. Morphs exhibit right- and left-handedness in jaw morphology; the former removes scales from the prey's left side, while the latter removes scales from the prey's right side (63).

Amphibians

Studies investigating adaptive plasticity in amphibian metamorphosis have provided important information on the relative costs and benefits of alternative adaptive phenotypes (20, 101–104, 192). Amphibians may be polymorphic with respect to metamorphosis, with some populations exhibiting both aquatic and metamorphosed adults (3, 17, 102, 144) (Table 1). Larval morphs may differ in numerous cranial and postcranial structures such as teeth, jaw musculature, and body size (11, 109, 197), intestine length (115), and age and size at metamorphosis (22, 115), some relating to whether they are carnivorous (often cannibalistic) or omnivorous.

In the New Mexico spadefoot toad (*Scaphiopus multiplicatus*), rapidly developing cannibalistic morphs grow larger than more slowly developing omnivorous morphs. Carnivorous tadpoles differ from typical omnivorous larvae in their hypertrophied jaw musculature, fewer teeth, decreased melanization, and shorter intestine (109, 119). All these characteristics may be induced during metamorphosis in anurans by exposing them to thyroid hormone (31, 36, 50, 52).

Polymorphisms in trophic structures also occur in some subspecies of larval and adult tiger salamanders (*Ambystoma tigrinum*) (18–20, 22, 118, 120, 132). Morphs may differ in maturation patterns; these may include retention of larval characteristics at sexual maturity in aquatic habitats (paedomorphosis via neoteny), or metamorphosis and sexual maturation in terrestrial habitats. Larval forms include "typical" morphs, which feed on zooplankton and other invertebrates, and cannibalistic morphs, which feed on both invertebrates and conspecifics (20, 22, 60, 77, 118). Relative to the typical morph, cannibalistic morphs are characterized by broader heads and mouths with enlarged vomerine teeth, adaptations apparently evolved for feeding on conspecifics (22, 118). Three morphs occur in adults, including metamorphosed, typical branchiate, and cannibal branchiate (21, 22), but there are few studies showing ecological differences among these morphs. Many species of *Ambystoma* vary in their propensity to metamorphose (Table 1), a variation resulting in obvious differences in resource use (terrestrial vs aquatic). For example, most *A. talpoideum* larvae metamorphose into terrestrial juveniles or adults, while in other populations individuals retain a larval morphology and reproduce as paedomorphs (112, 142).

That some color polymorphisms include a resource component has also been documented. These involve predation avoidance and use of differing microhabitats by being cryptic, although such examples are somewhat more difficult to demonstrate (100, 121, 180). For example, microhabitat selection by green and brown color morphs of the Pacific tree frog (*Pseudacris regilla*) appears to occur in response to predation (97).

Birds

Trophic morphs in birds may show differences in morphology, behavior, or a combination of both (Table 1). Bill-size morphs of the African finch *Pyrenestes ostrinus* feed on sedge seeds which differ in hardness. Morphs with small bills feed more efficiently on soft seeds, while the large-billed morphs do so on hard seeds (157–163). Studies of reproductive behavior indicate that finches mate randomly with respect to bill size (156, 157, 163). Different bill types of the hook-billed kite (*Chondrohierax uncinatus*) appear to be related to feeding on different size/age classes of tree snails (164), while in the oystercatcher (*Haematopus ostralegus*), differences in bill morphology are correlated with differences in feeding behavior and arise by differential wear of the bill. "Stabbers" have pointed bills and feed on mussels by inserting their bill between the valves, while "hammerers" exhibit blunt bills and break shells open by pounding (40, 107, 169). The frequency of morphs may also change seasonally as a function of differential wear imposed by dietary switches (64). An example of a behavioral polymorphism that has given rise to differential discrete niche use is seen in the Cocos Island finch (*Pinaroloxias inornata*). While the population exhibits little morphological variation, individuals show a diverse array of feeding behaviors, equivalent to those of several families of birds (189). The common color polymorphisms in herons may be maintained by differences in morph crypticity related to foraging mode (61, 98). Differences in habitat and foraging mode have been best documented between morphs of the Pacific reef heron (*Egretta sacra*), in which dark morphs employ a "standing and waiting" or "running mode" in shallow water, while the white phase employs a "flight-freeze" in breaking surf. It is believed that the white phase is more cryptic to prey in breaking surf and the dark phase more so in shallow, calmer waters (130). Experimental evidence for differential hunting success of color morphs is presented by Mock (96). Mock found that more fish were captured near models of white herons than around dark herons when the models were placed in shallow water on sunny days where herons typically forage. This experiment supports the assertion that the white form is more cryptic to fish when viewed against a clear sky. In general, however, although color polymorphisms are widespread in birds (15, 65), relatively few examples exist in which color morphs use resources differently.

Behavior polymorphisms with a demonstrated genetic basis include migration tendency in populations of two European birds, the blackcap (*Sylvia atricapilla*) and the robin (*Erithacus rubecula*) (6). Some populations show bimodality in migration behavior, in which some individuals are resident and others are migratory. While the evolutionary significance is unclear, it is likely to be resource-based (6). Some resource polymorphisms may be ephemeral in nature, such as in the case of the Darwin's finch, *Geospiza conirostris*, on the island of Genovesa in the Galápagos Islands (41, 44, 45). Males during one season were found to exhibit two discrete song types, while the male of each type differed in bill length and foraging mode. Longer billed, song A males, fed on *Opuntia* cactus flowers, whereas shorter billed, song B males, spent more time feeding on rotting *Opuntia* pads where they obtained larvae and pupae. While some initial evidence suggested assortative mating, morphological differences soon disappeared through random mating (41, 42).

ECOLOGICAL CIRCUMSTANCES THAT PROMOTE RESOURCE POLYMORPHISMS

Open Niches, Habitat Variability and the Relaxation of Interspecific Competition

From the examples discussed thus far, two circumstances appear fundamental in promoting resource polymorphisms: the existence of "open niches" or underutilized resources, and a relaxation of interspecific competition. Resources in many young lakes in the northern hemisphere are discrete, and fish tend to occur in either benthic or limnetic habitats (127, 141). For example, pumpkinseed and bluegill sunfish occupy distinct ecological niches where they co-occur; however, where only the pumpkinseed occurs, it has differentiated into two morphs (128). In whitefish the presence of competition may be inferred from morphology (76). Limnetic morphs of *Coregonus* with high gillraker counts are not found where the highly specialized cisco (*Leucichthys*) is present, and limnetic forms of *Prosopium* are not found where both of the other genera are present (76). Similarly, morphs of arctic charr are found where competing species are few or absent (78, 122). For instance, in Iceland, where only three species of fish are common in lakes, unusually diverse arctic charr morphs have evolved, taking advantage of most available habitats and resources (79, 127, 148, 152, 153, 167). Similarly, typical and cannibal morphs of tiger salamanders occur primarily where resource competitors and potential predators (fish) are relatively few (22). The lack of interspecific competition and the occurrence of an "empty niche" is also likely operating in the endemic Cocos island finch. Existing on an isolated oceanic island, much like a species of fish existing in a landlocked lake, and lacking interspecific competitors, this finch has diversified intraspecifically, exhibiting an array of distinct foraging behaviors typical of different species (189).

Habitat diversity and distinctness of resources can play an important role in fostering resource polymorphisms. For instance, volcanic lakes offer diverse and complex benthic habitats, often associated with subbenthic spaces, fissures, and caves rich in invertebrate prey for fish to exploit. Trophic adaptations in benthic charr morphs are clearly associated with subbenthic volcanic habitats (135, 148, 167). It is probably therefore not a coincidence that fish morphs and recently evolved species flocks are often found in volcanic-, rift-, and crater lakes as well as in recently deglaciated lakes (27, 67, 72, 93, 95, 137, 155, 167).

The Role of Specialization

Resource polymorphisms may also arise in some species-rich environments if resources are unique and require specialized traits to use. This appears to be the case in the finch *Pyrenestes ostrinus,* in which each morph specializes on sedge seeds differing in hardness, even though they occur in equatorial Africa, a region rich in granivorous birds species (156, 160–163). In this case, the resource on which these finches specialize is not used by other species. Cracking hard sedge seeds requires very specialized, broad, stout bills that sympatric granivorous species lack (162, 163). Because the most closely related species of finches have small bills, similar in size and shape to the small morph, it is believed that larger billed morphs have evolved from small billed forms by specializing on harder seeded sedge species (162). This is similar to the situation in the oystercatcher in which initial specialization on intertidal mussels seems to have given rise to even greater specialization, resulting in discrete feeding modes and a dimorphism in bill morphology (40). Another example in which specialization may lead to utilizing new resources is seen in the Lake Tanganyikan scale-eating cichlid fish (62, 63). Morphs are either left-handed or right-handed in the direction of mouth opening. In all three instances, specialization is associated with utilizing new, previously underexploited, resources.

Behavioral specialization of individual fish (cf 25), even at a very early age, may play an important first step in segregation. For example, variability in early behavior in the Atlantic salmon influences foraging and thus growth and later life history (88, 178); varying levels of flexibility in foraging behavior have been identified as key features in some polymorphic systems (7, 28, 29, 43, 47, 79, 84, 89, 90, 153, 197).

MECHANISMS MAINTAINING RESOURCE POLYMORPHISMS

Modes of Selection

Disruptive and frequency-dependent selection may play important roles in maintaining resource polymorphisms. A dramatic example of frequency-dependent selection is found in the Lake Tanganyika scale-eating cichlid fish. The frequency of right-mouthed and left-mouthed morphs fluctuated around a ratio of 1:1 over a ten-year period. Apparently individuals of the rarer morph are at a selective advantage because they are more successful in snatching scales from the flanks of prey (63). In New Mexico, omnivorous and carnivorous larval morphs of the spadefoot toad coexist in ephemeral ponds. Because of its faster developmental rate, the carnivorous morph is favored in short-duration ponds, but in longer-duration ponds, the slower-developing omnivorous morph is favored because its larger fat reserves enhance postmetamorphic survival. In ponds of intermediate duration, the abundance of each morph is frequency dependent (113, 115, 119).

In the African finch, *Pyrenestes ostrinus,* disruptive selection is most intense in juveniles (158, 161). Fitness peaks are associated with small and large morphs and correspond to performance peaks on soft- and hard-seeded sedges (162). In these finches, individuals at the extremes of each morph exhibit lower feeding performance and survival. Selection appears to be most intense following the major dry season, when food availability is low (160, 161).

Studies on the limnetic and benthic pairs of threespine stickleback in British Columbia show that the different forms have relatively higher feeding performances and grow best in their respective habitats, while hybrids do poorly in either habitat, suggesting selection against hybrids (139). Similarly, studies of early development suggest that hybrids of sockeye and kokanee suffer higher mortality than do pure lines (198). Behavioral studies on other fish species almost exclusively show that performance of each morph is positively correlated with the resource it utilizes (28, 29, 78, 90, 153, 193). In general, divergent selection (selection against intermediates) in the evolution of sympatric morphs and/or new species of freshwater fish is likely the result of intraspecific competition for food between phenotypically similar individuals. Subsequent, increased phenotypic divergence of sympatric forms would lead to reduced competition between them (74). This has recently been thoroughly discussed in the context of character release and displacement (127, 138, 140, 141, 173).

Phenotypic Plasticity, Induction, and Genetic Basis

Distinct morphological phenotypes may result from phenotypic plasticity (136, 184, 191). For example, cannibalistic and paedomorphic morphs of tiger salamanders and spadefoot toads arise through a developmental response to varying densities of conspecifics and food type (19, 111, 113–115, 117, 119). In tiger salamanders, morphogenesis is also responsive to kinship, with mixed broods more likely to develop the cannibal morphology than full-sib groups (116). Similarly, food type and quality change trophic morphologies in species of cichlid fishes (89, 194–197) and in pumpkinseed sunfish (187). In fishes,varying degrees of plasticity in foraging behavior and technique within and among morphs (28, 29, 74, 78, 90, 153) may lead to greater morphological specialization, which in turn channel the array of behavioral possibilities (90, 197). Similarly for oystercatchers, behavioral flexibility in feeding on discrete resources results in a bill dimorphism (40, 169). Life history among morphs is often highly plastic in fish, depending on food and habitat (58, 106, 151, 165, 166, 186). Such differences may depend on variability in foraging and social behavior, even very early in life (47, 88, 178).

The relative contribution of heredity and environment to phenotypic differences in fishes seems to vary not only among species but also among populations. The relative contribution varies among lakes in arctic charr (58, 106, 149, 151, 153, 171) and, based on rearing experiments, among forms in whitefish and Mexican *Ilyodon* (179). In contrast, morphological differences in species pairs of threespine sticklebacks in British Columbia have a strong genetic component (85, 86), but environmentally induced effects on morphology may accentuate their segregation. In the wild, the limnetic form has a more variable diet than the benthic form, and this is associated with relatively greater morphological plasticity in the former (24). Similarly, it is likely that both foraging behavior and growth patterns are less plastic in small benthivorous charr than in planktivorous charr in Thingvallavatn (78, 151, 153). The latter occupies a more temporally unstable niche than does the former, and long-term life-history studies of planktivorous charr show great fluctuations among year classes in growth and maturation patterns (165, 166).

In general, the genetic basis behind most resource polymorphisms is poorly understood. Both the bill size polymorphism in *Pyrenestes ostrinus* and handedness in the scale-eating cichlids appear to be determined by one locus with two alleles (63, 162). Moreover, the simple genetic control of at least some polymorphisms suggests that reaching new adaptive peaks may occur through mutations of large effect (108). To what extent other resource polymorphisms are controlled by one or a few loci will require further study.

While most alternative morphs are conditional and nonreversible, a few show reversible plastic phenotypes (90, 91, 115). In the cichlid *Cichlasoma citrinellum,* individual morphology may change reversibly in different seasons, depending on the kind of food available (91), a situation similar to that found in the oystercatcher, discussed earlier. While switches based exclusively on phenotypic plasticity seem to have evolved in unstable environments (22, 106, 115), switches under strong genetic control (63, 156, 158, 160, 162, 163) seem to evolve under relatively more stable environmental conditions. However, the determination of the role and importance of environmental stability requires further work. For example, common garden experiments involving the salamander *Ambystoma talpoideum* suggest that some populations have evolved phenotypic plasticity with respect to the tendency to metamorphose as ponds dry, while others show a genetic polymorphism (54, 144).

The developmental patterns producing different morphs also need further study. In several cases, heterochrony (1) has been emphasized (4, 22, 89, 149). A release from developmental and/or functional constraints can partly explain why some species are polymorphic and others are not. It has been suggested that a release of functional constraints in the jaw structure of cichlids allowed diversification in their feeding behavior, and that this partly explains their extensive radiation (75, 89, 90). In this case, phenotypic plasticity could either increase the rate of speciation or buffer against extinction, both leading to a net increase in species over time (89).

Alternative Adaptive Phenotypes and Speciation

The potential role resource polymorphisms play in speciation has been debated for decades (13, 14, 32, 51, 82, 94, 156, 190, 191). Most models and much of the debate center on whether the ecological separation caused by resource polymorphisms is sufficient to promote assortative mating and reproductive isolation in sympatry (14, 32, 82). Most alleged examples of sympatric speciation are correlative in that sympatric speciation is inferred from the dispersion pattern of already discrete species or races, as in some insects (174). A convincing example in vertebrates is found in some cichlids from Cameroon, West Africa (181): Mitochondrial DNA analysis of cichlid species flocks endemic to two crater lakes strongly suggests that each lake contains a monophyletic group of species that originated sympatrically (137). Species within lakes are more closely related to each other than they are to riverine species or species from adjacent lakes, and there are no geographic features of the lakes that could have provided geographic isolation. While polymorphisms were not demonstrated in this instance, they are likely to have been an intermediate step, given the many lakes in East and West Africa that contain polymorphic populations of cichlids (93, 95).

The amount of gene flow among sympatric morphs is variable and, if restricted, may lead to divergence and speciation (10, 34, 35, 56, 59, 87, 175–177). This may occur through either postzygotic mechanisms such as reduced fitness of hybrids (90, 94, 139, 198) or prezygotic mechanisms such as spatial and temporal segregation in breeding (often promoted by philopatry) and differences in breeding behavior or mate choice (9, 26, 27, 33, 80, 86, 91, 94, 124, 128, 132, 146, 152, 155, 175). These kinds of isolating mechanisms may coevolve with the phenotypic attributes such as size and color and the ecological segregation that characterize adaptations of morphs in their respective subniches (27, 91, 93, 128, 146, 152, 155). Among freshwater fish populations, segregation has occurred repeatedly within the same freshwater system and even within the same lake (10, 35, 56, 59, 86, 175), and the degree of genetic divergence between sympatric morphs is highly variable. In some cases, gene flow may be unimpeded, while in others, sympatric types may appear partially or completely reproductively isolated (10, 34, 56, 59, 86).

Niche-specific adaptation, typical of resource polymorphisms, is a key element in the divergence with gene flow speciation model (30, 32, 123, 154). A recent review (123) of laboratory studies involving *Drosophila* finds considerable support for the model when positive assortative mating occurs as a by-product of pleiotropy and/or genetic hitchhiking. The model proposes that speciation may occur under a range of selection intensities and levels of gene flow lying along a continuum: At one extreme is a population in a homogeneous environment with selection for two opposing phenotypes, and at the other extreme are parapatric populations experiencing differing directional selective forces in each. Reproductive isolation occurs if traits important in isolation are correlated or if they are the same as the traits important in resource use. Rice & Hostert (123) refer to this as the single-variation model of divergence-with-gene-flow speciation. In the model, reproductive isolation evolves through pleiotropy and/or genetic hitchhiking (sampling error–induced linkage disequilibrium between alleles affecting positive assortative mating and alleles affecting divergently selected characters) (123). Particularly salient is that a complete barrier to gene flow is unnecessary for speciation to occur if selection is strong and the trait under selection is also important in reproductive isolation. One could imagine just such a situation if morphs tended to reproduce where they fed and discrete resources occurred in different habitats. We believe this model has particular merit in understanding the possible role of resource polymorphisms in speciation. Incorporating aspects (Figure 1) of this model, resource polymorphisms could lead to speciation in the following steps: (*a*), (*a*), . . . , 1) invasion or exploitation of new or unexploited resource ("open" niches), 2) a decrease in intraspecific competition, 3) multifarious (usually divergent)

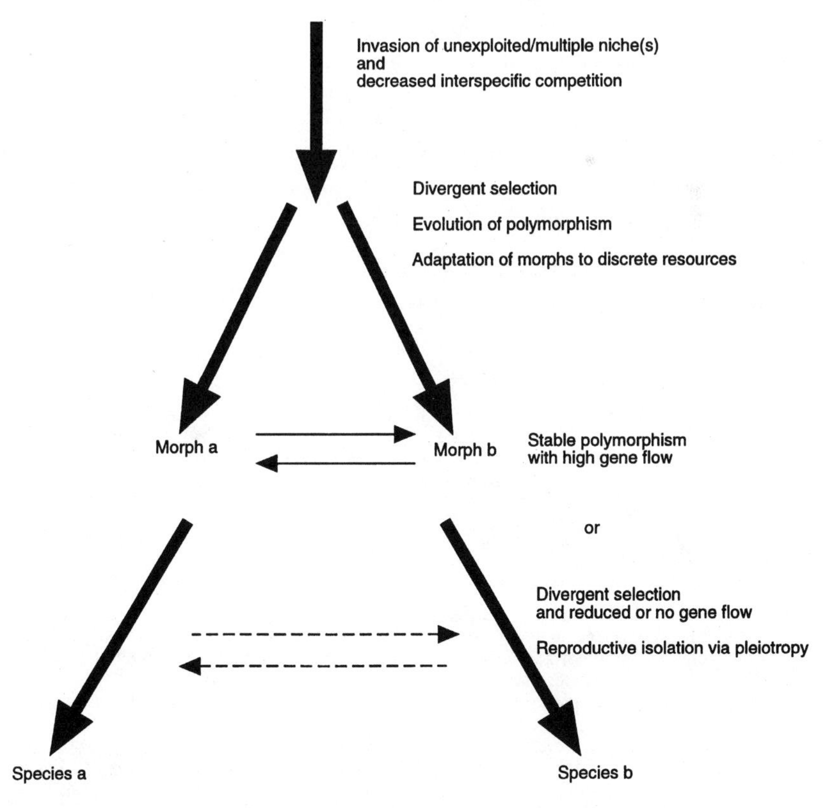

Figure 1 Generalized schematic showing possible steps and mechanisms leading to resource polymorphisms and speciation.

selection and evolution of a polymorphism, and 4) reduced gene flow and the evolution of prezygotic reproductive isolation mechanisms (150). If the model proves correct, as laboratory studies suggest, perhaps the reason resource polymorphisms not more common is that they more likely lead to speciation than to stable polymorphisms.

CONCLUDING REMARKS

Resource polymorphisms have been largely underestimated as a diversifying force and are likely more common and important in speciation than is currently appreciated. Many of the mechanisms that produce and maintain resource polymorphisms are similar among diverse taxa. Recognizing that many of the differences that separate morphs may be subtle, but nevertheless discrete, will likely lead to identification of more examples, especially in taxa where there is a tendency to assume that discontinuous morphotypes represent distinct species. The respective roles of development, phenotypic plasticity, genetics, natural selection, and ecology in maintaining and producing resource polymorphism, and a consideration of why polymorphisms appear more common in some taxa than others are fruitful areas for further investigation.

ACKNOWLEDGMENTS

We thank H Briem, T Eiriksson, H Malmquist, A Meyer, D Noakes, S Rohwer, E Routman, SS Snomason, D Stauffer, and E Taylor for helpful discussion and/or comments on the manuscript.

Literature Cited

1. Alberch P, Gould SJ, Oster GF, Wake DB. 1979. Size and shape in ontogeny and phylogeny. *Paleobiology* 5:296–317
2. Amundsen P. 1988. Habitat and food segregation of two sympatric populations of whitefish (*Coregonus lavaretus* L.) in Stuorajavri, Northern Norway. *Nordic J. Freshw. Res.* 64:67–73
3. Anderson JD, Worthington RD. 1971. The life history of the Mexican salamander *Ambystoma ordinarium* Taylor. *Herpetologica* 27:165–76
4. Balon EK. 1984. Life history of arctic charrs: an epigenetic explanation of their invading ability and evolution. In *Biology of the Arctic Charr,* ed. L Johnson, BL Burns, pp. 109–141. Winnipeg: Univ. Manitoba Press
5. Bernatchez L, Dodson JJ. 1990. Allopatric origin of sympatric populations of lake whitefish (*Coregonus clupeaformis*) as revealed by mitochondrial-DNA restriction analysis. *Evolution* 44:1263–71

6. Berthold P. 1991. Genetic control of migratory behaviour in birds. *Trends Ecol. Evol.* 6:254–57
7. Biro PA, Ridgway MS. 1995. Individual variation in foraging movements in a lake population of young-of-the-year brook charr (*Salvelinus fontinalis*). *Behaviour* 132:57–74
8. Birt TP, Green JM, Davidson WS. 1991. Mitochondrial DNA variation reveals genetically distinct sympatric populations of anadromous and nonanadromous Atlantic salmon, *Salmo salar. Can. J. Fish. Aquat. Sci.* 48:577–82
9. Bodaly RA, Clayton JW, Lindsey CC. 1988. Status of the Squanga Whitefish, *Coregonus* sp., in the Yukon Territory, Canada. *Can. Field Nat.* 102:114–25
10. Bodaly RA, Clayton JW, Lindsey CC, Vuorinen J. 1992. Evolution of the lake whitefish (*Coregonus clupeaformis*) in North America during the Pleistocene: genetic differentiation between sympatric populations. *Can. J. Fish. Aquat. Sci.* 49:769–79
11. Bragg AN, Hayes S. 1963. A study of labial teeth rows in tadpoles of Couch's spadefoot. *Wasmann J. Bio.* 21:149–54
12. Burlando B, Fadat C, Spano S. 1994. Hypothèse de travail sur le problème des becasses a bec court. *Proc. Fourth European Woodcock and Snipe Workshop.* Slimbridge, UK
13. Bush GL. 1975. Sympatric speciation in phytophagous parasitic insects. In *Evolutionary Strategies of Parasitic Insects,* ed. PW Price, pp. 187–206. New York: Plenum
14. Bush GL. 1994. Sympatric speciation in animals: new wine in old bottles. *Trends Ecol. Evol.* 9:285–88
15. Butcher GS, Rohwer S. 1989. The evolution of conspicuous and distinctive coloration for communication in birds. In *Current Ornithology,* ed. DM Power, pp. 51–108. New York: Plenum
16. Caldwell GS. 1986. Predation as a selective force on foraging herons: effects of plumage color and flocking. *Auk* 103:494–505
17. Collins JP. 1979. Sexually mature larvae of the salamanders *Ambystoma rosaceum* and *A. tigrinum* from Chihuahua, Mexico: taxonomic and ecological notes. *J. Herpetol.* 13:351–54
18. Collins JP. 1981. Distribution, habitats and life history variation in the tiger salamander, *Ambystoma tigrinum,* in east-central and southeast Arizona. *Copeia* 3:666–75
19. Collins JP, Cheek JE. 1983. Effect of food and density on development of typical and cannibalistic salamander larvae in *Ambystoma tigrinum nebulosum. Am. Zool.* 23:77–84
20. Collins JP, Holomuzki JR. 1984. Intraspecific variation in diet within and between trophic morphs in larval tiger salamander (*Ambystoma tigrinum nebulosum*). *Can. J. Zool.* 62:168–74
21. Collins JP, Mitton JB, Pierce BA. 1980. *Ambystoma tigrinum:* a multi-species conglomerate? *Copeia* 1980:938–41
22. Collins JP, Zerba KE, Sredl MJ. 1993. Shaping intraspecific variation: development, ecology and the evolution of morphology and life history variation in tiger salamanders. *Genetica* 89:167–83
23. Cresko WA, Baker JA. 1995. Two morphotypes of lacustrine threespine stickleback, *Gasterosteus aculeatus,* in Benka Lake, Alaska. *Environ. Biol. Fish.*
24. Day T, Pritchard J, Schluter D. 1994. A comparison of two sticklebacks. *Evolution* 48:1723–34
25. Dill LM. 1983. Adaptive flexibility in the foraging behavior of fishes. *Can. J. Fish Aquat. Sci.* 40:398–408
26. Dominey WJ. 1984. Effects of sexual selection and life history on speciation: species flocks in African cichlids and Hawaiian *Drosophila.* In *Evolution of Fish Species Flocks,* ed. AA Echelle, I Kornfield, pp. 231–250. Orono, Maine: Univ. Maine Press
27. Echelle AA, Kornfield I. 1984. *Evolution of Fish Species Flocks.* Orono: Univ. Maine Press
28. Ehlinger TJ. 1990. Habitat choice and phenotypic-limited feeding efficiency in bluegill: individual differences and trophic polymorphism. *Ecology* 71:886–96
29. Ehlinger TS, Wilson DS. 1988. Complex foraging polymorphism in bluegill sunfish. *Proc. Natl. Acad. Sci. USA* 85:1878–82
30. Endler JA. 1977. *Geographic Variation, Speciation, and Clines.* Princeton, NJ: Princeton Univ. Press
31. Etkin W. 1968. Hormonal control of amphibian metamorphosis. In *Metamorphosis: A Problem in Developmental Biology,* ed. W Etkin, LI Gilbert, pp. 131–348. New York: Appleton-Century-Crofts
32. Felsenstein J. 1981. Skepticism towards Santa Rosalia, or why are there so few kinds of animals? *Evolution* 35:124–38
33. Fenderson OC. 1964. Evidence of subpopulations of lake whitefish, *Coregonus*

clupeaformis involving a dwarfed form. *Trans. Am. Fish. Soc.* 93:77–94
34. Ferguson J, Taggart JB. 1991. Genetic differentiation among the sympatric brown trout (*Salmo trutta*) populations of Lough Melvin, Ireland. *Biol. J. Linn. Soc.* 43:221–37
35. Foote CJ, Wood CC, Withler RE. 1989. Biochemical genetic comparison of sockeye and kokanee, the anadromous and non-anadromous forms of *Oncorhynchus nerka*. *Can. J. Fish Aquat. Sci.* 46:149–58
36. Fox H. 1984. *Amphibian Morphogenesis*. Clifton, NJ: Humana Press, Wiley
37. Frandsen F, Malmquist HJ, Snorrason SS. 1989. Ecological parasitology of polymorphic Arctic charr, *Salvelinus alpinus* (L.) in Thingvallavatn, Iceland. *J. Fish Bio.* 34:281–97
38. Fryer G, Iles TD. 1972. *The Cichlid Fishes of the Great Lakes of Africa: Their Biology and Evolution*. Edinburgh: Oliver & Boyd
39. Galat DL, Vucinich N. 1983. Food partitioning between young of the year of two sympatric Tui Chub morphs. *Trans. Am. Fish. Soc.* 112:486–97
40. Goss-Custard JD, Le V, Dit Durell SEA. 1983. Individual and age differences in the feeding ecology of Oystercatchers *Haematopus ostralegus* wintering on the Exe Estuary, Devon. *Ibis* 125:155–71
41. Grant BR, Grant PR. 1983. Fission and fusion in a population of Darwin's finches: an example of the value of studying individuals in ecology. *Oikos* 41:530–47
42. Grant BR, Grant PR. 1989. *Evolutionary Dynamics of a Natural Population: The Large Cactus Finch of the Galapagos*. Chicago: Univ. Chicago Press
43. Grant JWA, Noakes DLG. 1988. Aggression and foraging mode of young-of-the-year brook charr, *Salvelinus fontinalis* (Pices, Salmonidae). *Behav. Ecol. Sociobiol.* 22:435–45
44. Grant PR, Grant BR. 1989. Sympatric speciation and Darwin's finches. In *Speciation and Its Consequences*, ed. D Otte, JA Endler, pp. 433–57. Sunderland, MA: Sinauer
45. Grant PR, Grant BR, Smith JNM, Abbott I, Abbott LK. 1976. Darwin's Finches: population variation and natural selection. *Proc. Natl. Acad. Sci. USA* 73:257–61
46. Greenwood PH. 1965. Environmental effects on the pharygeal gill of a cichlid fish, *Astatoreochromis alluaudi* and their taxonomic implications. *Proc. Biol. J. Linn. Soc.* 176:1–10
47. Griffiths D. 1994. The size structure of lacustrine arctic charr (Pisces: *Salmonidae*) populations. *Biol. J. Linn. Soc.* 43:221–37
48. Grudzien TA, Turner BJ. 1984. Direct evidence that the Ilydon morphs are a single biological species. *Evolution* 38:402–7
49. Gudkov PK. 1994. Sympatric charr of the genus *Salvelinus* from lakes of the Chukotsk Peninsula. *J. Ichthyology* 34:48–59
50. Hanken J, Hall BK. 1988. Skull development during anuran metamorphosis: II. Role of thyroid hormone in osteogenesis. *Anat. Embryol.* 178:219–27
51. Hanken J, Hall BK. 1993. Mechanisms of skull diversity and evolution. In *The Skull: Functional and Evolutionary Mechanisms*, ed. J Hanken, BK Hall, pp. 1–36. Chicago: Univ. Chicago Press
52. Hanken J, Summers CH. 1988. Skull development during anuran metamorphosis: III. Role of thyroid hormone in chondrogenesis. *J. Exp. Zoo.* 246:156–70
53. Harris RN. 1987. Density-dependent paedomorphosis in the salamander *Notophthalmus viridescens dorsalis*. *Ecology* 68:705–12
54. Harris RN, Semlitsch RD, Wilber HM, Fauth JE. 1990. Local variation in the genetic basis of paedomorphosis in the salamander *Ambystoma talpoideum*. *Evolution* 44:1588–603
55. Healy WR. 1974. Population consequences of alternative life histories in *Notophthalmus* v. *viridescens*. *Copeia* 1974:221–29
56. Hindar K. 1994. Alternative life histories and genetic conservation. In *Conservation Genetics*, ed. V Loeschke, J Tomink, SK Jain, pp. 323–336. Basel: Birkhäuser
57. Hindar K, Jonsson B. 1982. Habitat and food segregation of dwarf and normal arctic charr (*Salvelinus alpinus*) from Vangsvatnet Lake, western Norway. *Can. J. Fish. Aqatic. Sci.* 39:1030–45
58. Hindar K, Jonsson B. 1993. Ecological polymorphism in Arctic charr. *Biol. J. Linn. Soc.* 48:63–74
59. Hindar K, Ryman N, Stahl G. 1986. Genetic differentiation among local populations and morphotypes of Arctic charr, *Salvelinus alpinus*. *Biol. J. Linn. Soc.* 27:269–85
60. Holomuzki JR, Collins JP. 1987. Trophic dynamics of a top predator, *Ambystoma tigrinum nebulosum* (Caudata: *Ambystomatidae*), in a lentic community. *Copeia* 1987:949–57
61. Holyoak WG. 1973. Significance of colour dimorphism in Polynesian populations of *Egretta sacra*. *Ibis* 115:419–20

62. Hori M. 1991. Feeding relationships among cichlid fishes in Lake Tanganyika: effects of intra- and interspecific variations of feeding behavior on their coexistence. *Eco. Int. Bull.* 19:89–101
63. Hori M. 1993. Frequency-dependent natural selection in the handedness of scale-eating cichlid fish. *Science* 260:216–19
64. Hulscher JB. 1985. Growth and abrasion of the oystercatcher bill in relation to dietary switches. *Nether. J. Zool.* 35:124–54
65. Huxley JS. 1955. Morphism in birds. In *Acta XI Intern. Ornithol. Congr.*, pp. 303–328. Basel, Switzerland
66. Johnson L. 1980. The arctic charr, *Salvelinus alpinus*. In *Charrs, Salmonid Fishes of the Genus Salvelinus*, ed. EK Balon, pp. 15–98. Hague: W. Junk
67. Jonasson PM, ed. 1992. Ecology of oligotrophic, subarctic Thingvallavatn. *Oikos* 64:437
68. Jonsson B, Hindar K. 1982. Reproductive strategy of dwarf and normal Arctic charr (*Salvelinus alpinus*) from Vangsvatnet Lake, Western Norway. *Can. J. Fish. Aquat. Sci.* 39:1404–13
69. Jonsson B, Skúlason S, Snorrason SS, Sandlund OT, Malmquist HJ, et al. 1988. Life history variation of polymorphic Arctic charr (*Salvelinus alpinus*) in Thingvallavatn, Iceland. *Can. J. Fish. Aqatic. Sci.* 45:1537–47
70. Kirkpatrick M, Selander RK. 1978. Genetics of speciation in lake whitefishes in the Allegash Basin. *Evolution* 33:478–85
71. Kondrashov AS. 1986. Sympatric speciation: When is it possible? *Biol. J. Linn. Soc.* 27:201–16
72. Kornfield I, Smith DC, Gagnon PS, Taylor JN. 1982. The cichlid fish of Cuatro Ciénagas, Mexico: direct evidence of conspecificity among distinct trophic morphs. *Evolution* 36:658–64
73. Kurenkov SI. 1977. Two reproductively isolated groups of kokanee salmon, *Oncorhynchus nerka kennerlyi*, from Lake Kronotskiy. *J. Ichthyology* 17:526–34
74. Liem KF, Kaufman LS. 1984. Intraspecific macroevolution and functional biology of the polymorphic cichlid species *Cichlasoma minckleyi*. In *Evolution of Species Flocks*, ed. AA Echelle, I Kornfield, pp. 203–16. Orono: Univ. Maine Press
75. Liem KF, Osse JWM. 1975. Biological versatility, evolution, and food resource exploitation in African cichlid fishes. *Am. Zool.* 15:427–54
76. Lindsey CC. 1981. Stocks are chameleons: plasticity in gill rakers of coregonid fishes. *Can. J. Fish. Aquat. Sci.* 38:1497–506
77. Loeb MLG, Collins JP, Maret TJ. 1994. The role of prey in controlling expression of a trophic polymorphism in *Ambystoma trgrinum nebulosum. Functional Ecol.* 8:151–58
78. Malmquist HJ. 1992. Phenotype-specific feeding behaviour of two arctic charr *Salvenlinus alpinus* morphs. *Oecologia* 92:354–61
79. Malmquist HJ, Snorrason SS, Skúlason S, Jonsson B, Sandlund OT, Jonasson PM. 1992. Diet differentiation in polymorphic arctic charr in Thingvallavatn, Iceland. *J. Anim. Ecol.* 61:21–35
80. Mann GJ, McCart PJ. 1981. Comparison of sympatric dwarf and normal populations of least cisco (*Coregonus sardinella*) inhabiting Trout Lake, Yukon Territory. *Can. J. Fish. Aquat. Sci.* 38:240–44
81. Marangio MS. 1978. The occurrence of neotenic rough-skinned newts (*Taricha granulosa*) in montane lakes of southern Oregon. *NW Sci.* 52:343–50
82. Maynard Smith J. 1966. Sympatric speciation. *Am. Nat.* 100:637–50
83. McCart P. 1970. Evidence for the existence of sibling species of pygmy whitefish (*Prosopium coulteri*) in three Alaskan lakes. In *Biology of Coregonid Fishes*, ed. CC Lindsey, CS Woods, pp. 81–98. Winnipeg: Univ. Manitoba Press
84. McLaughlin RL, Grant JWA. 1994. Morphological and behavioural differences among recently-emerged brook charr, *Salvelinus fontinalis*, foraging in slow- vs. fast-running water. *Environ. Biol. Fishes* 39:289–300
85. McPhail JD. 1984. Ecology and evolution of sympatric sticklebacks (*Gasterosteus*): morphological and genetic evidence for a species pair in Enos Lake, British Columbia. *Can. J. Zool.* 62:1402–8
86. McPhail JD. 1994. Speciation and the evolution of reproductive isolation in sticklebacks (*Gasterosteus*) of southwestern British Columbia. In *The Evolutionary Biology of the Threespine Stickleback*, ed. MA Bell, SA Foster, pp. 399–437. Oxford: Oxford Univ. Press
87. McVeigh HP, Hynes RA, Ferguson A. 1995. Mitochondrial DNA differentiation of sympatric populations of brown trout, *Salmo trutta* L., from Lough Melvin, Ireland. *Can. J. Fish. Aquat. Sci.* 52:1617–22
88. Metcalfe N. 1993. Behavioural causes and consequences of life history variation in fish. *Mar. Behav. Physiol.* 23:205–17

89. Meyer A. 1987. Phenotypic plasticity and heterochrony in *Cichlasoma managuense* (Pisces, Cichlidae) and their implications for speciation in cichlid fishes. *Evolution* 41:1357–69
90. Meyer A. 1989. Cost of morphological specialization: feeding performance of the two morphs in the trophically polymorphic cichlid fish, *Cichlasoma citrinellum. Oecologia* 80:431–36
91. Meyer A. 1990. Ecological and evolutionary consequences of the trophic polymorphism in *Cichlasoma citrinellum* (Pisces: Cichlidae). *Biol. J. Linn. Soc.* 39:279–99
92. Meyer A. 1990. Morphometrics and allometry in the trophically polymorphic cichlid fish, *Cichlasoma citrinellum:* alternative adaptations and ontogenetic changes in shape. *J. Zoology* 221:237–60
93. Meyer A. 1993. Phylogenetic relationships and evolutionary processes in East African cichlid fishes. *Trends Ecol. Evol.* 8:279–84
94. Meyer A. 1993. Trophic polymorphisms in cichlid fish: Do they represent intermediate steps during sympatric speciation and explain their rapid radiation? In *Trends in Ichthyology,* ed. JH Schroder, J Bauer, M Schartl, pp. 257–66. GSF-Bericht: Blackwell
95. Meyer A, Kocher TD, Basasibwaki P, Wilson AC. 1990. Monophyletic origin of Lake Victoria cichlid fishes suggested by mitochondrial DNA sequences. *Nature* 347:550–53
96. Mock DW. 1981. White-dark polymorphism in herons. In *Proc. 1st Welder Wildlife Symp.,* pp. 145–161. Stinton, TX
97. Morey SR. 1990. Microhabitat selection and predation in the pacific treefrog, *Pseudacris regilla. J. Herpetol.* 24:292–96
98. Murton RK. 1971. Polymorphism in Ardeidae. *Ibis* 113:97–99
99. Nelson JA. 1968. Distribution and nomenclature of the North American kokanee, *Oncorhynchus nerka. J. Fish. Res. Board Can.* 25:415–20
100. Nevo E. 1973. Adaptive color polymorphism in cricket frogs. *Evolution* 27:353–67
101. Newman RA. 1989. Developmental plasticity of *Scaphiopus couchii* tadpoles in an unpredictable environment. *Ecology* 70:1775–87
102. Newman RA. 1992. Adaptive plasticity in amphibian metamorphosis. *BioScience* 42:671–78
103. Newman RA. 1994. Effects of changing density and food level on metamorphosis of a desert amphibian, *Scaphiopus couchii. Ecology* 75:1085–96
104. Newman RA. 1994. Genetic variation for phenotypic plasticity in the larval life history of spadefoot toads (*Scaphiopus couchii*). *Evolution* 48:1773–85
105. Noakes DLG, Skúlason S, Snorrason SS. 1989. Alternative life styles in salmonine fishes with emphasis on arctic charr *Salvelinus alpinus.* In *Alternative Life-History Styles of Animals,* ed. MN Bruton, pp. 329–46. Dordrecht: Kluwer Acad.
106. Nordeng H. 1983. Solution to the "charr problem" based on Arctic Char (*Salvelinus alpinus*) in Norway. *Can. J. Fish. Aquat. Sci.* 40:1372–87
107. Northon-Griffith M. 1967. Some ecological aspects of feeding behavior of the oystercatcher *Haematopus ostralegus* on the edible mussel *Mytilus edulis. Ibis* 109:412–24
108. Orr HA, Coyne JA. 1992. The genetics of adaptation: a reassessment. *Am. Nat.* 140:725–42
109. Orton GL. 1954. Dimorphism in larval mouthparts in spadefoot toads of the *Scaphiopus hammondi* group. *Copeia* 1954:97–100
110. Osinov AG, Ilyin II, Alexeev SS. 1990. Forms of Lenok (*Brachymystax* Salmoniformes, Salmonidae), delineated by genetic analysis. *J. Ichthyol.* 30:138–53
111. Packer EA. 1991. Kinship, cooperation and inbreeding in African lion: a molecular genetic analysis. *Nature* 351:563–65
112. Patterson KK. 1978. Life history aspects of paedogenic populations of the mole salamander, *Ambystoma talpoideum. Copeia* 1978:649–55
113. Pfennig DW. 1990. The adaptive significance of an environmentally cued developmental switch in an anuran tadpole. *Oecologia* 85:101–7
114. Pfennig DW. 1990. "Kin recognition" among spadefoot toad tadpoles: a side-effect of habitat selection? *Evolution* 44: 785–98
115. Pfennig DW. 1992. Polyphenism in spadefoot toad tadpoles as a locally adjusted evolutionarily stable strategy. *Evolution* 46:1408–20
116. Pfennig DW, Collins JP. 1993. Kinship affects morphogenesis in cannibalistic salamanders. *Nature* 362:836–38
117. Pfennig DW, Mabry A, Orange D. 1991. Environmental causes of correlations between age and size at metamorphosis in *Scaphiopus multiplicatus. Ecology* 72:2240–48
118. Pierce BA, Mitton JB, Jacobson L, Rose

FL. 1983. Head shape and size in cannibal and noncannibal larvae of the tiger salamader from west Texas. *Copeia* 1983:1006–12
119. Pomeroy LV. 1981. *The ecological and genetic basis of polymorphism in the larvae of spadefoot toads, Scaphiopus*. PhD thesis. Univ. Calif., Riverside
120. Powers JH. 1907. Morphological variation and its causes in *Ambystoma tigrinum*. *Stud. Univ. Neb.* 7:197–274
121. Pyburn WF. 1961. The inheritance and distribution of vertebrate stripe color in the cricket frog. In *Vertebrate Speciation*, ed. WF Blair, pp. 235–61. Austin: Univ. Texas
122. Reist JD, Gyselman E, Babaluk JA, Johnson JD, Wissink R. 1995. Evidence for two morphotypes of Arctic Char (*Salvelinus alpinus* L.) from Lake Hazen, Ellesmere Island, Northwest Territories, Canada. *Nordic J. Freshw. Res.* 71:396–410
123. Rice RR, Hostert EE. 1993. Laboratory experiments on speciation: What have we learned in 40 years? *Evolution* 47:1637–53
124. Ridgway MS, McPhail JD. 1984. Ecology and evolution of sympatric sticklebacks (*Gasterosteus*): mate choice and reproductive isolation in Enos Lake. *Can. J. Zool.* 62:1813–18
125. Riget A, Nygaard KH, Christensen B. 1986. Population structure, ecological segregation, and reproduction in a population of arctic charr (*Salvelinus alpinus*). *Can. J. Fish. Aquat. Sci.* 43:985–92
126. Roberts TR. 1974. Dental polymorphism and systematics in *Saccodon*, a neotropical genus of freshwater fishes (*Parodontidae, Characoidei*). *J. Zool.* 173:303–21
127. Robinson BW, Wilson DS. 1994. Character release and displacement in fishes: a neglected literature. *Am. Nat.* 144:596–627
128. Robinson BW, Wilson DS, Margosian AS, Lotito PT. 1993. Ecological and morphological differentiation of pumpkinseed sunfish in lakes without bluegill sunfish. *Evol. Ecol.* 7:451–64
129. Rohwer S. 1983. Formalizing the avoidance-image hypothesis: critique of an earlier prediction. *Auk* 100:971–74
130. Rohwer S. 1990. Foraging differences between white and dark morphs of the Pacific reef heron *Egretta sacra*. *Ibis* 132:21–26
131. Rohwer S, Paulson DR. 1987. The avoidance-image hypothesis and color polymorphism in Buteo hawks. *Ornis Scandinavica* 18:285–90
132. Rose FL, Armentrout D. 1976. Adaptive strategies of *Ambystoma tigrinum* Green inhabiting the Llano Estacado of West Texas. *J. Anim. Ecol.* 45:713–29
133. Safina C. 1990. Bluefish mediation of foraging competition between roseate and common terns. *Ecology* 71:1804–9
134. Sage RD, Selander RK. 1975. Trophic radiation through polymorphism in cichlid fishes. *Proc. Natl. Acad. Sci. USA* 74:4669–73
135. Sandlund OT, Jonsson B, Malmquist HJ, Gydemo R, Lindem T, et al. 1987. Habitat use of arctic charr *Salvelinus alpinus* in Thingvallavatn, Iceland. *Environ. Biol. Fish.* 20:263–74
136. Scheiner SM. 1993. Genetics and evolution of phenotypic plasticity. *Annu. Rev. Ecol. Syst.* 24:35–68
137. Schliewen UK, Tautz D, Pääbo S. 1994. Sympatric speciation suggested by monophyly of crater lake ciclids. *Nature* 368:629–32
138. Schluter D. 1994. Experimental evidence that competition promotes divergence in adaptive radiation. *Science* 266:798–801
139. Schluter D. 1995. Adaptive radiation in sticklebacks: trade-offs in feeding performance and growth. *Ecology* 76:82–90
140. Schluter D, McPhail JD. 1992. Ecological character displacement and speciation in sticklebacks. *Am. Nat.* 140:85–108
141. Schluter D, McPhail JD. 1993. Character displacement and replicate adaptive radiation. *Trends Ecol. Evol.* 8:197–200
142. Semlitsch RD. 1985. Reproductive strategy of a facultatively paedomorphic salamander *Ambystoma talpoideum*. *Oecologia* 65:305–13
143. Semlitsch RD. 1987. Paedomorphosis in *Ambystoma talpoideum:* effects of density, food and pond drying. *Ecology* 68:994–1002
144. Semlitsch RD, Harris RN, Wilber HM. 1990. Paedomorphosis in *Ambystoma talpoideum:* maintenance of population variation and alternative life-history pathways. *Evolution* 44:1604–13
145. Shaffer HB. 1984. Evolution in a paedomorphic lineage. I. An electrophoretic analysis of the Mexican ambystomatid salamaders. *Evolution* 38:1194–206
146. Sigurjonsdottir H, Gunnarsson K. 1989. Alternative mating tactics of arctic charr, *Salvelinus alpinus*, in Thingvallavatn, Iceland. *Environ. Biol. Fish.* 26:159–76
147. Skreslet S. 1973. Group segregation in landlocked Arctic Charr. *Salvelinus alpinus* of Jan Mayen Island in relation to the

charr problem. *Astarte* 6:55–58
148. Skúlason S, Antonsson TH, Gudbergsson G, Malmquist HJ, Snorrason SS. 1992. Variability in Icelandic arctic charr. *Iceland. Agric. Sci.* 6:143–53
149. Skúlason S, Noakes DLG, Snorrason SS. 1989. Ontogeny of trophic morphology in four sympatric morphs of arctic charr *Salvelinus alpinus* in Thingvallavatn, Iceland. *Biol. J. Linn. Soc.* 38:281–301
150. Skúlason S, Smith TB. 1995. Resource polymorphisms in vertebrates. *Trends Ecol. Evol.* 10:366–70
151. Skúlason S, Snorrason SS, Noakes DLG, Ferguson MM. 1996. Genetic basis of life history variations among sympatric morphs of arctic charr *Salvelinus alpinus. Can. J. Fish Aquat. Sci.* 00:00-
152. Skúlason S, Snorrason SS, Noakes DLG, Ferguson MM, Malmquist HJ. 1989. Segregation in spawning and early life history among polymorphic Arctic charr, *Salvelinus alpinus,* in Thingvallavatn, Iceland. *J. Fish Biol.* 35:225–32
153. Skúlason S, Snorrason SS, Ota D, Noakes DLG. 1993. Genetically based differences in foraging behaviour among sympatric morphs of arctic charr (Pisces: Salmonidae). *Anim. Behav.* 45:1179–92
154. Slatkin M. 1982. Pleiotropy and parapatric speciation. *Evolution* 36:263–70
155. Smith GR, Todd TN. 1984. Evolution of fish species flocks in north-temperate lakes. In *Evolution of Fish Species Flocks,* ed. AA Echelle, I Kornfield, pp. 47–68. Orono: Univ. Maine Press
156. Smith TB. 1987. Bill size polymorphism and intraspecific niche utilization in an African finch. *Nature* 329:717–19
157. Smith TB. 1990. Comparative breeding biology of the two bill morphs of the black-bellied seedcracker (*Pyrenestes ostrinus*). *Auk* 107:153–60
158. Smith TB. 1990. Natural selection on bill characters in the two bill morphs of the African finch *Pyrenestes ostrinus. Evolution* 44:832–42
159. Smith TB. 1990. Patterns of morphological and geographic variation in trophic bill morphs of the African finch *Pyrenestes. Biol. J. Linn. Soc.* 41:381–414
160. Smith TB. 1990. Resource use by bill morphs of an African finch: evidence for intraspecific competition. *Ecology* 71:1246–57
161. Smith TB. 1991. Inter- and intra-specific diet overlap during lean times between *Quelea erythrops* and bill morphs of *Pyrenestes ostrinus. Oikos* 60:76–82
162. Smith TB. 1993. Disruptive selection and the genetic basis of bill size polymorphism in the African finch, *Pyrenestes. Nature* 363:618–20
163. Smith TB. 1993. Ecological and evolutionary significance of a third bill form in the polymorphic finch *Pyrenestes ostrinus.* Presented at the *Birds and the African Environment: Proc. 8th Pan African Ornithological Congress,* Bujumbura, Burundi 1993
164. Smith TB, Temple SA. 1982. Feeding habits and bill polymorphism in hook-billed kites. *Auk* 99:197–207
165. Snorrason SS, Jonsson PM, Jonsson B, Lindem T, Magnasson KP, et al. 1992. Population dynamics of the planktivorous arctic charr, *Salvelinus alpinus* on Thingvallavatn. *Oikos* 64:352–64
166. Snorrason SS, Malmquist HJ, Jonsson B, Jonasson PM, Sandlund OT, Skúlason S. 1994. Modifications in life history charateristics of plantivorous arctic charr (*Salvelinus alpinus*) in Thingvallavatn, Iceland. *Verh. Int. Verein. Limnol.* 25:2108–12
167. Snorrason SS, Skúlason S, Jonsson B, Malmquist HJ, Jonasson PM, et al. 1994. Trophic specialization in Arctic charr *Salvelinus alpinus* (Pices: Salmonidae): morphological divergence and ontogenetic niche shifts. *Biol. J. Linn. Soc.* 52:1–18
168. Sparholdt H. 1985. The population, survival, growth, reproduction and food of arctic charr, *Salvelinus alpinus* (L.), in four unexploited lakes in Greenland. *J. Fish. Biol.* 26:313–30
169. Sutherland WJ. 1987. Why do animals specialize? *Nature* 325:483–84
170. Svärdson G. 1979. Speciation in Scandinavian Coregones. *Inst. Fresh Water Res. Drottningholm* 64:1–95
171. Svendang H. 1990. Genetic basis of life-history variation of dwarf and normal Arctic charr, *Salvelinus alpinus* (L.), in Stora Rosjon, central Sweden. *J. Fish Biol.* 36:917–32
172. Swain DP, Holtby LB. 1989. Differences in morphology and behavior juvenile coho salmon (*Oncorhynchus kisutch*) rearing in a lake and in its tributary stream. *J. Fish. Aquat. Sci.* 46:1406–14
173. Taper ML, Case TJ. 1992. Models of character displacement and the theoretical robustness of taxon cycles. *Evolution* 46:317–33
174. Tauber CA, Tauber MJ. 1989. Sympatric speciation in insects: perception and perspective. In *Speciation and its Consequences,* ed. D Otte, JA Endler, pp. 307–

175. Taylor EB, Bentzen P. 1993. Evidence for multiple origins and sympatric divergence of trophic ecotypes of smelt (*Osmerus*) in northeastern North America. *Evolution* 47:813–32
176. Taylor EB, Foote CJ, Wood CC. 1996. Molecular genetic evidence for parallel life-history evolution in a Pacific salmon: sockeye salmon and kokanee (*Oncorhynchus nerka*). *Evolution* 50:401–16
177. Taylor EB, Schluter D, McPhail JD. 1996. History of ecological selection pressures in sticklebacks: uniting phylogenetic and experimental approaches. In *Molecular Evolution and Adaptive Radiation*, ed. T Givnish, K Systema. Cambridge: Cambridge Univ. Press
178. Thorpe JE, Metcalfe NB, Huntingford FA. 1992. Behavioral influences on life-history variation in juvenile Atlantic salmon, *Salmo salar. Environ. Biol. Fish.* 33:331–40
179. Todd TN, Smith GR, Cable LE. 1981. Environmental and genetic contributions to morphological differentiation in ciscoes (Coregoninae) of the Great Lakes. *Can. J. Fish Aquat. Sci.* 38:58–67
180. Tordoff HW. 1980. Selective predation of gray jay, *Perisoreus canadensis*, upon boreal chorus frogs, *Pseudacris triseriata. Evolution* 34:1004–8
181. Trewavas E, Green J, Corbert SA. 1972. Ecological studies on crater lakes in West Cameroon, fishes of Barombi Mbo. *J. Zool. Lond.* 167:41–95
182. Turner BJ, Grosse DJ. 1980. Trophic differentiation in *Ilyodon*, a genus of stream-dwelling goodeid fishes: speciation versus ecological polymorphism. *Evolution* 34:259–70
183. Verspoor E, Cole LJ. 1989. Genetically distinct sympatric populations of resident and anadromous Atlantic salmon, *Salmo salar. Can. J. Zool.* 67:1453–61
184. Via S, Gomulkiewicz R, De Jong G, Scheiner SM, Schlichting CD, Van Tienderen PH. 1995. Adaptive phenotypic plasticity: consensus and controversy. *Trends Ecol. Evol.* 10:212–17
185. Vourinen AS, Bodaly RA, Reist JD, Bernatchez L, Dodson JJ. 1993. Genetic and morphological differentiation between dwarf and normal size forms of lake whitefish (*Coregonus clupeaformis*) in Como Lake, Ontario. *Can. J. Fish. Aquat. Sci.* 50:210–16
186. Vrijenhoek RC, Marteinsdottir G, Schenck R. 1987. Genotypic and phenotypic aspects of niche diversification in fishes. In *Community and Evolutionary Ecology of North American Stream Fishes*, ed. WJ Matthews, DC Heins, pp. 245–50. Norman OK: Univ. Oklahoma Press
187. Wainwright PC, Osenberg CW, Mittlebach GG. 1991. Trophic polymorphism in the pumpkinseed sunfish (*Lepomis gibbosus* Linnaeus): effects of environment on ontogeny. *Funct. Ecol.* 5:40–55
188. Walker AF, Greer RB, Gardner AS. 1988. Two ecologically distinct forms of artic charr *Salvelinus alpinus* (L.) in Lock Rannoch. *Biol. Conser.* 43:43–61
189. Werner TK, Sherry TW. 1987. Behavioral feeding specialization in *Pinaroloxias inornata*, the Darwin's finch of Cocos Island, Costa Rica. *Proc. Natl. Acad. Sci. USA* 84:5506–10
190. West-Eberhard MJ. 1986. Alternative adaptations, speciation, and phylogeny (a review). *Proc. Natl. Acad. Sci. USA* 83:1388–92
191. West-Eberhard MJ. 1989. Phenotypic plasticity and the origins of diversity. *Annu. Rev. Ecol. Syst.* 20:249–78
192. Wilbur HM, Collins JP. 1973. Ecological aspects of amphibian metamorphosis. *Science* 182:1305–14
193. Wilson DS. 1989. The diversification of single gene pools by density- and frequency-dependent selection. In *Speciation and its Consequences*, ed. D Otte, J Endler, pp. 366–85. Sunderland, MA: Sinauer
194. Wimberger PH. 1991. Plasticity of jaw and skull morphology in the neotropical cichlids *Geophagus brasiliensis* and *G. steindachneri. Evolution* 45:1545–63
195. Wimberger PH. 1992. Plasticity of fish body shape: the effects of diet, development, family and age in two species of *Geophagus* (Pisces: Cichlidae). *Biol. J. Linn. Soc.* 45:197–218
196. Wimberger PH. 1993. Effects of vitamin C deficiency on body shape and skull osteology in *Geophagus brasiliensis*: implications for interpretations of morphological plasticity. *Copeia* 1993:343–51
197. Wimberger PH. 1994. Trophic polymorphisms, plasticity, and speciation in vertebrates. In *Advances in Fish Foraging Theory and Ecology*, ed. DJ Stouder, K Fresh. Columbia, SC: Belle Baruch Press
198. Wood CC, Foote CJ. 1990. Genetic differences in early development and growth of sympatric sockeye salmon and Kokanee (*Oncorhynchus nerka*) and their hybrids. *Can. J. Fish. Aquat. Sci.* 47:2250–60

Annu. Rev. Ecol. Syst. 1996. 27:135–62

MANAGEMENT OF THE SPOTTED OWL: A Case History in Conservation Biology

Barry R. Noon and Kevin S. McKelvey
US Department of Agriculture, Forest Service, Redwood Sciences Laboratory, 1700 Bayview Drive, Arcata, California 95521

KEY WORDS: northern spotted owl, *Strix occidentalis*, forest management, population dynamics, endangered species, conservation biology

ABSTRACT

Official conservation efforts for the northern spotted owl began in the United States in 1975 when it was declared "threatened" in the state of Oregon; efforts continued in a sporadic and unsystematic way through the 1980s. In 1989 the Interagency Scientific Committee (ISC) was established by Congress and charged with the development of a scientifically defensible conservation strategy covering the entire range of the northern spotted owl, which includes parts of the states of Oregon, Washington, and California. The ISC collated all spotted owl research and approached questions concerning the need for a conservation strategy and the efficacy of potential reserve designs as testable hypotheses. Because the hypothesis tests were based on incomplete data and highly stylized population models, uncertainty concerning the conclusions of the ISC remained. Subsequent research focused on answering those uncertainties, and here we revisit the ISC's conclusions, asking which if any of them have been invalidated. The ISC's major conclusions have remained robust: The population of spotted owls is declining due to reductions in old growth habitat. Subsequent trend-analyses confirmed the levels of population decline calculated by the ISC and in addition concluded that the rate of decline was accelerating. The ISC's response to these conclusions was to recommend the establishment of an extensive network of large reserves. Subsequent research and more detailed computer modeling have confirmed the conceptual validity of this conservation plan but suggest that optimistic assumptions led the ISC to propose a minimal reserve structure. Current federal management plans in the Pacific Northwest propose more habitat than the ISC envisioned, providing a greater likelihood of persistence.

0066-4162/96/1120-0135$08.00

INTRODUCTION

The conservation saga of the northern spotted owl (*Strix occidentalis caurina*) is long and controversial (63, 85). Even today, despite the subspecies' listing as threatened under the Endangered Species Act (ESA), controversy continues, and some observers continue to assert that the owl's population is not in jeopardy, (e.g. 17). Numerous field studies indicate that the fate of the northern spotted owl is inextricably linked to the fate of large, old (>150 years) trees. By 1950 almost all old forest on private lands in western Washington and Oregon and northwestern California had been harvested. The remaining 10–15% of the original old forest was found almost exclusively on public lands administered by the Forest Service (FS) and Bureau of Land Management (BLM). After World War II, old forests on these lands began to be cut at the rate of 28,000–40,000 ha per year (85). Given the combined effects of forest fragmentation and habitat loss, listing the subspecies under the ESA was inevitable.

The vulnerability of the owl to continuing habitat loss was officially noted as early as 1975 when the state of Oregon designated the subspecies as "threatened." Thereafter, a series of public responses chronicled an increasing concern for its fate: In 1981 the Fish and Wildlife Service (FWS) recognized the threats to the owl but concluded that the subspecies did not yet warrant listing under the ESA (92). In 1983 the Forest Service (FS) designated the owl an "indicator species" of the integrity of old-growth forest ecosystems in the Pacific Northwest (PNW) (88), and in 1984 special management action was proposed in Oregon (habitat was protected around 375 owl pairs). In 1985 a team of scientists designated by the National Audubon Society concluded that the subspecies was headed toward listing and that immediate management intervention was justified (23). In 1987 the FWS conducted a second status review but again concluded that listing was not warranted. This decision was appealed, and the Federal District Court ruled that the decision not to list was "arbitrary and capricious," triggering the initiation of a third status review. In 1988 the FS adopted a special management plan for the owl but acknowledged that the species had a "poor" long-term chance of success. The plan was appealed by environmental groups, on the basis of failure to comply with existing environmental laws, and by the timber industry claiming unjustified economic hardship.

Against this increasingly hostile and polarized background, in 1989 Congress intervened and proposed an interim solution. By attaching a rider to the 1990 Interior Appropriations Bill (Section 318), Congress specified additional protection for the northern spotted owl, established a harvest-level acceptable to the timber industry, and exempted federal agencies (FS and BLM) from legal appeals. (Timber sales contracted under the Section 318 provision are still being disputed in 1996 as a consequence of the Recission Bill signed by

President Clinton.) A key provision of Section 318 was the establishment of an Interagency Scientific Committee (ISC) charged with the development of a long-term conservation strategy for the northern spotted owl on public lands. Jack Ward Thomas was selected as leader of that committee.

The ISC delivered its report to four agency heads (FS, BLM, National Park Service, and FWS) and Congress in April, 1990 (84). The conservation strategy called for the designation of approximately 2.4 million ha of federal lands, in addition to suitable habitat already in wilderness areas and national parks, to be arranged in a network of habitat conservation areas (HCAs), widely distributed throughout the range of the northern spotted owl.

Individuals active in the domains of economics, public policy, and politics were unnerved by the magnitude of the proposed conservation "solution." Ultimately, however, the most significant impact of the northern spotted owl conservation plan has been the increased scope of planning efforts for management of FS and BLM lands in the Pacific Northwest. The original focus on conservation of the northern spotted owl has expanded markedly to include viability concerns of other species, including aquatic organisms, invertebrates, plants (86), and indeed, the maintenance of the entire forest ecosystem (90). This most ambitious effort began in April 1993, when President Clinton established the Forest Ecosystem Management Assessment Team (FEMAT) to develop management options for Pacific Northwest forests within the range of the northern spotted owl. Despite a much broader focus, conservation planning for the northern spotted owl remains central to these additional planning efforts.

Since 1990, research on the northern spotted owl has continued and, in some cases, accelerated. The new information has increased our understanding of the ecology and life history of the owl, including new insights into its demography, distribution, habitat relations, behavior, and associations with prey. Significant advances have also taken place in conservation biology in general, and specifically in the principles of reserve design and the dynamics of spatially structured populations in the context of real landscapes (29, 54, 65, 66).

In this review, we do not attempt to discuss all facets of northern spotted owl biology and the challenge to conserve this subspecies. First, comprehensive reviews of new sources of information on northern spotted owl biology have recently been published (37, 86). Second, the scope of the conservation issue, in terms of its many economic, social, and legal ramifications is simply too broad. Instead we focus on the scientific foundation of the recommendations of the ISC and subsequent conservation strategies. We review the methods, data, and logic brought to bear to find a scientifically credible solution to the challenge of northern spotted owl conservation. Our approach is to revisit three fundamental hypotheses tested by the ISC (58, 84) and to re-evaluate the decisions based on those hypotheses in the context of more recent information and continuing advances in our understandings of the dynamics of wild populations. Similarly, we revisit the five basic principles of reserve design invoked by the ISC (58, 84, 97) and reassess their relevance to current understandings of conservation planning and reserve design.

FUNDAMENTAL HYPOTHESES TESTED BY THE ISC

To develop a scientifically credible conservation strategy, the ISC used the hypothetico-deductive methods of hypothesis testing (57, 58, 74, 75). Three hypotheses are generally applicable to species whose populations are threatened by the loss and fragmentation of their habitat. These hypotheses, framed in the context of the northern spotted owl conservation plan, were:

1. H_0: The finite rate of change ($\lambda \propto N_{t+1}/N_t$) of the northern spotted owl population is ≥ 1.0.
2. H_0: Northern spotted owls do not differentiate among habitats on the basis of forest age, structure, or composition.
3. H_0: No decline has occurred in the areal extent of habitat types selected by northern spotted owls for foraging, roosting, or nesting.

Null Hypothesis 1: Declining Populations

Estimates of λ are based on the eigenanalysis of stage projection matrices (22), parameterized from extensive field studies using capture-recapture and other standardized methods (35). The first null hypothesis was originally rejected based on the observation that λ was significantly less than 1.0 at two long-term demographic study sites (84). More rigorous tests of this hypothesis are now possible, based on data from several additional study sites widely spaced across the owl's range, and additional years of data from the original sites (32). A recent reanalysis of the demographic data from 11 study sites resulted in the overwhelming rejection of this hypothesis (16). Current estimates of λ, adjusted for bias (see below), indicate that populations of resident, territorial females have declined at an estimated rate of 4.5% per year from 1985 through 1993. Currently, only one study area reports a λ-estimate that suggests a stable population (41).

Greater availability of banding data and advances in the methods of analysis (e.g. 14, 48) have allowed more insight into a possible time dependency in the survival rates. Importantly, adult females show a significant, negative time-trend in annual survival rate over the period 1985–1993 (15), the parameter that most influences the value of λ (47, 60). Thus, as the USFWS Recovery Team

(93) and Burnham et al (15, 16) concluded, the rate of population decline has probably accelerated since the ISC's original analyses.

Based on statistical criteria and data availability, the analysis of Burnham et al (15, 16) was more rigorous than that by the ISC; Burnham et al used more study sites, years, capture histories, and sophisticated modeling techniques. Recently, however, concerns have been raised that these analyses may be based on negatively biased parameter estimates that lead to underestimates of λ (7). As discussed below, parameter estimates were adjusted for bias (15, 16), but Bart (7) argues that additional biases exist, increasing the likelihood of type I errors when testing hypothesis 1.

BIASES IN THE ESTIMATES OF λ Reliable estimates of rates of population change (λ) require accurate and precise estimates of the age- or stage-specific birth and death rates (22). It is important, therefore, to examine thoroughly possible sources of bias in the parameter estimates and to adjust for those biases whenever possible. For the northern spotted owl, concern over possible biased estimates has focused on birth and first-year survival rates.

Bias in the estimation of birth rate (b) The two possible sources of bias in the estimation of b, one positive and one negative, both arise from limitations in sampling efficiency. If breeding pairs are more readily detected than nonbreeding pairs, they would be overrepresented in the sample, introducing a positive bias to b. On the other hand, death of some newborns prior to detection would lead to an underestimate of productivity and a negative bias in b. The consensus among owl biologists is that these sources of bias are minimal, and no bias-adjustment has been developed.

Bias in the estimation of juvenile survival rate (s_0) Owl biologists have been most concerned about a negative bias in the estimation of s_0. Bias may arise because juvenile birds emigrate from the study area, survive at least one year, and are never reobserved. The consequence is that λ is underestimated. To adjust for this source of bias, an estimate of juvenile emigration rate (E) is needed. Assuming that emigrating juveniles survive at the same rate as nonemigrants, data derived from the radio-tracking of juvenile owls allowed Burnham et al (15, 16) to estimate E and adjust the Jolly-Seber (40, 80) estimate of s_0. The upward adjustment of s_0 increased the estimate of λ, but the null hypothesis of a stable or growing population was still consistently rejected (16).

Limits to inference from estimates of λ Estimates of λ pertain to rates of change over the period of study only for the population of territorial owls, not the floaters (see 7 for a counter opinion). Moreover, it is indefensible to use these estimates to compute past population sizes or future trends. To do so, one must assume constant birth and death rates, an untenable assumption for northern spotted owls (16). For example, fecundity varies annually, possibly because of winter/early spring weather patterns (99).

Additional variation in the parameters can arise due to sampling errors. Sampling variation arises from the fact that relatively few individuals in a population produce most of the offspring (34, 94), a pattern observed in virtually all extensive studies of lifetime reproductive success among birds (59). An additional limitation may arise as an artifact of sampling. When λ is estimated from a heterogeneous population consisting of a mixture of source and sink territories (64), inferences about population trend are uncertain. The overall λ-estimate from a heterogeneous population is a weighted average—with weights determined by the abundance of the two types of territories in the sample. Weighted λ-estimates <1.0 could arise even though the population as a whole were stable or growing.

Given these understandings, estimates of persistence likelihood for a population based on a heterogeneous sample must incorporate other relevant factors—for example, the smallest source population needed for local stability. Source populations are determined by the constraints of demographic and environmental stochasticity and the spatial distribution of breeding pairs (e.g. 45). For more global considerations, overall persistence of a metapopulation is determined by the number and spatial distribution of locally stable source populations. Rare, catastrophic events require that local populations be widely dispersed so that adverse effects are not experienced simultaneously by all local populations (24).

Vigorous debate over the shortcomings of demographic studies (70), complexities of the sampling process, limitations of the capture/recapture models (e.g. 48), and difficulties in the direct interpretation of λ-estimates (7) have led to more cautious interpretations of rates of population change. Despite these cautions, however, the weight of evidence still leads to a rejection of hypothesis 1 (16, 32).

Null Hypothesis 2: Habitat Specialization

All studies of northern spotted owl habitat use reviewed by the ISC led to rejection of the second null hypothesis and concluded that owls select old forests, or younger forests that have retained characteristics of old forests (84). An apparent exception occurred in the coastal redwood forests of northern California, where owls were also found in younger stands (30, 81). At an early age, redwood forests (representing $< 7\%$ of the owl's range) exhibit stand characteristics similar to those of old-growth forests elsewhere. However, even in these forests, owls nest in stands with a residual, large-diameter, old tree component (30). Owls may persist in these highly managed landscapes because of abundant prey, such as dusky-footed woodrats, which thrive in early seral habitats (18, 78, 84).

Studies published since Thomas et al (84) have provided additional falsification of hypothesis 2 (i.e. 5, 9, 12, 18, 20, 39, 49, 69, 83). Most studies of habitat selection relating to tests of hypothesis 2 have recently been reviewed (4, 86).

The accumulated knowledge from habitat studies allows a discussion of habitat use and the validity of hypothesis 2 at several spatial scales, including within and among forest stands, and home range and landscape scales. In the Oregon Coast Range (19), the Klamath Province in California (83), and the eastern slopes of the Washington Cascades (12), further study has confirmed selection for conifer-dominated stands characterized by large, old-tree components and closed canopies. However, patterns can vary geographically, with selection for old-growth forest at the stand level more pronounced west of the Cascade Mountain crest (43) than to the east (13). A stronger geographic contrast in habitat use occurs in northwestern California and southwestern Oregon where selected nest sites often had a well-developed hardwood understory (9, 42, 73, 83), a vegetation type absent from sites used elsewhere in the owl's range. Nest tree selection and nest substrate, however, showed even greater geographic variability—in the eastern Washington Cascades (12) nests were in clumps of mistletoe (*Arceuthobium douglassi*), and nest trees were generally smaller and younger than in Oregon (31) and California (43), where nests were in stick platforms or side cavities in large trees.

Most habitat-use information is provided by studies in which the owls are fitted with radio transmitters. The study by Blakesley et al (9), based on a sample from a contiguously distributed, local population in northwestern California, is an exception. Site selection within this local geographic distribution was significantly nonrandom, with daytime use concentrated in mature and old-growth forest. The selection of stands with old-growth components was most pronounced for nest sites.

At a landscape scale, the most extensive work is Carey et al (20). Based on data collected from owls fitted with transmitters in southwestern Oregon and the Olympic Peninsula, Carey et al (20) found significant landscape-level effects on home-range size and overlap, adult behavior, and the age composition of breeding pairs. Specifically, home-range size and overlap increased significantly in heavily fragmented landscapes, an effect that was more pronounced in mixed-conifer than in Douglas-fir forest. In general, twice as much old-growth forest was included within pair home ranges in landscapes dominated by Douglas-fir than in landscapes dominated by mixed-conifer forest. Particularly relevant to hypothesis 2, Carey et al (20) found selection for old forest to be significant at three spatial scales—landscape, annual home range of pairs, and foraging and roosting site selection by individuals within home ranges.

A number of studies have investigated habitat selection at multiple spatial scales (42, 49, 56, 72, 73). The general conclusion from these studies is that northern spotted owls select nest and roost sites in mature and old growth forest in areas that are less fragmented than the surrounding landscape. The spatial scale of selection around nest and roost sites (territory size) varies geographically from approximately 450 ha in northwestern California (42) to >1800 ha in Oregon (73) and >3200 ha in Washington (49).

In summary, northern spotted owls show extensive geographic variation in patterns of habitat use, but consistent associations between owl nesting and roosting locations and old-growth forest components have been found in all studies throughout the range of the owl. The weight of evidence, therefore, leads to a rejection of hypothesis 2 from a local to landscape scale. In general, the removal of old forest and its components results in reduced owl habitat.

CURRENT UNDERSTANDING OF HABITAT SELECTION Recent habitat studies have been less descriptive and more explanatory. Geographic variation in habitat use and area requirements has been most readily explained by variation in the abundance and composition of the prey base (18, 20, 98). For example, most studies continue to demonstrate strong selection of old forest stands by foraging owls; use of younger forests seems to be related to both the presence of hardwoods and the abundance of dusky-footed and bushy-tailed woodrats [*Neotoma fuscipes* and *N. cinera,* respectively; (98), (20), (18)]. In northwestern California and southwestern Oregon, conifer-dominated forests below 1250 m often have hardwood understories, and dusky-footed wood rats are the major prey species (2, 31, 82, 95, 98). In contrast, at higher elevations, and throughout the rest of the owl's range, flying squirrels (*Glaucomys sabrinus*) are the primary prey (18, 20, 31, 84). Woodrats respond positively to forest harvest and fragmentation—they reach their highest densities in young, regenerating clear-cut with abundant brush (67, 68, 78, 84). Though most woodrats live in stands too dense for foraging owls (78), woodrats regularly move from these stands into adjacent old forest (79). Flying squirrels are generally found in their highest densities in mature and old-growth forest, and they decrease in density in logged areas (20, 96; but see 76).

Variation in home-range size also seems related to prey base. Home range size decreases with increased abundance of medium-sized prey (18), and with increased abundance of woodrats in the diet (98). In areas where flying squirrels are the primary prey, home-range size decreases as the amount of old forest within the home-range increases (19, 20).

Though the prey abundance hypothesis currently has the most support, other factors such as differences in stand structure and microclimate may contribute to the selection of old forest (77). For example, northern spotted owls have a narrow thermoneutral zone (36), and selection for older forests may facilitate thermoregulation (3).

Null Hypothesis 3: Habitat Trends

The rejection of hypothesis 2 led the ISC to test hypothesis 3. Based on data from National Forest lands in Oregon and Washington (which provide >70% of all remaining owl habitat), the ISC found significant declines since 1940 in the extent of owl habitat, a trend projected to be continued into the future (58). Additional evidence since the ISC Report provides data for California (53), and more regionally specific estimates of habitat loss are found in the draft Recovery Plan (93). Figure 1 (93) is indicative of the magnitude and uniformity of habitat loss throughout the range of the northern spotted owl.

The most recent estimates of rates of habitat loss were made as part of the FEMAT process (86). Analysis of logging records on public lands within the demography study areas found that owl habitat was declining at a rate of 0.9 to 1.5% per year on FS study areas and 1.3 to 3.1% per year in study areas managed by the BLM. Analyses of habitat loss on three study areas using LANDSAT imagery, including both public and private lands, was 1.1 to 5.4% per year (86). The highest rate of habitat loss was on BLM's Medford District, in Oregon, which also recorded the highest rate of decline in the owl population (16).

In general, the relationship between the amount of habitat and population trend is poorly understood. A specific test, at the landscape scale, of the

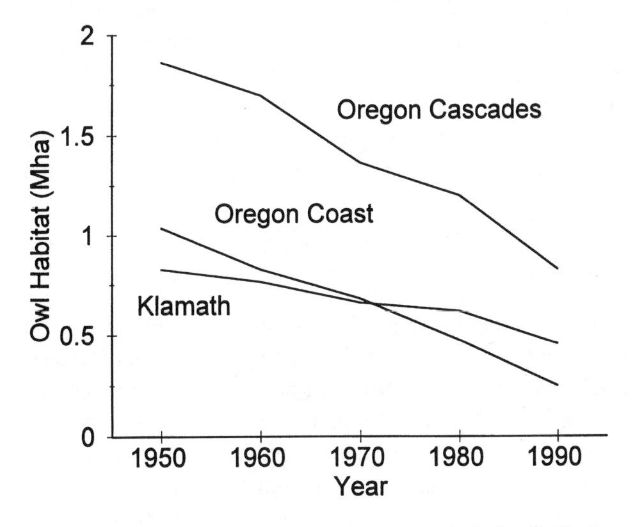

Figure 1 Trends in northern spotted owl habitat for three regions in the Pacific Northwest. Acreage trends include all ownerships. Data from Reference 93.

hypothesis that demography (survival, fecundity, and growth rate) was independent of the amount and pattern of habitat was not rejected (70). The absence of significant relationships is surprising given the extensive empirical evidence of selection for mature and old-growth forests. In the data used for this analysis (70), the amount of old forest surrounding pairs of owls was significantly greater than in the surrounding landscape for 10 of the 11 demographic studies (32). The most parsimonious explanation for the lack of pattern concerns spatial scale. Relationships between demography and habitat are more pronounced at the within-population scale with individual reproductive pairs the unit of analysis, rather than entire subpopulations (70). In support of this assertion, significant linear relationships between the proportion of suitable habitat within the home range and fecundity (5, 8, 73) and adult survival rate (5, 8) have been reported.

Conclusion

Based on studies reported since the ISC Report, decisions on the three null hypotheses remain the same. The observation that the resident, territorial population of northern spotted owls is in decline is most parsimoniously explained by the past and ongoing decline in the owl's habitat—mature and old-growth forest in the Pacific Northwest.

This downward trend in habitat is particularly relevant to territorial species with obligate juvenile dispersal, such as the spotted owl. Based on predictions from theoretical models, sharp thresholds to persistence are approached as habitat is lost and fragmented (44-46, 84). One extinction threshold arises directly from the loss of habitat: if the amount of suitable habitat is reduced to some small fraction of the landscape, the difficulty of an owl's finding a suitable territory leads to extinction (Figure 2). The second is due to an Allee effect (1)—if population numbers fall too low, the probability of finding a mate drops below the level required to maintain reproductive rates required for population stability.

THE RESERVE DESIGN PROCESS

Rejection of the three fundamental hypotheses justified management intervention to arrest the decline in the northern spotted owl population. Towards this goal, the ISC adopted a map-based planning approach that focused on the location, size, shape, spacing, and context of habitat patches planned for inclusion in a reserve system. The goal of this planning exercise was to propose a reserve design that would establish local, stable populations of owls, widely distributed throughout their historic range. Even though suitable habitat was projected to continue to decline outside of reserves for at least the next 50 years (89), habitat loss within reserves would stop and the process of renewal begin.

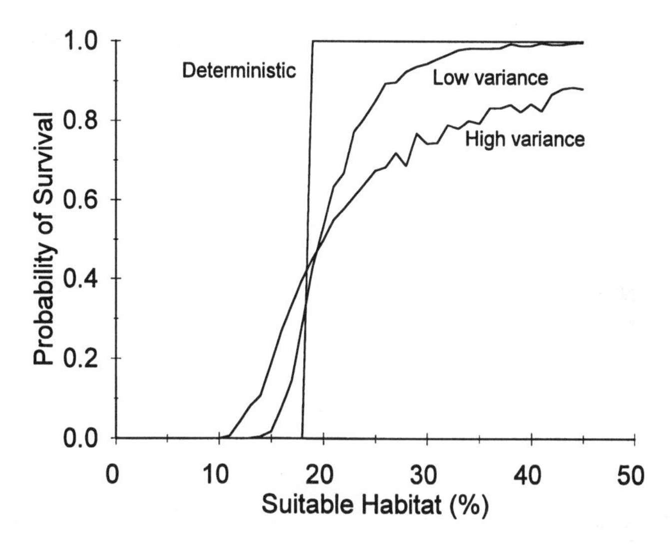

Figure 2 250 year survival probability as function of the proportion of the landscape suitable for nesting. Probability functions were generated using a stochastic model in which owls are forced to search the landscape in order to find both mates and suitable habitat (84, Appendix M; 44).

To provide scientific rigor, the reserve design was portrayed as a map with quantifiable properties that served as falsifiable hypotheses. That is, assertions of map properties were subject to testing with empirical data or theoretical predictions (58). The ISC used information from four map layers (current and historic northern spotted owl distributions; current and historic habitat distributions; known owl locations; land ownership) that, when overlaid, collectively provided the preliminary outline of a reserve system. The initial map represented the maximum size and number of habitat conservation areas (HCAs) available for planning purposes.

The distribution of HCAs was discontinuous (58, 84), suggesting that a feasible reserve design would result in a metapopulation—spatially distinct local populations "connected" by dispersal (38, 50, 51). In the Pacific Northwest, habitat discontinuities result from many factors including diverse topography and natural variation in vegetation structure and composition, and from clearcut timber harvest and permanent land conversion reflecting patterns of private ownership.

Working within the constraints of available habitat and land ownership, the system of HCAs had to provide sufficient habitat to assure a balance between local extinction and recolonization events. The best biological solution for the northern spotted owl was to designate all potential HCAs as part of the reserve system. This solution was not, however, politically, socially, or economically acceptable (85). Therefore, a process was needed to select a subset that would meet the conservation objective. Towards that goal, the ISC tested map-generated hypotheses in the context of five widely accepted concepts of optimal reserve design.

MAJOR RESERVE DESIGN PRINCIPLES INVOKED

Drawing from the fields of population viability analysis (10) and island biogeography (52), the ISC constructed its strategy around five principles that, in various forms, are central to conservation biology (24-26, 97):

1. Species that are well distributed across their range are less prone to extinction than species confined to small portions of their range.
2. Large blocks of habitat supporting many individuals are more likely to sustain those populations than are small blocks of habitat with only a few individuals.
3. Habitat blocks in close proximity facilitate dispersal and recolonization and are preferable to widely dispersed blocks.
4. Contiguous, unfragmented blocks of habitat are superior to highly fragmented blocks of habitat.
5. Habitat between protected areas is more easily traversed by dispersing individuals the more closely it resembles suitable habitat.

Subsequent refinement of the conservation strategy was based on the results of map-based tests of these five principles stated as falsifiable hypotheses (58, 84). To the extent possible, hypothesis tests were direct, based on empirical data collected from studies of the northern spotted owl. When this was not possible, indirect tests were based on ecological theory, data from similar species, or the predictions of simulation models (84). In an iterative fashion, test results were used to reject or retain various map-based elements of alternative reserve designs (58).

One area of scientific uncertainty relevant to reserve design for the northern spotted owl was the minimum size and spacing of HCAs for local population stability. Only general design rules were available from the conservation literature and existing biogeographic principles were too broad for specific application to the northern spotted owl problem. To refine the design rules, fundamental aspects of spatial structure and discontinuous habitat were incorporated into simple simulation models. These models extended the work of Lande (46) and further investigated the effects of habitat loss and fragmentation on territorial species. Based on information from field studies, these models were structured and parameterized specifically in terms of northern spotted owl life history. Using occupancy rate of home range–sized units of suitable habitat as the dependent variable, model simulations investigated the effects of reducing habitat amount and fragmentation, and the benefits of clustering habitat into blocks (HCAs) of various sizes (28, 44, 45, 84).

THE ROLE OF COMPUTER MODELS IN THE CONSERVATION STRATEGY

These initial models became the focus of considerable debate, both in Congressional hearings and in legal challenges to the ISC Report (63, 85). By changing model assumptions or parameter estimates, it was possible to suggest, for example, that the proposed reserve design was inadequate for the owl's persistence, or that the strategy set aside more habitat than was needed. These arguments, however, overly stressed the role of the models and downplayed the fundamental role of owl biology. For the most part, the ISC reserve design (HCA size, spacing, and number) was a logical consequence of the basic biology and life history of the northern spotted owl. The simple combination, for example, of home-range area requirements, habitat use patterns, and minimum population sizes for persistence (71) required the designation of large acreages of suitable habitat (mature and old-growth forests) dispersed throughout the owl's range.

A general finding was that arranging suitable habitat into large blocks asymptotically increased the mean occupancy rate of suitable sites (21, 28, 45; Figure 3). Higher levels of pair occupancy, in turn, translated into an increased likelihood of persistence. ISC model results were based on four critical assumptions that could not be directly tested: 35% of the forested landscape was within HCAs; on average, 60% of the area with an HCA would be suitable habitat in the future; each HCA was "connected" to adjacent HCAs by way of dispersal; and females had no problem in finding mates. Given these assumptions, model results suggested a minimum HCA size sufficient to provide 20 pair sites (territories/HCA) spaced not more than 19 km apart sufficed to provide for stable populations (Figure 3; 44, 45, 61, 84). Importantly, given a target equilibrium population size, arranging suitable habitat into blocks required less total area to be set aside than if habitat were randomly arranged as single-pair sites (45).

In summary, the ISC's conservation strategy, portrayed as a map, was described by the number, size, and distribution of its HCAs, and prescribed minimum habitat criteria between HCAs. In retrospect, it is clear that these rules were generated based on optimistic assumptions. Depensatory density-

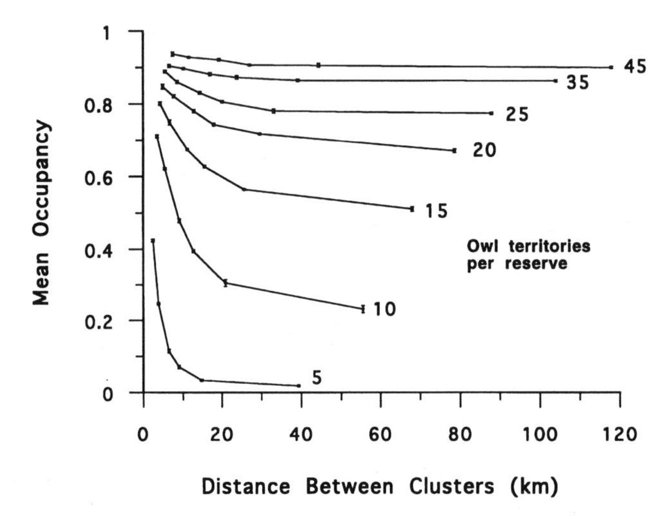

Figure 3 Average proportion of sites occupied by reproductive pairs as a function of reserve spacing and size. Occupancy levels were generated using a stochastic model which simulated habitat search within and between reserves (45, 61, 84).

dependent effects due to difficulties in finding mates at low densities (1,22) and environmental stochasticity were not modeled. Reserve shape and spacing were also optimal, and potential losses from source populations in the reserves to the exterior sink areas were probably underestimated. In addition, the habitat in many of the proposed reserves was < 60% , and in these areas habitat recovery was likely to be slow.

UNDERSTANDING GAINED THROUGH ADDITIONAL COMPUTER MODELING

Much of the modeling since the ISC report has addressed the consequences of these simplifying assumptions. The initial models were useful for demonstrating the interactions of populations with hypothetical landscape geometries, but they were unable to incorporate the complexities of real landscapes. One of the most influential recent themes in conservation biology is that the distribution of populations and their habitats—their geometry within the larger landscape context—strongly influence population trends and extinction likelihoods (29, 54, 65, 66, 87). For managers to benefit from this new understanding (6) and for the incorporation of the necessary level of detail, spatially explicit models interfaced with real landscape patterns would be required.

Effects of Reserve Shape

A circular HCA of a given size has the minimum perimeter to area ratio. Early owl models assumed circular reserves and did not explore the effects of HCA shape on population stability. Recent models allowing habitat to be clustered into different arrangements (45, 61), and a "stepping stone" model in which irregular habitat clusters could be directly represented (54), suggest that edge effects can be significant. The best shape for a reserve is round, the poorest are long and narrow or irregular. Irregularly shaped clusters have much higher losses to edge effects and lower occupancy rates than the round clusters assumed by the ISC (54). The potential losses associated with edges suggest that the high occupancy rate for clusters of 20–25 pair-sites (Figure 3) was an artifact of the ISC's model assumption of circular clusters; HCAs that were irregular in shape would need to support >25 pairs for comparable levels of occupancy.

Northern Spotted Owl Modeling and the Evolution of Land Management Plans

Even though the model results, in conjunction with empirical data from dispersal measurements and biogeographic theory, provided a basis for a reasonable set of rules controlling size and spacing of the reserves, the actual map was highly constrained by geography and land-ownership. Further, up-to-date vegetation maps for the National Forests and BLM lands were not available, so the habitat conditions of both the HCAs and the intervening landscape were approximations. These limitations, in combination with the uncertainty about current habitat conditions, precluded direct tests of the efficacy of the reserve design. During the latter stages of the ISC process, work began on a new owl model designed to directly incorporate vegetation maps, through a GIS (Geographic Information System) interface, into a habitat-based population dynamics model (54). Better habitat maps became available as need increased—a consequence of demand by the Spotted Owl Recovery Team (93) and President Clinton's charge to solve the "timber crisis" and avoid future "train wrecks" in the Pacific Northwest (see FEMAT introduction).

The first attempt to model the impacts of the conservation strategy on public lands was made by the BLM in conjunction with the FWS Recovery Team in 1992 (93). For timber harvest scheduling, specific forest stand data derived from GIS were stratified into productivity classifications, constrained by alternative management strategies, and evaluated to determine the pattern of harvest that maximized wood fiber production (27). A 1000 ha (approximate territory size for owls in this area of Oregon) hexagonal grid was placed over these simulated landscapes, and based on the fraction of older forest within the cell, the relative quality of each cell was evaluated. This process was repeated at

10-year intervals for each of six proposed management alternatives, creating dynamic landscape projections 100 years into the future (27).

Linking Vegetation to Demographics

To implement the process, habitat attributes were associated with birth and survival rates through a series of regression equations (5, 8, 39, 93). The attribute that best explained variation in survival, fecundity, and nest density was the amount of mature (>120 years old) forest within a home range–sized area surrounding an owl nest site (5, 8, 39, 73). Based on these data, functions relating survival, fecundity, and the probability of nesting to the amount of mature and old forest in a home-range–sized hexagon were estimated. These functions provided initial estimates of the vital rates associated with different amounts of mature forest. Finally, BLM biologists assessed the time required for stands to recover to owl habitat after a variety of silvicultural treatments (clear-cut, various partial cuts). The combination of rules controlling spatially explicit harvest scheduling, post-harvest recovery rates, and habitat quality provided the information needed for model parameterization and allowed comparison of land management plans.

Modeling FEMAT Alternatives

As the FEMAT team expanded its mandate to consider all species associated with old-growth forests, the draft BLM plan was subsumed into this larger effort. The proposed FWS Recovery Plan (93)—virtually identical to the ISC's strategy—was retained as one of 10 land management options for public lands [Option 7; (90)]. Because all other options in FEMAT addressed concerns for numerous species other than the northern spotted owl, Option 7 had the smallest reserve acreage and the least stream protection of the 10 options proposed. Option 1 was the most restrictive, with the preferred option (Option 9) fitting approximately midway between 1 and 7. A decision was made in late 1993 to model owl dynamics across the range of the northern spotted owl using methods developed to evaluate BLM management alternatives (27, 69) and a spatially explicit owl model (54, 55). The task was to compare the conservation benefits of options 1, 7, and 9, plus an option that retained all currently suitable owl habitat (69).

Given identical rules concerning vegetation quality and the owl's behavior, and assuming no regrowth of suitable habitat, the plans diverged greatly in terms of both the expected number of owls and their distribution over the landscape (Figures 4*a–d*). Option 9 was similar to Option 1 in both Washington and California but projected a smaller and more disjunct owl population for Oregon (cf Figures 4*b* and 4*d*). Option 9, however, retained large blocks of habitat within the Cascades, and the demographic stability of these blocks was expected to be high (Figure 4*d*). Option 7 (Figure 4*c*), on the other hand, provided significantly less protection for all areas with the exception of the Olympic Peninsula (where virtually all federal land is reserved under all options). Under rules that predict large, stable populations given current habitat conditions (Figure 4*a*), Option 7 produced few, widely spaced areas with stable populations (cf. Figures 4*a* and 4*c*).

Reanalysis of Owl Populations on the Olympic Peninsula

The most comprehensive attempt to update the ISC analysis of population stability occurred in 1994 when a team of modelers and owl biologists was formed to analyze the stability and degree of isolation of owl populations on the Olympic Peninsula in Washington (39). In its analyses, the team used the most current parameter estimates (33), explored owl population dynamics in the context of a spatially explicit model (54, 55), and forecast future population sizes and distributions on the basis of empirical relationships between demographics and habitat quality (5, 8, 93).

Habitat-demography relationships were estimated by regressing habitat condition (amount of old forest) on estimates of survival and fecundity (8, 39). Initial population estimates were derived from survey-based estimates of owl numbers within Olympic National Park (62). Territory size and the elevational range of northern spotted owls was determined from radiotelemetry studies (84). Survival rate estimates were based on the recent advances in capture/recapture modeling (48), and juvenile survival rate estimates were adjusted for emigration based on extensive data from radio-tagged juvenile owls (E Forsman, unpublished data).

Accurate maps of current owl habitat on public, state, and private lands were available, allowing reserve size and shape to be directly modeled (Figure 5). These maps also allowed the testing of assumed habitat demography relationships relative to the current distribution of owls. Assumed demographic rules produced sink habitat in areas with $< 30\%$ suitable habitat, source habitat in areas with $> 40\%$ habitat, and $\lambda \simeq 1.0$ for habitat classes with 30–40% suitable habitat (8, 39)—model results that conformed well to habitat distributions around known owl sites on the Olympic Peninsula.

TRANSITION DYNAMICS AND EQUILIBRIUM ANALYSIS

Given full implementation of their strategy, Thomas et al (84) argued that the northern spotted owl population should reach a positive equilibrium size sometime within the next 100 years. In making this argument they assumed three conditions: the amount and distribution of habitat would become fixed and sufficient to support a stable population, and the population would quickly

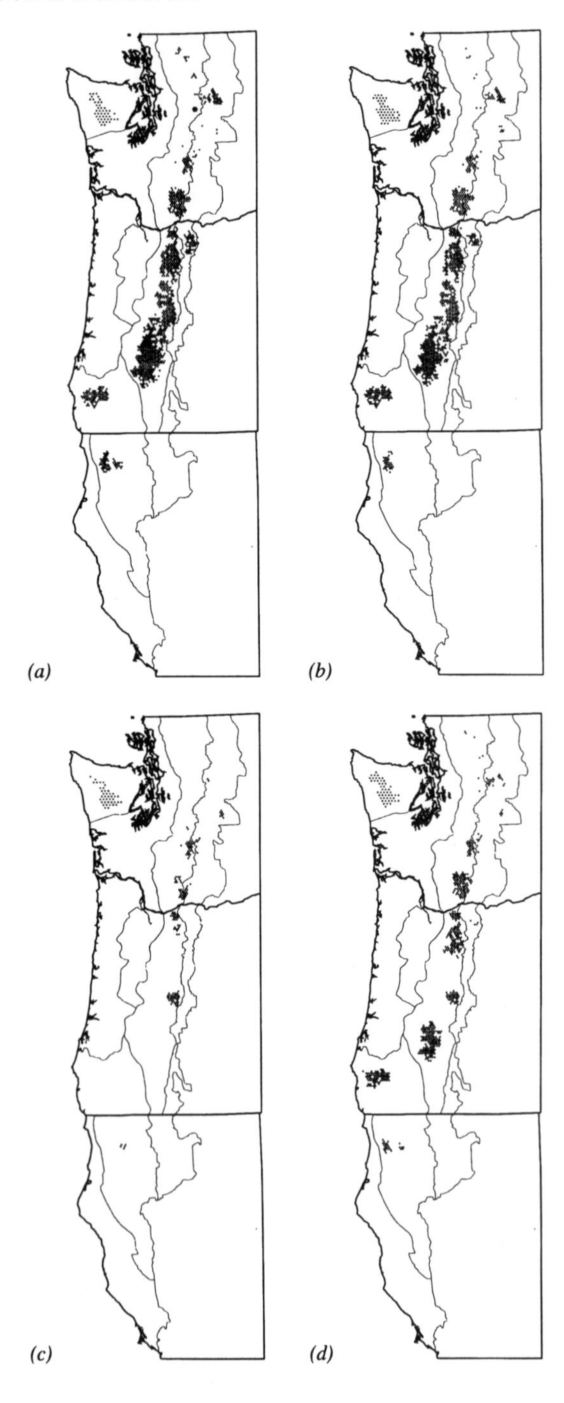
(a)
(b)
(c)
(d)

converge to its habitat-based carrying capacity. These understandings were theoretically formalized (46, 44, 84). If the combination of critical habitat amount and territory occupancy rate fell below a critical landscape threshold (labeled G* in Figure 6), the population would follow a trajectory to extinction. Once on this trajectory, even large increases in habitat proportion are unlikely to stabilize the population unless juvenile search ability is very high (Figure 6). For these reasons, and because habitat was far below modeled levels in many areas (particularly on BLM lands in Oregon), transition dynamics were particularly important to any assumptions of long-term stability of the reserve system.

Analysis of Transient Dynamics in the ISC

The existence of threshold points, arising from habitat loss and distribution, has been demonstrated in a number of the owl models (44, 46, 84). If these models forecast reality, the success of the conservation strategy depended on reaching a condition of no-net-loss in suitable habitat before the population encountered an extinction threshold. The ISC concluded (84), "Our knowledge of the model structure and of spotted owl dispersal and search capabilities is incomplete, however, and we cannot accurately predict the population size, suitable habitat, or amount of habitat fragmentation thresholds that, once crossed, would lead to a population crash." Since that time no further insights have arisen relative to the locus of the thresholds.

Despite the possible existence of thresholds, the ISC was confident its strategy would eventually arrest the population decline and stabilize its dynamics. That this equilibrium would probably be far below the current population levels was fully understood by the ISC: "In the worst case scenario, we estimate that the strategy could result in a 50 to 60% reduction in current owl numbers. This figure assumes that all pair sites outside of the HCAs will eventually be lost through habitat removal or become permanently vacant because of demographic factors resulting from increasing isolation" (84).

The ISC assumed that habitat within the HCAs would recover to the extent specified and subsequently would retain nearly full occupancy by owls. For the reserve in total, this was an optimistic scenario. In fact, little likelihood existed

←

Figure 4 Modeled pair occupancy (dots represent locations occupied at a rate ≥ 70% during simulation runs of 100 years) if all current habitat were retained (4*a*), and assuming implementation of FEMAT (90) Options 1, 7, and 9 (4*b*, *c*, and *d* respectively). Mapped areas include western Washington, Oregon, and northwestern California and assumed no habitat regrowth for 100 years into the future (69).

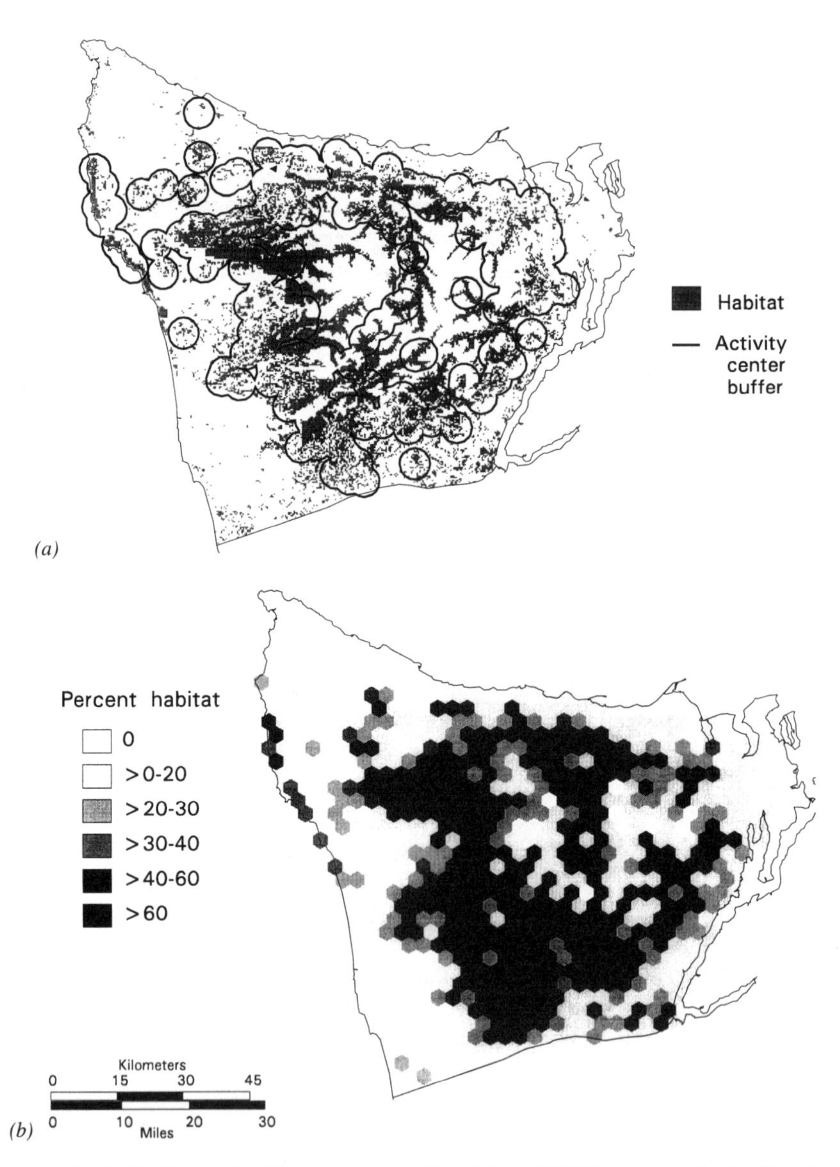

Figure 5 Owl habitat on the Olympic Peninsula (*a*) and derived 1500 ha hexagonal cells (*b*) used for model simulation (39). Known owl activity centers are circled on map *a*. Shading (*b*) indicates the percentage of habitat within hexagonal cells.

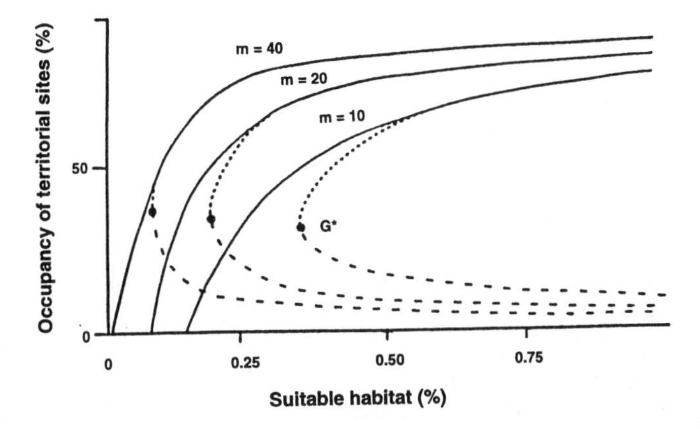

Figure 6 Model output showing the locations of stable and unstable equilibria based on both habitat occupancy and proportion of the landscape in suitable habitat. The solid lines represent no mate search, hence no Allee effect. The dotted and dashed curves show a stable and unstable Allee effect, respectively. The three M values represent different search efficiencies, with 40 the most vagile. If occupancy is too low, the population will follow the unstable (*dashed*) equilibrium. G* is the minimum proportion of suitable habitat for which stable positive equilibria can be achieved. If habitat proportions fall below G*, the population will go extinct regardless of initial occupancy (44).

that many of the HCAs would recover to their potential. Upon reexamination of HCAs on BLM lands in Oregon, reserve acres were fewer than originally estimated. This area of the owl's range is characterized by a "checkerboard" of ownerships with little likelihood of achieving > 50% owl habitat without mandating contributions from private property. Recent modeling that explicitly looked at harvest and regrowth in the next 50 years forecast the ISC's worst-case scenario. Assuming optimistic modeling rules that allow a stable or increasing population given current habitat levels, the models projected 50–60% declines in owl populations in Oregon and California if the ISC strategy (renamed Option 7 in FEMAT) was implemented (69; cf Figures 4a and 4c). Thus, the combination of continued harvest of old forest and the slow renewal rate of owl habitat suggested a substantial population decline before a new equilibrium was reached (69).

Population Stability Due to Regional Variability

Given the wide geographic distribution of the northern spotted owl, and natural variation in both forest condition and the owls' response to disturbance, it is unlikely that a global extinction threshold occurs at a specific point. Rather, a range of suitable habitat proportions likely exists in which the population trajectory becomes strongly negative (Figure 2). Further, the range of habitat proportions that defines this region varies in both space and time. Temporal and spatial variations in the threshold region arise from the inherent stochastic nature of the environment and the geographic variation in the species' ecological relations (for example, geographic variation in the owl's habitat associations and prey relationships).

It is therefore unlikely that the entire northern spotted owl population would simultaneously move past an extinction threshold. For example, we suspect that northern spotted owls in the Coast Range of Oregon may occupy habitats within or below the threshold region, but those in the Klamath Mountains in California probably do not (Figures 1 and 4). Both the amount and distribution of suitable habitat (86) and the population trend (15, 16, 32) vary geographically throughout the species' range.

Geographical variability will provide some degree of collective stability during the transition period. For example, crossing a habitat threshold in one part of the species' range may have little or no effect on the species' population dynamics in other parts of its range. Thus, assuming the existence of self-sustaining populations in parts of the species' range, growth of suitable habitat where the owl population has collapsed should lead eventually to recolonization of the collapsed areas by immigration from other populations (11). Both these assumptions were critical to the overall success of the strategy, and well within accepted ecological theory and empirical observations from a variety of species.

Habitat recovery will take longer in some parts of the owl's range than in others (69; Figure 4). Thus the transient behavior of local owl populations will vary depending on local rates of habitat renewal. Populations currently in precipitous decline and occupying areas deficient in habitat may continue to decline at least until a condition of no-net-loss of habitat occurs. During the interim, it is possible that some populations may collapse.

VARIABILITY IN DEMOGRAPHIC RATES Although the available demographic data indicate that the northern spotted owl's population is declining (15, 16), interpretation of these trends, relative to proximity to a threshold point, is unclear. Based on a meta-analysis, there is universal evidence for accelerating rates of decline in adult female survivorship, but individual study areas vary in the degree to which they show this trend (15, 16, 32). Given geographic variation in the location of threshold regions, the proximity of local owl populations to such regions, and potential confounding effects of increased intraspecific competition among owls as a result of packing by displaced birds, no general inference to threshold points can yet be drawn. Nevertheless, the demographic results are troubling and suggest some prudent modifications to the original conservation strategy.

IS THE CURRENT CONSERVATION STRATEGY ADEQUATE?

Since 1990, insights provided by the study of habitat relations and demography have confirmed the fundamental premises of the ISC—the owl requires large areas of late seral forest within its home range, and the loss and fragmentation of suitable habitat leads to population declines. Because the basic biological understandings have not changed, the ISC's reserve design has remained conceptually robust. Large reserves, widely distributed across the landscape, are needed to retain the owl's current range.

Will Owl Populations Reach a New Equilibrium?

Although supporting examples were drawn from empirical studies (mostly island studies), model results played an important part in the ISC's assumption of a future, stable equilibrium. Retrospectively, it is noteworthy that the questions of long-term stability within the modeling context were never rigorously pursued by the ISC. Although Thomas et al (84) noted that short-term patterns in population persistence could lead to very different results in the long term, the model used to test potential reserve designs was not extended beyond 100 years. The decision to truncate the models runs was based on the pragmatic argument that decision makers could not plan beyond this time horizon.

Subsequent modeling suggests that reserves with a carrying capacity of 20 pairs are stable only if juvenile search efficiency is high and edge effects are minimal (54, 55). To achieve local stability within the constraints of real landscapes, more recent modeling suggests that carrying capacities of perhaps 30–40 pairs per HCA are needed. In addition, a few large reserves (>100 pairs) significantly safeguard against population extinction. For these reasons, the original reserve design proposed by the ISC represents a minimum system, with greater risks to persistence then initially envisioned.

More recent, multispecies plans have also viewed the ISC as minimally acceptable. For example, in alternative plans for management of northwestern forests within the range of the northern spotted owl, the ISC's strategy [Option 7, (90)] provided the smallest acreage in reserves. All other options started with the ISC's reserve design and added additional reserve acreage. The adopted alternative (91), Option 9, added > 800,000 ha to the reserve system in areas adjacent to HCAs and of similar habitat quality. Some of these additions resulted in large reserve areas capable of supporting local populations of 40 to >170 owl pairs (39, 69). Simulation results suggest that these large reserves should support stable populations under a wide variety of demographic assumptions

(69; Figure 4d). While range reductions are possible, global extinction appears to be much less likely under Option 9.

The long-term dynamics of owls in the smaller HCAs and the matrix lands between HCAs are less clear. Here too, however, Option 9 provides a variety of restrictions on land use that exceed those proposed by the ISC. The ISC strategy assumed no suitable nesting habitat in the forest matrix—rather, it stipulated a minimal habitat structure to facilitate dispersal between HCAs [50% of the land with tree diameter > 28 cm and with canopy closure > 40% (84)]. Option 9 provides for 40 ha reserves around known owl sites in the matrix, and for 46–91 m riparian buffers along all streams (approximately 40% of the matrix lands lie within these "riparian reserves"). In addition, 15% of the trees must be retained within each cutting unit, preferably in groups, and these trees will be exempted from future cutting (91). These are significant restrictions, and greatly increase the likelihood that owls will nest within matrix areas as well. Thus, given full implementation of the management standards in guidelines of Option 9 (91), we believe the prognosis for the long-term (>100 years) persistence of the northern spotted owl appears to be very favorable.

In conclusion, the process used by the ISC to develop a scientifically credible reserve design—the hypothesis-testing framework, and the generation of maps with testable properties—is tenable. The scientific methods that evolved and were advanced during the northern spotted owl research program provide an exemplary model of how to acquire a defensible foundation for conservation planning. In the context of the contentious arena in which they have sought solutions to complex problems, we believe the northern spotted owl researchers and their colleagues have made significant contributions to the conservation sciences.

ACKNOWLEDGMENTS

We greatly admire and value the hard work and creative energy expressed by the many biologists who have studied northern spotted owls in the wild. Their passion for the owl and the forests of the Pacific Northwest are the foundation of our conservation efforts. In addition, we thank Annette Albert for her editorial expertise on this, and many other manuscripts—she makes our writing much easier and makes sure that we always "color between the lines."

Literature Cited

1. Allee WC. 1931. *Animal Aggregations: A Study in General Sociology*. Chicago: Univ. Chicago Press
2. Barrows CW. 1980. Feeding ecology of the spotted owl in California. *J. Raptor Res.* 14:73–85
3. Barrows CW. 1981. Roost selection by spotted owls: an adaption to heat stress. *Condor* 83:302–09
4. Bart J, Earnst S. 1992. Suitable habitat for northern spotted owls: an update. *U.S. Dep. Interior, Final Draft Recovery Plan for the Northern Spotted Owl, Appendix B. Fish Wildlife Serv. 2001.* Portland, OR
5. Bart J, Forsman ED. 1992. Dependence of northern spotted owls (*Strix occidentalis caurina*) on old-growth forest in the western USA. *Biol. Conserv.* 62:95–100
6. Bart J. 1995. Acceptance criteria for using individual-based models to make management decisions. *Ecol. Appl.* 5:411–20
7. Bart J. 1995. Evaluation of population trend estimates calculated using capture-recapture and population methods. *Ecol. Appl.* 5:662–71
8. Bart J. 1995. Amount of suitable habitat and viability of northern spotted owls. *Conserv. Biol.* 9:943–46
9. Blakesley JA, Franklin AB, Gutiérrez RJ. 1992. Spotted owl roost and nest site selection in northwestern California. *J. Wildl. Manage.* 56:388–92
10. Boyce MS. 1992. Population viability analysis. *Annu. Rev. Ecol. Syst.* 23:481–506
11. Brown JH, Kodric-Brown A. 1977. Turnover rates in insular biogeography: effect of immigration on extinction. *Ecology* 58:445–49
12. Buchanan JB, Irwin LL, McCutchen EL. 1993. Characteristics of spotted owl nest trees in the Wenatchee National Forest. *J. Raptor Res.* 27:1–7
13. Buchanan JB, Irwin LL, McCutchen EL. 1995. Within-stand nest site selection by spotted owls in the eastern Washington Cascades. *J. Wildl. Mange.* 59:301–10
14. Burnham KP, Anderson DR. 1992. Demographic analysis of northern spotted owl populations. *Recovery Plan for the Northern Spotted Owl, Appendix C.* US Fish & Wildl. Serv., Portland, OR
15. Burnham KP, Anderson DR, White GC. 1994. Estimation of vital rates of the northern spotted owl. *USDA Forest Service, Final Supplemental Environmental Impact Statement,* Vol. II. *Appendix J.* pp. 1–26. Portland, OR
16. Burnham KP, Anderson DR, White GC. 1996. Meta-analysis of vital rates of the northern spotted owl. *Stud. Avian Biol.* 17: In press
17. California Forestry Association. 1993. *A petition to remove the northern spotted owl in California from the list of threatened species.* Unpublished document, Calif. For. Assoc., Sacramento, CA
18. Carey AB, Peeler KC. 1995. Spotted owls: resource and space use in mosaic landscapes. *J. Raptor Res.* 29:223–39
19. Carey AB, Reid JA, Horton SP. 1990. Spotted owl home range and habitat use in southern Oregon Coast Ranges. *J. Wildl. Manage.* 54:11–17
20. Carey AB, Horton SP, Biswell BL. 1992. Northern spotted owls: influence of prey base and landscape character. *Ecol. Monogr.* 62:223–50
21. Carroll JE, Lamberson RH. 1993. The owl's odyssey: a continuous model for the dispersal of territorial species. *Siam J. Appl. Math.* 53:205–18
22. Caswell H. 1989. *Matrix Population Models.* Sunderland, MA: Sinauer. 328 pp.
23. Dawson WR, Ligon JD, Murphy JR, Meyers JP, Simberloff D, Verner J. 1986. Report of the advisory panel on the spotted owl. *Audubon Conserv. Rep., 7.* Natl. Audubon Soc., New York
24. Den Boer PJ. 1981. On the survival of populations in a heterogeneous and variable environment. *Oecologia* 50:39–53
25. Diamond JM. 1975. The island dilemma: lessons of modern biogeographic studies for the design of natural reserves. *Biol. Conserv.* 7:129–46
26. Diamond JM. 1976. Island biogeography and conservation: strategy and limitations. *Science* 193:1027–29
27. Dippon D, Cadwell C, Nighbert J, McKelvey K. 1992. Linking a spatially-oriented northern spotted owl population dynamics model with the Western Oregon Digital Database. *GIS '92 Symp.* Vancouver, British Columbia, Canada
28. Doak D. 1989. Spotted owls and old-growth logging in the Pacific Northwest. *Conserv. Biol.* 3:389–96
29. Dunning JB Jr, Stewart DJ, Danielson BJ, Noon BR, Root TL, et al. 1995. Spatially explicit population models current forms and future uses. *Ecol. Appl.* 5:3–11
30. Folliard L. 1993. *Nest site characteristics of northern spotted owls in managed forests of northwest California.* MS thesis. Univ.

Idaho, Moscow

31. Forsman ED, Meslow EC, Wight HM. 1984. Distribution and biology of the spotted owl in Oregon. *Wildl. Monog.* 87:1–64
32. Forsman ED, DeStefano S, Gutiérrez RJ, Raphael M, eds. 1996. Demography of the Northern Spotted Owl. *Stud. Avian Biol.* 17: In press
33. Forsman ED, Sovern SG, Maurice KJ, Taylor M, Zisa JJ, Seaman ED. 1996. Demography of the northern spotted owl on the Olympic Peninsula and Cle Elum Ranger District, Washington. *Stud. Avian Biol.* 17: In press
34. Franklin AB. 1992. Population regulation in northern spotted owls: theoretical implications for management. In *Wildlife 2001: Populations.* ed. DR McCullough, RH Barrett Jr, pp. 815–30. New York: Elsevier
35. Franklin AB, Anderson DR, Forsman ED, Burnham KP, Wagner FF. 1996. Methods for collecting and analyzing demographic data on the northern spotted owl. *Stud. Avian Biol.* 17: In press
36. Ganey JL, Balda RP, King RM. 1993. Metabolic rate and evaporative water loss of Mexican spotted and great horned owls. *Wilson Bull.* 105:645–56
37. Gutiérrez RJ, Franklin AB, LaHaye WS. 1995. Spotted Owl (*Strix occidentalis*). In *The Birds of North America, No. 17.* ed. A Poole, F Gill. Acad. Nat. Sci., Philadelphia, and Am. Ornithologist's Union, Washington, DC
38. Harrison S. 1993. Metapopulations and conservation. In *Large-Scale Ecology and Conservation Biology,* ed. PJ Edwards, RM May, NR Webb, pp. 111–28. Boston, MA: Blackwell Sci.
39. Holthausen RS, Raphael MG, McKelvey KS, Forsman ED, Starkey EE, Seaman DE. 1995. The contribution of federal and nonfederal habitat to persistence of the northern spotted owl on the Olympic Peninsula, Washington. *Rep. of the Reanalysis Team. USDA For. Serv., Gen. Tech. Rep. PNW-GTR-352.* Pacific NW Res. Station, Portland, OR. 68 pp.
40. Jolly GM. 1965. Explicit estimates from capture-recapture data with both death and dilution—stochastic model. *Biometrika* 52:225–47
41. Hopkins S, Forsman ED, Logan WD. 1996. Demography of northern spotted owls on the Salem District of the Bureau of Land Management in northwestern Oregon. *Stud. Avian Biol.* 17. In press
42. Hunter JE, Gutiérrez RJ, Franklin AB. 1995. Habitat configuration around spotted owl sites in northwestern California. *Condor* 97:684–93
43. LaHaye WS. 1988. *Nest site selection and nesting habitat of the northern spotted owl (*Strix occidentalis caurina*) in northwestern California.* MS thesis. Humboldt State Univ., Arcata, CA
44. Lamberson RH, McKelvey R, Noon BR, Voss C. 1992. A dynamic analysis of northern spotted owl viability in a fragmented forest landscape. *Conserv. Biol.* 6:505–12
45. Lamberson RH, Noon BR, Voss C, McKelvey KS. 1994. Reserve design for territorial species: the effects of patch size and spacing on the viability of the northern spotted owl. *Conserv. Biol.* 8:185–95
46. Lande R. 1987. Extinction thresholds in demographic models of territorial populations. *Am. Nat.* 130:624–35
47. Lande R. 1988. Demographic models of the northern spotted owl (*Strix occidentalis caurina*). *Oecologia* 75:601–07
48. Lebreton JD, Burnham KP, Clobert J, Anderson DR. 1992. Modeling survival and testing biological hypotheses using marked animals: a unified approach with case studies. *Ecol. Monogr.* 62:67–118
49. Lehmkuhl JF, Raphael MG. 1993. Habitat pattern around northern spotted owl locations on the Olympic Peninsula, Washington. *J. Wildl. Manage.* 57:302–14
50. Levins R. 1969. The effects of random variation of different types on population growth. *Proc. Natl. Acad. Sci. USA* 62:1061–65
51. Levins R. 1970. Extinction. *Lectures on Math. in Life Sci.* 2:75–107
52. MacArthur RH, Wilson EO. 1967. *The Theory of Island Biogeography.* Vol. 1. *Monographs in Population Biology.* Princeton, NJ: Princeton Univ. Press
53. McKelvey KS, Johnston JD. 1992. Historical perspectives on forests of the Sierra Nevada and the Transverse Ranges of southern California: forest conditions at the turn of the century. In *The California Spotted Owl: A Technical Assessment of its Current Status,* tech coords. J Verner, KS McKelvey, BR Noon, RJ Gutiérrez, GI Gould Jr, TW Beck, pp. 225–46. *USDA For. Serv., Gen. Tech. Rep. PSW-GTR-133.* Pacific SW Res. Station, Albany, CA
54. McKelvey K, Noon BR, Lamberson RH. 1993. Conservation planning for species occupying fragmented landscapes: the case of the northern spotted owl. In *Biotic Interactions and Global Changes*, ed. PM Karieva, JG Kingsolver, RB Huey, pp. 424–50. Sunderland, MA: Sinauer Assoc.
55. McKelvey KS, Crocker J, Noon BR. 1996.

A spatially explicit life-history simulator for the northern spotted owl. In review
56. Meyer JS. 1996. Influence of habitat fragmentation on spotted owls in western Oregon. *Wildl. Monogr.* In press
57. Murphy DD, Noon BR. 1991. Coping with uncertainty in wildlife biology. *J. Wildl. Manage.* 55:773–82
58. Murphy DD, Noon BR. 1992. Integrating scientific methods with habitat conservation planning: reserve design for the northern spotted owl. *Ecol. Appl.* 2:3–17
59. Newton I, ed. 1989. *Lifetime Reproduction in Birds.* London/New York: Academic
60. Noon BR, Biles CM. 1990. Mathematical demography of spotted owls in the Pacific Northwest. *J. Wildl. Manage.* 54:18–27
61. Noon BR, McKelvey KS. 1992. Stability properties of the spotted owl metapopulation in southern California. In *The California Spotted Owl: A Technical Assessment of Its Current Status*, tech coords. J Verner, KS McKelvey, BR Noon, RJ Gutiérrez, GI Gould Jr, TW Beck, pp. 187–206. *USDA Forest Service, Gen. Tech. Rep. PSW-GTR-133.* Pacific SW Res. Station, Albany, CA
62. Noon BR, Eberhardt LL, Franklin AB, Nichols JD, Pollock KH. 1993. *A survey design to estimate the number of territorial spotted owls in Olympic National Park. Final Report submitted to Olympic National Park.* 31 pp.
63. Noon BR, Murphy DD. 1994. Management of the spotted owl: the interaction of science, policy, politics, and litigation. In *Principles of Conservation Biology*, ed. GK Meffe, CR Carroll, pp. 380–88. Sunderlund, MA: Sinauer
64. Pulliam HR. 1988. Sources, sinks, and population regulation. *Am. Nat.* 132:652–61
65. Pulliam HR, Dunning JB Jr, Liu J. 1992. Population dynamics in complex landscapes: a case study. *Ecol. Appl.* 2:165–77
66. Pulliam HR, Liu J, Dunning JB Jr, Stewart DJ, Bishop TD. 1995. Modelling animal populations in changing landscapes. *Ibis* 137:5120–26
67. Raphael MG. 1984. Wildlife populations in relation to stand age and area in Douglas-fir forests of northwestern California. In *Proc. Symp. of Fish and Wildlife Relationships in Old-Growth Forests,* ed. WR Meehan, TR Merrell Jr, TA Hanley, pp. 259–74. Juneau, Alaska
68. Raphael MG. 1988. Long-term trends in abundance of amphibians, reptiles, and mammals in Douglas-fir forests of northwestern California. In *Management of Amphibians, Reptiles, and Small Mammals in North America*, tech. coords. RC Szaro, KE Severson, DR Patton, pp. 23–31. *General Tech, Rep. RM-166.* USDA For. Serv.
69. Raphael MG, Young JA, McKelvey KS, Galleher BM, Peeler KC. 1994. A simulation analysis of population dynamics of the northern spotted owl in relation to forest management alternatives. *Final Environmental Impact Statement on Management of Habitat for Late-Successional and Old-Growth Forest Related Species Within the Range of the Northern Spotted Owl.* Vol. II. *Appendix J-3.*
70. Raphael MG, Young JA, Forsman ED. 1996. Distribution and trend of habitat of the northern spotted owl. *J. Wildl. Manage.* In review
71. Richter-Dyn N, Goel NS. 1972. On the extinction of colonizing species. *Theor. Pop. Biol.* 3:406–33
72. Ripple WJ, Johnson DH, Hershey KT, Meslow EC. 1991. Old-growth and mature forest near spotted owl nests in western Oregon. *J. Wildl. Manage.* 55:316–18
73. Ripple WJ, Lattin PD, Hershey KT, Wagner FF, Meslow EC. 1996. Landscape patterns around northern spotted owl nest sites in southwest Oregon. *J. Wildl. Manage.* In press
74. Romesburg HC. 1981. Wildlife science: gaining reliable knowledge. *J. Wildl. Manage.* 45:293–313
75. Romesburg HC. 1989. More on gaining reliable knowledge: a reply. *J. Wildl. Manage.* 53:1177–80
76. Rosenberg DK, Anthony RG. 1992. Characteristics of northern flying squirrel populations in second- and old-growth forests in western Oregon. *Can. J. Zool.* 70:161–66
77. Rosenberg DK, Zabel CJ, Noon BR, Meslow EC. 1994. Northern spotted owls: influence of prey base—a comment. *Ecology* 75:1512–15
78. Sakai HF, Noon BR. 1993. Dusky-footed woodrat abundance in different-aged forests in northwestern California. *J. Wildl. Manage.* 57:373–81
79. Sakai HF, Noon BR. 1996. Between-habitat movement of dusky-footed woodrats and vulnerability to predation by spotted owls. *J. Mammal.* In review **!]
80. Seber GAF. 1965. A note on the multiple recapture census. *Biometrika* 52:249–59
81. Simpson Timber Company. 1992. *Habitat conservation plan for the northern spotted owl on the California Timberlands of Simpson Timber Company.* Arcata, CA: Published by Simpson Timber Company, PO Box 1169, Arcata, CA 95521-1169
82. Solis DM Jr. 1983. *Summer habitat ecology of spotted owls in northwestern California.*

MS thesis, Humboldt State Univ., Arcata, Calif. 230 pp.

83. Solis DM Jr, Gutiérrez RJ. 1990. Summer habitat ecology of northern spotted owls in northwestern California. *Condor* 92:739–48
84. Thomas JW, Forsman ED, Lint JB, Meslow EC, Noon BR, Verner J. 1990. *A conservation strategy for the northern spotted owl: Report of the Interagency Scientific Committee to address the conservation of the northern spotted owl. USDA Forest Service, Bureau of Land Management, US Fish & Wildl. Serv., and Natl. Park Serv.* Portland, OR.
85. Thomas JW, Verner J. 1992. Accommodation with socio-economic factors under the endangered species act—more than meets the eye. *Trans. 57th N Am. Wildl. & Nat. Res. Conf., Special Session 9. Int. Resource Issues and Opportunities.* Pp. 627–41.
86. Thomas JW, Raphael MG, Anthony RG, Forsman ED, Gunderson AG, et al. 1993. *Viability assessments and management considerations for species associated with late-successional and old-growth forests of the Pacific Northwest. The report of the Scientific Analysis Team. USDA For. Serv., Natl. For. System, For, Serv. Res.* Portland, OR
87. Turner MG, Arthaud GJ, Engstrom RT, Hejl SJ, Liu J, et al. 1995. Usefulness of spatially explicit population models in land management. *Ecol. Appl.* 5:12–16
88. US Department of Agriculture, Forest Service. 1983. *Internal letter of instruction, dated 9 February 1983, from the Regional Forester instituting guidelines for incorporating minimum management requirements in forest plans.* Pacific Northwest Region, Portland, OR
89. US Department of Agriculture, Forest Service. 1992. *Final environmental impact statement on management for the northern spotted owl in the national forests.* Portland, OR
90. US Department of Agriculture, Forest Service, US Fish and Wildlife Service, National Oceanic and Atmospheric Administration, National Marine Fisheries Service, National Park Service, Bureau of Land Management, and Environmental Protection Agency. 1993. *Forest ecosystem management: an ecological, economic, and social assessment: report of the Forest Ecosystem Management Assessment Team.* Portland, OR
91. US Department of Agriculture, Forest Service, US Department of Interior. 1994. *Record of decision for admendments to Forest Service and Bureau of Land Management planning documents within the range of the northern spotted owl.* Portland, OR
92. US Department of Interior. 1982. *The northern spotted owl: a status review.* US Fish Wildl. Serv., Portland, OR
93. US Department of Interior. 1992. *Final Draft Recovery Plan for the Northern Spotted Owl.* Vol. 2. US Fish Wildl. Serv., Portland, OR
94. Verner J, Gutiérrez RJ, Gould GI Jr. 1992. The California spotted owl: general biology and ecological relations. In *The California Spotted Owl: A Technical Assessment of its Current Status,* tech. coords. J Verner, KS McKelvey, BR Noon, RJ Gutiérrez, GI Gould Jr, TW Beck, pp. 55–77. *Gen. Tech. Rep. PSW-GTR-133.* Albany, CA: Pac. SW Res. Station, For. Serv., US Dep. Agric.
95. Ward JP. 1990. *Spotted owl reproduction, diet, and prey abundance in northwestern California.* MS thesis, Humboldt State Univ., Arcata, CA. 70 pp.
96. Waters JR, Zabel CJ. 1995. Northern flying squirrel densities in fir forests of northeastern California. *J. Wildl. Manage.* 59:858–66
97. Wilcove D, Murphy D. 1991. The spotted owl controversy and conservation biology. *Conserv. Biol.* 5:261–62
98. Zabel CJ, McKelvey KS, Ward JP. 1995. Influence of primary prey on home-range size and habitat-use patterns of Northern Spotted Owls (*Strix occidentalis caurina*). *Can. J. Zool.* 73:433–39
99. Zabel CJ, Salmons SE, Brown M. 1996. Demography of northern spotted owls in southwestern Oregon. *Stud. Avian Biol.* 17. In press

Annu. Rev. Ecol. Syst. 1996. 27:163–96

HISTORICAL BIOGEOGRAPHY OF WEST INDIAN VERTEBRATES

S. Blair Hedges
Department of Biology and Institute of Molecular Evolutionary Genetics, 208 Mueller Laboratory, Pennsylvania State University, University Park, Pennsylvania 16802

KEY WORDS: Caribbean, Antilles, dispersal, vicariance, evolution

ABSTRACT

The vertebrate fauna of the West Indies (1262 species) exhibits high levels of endemism and has a taxonomic composition characteristic of more isolated oceanic islands. Many groups that are widespread on the mainland are absent in the islands, and some of those present are characterized by large adaptive radiations. The growing fossil record of West Indian vertebrates, including mid-Tertiary amber fossils (considered here to be 20–30 million years old), indicates that this pattern of reduced higher-taxon diversity has persisted for a long period of time. Phylogenetic relationships of nonvolant groups display a strong South American influence, whereas volant groups (birds and bats) and freshwater fish show closer ties with Central and North America. Molecular estimates of divergence times between island taxa and their mainland counterparts indicate a Cenozoic origin (within the last 65 million years) for nearly all groups examined. Together, data from different sources point to an origin by overwater dispersal for a large majority of the vertebrate fauna. The prevailing current direction, from southeast to northwest, and the wide scattering of estimated times of origin suggest that much of the nonvolant fauna arrived by flotsam carried from the mouths of major rivers in northeastern South America. Spatial relationships, especially considering low sea levels during the Pleistocene, appear to better explain the routes of colonization taken by the volant fauna and freshwater fish. Caribbean geologic history does not preclude an origin by late Mesozoic vicariance for several possibly ancient groups, although an early Cenozoic arrival by dispersal also cannot be discounted. An integrative approach to historical biogeography is shown to be more insightful than the current trend in the field, cladistic biogeography, which places prime emphasis only on phylogenetic relationships.

0066-4162/96/1120-0163$08.00

INTRODUCTION

From a biogeographic standpoint, the West Indies includes the Greater Antilles (Cuba, Jamaica, Hispaniola, Puerto Rico), Lesser Antilles, Bahamas, and some peripheral islands. Trinidad, Tobago, and the islands adjacent to Venezuela usually are excluded from this definition because they have a biota more characteristic of South America (11). Among the plants and animals of the West Indies, the vertebrates (1262 species) exhibit some of the highest levels of endemism and therefore have been of considerable interest in biogeographic studies of the Caribbean region. Compared with Madagascar, which has three times the area, there are 60% more species in the West Indies, representing about 5% of all known extant terrestrial vertebrates (Table 1; 197). Sharply rising discovery curves for some West Indian groups indicate that more species remain to be discovered (74). Despite this high species diversity, many major groups are absent when compared with the adjacent mainland, including primary division freshwater fishes, salamanders, caecilians, marsupials, carnivores, ungulates, lagomorphs, and most families of frogs, turtles, and snakes. Instead, exceptionally large radiations characterize some of the groups present, such as eleutherodactyline frogs, anoline and sphaerodactyline lizards, capromyid rodents, and megalonychid edentates.

The focus of this review is the origin of the West Indian vertebrate fauna. Although not an exhaustive survey, aspects of the complex geologic history, fossil record, and biogeographic mechanisms are briefly discussed, and taxon-specific patterns are reviewed. Finally, these data are brought together to elucidate general patterns of historical biogeography for the West Indian vertebrate fauna as a whole.

Table 1 Numbers of orders, families, genera, and species of native West Indian vertebrates[a]

			Genera			Species		
Group	Orders	Families[b]	Total	Endemic	% Endemic	Total	Endemic	% Endemic
Freshwater fishes	6	9	14	6	43	74	71	96
Amphibians	1	4	6	1	17	166	164	99
Reptiles	3	19	50	9	18	449	418	93
Birds	15	49	204	38	19	425	150	35
Mammals:								
Bats	1	7	32	8	25	58	29	50
Other	4	9	36	33	92	90	90	100
TOTAL	30	97	342	95	28	1262	922	73

[a]Following sources listed in subsequent tables.
[b]Including one endemic family of birds and four of mammals.

GEOLOGIC HISTORY

Tectonic Development of the Caribbean

Only general aspects of the complex geologic history of the Caribbean region are mentioned here. For those interested in more detailed coverage, geologists provide several recent reviews (e.g. 36, 43).

The complexity of Caribbean geology is due in part to the position of the region, wedged between two major continental and oceanic plates (North and South America), and bounded on the west by Pacific plates. Although rocks as old as Precambrian occur in the Greater Antilles (44), the origin of the Caribbean region is tied to the breakup of Pangaea (mid-Jurassic), when Laurasia began to separate from Gondwana. The initial ocean floor that formed the gap between the two continents subsequently disappeared through subduction. Later, in the mid-Cretaceous, the Caribbean Plate formed in the eastern Pacific and has since moved eastward relative to the North and South American Plates. Volcanic islands formed along the northern and eastern margins of this plate as it moved, due to subduction of the North American Plate beneath the lighter Caribbean Plate, creating the proto-Antilles. In the early Tertiary, the northeastern boundary of the Caribbean Plate (Cuba, Hispaniola, Puerto Rico) collided with the Bahamas Platform (on the North American Plate) and essentially "plugged" the subduction in that region. Subsequently, the plate began to move in a more easterly direction, and a major fault developed to the south of Cuba, along with a small spreading center and associated fault zone (Cayman Trough). Jamaica and the southern portion of Hispaniola moved eastward along the northern edge of the Caribbean plate; eventually southern Hispaniola collided with northern Hispaniola in the Miocene (85). Southeastern Cuba and northern Hispaniola may have been connected in the early Tertiary and separated in the middle to late Eocene (44) or Oligocene (87a). The Bahamas Bank has remained as a carbonate platform fixed to the North American Plate, and the Lesser Antilles have remained as a classical island arc, moving eastward along the leading edge of the Caribbean Plate (36, 43).

Recently, it was suggested that the southern and northern Lesser Antilles have had separate geologic histories that have influenced the evolution of anoline lizards (158). However, a closer examination of the geologic evidence does not support that suggestion. For example, the major evidence was a purported fault zone in the central portion of the Lesser Antilles proposed in the late 1970s and early 1980s (12), but geologists (and others) abandoned that hypothesis in the late 1980s for lack of evidence (13, 84, 121). Also, other geologic studies cited (e.g. 174) do not support a geologic break between Dominica and Martinique.

Instead, the Lesser Antilles would appear to form a classic island arc of volcanic islands built above a subduction zone (184).

PALEOGEOGRAPHY The aspect of Caribbean geologic history of greatest interest to biogeographers, the relationships of emergent land areas, unfortunately is the one that is most poorly understood. There is no place in the West Indies that is known by the presence of a continuous sequence of sediments to have been emergent since the late Cretaceous, although some areas of Cuba, northern Hispaniola, and possibly Puerto Rico, may have been. Current geologic evidence is inconclusive but suggests that the proto-Antilles did not form a continuous dry land connection similar to the present-day Isthmus of Panama; instead it was probably a chain of islands (41, 42, 138). However, the recent suggestion that there were no permanently subaerial landmasses in the Greater Antilles prior to 42 million years ago (mya) (111a, 112a) is speculative; it can be neither refuted nor supported with current evidence.

Karst topography and exposed limestone rock are abundant in the West Indies and reflect the widespread mid-Tertiary inundation of most land areas. Many of the present high mountain ranges are the result of relatively recent (late Tertiary and Quaternary) orogenic activity and cannot be used as a guide to the early or mid-Tertiary physiography of the West Indies. For example, probably no part of Jamaica was more than a few meters above sea level from the middle Eocene to the middle Miocene (2, 20, 155), although subsequent orogenic activity has obliterated part of the evidence (widespread limestone formations). The Blue Mountains of eastern Jamaica, rising to over 2200 m, were uplifted only 5–10 mya (30, 185).

In Cuba, paleogeographic reconstructions suggest that some land areas were emergent throughout the Cenozoic (65–0 mya), possibly forming small islands. Beginning in the early Miocene, the present major upland areas of Cuba probably were large islands, later coalescing in the late Miocene and Pliocene as the present island shape emerged (44, 87).

In Hispaniola, the emergent land areas during the mid-Tertiary quiescent period were the Cordillera Central, Cordillera Oriental, and possibly a portion of the La Selle-Baoruco Range in the south (14, 104, 118, 126). Even in the Cordillera Central, the recent major uplift of that mountain range took place only 3–4 mya (105). Initial uplift of the Massif de La Hotte on the southwestern peninsula of Hispaniola is correlated with the collision between the South Island and northern Hispaniola; this began in the middle Miocene (15 mya), and the entire peninsula emerged in the Pliocene (5 mya) and Pleistocene (105). The major uplift of the Massif de La Selle and Sierra de Baoruco also began in the middle Miocene, although some small portions may have been emergent throughout the mid-Tertiary (118). The Sierra de Neiba and the Sierra Martin

Garcia, which are areas of endemism for vertebrates (164), apparently were not emergent until the late Miocene or Pliocene (5–10 mya; 126).

Most of Puerto Rico, and probably the Virgin Islands, was submerged from the late Eocene to the Pliocene (105). However, several periods of localized uplift occurred, from the late Cretaceous to the late Oligocene, on the Puerto Rican Bank. During one such event (late Eocene to mid-Oligocene), uplift of several kilometers occurred (99), and palynological evidence suggests a high altitude flora at that time (57, 58). These data indicate that some emergent land areas may have persisted on the Puerto Rican Bank throughout the Cenozoic.

The Bahamas Bank has been a carbonate platform since the mid-Mesozoic, gradually subsiding but maintaining near sea-level elevations as carbonate reef-deposits accumulated (39, 131). As such, it has been subject to periodic submergence and emergence, the latter most recently in the Pleistocene. Basement rocks in the northern Lesser Antilles are as old as Jurassic, but the oldest rocks in the southern Lesser Antilles are middle Eocene (184). The Cretaceous date for rocks on Union Island (Grenadines), mentioned in the earlier geologic literature and recently cited for biogeographic purposes (158), was in error (12, 175). The degree to which Lesser Antillean volcanoes have been emergent during the history of the island arc is unknown.

FOSSIL RECORD

Tertiary

AMBER FOSSILS Amber deposits in the Dominican Republic provide dramatic documentation of a mid-Tertiary biota, including (among vertebrates) frogs (*Eleutherodactylus*, seven specimens), lizards (*Anolis*, five specimens; *Sphaerodactylus*, six specimens), mammals (hair and bones), and a bird (feathers) (9, 99a, 111a, 141–143, 154). The mammal hair is believed to be from a rodent, the mammal bones from an insectivore, and one bird feather has been identified as belonging to a woodpecker or relative (Picidae). Critical to understanding the importance and relevance of these amber fossils to Caribbean biogeography is the establishment of their age.

Recent tabulations of dates for Tertiary fossils of West Indian vertebrates (112, 112a) listed two of the lizard fossils as Miocene and a frog fossil as Late Eocene. However, those and several other important vertebrate fossils all came from the same mine (La Toca) in the Cordillera Septentrional and therefore are considered to be of the same age (143). That age was determined to be late Eocene (40 mya) based on microfossils from a locality (El Mamey) west of La Toca in the Altimira Formation (47) and on correlative evidence from nuclear magnetic resonance spectroscopy (98, 143). Since then, further studies on the geology of the Cordillera Septentrional have distinguished the formation containing the amber mine as the La Toca Formation (38, 40). This formation ranges from Lower Oligocene to lower Middle Miocene and contains massive conglomerates in the lowermost portion and turbiditic sandstones and mudstones in the middle and upper portions (40). Because Dominican amber is found in turbiditic sandstones (45), and because those rocks, within the La Toca Formation, have been dated by nanofossils as upper Oligocene to lower Middle Miocene (40), the age range (minimum to maximum) of the La Toca amber fossils is 15–30 mya. Indirect evidence from nuclear magnetic resonance spectroscopy and hardness (98) would suggest that the older portion of that time interval (20–30 mya) contains the actual age of the fossils, concordant with an upper Oligocene time assigned to amber from the Cordillera Septentrional in a stratigraphic correlation (119).

NONAMBER FOSSILS There are several other pre-Quaternary fossils of vertebrates: a cichlid fish (*Cichlasoma*) from Haiti (28), ground sloths from Cuba (112) and Puerto Rico (112a), and a capromyid rodent and platyrrhine primate from Cuba (112a). All are early Miocene except the Cuban sloth (early Oligocene). Two reptilian vertebrae are known from Miocene deposits on Puerto Rico and are believed to belong to a boid snake and an iguanid lizard (114). All of these nonamber fossils belong to extant (or recently extinct) Antillean families.

Quaternary

The best fossil record for West Indian vertebrates comes from Pleistocene and Holocene cave and fissure deposits. During the glacial maxima, the West Indies experienced much drier conditions than they do now, and arid-adapted species flourished (149). The reason for the relatively large number of vertebrate extinctions since the Pleistocene is not well established, but it is likely a combination of several factors: reduction in arid habitats, reduction of land area by elevated sea levels, and human-associated causes (128, 129, 134, 148, 149, 176). Humans colonized the West Indies about 7000 years ago (159), and it is probable that many Holocene extinctions were the result of contact by Amerindians and Europeans (129).

BIOGEOGRAPHIC MECHANISMS

Early Ideas

Darwin (34) used the West Indian mammal fauna as an example of how islands surrounded by deeper water exhibit greater endemism, but otherwise he offered no speculation as to the origin of the fauna. Wallace (186), however, was one of the first to discuss, albeit briefly, the zoogeography of West Indian vertebrates. He made special note of the impoverished nature of the fauna (at higher taxonomic levels), an observation that would be repeated often (33, 117, 172) and one that has been appropriately termed "the central problem" (189). Because such a pattern of taxonomic composition is exactly what one sees on remote, oceanic islands (137a), this has been interpreted as evidence of overwater dispersal. However, land bridges between the islands and the continents also were proposed to explain the same patterns of distribution and faunal composition (5, 162, 163), launching a lengthy debate (33). Deep water now is known to surround many of the islands in the Greater and Lesser Antilles, precluding dry land connections due to recent sea-level changes, and therefore the "land bridge hypothesis" is no longer viable. However, vicariance, through the mechanism of plate tectonics, essentially has replaced the land bridge hypothesis as one of the two primary theories (with overwater dispersal) for the origin of the West Indian biota.

Vicariance and Dispersal

PHYLOGENY The vicariance theory of Caribbean biogeography suggests that the present West Indian biota represents the fragmented remnants of an ancient biota that was continuous with those of North and South America in the late Cretaceous (156, 157). Plate tectonic reconstructions still have not "stabilized" to a single well-supported scenario for the Caribbean region (71, 139, 140), but the proto-Antillean island arc required in the vicariance hypothesis is, nonetheless, a common theme in most reconstructions. Thus, proto-Antillean vicariance cannot be eliminated on geological grounds.

The primary evidence used to support the vicariance theory has come from cladistic (vicariance) biogeography (86, 133). The basic premise of this approach is that congruence among organismal phylogenies, and between those phylogenies and area relationships, supports vicariance. The early proponents of cladistic biogeography considered dispersal to be untestable and unscientific (133), a viewpoint that has not changed significantly (130). Even the less extreme viewpoints place dispersal in a secondary role: "dispersal should be a last resort for explaining modern distributions and used only after all vicariance possibilities have been considered" (60). Following this approach, cladistic biogeographers have claimed considerable support for a vicariant origin for West Indian vertebrates (32, 62, 63, 93, 94, 135, 156, 157, 160, 161).

However, this fundamental tenet of cladistic biogeography recently was called into question (79). The claim that phylogenetic evidence alone can provide such support of vicariance is rejected for the simple reason that dispersal can produce the same phylogenetic patterns (173a). Fully congruent phylogenies, even those congruent with a geological scenario, can be produced through concordant dispersal. Geographic proximity, air currents, water currents, and other factors all combine to produce higher probabilities of dispersal from some areas than others. In the West Indies, this is especially true because ocean currents flow almost unidirectionally from southeast to northwest (see Figure 1). Therefore, dispersal to the West Indies by rafting on currents is much more likely from South America than from Central or North America (74).

Because many West Indian vertebrate groups have their closest relatives in South America (see below), the methodology of cladistic biogeography would dictate that a vicariant event must have separated South America from the West Indies. However, concordant dispersal provides an equally valid explanation. The data that can distinguish between these two possibilities are the times of divergence of West Indian taxa from their closest relatives on the mainland (74, 78, 79). Groups that diverged at the same time as the geologic separation can be inferred to have arisen by vicariance. Those groups that arose after the geologic separation can be inferred to have arisen by dispersal. Thus, phylogeny can help to establish whether a pattern exists and to identify the source area, but it is the timing of the divergence that allows one to distinguish between these alternative explanations (74, 79).

TIMING Information on the time of divergence of West Indian vertebrate groups from their mainland relatives has come from the fossil record and from the use of molecular clocks. The presence of mid-Tertiary fossils has been used both for (143) and against (147) the vicariance theory. In the former case, an Eocene (40 mya) date for the amber fossils was used to suggest that a diverse fauna was present in the Antilles at an earlier date than generally proposed (143). However, those fossils now are considered to be younger (15–30 mya; see above), and dispersal could have occurred at any time (189). Although the fossil record continually is improving, it is at present of limited value in providing the crucial times of origin for West Indian lineages.

More than 65 studies have been published involving protein electrophoretic, immunological, and DNA sequence divergence among West Indian vertebrates, mostly amphibians and reptiles (145). Of those, the data most frequently used to obtain times of divergence are immunological estimates of amino acid sequence divergence of a protein, serum albumin (23, 56, 64, 66, 69, 78, 169). The rate of evolution in this gene is remarkably constant in studies where geological and paleontological calibration was possible (122).

To draw some general conclusions regarding the origin of the West Indian vertebrate fauna, albumin immunological data from diverse lineages of amphibians and reptiles were assembled (78). In all 13 comparisons between West Indian

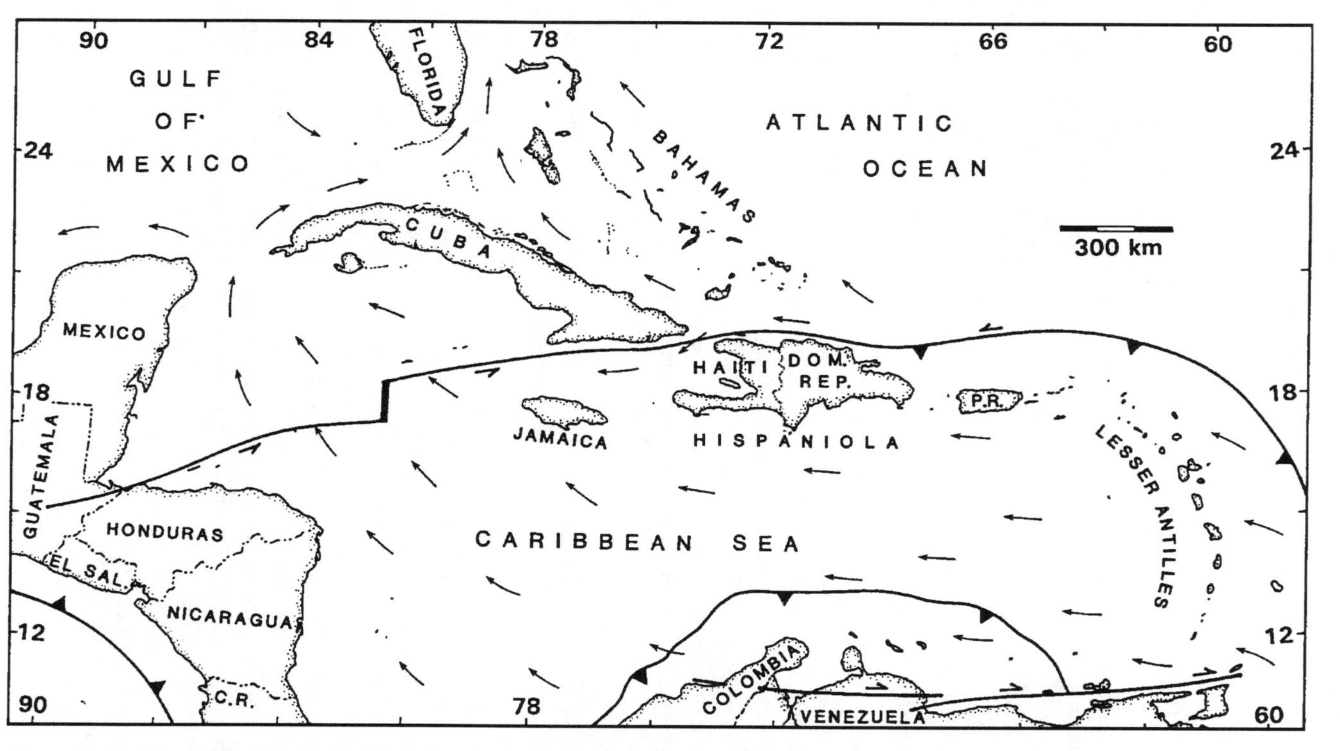

Figure 1 The West Indies, showing direction of water currents (long arrows). The geologic features, indicated by thick lines, are faults (short arrows indicate direction of plate movement), subduction zones (teeth indicate direction of subduction), and a spreading center (double-thick line).

lineages and their closest relatives on the mainland, the estimated divergence times were later than predicted by vicariance, indicating mid-Tertiary rather than late Cretaceous origins (78, 79). Since then, other molecular and non-molecular data were used to examine the remaining 64 independent lineages of West Indian amphibians and reptiles (74). With the possible exception of the frog genus *Eleutherodactylus* and the xantusiid lizard *Cricosaura typica,* all lineages appear to have originated in the Cenozoic. Phylogenetic analyses indicate that most West Indian groups have affinities with South American taxa, suggesting overwater dispersal from that continent. It was proposed that the nearly unidirectional (southeast to northwest) ocean currents have carried flotsam (e.g. 70, 91) from the mouths of major rivers in South America to islands of the West Indies throughout the Cenozoic era (74).

The Bolide Impact

Given the geologic possibility for vicariance, the virtual absence of an ancient West Indian vertebrate fauna suggests that numerous extinctions may have occurred since the formation of the proto-Antilles in the late Cretaceous. Because most of the mid-Tertiary fossils are of groups extant in the Holocene, an ancient West Indian biota—if it existed—would appear to have become extinct earlier in the Tertiary. To explain this, it was proposed that the impact of the Cretaceous-Tertiary (K-T) asteroid or comet (bolide) at 65 mya probably devastated the ancient West Indian biota because of its very close proximity to that region (78).

A large crater (Chicxulub) about 1 km beneath the surface of the Yucatán peninsula is believed by most geologists to be the impact site for the K-T bolide (81, 96, 120). The size of the crater is not agreed upon, but it is between 170 km and 320 km in diameter, making it the largest known impact structure in the inner solar system during the last 4 billion years (82, 166, 167). At the time of impact, the Greater Antilles were the closest land masses, located only 1–3 crater diameters away (139). It was, in fact, the local effects of this impact that helped locate the crater (81). Giant wave deposits found in Haiti and Cuba indicated disturbances in sediments that were at least 2 km below sea level at that time (120). "Megawave" deposits also have been found in deep water (> 400 m) sediments in northeastern Mexico, containing leaves, wood, and terrestrial debris (173), and in the southern United States (81) and the Gulf of Mexico (1). Gigantic hurricanes (hypercanes), resulting from local heating of sea water, also have been proposed (48). Besides the devastating global effects of the impact, these catastrophic local effects would almost certainly have caused widespread extinctions of many organisms that might have existed in the Caribbean region at that time.

Table 2 Numbers of families, genera, and species of native West Indian freshwater fishes[a]

Order and family[b]	Genera			Species		
	Total	Endemic	% Endemic	Total	Endemic	% Endemic
Luganoiiformes						
Lepisosteidae	1	0	0	1	1	100
Ophidiiformes						
Bythitidae	1	1	100	4	4	100
Cyprinodontiformes						
Poeciliidae	5	3	60	46	45	98
Cyprinodontidae	2	1	50	8	8	100
Fundulidae	1	0	0	1	0	0
Rivulidae	1	0	0	7	7	100
Atheriniformes						
Atherinidae	1	1	100	1	1	100
Synbranchiformes						
Synbranchidae	1	0	0	1	0	0
Perciformes						
Cichlidae	1	0	0	5	5	100
TOTAL	14	6	43	74	71	96

[a]After Burgess & Franz (17); but including the Miocene cichlid (28) and some recently described taxa.
[b]There are no endemic families.

TAXON-SPECIFIC PATTERNS

Freshwater Fishes

There are six orders and nine families of freshwater fishes in the West Indies (Table 2). Although no families are endemic, 6 of the 14 genera (43%) and 71 of the 74 species (96%) are endemic to the West Indies. Most of those species are restricted to a single island, portion of an island, lake, or even sinkhole. Species diversity is correlated with island area; Cuba (28 species) and Hispaniola (35 species) have the most species, with relatively few on Jamaica (6 species), the Bahamas (5 species), Cayman Islands (4 species), and Martinique (1 species). There are no endemic species on Puerto Rico. Relationships still are not well known for many groups, but approximately 17 independent lineages of West Indian freshwater fishes can be identified: 9 with North (or Central) American affinities, 6 with relationships to South American taxa, and 2 of marine origin (17).

Discussion of the historical biogeography of West Indian fishes has included both vicariance and dispersal as mechanisms for the origin of the fauna (16, 17, 156). However, the Tertiary fossil record consists of only one fossil cichlid from the early Miocene of Haiti (28), and there are no molecular data that would permit dating of divergences. While it is a reasonable assumption that the few non-endemic species arrived by dispersal in the Quaternary, the time of origin for the remainder of the species cannot be inferred. All of the species are secondary not primary freshwater species, which means that they are to some degree tolerant of salt water. This fact has led previous workers (e.g. 132) to consider secondary freshwater species as inadequate indicators of biogeographic patterns. With the possible exception of the Cuban gar, there is no evidence to indicate that any of the West Indian lineages of freshwater fishes are of ancient (Cretaceous) origin.

The Cuban gar, *Atractosteus tristoechus,* is a lepisosteid and the only luganoiiform in the West Indies. It is believed to be most closely related to species in the same genus inhabiting North and Central America (17). The oldest lepisosteid fossil is early Cretaceous, and fossils of *Lepisosteus* range back to the late Cretaceous (80 mya) in North America and Europe (54). Four species of brotulas or cusk-eels (Ophidiiformes) occur in the West Indies and are placed in an endemic genus, *Lucifuga.* They inhabit freshwater sinkholes on Cuba and a brackish water cave system in the Bahamas, and they are believed to have had a marine origin (17, 101).

The largest component of the West Indian freshwater ichthyofauna (84% of species) belongs to the order Cyprinodontiformes. Most of those species are in the family Poeciliidae. There is disagreement over the relationships of West Indian *Gambusia* (10 species), although there appear to be at least three species groups representing three separate colonizations, most likely from North or Central America (17, 106, 150, 151). Species in the endemic Cuban genera *Girardinus* (8 species) and *Quintana* (1 species) together form a monophyletic group with affinities to *Carlhubbsia* of Guatemala (151, 156). The endemic genus *Limia* (24 species) and the three species of *Poecilia* (monophyletic subgroup) are each believed to represent a single colonization from South America (17, 150). There are no documented pre-Quaternary fossil poeciliids, although there is an undocumented report of an early Tertiary (Paleocene) fossil from Argentina in the literature (137).

Eight species of killifishes (Cyprinodontidae) are found in the West Indies (17; and those subsequently described). *Cubanichthys* (3 species), a West Indian endemic, is believed to have arrived by dispersal from North or Central America (17). At least two dispersals from North or Central America, probably in the Quaternary, are required to explain the origin of West Indian *Cyprinodon* (5 species). The oldest fossil cyprinodontid is mid-Tertiary (Oligocene), although some early Tertiary (Paleocene) scales, possibly belonging to this family, exist (137). A single, non-endemic, species of fundulid (*Fundulus grandis*) occurs in northern Cuba; it has affinities to populations in southern

Florida. Dispersal from North America during the Quaternary has been postulated to explain its origin (17). Seven species of rivulids (*Rivulus*) are known from the West Indies. Their relationships are not well understood, and therefore the number of independent lineages is unknown, although at least two are indicated. The rivulids may have arrived by dispersal from South America (17). The order Atheriniformes is represented by a single endemic, monotypic genus of silverside, *Alepidomus,* occurring in western Cuba. It is believed to be of marine origin (17). A single, non-endemic, species of swamp eel (synbranchiform), *Ophisternon aenigmaticum,* occurs in Cuba. Its distribution in South and Central America suggests relatively recent (Quaternary) dispersal. The order Perciformes is represented by five endemic species of cichlids in the genus *Cichlasoma.* One of those species is the Miocene *C. woodringi* from Haiti (28); the remaining four are extant. The number of independent lineages is not known, but they are believed to have arisen by dispersal from South America (17). The oldest cichlid fossil is mid-Tertiary (Oligocene) and is from Somalia (137).

Amphibians

All West Indian amphibians are anurans (frogs and toads), and they represent 3.7% of all extant amphibian species. Of the four families and six genera, only one genus (*Osteopilus*) is endemic (Table 3). However, nearly all (99%) of the 166 native species are endemic, and most (138 species) are in the enormous Neotropical leptodactylid genus *Eleutherodactylus* (> 520 species). With few exceptions, each species is restricted to a single island and often to a small area within an island (< 100 km^2; 75). Species diversity is concentrated in the Greater Antilles; there are no endemic species in the Bahamas, and only nine species are endemic to the Lesser Antilles. The nine independent lineages of West Indian amphibians are believed to represent one origin by late Cretaceous vicariance and eight independent Cenozoic dispersals from the mainland (74);

Table 3 Numbers of genera, and species of native West Indian amphibians (all, order Anura)[a]

Family[b]	Genera			Species		
	Total	Endemic	% Endemic	Total	Endemic	% Endemic
Bufonidae	1	0	0	11	11	100
Dendrobatidae	1	0	0	1	1	100
Hylidae	2	1	50	10	10	100
Leptodactylidae	2	0	0	144	142	99
TOTAL	6	1	17	166	164	99

[a]After Hedges (74).
[b]There are no endemic families.

the source area for seven lineages is South America, and the other two lineages are from Central or South America.

The West Indian bufonids (*Bufo peltocephalus* group) represent a monophyletic group (146) among Neotropical species of the cosmopolitan genus *Bufo,* with probable affinities to the *Bufo granulosus* group of South America (27). Albumin immunological distance (ID) data (78) suggested an early Cenozoic origin for this group, probably by dispersal from South America (74). The single dendrobatid from Martinique (89) belongs to a diverse genus distributed almost entirely in South America (including Trinidad and Tobago). This fact, its location in the southern Lesser Antilles, and consideration of albumin ID data among dendrobatids (124) together suggest that it arrived by dispersal from South America during the Cenozoic (74).

Several hypotheses for the origin of West Indian hylid frogs have been proposed. Albumin ID data (78) largely supported an earlier hypothesis (46) that suggested that the West Indian species (except for one) are monophyletic and represent an early Cenozoic origin by dispersal from South America (74) rather than six independent colonizations (183). Those data also indicated that a single species from Hispaniola, *Hyla heilprini,* represents a separate invasion (dispersal) from South America, but the timing of that event is not known.

The leptodactylid genus *Eleutherodactylus* is the largest genus of vertebrates, and resolving the origin of the 138 known West Indian species has been a formidable challenge. Infrageneric classification largely has been based on several key morphological and allozyme characters (73, 107, 108), and much of the internal phylogenetic structure of the genus remains to be determined. A combination of slow-evolving allozyme loci, albumin ID data, and some unconventional morphological characters defined a western Caribbean clade (subgenus *Euhyas*) of 82 mostly terrestrial species and an eastern Caribbean clade (subgenus *Eleutherodactylus*) of 50 mostly arboreal taxa, with a third group of six large species (subgenus *Pelorius*) confined to Hispaniola (66, 73). The time of separation between the subgenera *Euhyas* and *Eleutherodactylus* was estimated to be late Cretaceous (70 ± 6.8 mya), thus agreeing well with a proto-Antillean vicariance model (66, 73). However, additional phylogenetic data are needed to clarify the origin of the West Indian lineage or lineages (74). Aside from two species of South American origin in the southern Lesser Antilles (90), there is no evidence that any dispersal events took place from the mainland to the West Indies during the Cenozoic. However, a single dispersal back to the mainland (probably from Cuba in the mid-Cenozoic) likely occurred and led to the subgenus *Syrrhophus* in Central America and southern North America (66, 73). The 17 native Jamaican species, all *Euhyas,* are a monophyletic group that resulted from a single colonization (probably from

Cuba) in the Miocene (8, 66, 72). Most species in the southern portion ("South Island") of Hispaniola, which was a separate island until the late Miocene, also belong to the subgenus *Euhyas* and apparently had a similar history as the Jamaican species (73). After collision with the remainder of Hispaniola ("North Island"), there was a limited exchange of species between the two regions.

Four species of *Leptodactylus* represent four independent dispersals to the West Indies from South America (or possibly Central America in the case of *L. insularum*). Three of these represent relatively recent (Pliocene or Quaternary) arrivals, but the Puerto Rican endemic (*L. albilabris*) is estimated to have arrived 40 mya (74, 123).

Reptiles

With 449 species (93% endemic), this is the largest component of the West Indian vertebrate fauna (Table 4), representing 7.4% of all extant reptiles. There are no endemic orders or families, although 9 of the 50 genera are endemic. Two genera of lizards, *Anolis* (138 species) and *Sphaerodactylus* (80 species), account for about one half of the total species diversity. With the possible exception of one Cuban species (*Cricosaura typica*) that may represent an ancient vicariant relict, all 68 independent lineages of West Indian reptiles are believed to have arrived by overwater dispersal from the following source areas: North America (3 lineages), Central America (8 lineages), Central or South America (14 lineages), South America (35 lineages), Africa (4 lineages), and "New World" (4 lineages) (74).

The three species of crocodilians represent three separate dispersals in the late Tertiary or the Quaternary, apparently from Central and South America (74). The relationships of amphisbaenians still are poorly known, although albumin ID data (78) together with other data (e.g. 29) suggest the 14 West Indian species may comprise a single radiation (74). The origin of that lineage or lineages probably was from South America, and the dispersal was estimated to have occurred in the early Cenozoic (78).

West Indian anguid lizards belong to the Neotropical subfamily diploglossinae. Until recently, five genera were recognized, based on osteoderm structure (177); four of these were distributed on the islands. Although two of the genera (*Sauresia* and *Wetmorena*) were restricted to Hispaniola, they were considered to be derived from Hispaniolan *Celestus* (177). Recently, an earlier generic arrangement was resurrected and a classical proto-Antillean vicariance scenario postulated to explain the origin of the West Indian species (161). In it, the two Hispaniolan endemic genera were considered to be early derivatives of *Diploglossus*. However, albumin ID data (74, 78) support the closer relationship of those genera to Hispaniolan *Celestus*. Moreover, albumin ID data

Table 4 Numbers of genera and species of native West Indian reptiles[a]

Order and family[b]	Genera			Species		
	Total	Endemic	% Endemic	Total	Endemic	% Endemic
Crocodylia						
Crocodylidae	1	0	0	3	1	33
Squamata						
Amphisbaenians						
Amphisbaenidae	1	0	0	14	14	100
Lizards						
Anguidae	2	0	0	22	22	100
Gymnophthalmidae	2	0	0	3	1	33
Iguanidae	5	2	40	172	169	98
Gekkonidae	7	0	0	97	89	92
Scincidae	1	0	0	2	1	50
Teiidae	4	0	0	24	20	83
Xantusiidae	1	1	100	1	1	100
Snakes						
Boidae	3	0	0	11	9	82
Colubridae	15	6	40	46	42	91
Elapidae	1	0	0	1	0	0
Leptotyphlopidae	1	0	0	8	6	75
Tropidophiidae	1	0	0	13	13	100
Typhlopidae	1	0	0	24	24	100
Viperidae	1	0	0	2	2	100
Testudines						
Emydidae	1	0	0	4	4	100
Kinosternidae	1	0	0	1	0	0
Testudinidae	1	0	0	1	0	0
TOTAL	50	9	18	449	418	93

[a]After Hedges (74).
[b]There are no endemic families.

indicate a mid-Cenozoic origin, by dispersal, for the West Indian species (74, 78). Three species of gymnophthalmids represent independent colonizations of the West Indies from South America. Two of those (*Bachia heteropus* and *Gymnophthalmus underwoodi*) almost certainly dispersed in the Quaternary, whereas the time of arrival for the third lineage (*G. pleei*) is unknown, although it was most likely during the Cenozoic (74). Iguanid lizards have inhabited the West Indies at least since the early Miocene, but the exact number of colonizations is not known. There is broad agreement over the definition of species groups and series of anoline lizards, but the higher-level relationships have not been determined despite considerable study with morphological, chromosomal, and molecular data (e.g. 19, 68, 169, 187, 188, 198). An attempt at synthesis of disparate sources of published data for anoline lizards (62) was unsuccessful (26, 63, 190).

Despite the current lack of consensus regarding anoline relationships, evidence for the timing of their arrival to the West Indies, from albumin ID data, supports an arrival by dispersal in the mid-Cenozoic (16–36 mya) from Central or South America (74). This timing is concordant with the fossil evidence noted above and contradicts speculation (62, 158) that their origin was the result of proto-Antillean vicariance. Future resolution of relationships will help determine the number of independent colonizations that occurred. Besides the anoline lizards, eight additional colonizations by iguanids to the West Indies can be identified, occurring in the Quaternary. Two exceptions are endemic genera that apparently arrived in the mid-Cenozoic from Central or South America (*Cyclura*) and North America (*Leiocephalus*), respectively (74).

The West Indian gekkonid lizards represent 11 independent colonizations, primarily from South America and Africa and mostly in the Quaternary (74). Of special note is the large genus *Sphaerodactylus* (80 species). Analysis of allozyme and albumin ID data led Hass (64) to postulate a South American origin for the West Indian species by dispersal in the mid-Cenozoic. This timing is in agreement with the presence of the genus in Dominican amber (9, 142). A recent DNA sequence analysis of relationships among *Sphaerodactylus* (65) identified several well-supported monophyletic groups in the West Indies and refined the classification of the genus but did not alter that hypothesis for the origin of the West Indian species. However, additional data are needed for all species, including those in the mainland *lineolatus* section (10 species), before the number or direction of dispersal events can be accurately determined. An origin for the mainland species by dispersal from the West Indies cannot yet be ruled out.

Only a single endemic scincid lizard is present in the West Indies. A possible relationship with a South American species in the same genus (*Mabuya*) suggests an origin by dispersal from that continent. At least eight independent lineages of teiid lizards are present, and most arose by dispersal from South America in the Quaternary. The relationships of the 20 endemic species of *Ameiva* (considered to be one lineage) are not yet established, but albumin ID data suggest a mid-Cenozoic origin by dispersal from Central or South America (74, 78).

The single endemic xantusiid lizard, *Cricosaura typica*, represents a biogeographic enigma because of its restricted distribution in an unexpected location: the Cabo Cruz region of eastern Cuba. Although analysis of some published morphological data supported a close relationship with *Lepidophyma* of Central America (32), several key morphological characters in that study were found to be incorrectly scored, thus putting that conclusion into question (76). Instead, DNA sequence data provided statistical support for the basal position of *Cricosaura* within the family (76, 77), and this was further supported by chromosome evidence (67). The presence of Middle Paleocene xantusiid fossils from North America (50) suggests an early isolation, perhaps by proto-Antillean vicariance, for *Cricosaura*. However, a more recent dispersal to the West Indies, and subsequent extinction of the mainland source population, cannot be ruled out due to the relictual nature of xantusiid lizard distribution (6, 74, 77). No albumin ID data are available for these xantusiid lizards.

A single vertebra records the presence of boid snakes in the West Indies by the early Miocene (114). Of the four independent lineages now present, only one (*Epicrates*) contains endemic species (74). There is one mainland species in that genus, and the nine West Indian species are believed to form a monophyletic group (93, 182). Although proto-Antillean vicariance was postulated to explain the origin of the West Indian species (94), a low albumin ID between *Epicrates* and *Boa* (37), corresponding to a divergence time of about 22 mya, argues instead for a mid-Cenozoic origin by dispersal (74).

Colubrid snakes colonized the West Indies at least 11 times, mostly from South America and mostly in the late Cenozoic (74). Of special note is the large assemblage (33 species) of alsophines, including six endemic genera. Three dispersals from the mainland were proposed to explain their origin, based on a morphological analysis (115). However, albumin ID data are more suggestive of a single, monophyletic group (22, 78). Those same data, and others (23), also indicate a mid-Cenozoic origin by dispersal from South America (74). The recent suggestion (31) that the North American genus *Farancia* is "internested" among West Indian alsophines is not supported by the strict consensus tree of that same study, and it is contradicted by albumin ID data (24), indicating that *Farancia* is not closely related to alsophines.

A single (nonendemic) species of elapid snake is recorded from Isla de Providencia off the coast of Nicaragua. Its origin likely was by dispersal from Central America in the Quaternary (74). Four independent colonizations of the West Indies by leptotyphlopid snakes are believed to have occurred (74). Relationships of species in this family are poorly known and in need of study, but the five endemic species of the *bilineata* group are believed to represent a single colonization and radiation (181). A single Bahamian species, *L. columbi*, may represent a separate colonization, perhaps from South America in the late Tertiary or Quaternary (74). The snake family Tropidophiidae is Neotropical, and the genus *Tropidophis* is primarily West Indian. An origin by dispersal from South America in the early to mid-Tertiary is indicated by albumin ID data (74, 78). Most West Indian typhlopid snakes are believed to form a monophyletic group (20 species) with African affinities, whereas the remaining four species are believed to represent a single group with New World affinities (180). However, albumin ID data suggest a closer relationship between those two lineages and an origin by dispersal in the Cenozoic (74, 78). Two species of viperid snakes are endemic to islands in the southern Lesser Antilles. Consideration of their distribution, affinities (100), and albumin ID data among related species (25) suggests that they arrived by dispersal from South America in the late Tertiary or Quaternary (74).

The Antillean genus of emydid turtles (*Trachemys*) has a single non-West Indian species, *T. scripta* (165). There are no albumin ID data, but consideration of the fossil record, distribution, and relationships suggests a single dispersal from Middle or North America in the mid- to late Cenozoic, with a reverse dispersal leading to *T. scripta* on the mainland (74, 165, 165a). The single, non-endemic kinosternid turtle species occurring on islands adjacent to Central America probably arrived by dispersal in the Quaternary. It is unknown whether the Miocene pelomedusid turtles inhabited freshwater or saltwater, and little is known of their origin (74, 112a, 114). The single extant testudinid turtle is not endemic; it likely arrived by dispersal in the Quaternary. Fossils of giant species in this family are known from the West Indies, but their time of origin and source area are not known.

Birds

There are 15 orders and 49 families of native West Indian birds representing about 4.4% of the world's bird fauna, although endemism is relatively low (Table 5). One family (Todidae) out of 49 (2%), 38 of the 204 genera (19%) and 150 of the 425 species (35%) are endemic to the West Indies. There never has been a comprehensive treatment of the historical biogeography of West Indian birds, although some general patterns have been discussed (10, 97, 134). Also, the phylogenetic relationships of many West Indian endemic groups remain poorly known, limiting biogeographic inferences (92, 125, 166) or tests of ecological models (51, 152, 153, 179).

Notwithstanding the limitations of the avian database, it has been postulated that the origin of the entire West Indian avifauna was by dispersal, largely from North America (10, 134), except for the avifauna of the southern Lesser Antilles, which appears to be mostly South American in derivation (49). The single endemic family, the Todidae, is believed to be the oldest lineage in the West Indies, having arrived from North America in the Oligocene; this theory is based on mainland fossils (134), but also some known from France. Each of the 275 non-endemic species represents at least one separate dispersal from the mainland, most likely in the Quaternary. However, that number is almost certainly an underestimate based on the finding of complex and multiple colonization patterns in some well-studied species (92, 166). The remaining 150 endemic species represent fewer than that number of colonizations, but the actual number of independent lineages presently is unknown.

Mammals

There are 5 orders and 16 families of native West Indian mammals representing about 3.1% of the world's mammal fauna (Table 6). Four families, 38 of the 65 genera (58%), and 116 of the 145 species (80%) are endemic to the West Indies. Although all West Indian vertebrate groups have suffered at least some human-caused extinctions, mammals have, by far, suffered the most. Almost 90% of the known species of nonvolant West Indian mammals went extinct during the last 20,000 years (129). Climatic changes related to the Pleistocene glaciations are believed to have caused some extinctions (149), while Amerindians and Europeans both have been implicated in other extinctions (129). Phylogenetic relationships are still not well established, but at least 51 independent lineages can be identified: 9 for nonvolant species and 42 for bats (129). Of those, the nonvolant mammals show a strong South American influence (7 lineages) with only one lineage showing affinities to Central America and one to Central or North America. Bats, on the other hand, show a greater influence from the

Table 5 Numbers of families, genera, and species of native West Indian birds[a]

		Genera			Species		
Order	Families[b]	Total	Endemic	% Endemic	Total	Endemic	% Endemic
Craciformes	1	1	0	0	1	0	0
Galliformes	1	1	0	0	1	0	0
Anseriformes	2	6	0	0	20	1	5
Piciformes	1	7	2	29	12	9	75
Trogoniformes	1	1	1	100	2	2	100
Coraciformes	2	2	1	50	7	5	71
Cuculiformes	2	4	2	50	10	6	60
Psittaciformes	1	3	0	0	13	11	85
Apodiformes	1	4	0	0	7	2	29
Trochiliformes	1	10	5	50	17	14	82
Strigiformes	4	10	2	20	16	6	38
Columbiformes	1	6	1	17	17	10	59
Gruiformes	3	3	1	33	13	1	8
Ciconiiformes	18	59	0	0	104	3	3
Passeriformes	10	87	23	26	185	80	43
TOTAL	49	204	38	19	425	150	35

[a]After Bond (11); with subsequent taxonomic changes and classification of Sibley and Ahlquist (170) and Sibley and Munroe (171).

[b]The Todidae (Coraciiformes) is the only endemic family.

Table 6 Numbers of genera and species of native West Indian mammals[a]

	Genera			Species			
Order and family	Total	Endemic	% Endemic	Total	Endemic	% Endemic	% Extinct
Edentata							
Megalonychidae	11	11	100	18	18	100	100
Insectivora							
Solenodontidae[b]	1	1	100	4	4	100	25
Nesophontidae[b]	1	1	100	8	8	100	100
Chiroptera							
Emballonuridae	1	0	0	1	0	0	0
Noctilionidae	1	0	0	1	0	0	0
Mormoopidae	2	0	0	8	6	75	25
Phyllostomidae	18	8	44	27	16	59	11
Natalidae	1	0	0	4	2	50	0
Vespertilionidae	5	0	0	9	4	44	0
Molossidae	4	0	0	8	1	13	0
Primates							
Cebidae	4	2	50	4	4	100	100
Callitrichidae	1	1	100	2	2	100	100
Rodentia							
Echimyidae	4	4	100	7	7	100	100
Capromyidae[b]	8	8	100	34	34	100	61
Heptaxodontidae[b]	4	4	100	5	5	100	100
Muridae	2	1	50	8	8	100	100
TOTAL	68	41	60	148	119	80	54

[a]After Baker and Genoways (4), Jones (88), Koopman (95), MacPhee and Iturralde-Vinent (112a), Wilson and Reeder (192), and Woods (195).
[b]Endemic family.

west: 18 lineages from Central America, 14 from South America, and 2 from North America.

All West Indian (and other) ground sloths are extinct, although some apparently were contemporaneous with Amerindians as recently as 3715 years ago (129). The West Indian species all belonged to the family Megalonychidae, which was distributed in North and South America and first appears in the fossil record in the Lower Oligocene of Puerto Rico (112a). Relationships among the species of megalonychids are not well known, although recent studies (112, 112a, 195) have supported an earlier suggestion (117) that the West Indian species likely arose by a single dispersal from South America in the mid-Tertiary. A continuously exposed Oligocene land bridge (Aves Ridge) was proposed to explain that dispersal event (112). However, geologic evidence favors only a small chain of islands, if up to 1000 m of subsidence is taken into account (83). Also, if a dryland connection occurred, it must be explained why groups that were present in South America at that time (e.g. marsupials, notoungulates, and astrapotherians among mammals; and many groups of amphibians and reptiles) did not come across and are not represented in the later fossil record or among extant fauna. Moreover, the fossil discovery of a marine ground sloth (35) indicates that saltwater dispersal probably would not have been a problem for these animals. Thus, the origin of West Indian ground sloths probably was the result of a single overwater dispersal from South America in the mid-Tertiary, possibly using the Aves island arc as "stepping stones."

Both families of West Indian insectivores, Solenodontidae and Nesophontidae, are endemic. Although none of the eight known species of nesophontids is extant, most became extinct in post-Columbian times (129). The surviving solenodontids are reduced in numbers and approaching extinction (196). Reconstructing the biogeographic origin of insectivores in the West Indies has proven to be difficult because of the present lack of consensus regarding relationships of the families of insectivores (21, 113), although the two West Indian families often are considered closest relatives (109) and the result of a single colonization event (129). A late Mesozoic origin by Proto-Antillean vicariance has been suggested (109, 110) and is compatible with the age (late Cretaceous) of insectivore fossils from North America (178). However, it also has been noted that the presence of insectivores on Ile Tortue (separated from Hispaniola by deep water) and in the Cayman Islands is evidence that they were capable of dispersing over salt water (195). No molecular data are available that would permit time estimations for the origin of these two families in the West Indies. However, some amber-encased bones believed to be of an insectivore recently were reported from the Dominican Republic (111a), which would establish their presence in the West Indies by 20–30 mya.

Seven families of bats are known from the West Indies (Table 6). There are no endemic families, and only 8 of the 32 genera (25%) and 29 of the 58 species (58%) are endemic. The number of species on each island is partly correlated with island area (59), but it also is related to distance from mainland source areas (4). Bahamian species are derived entirely from the West Indian fauna (almost all from Cuba) rather than from Florida (127), and those in the southern Lesser Antilles (e.g. Grenada) are derived from South America. It has been emphasized in the past that much of the West Indian bat fauna was derived from Central America (e.g. 95, 195). This is true for the non-endemic Jamaican and Cuban species (4, 95) but the Lesser Antillean bat fauna is largely derived from South America (15). The total number of known colonizations (N = 42) for all West Indian bats, including those from Grenada, reflects the different origins for the Greater versus Lesser Antillean bat faunas: 18 from Central America, 14 from South America, 2 from North America, and 8 of undetermined source (88, 95, 129, 192). However, the origin of the lineages leading to the endemic Antillean genera has yet to be determined (4, 59).

The low level of endemism in West Indian bats is almost certainly the result of their enhanced dispersal abilities compared with nonvolant vertebrates: "most bats hardly need a raft" (4). Late Pliocene or Quaternary dispersal can be assumed for all non-endemic species (4). The time of origin for the remainder of the bat fauna is difficult to determine, because no pre-Quaternary West Indian fossils exist, and the bat fossil record, in general, is poor (178). No pre-Quaternary fossils exist anywhere for noctilionids, mormoopids, or natalids, whereas the earliest phyllostomid is Middle Miocene, and vespertilionids and molossids first appear in the mid-Eocene (178). Although drawing inferences concerning divergence times from the fossil record is subject to sampling biases (116), there is no evidence (e.g. an endemic family) of ancient lineages of bats in the West Indies that may have arisen by late Cretaceous vicariance. Thus, the best explanation for the origin of the endemic West Indian bat fauna is by dispersal during the middle or late Cenozoic.

Two families of primates are known from the Tertiary and Quaternary fossil record of the West Indies, but there are no extant species (112a, 195). The Hispaniolan cebid is thought to have affinities with South American species in the genus *Cebus* (196). One of the two Cuban cebids, *Ateles fusciseps*, probably was introduced in historical times (112a, 129). The species *Xenothrix mcgregori* occurred on Jamaica and is believed to have had South American affinities; it has been placed both in the Callitrichidae (53, 191) and its own monotypic family Xenotrichidae (111, 128). The remaining West Indian primate material is insufficient to clearly infer taxonomic position or affinities (52, 128, 129, 195).

The oldest cebid and callitrichid fossils are Lower Miocene and Middle Miocene, respectively (112a, 178). If the divergence between Old World and New World monkeys occurred about 55 mya, considering sampling biases (116), then the split between the cebids and callitrichids must have occurred more recently (55–20 mya); this would provide an upper bound on the time of origin for lineages within either family. Considering these constraints, the origin of West Indian primates is best explained by at least two independent overwater dispersals from South America during the middle or late Cenozoic.

Of the four families of native rodents known to have existed in the West Indies, only one (Capromyidae) is extant (194). The greatest generic and species diversity is seen in the three hystricognath families: Echymyidae, Capromyidae, and Heptaxodontidae. Echymyids show their greatest generic diversity on

Puerto Rico, but they also are known from Hispaniola and Cuba. Capromyids are not known from Puerto Rico but occurred on Hispaniola (where they show their greatest generic diversity), Cuba, the Cayman Islands, Jamaica, and the Bahamas. The giant hutias (heptaxodontids), some as large as 200 kg (7), are known from the northern Lesser Antilles, Puerto Rico, Hispaniola, and Jamaica.

Based on phylogenetic relationships and distribution, it has been proposed that all West Indian hystricognath rodents form a monophyletic group and owe their origin to a single overwater dispersal from South America in the late Oligocene or early Miocene (194). The initial disperser was believed to be a heteropsomyine echymyid, which colonized and diversified on Puerto Rico, later dispersing to Hispaniola and Cuba. A lineage of that initial Antillean radiation on Hispaniola, in turn, is thought to have led to the radiation of capromyids. Finally, the heptaxodontids are believed to have arisen from a capromyid ancestor on Hispaniola (194). If that hypothesis is correct, then perhaps the West Indian echymyids (Heteropsomyinae) and heptaxodontids should be placed in the Capromyidae. A different scenario involving more than one dispersal of hystricognaths to the West Indies also has been proposed (136).

West Indian sciurognaths are represented by two genera and eight species of murids, all extinct. Except for a single species from Jamaica, they are known only from the Lesser Antilles (194). The several species of *Oryzomys* are thought to represent two dispersals to the West Indies in the late Pliocene or Pleistocene: one from Central America leading to the Jamaican species *O. antillarum*, and the other from South America leading to the Lesser Antillean species. The several species of *Megalomys* are believed to represent one dispersal from South America in the late Pliocene or Pleistocene (194).

GENERAL PATTERNS

Knowledge of the diversity and phylogeny of West Indian vertebrates remains incomplete, but sufficient data are available now to draw some conclusions regarding the origin of the fauna. The general pattern that emerges is an origin by dispersal during the Cenozoic for an overwhelming majority (99%) of the independent lineages. In addition, the source area for a large fraction (66%) of the nonvolant vertebrate fauna is South America rather than the closer mainland areas of North America (16%) and Central America (11%) (Table 7). If fish are removed, this pattern is even more pronounced. Such a dispersal pattern can be explained by the nearly unidirectional current flow from the southeast to the northwest (18, 61), bringing flotsam from the mouths of South American rivers (e.g. Amazon, Orinoco) to the islands of the West Indies.

In contrast, the primary source areas for the freshwater fish and volant groups are North and Central America, suggesting that this difference lies in the mode of dispersal: passive (flotsam) versus active (swimming and flying). Passive dispersers must rely on the surface currents to transport them, whereas active dispersers such as fish, bats, and birds have more control over their direction and speed of movement. Although air currents reaching the West Indies are mostly from the northeast, which might explain an origin from peninsular Florida for the volant fauna, this would not explain the large number of bat and bird lineages derived from Central America, or the Antillean derivation of the Bahamian bird and bat faunas. A more likely explanation for the origin of the fish and volant faunas involves a simple distance effect, with dispersal over shorter distances being favored. During Pleistocene sea-level lows, Cuba was nearly in contact with the exposed Great Bahama Bank, and Jamaica was much closer to Central America via the exposed Nicaraguan Rise, facilitating active dispersal. The other Bahamian vertebrate groups also show a derivation from the Antilles rather than from North America, which may be the result of both short distance and northward flowing water currents. However, the relatively low levels of endemism in the Bahamas probably reflect an origin following the Pleistocene high sea levels, when most or all of the Bank was submerged.

Evidence for this general pattern comes from a diversity of sources. The unusual taxonomic composition of the West Indian vertebrate fauna, with reduced higher-taxon diversity, always has favored overwater dispersal, and the growing Tertiary fossil record has yet to alter that conclusion. Evidence that this "unbalanced" fauna is not an artifact of an incomplete fossil record is found in the morphologies and ecologies of lineages that have radiated in the West

Table 7 The origin of West Indian vertebrates[a]

	Fish	Amphibians	Reptiles	Birds[b]	Mammals		Total
					Bats	Other	
Mechanism:							
Dispersal	16	8	67	425	42	8	566
Vicariance	0	1	0	0	0	0	1
Undetermined	1	0	1	0	0	1	3
Source							
South America	6	7	35	—	14	7	69
Central America	0	0	8	—	18	1	27
North America	9[c]	0	3	—	2	1	15
Other	2	0	4	—	0	0	6
Undetermined	0	2	18	—	0	0	20

[a]Shown are the numbers of independent lineages.

[b]The exact number of lineages is not known for birds; there are at least 300 and probably more than 500 independent colonizations (see text). The predominate source area for West Indian birds is North America, but the specific number of lineages from each source area is not known.

[c]Some of these lineages may have arrived from Central America (see text).

Indies. For example, the ground sloths and hystricognath rodents underwent unusually large radiations, filling niches normally occupied by primates, squirrels, porcupines, and ungulates (129, 195). And the absence of carnivores is believed to be responsible for the evolution of giant raptorial birds, now extinct, in the West Indies (134). Additional examples are found among the amphibians and reptiles. The primary source area for nonvolant colonists, South America, agrees with long-established current patterns, and the widely scattered Cenozoic dates of origin estimated by molecular data (74, 78) are concordant with a random mechanism such as overwater dispersal.

The large number of claims in the literature of an origin by vicariance is remarkable considering that geologists have not been able to establish a clear pattern of area relationships for the proto-Antilles (42), and that congruence of multiple phylogenies also can be attributed to concordant dispersal (79). In fact, the general pattern proposed here, dispersal on currents coming from South America, is such an example of concordant dispersal.

Some ancient lineages of West Indian vertebrates nonetheless may be present, and several candidates are the Cuban gar, frogs of the genus *Eleutherodactylus* (73), the xantusiid lizard *Cricosaura* (77), and the insectivores (109). The proto-Antillean land mass required by the vicariance theory cannot be eliminated on geological grounds, and indirect evidence suggests that some land areas in the Greater Antilles have been above water throughout the Cenozoic. For these reasons, vicariance may explain the origin of some lineages of vertebrates. However, catastrophic local effects of the K-T bolide impact, especially the giant tsunamis, must have resulted in widespread extinctions on any Antillean islands that were emergent at the time. For this reason, dispersal in the early Tertiary, immediately following the impact, also may explain the presence of ancient lineages in the West Indies.

The idea of the Lesser Antilles being a classic dispersal filter is well supported by the different distances that South American groups have extended up the chain (102, 103). For example, the faunal break for eleutherodactyline frogs occurs between St. Lucia and St. Vincent (90), whereas the break for anoline lizards is between Dominica and Martinique (55). Geologic evidence (see above) and the position of faunal breaks for different groups do not support the recent suggestion that the northern and southern Lesser Antilles were separated by a fault between Dominica and Martinique with major biogeographic consequences (158).

Future molecular phylogenetic studies of West Indian vertebrates should help to refine time estimates for the origin of independent lineages and more accurately determine source areas. Also, as the geological evolution of the Caribbean region becomes better known, it should be possible to examine the influence of intra-Caribbean tectonic events on organismal evolution. At least one such event, the fusion that resulted in the present-day island of Hispaniola, appears to have had an impact on some of the fauna (73). Additional fossils, especially from the Tertiary, will give a better estimate of the taxonomic composition of the early vertebrate fauna and its bearing on biogeographic models. Although the recent trend in historical biogeography has been to focus on only one element of information, phylogeny, the integrative approach provides a better explanation of the geographic distribution of organisms through time.

ACKNOWLEDGMENTS

I thank Carla Hass, Robert Henderson, and Robert Powell for providing comments on the manuscript. This work was supported by grants from the National Science Foundation.

Literature Cited

1. Alvarez W, Smit J, Lowrie W, Asaro F, Margolis SV, et al. 1992. Proximal impact deposits at the Cretaceous-Tertiary boundary in the Gulf of Mexico: a restudy of DSDP Leg 77 Sites 536 and 540. *Geology* 20:697–700
2. Arden DD. 1975. The geology of Jamaica and the Nicaraguan Rise. In *Ocean Basins and Margins,* Vol. 3, *Gulf Coast, Mexico, and the Caribbean,* ed. AEM Nairn, FG Stehli, pp. 617–61. New York: Plenum
3. Azzaroli A, ed. 1990. *Biogeographical Aspects of Insularity.* Rome: Accadem. Nazionale dei Lincei
4. Baker RJ, Genoways HH. 1978. Zoogeography of Antillean bats. *Proc. Acad. Nat. Sci. Philadelphia, Spec. Publ.* 13: 53–97
5. Barbour T. 1916. Some remarks upon Matthew's "Climate and Evolution." *Ann. NY Acad. Sci.* 27:1–15
6. Bezy RL. 1972. Karyotypic variation and evolution of the lizards in the family Xantusiidae. Nat. Hist. Mus. Los Angeles Cty. *Contr. Sci.* 227:1–29
7. Biknevicius AR, McFarlane DA, MacPhee RDE. 1993. Body size in *Amblyrhiza inundata* (Rodentia: Caviomorpha), and extinct megafaunal rodent from the Anguilla Bank, West Indies: estimates and implications. *Am. Mus. Novit.* 3079:1–25
8. Bogart JP, Hedges SB. 1995. Rapid chromosome evolution in Jamaican frogs of the genus *Eleutherodactylus* (Leptodactylidae). *J. Zool.* 235:9–31
9. Böhme W. 1984. Erstfund eines fossilen Kugelfingergeckos (Sauria: Gekkonidae: Sphaerodactylinae) aus Dominikanischem Bernstein (Oligozän von Hispaniola, Antillen). *Salamandra* 20:212–20
10. Bond J. 1978. Derivations and continental affinities of Antillean birds. *Proc. Acad. Nat. Sci. Philadelphia, Spec. Publ.* 13:119–28
11. Bond J. 1979. *Birds of the West Indies.* London: Collins
12. Bouyesse P. 1984. The Lesser Antilles arc: structure and geodynamic evolution. *Initial Reports of the Deep Sea Drilling Project,* Vol. 78A, ed. B Biju-Duval, et al, pp. 83–103. Washington, DC: US Govt. Printing Off.
13. Bouyesse P. 1988. Opening of the

Grenada back-arc basin and evolution of the Caribbean during the Mesozoic and early Paleogene. *Tectonophysics* 149:121–43
14. Bowin C. 1975. The geology of Hispaniola. In *Ocean Basins and Margins,* Vol. 3, *Gulf Coast, Mexico, and the Caribbean,* ed. AEM Nairn, FG Stehli, pp. 501–22. New York: Plenum
15. Breuil M, Masson D. 1991. Some remarks on Lesser Antillean bat biogeography. *CR Seances Soc. Biogeogr.* 67:25–39
16. Briggs JC. 1984. Freshwater fishes and biogeography of Central America and the Antilles. *Syst. Zool.* 33:428–34
17. Burgess GH, Franz R. 1989. Zoogeography of the Antillean freshwater fish fauna. See Ref. 193, pp. 263–304
18. Burkov VA. 1993. *General Circulation of the World Ocean.* Rotterdam: AA Balkema
19. Burnell KL, Hedges SB. 1990. Relationships of West Indian *Anolis* (Sauria: Iguanidae): an approach using slow-evolving protein loci. *Caribb. J. Sci.* 26:7–30
20. Buskirk R. 1985. Zoogeographic patterns and tectonic history of Jamaica and the northern Caribbean. *J. Biogeogr.* 12:445–61
21. Butler PM. 1988. Phylogeny of the insectivores. In *The Phylogeny and Classification of the Tetrapods,* Volume II, ed. MJ Benton, pp. 117–41. Oxford: Clarendon
22. Cadle JS. 1984. Molecular systematics of neotropical xenodontide snakes. I. South American xenodontines. *Herpetologica* 40:8–20
23. Cadle JS. 1985. The Neotropical colubrid snake fauna (Serpentes: Colubridae): lineage components and biogeography. *Syst. Zool.* 34:1–20
24. Cadle JS. 1988. Phylogenetic relationships among advanced snakes: a molecular perspective. *Univ. Calif. Berkeley Publ. Zool.* 119:1–77
25. Cadle JS. 1992. Phylogenetic relationships among pitvipers: immunological evidence. In *Biology of the Pitvipers,* ed. JA Campbell, ED Brodie Jr, pp. 41–47. Tyler, TX: Selva
26. Cannatella DC, de Queiroz K. 1989. Phylogenetic systematics of the anoles: Is a new taxonomy warranted? *Syst. Zool.* 38:57–69
27. Cei JM. 1972. Bufo of South America. In *Evolution in the Genus Bufo,* ed. WF Blair, pp. 82–92. Austin, TX: Univ. Texas Press
28. Cockerell TDA. 1924. A fossil cichlid fish from the Republic of Haiti. *Proc. US Nat. Mus.* 63:1–3
29. Cole CJ, Gans C. 1987. Chromosomes of *Bipes, Mesobaena,* and other amphisbaeaians (Reptilia), with comments on their evolution. *Am. Mus. Novit.* 2869:1–9.
30. Comer JB. 1974. Genesis of Jamaican bauxite. *Econ. Geol.* 69:1251–64
31. Crother BI, Hillis DM. 1995. Nuclear ribosomal DNA restriction sites, phylogenetic information, and the phylogeny of some xenodontine (Colubridae) snakes. *J. Herpetol.* 29:316–20
32. Crother BI, Miyamoto MM, Presch WF. 1986. Phylogeny and biogeography of the lizard family Xantusiidae. *Syst. Zool.* 35:37–45
33. Darlington PJ, Jr. 1957. *Zoogeography: The Geographical Distribution of Animals.* New York: Wiley
34. Darwin C. 1859. *The Origin of Species.* London: John Murray
35. de Muizon C, McDonald HG. 1995. An aquatic sloth from the Pliocene of Peru. *Nature* 375:224–27
36. Dengo G, Case JE, ed. 1990. *The Geology of North America.* Vol. H. *The Caribbean Region.* Boulder, CO: Geol. Soc. Am.
37. Dessauer HC, Cadle JE, Lawson R. 1987. Patterns of snake evolution suggested by their proteins. *Fieldiana Zool.* 34:1–34
38. de Zoeten R, Mann P. 1991. Structural geology and Cenozoic tectonic history of the Central Cordillera Septentrional, Dominican Republic. *Geol. Soc. Am., Spec. Pap.* 262:265–79
39. Dietz RS, Holden JC, Sproll WP. 1970. Geotectonic evolution and subsidence of Bahama Platform. *Geol. Soc. Am. Bull.* 81:1915–28
40. Dolan J, Mann P, de Zoeten R, Heubeck C, Shiroma J, Monechi S. 1991. Sedimentologic, stratigraphic, and tectonic synthesis of Eocene-Miocene sedimentary basins, Hispaniola and Puerto Rico. *Geol. Soc. Am. Spec. Pap.* 262:217–63
41. Donnelly TW. 1989. History of marine barriers and terrestrial connections: Caribbean paleogeographic inference from pelagic sediment analysis. See Ref. 193, pp. 103–18
42. Donnelly TW. 1990. Caribbean biogeography: geological considerations bearing on the problem of vicariance vs. dispersal. See Ref. 3, pp. 595–609

43. Donovan SK, Jackson TA, ed. 1994. *Caribbean Geology: An Introduction*. Kingston, Jamaica: Univ. West Indies Publishers' Assoc.
44. Draper G, Barros JA. 1994. Cuba. See Ref. 43, pp. 65–86
45. Draper G, Mann P, Lewis JF. 1994. Hispaniola. See Ref. 43, pp. 129–50
46. Dunn ER. 1926. The frogs of Jamaica. *Proc. Boston Soc. Nat. Hist.* 38:11–130
47. Eberle W, Hirdes W, Muff R, Pelaez M. 1980. The geology of the Cordillera Septentrional. In *Transactions of the Ninth Caribbean Geologic Conference*, ed. Anonymous, pp. 619–32. Santo Domingo, Dominican Republic: Amigo del Hogar
48. Emanuel KA, Speer K, Rotunno R, Srivastava R, Molina M. 1994. Hypercanes: a possible link in global extinction scenarios. *Eos* 75 (44, suppl.):409 (Abstr.)
49. Erard C. 1991. Landbirds of the Lesser Antilles. *CR Seances Soc. Biogeogr.* 67:3–23
50. Estes R. 1976. Middle Paleocene lower vertebrates from the Tongue River Formation, southeastern Montana. *J. Paleontol.* 50:500:20
51. Faaborg J. 1985. Ecological constraints on West Indian bird distributions. *Ornithol. Monogr.* 35:621–53
52. Ford SM. 1990. Platyrrhine evolution in the West Indies. *J. Hum. Evol.* 19:237–54
53. Ford SM, Morgan GS. 1986. A new ceboid femur from the late Pleistocene of Jamaica. *J. Vert. Paleontol.* 6:281–89
54. Gardiner BG. 1993. Osteichthyes: basal actinopterygians. In *The Fossil Record 2*, ed MJ Benton, pp. 611–19. London: Chapman & Hall
55. Gorman GC, Atkins L. 1969. The zoogeography of Lesser Antillean *Anolis* lizards—an analysis based upon chromosomes and lactic dehydrogenases. *Mus. Comp. Zool. (Harv. Univ.) Bull.* 138:53–80
56. Gorman GC, Wilson AC, Nakanishi M. 1971. A biochemical approach towards the study of reptilian phylogeny: evolution of serum albumin and lactic dehydrogenase. *Syst. Zool.* 20:167–85
57. Graham A. 1993. Contribution toward a Tertiary palynostratigraphy for Jamaica: the status of Tertiary paleobotanical studies in northern Latin America and preliminary analysis of the Guys Hill Member (Chapelton formation, Middle Eocene) of Jamaica. In *Biostratigraphy of Jamaica*, ed. RM Wright, E Robinson, pp. 443–61. Boulder, CO: Geol. Soc. Am.
58. Graham A, Jarzen DM. 1969. Studies in neotropical paleobatany. I. The Oligocene Communities of Puerto Rico. *Ann. Mo. Bot. Garden* 56:308–57
59. Griffiths TA, Klingener D. 1988. On the distribution of Greater Antillean bats. *Biotropica* 20:240–51
60. Grimaldi DA. 1988. Relicts in the Drosophilidae (Diptera). In *Zoogeography of Caribbean Insects*, ed. JK Liebherr, pp. 183–213. Ithaca, NY: Cornell Univ. Press
61. Guppy HB. 1917. *Plants, Seeds, and Currents in the West Indies and Azores*. London: Williams & Norgate
62. Guyer C, Savage JM. 1986. Cladistic relationships among anoles (Sauria:Iguanidae). *Syst. Zool.* 35:509–31
63. Guyer C, Savage JM. 1992. Anole systematics revisited. *Syst. Zool.* 41:89–110
64. Hass CA. 1991. Evolution and biogeography of West Indian *Sphaerodactylus* (Sauria: Gekkonidae): a molecular approach. *J. Zool.* 225:525–61
65. Hass CA. 1996. Relationships among West Indian geckos of the genus *Sphaerodactylus:* a preliminary analysis of mitochondrial 16S ribosomal RNA sequences. See Ref. 144, pp. 175–94
66. Hass CA, Hedges SB. 1991. Albumin evolution in West Indian frogs of the genus *Eleutherodactylus* (Leptodactylidae): Caribbean biogeography and a calibration of the albumin immunological clock. *J. Zool.* 225:413–26
67. Hass CA, Hedges SB. 1992. Karyotype of the Cuban lizard *Cricosaura typica* and its implications for xantusiid phylogeny. *Copeia* 1992:563–65
68. Hass CA, Hedges SB, Maxson LR. 1993. Molecular insights into the relationships and biogeography of West Indian anoline lizards. *Biochem. Syst. Ecol.* 21:97–114
69. Hass CA, Hoffman MA, Densmore LD III, Maxson LR. 1992. Crocodilian evolution: insights from immunological data. *Mol. Phylogenet. Evol.* 1:193–201
70. Heatwole H, Levins R. 1972. Biogeography of the Puerto Rican Bank: flotsam transport of terrestrial animals. *Ecology* 53:112–17
71. Hedges SB. 1982. Caribbean biogeography: implications of the recent plate tectonic studies. *Syst. Zool.* 31:518–22
72. Hedges SB. 1989. An island radiation: allozyme evolution in Jamaican

frogs of the genus *Eleutherodactylus* (Anura, Leptodactylidae). *Caribb. J. Sci.* 25:123–47

73. Hedges SB. 1989. Evolution and biogeography of West Indian frogs of the genus *Eleutherodactylus*: slow-evolving loci and the major groups. See Ref. 193, pp. 305–70
74. Hedges SB. 1996. The origin of West Indian amphibians and reptiles. See Ref. 144, pp. 95–128
75. Hedges SB. 1996. Distribution patterns of amphibians in the West Indies. In *Regional Patterns of Amphibian Distribution: A Global Perspective,* ed. WE Duellman. Lawrence, KS: Univ. KS Mus. Nat. Hist. Spec. Publ. In press
76. Hedges SB, Bezy RL. 1993. Phylogeny of xantusiid lizards: concern for data and analysis. *Mol. Phylogenet. Evol.* 2:76–87
77. Hedges SB, Bezy RL, Maxson LR. 1991. Phylogenetic relationships and biogeography of xantusiid lizards inferred from mitochondrial DNA sequences. *Mol. Biol. Evol.* 8:767–80
78. Hedges SB, Hass CA, Maxson LR. 1992. Caribbean biogeography: molecular evidence for dispersal in West Indian terrestrial vertebrates. *Proc. Nat. Acad. Sci. USA* 89:1909–13
79. Hedges SB, Hass CA, Maxson LR. 1994. Towards a biogeography of the Caribbean. *Cladistics* 10:43–55
80. Henderson RW, Hedges SB. 1995. Origin of West Indian populations of the geographically widespread boa *Corallus enydris* inferred from mitochondrial DNA sequences. *Mol. Phylogenet. Evol.* 4:88–92
81. Hildebrand AR, Boynton WV. 1990. Proximal Cretaceous-Tertiary boundary impact deposits in the Caribbean. *Science* 248:843–47
82. Hildebrand AR, Pilkington M, Ortiz-Aleman C, Chavez RE, Conners M. 1994. The Chicxulub crater: size, structure, and hydrogeology. *Eos 75 (No. 44,* suppl.):409
83. Holcombe TL, Edgar NT. 1990. Late Cretaceous and Cenozoic evolution of Caribbean ridges and rises with special reference to paleogeography. See Ref. 3, pp. 610–26
84. Holcombe TL, Ladd JW, Westbrook G, Edgar NT, Bowland CL. 1990. Caribbean marine geology; ridges and basins of the plate interior. See Ref. 36, pp. 231–60
85. Huebeck C, Mann P. 1991. Structural geology and Cenozoic tectonic history of the southeastern termination of the Cordillera Central, Dominican Republic. *Geol. Soc. Am., Spec. Pap.* 262:315–36
86. Humphries CJ. 1992. Cladistic biogeography. In *Cladistics: A Practical Course in Systematics,* PL Forey et al, pp. 137–59. Oxford: Clarendon
87. Iturralde-Vinent MA. 1988. *Naturaleza Geológica de Cuba.* Havana: Editorial Científico-Técnica

87a. Iturralde-Vinent MA. 1994. Cuban geology: a new plate-tectonic synthesis. *J. Petrol. Geol.* 17:39–70

88. Jones JK. 1989. Distribution and systematics of bats in the Lesser Antilles. See Ref. 193, pp. 645–60
89. Kaiser H, Coloma LA, Gray HM. 1994. A new species of *Colostethus* (Anura: Dendrobatidae) from Martinique, French Antilles. *Herpetologica* 50:23–32
90. Kaiser H, Sharbel TF, Green DM. 1994. Systematics and biogeography of eastern Caribbean *Eleutherodactylus* (Anura: Leptodactylidae): evidence from allozymes. *Amphibia-Reptilia* 15:375–94
91. King FW. 1962. The occurrence of rafts for dispersal of land animals into the West Indies. *Q. J. Fla. Acad. Sci.* 25:45–52
92. Klein NK, Brown WM. 1994. Intraspecific molecular phylogeny in the Yellow Warbler (*Dendroica petechia*), and implications for avian biogeography in the West Indies. *Evolution* 48:1914–32
93. Kluge AG. 1988. Parsimony in vicariance biogeography: a quantitative method and a Greater Antillean example. *Syst. Zool.* 37:315–28
94. Kluge AG. 1989. A concern for evidence and a phylogenetic hypothesis of relationships among *Epicrates* (Boidae, Serpentes). *Syst. Zool.* 38:7–25
95. Koopman KF. 1989. A review and analysis of the bats of the West Indies. See Ref. 193, pp. 635–44
96. Kring DA, Boynton WV. 1992. Petrogenesis of an augite-bearing melt rock in the Chicxulub structure and its relationship to K/T impact spherules in Haiti. *Nature* 358:141–44
97. Lack D. 1976. *Island Biology.* Los Angeles: Univ. Calif. Press
98. Lambert JB, Frye JS, Poinar GO, Jr. 1985. Amber from the Dominican Republic: analysis of nuclear magnetic resonance spectroscopy. *Archaeometry*

27:43–51
99. Larue DK. 1994. Puerto Rico and the Virgin Islands. See Ref. 43, pp. 151–66
99a. Laybourne RC, Deedrick DW, Hueber FM. 1994. Feather in amber is earliest New World fossil of Picidae. *Wilson Bull.* 106:18–25
100. Lazell JD, Jr. 1964. The Lesser Antillean representatives of *Bothrops* and *Constrictor. Mus. Comp. Zool. (Harv. Univ.) Bull.* 132:245–73
101. Lee DS, Platania SP, Burgess GH, ed. 1983. *Atlas of North American Freshwater Fishes. 1983 Supplement.* Raleigh, NC: NC State Mus. Nat. Hist.
102. Lescure J. 1987. Le peuplement en reptiles et amphibiens des Petites Antilles. *Soc. Zool. Fr. Bull.* 112:327–42
103. Lescure J, Jeremie J, Lourenco W, Mauries JP, Pierre J, et al. 1991. Biogéographie et insularité: l'example des Petites Antilles. *CR Seances Soc. Biogeogr.* 67:41–59
104. Lewis JF. 1980. Cenozoic tectonic evolution and sedimentation in Hispaniola. In *Transactions of the Ninth Caribbean Geologic Conference,* pp. 65–73. Santo Domingo, Dominican Republic: Amigo del Hogar
105. Lewis JF, Draper G. 1990. Geology and tectonic evolution of the northern Caribbean margin. See Ref. 36, pp. 77–140
106. Lydeard C, Wooten MC, Meyer A. 1995. Molecules, morphology, and area cladograms: a cladistic and biogeographic analysis of *Gambusia* (Teleostei: Poeciliidae). *Syst. Biol.* 44:221–36
107. Lynch JD. 1976. The species groups of South American frogs of the genus *Eleutherodactylus* (Leptodactylidae). *Occas. Pap. Mus. Nat. Hist. Univ. Kans.* 61:1–24
108. Lynch JD. 1986. The definition of the Middle American clade of *Eleutherodactylus* based on jaw musculature (Amphibia: Leptodactylidae). *Herpetologica* 42:248–58
109. MacFadden B. 1980. Rafting mammals or drifting islands? Biogeography of the Greater Antillean insectivores *Nesophontes* and *Solenodon. J. Biogeogr.* 7:11–22
110. MacFadden B. 1981. Comments on Pregill's appraisal of historical biogeography of Caribbean vertebrates: vicariance, dispersal, or both? *Syst. Zool.* 30:370–72
111. MacPhee RDE, Fleagle JG. 1991. Postcranial remains of *Xenothrix mcgregori* (Primates, Xenotrichidae) and other late Quaternary mammals from Long Mile Cave, Jamaica. *Am. Mus. Nat. Hist. Bull.* 206:287–321
111a. MacPhee RDE, Grimaldi DA. 1996. Mammal bones in Dominican amber. *Nature* 380:489–90
112. MacPhee RDE, Iturralde-Vinent MA. 1994. First tertiary land mammal from Greater Antilles: an early Miocene Sloth (Xenarthra, Megalonchidae) from Cuba. *Am. Mus. Novit.* 3094:1–13
112a. MacPhee RDE, Iturralde-Vinent MA. 1995. Origin of the Greater Antillean land mammal fauna. 1. New Tertiary fossils from Cuba and Puerto Rico. *Am. Mus. Novit.* 3141:1–30
113. MacPhee RDE, Novacek MJ. 1993. Definition and relationships of the Lipotyphla. In *Mammalian Phylogeny,* Vol. 2, *Placentals,* ed. FS Soulé, MJ Novacek, MC McKenna, pp. 13–31. New York: Springer-Verlag
114. MacPhee RDE, Wyss AR. 1990. Oligo-Miocene vertebrates from Puerto Rico, with a catalog of localities. *Am. Mus. Novit.* 2965:1–45
115. Maglio VJ. 1970. West Indian xenodontine colubrid snakes: their probable origin, phylogeny, and zoogeography. *Mus. Comp. Zool. (Harv. Univ.) Bull.* 141:1–54
116. Martin RD. 1993. Primate origins: plugging the gaps. *Nature* 363:223–34
117. Matthew WD. 1918. Affinities and origin of the Antillean mammals. *Geol. Soc. Am. Bull.* 29:657–66
118. Maurrassee FJ-MR. 1982. *Survey of the Geology of Haiti.* Miami: Miami Geol. Soc.
119. Maurrassee FJ-MR. 1990. Stratigraphic correlation for the circum-Caribbean region. See Ref. 36, plates 4–5
120. Maurrassee FJ-MR, Sen G. 1991. Impacts, tsunamis and the Haitian Cretaceous-Tertiary boundary layer. *Science* 252:1690–93
121. Maury RC, Westbrook GK, Baker PE, Bouysse P, Westercamp D. 1990. Geology of the Lesser Antilles. See Ref. 36, pp. 141–66
122. Maxson LR. 1992. Tempo and pattern in anuran speciation and phylogeny: an albumin perspective. In *Herpetology: Current Research on the Biology of Amphibians and Reptiles,* ed. K Adler, pp. 41–57. Oxford, OH: Soc. Study Amphib. Rept.
123. Maxson LR, Heyer WR. 1988. Molecular systematics of the frog genus *Lepto-*

dactylus (Amphibia: Leptodactylidae). *Fieldiana, Zool.* 41:1–13
124. Maxson LR, Myers CW. 1985. Albumin evolution in tropical poison frogs (Dendrobatidae): a preliminary report. *Biotropica* 17:50–56
125. McDonald MA, Smith MH. 1990. Speciation, heterochrony, and genetic variation in Hispaniolan Palm-tanagers. *Auk* 107:707–17
126. McLaughlin PP, van den Bold WA, Mann P. 1991. Geology of the Azua and Enriquillo basins, Dominican Republic; 1, Neogene lithofascies, biostratigraphy, biofacies, and paleogeography. *Geol. Soc. Am., Spec. Pap.* 262:337–66
127. Morgan GS. 1989. Fossil Chiroptera and Rodentia from the Bahamas, and the historical biogeography of the Bahamian mammal fauna. See Ref. 193, pp. 685–740
128. Morgan GS. 1993. Quaternary land vertebrates of Jamaica. In *Biostratigraphy of Jamaica,* ed. RM Wright, E Robinson, pp. 417–42. Boulder, CO: Geol. Soc. Am.
129. Morgan GS, Woods CA. 1986. Extinction and the zoogeography of West Indian land mammals. *Biol. J. Linn. Soc.* 28:167–203
130. Morrone JJ, Crisci JV. 1995. Historical biogeography: introduction to methods. *Annu. Rev. Ecol. Syst.* 26:373–401
130a. Mourer-Chauviré C. 1985. Les Todidae (Aves, Coraciiformes) des phosphorites du Querrcy (France). *Proc. Koninklije Nederlandse Akad. van Wetenschappen.* Ser. E Vol. 88(4)
131. Mullins HT, Lynts GW. 1977. Origin of the northwestern Bahama Platform: review and interpretation. *Geol. Soc. Am. Bull.* 88:1447–61
132. Myers GS. 1938. Freshwater fishes and West Indian zoogeography. *Smithson. Year Annu. Rep. Smithson. Inst.* 1937:339–64
133. Nelson G, Platnick NI. 1981. *Systematics and Biogeography: Cladistics and Vicariance*. New York: Columbia Univ. Press
134. Olson SL. 1978. A paleontological perspective of West Indian birds and mammals. *Proc. Acad. Nat. Sci. Philadelphia, Spec. Publ.* 13:99–117
135. Page RDM, Lydeard C. 1994. Towards a cladistic biogeography of the Caribbean. *Cladistics* 10:21–41
136. Pascual R, Vucetich MG, Scillato-Yané GJ. 1990. Extinct and recent South American and Caribbean Megalonichyidae edentates and Hystricognathi rodents: outstanding examples of isolation. See Ref. 3, pp. 627–39
137. Patterson C. 1993. Osteichthyes: basal actinopterygians. In *The Fossil Record 2,* ed. MJ Benton, pp. 621–56. London: Chapman & Hall
137a. Paulay G. 1994. Biodiversity on oceanic islands: its origin and extinction. *Am. Zool.* 34:134–44
138. Perfit MR, Williams EE. 1989. Geological constraints and biological retrodictions in the evolution of the Caribbean Sea and its islands. See Ref. 193, pp. 47–102
139. Pindell J. 1994. Evolution of the Gulf of Mexico and the Caribbean. See Ref. 43, pp. 13–39
140. Pindell J, Barrett SF. 1990. Geological evolution of the Caribbean region: a plate tectonic perspective. See Ref. 36, pp. 405–32
141. Poinar GO Jr. 1988. Hair in Dominican amber: evidence for tertiary land mammals in the Antilles. *Experienta* 44:88–89
142. Poinar GO Jr. 1992. *Life in Amber*. Stanford, CA: Stanford Univ. Press
143. Poinar GO Jr, Cannatella DC. 1987. An Upper Eocene frog from the Dominican Republic and its implication for Caribbean biogeography. *Science* 237:1215–16
144. Powell R, Henderson RW, eds. 1996. *Contributions to West Indian Herpetology: A Tribute to Albert Schwartz.* Ithaca, New York: Soc. Study Amphib. Rept.
145. Powell R, Henderson RW. 1996. A brief history of West Indian herpetology. See Ref. 144, pp. 29–50
146. Pregill GK. 1981. Cranial morphology and the evolution of West Indian toads (Salientia: Bufonidae): resurrection of the genus *Peltophryne* Fitzinger. *Copeia* 1981:273–85
147. Pregill GK. 1981. An appraisal of the vicariance hypothesis of Caribbean biogeography and its application to West Indian terrestrial vertebrates. *Syst. Zool.* 30:147–55
148. Pregill GK, Crombie RI, Steadman DW, Gordon LK, Davis FW, Hilgartner WB. 1992. Living and late holocene fossil vertebrates, and the vegetation of the Cockpit Country, Jamaica. *Atoll Res. Bull.* 353:1–19
149. Pregill GK, Olson SL. 1981. Zoogeography of West Indian vertebrates in relation to Pleistocene climatic cycles. *Annu.*

Rev. Ecol. Syst. 12:75–98
150. Rauchenberger M. 1988. Historical biogeography of poecilid fishes in the Caribbean. *Syst. Zool.* 37:356–65
151. Rauchenberger M. 1989. Systematics and biogeography of the genus *Gambusia* (Cyprinodontifomres: Poecilidae). *Am. Mus. Novit.* 2951:1–74
152. Ricklefs RE, Cox GW. 1972. Taxon cycles in the West Indian avifauna. *Am. Nat.* 106:195–219
153. Ricklefs RE, Cox GW. 1978. Stage of the taxon cycle, habitat distribution, and population density in the avifauna of the West Indies. *Am. Nat.* 112:875–95
154. Rieppel O. 1980. Green anole in Dominican amber. *Nature* 286:486–87
155. Robinson E. 1994. Jamaica. See Ref. 43, pp. 111–27
156. Rosen DE. 1975. A vicariance model of Caribbean biogeography. *Syst. Zool.* 24:431–64
157. Rosen DE. 1985. Geological hierarchies and biogeographic congruence in the Caribbean. *Ann. Mo. Bot. Garden* 72:636–59
158. Roughgarden J. 1995. *Anolis Lizards of the Caribbean*. New York: Oxford
159. Rouse I. 1989. Peopling and repeopling of the West Indies. See Ref. 193, pp. 119–36
160. Savage JM. 1982. The enigma of the Central American herpetofauna: dispersals or vicariance? *Ann. Mo. Bot. Garden* 69:464–547
161. Savage JM, Lips KR. 1993. A review of the status and biogeography of the lizard genera *Celestus* and *Diploglossus* (Squamata: Anguidae), with description of two new species from Costa Rica. *Rev. Biol. Trop.* 41:817–42
162. Scharff RF. 1912. *Distribution and Origin of Life in America*. New York: MacMillan
163. Schuchert C. 1935. *Historical Geology of the Antillean-Caribbean Region*. New York: Wiley
164. Schwartz A, Henderson RW. 1991. *Amphibians and Reptiles of the West Indies: Descriptions, Distributions, and Natural History*. Gainesville, FL: Univ. FL Press
165. Seidel ME. 1988. Revision of the West Indian emydid turtles (Testudines). *Am. Mus. Novit.* 2918:1–41
165a. Seidel ME. 1996. Current status of biogeography of the West Indian turtles in the genus Trachemys (Emydidae). See Ref. 144, pp. 169–74
166. Seutin G, Klein NK, Ricklefs RE, Bermingham E. 1994. Historical biogeography of the bananaquit (*Coereba flaveola*) in the Caribbean region: a mitochondrial DNA assessment. *Evolution* 48:1041–61
167. Sharpton VL, Burke K, Zanoguera AC, Hall SA, Lee DS, et al. 1993. Chicxulub multiring impact basin: size and other characteristics derived from gravity analysis. *Science* 261:1564–67
168. Sharpton VL, Marin LE. 1994. How big is the Chicxulub impact basin, Yucatán, Mexico? *Eos* 75 (44, suppl.):409 (Abstr.)
169. Shochat D, Dessauer HC. 1981. Comparative immunological study of albumins of *Anolis* lizards of the Caribbean islands. *Comp. Biochem. Physiol.* 68A:67–73
170. Sibley CG, Ahlquist JE. 1990. *Phylogeny and Classification of Birds*. New Haven: Yale Univ. Press.
171. Sibley CG, Munroe BL Jr. 1990. *Distribution and Taxonomy of Birds of the World*. New Haven: Yale Univ. Press.
172. Simpson GG. 1956. Zoogeography of West Indian land mammals. *Am. Mus. Novit.* 1759:1–28
173. Smit J, Montanari A, Swinburne NHM, Alvarez W, Hildebrand AR, et al. 1992. Tektite-bearing, deep-water classic unit at the Cretaceous-Tertiary boundary in northeastern Mexico. *Geology* 20:99–103
173a. Sober E. 1988. The conceptual relationship of cladistic phylogenetics and vicariance biogeography. *Syst. Zool.* 37:245–53
174. Speed RC. 1985. Cenozoic collision of the Lesser Antilles Arc and continental South America and the origin of the El Pilar Fault. *Tectonophysics* 4:41–69
175. Speed RC, Smith-Horowitz PL, Perch-Nielsen KvS, Saunders JB, Sanfilippo AB. 1993. Southern Lesser Antilles arc platform: pre-Late Miocene stratigraphy, structure, and tectonic evolution. *Geol. Soc. Am., Spec. Pap.* 277:1–98
176. Steadman DW, Pregill GK, Olson SL. 1984. Fossil vertebrates from Antigua, Lesser Antilles: evidence for late Holocene human-caused extinctions in the West Indies. *Proc. Natl. Acad. Sci. USA* 81:4448–51
177. Strahm MH, Schwartz A. 1977. Osteoderms in the anguid lizard subfamily Diploglossinae and their taxonomic importance. *Biotropica* 9:58–72
178. Stucky RK, McKenna MC. 1993. Mammalia. In *The Fossil Record 2,* ed. MJ

Benton, pp. 739–71. London: Chapman & Hall

179. Terborgh JW, Faaborg J, Brockmann HJ. 1978. Island colonization by Lesser Antillean birds. *Auk* 95:59–72
180. Thomas R. 1989. The relationships of Antillean *Typhlops* (Serpentes: Typhlopidae) and the description of three new Hispaniolan species. See Ref. 193, pp. 409–32
181. Thomas R, McDiarmid RW, Thompson FG. 1985. Three new species of thread snakes (Serpentes: Leptotyphlopidae) from Hispaniola. *Proc. Biol. Soc. Wash.* 98:204–20
182. Tolson PJ. 1987. Phylogenetics of the boid snake genus *Epicrates* and Caribbean vicariance theory. *Occas. Pap. Mus. Zool. Univ. Mich.* 715:1–68
183. Trueb L, Tyler MJ. 1974. Systematics and evolution of Greater Antillean hylid frogs. *Occas. Pap. Mus. Nat. Hist. Univ. Kans.* 24:1–60
184. Wadge G. 1994. The Lesser Antilles. See Ref. 43, pp. 167–78
185. Wadge G, Dixon TH. 1984. A geological interpretation of SEASAT-SAR imagery of Jamaica. *J. Geol.* 92:561–81
186. Wallace AR. 1881. *Island Life.* New York: Harper
187. Williams EE. 1969. The ecology of colonization as seen in the zoogeography of anoline lizards on small islands. *Q. Rev. Biol.* 44:345–89
188. Williams EE. 1976. West Indian anoles: a taxonomic and evolutionary summary. I. Introduction and a species list. *Breviora* 440:1–21
189. Williams EE. 1989. Old problems and new opportunities in West Indian biogeography. See Ref. 193, pp. 1–46
190. Williams EE. 1989. A critique of Guyer and Savage (1986): cladistic relationships among anoles (Sauria: Iguanidae): Are the data available to reclassify the anoles? See Ref. 193, pp. 433–78
191. Williams EE, Koopman KF. 1952. West Indian fossil monkeys. *Am. Mus. Novit.* 1546:1–16
192. Wilson DE, Reeder DM, ed. 1993. *Mammal Species of the World.* Washington, DC: Smithsonian Inst. 2nd ed.
193. Woods CA, ed. 1989. *Biogeography of the West Indies: Past, Present, and Future.* Gainesville, FL: Sandhill Crane
194. Woods CA. 1989. The biogeography of the West Indian rodents. See Ref. 193, pp. 741–98
195. Woods CA. 1990. The fossil and recent land mammals of the West Indies: an analysis of the origin, evolution, and extinction of an insular fauna. See Ref. 3, pp. 642–80
196. Woods CA, Ottenwalder JA. 1992. *The Natural History of Southern Haiti.* Gainesvile, FL: FL Mus. Nat. Hist.
197. World Resources Institute. 1994. *World Resources 1994–95.* New York: Oxford
198. Wyles JS, Gorman GC. 1980. The albumin immunological and Nei electrophoretic distance correlation: a calibration for the saurian genus *Anolis* (Iguanidae). *Copeia* 1980:66–71

Annu. Rev. Ecol. Syst. 1996. 27:197–235

TROUBLE ON OILED WATERS: Lessons from the *Exxon Valdez* Oil Spill

R. T. Paine, Jennifer L. Ruesink, Adrian Sun, Elaine L. Soulanille, Marjorie J. Wonham, Christopher D. G. Harley, Daniel R. Brumbaugh, and David L. Secord

Department of Zoology, University of Washington, Seattle, Washington 98195-1800

KEY WORDS: baseline, damage assessment, disturbance, experiments, Prince William Sound, restoration

ABSTRACT

The *Exxon Valdez* oil spill was the largest in US maritime history. We review post-spill research and set it in its legal context. The Exxon Corporation, obviously responsible for the spill, focused on restoration, whereas the Trustees, a coalition of state and federal entities, focused on damage and its assessment. Despite billions of dollars expended, little new understanding was gained about the recovery dynamics of a high latitude marine ecosystem subject to an anthropogenic pulse perturbation. We discuss a variety of case studies that highlight the limitations to and shortcomings of the research effort. Given that more spills are inevitable, we recommend that future studies address spatial patterns in the intertidal, and focus on the abundances of long-lived species and on organisms that preserve a chronological record of growth. Oil spills, while tragic, represent opportunities to gain insight into the dynamics of marine ecosystems and should not be wasted.

"You get a guy with four PhDs saying no fish were hurt, then you get a guy with four PhDs saying, yeah, a lot of fish were hurt.... They just kind of delete each other out." (Barker 1994, p. 74)

INTRODUCTION

The *Exxon Valdez* oil spill (EVOS) is likely to be remembered as one of the great environmental tragedies of North America in the late twentieth century. Through pictures of dead birds, struggling baby seals, and great dark stains across pristine beaches framed by snowcapped mountains, the spill imposed

0066-4162/96/1120-0197$08.00

itself upon the public consciousness. Collective outrage provoked costly cleanup new federal legislation, and the most expensive settlement for oil spill damages ever, though the case is still in litigation. This paper represents an analysis of some of the ecological research performed after EVOS. Enormous effort has gone into post-spill research, with many results yet unpublished or buried in gray literature. Clearly we cannot summarize all that has been done. Thus, we have focused our efforts on three areas:

1. to set the stage for the scientific response to EVOS, we detail the limitations on research imposed by US environmental legislation;
2. we discuss EVOS research that is particularly representative of the results, problems, or conflicts associated with the studies; and
3. we describe the sorts of research and monitoring that we believe might be more useful when future spills occur, as they surely will.[1]

We take the perspective that oil spills and subsequent cleanup activities are pulse perturbations to communities. Pulse perturbations, which are one-time, short-term alterations of some component of an ecosystem, are commonly used by ecologists (usually at smaller scales and by manipulating factors other than hydrocarbon concentration) to explore the dynamic interconnectedness of biotic systems. The few scientific conclusions that can be reached after six years of study and the expenditure of hundreds of millions of dollars suggest the inefficiency of the research effort and the squandering of a rare albeit unfortunate opportunity.

QUANTITIES AND COSTS

Grounding of the T/V Exxon Valdez

On 24 March 1989, shortly after midnight, the single-hulled bulk oil carrier T/V *Exxon Valdez* ran aground on Bligh Reef in the eastern part of Prince William Sound, Alaska (PWS 61°02′N, 146°05′W), spilling approximately 36,000 metric tons (10.8 million gallons) of North Slope crude oil onto a topographically varied, biologically rich, and poorly known, high latitude marine ecosystem (112). The spill, the largest to date in US maritime history, had a number of immediate repercussions: It shocked the American public, dominating the news

[1]As we write, efforts are underway to control the 15 February 1996 spill from the *Sea Empress*, a single hull supertanker carrying 132,000 metric tons (36.7 million gallons) of North Sea light crude oil. The tanker ran aground off the coast of Wales (UK), spilling to date at least 88,000 metric tons (20 million gallons).

media for weeks; individuals and organizations were galvanized into action, generating an economic boom for southeast Alaska; the Oil Pollution Act of 1990 was passed by Congress; the spill thwarted then-President Bush's intent to explore for and exploit known and suspected oil reserves in the Alaska National Wildlife Refuge; and it generated the largest corporate response and subsequent fine in US financial history.

EVOS Compared to Prior Spills

What is lost among these superlatives is the fact that the spill was hardly exceptional: Of spills occurring between 1967 and 1994 in excess of 1000 metric tons, EVOS ranked 40th (4, 116) (Figure 1). By comparison, over 6500 times more oil (240 million metric tons) was released from January to June 1991 in the Persian Gulf. On the other hand, EVOS shares numerous similarities with other large spills: It was caused by human error; its occurrence was inevitable, although the magnitude, timing, and position were not; and it affected a site characterized by scant pre-existing environmental data (64, 78, 157).

EVOS Oil Trajectory

After the grounding, oil was blown onto southwestern islands (Figure 2) by storm winds coming from the northeast three days after the spill. Within three weeks, about 40% of the spilled oil landed on Prince William Sound beaches, and one fourth of the oil moved into the Gulf of Alaska (56, 57). Oil traveled as far as 750 km from the spill site, contacting 1750 km of shoreline along the way (900 km in PWS) (76, 100). Some beached oil was collected by cleanup crews, but much eventually ended up back in the water column, lifted off beaches when they were washed by humans or pummeled by winter waves (67, 82, 107, 172).

Oil that was not collected during cleanup disappeared from the environment in two ways: through evaporation and through degradation into other carbon-based compounds. The lightest, most toxic fractions probably evaporated within the first 10 days and constituted no more than 20% of the total amount (172). Degradation of heavier oil fractions takes a great deal longer and is often assumed to follow an exponential decline, as hydrocarbons are broken down by light and by microbial consumers (172). Five years after the spill, about 2% of the oil remained on beaches and 13% in sediments, with only a tiny fraction still dispersed in water (172).

Cleanup Operations

Cleanup of oil following previous high-latitude coastal spills often met with limited success. Booms to cordon off stretches of shoreline become ineffective in stormy seas; boats equipped to skim oil off the water surface prove fruitless when entangled in macroalgae [e.g. Santa Barbara oil platform spill in 1969

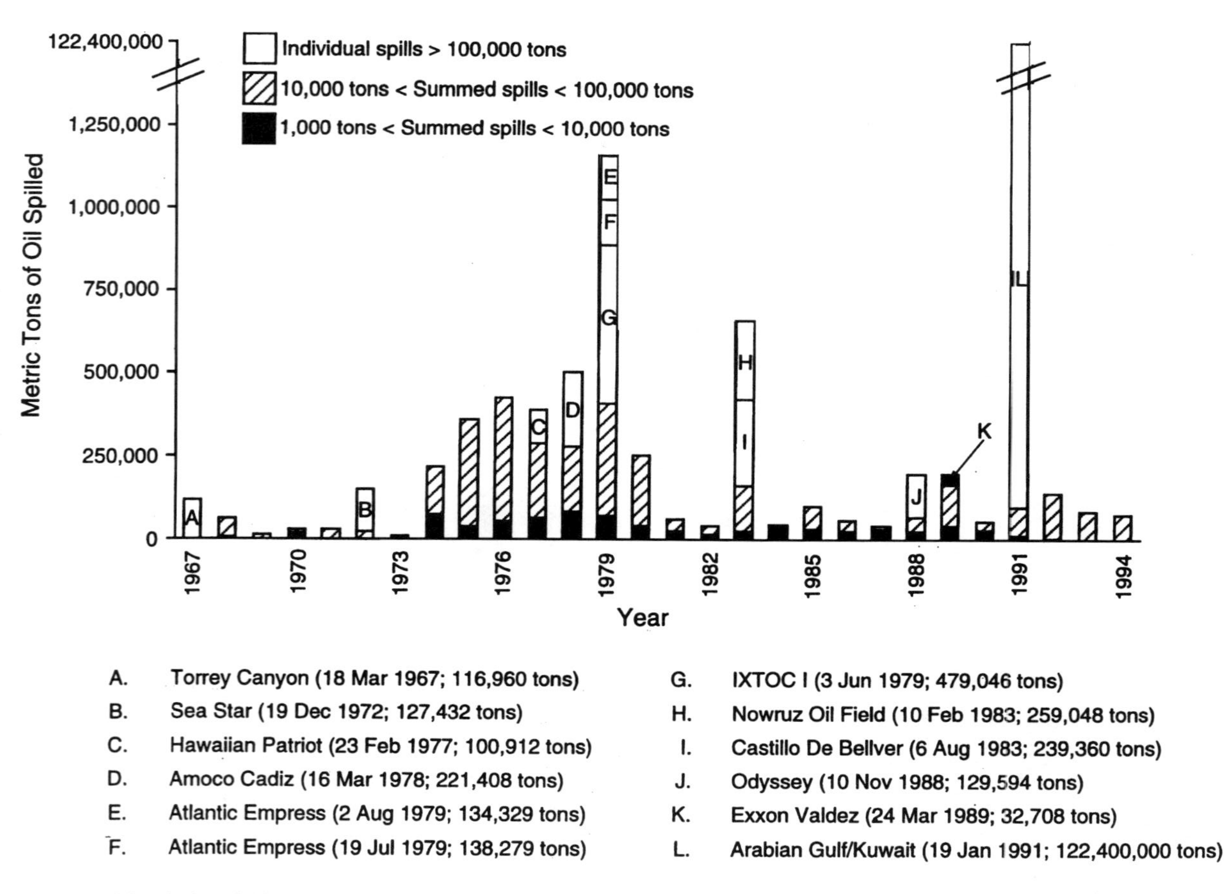

Figure 1 Magnitudes of oil spilled into the ocean since 1967. From (4, 116) and *Oil and Gas Journal*, Vol. 91–92 (1993–1994)

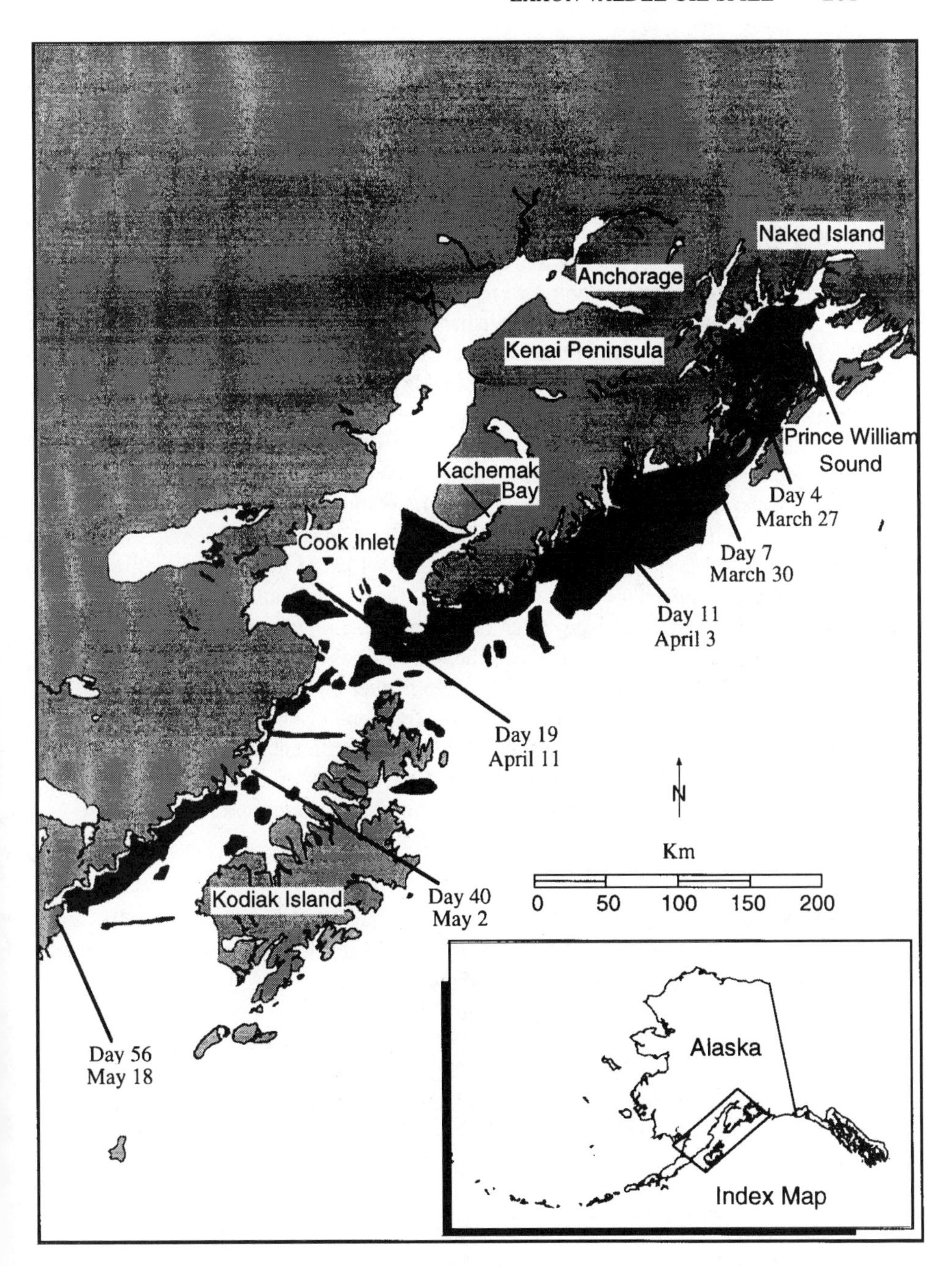

Figure 2 Spread of oil spilled from T/V *Exxon Valdez* during the initial 56 days. Courtesy USDA Forest Service (from 27).

(115, 149)]; dispersants, detergents, and hot water cleaning of shoreline cause substantially more mortality than oil itself, with extended effects on intertidal dynamics [e.g. *Torrey Canyon* tanker in 1967, *Amoco Cadiz* tanker in 1978 (12, 18, 62, 65, 70, 145)]. The acute toxicity of oil in terms of birds and mammals killed has become legendary. In contrast, on shorelines exposed to high wave energy, the developing wisdom is that the cleanup "cure" does more environmental damage than the spill (50, 59, 65, 118).

After EVOS, federal legislation and public outrage required a variety of immediate remediations. Booms were deployed selectively to keep oil from contaminating salmon hatcheries, while fishing in oiled areas was prohibited in 1989 (140). Oiled birds and mammals were retrieved and brought to rehabilitation centers (73, 109). Exxon used the world supply of booms and skimmers, involved 1500 boats in assessment, cleanup, and wildlife rescue, and organized some 12,000 people to remove oil manually from beaches (61, 100). Most of the moderately to heavily oiled shorelines in PWS (>500 km) were sprayed with seawater, often at high pressures and temperatures, and about 110 km of shoreline were treated with fertilizer in 1989 to stimulate natural biodegradation (6, 20, 76, 105, 137). These efforts removed 10–14% of the total oil spilled (61, 172), and required 40 times that much oil to run the boats and equipment involved in cleaning (61). A day after the grounding, a small portion (< 0.27%) (87, 172) of the spill was used to test the effectiveness of burning, and up to 98% of the trapped oil burned away (3). No further burning was carried out, either because too many of the volatile oil fractions evaporated and the oil would not ignite (87), or because smoke from the experimental burn distressed unprepared and unwarned Alaskans, and the Coast Guard forbade larger efforts (166). Some estimates suggest that 50% of the *Exxon Valdez* spill could have been burned without any danger to the ship itself (42).

Costs of EVOS

EVOS provided southeast Alaska with a short economic boom, as the media converged to cover the disaster and funds from Exxon for cleanup and compensation poured in. Exxon paid immediate post-spill costs to fishermen and cleanup workers ($2300 million) and eventually reached an out-of-court settlement with Federal and State officials to pay $900 million more over 10 years to "restore the resources injured by the spill, and the reduced or lost services (human uses) they provide" (155) (Table 1). The money, which supports research and restoration projects, is administered by the six-member EVOS Trustee Council.[2]

[2]Attorney General of Alaska; State Commissioner of Environmental Conservation; State Commissioner of Fish and Game; Assistant Secretary of Interior; Director of the Alaska Region of National Marine Fisheries Service; Alaska Regional Forester for Department of Agriculture.

Table 1 Money spent by Exxon Corporation subsequent to EVOS (in millions of dollars) (11, 155)

Immediate Costs (1989, 1990)	
Cleanup	$2,000
Fishermen	300
Out-of-Court Settlement (1991–2001)	
Damage assessment	214
Habitat protection	375
Administrative costs	35
Research, monitoring and general restoration	180
Restoration reserve	108
Accumulated interest less Court fees	12
TOTAL	$3,224
Civil Trial (1995)	
Compensation to fishermen	$287
Punitive compensation (under appeal)	5000

A $5 billion punitive award has been assessed for Exxon but is currently under appeal (162). The legal costs, ongoing and difficult to ascertain, must also be enormous. For instance, in the first two years after the spill, the State of Alaska alone spent nearly $20 million on litigation (107a).

Although much of the oil is now gone, EVOS remains by far the largest spill in US coastal waters and is unquestionably the most expensive. From Exxon's perspective, the $3.2 billion expended to date on all phases of cleanup, damage assessment, and restoration has increased the value of a spilled barrel from $15 to $12,000 (164). For scale, the same $3.2 billion would, at FY 1995 rates, support the National Science Foundation's programs in General Ecology for 227 years, Ecosystems for 192 years, or Biological Oceanography for 128 years. We believe basic science was shortchanged, and that remediation, restoration, or management protocols will not have been substantially improved for any high-latitude rocky shore coastal systems.

POLITICAL AND LEGAL CONSTRAINTS ON US OIL SPILL RESPONSE

Current legislation discourages research into the consequences of oil as a pulse perturbation for two reasons. First, laws demand that effort be channeled into wildlife rehabilitation, while more effective means of mitigating or preventing spills are not required. Second, legislation requires that oil corporations pay for any damages caused by a spill, which created a situation in which government scientists had incentive to prove injury occurred, while Exxon scientists looked for countervailing evidence.

Major legislation, especially the Federal Clean Water Act (CWA) and the Comprehensive Environmental Response, Compensation, and Liability Act (CERCLA) dictate federal guidelines and responsibilities for damage assessment and potential restoration subsequent to oil spills. Of particular note are the regulations requiring the Departments of Interior and Commerce to arrange for the collection and rehabilitation of affected wildlife. Alaska state legislation exists in parallel with, and is compatible with, the federal acts (173). At the time of the spill the Environmental Protection Agency (EPA) had in place, as required, a contingency plan giving details of how scientists, response teams, and administrators should interact. While Federal law assigns supervision of an oil spill in maritime waters to the US Coast Guard (USCG), the option exists for the responsible party (Exxon) to direct the cleanup response. Exxon assumed this responsibility, with the USCG and the Alaska Department of Environmental Conservation (ADEC) overseeing the Exxon response and cleanup effort (113, 136). By 31 March, the heads of all Trustee agencies had met and agreed on a general course of action, including an initial payment by Exxon. On 28 April a memorandum of agreement was signed by the heads of the major Federal agencies. The State of Alaska refused to sign because "the state felt that it should be the lead trustee" (173). Nevertheless, formal damage assessment began with this action.

At the outset there appears to have been a spirit of cooperation among corporate and State and Federal agencies, as indicated by shared overflights, maps of the oil trajectory, and mammal and bird rehabilitation. However, conflicts within the agreed-upon administrative infrastructure developed rapidly. A major contributor was a confidentiality restriction placed on all government scientists by the Department of Justice (173). Exxon scientists were similarly gagged by the Corporation (166, p. 118). All accounts agree that the impending litigation both directed research and hampered the development of cooperative efforts, and that the activities rapidly polarized themselves into cleanup/restoration (Exxon) and damage assessment (Trustees). The research performed to test for spill effects was driven by legal definitions of injury, baseline, and recovery.

Injury

According to the Code of Federal Regulations (CFR), injury occurs when there is a "measurable adverse change, either long- or short-term, in the chemical or physical quality or the viability of a natural resource resulting either directly or indirectly from exposure to a discharge of oil..." (43 CFR §11.14 (v)). In the event of injury, the responsible party must pay compensatory damages. Establishing that injury has occurred depends on knowledge of the status of the resource just prior to the spill.

Baseline

In the legal context of damage assessment, baseline is defined as the condition or conditions that would have existed at the assessment area had the discharge of oil not occurred (43 CFR §11.14 (e)). In practice, though, legal proceedings have recognized that these ideal baseline data are unlikely to exist—precisely because some anthropogenic change has occurred—and that various approximations will have to be used. Additionally, as all damage assessments have been settled out of court, no judicial precedents exist on the appropriate or standard uses of baseline approximations or inferences. Rather, in each case, baseline data have been used to influence the outcomes of negotiated settlements.

Recovery

The Code of Federal Regulations defines recovery as a return to baseline conditions (43 CFR §11.14 (gg)). However, two distinct perspectives on the definition of recovery following EVOS developed: that of the Trustees and that of Exxon. In their 1993 Draft Restoration Plan, the Trustees state that a resource will be restored once it has recovered to where it would have been had no oil spill occurred (152). Their 1994 restoration plan does not give an explicit general definition but describes recovery objectives for each injured resource (154). Criteria can include one or more of the following: a return to pre-spill conditions; establishment of conditions comparable to those within unoiled areas; establishment of conditions that would have prevailed had the spill not occurred (i.e. baseline); stable or increasing populations (though in some cases populations were decreasing prior to the spill); or the public perception that the resource is restored. For example, the criteria for the recovery of the intertidal are the return to baseline conditions of community composition, population abundances, age-class distributions, and ecosystem function and services.

The above perspective contrasts that of an Exxon-funded review: "Recovery is marked by the re-establishment of a healthy biological community in which the plants and animals characteristic of that community are present and functioning normally" (9). Explicit caveats are added that age structure or species composition may be different from what it was before the damage, and that it is impossible to determine whether a recovered community is the same as one that would have existed had the spill not occurred.

The Trustees' requirements seem idealistic and unattainable, because, as recognized by Exxon, natural variability makes it impossible to know what a population or community would have been like in the absence of the spill. Nonetheless, judging recovery solely by criteria of ecosystem function minimizes the significance of specific biological detail such as species density and age structure.

POST-SPILL RESEARCH: GENERAL PROBLEMS

Spilled oil is a pulse perturbation. In ecological experiments, pulse perturbations involve altering an extrinsic factor or the density of a species. Subsequent observations serve to evaluate the dynamics of a complicated system. Thus, the *Exxon Valdez* Oil Spill was a sort of unnatural experiment in Prince William Sound. Experiments designed to test hypotheses have the following qualities: (*a*) samples taken before and after the perturbation; (*b*) manipulated (experimental) and unmanipulated (control) treatments; and (*c*) sufficient replication. EVOS generally violated all these requirements.

EVOS: The Imperfect Experiment

It is easy to imagine why EVOS fits poorly into the framework of a rigorous experiment. For most species, there were no "before" samples. If samples did exist, they often were collected in the distant past, suffered from low sample size or high natural variation, or showed populations in decline prior to the spill. The prior studies that did exist can be grouped into (*a*) irregular counts of some of the more conspicuous species, (*b*) annual information about fisheries, (*c*) a few ongoing research projects on orcas and birds, and (*d*) rapid assessment of intertidal areas in PWS and along the Kenai Peninsula where oil was predicted to land.

Research after EVOS also involved comparisons of oiled and unoiled sites. However, since the oil primarily affected the southwestern portion of PWS, most oiled sites were in one region. Because control and experimental sites were not interspersed or randomized, shared conditions other than oil might spuriously distinguish them. The only chance to do a proper experiment would have involved the application of cleanup technologies such as hot water washing or bioremediation, with similar untreated beaches as controls. However, this opportunity was lost early on due to ubiquitous but often unrecorded cleanup activities sponsored by Exxon (166).

Compounding the design problems of this unnatural experiment are several statistical issues. Many tests looked for effects of oil on multiple dependent variables (e.g. population changes of many bird taxa, or lesions in a dozen seal tissues). In general, a treatment effect is termed "significant" when the likelihood is less than 5% that a given difference is due to chance. If multiple dependent variables are examined, 5% will likely show "significant" change even in the absence of a perturbation, and the hypothesis of "no effect" will be rejected when in fact it should be accepted (Type 1 error). For example, growth of 86 bird populations was compared in oiled and unoiled parts of PWS. Two of these populations (golden eye and merganser), i.e. 2.3% of the cases, showed slower growth rates at oiled sites—it is difficult to tell if this was caused by chance, or by the oil. Exxon scientists made similar counts of many bird species to assess whether birds tended to avoid oiled areas using an extremely liberal measure ($p < 0.2$) of "oil effect" (30).

Type 2 errors, in which a hypothesis that is false is accepted, are just as misleading. For those parties interested in proving "no effect" of oil, all that is required is low replication of samples. Bayesian statistics (72) and power analyses (26) reduce the possibility of committing type 1 or type 2 errors, although they cannot overcome the limitations of small sample size. In Bayesian analyses, data are used to distinguish among several alternative hypotheses. If "significant effect of oil" and "insignificant effect" are equally likely, then insufficient data have been collected.

Second, difficulties arose in separating correlation from causation, that is, in determining whether post-spill changes were due to oil or to some other factor. We provide two examples: 1. Some population trends in Prince William Sound are too dramatic to question. From 1983–1988 herring spawned along 106–273 km of PWS shoreline. These values did not change immediately after EVOS (1989: 158 km; 1990: 182 km) (164). In 1994, only 12 km of shoreline received herring spawn, signaling a crash in stocks that prompted complete closure of the fishery (55). There may indeed be latent effects of oil now preventing recruitment and reducing survival, but an alternative explanation, involving competition and predation from hatchery salmon, is also plausible. 2. Few studies (but see 166) have mentioned the possible consequences of a weeklong freeze in late January 1989, when Valdez recorded its coldest day ever measured (156). The freeze was part of a record-setting cold that appeared general across the south coast of Alaska and could have caused substantial mortality prior to EVOS; thus possible oil effects are confounded by a natural calamity. In general, a strong case that correlation is in fact causation requires likely biological explanations for the timing of events (e.g. 22) and experiments demonstrating a mechanism.

Perhaps the most insidious problem associated with detecting effects stems from the natural variability of biotic assemblages. Below we discuss two specific illustrations of the generic difficulty with detecting "injury" in natural communities in the face of tremendous background temporal and spatial variation.

Resampling After Long Intervals

Many of the pre-spill data available for PWS stemmed from studies done in the 1970s and 1980s. In species such as marbled murrelets, Steller sea lions, and harbor seals, it was well accepted that populations had likely declined by 50% in the decades prior to EVOS. It would have been possible, in the absence of intervening information, to ascribe all of the decline to oil effects.

Without baseline studies that focus on potential spill sites, target appropriate species, and involve extensive replication and continuous data collection, it is very difficult to document the effects of a perturbation. To examine these difficulties, we resampled after nearly 20 years a baseline study site in Washington state (120, 121), returning to the same area (within a few meters) and tidal level, at a comparable season, and employing identical procedures (four 0.5 × 0.5 m quadrats). The resident biota overlaps substantially with that characteristic of Prince William Sound, and because the area is in a vigorously protected nature reserve, we expected little change. In fact, only 2 of 15 common taxa showed significantly different abundances in 1995 compared to the 1970s, based on a 1-way ANOVA in which data were partitioned into past (with nested sample periods) and current times. One of these taxa, ulvoid algae, is highly opportunistic, fast growing, and ephemeral, and its life history makes it a poor candidate for judging biotic change (Figure 3). However, those species showing no statistical change in abundance were not necessarily invariant. Rather, the low sample size, combined with high spatial (within years, e.g. *Balanus glandula*) and temporal (between years, e.g. *Fucus gardneri*) variation caused estimates of abundance to embrace possible extinction in 10 of the 15 species. It would be extremely difficult under these circumstances to demonstrate harm—to show that a sample is an outlier and differs from what would have been expected without the perturbation. Clearly, detecting population change requires numerous samples, distributed through time, focusing on long-lived species. The 5–10 replicates per tide height per site used by most intertidal research teams after EVOS may characterize the intrinsic variability of intertidal biota insufficiently (60, 71, 76).

Modelling Effects of Seasonality and Organism Interactions

Many intertidal organisms in Prince William Sound including mussels, barnacles, and algae exhibit strong seasonal patterns of recruitment, growth, and mortality. In addition, such organisms show marked age/size structure and susceptibility to predation. Given this, what difficulties might arise in determining the population trajectory through percent cover estimates made in different seasons? In a model designed to explore this question, we assume an initial recruitment of mussels onto bare substrate. We obtained size-specific growth parameters from the literature (142, 143) and examined changes in percent cover over a range of size-dependent mortality estimates representing either natural or spill-related changes in predatory dogwhelk numbers (133). Figure 4 suggests that population decline (as an index of "recovery") could result from normal dynamics and not chronic oiling effects. In addition, monitoring studies focused on percentage of cover must be sensitive to the difference that can be produced by just a few months seasonal displacement in sampling. For example, percent cover estimates in April (month 0) immediately after mussel settlement would reveal no differences between sites with the maximum and

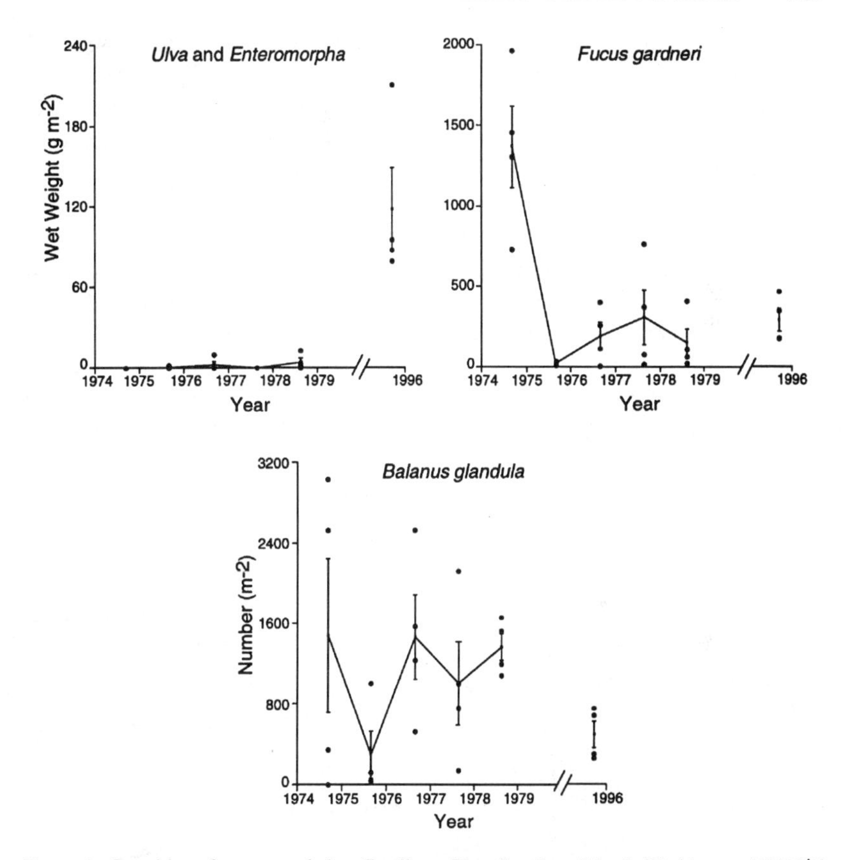

Figure 3 Densities of taxa sampled at Cantilever Pier, San Juan Island, Washington (48°33′N 123°W), for five years in the mid-1970s and again in September 1995. (*a*) Ulvoid green algae (i.e. *Ulva* and *Enteromorpha*), (*b*) *Fucus gardneri* (perennial brown alga), (*c*) *Balanus glandula* (barnacle). Circles represent actual samples. Bars represent standard errors of the means.

minimum predator density; samples six months later (October) would find a 28% difference, and nine months later (January) a 47% difference. Finally, the same proportion of cover at different times or sites may reflect very different underlying predator-prey dynamics. As in all monitoring studies, site-specific sampling data, and the conclusions drawn from them, will be rendered more variable by interactions.

POST-SPILL RESEARCH: CASE STUDIES

The various interested parties were able to agree on only a few findings: The path of the spill, for example, is substantiated by detailed maps (56) (Figure 2).

However, even fundamental "facts" are disputed. Exxon estimated that roughly 36,000 metric tons (10.8 million gallons) were spilled; ADEC's official estimates were 37,333 metric tons (11.2 million gallons), an 11% discrepancy (96). Similarly, estimates of the proportion of the region's shoreline impacted ranged from 10% to 18%, variation that enters from disagreement over the spatial boundaries of the region and the fractal nature of shorelines (166). The biological effects of the spill met with even less agreement.

Table 2 summarizes research done on a variety of species. The morgue counts represent the number of corpses found and attributed to oil, although caution is warranted in accepting either species identifications or cause of death. For example, only 4 of 12 corpses identified as yellow-billed loons and later reexamined, turned out to be that rare species (S Rohwer personal communication). Extrapolations to total deaths were based upon the proportion of tagged and released carcasses found again. From previous studies, it is clear that retrieval generally increases with the number of corpses released, while it

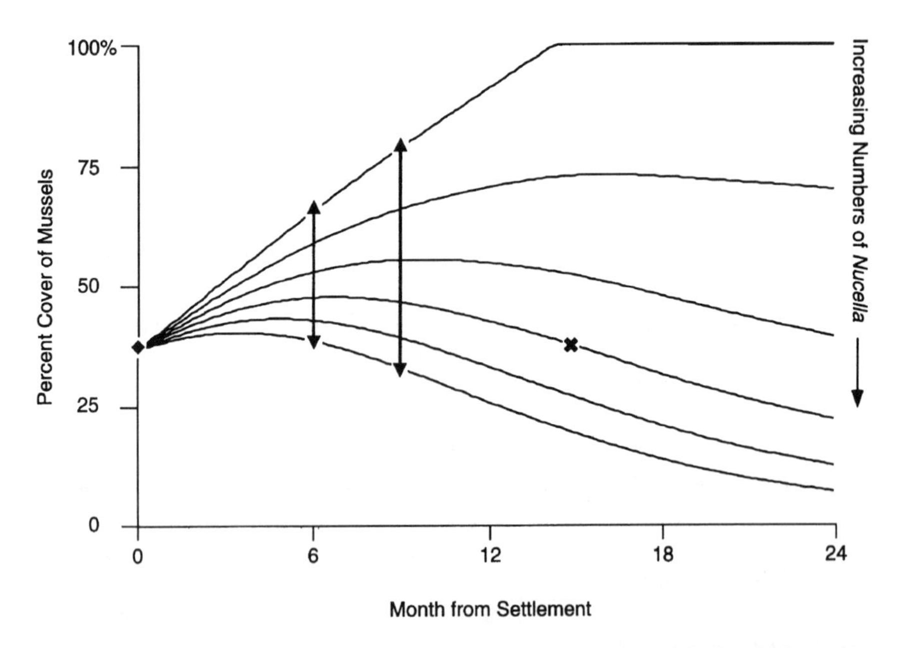

Figure 4 Model results for a scenario in which mussels recruit annually and die from both natural mortality and predation by dogwhelks (*Nucella*). Vertical arrows show that percent cover of mussels 6–9 months after settlement depends upon the density of dogwhelks present. For any given density of dogwhelks, mussel cover varies over the course of the year, although samples obtained at particular times may indicate unchanging cover ♦ and ×.

remains variable (0–59% recovery) due to local conditions (135). At least three corpse-drift studies were performed after EVOS (of 100, 94, and 144 birds), with recovery rates of just 2–3% (41, 135). Corpses also disappeared, once beached, through sand burial and scavenging (41). Radio-tagged carcasses had an average beach tenure of just 24 h (54), and 0–100% of (chicken) corpses were gone within 10 days (171). While the fates of corpses confounded estimates of total mortality, post-spill injury was also examined through censuses of population change and observations of reproductive success. Tissue samples were collected from numerous species to assess toxicological effects of oil.

Many organisms died as a consequence of oil, due to toxicity, hypothermia, or smothering. Their corpses stand as an incontrovertible sign of damage, although separating natural from oil-related mortality and translating corpses into total organisms killed were clearly difficult. Sublethal effects of oil were also evident after the spill. Some organisms such as mussels and seals stored hydrocarbons directly. Vertebrates were able to detoxify oil, but an oil signal could still be detected in bile (oil metabolites) and liver and brain (lesions). Few of the studies done in Prince William Sound to assess changes in population density or demography gave incontrovertible evidence of an oil effect. For example, the productivity of black oystercatchers was estimated to have been reduced by 6% throughout the Gulf of Alaska during the year of the spill, but this calculation was based on 9 corpses, and on poor nestling survivorship in oiled areas of one island (144). The conclusions seem particularly uncertain in light of the fact that nest failure due to predation varied by 20% among oiled and unoiled areas—perhaps the oiled beaches actually kept predators away. In general, spill effects could not be separated from the consequences of cleanup activities and tended to dissipate after a few years.

We have chosen to examine in greater detail the research conducted on a few species to illustrate both problems with the data and the conflicting interpretations of Exxon and Trustee scientists.

*Orcas, or Killer Whales (*Orcinus orca*)*

Orcas highlight the fact that even when exceptional baseline data are available about population size, demography, and individual traits, it may be difficult to show that any changes after EVOS are due to that particular perturbation. Cetaceans are not uncommon in PWS, and orcas have been studied there since at least 1977. By 1987, 221 individuals in seven resident pods or matriarchal groups had been photographically identified by distinctive color patterns or scarring (103). The post-spill study was apparently funded only by the Trustees, and not by Exxon, in contrast to research on many other taxa.

Resident orca pod AB contained 35 individuals in 1984. Between 1984 and 1989 there were 8 suspected mortalities (that is, disappearances with no

Table 2 Highlights of organismal research after the Exxon Valdez oil spill. In 1989, over 50 Natural Resource Damage Assessments were funded for research on birds, mammals, fish and shellfish, and coastal habitats. Some funded studies were not performed (e.g. passerines, migratory birds); some found few individuals to analyze (e.g. sea urchins, cetaceans); some were discontinued after a year, while priorities shifted from damage assessment to restoration. Data are also available on other intertidal (e.g. limpets, *Nucella* spp., barnacles) and subtidal (e.g. helmet crabs, leather and sun stars, eelgrass, kelps) species (31–33, 71, 76, 84). Standard font = Trustee-sponsored research; *italics = Exxon-sponsored research.* n = Number of sites (or individuals if noted by "indiv."). B = data before spill; A = data after spill; U = data from unoiled site; O = data from oiled site; T = data from treated (e.g. cleaning attempted) site; L = data from lightly-oiled site; M = data from moderate to heavily-oiled site; H = hydrocarbons. Morgue counts refer to total numbers of carcasses of each taxon found after EVOS in the entire spill area (or in PWS if noted). For multiple comparisons, only significant differences have been included. > and < signs represent significant differences that may or may not be due to oil.

SPECIES	MORGUE COUNT	ESTIMATED DEATHS #{% of exposed population}	POPULATION CHANGE	DEMOGRAPHY	TOXICOLOGY	REF.
All whales	37 strandings in '89 (A = B)				tissue ('89): no H (n = 7 stranded indiv.)	98
Humpback	no strandings in '89		density ('89–90): A < B ('88), due to wider search	reproductive rate ('89, 90): A = B ('80–88)		161
Orca	strandings ('90): 3; ('92): 1		pod AB ('88–90): 13 of 36 gone; other pods: 0 gone; transients: 9 gone			28, 103
Steller Sea Lion	12; 16 killed for tissue samples	not oil-related	population decline ('60–'90): 70%; ('89): A = B	# pups: O = U; '89 < '90; premature births: O > U	bile H metabolites ('89): high; lesions: none	26
Harbor Seal	18; 27 killed for tissue samples	300 {36%}	population decline ('75–'90): > 10%/yr; ('89): O (n = 7) > U (n = 18)	pups / total: (O) '89 < '90; (U) '89 = '90	bile, brain, blubber H ('89): O (n = 13) > U (n = 14 indiv.); brain lesions: 35 d post-spill (n = 1) > 3 mo. (n = 11 indiv.)	51–53, 147

Sea Otter	1011; 493 in PWS	2800–4028	census ('89–'91): (O) A < B ('85), (U) A > B; *census ('91): A = B ('85) (n = 3)*	prime age animals as proportion of deaths ('89–91): A > B; *reproduction ('90, '91): (O) A = B*	tissue H ('89): O (n = 10) > U (n = 12 indiv.)	10, 16, 24, 44, 58, 85, 114
River Otter	few		latrine site abandonment: O > U; scat decline ('90): O > U	weight ('90, '91): O < U; ('92): O = U	blood haptoglobins : ('90, '91) O > U; '90 > '91 > '92	37, 38, 166
Bald Eagle	151	902 {11%}	census ('89–91): A > B ('82); *habitat use (to '90):* O < U	nest failure ('89): O (cleanup?) > U; survival of tagged birds (8/89): O = U; *density ('90, '91): O > U; nest success ('90): O = U, ('91): O > U; young/nest ('90): O = U; ('91): U > O*	blood ('89, '91): H low; uric acid: H depend on recent food	19, 30, 167
Peregrine Falcon	no known deaths		40–60 prs in spill path	nest productivity ('89): O (n = 13) < AK in general		79
Marbled Murrelet	289 in PWS; 334 total murrelets	5000 {1.25%}	density ('73–'85): decline in PWS; census ('89): A < B ('78–80); ('90, '91): A = B; negative correlation between boat/air traffic and density	juvenile-to-adult ratio ('89): O (n = 3) < U (n = 1)		27, 90, 135
Kittlitz's Murrelet	72 ≥ 100	1000–2000 {5–10% of species}				159, 160

(*Continued*)

Table 2 (*Continued*)

SPECIES	MORGUE COUNT	ESTIMATED DEATHS #{% of exposed population}	POPULATION CHANGE	DEMOGRAPHY	TOXICOLOGY	REF .
Common Murre	21,604; 438 in PWS	375,000–435,000 {50–90%}	census ('89–91): A > B; *density:* *A = B (Barren Isl.)*	breeding date ('89): A later than B; *attendance ('91):* *A = B*		43, 49, 68, 122, 129, 134, 135
Pigeon Guillemot	135 in PWS	{% of 5000}	density ('72–'85): decline by 50%; decline ('89 vs. '85): O > U; *habitat use:* *('89–91): O = U*	hatching, fledging, feeding rates & fledging weights ('89): O = U (n = 1); chick weight, hatching success: '89 > '90		30, 66, 91, 123, 135
Storm Petrel	12 in PWS; die-off in Aug. 89		density ('89): A = B (Barren Isl.)	reproductive success ('89): A = B		119, 135
Black-legged Kittiwake	123 on beaches		*habitat use: ('89–91):* *O = U*	chicks / nest ('89): A < B ('85–88); ('90–92) O (n = 10) & U (n = 14) decline	oiled feathers ('89): ≤37% of birds; liver ('89): H in of 10 indiv.; ('90) H in 0 of 5	30, 80, 171
Glaucous-Winged Gull	52 total gulls in PWS		*habitat use (to '90):* *O < U*	breeding pairs, nest density, clutch size, hatching success, fledging success ('89): A = B on island 25 km from spill		30, 130, 134
Harlequin Duck	148 in PWS	423 {of ≫ 2000 total }	*habitat use (to '91):* *O < U*	pairs/stream ('91, '92): O (n = 16, 37) < U (n = 12, 20)	oil metabolites in bile ('89): O > U	30, 131

Goldeneye	64 total goldeneyes			population increase ('90–94): U>O		2, 131
Merganser			*habitat use (to '89 or '91):* O < U	population increase ('90–94): U>O		2, 30
Oystercatcher	9	120	census: O < U; *habitat use (to '91): O < U*	chick growth, survival, feeding rate ('89): O<U; hatching success: O = U; nest predation: O < U (n = 1 pr)		30, 144
Cormorant	461 in PWS					131
Yellow-Billed Loons	87; 395 total loons in PWS	300–900 {6–18% of AK population}	density ('89): A = B ('83, '84) (n = 1)	% nesting ('89): A < B		48
Pink Salmon	no known deaths		hatchery stocks at record levels; many wild stocks low	egg mortality ('89–91): O > U; egg to fry survival: O = U; growth: O (low temperature) < U; *egg viability ('89–91): O = U in lab; adult return: O = U*	lesions in larvae ('89–91): O > U; H in bile (n = 5 individuals)	21, 23, 89, 101, 139, 141, 168, 169
Dolly Varden fish	no known deaths			survival ('89, 90): O (n = 2)<U (n = 3); growth: O (n = 2) < U (n = 3); uncertain because 47% "evaded recapture"	liver lesions: O (n = 1) > U (n = 3)	69, 102
Herring	no known deaths		length of spawn ('89–90): A = B; ('94) A ≪ B ('83–88)	40% (*4%*) of spawn area oiled; egg mortality: O > U; hatching rate ('92): O (n = 1)<U (n = 1); *larval hatching: no relation to H*	egg, larval abnormalities ('92): O > U; *larval abnormalities: no relation to H*	14, 55, 88, 132
Rockfish	5				lesions ('89, by species): O > U	102, 104

(Continued)

Table 2 *(Continued)*

SPECIES	MORGUE COUNT	ESTIMATED DEATHS #{% of exposed population}	POPULATION CHANGE	DEMOGRAPHY	TOXICOLOGY	REF.
Shrimp (*Pandalus*)		*catch ('89–91): A < B ('80–88); larvae: '89 > '90*	female density ('89): O (n = 3) < U (n = 3)	dead eggs in females ('89): O > U; *fecundity* ('89–91): *O = U*		5, 34
Clam (*Protothaca*)	'89 counts (n = 9 in PWS): few dead clams		density ('90–92): T (n = 4) < O (n = 5) = U (n = 3)	growth ('89): T = O = U; growth ('91): high sediment H < low H		35, 76
Snail (*Littorina*)			density ('90): O < U; ('92): T > O = U	size ('90): T (n = 4) = O (n = 5) < U (n = 5)	tissue H: T (n = 9) > O (n = 7) > U (n = 7)	71, 76, 77
Mussel (*Mytilus*)			density: O = U; O = U = T; *density: L = M = U*	growth ('90): T = O > U; mortality ('90): T = O < U	tissue H: T > O > U; *tissue H:* '89 > '90 > '91	17, 60, 71, 76, 77, 105
Rockweed (*Fucus*)			large plant density ('90): U > O; ('93): O > U; density ('90): T < O = U; ('92): T = O = U; *density ('90): L < U = M or M > L = U*	reproduction: O < U		60, 71, 76

subsequent resurrections) and 9 calves survived to an age of at least 6 months. Thus in September 1988 there were 36 whales in this pod; on 31 March 1989 only 29 were observed and an additional 6 whales disappeared between then and June 1990. Calculated annual mortality for these two years was 19% and 21%. There were no births during this interval.

Because orcas survive to an age of 80 or more years and live in small family groups, pod-specific mortality rate estimates can be vexed by small sample sizes. A larger data set garnered over a 17-year interval in Puget Sound, Washington, and British Columbia suggests that a pod of 29 members has only a 2.3% chance of losing >6 members in a year (28, 124). Because six other resident PWS orca pods show no evidence for accelerated mortality during this interval, and because no corpses were recovered, attributing the disappearances to oil remains speculative (28).

Another source of uncertainty is the hostile interaction between a sablefish fishery and these whales. Many whales have been shot, and bullet wounds have been identified in both photographs and corpses. Such an alternative potential mortality source also makes any conclusion about an EVOS effect scientifically insupportable despite the unusually detailed prior database.

*Sea Otters (*Enhydra lutris*)*

Both Exxon- and Trustee-biased accounts agree that sea otter populations were numerically and visibly impacted. Because they represent prototypical charismatic megafauna, they became immediately symbolic of the destructive effects of a massive oil spill. Money flowed towards them: $18.3 million for capture and rehabilitation (44) and $3–4 million for damage assessment and research (148).

Commercial exploitation from 1741 to 1911 drove otters close to extinction before they were granted federal protection. The PWS population probably represented a remnant (93): It is estimated to have grown 9% per year until 1989, when 6500 otters were thought to have inhabited the oiled area (58). They thus represent a species for which pre-spill data suggested a growing local population.

The impact of EVOS on otters is evident from morgue counts. Intensive beach surveys collected 871 otters dead from oil or natural causes (493 in PWS) (10, 148). Another 357 were captured and delivered to rehabilitation centers: Of these, 197 were eventually released, and 25 were transferred to permanent holding facilities. Estes (44) calculated that the cost per saved otter exceeded $80,000, and he concluded that great expense, high mortality of oiled individuals eventually released, and a 5–10% mortality due to the stress of capture of even healthy otters renders such humanitarian efforts ecologically "unrealistic."

In the case of otters, prior data provide convincing evidence of population decline due to oil. A pre-spill census conducted by boat in 1984–1985 involved 742 shoreline transects (822 km^2) (81). This heroic effort estimated a density of 5.12 ($\pm$0.12 SE) otters/ km shoreline. Comparable post-spill sampling in 1989–1991 yielded significantly lower densities with no sign of recovery (July 1989: 3.30 [($\pm$0.42 SE); July 1990: 2.76 ($\pm$0.47 SE); July 1991: 2.91 ($\pm$0.34 SE)] (24). Otter abundances remained depressed at least through 1993 (1).

In contrast, Exxon scientists found that otter densities at two of three sites were indistinguishable from prior censuses, and at the third site otters recovered to pre-spill densities by 1991. In order to resolve unchanged population densities with known deaths, the otter population was hypothesized to have increased in the time between pre-spill censuses and EVOS. As further indications of complete recovery, Exxon scientists found that otters produced just as many pups one to two years after the spill as they had before, even at heavily oiled sites, and otters used oiled areas just as frequently as similar unoiled habitats (85).

Within PWS as a whole, sea otter populations are probably increasing today, although doubt may remain about the time course to recovery because of both chronic effects of oil (111) and possible disease introduction from the intensive rehabilitation efforts (110). However, many millions of dollars appear to have been squandered as the antagonists jockeyed for legal advantage, and little seems to have been learned of significance for the conservation, restoration, and especially management of this ecologically conspicuous species.

*Bald Eagles (*Haliaeetus leucocephalus*)*

Both Trustee (19) and Exxon (167) biologists agree on some basic facts: About 150 eagle carcasses were recovered in 1989; most of the 1989 damage was due to a negative effect of the massive cleanup effort on breeding birds; no post-spill oil-related effects were visible in 1990 and 1991. Another 137 birds were captured, cleaned, and released, at a cost of approximately $10,000 per bird (166). We focus on the body count to make the following general point. Mortality from the spill should add to that occurring naturally. This was the most convincing argument presented by orca biologists that pod AB was probably damaged by oil: Observed post-spill mortality was greater than expected under normal conditions. For eagles, however, this type of reasoning was not applied. Spies (146) refers to the 150 carcasses as "unequivocal evidence of the impact." Other Trustee biologists (19) have applied standard factors to adjust for unrecovered corpses and increased the estimated mortality to 614 to 1871 eagles, with a best estimate of 902 deaths or 11% of the 8000 eagles thought to live in the vicinity of the spill. No one developed a model of null expectation: How many eagle corpses might have been found in an intensive search of the area in any other year, especially after an exceptionally cold winter? Maximum known longevity for this species is 22 years (15), which generates a minimal death rate for adult eagles of 5% yr^{-1}, or 400 birds in the spill trajectory. A more reasonable life span of 10–15 years would result in a higher natural mortality rate. Since eagles feed and nest in the marine near-shore environment, dead eagles are likely to be deposited along coastlines. We may not need to invoke EVOS to explain a retrieval of 150 corpses from 800 km of searched shore, or 328 extrapolated to the entire spill trajectory of 1750 km.

*Murres (*Uria *spp.)*

Murres are abundant, colonial-nesting seabirds in PWS and waters adjacent to the oil trajectory. They accounted for about 74% of the vertebrate carcasses recovered up to August 1989. This body count (21,500 birds) is the only reliable fact underlying subsequent reconstructions of mortality (135). Controversy centers on how to estimate both the actual number of murres killed and the post-spill impacts on colony breeding success (129, 134, 135). Trustee biologists have extended the minimal mortality estimate by considering whether carcasses sink or float, whether oiled murres swim toward or away from land, and the potential role of scavengers. One recent estimate was that 375,000 murres died, of which approximately 6% were initially recovered (134). Exxon scientists argue for interpretive caution since no reliable estimates exist for pre-spill murre population numbers in the impacted region, and even in the best-studied site (Barren Islands) "the various estimates varied by 80,000 murres within a two-year period" (129). Certainly the body count represents the absolute minimum mortality, but the percent of murres killed by oil, the breeding colony to which they could be assigned, and even the possibility of a 1 to 2 year decrease in reproductive effort and colony success all remain conjectural. The continuing debate about the "facts" strains the credibility of all biological inferences. At the worst, sensationalized accounts of the event, tending to err on the high side, have made their way into textbooks (108), adding yet another layer of misinformation.

*Kittlitz's Murrelet (*Brachyramphus brevirostris*)*

Our concern here is an inability or unwillingness of agencies to focus on rare or seriously threatened species. Rarity of a species suggests that a priori data are few, and that quantitative assessment will be difficult; it also implies a high susceptibility to natural or human-caused catastrophe—all traits that predispose the species for federal listing. Kittlitz's murrelet is a small seabird whose world population is estimated at 18,300, 95% of which survive in Alaskan waters (159). Its biology is tied to coastal glaciers by breeding in the high alpine zone and foraging in glacially silted freshwater plumes. The 30,000 km^2 area impacted by the oil represented a core area for its breeding, molting, and feeding, and this area contained approximately one half of its known population (160). Thus Kittlitz's Murrelet seems especially susceptible to high latitude coastal spills such as EVOS.

Of the 34,977 carcasses logged in a US Fish and Wildlife Service database, only 67 were of this species (41). However, when impact is defined as the proportion of the total world population killed by, or disappearing after, some catastrophic event, direct mortality for this species may have been as high as 5–10% . In general, rare species and most transients, e.g. migratory waterfowl, shorebirds, and land birds foraging in the intertidal zone, seem to have been ignored.

*Pink Salmon (*Oncorhynchus gorbuscha*)*

Pink, or humpback, salmon, a commercially valuable species worth well over $50 million annually, has been carefully monitored for many years in Prince William Sound (100). Highly variable annual catches ranged between 10 and 30 million fish in the eight years prior to EVOS (139). Population fluctuations, migration, growth, and survival are tracked by marking hatchery fish with coded wire tags before their release as juveniles.

Pink salmon have a 2-yr life cycle, where the first year is spent developing in the intertidal and nearshore environment, and the second year is spent in the northern Pacific, with a final return to the shore. There are no data on immediate oil-induced mortalities, but several generations were exposed to oil. Exxon scientists showed that eggs collected from oiled and unoiled streams hatched equally well in the lab [91.1% vs. 90.6% (21)], whereas a Trustee-sponsored field survey of 31 streams found significantly higher egg mortality in oiled than in unoiled streams in 1989–91 (23). Coded wire tag data indicate that juvenile growth rates were significantly lower in moderately oiled hatchery areas than in lightly oiled areas in 1989, although low water temperatures confounded oiling (170).

These lower growth rates were calculated to reduce juvenile survival in 1989 by about 2% at three hatcheries. Nevertheless, the 1990 harvest set a PWS catch record with over 44 million fish. This dramatically high catch has been attributed to unusually favorable weather conditions and a record-setting plankton bloom that served as food for the juveniles in 1989 (140). The 1991 catch, 37.3 million salmon, was the second largest catch in PWS (139). In 1992 and 1993 the catch fell to below 10 million fish and then bounced back up to over 30 million in 1994 (141). Despite the wealth of long-term (several decades) baseline data available, it remains difficult to identify population effects of EVOS, or delayed changes mediated through indirect effects, because of the high population variability that exists naturally.

*Rockweed (*Fucus gardneri*)*

This perennial brown alga forms a major, biologically important zone in PWS and throughout its broad geographic distribution; it may be the single most significant species at mid to high levels on rocky intertidal shores in terms of biomass (71). *Fucus* was severely impacted by both the oil from EVOS and the high pressure hot water cleanup (92). It was studied using funds from Exxon, Trustees, and the National Oceanic and Atmospheric Administration (NOAA).

The response of *Fucus* to oil on other high latitude shores has been followed closely (65, 145). *Fucus* colonized bare rock about six months after the 1967 *Torrey Canyon* spill and persisted for five years before a combination of natural senescence and unusually high densities of limpets eliminated almost all algal cover. Populations of limpets subsequently crashed from lack of food. Cycles of algal recruitment and disappearance persisted for at least a decade (65).

In Prince William Sound the fate of *Fucus* was used as an index of intertidal recovery. Exxon scientists, focusing on plant biomass, claimed that most of the *Fucus* zone on oiled shorelines had recovered by summer 1990, 15–18 months after the spill (60). In stark contrast, a Trustee-sponsored team concluded that "full recovery has not yet occurred more than 4 years after the spill" (71). This conclusion was based, not on mass, but on population age/size structure, density of reproductive plants, and the number of eggs settling on the shoreline. Additional field experiments (71) and longer term studies (75) in PWS suggest that as the initial *Fucus* cohort senesces after about 6 years, percentage of cover decreases dramatically. Subsequent regrowth of *Fucus* permits the return of grazers. The pattern appears strikingly similar to the grazer-mediated cycles observed after the Torrey Canyon spill (65), and it may exemplify the sorts of dynamics likely after oil spills.

Exxon's assessment of *Fucus* recovery, based on comparative studies of plant mass in oiled and control sites, is statistically correct yet biologically flawed. Prior consideration of the *Torrey Canyon* results should have fostered a more conservative, longer-term perspective, in which re-establishment of age structure and reduction in biomass oscillations are critical components of recovery. Re-establishment of age structure is admittedly difficult to judge in *Fucus*, as age can be estimated only by size. However, both growth rates and age can be determined in species, from algae to fish, with distinct annual growth increments. Even after a spill, these species can be used to assess whether oil causes unexpectedly high mortality in certain age classes, whether growth rates change in concert with oiling, and the extent of natural variability in species' growth rates or age structures.

OIL REMOVAL

Although emergency response plans are legislated for all areas with major oil tanker traffic, current containment and recovery techniques are inadequate for large spills (87, 165). Thus, shoreline cleanup and remediation have been important components of oil spill response, and numerous options for cleanup existed at the time of EVOS: burning, washing, scrubbing, booms, dispersants, and detergents. Most of these cleanup techniques, however, have been employed in essentially unchanged fashion for decades, as spill after spill has touched shore. This shallow learning curve can be attributed in part to federal constraints on developing and field-testing novel cleanup technologies (117; but see 42, 47, 138). Although detergents with well-known toxic properties were applied sparingly on PWS shores (65, 118), other high impact techniques, including high-pressure hot-water washing and heavy equipment in marshes, were employed despite warnings from science advisers (95). Sadly, the State of Alaska mandated that oil be removed, rather than taking a scientifically favored focus of minimizing total ecological effects (95). Two of the most conspicuous attempts at cleaning oiled beaches in PWS involved washing and bioremediation.

High-Pressure Hot-Water Washing

High-pressure hot sea water (71°C) was sprayed onto many shorelines from omni-barges even though American Petroleum Institute guidelines recommend "natural cleansing" as the preferred method to remove oil from exposed rocky shores. In fact, high pressure flushing is "not advisable" as it tends to remove organisms and substrate as well as oil (50). Such an outcome was manifest in PWS, whenever both short- and long-term impacts were assessed. The immediate consequence was the removal of an estimated 15% to 27% of the oil from rocks, but flushing also increased the number of dead mussels by 20 times and reduced the mean number of species in 0.25 m^2 quadrats from 8 to 3 (105). The addition of an oil dispersant along with hot water resulted in 75% mortality of clams (*Protothaca staminea*) subtidal to cleaned areas (105). Researchers speculate that the sediment washed out and transported by the high pressure cleaning might have smothered organisms even deeper in the subtidal zone (76). While high-pressure hot water effectively removes oil, it also increases the mortality of organisms that manage to survive the initial spill.

Longer-term consequences were examined by comparing a few (nine) oiled areas that were set aside and not cleaned, with areas that were oiled and cleaned, and areas that were not oiled. Percent cover of rockweed, which reaches 50% on unoiled areas, returned to normal values by 1991 on oiled sites but not until 1992 if those sites were also cleaned (76) (however, remember the age/size caveats above). Cleaning also apparently reduced the diversity of species found in soft-sediment cores for at least 2 years after the spill, while oiled sites that had not been cleaned showed reductions in species abundances but not in diversity (76, 77, 105). In general, cleaning produced more biological differences than oiling, in comparison to pristine sites.

Bioremediation

One of the more promising techniques for post-spill cleanup involves the degradation of petroleum products by microorganisms (6). Hydrocarbon-degrading microbes occur naturally in PWS, but oil disappears more rapidly if these resident consumers can be stimulated to break down oil. Fertilizers were added to oiled shorelines to test whether biodegradation could be accelerated without causing nutrient toxicity or eutrophication of coastal sites. Early results showed clear "windows" of fertilized beach surrounded by oil, and bioremediation was termed a success (137). However, microorganism counts were not significantly elevated by fertilization, nor could any significant declines in oil be detected by gravimetric techniques (e.g. oil per area or volume) (150).

Oil degradation has traditionally been measured using large hydrocarbons as a persistent background against which to judge disappearance of other oil fractions. Because the resident microbes in PWS break down large hydrocarbon molecules, this technique failed until still larger, undegraded compounds were identified. Recent reanalyses show that fertilization accelerated the loss of hydrocarbons relative to unfertilized control areas at one of three sites; none of the sites showed signs of eutrophication (20). The successful fertilization occurred at a site with a high nitrogen concentration to oil load (nitrogen per oil per sediment volume), in which degradable oil fractions were still prevalent. If fertilizer is applied early, bioremediation may reduce oil, but the utility of this approach (i.e. how much additional oil microorganisms remove in comparison to wave action, and whether any intertidal species suffer less as a consequence) is equivocal.

NO USE CRYING OVER SPILLED OIL: FUTURE POST-SPILL RESEARCH

In our estimation, the research initiated after EVOS failed in three ways. First, much of it was carried out to assess injury in terms of changes in population size, using species lacking adequate baseline information. Few significant, unambiguous changes were found, but this is likely due to an inability to detect change rather than to an absence of mortality: Tremendous declines would have been required as incontrovertible evidence of injury when assessed as departures from baselines. Second, opportunities for controlled experimental examination of the effects of various cleanup techniques were lost. The blame for this omission should be shared: public outcry to "do something," NOAA's decision to set aside 9 rather than 61 uncleaned sites (164), and Exxon's failure to keep track of shoreline treatments (166) all contributed.

The third failure involves outlays for restoration projects. Restoration has become the major beneficiary of money still available from Exxon's $900 million out-of-court settlement, but the money seems misdirected in light of available evidence about what components of the PWS ecosystem suffered injury. For example, despite the fact that hatchery-raised pink salmon have reached record returns, and that evidence of negative effects of oil on salmon is disputed, nearly 20% of all restoration funds are being expended on salmon-related projects in 1996 (155). Even at the outset, although partly required by law, much effort went into the rehabilitation of oiled birds and otters, which did little to aid populations, and into beach-cleaning, which arguably did more harm than good. If oil shippers pay for restoration after a spill (either to compensate for damages or as punishment for littering), then these funds would seem best directed toward improving the status of the most rare and endangered species within a system (e.g. Kittlitz's murrelet) and identifying the causes underlying population declines unrelated to spill effects (e.g. sea lions and harbor seals). Foxes introduced to islands as a source of fur have done greater harm to Alaskan bird populations than an oil spill possibly could (8, 25), and moderating this and other anthropogenic onslaughts such as dredging or clear-cutting should constitute a viable "restoration" alternative. In that way, the inevitable next oil spill will be less likely to cause extinctions.

Future oil spill research should (*a*) allow injury to be assessed in a more convincing way, (*b*) incorporate studies of both direct and indirect effects of oil, and (*c*) use the spill as a pulse perturbation to study the dynamics of coastal species assemblages. Our recommendations reflect these goals.

Better Baselines

Although often tedious to collect, baseline data are essential to understand whether an environmental disaster has affected species. The effectiveness of baseline data depends on knowing a species' status immediately prior to the spill and on the ability to detect departures from natural variation. Variability in density measures can be reduced by paying attention to season (e.g. densities vary within a year but may be consistent each summer), species (e.g. long-lived species and local dispersers tend to show less natural variability), and replication, which should be sufficient to overpower statistical variations in local distribution and abundance. In practice, baselines will concentrate on two categories of species: those that people care about, and those that are reliable indicators of change. Baselines need not be simply measures of species abundance, nor are all species equally suitable. In the following sections we discuss the value of understanding natural mortality, demography, age structure, recruitment, zonation, and patch structure.

Better Estimates of Acute Mortality

Deaths are an unarguable sign of injury, but corpses do not reflect the exact number of individuals that died due to oil. Two components are involved: determining what proportion of the corpses represent oil-related mortality, and extrapolating from these corpses to the total death toll. Especially needed are field measurements on the propensity for corpses to sink, float away, or be scavenged beyond recognition. Considering how critical this information is to both damage assessment and litigation, there remains no scientifically justifiable reason to avoid using retrieved corpses to generate data on the fate of organisms killed by oil.

Rapid Assessment

With the development of increasingly sophisticated models predicting oil spill trajectories, it should be possible to mobilize research teams to acquire pre-spill data (50). These data, which might include photographs, species abundances, or marked quadrats, give a view of the shoreline immediately prior to oiling that can be compared to samples after the spill. In the EVOS case, between three days and a week were available for these sorts of rapid assessments, but few were carried out in PWS (but see 135, 166). An assessment attempt along the Kenai Peninsula was foiled when oil failed to land on the stretches of intertidal that had been marked and inventoried (D Duggins, personal communication).

Demography

Whether or not effects show up as changes in species abundance, oil may affect demography, especially age structure, birth rates, and individual growth rates, and consequently a spill could alter population trajectories. For some species, these population metrics may be much more spatially and temporally consistent than density and thus provide a better indication of injury. Data might include pupping rates, adult-to-juvenile ratios, clutch size variation, or even dispersal tendencies. For example, after EVOS, black-legged kittiwakes, harbor seals, and *Fucus* all showed lower offspring production at oiled compared to unoiled sites, although hatching success in black oystercatchers was higher at oiled sites (Table 2).

In many invertebrate taxa, the presence of hard mineralized skeletons allows for assessments of pre- and post-spill demography. Candidate intertidal species in which it is possible to determine the age of individuals include sea urchins [rings in the ambulacral plate of *Strongylocentrotus droebachiensis*

(83)], gastropods [annual growth rings in the shell of Tegula (125) or Protothaca (76)], barnacles, and bryozoans (40). Some algae also show growth rings [in the stipe of *Laminaria* spp. (86); as remnants of each year's blade in *Constantinea* spp. (29)]. In all these species, individuals collected some time after the spill can be backdated to compare pre- and post-spill population age structure and growth rates. Prior reproductive effort may also be accessible for comparison if, as in some bryozoans, external brood chambers persist (40). By experimentally treating individual colonies with vital stains such as tetracycline immediately after a spill, it should be possible to compare pre- and post-spill reproductive efforts.

Recruitment

Demographic data indicate population changes of locally dispersing species, but for many intertidal organisms with long-distance planktonic larval stages, recruitment rather than reproduction has the greatest influence on recovery. A standard method for assessing recruitment is to clear the substrate or attach settling plates and observe what appears on these bare areas. The critical comparison is between oiled and unoiled substrates (plates) placed into oiled, cleaned, and unoiled sites (39, 74). If recruitment differs among sites, then either the proximity of oil or some confounding factor (e.g. currents) is responsible. If plates within sites differ, then oiled surfaces affect recruitment independently of site characteristics.

This technique becomes particularly powerful when combined with grazer-exclusion treatments because most intertidal studies show that limpets, chitons, crabs, and snails determine the algal species that appear in bare spaces (99, 127). An absence of a treatment effect in such a grazer-exclusion experiment would indicate an anomalously low level of herbivory.

Intertidal Zonation

Zonation, the conspicuous layer-cake arrangement of species, might well change after a spill (94). It integrates a wide variety of physical and biological processes into a relatively stable pattern that can be easily tracked photographically (126). In the aftermath of an oil spill, dramatic changes in zonation boundaries would be expected to occur as oil or cleanup kills mobile consumers, sessile invertebrates, and algae. Measuring the time to re-establishment of stable zonal boundaries provides an estimate of recovery time.

Patch Structure

Intertidal mussel beds often display gaps of a characteristic size-frequency and distribution (128). These gaps, or patches, result from disturbance by waves, logs, weather, and predators. A change in the disturbance regime might be expected to change patch structure and thus exert indirect effects on a number of species, since the heterogeneity afforded by patches promotes species coexistence. Oiling kills mussels and therefore creates large patches. Oil effects might also be mediated indirectly through sea otters that consume mussels, since a reduction in otter densities could cause the small patches characteristic of their intertidal foraging to decline in frequency (158).

Replicated Tests of Cleanup Techniques

Thirty years of experience with oil spills on rocky shores have shown that dispersants and hot water cleaning, while removing oil, also kill many surviving plants and animals (59, 65, 118). Nevertheless, there are situations in which cleaning will continue to be carried out: under intense public pressure; in low-energy estuaries and bays where natural oil removal is slow (151); and as chemical companies develop putatively safer cleaners. To determine the effects of various cleaning methods, it is necessary to perform replicated tests, applying the technique to half of a beach and maintaining the rest as a control. Furthermore, both oiled and unoiled beaches should be included in the cleaning, to separate oil effects from cleaning effects. To this end, EPA needs to commit itself to aid the development and testing of cleanup technology, even if this means encouraging carefully planned and executed experimental oil spills.

Food Chain and Indirect Effects

Oil effects beyond direct mortality can be manifest when hydrocarbons are passed from prey to predators, or when oil-induced alterations in the density of one species affect others. A developing controversy involves the passage of EVOS hydrocarbons through marine food chains (7, 13, 36, 63). Since mussels harbor patches of unweathered oil in anoxic sediments around their byssal threads, they can filter these slowly released hydrocarbons for years (7). Mussels are eaten by sea ducks, sea stars, and sea otters, so these and other intertidal foragers remain exposed to oil. Trustee-sponsored scientists have found at least 59 mussel beds around Knight Island that in 1993 still showed elevated hydrocarbons in sediment and mussels (7), although Exxon-sponsored scientists argue that levels are not high enough to make a difference to higher trophic levels (164).

Oil effects may cause—or be obscured by—longer-term indirect effects that propagate through assemblages. In fact, a high proportion of ecological impacts (up to 60%) are transmitted indirectly (106). For instance, otters are known to eat predatory invertebrates such as crabs and sea stars. These species in turn have dramatic effects on barnacles, limpets, and other species not normally consumed by otters, which subsequently affect algae that would never be otter food. Sea otters provided early convincing evidence of these sorts of trophic cascades, where fluctuations are transferred through feeding links (45, 46), and the substantial pulse perturbation of otters in PWS likely generated changes beyond those in otter number alone.

SUMMARY

The evidence that EVOS harmed individual organisms brooks little argument. Despite uncertainty in precise cause of death and the difficulty of determining total mortality from counts of corpses, numerous birds, mammals, and intertidal biota died from oil toxicity, hypothermia, smothering, or cleanup. Furthermore, tissue chemistry and histopathology indicated that individuals living in oiled parts of Prince William Sound suffered sublethal effects, although the manifestation of oil exposure varied greatly among species.

The population consequences of the oil spill have been more disputable, stemming from the difficulty of determining how abundant a species was before the spill, how many individuals actually died, and how these deaths and persistent but unquantified chronic injury would affect future population growth. More consistent baseline monitoring of a few indicator species could allow better determination of whether oil causes fluctuations outside the range of natural variation. Monitoring will, however, be costly. The missed opportunity of EVOS is that little was learned about long-term population response to, or interspecific interactions after, the dramatic pulse perturbation.

Perhaps the harshest lesson of EVOS is that basic science motivated by litigation served no master well. We have documented some of the conflicts stemming from biological research. A more dismal interpretation of EVOS research comes from Barker (11): Scientific evidence became discredited in the jury's eyes as expert witnesses with conflicting testimony canceled each other out.

In a practical sense, we hope that the lesson from EVOS is that preventive measures such as well-rested pilots and double-hulled ships, and a rapid response of burning or skimming spilled oil may eliminate future swaths of destruction. But the unending stream of ugly and toxic spills since 1989 belies any such education. Repeated spills cannot help but turn a shoreline into an asphalt parking lot. When another spill occurs that galvanizes the American public and creates a research agenda, perhaps the punishment can be meted out expediently, and the focus turned from arguing about whether feathered, furry, or edible species have been irreparably harmed to better understanding how processes underlying coastal systems are altered in the face of a massive perturbation.

ACKNOWLEDGMENTS

We thank Megan Dethier and DO Duggins for discussion and P Kareiva for a perceptive and needed review. C Holba at the Oil Spill Public Information

Center, Anchorage AK, sent us rare documents. (The Center is a useful source of further information.) NOAA provided RTP with a glimpse of Prince William Sound in 1994. Additional support has come from the Andrew W. Mellon Foundation. NSF predoctoral fellowships to JLR, AS, DRB, and DLS, as well as long-term support from the NSF program in Biological Oceanography to RTP, have generated our own perceptions of how assemblages on exposed rocky shores are organized. Without that experience, this review would have been impossible.

Literature Cited

1. Agler BA, Seiser PE, Kendall SJ, Irons DB. 1994. Marine bird and sea otter population abundance of Prince William Sound, Alaska: Trends following the T/V *Exxon Valdez* oil spill, 1989–93. *Exxon Valdez Oil Spill Restoration Project 93045 Final Rep.* US Fish & Wildl. Serv.
2. Agler BA, Seiser PE, Kendall SJ, Irons DB. 1995. Winter marine bird and sea otter abundance of Prince William Sound, Alaska: Trends following the T/V *Exxon Valdez* oil spill from 1990–1994. *Exxon Valdez* Oil Spill Restoration Proj. 94159 Final Rep. US Fish & Wildl. Serv.
3. Allen AA. 1991. Controlled burning of crude oil on water following the grounding of the *Exxon Valdez*. *Proc. 1991 Int. Oil Spill Conf.*:213–16
4. Anderson CM, Lear EM. 1994. MMS Worldwide Tanker Spill Database: An overview. *OCS Report MMS 94–0002*. US Dept. Interior, Minerals Manage. Serv.
5. Armstrong DA, Dinnel PA, Orensanz JM, Armstrong JL, McDonald TL, Cusimano RF, et al. 1995. Status of selected bottomfish and crustacean species in Prince William Sound following the *Exxon Valdez* oil spill. See Ref. 163, pp. 485–547
6. Atlas RM, Cerniglia CE. 1995. Bioremediation of petroleum pollutants: Diversity and environmental aspects of hydrocarbon biodegradation. *BioScience* 45:332–38
7. Babcock M, Irvine G, Rice S, Rounds P, Cusick J, Brodersen C. 1993. Oiled mussel beds two and three years after the *Exxon Valdez* oil spill. See Ref. 153, pp. 184–85 (Abstr.)
8. Bailey EP. 1992. Red foxes, *Vulpes vulpes*, as biological control agents for introduced Arctic foxes, Alopex lagopus, on Alaskan islands. *Can. Field-Nat.* 106:201–05
9. Baker JM, Clark RB, Kingston PF. 1990. *Environmental recovery in Prince William Sound and the Gulf of Alaska.* Edinburgh: Inst. Offshore Engin.
10. Ballachey BE, Bodkin JL, DeGange AR. 1994. An overview of sea otter studies. See Ref. 97, pp. 47–60
11. Barker E. 1994. The *Exxon* trial: a do it yourself jury. *Am. Lawyer* Nov:68–77
12. Bellamy DJ, Clarke PH, John DM, Jones D, Whittick A, Darke T. 1967. Effects of pollution from the Torrey Canyon on littoral and sublittoral ecosystems. *Nature* 216:1170–73
13. Bence AE, Burns WA. 1995. Fingerprinting hydrocarbons in the biological resources of the *Exxon Valdez* spill area. See Ref. 163, pp. 84–140
14. Biggs ED, Baker TT. 1993. Summary of known effects of the *Exxon Valdez* oil spill on herring in Prince William Sound, and recommendations for future inquiries. See Ref. 153, pp. 264–67 (Abstr.)
15. Bildstein K. 1995. Field notes: Redtail 0877-17127. *Hawk Mountain News* Fall,

No. 83:22–23
16. Bodkin JL, Udevitz MS. 1994. An intersection model for estimating sea otter mortality along the Kenai Peninsula. See Ref. 97, pp. 81–96
17. Boehm PD, Page DS, Gilfillan ES, Stubblefield WA, Harner EJ. 1995. Shoreline ecology program for Prince William Sound, Alaska, following the *Exxon Valdez* oil spill: Part 2. Chemistry and toxicology. See Ref. 163, pp. 347–97
18. Boucher G. 1980. Impact of Amoco *Cadiz* oil spill on intertidal and sublittoral meiofauna. *Mar. Poll. Bull.* 11:95–101
19. Bowman TD, Schempf PF. 1993. Effects of the *Exxon Valdez* oil spill on bald eagles. See Ref. 153, pp. 142–43 (Abstr.)
20. Bragg JR, Prince RC, Harner EJ, Atlas RM. 1994. Effectiveness of bioremediation for the *Exxon Valdez* oil spill. *Nature* 368:413–18
21. Brannon EL, Moulton LL, Gilbertson LG, Maki AW, Skalski JR. 1995. An assessment of oil spill effects on pink salmon populations following the *Exxon Valdez* oil spill. Part 1: Early life history. See Ref. 163, pp. 548–87
22. Buchanan JB. 1993. Evidence of benthic pelagic coupling at a station off the Northumberland coast. *J. Exp. Mar. Biol. Ecol.* 172:1–10
23. Bue BG, Sharr S, Moffitt SD, Craig A. 1993. Assessment of injury to pink salmon eggs and fry. See Ref. 153, pp. 101–03 (Abstr.)
24. Burn DM. 1994. Boat-based population surveys of sea otters in Prince William Sound. See Ref. 97, pp. 61–80
25. Byrd GV, Trapp JL, Zeillemaker CF. 1994. Removal of introduced foxes: a case study in restoration of native birds. *Trans. 59th No. Am. Wildl. & Natur. Resour. Conf.* :317–21
26. Calkins DG, Becker E, Spraker TR, Loughlin TR. 1994. Impacts on Steller sea lions. See Ref. 97, pp. 119–40
27. Carter HR, Kuletz KJ. 1995. Mortality of marbled murrelets due to oil pollution in North America. In *Ecology and Conservation of the Marbled Murrelet*, ed. CJ Ralph, GL Hunt Jr, MG Raphael, JR Piatt, pp. 261–69. Albany, CA: Pac. SW Res. Station, For. Serv., US Dept. Agric.
28. Dahlheim ME, Matkin CO. 1994. Assessment of injuries to Prince William Sound killer whales. See Ref. 97, pp. 163–71
29. Dawson EY. 1966. *Marine Botany*. New York: Holt, Rinehart & Winston
30. Day RH, Murphy SM, Wiens JA, Hayward GD, Harner EJ, Smith LN. 1995. Use of oil-affected habitats by birds after the *Exxon Valdez* oil spill. See Ref. 163, pp. 726–61
31. Dean TA, Jewett S. 1993. The effects of the *Exxon Valdez* oil spill on epibenthic invertebrates in the shallow subtidal. See Ref. 153, pp. 91–93 (Abstr.)
32. Dean TA, McDonald L, Stekoll MS, Rosenthal RR. 1993. Damage assessment in coastal habitats: lessons learned from *Exxon Valdez*. In *Proc. 1993 Int. Oil Spill Conf.* pp. 695–97
33. Dean TA, Stekoll M, Jewett S. 1993. The effects of the *Exxon Valdez* oil spill on eelgrass and subtidal algae. See Ref. 153, pp. 94–96 (Abstr.)
34. Donaldson W, Ackley D. 1990. Injury to Prince William Sound spot shrimp. *Fish/Shellfish Study No. 15, Preliminary Status Rep.* Alaska Dept. of Fish and Game, Juneau, AK
35. Donaldson W, Trowbridge C, Davis A, Ackley D. 1990. Injury to Prince William Sound clams; injury to clams outside Prince William Sound. *Fish/Shellfish Study Nos. 13 & 21 Preliminary Status Report.* Alaska Dept. Fish & Game, Juneau, AK
36. Doroff AM, Bodkin JL. 1994. Sea otter foraging behavior and hydrocarbon levels in prey. See Ref. 153, pp. 193–208 (Abstr.)
37. Duffy LK, Bowyer RT, Testa JW, Faro JB. 1994. Chronic effects of the *Exxon Valdez* oil spill on blood and enzyme chemistry of river otters. *Environ. Toxicol. Chem.* 13:643–47
38. Duffy LK, Bowyer RT, Testa JW, Faro JB. 1994. Evidence for recovery of body mass and haptoglobin values of river otters following the *Exxon Valdez* oil spill. *J. Wildl. Dis.* 30:421–25
39. Duncan PB, Hooten AJ, Highsmith RC. 1993. Influence of the *Exxon Valdez* oil spill on intertidal algae: tests of the effect of residual oil on algal colonization. See Ref. 153, pp. 179–81 (Abstr.)
40. Dyrynda PEJ, Ryland JS. 1982. Reproductive strategies and life histories in the cheilostome marine bryozoans *Chartella papyracea* and *Bugula flabellata*. *Mar. Biol.* 71:241–56
41. Ecological Consulting Inc. 1991. Assessment of direct seabird mortality in Prince William Sound and the Western Gulf of Alaska resulting from the *Exxon Valdez* oil spill. US Dept. Justice, US

Dept. Interior

42. Environment Canada, US Minerals Management Service, Canadian Coast Guard, American Petroleum Institute, et al. 1993. *NOBE: Newfoundland Offshore Burn Experiment.* Ottawa, Ontario
43. Erikson DE. 1995. Surveys of murre colony attendance in the northern Gulf of Alaska following the *Exxon Valdez* oil spill. See Ref. 163, pp. 780–819
44. Estes JA. 1991. Catastrophes and conservation: Lessons from sea otters and the *Exxon Valdez*. *Science* 254:1596
45. Estes JA, Duggins DO. 1995. Sea otters and kelp forests in Alaska: generality and variation in a community ecological paradigm. *Ecol. Monogr.* 65:75–100
46. Estes JA, Palmisano JF. 1974. Sea otters: Their role in structuring nearshore communities. *Science* 185:1058–60
47. Feder HM, Naidu AS, Paul AJ. 1990. Trace element and biotic changes following a simulated oil spill on a mudflat in Port Valdez, Alaska. *Mar. Pollution Bull.* 21:131–37
48. Field R, North MR, Wells J. 1993. Nesting activity of yellow-billed loons on the Colville River delta, Alaska, after the *Exxon Valdez* oil spill. *Wilson Bull.* 105:325–32
49. Ford RG, Bonnell ML, Varoujean DH, Page GW, Sharp BE, Heinemann D, et al. 1991. Assessment of direct seabird mortality in Prince William Sound and the western Gulf of Alaska resulting from the *Exxon Valdez* oil spill. Portland, OR; Ecological Consulting Inc.
50. Foster MS, Tarpley JA, Dearn SL. 1990. To clean or not to clean: the rationale, methods, and consequences of removing oil from temperate shores. *Northwest Environ. J.* 6:105–20
51. Frost KJ, Lowry LF. 1994. Assessment of injury to harbor seals in Prince William Sound, Alaska, and adjacent areas following the *Exxon Valdez* oil spill. *Marine Mammal Study #5, Restoration Study #73 Final Report.* Alaska Dept. Fish & Game, Wildl. Conserv. Div.
52. Frost KJ, Lowry LF, Sinclair EH, Hoef JV, McAllister DC. 1994. Impacts on distribution, abundance, and productivity of harbor seals. See Ref. 97, pp. 97–118
53. Frost KJ, Manen C-A, Wade TL. 1994. Petroleum hydrocarbons in tissues of harbor seals from Prince William Sound and the Gulf of Alaska. See Ref. 97, pp. 331–58
54. Fry DM. 1993. How do you fix the loss of half a million birds? See Ref. 153, pp. 30–33 (Abstr.)
55. Funk F. 1994. Preliminary summary of 1994 Alaska sac roe herring fisheries. Alaska Dept. Fish & Game, Juneau, AK
56. Galt JA, Lehr WJ, Payton DL. 1991. Fate and transport of the *Exxon Valdez* oil spill. *Environ. Sci. Technol.* 25:202–09
57. Galt JA, Watabayashi GY, Payton DL, Petersen JC. 1991. Trajectory analysis for the *Exxon Valdez*: a hindcast study. *Proc. 1991 Int. Oil Spill Conf.* pp. 629–34 Washington, DC: Am. Petrol. Inst.
58. Garrott RA, Eberhardt LL, Burn DM. 1993. Mortality of sea otters in Prince William Sound following the *Exxon Valdez* oil spill. *Mar. Mammal Sci.* 9:343–59
59. George M. 1961. Oil pollution of marine organisms. *Nature* 192:1209
60. Gilfillan ES, Page DS, Harner EJ, Boehm PD. 1995. Shoreline ecology program for Prince William Sound, Alaska, following the *Exxon Valdez* oil spill. Part 3. Biology. See Ref. 163, pp. 398–443
61. Goldstein B. 1991. The folly of the *Exxon Valdez* cleanup. *Earth Island J.* Winter:30–31
62. Gundlach ER, Boehm PD, Marchand M, Atlas RM, Ward DM, Wolfe DA. 1983. The fate of *Amoco Cadiz* oil. *Science* 221:122–29
63. Hartung R. 1995. Assessment of the potential for long-term toxicological effects of the *Exxon Valdez* oil spill on birds and mammals. See Ref. 163, pp. 693–725
64. Haven SB. 1971. Effects of land-level changes on intertidal invertebrates, with discussion of postearthquake ecological succession. *The Great Alaska Earthquake of 1964,* Washington, DC: Natl. Res. Coun., pp. 82–126.
65. Hawkins SJ, Southward AJ. 1992. The Torrey Canyon oil spill: Recovery of rocky shore communities. In *Restoring the Nation's Marine Environment,* ed. GW Thayer, pp. 583–631. College Park, MD: Maryland Sea Grand Coll.
66. Hayes DL. 1995. Recovery monitoring of Pigeon Guillemot populations in Prince William Sound, Alaska. *Restoration Project 94173 Final Report.* US Fish & Wildl. Serv.
67. Hayes MO, Michel J, Noe DC. 1991. Factors controlling initial deposition and long-term fate of spilled oil on gravel beaches. *Proc. 1991 Int. Oil Spill Conf.* pp. 453–65. Washington, DC: Am.

Petrol. Inst.
68. Heinemann D. 1993. How long to recovery for murre populations, and will some colonies fail to make the comeback? See Ref. 153, pp. 139–41 (Abstr.)
69. Hepler KR, Hansen PA, Bernard DR. 1993. Impact of oil spilled from the *Exxon Valdez* on survival and growth of Dolly Varden and Cutthroat trout in Prince William Sound, Alaska. See Ref. 153, pp. 239–40 (Abstr.)
70. Hess WN. 1978. *The Amoco Cadiz Oil Spill: A Preliminary Scientific Report.* Washington, DC: US Govt. Printing Off.
71. Highsmith RC, Stekoll MS, van Tamelen P, Hooten AJ, Saupe SM, Deysher L, et al. 1995. Herring Bay monitoring and restoration studies. *Monitoring and Restoration Project No. 93039.* Inst. Mar. Sci., Univ. Alaska, Fairbanks
72. Hilborn R. 1993. Detecting population impacts from oil spills: a comparison of methodologies. See Ref. 153, pp. 231–32 (Abstr.)
73. Holcomb J. 1991. Overview of bird search and rescue and response efforts during the *Exxon Valdez* oil spill. *Proc. 1991 Int. Oil Spill Conf.* pp. 225–28. Washington, DC: Am. Petrol. Inst.
74. Hooten AJ, Highsmith RC. 1993. *Exxon Valdez* oil spill: recruitment on oiled and non-oiled substrates. See Ref. 153, pp. 195–98 (Abstr.)
75. Houghton JP, Fukuyama AK, Driskell WB, Lees DC, Shigenaka G, Mearns AJ. 1993. Recovery of Prince William Sound intertidal epibiota from *Exxon Valdez* spill and treatments—1990–1992. See Ref. 153, pp. 79–82 (Abstr.)
76. Houghton JP, Fukuyama AK, Lees DC, Hague PJ, Cumberland HL, Harper PM, et al. 1993. *Evaluation of the condition of Prince William Sound shorelines following the Exxon Valdez oil spill and subsequent shoreline treatment.* Vol. II. *1992 Biological monitoring survey. NOAA Technical Memorandum NOS ORCA 73.* Natl. Oceanic and Atmos. Admin., US Dept. Commerce
77. Houghton JP, Lees DC, Ebert TA. 1991. *Evaluation of the condition of intertidal and shallow subtidal biota in Prince William Sound following the Exxon Valdez oil spill and subsequent shoreline treatment.* Vol. 1. *HMRB 91-1.* NOAA WASC Contract Nos. 50ABNC-0–00121 and 50ABNC-0–00122
78. Hubbard JD. 1971. Distribution and abundance of intertidal invertebrates at Olsen Bay, Prince William Sound, Alaska, one year after the 1964 earthquake. In *The Great Alaska Earthquake of 1964,* pp. 137–157. Washington DC: Natl. Res. Council
79. Hughes JH. 1990. *Impact assessment of the Exxon Valdez oil spill on Peale's Peregrine Falcons. Bird Study No. 5.* Juneau, AK: Alaska Dept. Fish & Game
80. Irons DB. 1993. Effects of the *Exxon Valdez* oil spill on black-legged kittiwakes. See Ref. 153, pp. 159 (Abstr.)
81. Irons DB, Nysewander DR, Trapp JL. 1988. Prince William Sound sea otter distribution in relation to population growth and habitat type. US Fish & Wildl. Serv., Anchorage, Alaska
82. Jahns HO, Bragg JR, Dash LC, Owens EH. 1991. Natural cleaning of shorelines following the *Exxon Valdez* spill. *Proc. 1991 Int. Oil Spill Conf.* pp. 167–79. Washington, DC: Am. Petrol. Inst.
83. Jensen M. 1969. Age determination of echinoids. *Sarsia* 37:41–44
84. Jewett SC, Dean TA. 1993. The effects of the *Exxon Valdez* oil spill on infaunal invertebrates in the eelgrass habitat of Prince William Sound. See Ref. 153, pp. 97–99 (Abstr.)
85. Johnson CB, Garshelis DL. 1995. Sea otter abundance, distribution, and pup production in Prince William Sound following the *Exxon Valdez* oil spill. See Ref. 163, pp. 894–929
86. Kain JM. 1963. Aspects of the biology of *Laminaria hyperborea.* 2. Age, weight and length. *J. Mar. Biol. Assoc. UK* 43:129–51
87. Kelso DD, Kendziorek M. 1991. Alaska's response to the *Exxon Valdez* oil spill. *Environ. Sci. Technol.* 25:16–23
88. Kocan RM, Baker T, Biggs E. 1993. Adult herring reproductive impairment following the *Exxon Valdez* oil spill. See Ref. 153, pp. 262–63 (Abstr.)
89. Krahn MM, Burrows DG, Ylitalo GM, Brown DW, Wigren DA, Collier TK, et al. 1992. Mass spectrographic analysis for aromatic compounds in bile of fish sampled after the *Exxon Valdez* oil spill. *Environ. Sci. Technol.* 26:116–26
90. Kuletz KJ. 1993. Effects of the *Exxon Valdez* oil spill on marbled murrelets. See Ref. 153, pp. 148–50 (Abstr.)
91. Laing KK, Klosiewski SP. 1993. Marine bird populations of Prince William Sound, Alaska, before and after the *Exxon Valdez* oil spill. See Ref. 153, pp. 160–62 (Abstr.)
92. Lees D, Driskell W, Houghton J. 1993.

Short-term biological effects of shoreline treatment on intertidal biota exposed to the *Exxon Valdez* oil spill. See Ref. 153, pp. 69–72 (Abstr.)

93. Lensink CJ. 1962. *The history and status of sea otters in Alaska.* PhD thesis. Purdue Univ.
94. Lewis JR. 1964. *The Ecology of Rocky Shores.* London: English Univ. Press
95. Lindstedt-Siva J. 1991. U.S. oil spill policy hampers response and hurts science. *Am. Petrol. Inst.* 4529:349–52
96. Loughlin TR. 1994. Appendix II: Oil tanker accidents. See Ref. 97, pp. 383–84
97. Loughlin TR. 1994. *Marine Mammals and the* Exxon Valdez. San Diego, CA: Academic. 395 pp.
98. Loughlin TR. 1994. Tissue hydrocarbon levels and the number of cetaceans found dead after the spill. See Ref. 97, pp. 359–70
99. Lubchenco J, Gaines SD. 1981. A unified approach to marine plant-herbivore interactions 1. Populations and communities. *Annu. Rev. Ecol. Syst.* 12:405–37
100. Maki AW. 1991. The *Exxon Valdez* oil spill: Initial environmental impact assessment. *Environ. Sci. Technol.* 25:24–29
101. Maki AW, Brannon EJ, Gilbertson LG, Moulton LL, Skalski JR. 1995. An assessment of oil spill effects on pink salmon populations following the *Exxon Valdez* oil spill. Part 2. Adults and escapement. See Ref. 162, pp. 585–625
102. Marty GD, Okihiro MS, Hinton DE. 1993. Histopathologic analysis of chronic effects of the *Exxon Valdez* oil spill on Alaska fisheries. See Ref. 153, pp. 243–46 (Abstr.)
103. Matkin CO, Ellis GM, Dahlheim ME, Zeh J. 1994. Status of killer whales in Prince William Sound, 1985–1992. See Ref. 97, pp. 163–72
104. Meacham CP, Sullivan JR. 1993. Summary of injuries to fish and shellfish associated with the *Exxon Valdez* oil spill. See Ref. 153, pp. 27–29 (Abstr.)
105. Mearns AJ. 1993. Recovery of shoreline ecosystems following the *Exxon Valdez* oil spill and subsequent treatment. *Coastal Zone '93: Proc., 8th Symp. on Coastal and Ocean Manage.* pp.466–79. New Orleans, LA
106. Menge BA. 1995. Indirect effects in marine rocky intertidal interaction webs: patterns and importance. Ecol. Monogr. 65:21–74
107. Michel J, Hayes MO, Sexton WJ, Gibeaut JC, Henry C. 1991. Trends in natural removal of the *Exxon Valdez* oil spill in Prince William Sound from September 1989 to May 1990. *Proc. 1991 Int. Oil Spill Conf.* pp. 181–87. Washington, DC: Am. Petrol. Inst.

107a. Miller, D.1991. "Deal OK could halt studies." Anchorage Times, April 6, 1991.p. B1, B7.

108. Miller GT. 1994. *Living in the Environment.* Belmont, CA: Wadsworth
109. Monahan TP, Maki AW. 1991. The *Exxon Valdez* 1989 wildlife rescue and rehabilitation program. *Proc. 1991 Int. Oil Spill Conf.* pp. 131–36. Washington, DC: Am. Petrol. Inst.
110. Monnett C, Rotterman LM. 1993. The efficacy of the *Exxon Valdez* oil spill sea otter rehabilitation program and the possibility of disease introduction into recipient otter populations. See Ref. 153, pp. 273 (Abstr.)
111. Monson DH, Ballachey BE. 1993. Age distributions of sea otters dying in Prince William Sound, Alaska after the *Exxon Valdez* oil spill. See Ref. 15, pp. 282–84 (Abstr.)
112. Moore WH. 1994. The grounding of *Exxon Valdez*: An examination of the human and organizational factors. *Mar. Technol.* 31:41–51
113. Morris BF, Loughlin TR. 1994. Overview of the *Exxon Valdez* oil spill, 1989–1992. See Ref. 97, pp. 1–22
114. Mulcahy DM, Ballachey BE. 1994. Hydrocarbon residues in sea otter tissues. See Ref. 97, pp. 313–330
115. National Academy of Science. 1985. Appendix A: Impact of some major spills (Spill case histories). In *Oil in the Sea: Inputs, Fates, and Effects,* pp. 549–82. Washington, DC: Natl. Acad. Sci.
116. National Oceanic and Atmospheric Association, Hazardous Materials Response and Assessment Division. 1992. *Oil Spill Case Histories, 1967–1991: Summaries of significant U.S. and international spills. HMRAD 92–11, NOAA*
117. National Research Council. 1993. *Environmental Information for Outer Continental Shelf Oil and Gas Decisions in Alaska.* Washington, DC: Natl. Acad. Sci.
118. Nelson-Smith A. 1968. The effects of oil pollution and emulsifier cleansing on shore life in south-west Britain. *J. Appl. Ecol.* 5:97–107
119. Nishimoto M. 1990. *Assessment of injury to waterbirds from the Exxon Valdez oil spill: effects of petroleum hydro-*

carbon on fork-tailed storm-petrel reproductive success. Bird Study No. 7. Juneau, AK: US Fish & Wildl. Serv. Anchorage, AK

120. Nyblade CF. 1977. *North Puget Sound Intertidal Study, Baseline Study Program.* Contract #76-043, Univ. Wash. for Wash. State Dept. Ecology
121. Nyblade CF. 1979. *Five year intertidal community change, San Juan Islands, 1974–197. Appendix F (update).* Univ. Wash. for Wash. State Dept. Ecology
122. Nysewander DR, Dippel C, Byrd GV, Knudtson EP. 1993. Effects of the T/V *Exxon Valdez* oil spill on murres: a perspective from observations at breeding colonies. See Ref. 153, pp. 135–38 (Abstr.)
123. Oakley KL, Kuletz KJ. 1993. Effects of the *Exxon Valdez* oil spill on pigeon guillemots (*Cepphus columba*) in Prince William Sound, Alaska. See Ref. 153, pp. 144–46 (Abstr.)
124. Olesiuk PF, Bigg MA, Ellis GM. 1990. Life history and population dynamics of resident killer whales (*Orcinus orca*) in the coastal waters of British Columbia and Washington State. In *Individual Recognition of Cetaceans: Use of Photo-Identification and Other Techniques to Estimate Population Parameters,* Special Issue 12, ed. PS Hammond, SA Mizroch, GP Donovan, pp. 109–243. Rep. Int. Whaling Commiss.
125. Paine RT. 1969. The Pisaster-Tegula interaction: Prey patches, predator food preference, and intertidal community structure. *Ecology* 50:950–61
126. Paine RT. 1974. Intertidal community structure. Experimental studies on the relationship between a dominant competitor and its principal predator. *Oecologia* 15:93–120
127. Paine RT. 1994. *Marine Rocky Shores and Community Ecology: An Experimentalist's Perspective.* Oldendorf /Luhe: Ecol. Inst. 152 pp.
128. Paine RT, Levin SA. 1981. Intertidal landscapes: disturbance and the dynamics of pattern. *Ecol. Monogr.* 51:145–78
129. Parrish JK, Boersma PD. 1995. Muddy waters. *Am. Sci.* 83:112–15
130. Patten SM. 1990. *Assessment of injury to Glaucous-winged Gulls using Prince William Sound. Bird Study No. 10.* Juneau, AK: Alaska Dept. Fish & Game
131. Patten SM. 1993. Acute and sublethal effects of the *Exxon Valdez* oil spill on harlequins and other seaducks. See Ref. 153, pp. 151–54 (Abstr.)
132. Pearson WH, Moksness E, Skalski JR. 1995. A field and laboratory assessment of oil spill effects on survival and reproductions of Pacific herring following the *Exxon Valdez* oil spill. See Ref. 163, pp. 626–61
133. Petraitis PS. 1995. The role of growth in maintaining spatial dominance by mussels (*Mytilus edulis*). *Ecology* 76:1337–46
134. Piatt J. 1995. Water over the bridge. *Am. Sci.* 83:396–98
135. Piatt JF, Lensink CJ, Butler W, Kendziorek M, Nysewander DR. 1990. Immediate impact of the *Exxon Valdez* oil spill on marine birds. *The Auk* 107:387–97
136. Piper E. 1993. *The* Exxon Valdez *oil spill, final report, State of Alaska response.* Alaska Dept. of Environ. Conserv., Juneau, Alaska
137. Pritchard PJ, Costa CF. 1991. EPA's Alaska oil spill bioremediation project. *Environ. Sci. Tech.* 25:372–79
138. Raloff J. 1993. Burning issues: Is torching the most benign way to clear oil spilled at sea? *Sci. News* 144:220–23
139. Rigby P, McConnaughey J, Savikko H. 1991. *Alaska Commercial Salmon Catches 1878–1991. Regional Information Report No. 5J91–16.* Alaska Dept. Fish & Game
140. Royce WF, Schroeder TR, Olsen AA, Allender WJ. 1991. *Alaskan Fisheries Two Years After the Spill.* Cook Inlet Fisheries Consult.
141. Savikko H. 1991–1995. *Preliminary Commercial Salmon Catches—"Blue Sheet."* Juneau, AK: Alaska Dept. Fish & Game
142. Seed R. 1968. Factors influencing shell shape in the mussel *Mytilus edulis. J. Mar. Biol. Assoc. UK* 48:561–84
143. Seed R. 1969. The ecology of *Mytilus edulis* L. (*Lamellibranciata*) on exposed rocky shores II. Growth and mortality. *Oecologia* 3:317–50
144. Sharp BE, Cody M. 1993. Black oystercatchers in Prince William Sound: oil spill effects on reproduction and behavior. See Ref. 153, pp. 155–58 (Abstr.)
145. Southward AJ, Southward EC. 1978. Recolonization of rocky shores in Cornwall after use of toxic dispersants to clean up the Torrey Canyon spill. *J. Fish. Res. Board Can.* 35:682–706
146. Spies RB. 1993. So why can't science tell us more about the effects of the *Exxon Valdez* oil spill? See Ref. 153, pp. 1–5 (Abstr.)
147. Spraker TR, Lowry LF, Frost KJ. 1994.

Gross necropsy and histopathological lesions found in harbor seals. See Ref. 97, pp. 281–311

148. St. Aubin DJ, Geraci JR. 1993. Summary and Conclusions. See Ref. 97, pp. 371–76
149. Steinhart CE, Steinhart JS. 1972. *Blowout.* North Scituate, MA: Duxbury
150. Stone R. 1992. Oil-cleanup method questioned. *Science* 257:320–21
151. Teal JM, Farrington JW, Burns KA, Stegeman JJ, Tripp BW, Woodin B, et al. 1992. The West Falmouth oil spill after 20 years: Fate of fuel oil compounds and effects on animals. *Mar. Pollut. Bull.* 24:607–14
152. Trustee Council, *Exxon Valdez* Oil Spill. 1993. Exxon Valdez *oil spill draft restoration plan: summary of alternatives for public comment.* Anchorage, Alaska
153. Trustee Council, *Exxon Valdez* Oil Spill. 1993. *Exxon Valdez* Oil Spill Symposium Abstract Book. Anchorage, AK: February 2–5, 1993. 356 pp.
154. Trustee Council, *Exxon Valdez* Oil Spill. 1994. *Science for the Restoration Process.* Anchorage, Alaska
155. Trustee Council, *Exxon Valdez* Oil Spill. 1995. *Fiscal Year 1996 Work Plan.* Anchorage, Alaska
156. US Department of Commerce. 1975–1994. *Environmental Data Service Climatological Data Annual Summary. Alaska.* Vol. 61–80, US Dept. Commerce
157. US Department of Interior. 1972. *Final Environmental Impact Statement, Proposed Trans-Alaska Pipeline*. Vol. 3, *Environmental setting between Port Valdez, Alaska, and West Coast ports. PB-206 921–3,* Washington, DC: Special Interagency Task Force for the Federal Task Force on Alaskan Oil Development
158. van Blaricom GR. 1988. Effects of foraging by sea otters on mussel-dominated intertidal communities. In *The Community Ecology of Sea Otters,* ed. GR van Blaricom, JA Estes, pp. 48–91. Berlin: Springer-Verlag
159. van Vliet G. 1993. Status concerns for the "global" population of Kittlitz's Murrelet: Is the "glacier murrelet" receding? *Pac. Seabird Group Bull.* 20:15–16
160. van Vliet G. 1994. Kittlitz's Murrelet: the species most impacted by direct mortality from the *Exxon Valdez* oil spill? *Pac. Seabirds* 21:5–6
161. von Ziegesar O, Miller E, Dahlheim ME. 1994. Impacts on humpback whales in Prince William Sound. See Ref. 97, pp. 173–92
162. The Wall Street Journal. September 12, 1995. Exxon seeks a new trial in Valdez oil- spill case. pp. A16
163. Wells PG, Butler JN, Hughes JS. 1995. Exxon Valdez *Oil Spill: Fate and Effects in Alaskan Waters.* Philadelphia, PA: Am. Soc. for Testing & Materials. 955 pp.
164. Wells PG, Butler JN, Hughes JS. 1995. Introduction, overview, issues. See Ref. 163, pp. 3–38
165. Westermeyer WE. 1991. Oil spill response capabilities in the United States. *Environ. Sci. Technol.* 25:196–200
166. Wheelwright J. 1994. *Degrees of Disaster: Prince William Sound: How* Nature *Reels and Rebounds.* New York: Simon & Schuster. 348 pp.
167. White CM, Ritchie RJ, Cooper BA. 1995. Density and productivity of bald eagles in Prince William Sound, Alaska, after the *Exxon Valdez* oil spill. See Ref. 163, pp. 762–79
168. Wiedmer M, Fink MJ, Stegeman JJ. 1993. Cytochrome P450 induction and histopathology in pre-emergent pink salmon from oiled streams in Prince William Sound, Alaska. See Ref. 153, pp. 104–07 (Abstr.)
169. Willette M. 1993. Impacts of the *Exxon Valdez* oil spill on the migration, growth, and survival of juvenile pink salmon in Prince William Sound. See Ref. 153, pp. 112–14 (Abstr.)
170. Willette TM, Carpenter G, Shields P, Carlson SR. 1994. Early marine salmon injury assessment in Prince William Sound. Exxon Valdez *Oil Spill State/Federal Natural Resource Damage Assessment Fish/Shellfish Study Number 4A. Final Report.* Alaska Dept. Fish & Game, Commercial Fisheries Manage., and Dev. Div., Cordova, Alaska
171. Wohl K, Denlinger L. 1990. *Assessment of injury to waterbirds from the* Exxon Valdez *oil spill: beached bird surveys in Prince William Sound and the Gulf of Alaska. Bird Study Number 1,* Anchorage, AK: US Fish & Wildl. Serv.
172. Wolfe DA, Hameedi MJ, Galt JA, Watabayashi G, Short J, O'Claire C, et al. 1994. The fate of the oil spilled from the *Exxon Valdez*. *Environ. Sci. Technol.* 28:561–68
173. Zimmerman ST, Gorbics CS, Lowry LF. 1994. Response activities. See Ref. 97, pp. 23–45

Annu. Rev. Ecol. Syst. 1996. 27:237–77

EVOLUTIONARY SIGNIFICANCE OF LOCAL GENETIC DIFFERENTIATION IN PLANTS

Yan B. Linhart and Michael C. Grant
Department of Environmental, Population, and Organismic Biology, University of Colorado, Boulder, Colorado 80308-0334

KEY WORDS: selection, genetic variation, microdifferentiation, plants, evolutionary dynamics

ABSTRACT

The study of natural plant populations has provided some of the strongest and most convincing cases of the operation of natural selection currently known, partly because of amenability to reciprocal transplant experiments, common garden work, and long-term in situ manipulation. Genetic differentiation among plant populations over small scales (a few cm to a few hundred cm) has been documented and is reviewed here, in herbaceous annuals and perennials, woody perennials, aquatics, terrestrials, narrow endemics, and widely distributed species. Character differentiation has been documented for most important features of plant structure and function. Examples are known for seed characters, leaf traits, phenology, physiological and biochemical activities, heavy metal tolerance, herbicide resistance, parasite resistance, competitive ability, organellar characters, breeding systems, and life history. Among the forces that have shaped these patterns of differentiation are toxic soils, fertilizers, mowing and grazing, soil moisture, temperature, light intensity, pollinating vectors, parasitism, gene flow, and natural dynamics. The breadth and depth of the evidence reviewed here strongly support the idea that natural selection is the principal force shaping genetic architecture in natural plant populations; that view needs to be more widely appreciated than it is at present.

INTRODUCTION

Genetic patterns in natural plant populations have fascinated evolutionary biologists and ecologists for two hundred years and continue to do so today. Reasons for this interest include the economic prominence of plants in agriculture

0066-4162/96/1120-0237$08.00

and forestry, their biological importance as the base of nearly all food chains, their significance in restoration programs, and their experimental tractability. The fundamental biology of plants—and we interpret this old-fashioned term broadly to include eukaryotic algae, ferns, bryophytes, and all seed plants—leads directly to particular and predictable patterns of genetic architecture. The principal "consequences of being a plant" (25, 150) that bear on the subject of this review are: 1. their sessile growth form, which generates high dependence upon extrinsic, small-scale patterns of soil chemistry, water availability, and light conditions; 2. their basal position in food chains, which subjects them to vigorous attack as a resource base; 3. their propensity to show considerable chromosomal flexibility, including both polyploid and aneuploid variation; 4. flexible breeding systems that include facultative sexual and asexual forms such as the ability to reproduce clonally; 5. the ability to hybridize both subspecifically and interspecifically; and 6. remarkably variable levels of phenotypic plasticity.

The literature on large-scale geographic variation is vast and continues to accumulate. Given the general appreciation for variation at the scale of the whole species, we choose not to revisit the topic in detail. However, we do refer to some of the best-studied early examples of geographic variation, because these studies alerted biologists to the association between the heterogeneity in physical landscapes and the matching genetic variability in plants distributed over those landscapes. Such findings spurred experimental analyses at gradually smaller and smaller scales, partly to ascertain the minimum distances at which one could detect the often conflicting interactions between selection and gene flow and partly to determine the finest temporal and spatial scales at which evolution operates.

This essay seeks to describe and interpret patterns of variation and differentiation at small scales both spatially (one to a few hectares, less than 1 m to 1000 m) and temporally (1 to 20 generations). We provide compelling illustrations of natural selection acting at small temporal and spatial scales. We also discuss the roles of gene flow and neutral dynamics, which include all behaviors of traits and loci that operate independently of natural selection. Given the extensive literature in these areas, we do not claim comprehensiveness in citations; we hope to be reasonably inclusive of the ideas.

In what follows, we emphasize repeatedly the interplay between the ecological setting and the genetic characteristics of the species in question. Specifically, we develop the following topics:

1. Environmental heterogeneity causally generates genetic heterogeneity.

2. Natural selection is of primary importance in shaping gene pools, often at

scales of 10–100 m and for all types of characters but especially polygenic ones.

3. Gene flow strongly influences patterns of genetic differentiation.

4. The dynamics of neutral loci, including isolation by distance, genetic drift, and neutral mutation, may best explain certain natural genetic patterns. We conjecture that most of these explanations are accepted in deference to the default position (null hypothesis) rather than as a result of positive demonstration that neutral forces actually dominate selective ones.

DESCRIPTIONS OF PATTERNS IN NATURE

The biological properties and economic importance of plants were so vital to human welfare that careful studies of intraspecific variation in morphology and development were initiated over 200 years ago. A thorough review of this early work was made by Langlet (141).

The sessile nature of most plants allowed some of the earliest and tidiest visual demonstrations of the hereditary basis of intraspecific differentiation in morphology and size. These studies centered on comparisons of plants from diverse environments grown in common gardens. The sampling scale was often continent-wide. The motivations for the research usually included a practical economic component. For example, in the late eighteenth century, the French navy was concerned about having adequate supplies of timber for its ships. The inspector-general of the navy, Duhamel de Monceau, was a botanist. He established common gardens of *Pinus sylvestris* from Scotland, Russia, Scandinavia, the Baltic, and Central Europe. Additionally, the agricultural importance of various cereal crops led to large-scale agronomic transfer of seeds which, in turn, allowed observant botanists, including Linnaeus, to recognize the rapid phenological adaptation of these annuals to their new environments. These studies of differentiation in both space and time were done well before Darwin's time; he used the results to support his argument about the interactions between inherent variation and natural selection (50). Nearly 100 years ago, Bonnier (43) and other botanists recognized the experimental advantages derived from the facts that plants can be moved readily, subdivided by experimenters, and then grown under highly controlled conditions.

Agronomists and foresters such as Vavilov, Nilsson-Ehle, Cieslar, Langlet, and others were at the forefront of large-scale studies in the early twentieth century. Reviews of this early work are available in Bennett (18), Heslop-Harrison (108), and Langlet (18, 108, 141). Recent reviews of intraspecific variation focus on the interplay of ecological settings, life history, and the evolutionary implications of these complexities in natural plant populations (22, 25, 31, 57, 78, 79, 120).

The best classical studies of geographic variation in plants were initiated by Clausen, Keck, Hiesey, and their colleagues. They stand out because they produced the most thorough analyses of morphological variation at that time. The taxa studied included *Achillea lanulosa, Potentilla glandulosa,* and several species of *Viola* and *Mimulus* (46, 47). Later, these studies were expanded to include physiological characters (19–21, 114).

These analyses demonstrated that differentiation can encompass significant segments of the genome, because it influences many characteristics associated in some way with fitness. For example, in *Potentilla glandulosa,* the following characters (and estimated numbers of loci involved) showed significant differentiation between populations from different localities: winter dormancy (3); seed weight (6); seed color (4); petal length (4), width (2), and color (1); pubescence (5); anthocyanin (4); flowering time (many); stem length (many); leaf length (many) (46). In addition, physiologically relevant suites of characters (water uptake, transpiration, and photosynthesis) were also shown to be quantitatively inherited (19–21, 114). Recent extensions of these studies testify to the enduring importance of this research (96).

Both continuous and discontinuous character gradients have been documented for single-gene polymorphisms and for quantitatively varying characters. Nearly all of the many dozens of species studied are quite variable and extensively differentiated on a large geographic scale, whether it be for morphological (17, 28, 46, 89, 141, 259), physiological (19–21, 53, 191, 193), isozymic (97, 102, 113, 155, 165, 168, 208, 283), or nucleic acid (118, 243) traits. The structuring of this variation at small scales is a principal theme of what follows.

Gradually, analyses of phenotypic differentiation came to be carried out at smaller and smaller scales, and the results led to the recognition that genetic differentiation could be observed over distances of even a few meters and, whenever localized selection was sufficiently intense, even a few centimeters (8). As documented below, such differentiation is commonly induced either by physical or biotic features of the environment or combinations of these.

In early studies of geographic variation, one major focus of the analyses was to determine whether or not the variation showed a continuous clinal pattern or a discontinuous ecotypic one. The same results were sometimes interpreted as clinal or ecotypic by different groups of researchers. This dichotomy of views reflected the fact that, for taxonomically inclined botanists operating within the confining structures of formal classification, the categorical nature of ecotype designation was far easier to interpret and deal with than were clines. Now, detailed analyses of variation show that, within the same species, some characters can vary gradually, others discontinuously, depending on, for example, gene flow, intensity of selection, number of genes involved, and terrain configuration. The cline versus ecotype controversy has not proved particularly useful and it has mostly faded; because of its varied usage, some argue the term "ecotype" is so imprecise that it should be discarded (141, 212).

Empirical microscale differentiation studies rely on the workhorses of population genetic studies: isozymes and morphology (196, 210, 211). New characters such as rflp traits (284), DNA fingerprinting (4, 174), and analysis of non-nuclear genotypes (179) have strengthened and broadened the data base. Some of these studies show high levels of concordance among types of traits studied, and some show considerable discordance (115, 187). It is still too early to generalize about those cases where there are disagreements between different types of characters, but as Hillis clearly pointed out (115), evolutionary systematists have always been faced with discordance among data sets even if, for example, two or more data sets included only morphological characters. On the other hand, there are good reasons to predict that neutral versus selectively responsive characters should often show different patterns; in addition, single gene characters are likely to behave quite differently from polygenic traits. We anticipate much progress in synthesis of these perspectives over the next few years.

Environmental Heterogeneity Generates Genetic Heterogeneity

The empirical association between environmental heterogeneity—biotic and abiotic—and genetic heterogeneity is so strong and so well documented (7, 106) that it seems almost unnecessary to emphasize it once again in this review. Routinely and consistently, when investigations of natural populations have been conducted, there tends to be a strong correlation between significant environmental heterogeneity and genetic heterogeneity. This topic has also been investigated several times theoretically. A recent review concludes that the empirical positive association accords with theory (106). We note, however, that association or correlation does not, in general, provide adequate demonstration of a cause and effect relationship. And so we pose the question, "Is there such a relationship?" We believe the answer is an unequivocal "yes." We reach this conclusion largely because of the sheer preponderance of the association in all types of environments across all types of organisms and for all types of characteristics. These characteristics span many traits of plant anatomy, morphology, function, development, chromosomal makeup, life history, and biochemistry.

We also believe we have a reasonably good understanding of the underlying mechanisms that produce the association. First and most importantly, different environments generate different selection pressures, and these, in turn, lead to genetic heterogeneity. Secondly, environmental heterogeneity such as elevation, exposure, and moisture availability quite often generates significant barriers to gene flow, e.g. via effects on phenology, and thus, it enhances genetic differentiation among semi-isolated or isolated populations. These two forces work synergistically, producing genetic heterogeneity in natural populations; this is one of the strongest generalizations currently known about differentiation at either large or small temporal and spatial scales in plants. Secondary or chance associations can also be produced by genetic drift and isolation by distance (39, 110, 111), but we consider these forces to be much less important in producing the closely matching associations between environmental and genetic heterogeneity across small spatial and temporal scales.

Exceptions to this general pattern of extensive differentiation do exist. They tend to occur in species that (*a*) have undergone special historical events, (*b*) show little genetic variability, or (*c*) live in aquatic environments. It is also true that plants with very high levels of phenotypic plasticity often show low levels of genetic variability, a point we cover in the discussion below. One of the most striking exceptions to the general pattern of differentiation is that of *Bromus tectorum* (199), which is now a widespread weed colonizing western US grasslands extensively. Despite the considerable range of this taxon on its newly colonized continent and the heterogeneity of local environments in this region, *B. tectorum* shows a remarkably low level of genetic heterogeneity. Similarly, the evolution of heavy metal tolerance has been known to be a powerful selective force (9) for several decades, yet Tonsor (250) was unable to find any evidence of genetic differentiation for lead tolerance in a small-scale study of *Plantago*, although the situation was one in which such differentiation could have been reasonably expected. A second important exception to the general pattern of positive association can be seen in the extremely low levels of genetic variation in some forest tree species, including *Pinus resinosa* (76) and *P. torreyana* (147). These particular exceptions stand out even more when considered against the backdrop of high levels of genetic diversity generally found in forest trees (100). Although we understand some of the circumstances that may lead to a lack of genetic differentiation, Barret & Kohn (15) argue persuasively that we still do not understand the impact of various historical influences.

The third situation in which surprisingly low levels of genetic variability have been documented is among aquatic taxa. They tend to have very low levels of differentiation which are typically detected only at large geographic scales, for example, in comparisons between Baltic Sea and Mediterranean populations or between extremely different habitats such as fresh versus brackish water. In a recent review, Triest (252) suggests that this may partly reflect the generally more homogenous substrates in water and wetlands as compared to strictly terrestrial habitats.

Selection Shapes Gene Pools

This section deals with selection-mediated genetic differentiation on small scales: tens to hundreds of meters. The section is organized to emphasize the various features of the physical and biotic environments known to generate differentiation via selection. These studies have dealt either with herbaceous species, whose populations can show genetic heterogeneity on a scale of tens of meters or less, and with larger woody species, which often show such heterogeneity on a scale of 100–300 m.

In his impressive review of natural selection, Endler (66) specifies some important criteria directly relevant to the subject of this review. In particular, he points out that selection can operate if specific traits, demonstrably under at least partial genetic control, are associated with fitness differences. Then, he argues, selection does operate if trait frequencies or distributions vary between parent and offspring or among age or life history classes, provided such differences go beyond those expected from ontogeny. Empirical studies that aim to demonstrate the operation of natural selection should (*a*) identify which traits are amenable to selection and the corresponding environmental factors that may affect them, (*b*) demonstrate that selection directly affects those traits, and (*c*) identify which environmental factors are actually causal, selectively. He also cautions that those studies are inadequate that rely exclusively on the presence of correlations between the frequencies of certain traits and certain environmental variables or on comparisons of variation patterns in specific traits as a function of specific habitats, in either related or unrelated species.

We consider that the studies reviewed herein adequately demonstrate natural selection when viewed as a large, collective body of work. One reason is that many utilize field observations coupled to experimental manipulation, e.g. reciprocal transplantation of either plants or seeds between contrasting environments. Furthermore, other studies demonstrate that heritable demographic parameters vary in predictable frequencies in the presence of selective agents (e.g. moisture gradients, parasites) but not in their absence. In addition to demonstrating natural selection, selection coefficients can be calculated in some experimental designs. Reviews of these studies indicate that selection coefficients in the range of 0.1 to 0.5 are common (66, 127, 151, 157, 237) (Table 1). However, such high values can be sustained only over the short term for any particular character and population.

Physical and biotic components of the environment often differ from each other in the patterns of differentiation they generate. Physical features (e.g. moisture, soil conditions, exposure) are typically contiguous in space and time.

They may vary either as smooth gradients or abruptly. They then produce corresponding patterns of differentiation between adjacent populations or generations. In contrast, biotic components typically vary much more dynamically, because the elements providing the selection, i.e. competitors, herbivores, and parasites, can move about within a given area. Plant competitors or parasites can move about via seed or spores from one generation to the next and, equally importantly, the biotic factors themselves may undergo evolutionary change, population flushes or crashes, etc, each of which can impact the plant population structure of interest. Consequently, microscale differentiation patterns generated and maintained by biotic forces tend to be less sharply defined than those associated with changes in the physical environment. Biotic components can and do produce differentiation within populations but typically in a mosaic, unsegregated fashion (79, 104, 163, 263). Thus, individuals with higher fitness, e.g. those that are better competitors or more parasite resistant, grow intermixed with individuals of lower fitness.

AGENTS OF SELECTION IN SPACE The analysis of spatial differentiation on a microscale occupies the largest and most important niche in the study of natural selection in plants.

Toxic Soils Mining activities have produced refuse consisting of soils mixed with high concentrations of toxic metals. These mine tailings are eventually colonized by plants, which usually evolve the ability to tolerate the toxic effects of the metals. Studies of the evolution of such metal tolerance in plants are among the most detailed and elegant in evolutionary biology. Some of the most important results are as follows.

Selection can be very intense. Selection coefficients (s) > 0.3 have been documented leading to the evolution of metal tolerant "races" within tens to hundreds of years (25, 123, 127). Because selection can produce sharply defined differentiation where the boundaries of mine tailings are abrupt, plants on either side of such a boundary, only 1 or 2 m apart, can differ markedly in their metal tolerance. In the smallest spatial scale studies in natural populations, several workers have demonstrated genetic differentiation with respect to metal tolerance over distances of only a few centimeters (3, 8).

Evolution of metal tolerance has occurred in a wide variety of plant species with many different life histories, pollination systems, and lifespans. Most are herbaceous, but such tolerance has evolved in at least one tree (*Betula*) (29). Herbaceous species include *Aegilops peregrina* (196), *Armeria maritima* (148), *Agrostis tenuis* (182), *Anthoxanthum odoratum* (7, 207), *Agrostis stolonifera* (280), *Arrhenatherum elatius* (60), *Silene cucubalus* (262), *Mimulus guttatus* (5), and the legume *Lotus purshianus* and its symbiont *Rhizobium loti* (281).

Several mosses have also been studied and differentiation in metal tolerance demonstrated (127). However, the speed and strength of such local differentiation appears to be less than that for angiosperms (230, 231). Soils also become toxic along roadsides, and evolutionary changes have also occurred in plant populations exposed either to lead from gasoline (279) or to deicing salts (132). We also know of one study that demonstrated metal tolerance in an alga *Ectocarpus siliculosus* (219).

Adaptation to heavy metals evolves at a physiological cost. Metal-tolerant individuals, when grown in metal-free soils, are generally competitively inferior to nontolerant individuals from adjacent, nonmetalliferous soils (26, 112). The same pattern has been documented in plants that are tolerant of serpentine soils (127, 137, 196, 231). This pattern seems to be a generally applicable rule.

Whenever edaphic conditions are extreme in terms of pH, mineral contents, or other features, they generate selection pressures of the sort documented in the context of human-modified soils discussed above. In the western United States, the biology of closely adjacent plant populations that span dramatically different soil chemistries has been the subject of much attention for several decades (136, 138). The presence of serpentine or ultramafic outcrops, soils generally characterized by high levels of magnesium and low levels of calcium, commonly produce strongly differentiated populations (137, 138, 176). The forces producing this differentiation in plants of these regions produce an entire syndrome of effects including specialized morphology (138), physiology (209), development, etc (125).

Kruckeberg (138) reports the regular and predictable existence of intraspecific, genetically based differentiation in resistance to ultramafic soils across many plant taxa that include pines (80), herbaceous dicots, and grass species (173). Nickel tolerance, for example, either via hyperaccumulation or exclusion, commonly occurs as one component of the serpentine syndrome in a taxonomically broad range of plants (138). Growth rates of serpentine plants in ultramafic soils are typically greater than those of their nonultramafic counterparts, a difference apparently related to calcium and magnesium balances. This response appears to be a widespread, common pattern in many phylogenetically unrelated, serpentine taxa. These syndromes are highly reminiscent of the examples of heavy metal tolerance.

These unusual soils are especially interesting, because they form sites for potential chemical endemics. For example, there are serpentine-associated endemic species in Zimbabwe, New Caledonia (where 2 families, over 30 genera, and 900 species are restricted to serpentine outcrops), Yugoslavia, and California. Edaphic endemics also exist on other unusual soil formations. *Astragalus phoenix* is restricted to calcareous alkaline soils in Nevada; *Hudsonia montana* to quartzite ledges in Burke County, NC (139). These endemics illustrate one major reason why the study of small-scale genetic differentiation can be important: It provides strong evidence about the formation of new species via selection. Kruckeberg (139) argues persuasively that the acquisition of an ultramafic syndrome by certain plant populations has been of such force as to form the initial basis of differentiation from which speciation has taken place in the genera *Ceanothus, Clarkia, Gilia, Phacelia*, and *Streptanthus*.

The repeated similarities of biologically important functional properties found in serpentine plants appear to constitute overwhelming evidence that such patterns can be adequately explained only by natural selection. Drift and isolation by distance are excluded because of the repeatability of the patterns in many different locations around the world; shared common ancestry can be excluded given the low levels of phylogenetic relatedness of the various taxa in which the syndrome has been identified.

Less extreme variation in soil conditions occurs more commonly in nature than do serpentine outcrops, but they too lead to documented differentiation. In the Swiss Alps, variation in frequencies of acyanogenic and cyanogenic morphs of *Lotus alpinus* (257) and *Ranunculus montanus* (56) was compared in adjacent populations occupying soils. These were characterized either as carbonate soils, which have unfavorable water regimes and nitrogen in nitrate form, or as acidic silicate soils, which have relatively well-balanced water regimes and nitrogen mostly in the form of ammonia. In both species studied, cyanogenic morphs occur with high frequencies in carbonate soils, but at much lower frequencies in acidic silicate soils. Cyanogenesis has been recognized as an important antiherbivore defense in several species (57), but it may also be important in the nitrogen economy of plants. Specifically, cyanogenic glucosides may serve as nitrogen storage structures, accumulated in times favorable for rapid nitrogen cycling, to be used later under more stringent nitrogen conditions. For this reason, cyanogenic plants may be favored under the physiological conditions associated with carbonate soils (257), which tend to be nitrogen deficient. In some species, as above, the differentiation observed can be associated with specific physiological features of the plants; other studies have detected differences in subpopulations on different soil types at the allozyme level (194, 196).

Soil moisture, nutrients, and pH can also affect distribution of resident microorganisms such as mycorrhizal fungi. As a result, plant populations spanning gradients of soil conditions may also be associated with different communities of soil microorganisms; the detailed consequences of such a scenario have apparently not been adequately studied yet but certainly deserve to be (34, 41).

Fertilizers The regular application of fertilizers over a number of years changes soil properties and therefore plant growing conditions. Many of the same conclusions obtained from studies of metal tolerance and its evolution are applicable in these cases. Once again, tolerance to or dependence upon high fertilizer levels evolved in less than 100 years. Most of these studies have been done at the Rothamsted Experimental Station, in plots that were originally (in 1856) set up to study the effects of various fertilizers. However, an interesting complexity was added indirectly: Fertilized plots produced a different biotic environment, e.g. taller vegetation. As a result, selection patterns were also changed and, consequently, there were specific evolutionary responses to these changes. More recent small-scale experiments also suggest that evolution can occur very rapidly (even after only 1 or 2 generations) because specific fertilizer regimes on certain plots in 1965 were associated with heritable changes by 1972 (237, 238).

In this context, we point out the fascinating but still understudied phenomenon by which apparently permanent (at least relatively long-term) changes in plant genomes have been brought about by the use of high levels of soil fertilizers. This has been documented in *Linum* and *Nicotiana* (61) and certainly deserves more focused, detailed study, given the intriguing results—genomic changes induced by a wide variety of environmental stresses (140, 180).

Herbicides Just as antibiotics, insecticides, and rodenticides have been effective agents of selection; herbicides too have led to the evolution of plants with enhanced ability to tolerate these toxins. As of 1991, over 80 species had shown evolution of herbicide resistance (144). In addition, other groups, e.g. the alga *Chlamydomonas reinhardi,* also show these types of responses. Most of these taxa are typical weedy plants such as *Convolvulus arvensis, Tripleurospermum inodorum, Daucus carota, Alopecurus myosuroides, Echinocloa crusgalli, Senecio vulgaris* and *Amaranthus retroflexus* (43), but they include shrubs and trees such as *Prunus persica*, *Malus silvestris*, *Citrus sinensis*, and *Quercus marylandica* (145). Resistance has been found to most important herbicides, including 2,4-di-nitro-toluene, triazines, carbamates, benzoic acids, and amides. Resistant populations of many species are posing serious problems in many regions, particularly in North America and Europe (43, 116, 117, 144, 145).

Lawns and grazed areas Repeated mowing, clipping, and grazing obviously produce strong directional selection: Anything that stands tall is cut down. For this reason, lawns and other public grasslands, grazed meadows, and pastures have attracted students of morphological differentiation in human-dominated habitats. The evolution of prostrate life forms that can flower and set seed very close to the ground has been documented in a number of British lawn weeds and other species including *Poa annua, Bellis perennis, Plantago lanceolata,*

P. major, and *Prunella vulgaris* (28, 264). However, similar comparisons with *Achillea millefolium,* which included reciprocal transplanting, indicated no evidence of genetically based differentiation (265).

Maritime exposure Plants growing on cliffs, dunes, and other seaside habitats are exposed to such extremes of light, wind, salt deposition, wave action, and other disturbance that they can be expected to show adaptation to these conditions. They were studied in some of the earliest investigations of intraspecific differentiation. Turesson (253) developed his concept of the ecotype with careful demonstrations of the constancy of morphological differences among plants from seaside and inland habitats. Gregor was the first to take this approach to a fine scale (94), and the first to demonstrate genetic differentiation between adjacent populations of *Plantago maritima.*

Moisture, temperature, and elevation These factors often vary in concert. Consequently, studies of differentiation that have been done along such habitat gradients consider them jointly; in these cases, the effects of single factors cannot be isolated. For example, genetic differentiation has been demonstrated in frequencies of allozyme alleles between spire-shaped trees within closed-canopy forests and their conspecifics growing nearby as krummholtz in both *Abies lasiocarpa* and *Picea engelmanii* (90). Differences between these habitats include temperature extremes, snow accumulation, insolation, competition and herbivory on seedlings, wind, and probably soil factors such as fertility and mycorrhizae. We cannot determine which of these factors is most important in producing the genetic differences observed. Nonetheless, we can at least conclude that there are significant genetic differences associated with growth morphology, and detectable at the scale of 100 m or less.

On a similar scale, there is significant genetic differentiation in *Pinus ponderosa* occupying north- and south-facing slopes in the Colorado Rocky Mountains. This differentiation is observed at a peroxidase locus that has two common alleles. One allele is most common at higher, moister, cooler elevations, the other at lower, drier, warmer elevations. The cool environment allele also occurs at significantly higher frequencies on north-facing slopes, whereas the warm environment allele is most common on south-facing slopes of comparable elevations (189). In the same setting, differentiation has also been observed in *Pseudotsuga menziesii,* where roots of plants from south-facing slopes show greater genetically based growth rates than those from adjacent north-facing slopes with more plentiful soil moisture (107). Similar patterns occur in other forest trees (13, 14, 95, 232).

In herbaceous species, the differentiation observed typically occurs on a smaller scale, probably because of the smaller size of the plants. *Avena barbata* is an introduced annual plant in California that shows significant genetic differentiation both in allozyme patterns and morphology, between cool, mesic northern California and the hot, xeric southern parts of the state. When analyzed on the scale of a single hillside, genetic differentiation was also demonstrated between locations 5–50 m apart, consistent with the large-scale, state-wide analyses. Genotypes and allele frequencies characteristic of mesic regions were also most common in the mesic sections of the hillside bottom. Conversely, the genotypes and allele frequencies characteristic of drier southern California were also found in the xeric hilltop (98, 101). Note that in both *A. barbata* and *P. ponderosa,* large-scale patterns and small-scale patterns are consistent with each other. This provides strong inferential evidence that selection produced the differentiation observed. Hypotheses of isolation by distance, genetic drift, and gene flow are all falsified in these particular examples because none predict these patterns of similarity across the different spatial scales just described.

Small-scale gradients also occur on the sides of small depressions that fill with water; as the water evaporates and recedes, microhabitat zones appear that vary from one another in many characteristics including moisture availability, soil pH, temperatures, soil aeration, and vegetation composition. Differentiation associated with environmental heterogeneity occurs in several species of annual dicots along these 5 to 10 m wide microgradients (156). *Veronica peregrina* has been studied in the most detail. Differentiation was significant between plants occupying the drier periphery (where *Veronica* are few, intraspecific competition is minimal, taller grasses are common, and interspecific competition is severe) and plants occupying the wet center (where *Veronica* are dense, and competition is primarily intraspecific). Genetic differentiation was demonstrated in electrophoretic variability (130) and in response to water-logging, phenology, reproductive output, plant size, seed size and seed number (157).

Light intensity Shade is often provided by adjacent plants, so it is, in part, a biotic effect with which plants must cope, but its primary impact is the physical reduction of available light. Once again, evidence suggests that if adjacent habitats are characterized by predictable conditions that create different selective pressures, plant populations will show evolutionary divergence. For example in *Plantago lanceolata,* there occur genetically differentiated sun and shade populations. These populations differ from each other in photosynthetic rates (245a). More recent work on tropical plants has shown differential abilities to respond to sunflecks over tiny spatial and temporal scales (44).

Competition Plants compete with one another for light, water, nutrients, space, pollinators, and other necessities. Therefore, competition entails many different kinds of competitive interactions (77). Because of interspecific differences in morphology, development, physiological requirements, and tolerances, intraspecific and interspecific competition are likely to differ. Interspecific competition is also likely to be species-specific: Competing against a shallow-rooted but shade-producing neighbor will be very different from competing against a deep-rooted, rosette-producing one. Unfortunately, most studies of competition have not dealt in any detail with the specifics of such differences, although competition-induced selection has been shown in *Veronica peregrina.* Subpopulations exposed to either intra- or interspecific competition showed adaptive differentiation in seed size, timing of germination, growth rate, branching patterns, and overall plant size as a result (157). Different kinds of interspecific competition have also been compared in *Trifolium repens*, which in competition with *Lolium, Agrostis, Phleum*, and *Dactylis* showed differentiation in response to long-term associations with these species (255). Originally, these results were interpreted as evidence of the selective roles of competition between *Trifolium* and the four grasses. However, it now appears that at least some of this differentiation reflects the consequences of specific underground interactions between *Rhizobium*, the bacterial symbiont of *Trifolium*, and the grass species (41).

Pollinating vectors Some evidence exists that adjacent populations of a given species can differ in various floral features that are associated with different modes of pollination. For example, *Polemonium viscosum* populations vary in floral odors: sweet versus stinky flowered populations are typically attended by bees and flies, respectively (82). *Plantago lanceolata* populations in windy habitats have dry, wind-dispersed pollen, whereas nearby populations in sheltered meadows had adhesive pollen dispersed primarily by syrphid flies (241a). The scale at which this differentiation originally occurred is unknown but in both species current populations do live near one another and the distinction appears to be maintained. At least in the case of *Plantago*, this differentiation can be interpreted in the context of relevant pollinating vectors.

Herbivory, predation, and parasitism Interactions between plants and their herbivores and parasites can lead to two broad classes of selection scenarios. One consists of two or more different species of plants that, themselves, are substrates to one species of parasite. Several studies of animal parasites have documented the existence of genetic differentiation between parasite races of specific plant hosts. The best known are those of *Rhagoletis* fruit flies that have differentiated along host systematic lines with apple and hawthorn specialists, showing significant divergence that appears to be leading to sympatric speciation (36). Studies with plant parasites are fewer but consistent with the overall pattern. There is evidence of genetic divergence along the lines of host resistance in the mistletoes *Viscum album* (201), *Phoradendron* (48), and *Arceuthobium* (198).

The more commonly studied pattern entails one host plant species interacting with one or more herbivores and parasites. In some cases, adjacent populations are either exposed to or not exposed to specific herbivores or parasites, and under these conditions there can be spatially identifiable patterns of genetic differentiation in response to fungi (202), mollusc herbivory (57), or insect parasitism (185, 240a). The important roles of fungi and arthropods in shaping evolutionary change in plants have also been demonstrated, usually in studies involving interactions between a specific plant species and a single fungus or insect species (35, 53, 79).

Most herbivores and parasites seldom have spatial distributions with sharp boundaries. For this reason, differentiation also seldom involves adjacent, contiguous plant populations, but rather it forms a mosaic of differentiated individuals within populations. In the presence of a disease organism or parasite, some individuals are susceptible while others can be either tolerant or resistant. These two alternatives sometimes involve different sets of plant characteristics and therefore tradeoffs that complicate the evolution of resistance (75).

Scenarios get especially complex and interesting when two or more parasites or herbivores operate within the same plant populations. For example, in the juvenile part of its life, a plant may be chewed by nematodes, rasped by mollusks, clipped by ants, invaded by fungal mycelia, colonized by bacteria, or it may play host to viral infection. With the onset of maturity, a plant can, in addition, be blighted, blasted, drilled, defoliated, grazed, shredded, sucked, skeletonized, or uprooted. If it reaches sexual maturity, its flowers, fruits, and seeds may be robbed, eaten, parasitized, or otherwise harmed. These are some of the biological consequences of being a plant and of having a basal position in the food-chain within the community in which they grow. In this position, plants often serve as hosts simultaneously or sequentially to a wide variety of disease organisms, parasites, herbivores, commensals, symbionts, and organisms with other life-styles. These species in the aggregate are referred to as dependent species. This term is useful because it includes both the species with negative impacts upon plants (the "pests" of agronomists and forest managers) and those with beneficial roles (such as mycorrhizae) (159). The number of dependent species that can be associated with one species of host plant can be in the dozens to hundreds (159) and include viruses, bacteria, fungi, insects, molluscs, birds, mammals, and parasitic angiosperms. Not all these dependent species will be active simultaneously, nor is it likely that every dependent species employs a unique set of criteria when settling upon a suitable host. Nevertheless, variation in the direction of selection can be expected when: 1. various dependent species differ in their preferences for individual hosts, and 2. the variable host characters associated with differential damage have a demonstrable genetic basis. For example, the brown alga *Dictyota dichotoma* contains diterpene alcohols that simultaneously deter feeding by the fish *Lagodon rhomboides* and stimulate feeding by the polychaete *Platynereis dumerilio* (104, 105). In *Trifolium repens,* the presence of cyanogenic glycosides deters feeding by various molluscs and vertebrates so that cyanogenic morphs are usually protected from such herbivory; simultaneously, however, they are more susceptible to parasitism by the fungus *Uromyces trifolia* (57).

Patterns of species-specific host selection are not always associated with a single compound. In *Pinus ponderosa,* the relative amounts of a whole suite of compounds in xylem and phloem determine whether a tree will be attacked by the beetle *Dendroctonus ponderosae*, the dwarf-mistletoe *Arceuthobium vaginatum*, or the squirrel *Sciurus aberti* within stands where all these species coexist. Trees are seldom attacked by all three (158, 163). Other well-documented examples support the existence of this pattern. For example, in wild *Brassica oleoracea*, differential herbivory by an array of herbivores has led to divergent selection that maintains a series of polymorphisms at four loci coding for defensive compounds (186). Most of these examples involve comparisons among host individuals within local populations, providing the opportunity for intrapopulation host selection and differentiation. The groups for which species-specific host selection has been documented include algae, herbaceous plants, and forest trees. In many cases, preference by dependent species, or susceptibility by a host, has not been related to precise features of the host phenotype (159, 234).

One potential consequence of multispecies challenges to a host plant is diversifying selection (66, 85). In its simplest formulation, diversifying selection will maintain two alleles at one locus if allele A1 confers some higher fitness than allele A2 in environment E1, whereas allele A2 confers higher fitness in E2 (52, 149). In the case of traits controlled by several loci, diversifying selection can operate when individual hosts with different phenotypes are favored under different conditions (40, 105, 162, 234, 242). In contrast to diversifying selection, stabilizing selection can also occur if different dependent species differentially attack individuals at the extremes of a continuously distributed trait. Directional selection can occur if two dependent species favor similar or positively correlated phenotypes, or no selection may be detected. These scenarios are detailed in Simms & Rauscher (235) and Phillips and Arnold (206). At this time too few studies of multi-species parasitism have been conducted to indicate which of these alternatives are to be expected most often.

Sometimes different dependent species select the same plant hosts to feed upon. These exceptions to differential host preferences appear to fall into one of two categories. The first category consists of situations where individual plants attacked by one species are weakened and therefore made prone to subsequent attacks by the same or other species (24, 53, 190). The second class of exceptions consists of plants attacked by several dependent species that are members of the same feeding guild and, in some but not all cases (172), are closely related. For example, two species of *Phyllotetra* show positive correlations in their distributions on individual *Brassica* hosts (129). The same is true for two species of *Pemphigus* gall aphids on *Populus* hosts (271). In these cases, it may be argued that certain groups of closely related species have similar physiologies or behaviors and thus similar desiderata. These exceptions do not disprove the existence of species-specific host selection but predict the conditions under which such selection can operate.

SELECTION OVER TIME A temporal component is relevant to microscale differentiation in two contexts. One is to determine rates of evolutionary change. The other is that rapid changes in environmental conditions can expose different cohorts within a population to different selection pressures, thereby generating temporal differentiation in genetic constitution. Certain studies have indeed permitted estimates of rates of evolutionary changes, e.g. analyses of populations residing on mine soils (123) or on fertilized plots created a few dozens or hundreds of years ago (237, 238). Rates of change for individual species can be directly calculated given generation times and the well-known time frames for the environmental changes. Among the most remarkable experimental approaches to this question were those of Harlan & Martini (103) with annual barley (*Hordeum*). In 1925, eleven agricultural varieties were mixed together in equal proportions and sown in 10 locations across the United States. Each year, at each location, seeds were collected for resowing. The identity and proportions of all varieties were known and tallied. The result was that different varieties became predominant in specific localities but were completely eliminated in others. This frequently happened within less than 10 generations. A recent dramatic example of rapid selective change was demonstrated in two species of weedy Compositae whose propagules (achenes plus pappus) have evolved reduced dispersal capacity on a series of small islands they colonized. This change occurred within 8 to 10 years (48a).

Cohort-specific temporal differentiation has been detected primarily in long-lived perennial plants (216). Within populations, individuals of different ages became established in different years, decades, and centuries. Because ecological conditions can change quickly over time, different cohorts can be exposed to different selection regimes at different times. This means that such age-structured populations can consist of groups occupying the same site but having very different genetic constitutions, despite the fact that the younger plants are descended from the older plants on the site and are therefore close relatives of these older plants. The theoretical aspects of this possibility have been explored by Charlesworth (42). Experimental support for this pattern has been obtained in several species. For example, metal-tolerance in parental plants is more pronounced than in their progeny (25, 182); these results demonstrate change in but a single generation. Temporal differentiation may also occur in the context of succession; early arriving colonists experience one set of environmental conditions, and their progeny, though establishing in the same geographic area, are faced by different conditions of light, moisture, competition, herbivory, and other factors, so different genotypes are selected (16, 93, 160, 251, 272). Temporal differences in selection and in genetic reshuffling associated with changes in the mating system can also be stored in seed banks. Such seed banks, in turn, can be stored either in soil (51, 152, 181, 215) or in closed fruits or cones kept on plants for several decades, as is known in genera such as *Pinus, Banksia, Hakea*, etc (83).

Gene Flow and Microdifferentation

The general view of the effectiveness of gene flow and its relevance to the genetic structuring of natural populations has swung across a considerable range of outlooks during the last 30 years. Articulating and strongly advocating the general view of most evolutionists for the previous several decades, Mayr (177) states that the real problem of speciation is not how to produce difference but rather how to escape from the cohesion of the gene complex. Although Mayr was explicitly referring to speciation in animals, the view that the cohesive force of gene flow represented a formidable barrier to genetic divergence certainly applies to plant population differentiation as well. A number of workers, stimulated by Ehrlich & Raven (63), challenged this view directly, arguing mostly on the basis of data about (*a*) animal movement, (*b*) pollen movement in natural populations, and (*c*) quantitative studies of gene flow measurements in crop plants that the apparent cohesiveness of most species could not be adequately accounted for by gene flow because it was entirely too localized and restricted. Partly as a consequence of this seminal paper, the study of gene flow levels became an item of major interest to evolutionary biologists. Gene flow remains today an area of considerable interest, but we do not attempt to review the large literature in this paper; rather we refer the interested reader to reviews (67, 152, 236), some empirical (62, 290) and some theoretical (1, 11, 37, 71, 73, 200, 223, 236, 256).

The task of empirically measuring actual gene flow in natural populations has been greatly aided by advances in several techniques, including dye marker procedures, development of a deeper understanding of genetic diversity indices and their utility in estimating gene flow, study of the distribution and movement of private alleles, and the application of maximum-likelihood statistical estimation concepts, especially in paternity analysis approaches (1, 33, 39, 183). Despite its relative newness, the literature is rather large, and outcomes show considerable diversity. In general, however, our reading of recent studies on natural plant populations is that gene flow occurs at rates considerably higher than those suggested by Ehrlich & Raven (63), due partly to several forces that have tended to cause serious underestimation of rates (54). Those effects include pollen carryover, multiple drops, reattachment to pollinators, and the difficulty of tracking long-distance movement. Some simulation work, on the other hand, suggests that gene flow calculated strictly on a gene frequency basis may produce an underestimate (72, 73). Gene flow seems especially great in forest trees (99, 100), where gene flow from immigrant pollen may account for half the successful paternities. High levels of gene flow have also been reported even in herbaceous taxa and across rather large geographic distances (33, 65, 152, 248, 278).

Although most plant population studies focus on gene flow via pollen, seed and spore movement are also significant in determining population structuring. In some cases they can be potent mechanisms of gene spread (11, 214, 223, 226, 239). In other cases, highly limited seed movement has been directly responsible for subdivision of populations into family groups (55, 162, 178, 227). Furthermore, direct comparisons between species that have different modes of seed dispersal show different scales of spatial differentiation. Greater potential seed-dispersal distance sometimes correlates positively with lower levels of differentiation (99, 274). On the other hand, we point out a recent elegant analysis of gene flow via colonization by spores that resulted in a considerable enhancement of local heterogeneity (which might easily be interpreted as a result of local selective forces or drift) (204, 214). Localized diversity may also be positively related to long-distance immigration and subsequent hybridization. The effects of pollen and seed dispersal may jointly be antagonistic, synergistic, or independent (135, 162). And in a manner directly contrary to its homogenizing influence, this type of gene flow also sometimes operates as an important source of local heterogeneity in certain restricted but biologically important circumstances. A recent theoretical analysis of long-distance dispersal suggests that if such dispersal into a given area occurs frequently enough, it can lead to the establishment of intermingled isolates that are highly differentiated from each other (197). Such results provide clear evidence of the importance of seed and spore dispersal mechanisms and, surprisingly, may affect differentiation patterns by generating either population divergence or convergence.

Recent comparisons of the relative rates of gene flow via pollen and seed (67) and developments in the prediction of population size (37) are relevant in this context.

Gene flow certainly operates as an important homogenizing force, even when strong selection might be expected (70, 250), and not all studies of genetic architecture produce evidence of local differentiation (6, 184) or even differentiation in a widespread, colonizing taxon (199). Nonetheless, the overwhelming proportion of small-scale genetic differentiation studies (see section above on selection) report important differences in at least some characters despite the presence of gene flow at high levels (278). We interpret the dominant pattern—existence of genetic differentiation—as providing strong evidence of the power of local forces—particularly selection—frequently but not always to override the generally homogenizing effects of gene flow.

Our knowledge about gene flow and its relationship to local genetic variation in nonvascular plants remains quite limited (2, 127, 214, 231, 239, 240, 282), but general conclusions about the levels and effectiveness based on vascular plants are not likely to be substantially different from those reached for seed plants.

Dynamics of Neutral Loci Influence Differentiation Patterns

Population bottlenecks and genetic drift are related mechanisms that also strongly influence the local genetic architecture of natural plant populations. The general expectation has been that bottlenecks tend to produce a drastic reduction in the variability of the gene pool (76, 146, 175), but such phenomena can, under certain circumstances, increase genetic diversity. If, for example, one local population crashes and undergoes a bottleneck constriction but another does not, then the two populations may actually end up more genetically differentiated than before the population crash. In addition, Willis & Orr (276) have shown that dominance can interact with population bottlenecks to generate an increase in heritable variation. The historical influence of bottlenecks can be quite important to contemporaneous genetic structure but difficult to assess empirically (15).

Genetic drift is a commonly postulated force with respect to local microgeographic differentiation (2, 111, 120, 169, 192, 269, 270), although the now widely used methods of estimating effective population sizes seem to have generally reduced the frequency with which this phenonemon is invoked as an explanatory mechanism because effective population size estimates tend to be larger than previously thought. Frequencies of neutral alleles tend to fluctuate in accordance only with mutation rates and population sizes (133)—not according to local conditions of heterogeneity. They will thus generally obscure or smear the matching patterns of genes and environments. Again, we interpret the overwhelming frequency of genetically differentiated characters, typically correlated with differentiation in environmental conditions, as providing strong evidence against neutral forces as the dominant shapers of local gene pools.

DISCUSSION

Plant populations show a remarkable degree of both spatial and temporal genetic differentiation (illustrated in Table 1). It is remarkable because (*a*) the scales at which it has been detected are very small, (*b*) it encompasses a diverse array of traits, singly or in combination, which affect all aspects of the genome and life history, and (*c*) most studies done to date provide strong evidence that this differentiation is driven primarily by selection. These features are discussed below in the form of general commentaries.

Spatial Scale of Differentiation and Its Consequences

Scales of differentiation depend primarily on four principal factors: (*a*) intensity of selection, (*b*) degree and distance of gene flow, (*c*) life history, and (*d*) plant stature. The finest scale of differentiation, 10 m or less, has been found in situations where environments generate strong selection pressures, either anthropogenic such as mine tailings (8, 25, 123, 230) or natural ones such as the sides of temporary pools (157) or ultramafic soils (138). Analyses of the interactions between selection coefficients (s) and gene flow for eight herbs shows that, for values of $s = 0.1$, differentiation generally occurs over 15–25 m, although limiting gene flow can reduce the distance to 7 m. When $s = 0.5$, the values are generally under 10 m. Such values have been recorded in several studies, all of which were on annual or short-lived plants of small stature (123, 151; Table 1).

Because of their stature and tendency to outcross, forest trees typically have much more extensive gene flow. In addition, they tend to be primarily outcrossing and therefore have a more open breeding system. For these reasons, the finest grained differentiation due to selection and documented to date between adjacent subpopulations is usually on the scale of 100–300 m (89, 90, 107, 134, 161, 205, 224, 225). Conversely, strong differentiation on a scale of 10–50 m can also be found, usually as a result of very localized seed dispersal, which generates distinct family groups (162, 227, 248). In addition, such genetic substructuring can happen after long-distance dispersal (several km) when new plants arise from seeds dispersed in groups of genetically related individuals (81, 248).

We have presented the case that the differentiation between adjacent subpopulations so often reported in the literature occurs primarily as a result of different selection regimes imposed by different environments. If that is the case, and

Table 1 Examples of small-scale genetic differentiation in plants.

Species (family)	Life form and breeding[a]	Scale of differentiation (space or time)	Selective factor(s)	Characters differentiated	s[b]	Reference
Toxic Soils						
Agrostis tenuis (Poaceae)	PG O	5 m	Cu in tailings	Root length	Mine 0.95 Past. 0.05	123
Agrostis tenuis (Poaceae)	PG O	21–31 yrs	Zn under pylons	Root growth		3
Mimulus guttatus (Scrophulariaceae)	PH O	6–500m	Cu in soils	Plant growth		5
Fertilizers						
Anthoxanthum odoratum (Poaceae)	PG O	10–20 m 6 yrs	$CaCo_3$ in soil and vegetation density	Size and growth		238
Herbicides						
Senecio vulgaris (Compositae)	AH M	6 yrs	Simazine	Survival		117
Mowed and grazed						
Poa annua (Poaceae)	AH I	5–10 m	Mowing	Dry weight	0.53–0.77	28
Bellis perennis (Compositae)	PH I	"adjacent to several km"	Mowing and grazing	Stature, reproduction	0.7–1.0	264
Maritime exposure						
Agrostis stolonifera (Poaceae)	PG O	15 m	Wind and salt	Stolon length	0.5–0.8	123
Plantago maritima (Plantaginaceae)	PG O	"adjacent"	Exposure and salt	Stature		94
Moisture, temperature, elevation						
Avena barbata (Poaceae)	AG I	5–50 m	Soil moisture	Allele freqs.		98
Cleome serrulata (Capparidaceae)	AH O	< 30 m	Soil mositure	Water potential		74a
Abies lasiocarpa and *Picea Engelmannii* (Pinaceae)	T O	100 m elev. 100–500 m dist.	Wind, temp.	Allele freqs.		90
Eucalyptus spp. (Myrtaceae)	T O	300–600 m dist. 100–150 m elev.	Frost resist.	Glaucousness (1—few genes)		13

Table 1 (*Continued*)

Species (family)	Life form and breeding[a]	Scale of differentiation (space or time)	Selective factor(s)	Characters differentiated	s[b]	Reference
Light intensity						
Plantago lanceolata (Plantaginaceae)	PH O	8 m	Light patterns	Photosynthetic rates and temperature response		245a
Anthoxanthum odoratum (Poaceae)	PG O	10–30 m	Light intensity	Morphology, survival	0.60–0.75	88
Competition						
Trifolium repens (Leguminosae)	PH O	10–50 m	Interspecific	Growth		254
Veronica peregrina (Scrophulariaceae)	AH	5–10 m	Intra- and interspecific	Morphology, life history	0.20–0.89	157
Pollinating vectors						
Polemonium viscosum (Polemoniaceae)	PH O	100 m elev. 500 m dist.	Krummholtz vs alpine	Flower odor		82
Plantago lanceolata (Plantaginaceae)	PH O	"adjacent"	Windy vs sheltered	Pollen stickiness		241a
Herbivory, predation, parasitism						
Podophyllum peltatum (Berberidaceae)	HP O	700 m	Fungal attack	Susceptibility to infection		202
Trifolium repens (Leguminosae)	HP O	<10 m	Mollusc herb, fungal parasite	Cyanogenic morphs		57
Phoradendron tomentosum (Viscaceae)	HP O	<1 km	Host tree spp.	Seedling estab.	0.32	48
Pinus edulis (Pinaceae)	T O	<10 m	Parasitism by Lepidoptera	Allele freqs.		194
Time						
Lactuca muralis (Compositae)	BH I	8–10 yrs	Dispersal reduction	Diaspore size		48a
Hypochaeris radiata (Compositae)	PH I	8–10 yrs	Dispersal reduction	Diaspore size		48a
Pinus ponderosa (Pinaceae)	T O	1 generation ~ 50 yrs	Shade and moisture	Allele freqs.		16

[a]Abbreviations are as follows: A, annual, P, perennial, G, grass, H, Herb, T, tree; I, inbreeding, M, mixed, O, outbreeding.
[b]Selection coefficient = s.

if this scenario leads to adaptation, one might expect that disruption of such adaptation would have seriously negative effects. Indeed, this hypothesis has been tested in the form of experiments in which plants from different environments were mated. This process breaks up the integrated gene complexes that formed an important part of the adaptation to a specific locale; consequently, the progeny of such matings were predicted to be adaptively inferior. That prediction has been borne out, and outbreeding depression, as it is now called, has been demonstrated in the herbaceous species *Ipomopsis aggregata, Delphinium nelsonii,* and *Impatiens capensis.* For these taxa, there is evidence that progeny of plants produced from outcrossing parents separated by about 100–200 m in some cases, and as little as 15–20 m in others, are poorly adapted to the environment of either parent. They tend to die off in higher frequencies or grow more slowly than progeny of crosses between parents 1–5 m apart (170, 222, 266, 267). Outbreeding depression clearly reduces the homogenizing effect of gene flow. If it turns out to be a widespread phenomenon, then the general effectiveness of gene flow as a cohesive factor, especially over long distances, may be less important than currently thought.

Biological Constraints

The concept of evolutionary constraint has taken a prominent role in current evolutionary literature. In this context, the word usually refers to the restricted set of evolutionary "choices" available within a population. Certain possible or imagined body plans, physiological characters, morphologies, or combinations of traits do not exist; evolution, obviously, is constrained to those choices that do exist.

LIFE HISTORY The primary constraints of life history such as life span, breeding system, and body size strongly influence patterns of differentiation, especially in scale and tempo. Species that are small in stature (<0.5 m) commonly have high levels of selfing and short life spans, and they tend to show differentiation at smaller scales (e.g. 10–50 m) than do taller, longer-lived, woody plants with more open recombination systems. The latter typically show selection-generated differentiation at minimum scales of a few hundred meters. Small herbaceous species commonly show rapid evolutionary responses (237, 238), whereas longer-lived species tend to respond more slowly.

Life spans and body sizes can also affect the extent of selection generated by multispecies herbivory. Smaller, short-lived plants harbor fewer parasites and herbivores (on the order of 10–30 taxa per species) than do larger, long-lived species, which can harbor over 200 taxa of dependent species (159). A reasonable conjecture from such observations is that the complexity of selection can be far greater in the larger, long-lived species (159).

Life history per se is not necessarily a constraint in that it is an admittedly complex suite of characters subject to selection (45). Such characters can and do vary within species. For example, *Poa annua* populations can vary from semelparous annuals to iteroparous short-lived perennials (142).

BREEDING SYSTEMS Plants are remarkably variable in their breeding systems. They vary from apomixis to complete self-pollination, to mixed mating systems that provide a combination of selfing and outcrossing, to complete dioecy; sometimes these traits vary facultatively. Certain genome segments are transmitted biparentally while some are passed on strictly maternally and still others, strictly paternally. In addition chromosome numbers range from 2 to over 1000. With respect to genetic recombination, this means that everything from a very conservative copying of parental genotypes via apomixis to complete genetic reshuffling every generation is possible; each of these systems generates different constraints upon differentiation.

From the point of view of this review, two matters are important about breeding systems. One is their impact upon the scale of gene flow and population divergence. Some generalizations based on flower size and shape, mating compatibilities, plant stature, and systematics are possible. For example, plants that are small and/or have small inconspicuous flowers often have limited gene flow and tend to be associated with self-pollination, unless they are self-incompatible. In contrast, species that are self-incompatible or have large, showy flowers tend to exhibit more extensive gene movement. For example, certain families tend to be predominantly outcrossed (e.g. many taxa in the Compositae, Pinaceae, Primulaceae, and Rubiaceae). In all these instances, genetic analyses suggest that differentiation between populations is less strongly marked than in taxa with high levels of self-pollination. As a result, there are relatively smaller amounts of genetic variability within, but relatively larger amounts among, populations in selfers (32, 99, 122, 143, 166). In addition, species with either pollen (65, 124) or seeds (135, 248) subject to dispersal at scales greater than 1 km are especially notable in that they have population structures that produce larger, landscape-scale homogeneity because of the exceptionally long distances traveled by their dispersers. These species are poorly studied and the long travel distances attributed to such propagule movement are largely anecdotal, but notable.

The second context in which breeding systems are critical is their place as integral parts of life history; those parts form a complex of traits subject to selection. It must be stressed that breeding systems are highly variable within taxa because they are also subject to local selection pressures; thus, overall patterns are seldom as clear as they seem. For example, certain *Taraxacum* spp. are triploid and generally considered to be strictly agamospermous, i.e. strictly asexual reproducers. Consequently, one would expect them to be rather uniform genetically, but in fact some species show some sexual reproduction and thus contain important amounts of genetic variation (119). In at least some cases, such variation can be observed between populations separated by distances of 100 m or less. Under these circumstances, breeding systems can be thought of as another suite of characters, subject to the effects of selection (30, 164, 217) rather than as a fixed constraint.

Armeria maritima is a classic example of a distylous species where pollen of pin morphs can only germinate on thrum stigmas and vice versa. This system ensures complete outcrossing, usually affected by insect vectors. However, in certain conditions, the distyly system breaks down and self-pollination occurs, illustrating that so-called breeding system constraints may be dramatically altered in response to local conditions (148, 217). The evolution of self-compatibility is known to contribute strongly to small-scale differentiation (8).

Differentiation in Clonally Reproducing Taxa

The facultative ability for many plant species to reproduce via clonal and sexual means provides enormous advantages over either mode alone (64, 84). Clonal reproduction allows successful genotypes to be replicated repeatedly, perhaps without a theoretical limit to the number of regenerations that could be produced via this method. Such reproduction often produces individual plants that can spread across many hectares and take up resources scattered over those spaces (91, 188), with the concomitant result that small-scale spatial heterogeneity can be either exploited or minimized and can certainly be averaged. Transfer of water and nutrients from a region where the supplies are abundant to an area of a clone where they are limited can enhance the viability of the entire genotype with demonstrable advantages to the clone as a whole (109, 188, 244, 273). Fragmentation of clones may produce physiologically independent individuals, which then more or less successfully exploit their own particular microenvironment. Clonal reproduction also spans the environments temporally, occasionally covering periods estimated to be 17 millenia (261) or possibly even a million years (87).

As a consequence of these propensities to experience environmental heterogeneity through both time and space by a single genotype, the physiological flexibility to deal with this heterogeneity appears to be generated in two different ways: 1. Physiological flexibility in underlying biochemical pathways often produces relatively stable phenotypes across dramatically different environments (171, 245, 246), and 2. enhanced individual genetic diversity, e.g. via polyploidy (92, 192, 214, 220). This chromosomal mechanism also enhances the significance of mutationally generated variation (84), which may be carried for long periods of time within an individual genome, perhaps as an exaptation. Both mechanisms can be viewed as selection for enhanced phenotypic plasticity in particular traits, i.e. biochemical in the first and chromosomal in the second (221, 245), enabling individual genotypes to operate successfully across many types of environments. Note that this adaptive pathway contrasts markedly with that of genetic differentiation of localized subpopulations that contain different distributions of genotypes (70, 126). The phenotypic plasticity enhancement paths typically operate antagonistically to local genetic differentiation in the traits exhibiting that plasticity. Nonetheless, considerable local genetic diversity has been found in many clonal taxa such as *Trifolium repens, Spartina,* and *Podophyllum peltatum* (4, 202, 203, 254), but not in others (12, 247).

Plant Characteristics Affected by Differentiation

Localized differentiation has been documented for most important features of plant structure and function. Examples can be found for seed size (157), leaf pubescence (138), seed germination (88), seed color (27), root morphology (107), plant stature (237), flowering phenology (88), leaf morphology (14), photosynthetic abilities (245a), heavy metal tolerance (9), herbicide resistance (144), herbivory resistance (233), pathogen resistance (203), response to competitors (255), pollinator availability (82), organelle traits (187), isozymes (98), DNA fingerprints (4), rflps (187), breeding systems (148), and life history (10, 128, 142).

Chromosome variability also plays an important role in genetic differentiation. Certain plants have significant levels of intraspecific chromosomal variability, both at the level of polyploidy (92, 131, 154, 167, 214, 249) and at the aneuploid levels (131, 154). Some of these taxa show differentiation over short distances, but the ecological and physiological contexts that are relevant to these chromosomal variants are far from clear. In a larger context, polyploidy has long been thought to be an important factor in the local origin of new taxa following hybridization (241), and some recent work continues to support that view (214). Additionally, Rabe & Haufler (213) provide experimental data indicating that polyploidy in *Adiantum* may come about via the mechanism of unreduced but fully functional polyploid spores without the accompanying process of hybridization.

Modes of Inheritance—Maternal, Paternal, and Biparental

Plants show a remarkable variety of inheritance modes, and, further, some of their reproductive patterns permit genetic study with means not available in other types of organisms. For example, the ability to identify maternal and paternal contributions to nuclear genotypes simultaneously in conifers has allowed detailed examination of several aspects of gene flow, differentiation, and responses to selection (187). Conifers are also particularly useful because chloroplasts are typically inherited paternally and mitochrondria maternally, contrary to most other organisms. Such diversity in modes of inheritance permits analysis of genetic structures from three different perspectives. In one such study, the closed-cone pines of California (*P. attenuata, P. muricata, P. radiata*) showed the commonly expected high amounts of isozymic variation between populations: 12–22% of the variation was found within populations. In contrast, chloroplast and mitochondrial DNA showed very different patterns. Very little variation was observed for cpDNA in *P. attenuata* and *P. radiata*, whereas *P. muricata* showed twice the level of differentiation among populations for cpDNA as that observed using isozymes. The patterns for mtDNA showed low within-population variation but Gst ranged from 75% in *P. radiata* to 96% in *P. muricata* (118, 243).

In the context of contrasts between the nuclear and organellar genomes, the concept of genetic flux should be mentioned. This term designates the exchange of genetic material between chloroplasts and mitochondria and both organelles and the nuclear genome. There is evidence for such movement in a number of different species. The evolutionary consequences of this genetic flux have not been examined in detail but are certainly relevant to phenomena such as soma-clonal variation and phenotypic plasticity (115a).

SUMMARY AND CONCLUSIONS

Natural plant populations routinely and consistently show small-scale genetic differentiation and do so in all types of characters studied to date, although important exceptions are known. More than 100 species have now been demonstrated to exhibit such differentiation; the vast majority of these are angiosperms. Several examples are known from conifers, but with few exceptions, the non-seed plants such as Pteridophytes, Bryophytes, and especially the algae are virtually *Plantae incognitae* from this perspective. Spatial differentiation has been studied most thoroughly and shows more dramatic effects than does temporal change. This may be due partly to the greater ease with which spatial patterns can be studied as compared to temporal ones, but fundamentally, evolution proceeds temporally; thus we foresee increased emphasis on temporal differentiation in future work. Genetic differentiation in response to physical environments typically occurs in a comparatively simple, contiguous fashion, whereas differentiation in response to biotic factors frequently shows very fine-scale mosaic patterns. When the work herein reviewed is considered as a collective whole, the inescapable conclusion, particularly for continuously variable traits of obvious functional significance, is that this genetic differentiation has been typically produced by natural selection in response to environmental heterogeneity, biotic and abiotic.

The most convincing cases for selectively driven differentiation across short spatial distances are those caused by dramatic soil differences, either human induced (e.g. mine tailings) or natural (e.g. ultramafic outcrops). Four aspects of these examples, taken jointly, make the strong case for natural selection: (*a*) the high frequency of similar biological responses to similar environmental differences among unrelated taxa, i.e. evolutionary convergence, (*b*) the consistency of change across many taxa in many different locations in several physiological traits which, together, form a recognizable response syndrome, (*c*) the direct relevance of these character changes to the agents of selection (i.e. soil chemistry and the plant processing connections to that chemistry), and (*d*) the demonstrated biological consequences of such differentiation on reproductive success when experimentally manipulated.

In the most extreme cases, genetic differentiation in response to serpentine outcrops appears to have extended to such a degree as to produce sympatric speciation in several different taxa. We do not believe that all of these cases can reasonably be explained by secondary invasions following development of reproductive isolation during allopatry, nor can they be convincingly explained by the operation of random processes alone. If our interpretations are correct, these examples provide strong arguments against hypotheses that entail a general decoupling of speciation from Darwinian processes. Rather, selection plays a central role in plant speciation.

It is our opinion that the zeitgeist of the evolutionary community at the present time can be fairly characterized as essentially pan-neutralist. In particular, most molecular variation observed in small-scale differentiation studies is presumed to be neutral or nearly neutral in character (133). In contrast, a significant minority of evolutionary biologists argue that many, if not most, protein variants (as opposed to most nucleic acid variants) are responsive to natural selection. Those students of evolution argue that much protein variation is inconsistent with the neutral view, e.g. (187, 195, 218, 277). As members of that minority we observe that there has been an historically consistent pattern of interpretation regarding the relative significance of selection versus neutral forces in newly observed traits. The pattern to which we refer is: 1. Interpret the data as consistent with neutralism (and, therefore, the explanatory hypothesis to be favored), then 2. shift interpretation toward the selective end of the spectrum when more detailed information has become available. One of the most relevant examples of this pattern can be observed in Th. Dobzhansky's writings in his remarkable series on the *Genetics of Natural Populations.* In the first paper of the series (59), the authors wrote about the different chromosomal inversion patterns of the third chromosome that they had studied as follows: "By far the most probable explanation of the observed differences between populations of the separate mountain ranges is that the frequency of a gene or a chromosome structure is subject to random fluctuations." Further, they wrote: "the supposition that these chromosome structures are subject to selection is extremely improbable...." Similarly, in the only study of plants in the series (68), Dobzhansky used the phrase he invented, "microgeographic races," to refer to blue and white flower morphs in *Linanthus Parryae*; the patterns of occurrence here too seemed best explained by neutral forces, particularly isolation by distance. Over the course of the next several years, he began to question the neutral explanation for chromosomal inversions and by 1948, he boldly titled paper XVII in the series, "Proof of operation of natural selection in wild populations of *Drosophila pseudoobscura*" (58a). He had, quite justifiably it turned out, completely reversed his position regarding the relative role of natural selection in ordering the variation in chromosomal arrangements for these natural populations. His conclusion that natural selection was the primary force has been repeatedly verified, most convincingly in experimental studies. Additional examples include human blood types (153), flower color morphs in *Linanthus Parryae* (69), color and banding patterns in snails (38), butterfly markings, etc.

Despite this historical pattern and extensive accumulation of literature, it is our impression that there now seems to be a whole generation of young scientists who apparently believe that natural selection, as an evolutionary force, is typically impotent, of secondary significance, or, at the very least, rarely demonstrated with convincing rigor. A similar sentiment was recently articulated by DePew & Webber (53a):

> "If there is one thing advocates of an expanded Darwinism and their new developmentalist counterparts have in common it is antipathy to adaptationism. So great is this shared antipathy that ... it spills over to the notions of adaptation and adaptedness themselves, and to natural selection...."

We believe there are several reasons for this intellectual diminution of the power and ubiquity of natural selection: 1. the pleasing, scientific formalism of considering neutralism as *the* null model against which observational data could be compared in mathematically and statistically rigorous ways, 2. the clear statistical predictability of allelic behavior over generations with finite population sizes, 3. the enormously seductive power of insight (apparent or real) gained by analysis of historical genetic events controlled only by statistical properties (e.g. systematic affinities, molecular clock divergences, etc), 4. the low power of most commonly used statistical procedures to detect significant deviation from neutral predictions, and 5. an overly effective swing of the intellectual, linguistic pendulum toward condemnation of "panselectionism," something of a straw characterization, albeit with a genuine kernel of truth. We believe the use of powerfully derisive language exemplified by phrases such as "idle Darwinizing," "Pan-Glossian selectionism," and "just-so" stories, while laudably intended to oppose unthinking, facile application of selective explanations (275), has resulted in an erroneous minimization of natural selection as a potent natural force in contemporary evolutionary thinking and writing. West-Eberhard (268) and Berry (18a) provide brief but excellent critiques of these issues.

It is our hope that this review partially redresses the imbalance we see between the current pan-neutralist outlook and a more defensible, intellectually inclusive location. In short, we argue that there is strong, unambiguous evidence that the heart of Darwinism—natural selection—dominates formation and maintenance of the genetic architectural patterns of natural plant populations, and we believe that the appreciation of this force has been unduly diminished.

ACKNOWLEDGMENTS

We thank Jeffry Mitton, Kathy Keeler, Robert Latta, Marc Snyder, JP Gibson, and two reviewers for constructive suggestions. This work was supported in part by NSF grant BSR-8911433 to Russell Monson and MCG; NSF grant BSR-8918478, a grant from the US National Park Service, and USDA grant 95-37101-1638 to YBL.

Literature Cited

1. Adams WT, Griffin AR, Moran GF. 1992. Using paternity analysis to measure effective pollen dispersal in plant populations. *Am. Nat.* 140:762–80
2. Akiyama H. 1994. Allozyme variability within and among populations of the epiphytic moss *Leucodon* (*Leucodontaceae: Musci*). *Am. J. Bot.* 81:1280–87
3. Al-Hiyaly SAK, McNeilly T, Bradshaw AD, Mortimer AM. 1993. The effect of zinc contamination from electricity pylons: genetic constraints on selection for zinc tolerance. *Heredity* 70:22–32
4. Alberte RS, Suba GK, Procaccini G, Zimmerman RC. 1994. Assessment of genetic diversity of seagrass populations using DNA fingerprinting: implications for population stability and management. *Proc. Natl. Acad. Sci. USA* 91:1049–53
5. Allen WR, Sheppard PM. 1971. Copper tolerance in some California populations of the monkey flower *Mimulus guttatus*. *Proc. R. Soc. London B* 177:177–96
6. Alvarez-Buylla ER, Garay AA. 1994. Population genetic structure of *Cecropia obtusifolia*, a tropical pioneer tree species. *Evolution* 48:437–53
7. Antonovics J. 1971. The effects of a heterogenous environment on the genetics

of natural populations. *Am. Sci.* 59:593–99
8. Antonovics J. 1978. The population genetics of mixtures. In *Plant Relations in Pastures,* ed. JR Wilson, pp. 233–52. East Melbourne: CSIRO. 425 pp.
9. Antonovics J, Bradshaw AD, Turner RG. 1971. Heavy metal tolerance in plants. *Adv. Ecol. Res.* 71:1–85
10. Argyres AZ, Schmitt A. 1991. Microgeographical genetic structure of morphological and life history traits in a natural population of *Impatiens capensis*. *Evolution.* 45:178–89
11. Asmussen MA, Schnabel A. 1991. Comparative effects of pollen and seed migration on the cytonuclear structure of plant populations. Maternal cytoplasmic inheritance. *Genetics* 128:639–54
12. Aspinwal N, Christian T. 1992. Clonal structure, genotypic diversity and seed production in populations of *Filipendula rubra, (Rosaceae)* from the North Central United States. *Am. J. Bot.* 79:294–99
13. Barber HN. 1955. Adaptive gene substitutions in Tasmanian Eucalypts: I. Genes controlling the development of glaucousness. *Evolution* 9:1–14
14. Barber HN. 1965. Selection in natural populations. *Heredity* 20:551–72
15. Barrett SCH, Kohn JR. 1991. Genetic and evolutionary consequences of small population size in plants: implications for conservation. In *Genetics and Conservation of Rare Plants,* ed. DA Falk, KE Holsinger, pp. 3–30. Oxford: Oxford Univ. Press, 283 pp.
16. Beckman JS, Mitton JB. 1984. Peroxidase allozyme differentiation among successional stands of ponderosa pine. *Am. Midl. Nat.* 112:43–49
17. Bell G, Lechowicz MJ. 1991. The ecology and genetics of fitness in forest plants: environmental heterogeneity measured by explant trials. *J. Ecol.* 79:663–85
18. Bennett E. 1964. Historical perspectives in genecology. *Scot. Pl. Breed. Stn. Rec.* 1964:49–115
18a. Berry A. 1996. Non-non-Darwinian evolution. *Evolution* 50:462–66
19. Björkman O. 1968. Carboxydismutase activity in shade-adapted and sun-adapted species of higher plants. *Physiol. Plant.* 21:1–10
20. Björkman O, Gauhl E, Nobs MA. 1969. Comparative studies of *Atriplex* species with and without B-carboxylation and their first generation hybrids. *Carnegie Inst. Yearbk.* 68:620–33
21. Björkman O, Holmgren P. 1963. Adaptation to light intensity in plants native to shaded and exposed habitats. *Physiol. Plant.* 19:854–55
22. Bock JH, Linhart YB. eds. 1989. *The Evolutionary Ecology of Plants*. Boulder, CO: Westview. 600 pp.
23. Bonnier G. 1895. Recherches experimentales sur l'adaptation des plantes au climat alpin. *Ann. Sci. Nat. Bot. Biol. Veg., 7th ser.* 20:217–358
24. Boyce JS. 1961. *Forest Pathology*. New York: McGraw–Hill. 572 pp. 3rd. ed.
25. Bradshaw AD. 1972. Some of the evolutionary consequences of being a plant. *Evol. Biol.* 5:25–47
26. Bradshaw AD, McNeilly TS, Gregory RPG. 1965. Industrialization, evolution, and development of heavy metal tolerance in plants. *Br. Ecol. Soc. Symp.* 6:327–43
27. Brayton RD, Capon B. 1980. Palatability, depletion and natural selection of *Salvia columbiarae* seeds. *Aliso* 9:581–87
28. Briggs D, Walters SM. 1984. *Plant Variation and Evolution.* Cambridge: Cambridge Univ. Press. 412 pp. 2nd ed.
29. Bronn MT, Wilkins DA. 1985. Zinc tolerance in a mycorrhizal *Betula. New Phytol.* 99:101–6
30. Brown AHD. 1990. Genetic characterization of plant mating systems. See Ref. 31, pp. 145–62
31. Brown AHD, Clegg MT, Kahler AL, Weir BS, eds. 1990. *Plant Population Genetics, Breeding, and Genetic Resources*. Sunderland, MA: Sinauer. 449 pp.
32. Brown CR, Jain SK. 1979. Reproductive system and pattern of genetic variation in two *Limnanthes* species. *Theor. Appl. Genet.* 54:181–90
33. Broyles SB, Schnabel A, Wyatt R. 1994. Evidence for long-distance pollen dispersal in milkweeds, (*Asclepias exaltata*). *Evolution* 48:1032–40
34. Brundrett M. 1991. Mycorrhizas in natural ecosystems. *Adv. Ecol. Res.* 21:171–213
35. Burdon J. 1987. *Diseases and Plant Population Biology*. Cambridge: Cambridge Univ. Press. 208 pp.
36. Bush GL. 1975. Modes of animal speciation. *Annu. Rev. Ecol. Syst.* 6:339–64
37. Caballero A. 1994. Developments in the prediction of effective population size. *Heredity* 73:656–79
38. Cain AJ, Sheppard PM. 1950. Selection in the polymorphic land snail *Cepea*

nemoralis. *Heredity* 4:275–94
39. Campbell DR, Dooley JL. 1992. The spatial scale of genetic differentiation in a hummingbird-pollinated plant: comparison with models of isolation by distance. *Am. Nat.* 139:735–48
40. Carroll CR, Hoffman CA. 1980. Chemical feeding deterrent mobilized in response to insect herbivory and counteradaptation by *Epilachna tredecimnotata*. *Science* 209:414–16
41. Chanway CP, Turkington R, Holl FB. 1991. Ecological implications of specificity between plants and rhizosphere micro-organisms. *Adv. Ecol. Res.* 21:122–69
42. Charlesworth B. 1994. *Evolution in Age Structured Populations*. Cambridge: Cambridge Univ. Press. 306 pp. 2nd ed.
43. Chauvel B, Gasquez J. 1994. Relationships between genetic polymorphism and herbicide resistance in *Alopecurus myosuroides* Huds. *Heredity* 72:336–44
44. Chazdon RL, Pearcy RW. 1991. The importance of sunflecks for forest understory plants: Photosynthetic machinery appears adapted to brief, unpredictable periods of radiation. *Bioscience* 41:760–66
45. Clark DA, Clark DB. 1992. Life history diversity of canopy and emergent trees in a neotropical rain forest (Costa Rica). *Ecol. Mono.* 62:315–44
46. Clausen J, Hiesey WM. 1958. Experimental studies on the nature of species. IV. Genetic structure of ecological races. *Carnegie Inst. Wash. Publ. 615*. Washington, DC
47. Clausen J, Keck DD, Hiesey WM. 1948. Experimental studies on nature of species. III. Environmental responses of climatic races of *Achillea*. *Carnegie Inst. Wash. Publ. 581* Washington, DC
48. Clay K, Dement D, Rejmaneck K. 1985. Experimental evidence for host races in mistletoe (*Phoradendron tomentosum*). *Am. J. Bot.* 72:1225–31
48a. Cody ML, Overton JM. 1996. Short-term evolution of reduced dispersal in island plant populations. *J. Ecol.* 84:53–61
49. Cox, PA. 1989. Baker's Law: plant breeding systems and island colonization. See Ref. 22, pp. 209–24
50. Darwin CR. 1872. *The Origin of Species*. Reprinted by Collier Books (1962). 512 pp. 6th ed.
51. Del Castillo RF. 1994. Factors influencing the genetic structure of *Phacelia dubia*, a species with a seed bank and large fluctuations in population size. *Heredity* 72:446–58
52. Dempster ER. 1955. Maintenance of genetic heterogeneity. *Cold Spring Harbor Symp. Quant. Biol.* 20:25–32
53. Denno RF, McClure MS, eds. 1983. *Variable Plants and Herbivores in Natural and Managed Systems*. New York: Academic. 717 pp.
53a. Depew DJ, Weber BH. 1995. *Darwinism Evolving*. Cambridge, Mass., MIT Press. 588 pp.
54. Devlin B, Ellstrand NC. 1990. The development and application of a refined method for estimating gene flow from angiosperm paternity analysis. *Evolution* 44:248–59
55. Dewey SE, Heywood JS. 1988. Spatial genetic structure in a population of *Psychotria nervosa*. I. Distribution of genotypes. *Evolution* 42:824–38
56. Dickenman R. 1982. Cyanogenesis in *Ranunculus montanus* from the Swiss alps. *Berlin Geobot. Inst. Zurich* 49:56–75
57. Dirzo R, Harper JL. 1982. Experimental studies on slug-plant interactions. IV. The performance of cyanogenic and acyanogenic morphs of *Trifolium repens* in the field. *J. Ecol.* 70:119–38
58. Dirzo R, Sarukhan J, eds. 1984. *Perspectives in Plant Population Ecology*. Sunderland, MA: Sinauer. 478 pp.
58a. Dobzhansky T, Levene H. 1948. Genetics of natural populations. XVII. Proof of operation of natural selection in wild populations of *Drosophila pseudoobscura*. *Genetics* 33:537–47
59. Dobzhansky Th, Queal ML. 1938. Genetics of natural populations. I. Chromosome variation in populations of *Drosophila pseudoobscura* inhabiting isolated mountain ranges. *Genetics* 23:239–51
60. Ducousso A, Petit D, Valero M, Vernet P. 1990. Genetic variation between and within populations of a perennial grass *Arrhenatherum elatius*. *Heredity* 65:178–88
61. Durrant A. 1971. Induction and growth of flax genotrophs. *Heredity* 27:277–98
62. Eguiarte LE, Burquex A, Rodriguez J, Martinez-Ramos M, Sarukhan J, Pinero D. 1993. Direct and indirect estimates of neighborhood and effective population size in a tropical palm *Astrocaryum mexicanum*. *Evolution* 47:75–87
63. Ehrlich PR, Raven PH. 1969. Differentiation of populations. *Science* 165:1228–32

64. Ellstrand NC. 1987. Patterns of genotypic diversity in clonal plant species. *Am. J. Bot.* 74:123–31
65. Emerson S. 1939. A preliminary survey of the *Oenothera organensis* population. *Genetics* 27:524–37
66. Endler JA. 1986. *Natural Selection in the Wild.* Princeton NJ: Princeton Univ. Press. 336 pp.
67. Ennos RA. 1994. Estimating the relative rates of pollen and seed migration among plant populations. *Heredity* 72:250–59
68. Epling C, Dobzhansky Th. 1942. Genetics of natural populations. VI. Microgeographic races in *Linanthus Parryae. Genetics* 27:317–32.
69. Epling C, Lewis H, Ball FM. 1960. The breeding group and seed storage: a study in population dynamics. *Evolution* 11:248–56
70. Epperson BK. 1990. Spatial autocorrelation of genotypes under directional selection. *Genetics* 124:757–71
71. Epperson BK. 1990. Spatial patterns of genetic variation within plant populations. See Ref. 31, pp. 229–53
72. Epperson BK. 1993. Recent advances in correlation studies of spatial patterns of genetic variation. *Evol. Biol.* 27:95–155
73. Epperson BK. 1995. Fine scale spatial structure: correlations for individual genotypes differ from those for local gene frequencies. *Evolution* 49:1022–26
74. Epperson BK. 1995. Spatial distributions of genotypes under isolation by distance. *Genetics* 140:1431–40

74a. Farris MA. 1987. Natural selection on the plant water relations of *Cleome serrulata* growing along natural moisture gradients. *Oecologia* 72:434–39

75. Fineblum WL, Rausher MD. 1995. Tradeoff between resistance and tolerance to herbivore damage in a morning glory. *Nature* 377:517–20
76. Fowler DP, Morris RW. 1977. Genetic diversity in red pine: evidence for low genetic heterozygosity. *Can. J. For. Res.* 7:343–47
77. Fowler NL. 1990. The effects of competition and environmental heterogeneity on three coexisting grasses. *J. Ecol.* 78:389–402
78. Freysen AHJ, Woldendorp JW, eds. 1978. *Structure and Functioning of Plant Populations.* Amsterdam: North Holland. 405 pp.
79. Fritz RS, Simms EL, eds. 1992. *Plant Resistance to Herbivores and Pathogens.* Chicago: Univ. Chicago Press. 590 pp.
80. Furnier GR, Adams WT. 1986. Geographic patterns of allozyme variation in Jeffrey pine. *Am. J. Bot.* 73:1009–15
81. Furnier GR, Knowles P, Clyde MA, Dancik BP. 1987. Effects of avian seed dispersal on the genetic structure of white bark pine populations. *Evolution* 41:607–12
82. Galen C, Kevan PG. 1980. Scent and color, floral polymorphisms and pollination biology in *Polemonium viscosum.* Nutt. *Am. Midl. Natur.* 104:281–89
83. Gibson JP, Hamrick JL. 1991. Heterogeneity in pollen allele frequencies among cones, whorls, and trees of Table Mountain pine (*Pinus pungen*). *Am. J. Bot.* 78:1244–51
84. Gill DE, Chao L, Perkins SL, Wolf JB. 1995. Genetic mosaicism in plants and clonal animals. *Annu. Rev. Syst. Ecol.* 26:423–44
85. Gillespie JM, Turelli M. 1989. Genotype-environment interactions and the maintenance of polygenic variation. *Genetics* 121:129–38
86. Gottlieb LD, Jain SK, eds. 1988. *Plant Evolutionary Biology.* New York: Chapman & Hall. 414 pp.
87. Grant MC. 1993. The trembling giant. *Discover* (Oct):83–88
88. Grant MC, Antonovics J. 1978. Biology of ecologically marginal populations of *Anthoxanthum odoratum.* I. Phenetics and dynamics. *Evolution* 32:822–38
89. Grant MC, Linhart YB, Monson RK. 1989. Experimental studies in ponderosa pine. II. Quantitative genetics of morphological traits. *Am. J. Bot.* 76:1033–40
90. Grant MC, Mitton JB. 1977. Genetic differentiation among growth forms of Engelmann spruce and subalpine fir at tree line. *Arctic Alpine Res.* 9:259–63
91. Grant MC, Mitton JB, Linhart YB. 1992. Even larger organisms. *Nature* 360:216
92. Grant MC, Proctor VW. 1980. Electrophoretic analysis of genetic variation in the *Charophyta.* I. Gene duplication via polyploidy. *J. Phycol.* 16:109–15
93. Gray AJ. 1987. Genetic change during succession in plants. In *Colonization, Succession and Stability,* ed. AJ Gray, MJ Crawley, PJ Edwards, pp. 273–94. Oxford: Blackwell. 482 pp.
94. Gregor JW. 1930. Experiments on the genetics of wild populations part I. *Plantago maritima. L. J. Genet.* 22:15–25
95. Gullberg U, Yazdanik R, Rudin D. 1982. Genetic differentiation between adjacent populations of *Pinus sylvestris. Silva*

Fennica 16:205–14
96. Gurevitch J. 1992. Sources of variation of leaf shape among two populations of *Achillea lanulosa. Genetics* 130:385–94
97. Guries RP, Ledig FT. 1982. Genetic diversity and population structure in pitch pine (*Pinus rigida* Mill.). *Evolution.* 36:387–99
98. Hamrick JL, Allard RW. 1972. Microgeographical variation in allozyme frequencies in *Avena barbata. Proc. Natl. Acad. Sci. USA* 69:2100–4
99. Hamrick JL, Godt MJ, Sherman-Broyles SL. 1992. Factors influencing levels of genetic diversity in woody plant species. *New For.* 6:95–124
100. Hamrick JL, Godt MJ. 1990. Allozyme diversity in plant species. See Ref. 31, pp. 43–63
101. Hamrick JL, Holden LR. 1979. Influence of microhabitat heterogeneity on gene frequency distribution and gametic phase disequilibrium in *Avena barbata. Evolution* 33:521–33
102. Hamrick JL, Linhart YB, Mitton JB. 1979. Relationships between life-history characteristics and electrophoretically detectable genetic variation in plants. *Annu. Rev. Ecol. Syst.* 10:173–200
103. Harlan HV, Martini ML. 1938. The effect of natural selection on a mixture of barley varieties. *J. Agric. Res.* 57:189–99
104. Hay ME, Fenical W. 1988. Marine plant-herbivore interactions: the ecology of chemical defense. *Annu. Rev. Ecol. Syst.* 19:111–45
105. Hay ME, Renaud PE, Fenical W. 1988. Large mobile versus small sedentary herbivores and their resistance to seaweed chemical defenses. *Oecologia* 75:246–52
106. Hedrick PW, Ginevan ME, Ewing EP. 1976. Genetic polymorphism in heterogenous environments. *Annu. Rev. Ecol. Syst.* 7:1–32
107. Herman RK, Lavender DP. 1968. Early growth of Douglas-fir from various altitudes and aspects in Southern Oregon. *Silvae Genet.* 17:143–51
108. Heslop-Harrison J. 1964. Forty years of genecology. *Adv. Ecol. Res.* 2:159–247
109. Hester MW, McKee KL, Burdick DM, Koch MS, Flynn KM, et al. 1994. Clonal integration of *Spartina patens* across a nitrogen and salinity gradient. *Can. J. Bot.* 72:767–70
110. Heywood JS. 1986. The effect of plant size variation on genetic drift in populations of annuals. *Am. Nat.* 127:851–61
111. Heywood JS. 1991. Spatial analysis of genetic variation in plant populations. *Annu. Rev. Ecol. Syst.* 22:335–56
112. Hickey DA, McNeilly T. 1975. Competition between metal tolerant and normal plant populations: a field experiment on normal soils. *Evolution* 29:458–64
113. Hiebert RD, Hamrick JL. 1983. Patterns and levels of genetic variation in Great Basin bristlecone pine, *Pinus longaeva. Evolution* 37:203–10
114. Hiesey WM, Milner HW. 1965. Physiology of ecological races and species. *Annu. Rev. Plant Physiol.* 16:203–16
115. Hillis DM. 1987. Molecular versus morphological approaches in systematics. *Annu. Rev. Ecol. Syst.* 18:23–42
115a. Hohn B, Dennis EM, eds. 1985. *Genetic Flux in Plants.* Springer Verlag. 253 pp.
116. Holliday RJ, Putwain PO. 1977. Evolution of resistance to simazine in *Senecio vulgaris. Weed Res.* 17:281–86
117. Holliday RJ, Putwain PO. 1980. Evolution of herbicide resistance in *Senecio vulgaris:* variation in susceptibility to simazine between and within populations. *J. Appl. Ecol.* 17:779–81
118. Hong Y-P, Hipkins VD, Strauss SH. 1993. Chloroplast DNA diversity among trees, populations and species in the California closed-cone pines (*Pinus radiata, Pinus muricata* and *Pinus attenuata*). *Genetics* 135:1187–96
119. Hughes J, Richards AJ. 1988. The genetic structure of populations of sexual and asexual *Taraxacum* dandelions. *Heredity* 60:161–72
120. Husband BC, Spencer SCH. 1992. Effective population size and genetic drift in tristylous *Eichornia paniculata* (*Pontederiaceae*). *Evolution* 46:1875–90
121. Jacquard P, Heim G, Antonovics J, eds. 1985. *Genetic Differentiation and Dispersal in Plants.* Berlin: Springer Verlag. 452 pp.
122. Jain SK. 1976. The evolution of inbreeding in plants. *Annu. Rev. Ecol. Syst.* 7:469–95
123. Jain SK, Bradshaw AD. 1966. Evolutionary divergence among adjacent plant populations. I. Evidence and its theoretical analysis. *Heredity* 21:407–41
124. Janzen D. 1971. Euglossine bees as long distance pollinators of tropical plants. *Science* 171:203–5
125. Jenny H. 1980. *The Soil Resource: Origin and Behavior.* New York: Springer Verlag. 431 pp.
126. Jordan N. 1992. Path analysis of local adaptation in two ecotypes of the annual

plant *Diodia teres* Walt. (*Rubiaceae*). *Am. Nat.* 140:149–65
127. Jules ES, Shaw AJ. 1994. Adaptation to metal-contaminated soils in populations of the moss, *Ceratodon purpureus:* vegetative growth and reproductive expression. *Am. J. Bot.* 81:791–97
128. Kalisz S, Wardle GM. 1994. Life history variation in *Campanula americana* (*Campanulaceae*): population differentiation. *Am. J. Bot.* 81:521–27
129. Kareiva P. 1986. Patchiness, dispersal and species interactions: consequences for communities of herbivorous insects. In *Community Ecology,* ed. J Diamond, T Case, pp. 192–206. New York: Harper & Row. 665 pp.
130. Keeler K. 1978. Intra-population differentiation in annual plants. I. Electrophoretic variation in *Veronica peregrina. Evolution* 32:638–45
131. Keeler K, Kwankin B. 1989. Polyploid polymorphism in the grasses of the North American prairie, See Ref. 22, pp. 99–128
132. Kiang YT. 1982. Local differentiation of *Anthoxanthum odoratum* L. populations on roadsides. *Am. Midl. Nat.* 107:340–50
133. Kimura M. 1983. *The Neutral Theory of Molecular Evolution.* Cambridge, UK: Cambridge Univ. Press. 367 pp.
134. Knowles P. 1990. Spatial genetic structure within two natural stands of black spruce (*Picea mariana* [Mill] B.S.P.) *Silvae Genet.* 40:13–19
135. Krauss S. 1994. Restricted gene flow within the morphologically complex species *Persoonia mollis* (*Proteaceae*): contrasting evidence from the mating system and pollen dispersal. *Heredity* 73:142–54
136. Kruckeberg AR. 1951. Response of plants to serpentine soils. *Am. J. Bot* 38:408–18
137. Kruckeberg AR. 1954. The ecology of serpentine soils. III. Plant species in relation to serpentine soils. *Ecology* 35:267–74
138. Kruckeberg AR. 1992. Plant life of western North American ultramafics. In *The Ecology of Areas With Serpentinized Rocks: A World View,* ed. BA Roberts, J Proctor, pp. 31–73. Dordrecht, Netherlands: Kluwer. 427 pp.
139. Kruckeberg AR, Rabinowitz D. 1985. Biological aspects of endemism in higher plants. *Annu. Rev. Ecol. Syst.* 16:447–79
140. Landman OE. 1991. The inheritance of acquired characteristics. *Annu. Rev. Genet.* 25:1–20.
141. Langlet O. 1971. Two hundred years of genecology. *Taxon* 20:653–722
142. Law R, Bradshaw AD, Putwain PD. 1977. Life history variation in *Poa annua. Evolution* 31:233–46
143. Layton CR, Ganders FR. 1984. The genetic consequences of contrasting breeding systems in *Plectritis* (*Valerianaceae*). *Evolution* 38:1308–25
144. Lebaron HM. 1991. Distribution and seriousness of herbicide resistant weed infestations world-wide. In *Herbicide Resistance in Weeds and Crops,* ed. JC Caseley, GW Cussans, AR Atkin, pp. 22–44. Oxford: Butterworth-Heineman. 513 pp.
145. Lebaron HM, Gressel J, eds. 1982. *Herbicide Resistance in Plants.* New York: Wiley. 401 pp.
146. Leberg PL. 1992. Effects of population bottlenecks on genetic diversity as measured by allozyme electrophoresis. *Evolution* 46:477–94
147. Ledig FT, Conkle MT. 1983. Gene diversity and genetic structure in a narrow endemic, Torrey pine (*Pinus torreyana* Parry. ex Carr.) *Evolution* 37:79–86
148. Lefebvre C, Vernet P. 1989. Microevolutionary processes on contaminated deposits. See Ref. 230, pp. 285–300
149. Levene H. 1953. Genetic equilibrium when more than one ecological niche is available. *Am. Nat.* 87:331–33
150. Levin DA. 1978. Some genetic consequences of being a plant. In *Ecological Genetics: The Interface,* ed. PF Brussard, pp. 189–212. New York: Springer-Verlag. 247 pp.
151. Levin DA. 1988. Local differentiation and the breeding structure of plant populations. See Ref 86, pp. 305–329
152. Levin DA, Kerster HW. 1974. Gene flow in seed plants. *Evol. Biol.* 7:139–220
153. Levine P. 1958. The influence of the ABO system on Rh hemolytic disease. *Human Biol.* 30:14–28
154. Lewis WH, ed. 1980. *Polyploidy: Biological Relevance.* New York: Plenum. 583 pp.
155. Li P, Adams WT. 1989. Range-wide patterns of allozyme variation in Douglas-fir (*Pseudotsuga menziesii*). *Can. J. For. Res.* 19:149–61
156. Linhart YB. 1976. Evolutionary studies of plant populations in vernal pools. In *Vernal Pools—Their Ecology and Conservation,* ed. S Jain, pp. 40–46. Davis, CA: Inst. Ecol. Univ. Calif. 93 pp.

157. Linhart YB. 1988. Intra-population differentiation in annual plants. III. The contrasting effects of intra- and inter-specific competition. *Evolution* 42:1047–64
158. Linhart YB. 1989. Interactions between genetic and ecological patchiness in forest trees and their dependent species. See Ref. 22, pp. 393–430
159. Linhart YB. 1991. Disease, parasitism and herbivory: multidimensional challenges in plant evolution. *Trends Ecol. Evol.* 6:392–96
160. Linhart YB, Davis ML. 1991. The importance of local genetic variability in Douglas-fir. In *Interior Douglas Fir: The Species and Its Management,* ed. DM Baumgartner, JE Lotan, pp. 63–72. Pullman, WA: Coop. Ext. Wash. State Univ. 301 pp.
161. Linhart YB, Mitton JB. 1985. Relationships among reproduction, growth rates and genetic protein heterozygosity in ponderosa pine. *Am. J. Bot.* 72:181–84
162. Linhart YB, Mitton JB, Sturgeon KB, Davis ML. 1981. Genetic variation in space and time in a population of ponderosa pine. *Heredity* 40:407–20
163. Linhart YB, Snyder MA, Gibson JP. 1994. Differential host utilization by two parasites in a population of ponderosa pine. *Oecologia* 98:117–20
164. Lloyd DG. 1975. Breeding systems in Cotula. III. Dioecious populations. *New Phytol.* 74:109–23
165. Loukas M, Vergini Y, Krimbas CB. 1983. Isozyme variation and heterozygosity in *Pinus halepensis* L. *Biochem. Genet.* 21:497–509
166. Loveless MD, Hamrick JL. 1984. Ecological determinants of genetic structure in plant populations. *Annu. Rev. Ecol. Syst.* 15:65–95
167. Lumaret R. 1984. The role of polyploidy in the adaptive significance of polymorphism at the GOT 1 locus in the *Dactylis glomerata* complex. *Heredity* 452:153–69
168. Lundkvist K. 1979. Allozyme frequency distributions in four Swedish populations of Norway spruce (*Picea abies* K.). I. Estimations of genetic variation within and among populations, genetic linkage and a mating system parameter. *Hereditas* 90:127–43
169. Lynch M. 1988. The divergence of neutral quantitative characters among partially isolated populations. *Evolution* 42:455–66
170. Lynch M. 1991. The genetic interpretation of inbreeding depression and outbreeding depression. *Evolution* 45:622–29
171. MacDonald SE, Chinnappa CC. 1989. Population differentiation for phenotypic plasticity in the *Stellaria longipes* complex. *Am. J. Bot.* 76:1627–37
172. Maddox GD, Root RB. 1990. Structure of the encounter between goldenrod (*Solidago altissima*) and its diverse insect fauna. *Ecology* 71:2115–24
173. Main JL. 1974. Differential responses to magnesium and calcium by natural populations of *Agropyron spicatum. Am. J. Bot.* 61:931–37
174. Maroof MAS, Biyashev RM, Yang GP, Zhang Q, Allard RW. 1994. Extraordinarily polymorphic microsatellite DNA in barley: species diversity, chromosomal locations and population dynamics. *Proc. Natl. Acad. Sci. USA* 91:5466–70
175. Maruyama T, Fuerst PA. 1984. Population bottlenecks and non-equilibrium models in population genetics. I. Allele numbers when populations evolve from zero variability. *Genetics* 108:745–63
176. Mayer MS, Soltis PS, Soltis DE. 1994. The evolution of the *Streptanthus glandulosus* complex (*Cruciferae*): genetic divergence and gene flow in serpentine endemics. *Am. J. Bot.* 81:1288–99
177. Mayr E. 1963. *Animal Species and Evolution.* Cambridge, MA: Belknap. 795 pp.
178. Mazzoni C, Gouyon PH. 1985. Horizontal structure of populations: migration, adaptation and drift: an experimental study of *Thymus vulgaris.* See Ref. 121, pp. 395–412
179. McCauley DE. 1994. Contrasting the distribution of chloroplast DNA and allozyme polymorphism among local populations of *Silene alba*: implications for studies of gene flow in plants. *Proc. Natl. Acad. Sci. USA* 91:8127–31
180. McClintock B. 1984. The significance of responses of the genome to challenge. *Science* 226:792–801
181. McGraw JB, Vavrek MC, Bennington CC. 1991. Ecological genetic variation in seed banks. I. Establishment of a time transect. *J. Ecol.* 79:617–25
182. McNeilly T. 1968. Evolution in closely adjacent plant populations. III. *Agrostis tenuis* on a small copper mine. *Heredity* 23:99–108
183. Meagher TR. 1991. Analysis of paternity within a natural population of *Chamaelirium luteum*: patterns of male

reproductive success. *Am. Nat.* 137:738–52
184. Merzeau D, Comps B, Thiebaut B, Cuguen J, Letouzey J. 1994. Genetic structure of natural stands of *Fagus sylvatica* L. (beech). *Heredity* 72:269–77
185. Michalakis Y, Sheppard AW, Noel V. 1993. Population structure of a herbivorous insect and its host plant on a microgeographic scale. *Evolution* 47:1611–16
186. Mithen R, Raybould AF, Giamoustaris A. 1995. Divergent selection for secondary metabolites between wild populations of *Brassica oleracea* and its implications for plant herbivore interactions. *Heredity* 75:472–84
187. Mitton JB. 1994. Molecular approaches to population biology. *Annu. Rev. Ecol. Syst.* 25:45–69
188. Mitton JB, Grant MC. 1996. Genetic variation and the natural history of quaking aspen. *Bioscience* 46:25–31
189. Mitton JB, Linhart YB, Hamrick JL, Beckman J. 1977. Population differentiation and mating systems in ponderosa pine in the Colorado Front Range. *Theor. Appl. Genet.* 51:5–14
190. Mitton JB, Sturgeon KB, eds. 1982. *Bark Beetles in North American Conifers.* Austin, TX: Univ. Texas Press. 527 pp.
191. Monson RK, Grant MC. 1989. Experimental studies of ponderosa pine. III. Differences in photosynthesis, stomatal conductance and water-use efficiency between two genetic lines. *Am. J. Bot.* 76:1041–47
192. Moody ME, Mueller LD, Soltis DE. 1993. Genetic variation and random drift in autotetraploid populations. *Genetics* 134:649–57
193. Mooney HA, Dunn EL. 1970. Convergent evolution of Mediterranean climate evergreen sclerophyllous shrubs. *Evolution* 24:282–303
194. Mopper S, Mitton JB, Whitham TG, Christensen KM. 1991. Genetic differentiation and heterozygosity in pinyon pine associated with herbivory and environmental stress. *Evolution* 45:989–99.
195. Nevo E. 1991. Genetic diversity and ecological heterogeneity in amphibian evolution. *Copeia* 1991:565–92
196. Nevo E, Krugman T, Beiles A. 1994. Edaphic natural selection of allozyme polymorphisms in *Aegilops peregrina* at a Galilee microsite in Israel. *Heredity* 72:109–12
197. Nichols RA, Hewitt GM. 1994. The genetic consequences of long distance dispersal during colonization. *Heredity* 72:312–17
198. Nickrent DL, Stell AL. 1990. Electrophoretic evidence for genetic differentiation in two host races of hemlock dwarf-mistletoe (*Arceuthobium tsugense*). *Biochem. Syst. Evol.* 18:267–75
199. Novak SJ, Mack RN, Soltis DE. 1991. Genetic variation in *Bromus tectorum Poaceae*: population differentiation in its North American range. *Am. J. Bot.* 78:1150–61
200. Ohsawa R, Furuya N, Ukai Y. 1993. Effect of spatially restricted pollen flow on spatial genetic structure of an animal-pollinated allogamous plant population. *Heredity* 71:65–73
201. Paine LE. 1950. The susceptibility of pear trees to penetration and toxic damage by mistletoe. *Phytopath. Zeitschrift.* 17:305–27
202. Parker MA. 1989. Disease impact and local genetic diversity in the clonal plant *Podophyllum peltatum. Evolution* 43:540–47
203. Parker KC, Hamrick JL. 1992. Genetic diversity and clonal structure in a columnar cactus *Lophocereus schottii. Am. J. Bot.* 79:86–96
204. Parks CR, Wendel JF, Sewell MM, Qiu Y-L. 1994. The significance of allozyme variation and introgression in the *Liriodendron tulipifera* complex (*Magnoliaceae*). *Am. J. Bot.* 81:878–89
205. Perry DJ, Knowles P. 1991. Spatial genetic structure within three sugar maple (*Acer saccharum* Marsh) stands. *Heredity* 66:137–42.
206. Phillips PC, Arnold SJ. 1989. Visualizing multivariate selection. *Evolution* 43:1209–22
207. Platenkamp GAJ, Shaw RG. 1992. Environmental and genetic constraints on adaptive population differentiation in *Anthoxanthum odoratum. Evolution* 46:341–52
208. Plessas ME, Strauss SH. 1986. Allozyme differentiation among populations, stands, and cohorts in Monterrey pine. *Can. J. For. Res.* 16:1155–64
209. Popp M. 1983. Genotypic differences in the mineral metabolism of plants adapted to extreme habitats. *Plant Soil* 72:201–73
210. Price SC, Schumaker KN, Kahler AL, Allard RW, Hill JE. 1984. Estimates of population differentiation obtained from enzyme polymorphisms and quantitative traits. *J. Hered.* 75:141–42
211. Qui Y-L, Parks CR. 1994. Disparity of allozyme variation levels in three Mag-

nolia (*Magnoliaceae*) species from the southeastern United States. *Am. J. Bot.* 81:1300–8
212. Quinn JA. 1987. Complex patterns of genetic differentiation and phenotypic plasticity versus an outmoded ecotype terminology. See Ref. 258, pp. 95–113
213. Rabe EW, Haufler CH. 1992. Incipient polyploid speciation in the maidenhair fern *Adiantum pedatum* (*Adiantaceae*). *Am. J. Bot.* 79:701–7
214. Ranker TA, Floyd SK, Trapp PG. 1994. Multiple colonizations of *Asplenium nigrum* into the Hawaiian archipelago. *Evolution* 48:1364–70
215. Reinartz JA. 1984. Life history variation of common mullein (*Verbascum thapsus*): differences among sequential cohorts. *J. Ecol.* 72:927–36
216. Reinartz JA. 1994. Bottleneck-induced dissolution of self-incompatibility and breeding system consequences in *Aster furcatus* (*Asteraceae*). *Am. J. Bot.* 81: 446–55
217. Richards AJ. 1986. *Plant Breeding Systems*. London: Allen & Unwin. 529 pp.
218. Ridley M. 1993. *Evolution*. Oxford: Blackwell. 670 pp.
219. Russell G, Morris P. 1970. Copper tolerance in the marine fouling alga *Ectocarpus siliculosus*. *Nature* 228:288–89
220. Samuel R, Pinsker W, Ehrendorfer F. 1990. Allozyme polymorphism in diploid and polyploid populations of *Galium*. *Heredity* 65:369–78
221. Schlichting CD. 1986. The evolution of phenotypic plasticity in plants. *Annu. Rev. Ecol. Syst.* 17:667–94
222. Schmitt J, Gamble SE. 1990. The effect of distance from the parental site on offspring performance and inbreeding depression in *Impatiens capensis:* a test of the local adaptation hypothesis. *Evolution* 44:2022–30
223. Schnabel A, Asmussen MA. 1992. Comparative effects of pollen and seed migration on the cytonuclear structure of plant populations: paternal cytoplasmic inheritance. *Genetics* 132:253–67
224. Schnabel A, Hamrick JL. 1990. Organization of genetic diversity within and among populations of *Gleditsia triacanthos* (*Leguminosae*). *Am. J. Bot.* 77:1060–69
225. Schnabel A, Lauschman RH, Hamrick JL. 1991. Comparative genetic structure of two co-occurring tree species: *Maclura pomifera* (*Moraceae*) and *Gleditsia triacanthos* (*Leguminosae*). *Heredity* 67:357–64
226. Schuster WS, Alles DL, Mitton JB. 1989. Gene flow in limber pine: evidence from pollination phenology and genetic differentiation along an elevational transect. *Am. J. Bot.* 76:1395–403
227. Shapcott A. 1995. The spatial genetic structure in natural populations of the Australian temperate rainforest tree *Atherosperma moschatum* (Labill) *Monimiaceae*. *Heredity* 74:28–38
228. Shaw J. 1987. Evolution of heavy metal tolerance in Bryophytes. II. An ecological and experimental investigation of the 'copper moss' *Scopelophila cataractae* (*Pottiaceae*). *Am. J. Bot.* 74:813–21
229. Shaw J. 1987. Effect of environmental pretreatment on tolerance to copper and zinc in the moss *Funaria hygrometrica*. *Am. J. Bot.* 74:1466–77
230. Shaw J, ed. 1990. *Heavy Metal Tolerance in Plants: Evolutionary Aspects*. Boca Raton, LA: CRC. 355 pp.
231. Shaw J, Antonovics J, Anderson LE. 1987. Inter- and intra specific variation of mosses in tolerance to copper and zinc. *Evolution* 41:1312–25
232. Shea KL. 1989. Genetic variation between and within populations of Engelmann spruce and subalpine fir. *Genome* 33:1–8
233. Simms EL, 1990. Examining selection on the multivariate phenotype: plant resistance to herbivores. *Evolution* 44:1177–88
234. Simms EL, Fritz RS. 1990. Ecology and evolution of host plant resistance to insects. *Trends Ecol. Evol.* 5:356–60
235. Simms EL, Rauscher MS. 1992. Use of quantitative genetics for studying the evolution of plant resistance. See Ref. 79, pp. 52–68.
236. Slatkin M. 1993. Isolation by distance in equilibrium and non-equilibrium populations. *Evolution* 47:264–79
237. Snaydon RW, Davies MS. 1976. Rapid population differentiation in a mosaic environment. II. Morphological variation in *Anthoxanthum odoratum*. *Evolution* 26:390–405
238. Snaydon RW, Davies MW. 1982. Rapid divergence of plant populations in response to recent changes in soil conditions. *Evolution* 36:289–97
239. Soltis PS, Soltis DE, Holsinger KE. 1988. Estimates of intragametophytic selfing and interpopulational gene flow in homosporous ferns. *Am. J. Bot.* 75:1765–70
240. Soltis PS, Soltis DE, Ness BD. 1989. Population genetic structure in *Cheilan-*

thes gracillima. Am. J. Bot. 76:1114–18

240a. Sork VL, Stowe KA, Hochwender C. 1993. Evidence for local adaptation in closely adjacent subpopulations of northern red oak (Quercus *rubra* L.) expressed as resistance to leaf herbivores. *Am. Nat.* 142:928–36

241. Stebbins GL. 1958. The role of hybridization in evolution. *Proc. Am. Phil. Soc.* 103:231–51

241a. Telleman PS. 1984. Reflections on the transition from wind pollination to ambophily. *Acta Bot. Neerl.* 33:497–508

242. Stephan BR. 1987. Differences in resistance of Douglas-fir provenances to the wooly aphid *Gilleteela cooleyi. Silvae Genet.* 36:76–79

243. Strauss SH, Hong Y-P, Hipkins VD. 1993. High levels of population differentiation for mitochondrial DNA haplotypes in *Pinus radiata, muricata* and *attenuata. Theor. Appl. Genet.* 86:605–11

244. Stuefer JF, During HJ, DeKroon H. 1994. High benefits of clonal integration in two stoloniferous species in response to heterogeneous light environments. *J. Ecol.* 82:511–18

245. Sultan SE, Bazzaz FA. 1993. Phenotypic plasticity in *Polygonum persicaria.* III. The evolution of ecological breadth for nutrient environment. *Evolution* 47:1050–71

245a. Teramura AH, Strain BR. 1979. Localized population differences in photosynthetic response to temperature and irradiance in *Plantago lanceolata. Can. J. Bot.* 57:2559–63

246. Thomas SC, Bazzaz FA. 1993. The genetic component in plant size hierarchies: norms of reaction to density in a *Polygonum* species. *Ecol. Mono.* 63:231–49

247. Thompson JD, McNeilly T, Gray AJ. 1991. Population variation in *Spartina anglica* CE Hubbard. I. Evidence from a common garden experiment. *New Phytol.* 117:115–28

248. Tomback DF, Linhart YB. 1990. The evolution of bird-dispersed pines. *Evol. Ecol.* 4:185–219

249. Tomekpe K, Lumaret R. 1991. Association between quantitative traits and allozyme heterozygosity in a tetrasomic species: *Dactylis glomerata. Evolution* 45:359–70

250. Tonsor SJ. 1990. Spatial patterns of differentiation for gene flow in *Plantago lanceolata. Evolution* 44:1373–78

251. Tonsor SJ, Kalisz S, Fisher J, Holtsford TP. 1993. A life-history based study of population genetic structure, seed bank to adults in *Plantago lanceolata. Evolution* 47:833–43

252. Triest L, ed. 1991. Isozymes in water plants. *Opera Botanica Belgica 4.* Natl. Bot. Gard. of Belgium. 264 pp.

253. Turesson G. 1922. The genotypical response of the plant species to the habitat. *Hereditas* 3:211–350

254. Turkington R. 1989. The growth, distribution, and neighbour relationships of *Trifolium repens* in a permanent pasture. V. The coevolution of competitors. *J. Ecol.* 77:717–33

255. Turkington RA, Aarssen LW. 1984. Local scale differentiation as a result of competitive interactions. See Ref. 58, pp. 107–27

256. Turner ME, Stephens JC, Anderson WW. 1982. Homozygosity and patch structure in plant populations as a result of nearest-neighbor pollination. *Proc. Natl. Acad. Sci. USA* 79:203–7

257. Urbanska KM. 1984. Polymorphism of cyanogenesis in *Lotus alpinus* from Switzerland. II. Phenotypic and allelic frequencies upon acidic silicate and carbonate. *Ber. Geobot. Inst. Zurich* 51:132–63

258. Urbanska KM. 1987. *Differentiation Patterns in Higher Plants.* London: Academic. 272 pp.

259. Van Tienderen PH, Van Der Toorn J. 1991. Genetic differentiation between populations of *Plantago lanceolata.* II. Phenotypic selection in a transplant experiment in three contrasting habitats. *J. Ecol.* 79:43–60

260. Van Treuran R, Bijlsma R, Ouborg NJ, Van Delden W. 1993. The effects of population size and plant density on outcrossing rates in a locally endangered *Salvia pratensis. Evolution* 47:1094–104

261. Vasek F. 1980. Creosote bush: long-lived clones in the Mojave desert. *Am. J. Bot.* 67:246–55

262. Verkleij JAC, Bast-Cramer WB, Levering H. 1985. Effect of heavy metal stress on the genetic structure of populations of *Silene cucubalus.* In *Structure and Functioning of Plant Populations,* ed. J Haeck, JW Woldendorp, 2:355–65. Amsterdam: North-Holland. 405 pp.

263. Via S. 1991. The genetic structure of host plant adaptation in a spatial patchwork: demographic variability among reciprocally transplanted pea aphid clones (*Acyrthosiphon pisum*). *Evolution* 45:827–52

264. Warwick SL Briggs I. 1979. The genecology of lawn weeds. III. Cultivation experiments with *Achillea millefolium* L., *Bellis perennis* L., *Plantago lanceolata* L., *Plantago major* L. and *Prunella vulgaris* L. collected from lawns and contrasting grassland habitats. *New Phytol.* 83:509–36
265. Warwick SL, Briggs I. 1980. The genecology of lawn weeds. VI. The adaptive significance of variation in *Achillea millefolium* L. as investigated by transplant experiments. *New Phytol.* 85:451–60
266. Waser N, Price MV. 1989. Optimal outcrossing in *Ipomopsis aggregata:* seed set and offspring fitness. *Evolution* 43:1097–109
267. Waser NM, Price MV. 1994. Crossing-distance effects in *Delphinium nelsonii:* outbreeding and inbreeding depression in progeny fitness. *Evolution* 48:842–52
268. West-Eberhard MJ. 1992. Adaptation: current usages. In *Keywords in Evolutionary Biology,* ed. EF Keller, EA Lloyd, pp.13–18, Cambridge, MA: Harvard Univ. Press. 414 pp.
269. Westerbergh A, Saura A. 1992. The effect of serpentine on the population structure of *Silene dioica, Caryophyllaceae. Evolution* 46:1537–48
270. Westerbergh A, Saura A. 1994. Genetic differentiation in endemic *Silene* (*Caryophyllaceae*) on the Hawaiian Islands. *Am. J. Bot.* 81:1487–93
271. Whitham TG. 1983. Host manipulation of parasites: within-plant variation as a defense against rapidly evolving pests. In *Variable Plants and Herbivores in Natural and Managed Systems,* ed. RF Denno, MS McClure, pp. 15–41. New York: Academic. 717 pp.
272. Whitlock MC. 1992. Temporal fluctuations in demographic parameters and the genetic variance among populations. *Evolution* 46:608–15
273. Wijesinghe DK, Handel SN. 1994. Advantages of clonal growth in heterogenous habitats: an experiment with *Potentilla simplex. J. Ecol.* 82:495–502
274. Williams CF, Guries RP. 1994. Genetic consequences of seed dispersal in three sympatric forest herbs. I. Hierarchical population–genetic structure. *Evolution* 48:791–805
275. Williams GC. 1966. *Adaptation and Natural Selection.* Princeton NJ: Princeton Univ. Press. 307 pp.
276. Willis JH, Orr HA. 1993. Increased heritable variation following population bottlenecks: the role of dominance. *Evolution* 47:949–57
277. Wilson AC, Carlson SS, White TJ. 1977. Biochemical evolution. *Annu. Rev. Biochem.* 46:573–639
278. Wolf PG, Soltis PS. 1992. Estimates of gene flow among populations, geographic races, and species in the *Ipomopsis aggregata* complex. *Genetics* 130:639–47
279. Wu L, Antonovics J. 1976. Experimental genetics of *Plantago.* II. Lead tolerance in *P. lanceolata* and *Cynodon dactylon* from a roadside. *Ecology* 37:205–8
280. Wu L, Bradshaw AD, Thurman DA. 1975. The potential for evolution of heavy metal tolerance in plants. III. The rapid evolution of copper tolerance in *Agrostis stolonifera. Heredity* 34:165–85
281. Wu L, Lin SL. 1990. Copper tolerance and copper uptake of *Lotus purshianus* and its symbiotic *Rhizobium loti* derived from a copper mine waste population. *New Phytol.* 116:531–38
282. Wyatt R, Odrzykowski KJ, Stoneburner A. 1989. High levels of genetic variability in the haploid moss *Plagiomnium ciliare. Evolution* 43:1085–96
283. Yeh FC, Cheliak WM, Dancik BP, Illingworth K, Trust DC, Pryhitka BA. 1985. Population differentiation in lodgepole pine, *Pinus contorta,* ssp. *latifolia*: a discriminant analysis of allozyme variation. *Can. J. Genet. Cytol.* 27:210–18
284. Zhang Q, Maroof MAS, Kleinhofs A. 1994. Comparative diversity analysis of rflps and isozymes within and among populations of *Hordeum vulgare* ssp *spontaneum. Genetics* 134:909–16

Annu. Rev. Ecol. Syst. 1996. 27:279–303

RATES OF MOLECULAR EVOLUTION: Phylogenetic Issues and Applications

David P. Mindell and Christine E. Thacker
Department of Biology and Museum of Zoology, University of Michigan, Ann Arbor, Michigan 48109-1079

KEY WORDS: rates of evolution, character weighting, constraints on character change, role of evolution

ABSTRACT

The proper relationship between systematics practice and our understanding of evolution has been long debated. Systematists seek to avoid assumptions about evolutionary process in their methods, yet a growing body of evidence indicates that patterns in rates of evolution can be used to reduce effects of homoplasy. We review variable evolutionary rates for molecular characters in the context of constraints on mutation and fixation. Some constraints, like the genetic code for protein-coding genes, are consistent in the direction of their effects, whereas others, like population size and cladogenesis frequency, are historically variable within and among lineages.

We review methods for assessing rate variability, and we estimate comparative absolute rates of change for five sets of mitochondrial DNAs in 12 vertebrates for application in phylogenetic analyses. Unequal weights for subsets of mitochondrial DNAs improved congruence with the most highly corroborated tree in many but not all cases. The largest data set (12,120 bases) yielded the same tree under all four weighting alternatives. This is consistent with the notion, echoing the law of large numbers, that as data sets increase in size, homoplasy will tend to cancel itself out. Even if this notion has validity, however, evolutionary biology will remain vital in systematics if we want to: match sets of taxa with characters likely to be historically informative (when data sets are not sufficiently large); avoid comparing characters with different histories due to reticulations, horizontal transfer, or lineage sorting; avoid assuming random distribution of homoplasy; be alerted to the possibility of long-branch attraction problems; and understand the cause of the hierarchy of taxa in nature as inheritance of genetic material and descent with modification.

0066-4162/96/1120-0279$08.00

INTRODUCTION

An appealing aspect of cladistics is its claim of relative freedom from assumptions about evolutionary processes (29, 86, 120). If we presume the existence of 1) a natural, divergent hierarchy of organisms based on common descent and 2) identifiable shared-derived characters (homologies) for taxa within the hierarchy, then discovery of monophyletic groups appears to be a straightforward task. Preference for a parsimony criterion in this discovery process derives from the general scientific practice of minimizing ad hoc assumptions (assumptions of homoplasy in the case of phylogenetic analyses) and the notion that parsimony correctly determines which phylogenetic hypothesis is best supported by the character evidence (32, 85). One need not invoke any specifics of evolutionary process to estimate genealogy. This enhances objectivity of phylogenetic analyses by separating them from preconceived notions of evolutionary process.

However, homologies can be difficult to identify. This may be attributed in part to the existence of a finite number of character states (four in the case of DNAs) and rates of change sufficient to yield independent expressions of the same state. Putative homologies, like putative relationships among taxa, are products of phylogenetic analysis (85); attempts to improve them must, therefore, involve refinements in phylogenetic analyses. Such refinement has long been sought in the use of conservative characters. By giving greater weight in phylogenetic analyses to characters changing less frequently, confounding effects of homoplastic similarity can be reduced.

Thus, improved understanding of rates of molecular evolution has the potential to improve phylogenetic analyses. However, for some this may appear to conflict with cladistic parsimony's appeal of making minimal assumptions about evolutionary process; and so several questions arise. Does character weighting based on relative rates of change conflict with a "total evidence" approach to character analysis (61, 111)? Can parsimony analyses recover genealogy without recourse to knowledge and theory regarding biological evolution? If not, what knowledge of evolution is needed? How is that knowledge gained, and how might it be applied? These questions focus on the relevance of biology for phylogenetic inference.

ASSESSING EVOLUTIONARY RATES

Organismal Rates

Absolute rates of molecular evolution have been estimated using fossil or biogeographic events together with pairwise distance measures, to provide a minimum age for the divergence of pairs of taxa. Such estimates must presume 1) that the rate is constant over time and across taxa for species involved in the pairwise distance comparisons and 2) that the fossils or biogeographic dates accurately reflect lineage divergence times. Unfortunately, these are not safe presumptions. Use of discrete character branch lengths, rather than distances, can obviate the first but not the second assumption.

A brief look at avian divergence time and rate estimates illustrates the difficulties. Sibley & Ahlquist's (106) calibration for DNA-DNA hybridization distances was based on assuming that a sister relationship existed between the ostrich and rhea lineages, and that the Atlantic Ocean created a dispersal barrier to their most recent common ancestor about 80 million years ago (mya). However, the sister relationship of ostriches and rheas as well as the Gondwana origin of ratites (including ostriches and rheas) has been contradicted by paleontological evidence (55) and mitochondrial DNA (mtDNA) analyses (18). Shields & Wilson (104) estimated the mtDNA divergence rate in geese at 1.8% to 2.3% per my, based on distances from RFLP data. This estimate is similar to absolute rate estimates of 2.0% per my for mtDNAs in mammals (122), suggesting similar rates in birds and mammals. However, the calculation for birds suffers from being based on a single divergence time estimate and has led to estimates of effective population sizes inconsistent with observed sizes (5, 90). This calculation also conflicts with evidence indicating slower rates of mitochondrial sequence evolution in birds relative to mammals (59, 76). Thus, the assumptions inherent in calibrating require that they be interpreted cautiously (52).

Relative rate tests for lineages are based on the expectation of equal amounts of character change in sister taxa relative to an outgroup, if rates of evolution in the sister taxa are equal. This follows from the fact that sister taxa have had equal amounts of time for accumulation of change relative to the outgroup, and it makes the test independent of fossil or biogeographic event dating. Relative rate tests based on distances (66, 98, 125) must assume equal amounts of homoplasy across taxa. Relative rate tests using branch lengths from phylogenetic trees to estimate amounts of unambiguous character state change and a binomial test to assess departure from the expected 50% of all change found in each of two ingroup taxa (75), together have the potential for recovering differential homoplasy amounts across taxa. The advantages of discrete character relative rate tests, compared to relative rate tests based on genetic distances, are that 1) different kinds of character change may readily be used in seeking to emphasize those with lower levels of homoplasy; 2) no assumptions of equal amounts of homoplasy across taxa are made, as homoplasious similarity may be recognized and tallied on phylogenetic tree branch lengths; and 3) internal branches shared by ingroup taxa may be excluded from analyses, thereby preventing one form of nonindependence (Figure 1). Alternative pairwise comparisons involving the same species, however, are not independent.

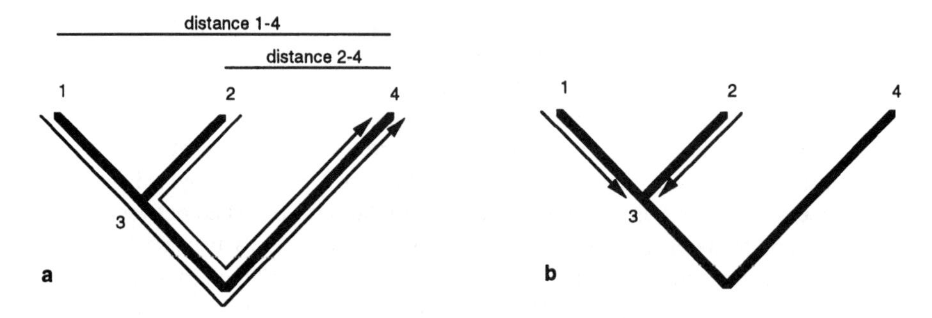

Figure 1 Comparison of distance versus parsimony-based relative rate tests (RRTs). *1a.* Distance-based RRTs have non-independence problems in counting changes twice that accumulate historically between 4 and 3. *1b.* Parsimony-based RRTs compare only autapomorphies for *1* and *2*. Further, assumptions of equal amounts of homoplasy across taxa are not made as in *1a* because homoplasious similarity may be recognized and tallied on phylogenetic tree branch lengths.

Evolutionary rate estimates may be biased in favor of finding faster rates in clades with more taxa, as overwritten character state changes can be discovered only by adding taxa to a clade (36, 45). Although less widely recognized, addition of taxa may also result in less of the actual change being detected due to convergent events (see 78, 97). Both potential biases stem from analyses of data sets with high levels of homoplasy. Focus on relatively conserved types of sequence change and data sets with low levels of homoplasy works to minimize these potential biases. Relative rate tests may lead to false conclusions if the two ingroup taxa are so recently diverged that actual differences in rate have not yet appeared, or if the designated outgroup is actually more closely related to one or the other of the ingroup taxa.

Though we focus on parsimony-based relative rate tests, least-squares and maximum-likelihood methods have also been developed and are designed for identifying sequences that differ in rate compared with an average rate for all sequences (35). Muse & Gaut (83) implement a likelihood approach in comparing relative rates of change across taxa for all characters as well as for transitions (TIs), transversions (TVs), synonymous and nonsynonymous nucleotide substitutions in their program CODRATES. Their evolutionary model uses the codon, rather than the nucleotide, as the unit of evolution, and it seeks to account for dependencies among nucleotides and multiple hits within a codon. F-ratio tests (96) have also been used to investigate rate variation across taxa by comparing length of sister branches from distance trees (103, 110), though these tests assume both additivity and independence (34).

Character Rates

Number of nucleotide substitution types found in pairwise comparisons of taxa may be plotted against estimated time since divergence of those taxa to estimate absolute rates of character change. Absolute rate estimates can be compared for various kinds of characters to determine their degree of saturation with change over time (Figure 2). Alternatively, the number of substitutions of different kinds can be plotted against each other to provide an estimate of relative rates of change (73, 116). If one kind of substitution decreases in frequency relative to another, or relative to absolute time, this suggests multiple changes at single nucleotide positions and some degree of saturation with change.

However, such interpretation of rate estimates entails assumptions and must be considered cautiously. Homoplasy can be recognized only in the context of a phylogeny and will be underestimated in pairwise comparisons (38). Comparisons averaging rates of change across sites will tend to underestimate biases in rates of change for specific nucleotide substitutions (116a). Application of pairwise comparisons also assumes a degree of rate constancy over time and across taxa. Increased sampling of taxa and of divergence time intervals will provide more robust rate comparisons, reducing the impact of any one taxon; however, the relative rate differences found remain tentative and should be reassessed as more data becomes available.

Apparent advantages of character rate comparisons based on branch lengths are as mentioned above (points 2 and 3), and it would be useful to know if the same patterns are seen in both types of comparison. We performed such comparisons and found that the relative rate estimates determined in a phylogenetic context show patterns similar to those from direct pairwise comparisons (Figures 2 and 3; see below for methods). For example, in the 13 mitochondrial protein-coding gene graphs, both estimates of change decrease in rate, suggesting multiple hits, after about 100 my. Estimates of evolutionary rate for categories of character change are generally not done in the context of a phylogeny because the phylogeny is unknown. Several methods have been used to estimate rates iteratively in phylogenetic analyses (31, 121); however, these focus on individual characters and do not pertain directly to learning about variable rates for particular categories of sequence change.

CONSTRAINTS AND PATTERNS FOR MOLECULAR RATES

Recognizing mutation and fixation as two fundamental steps in evolution, we summarize primary constraints influencing rates of change at these two levels (Table 1). Variation in mutation rate can stem from differences in replication

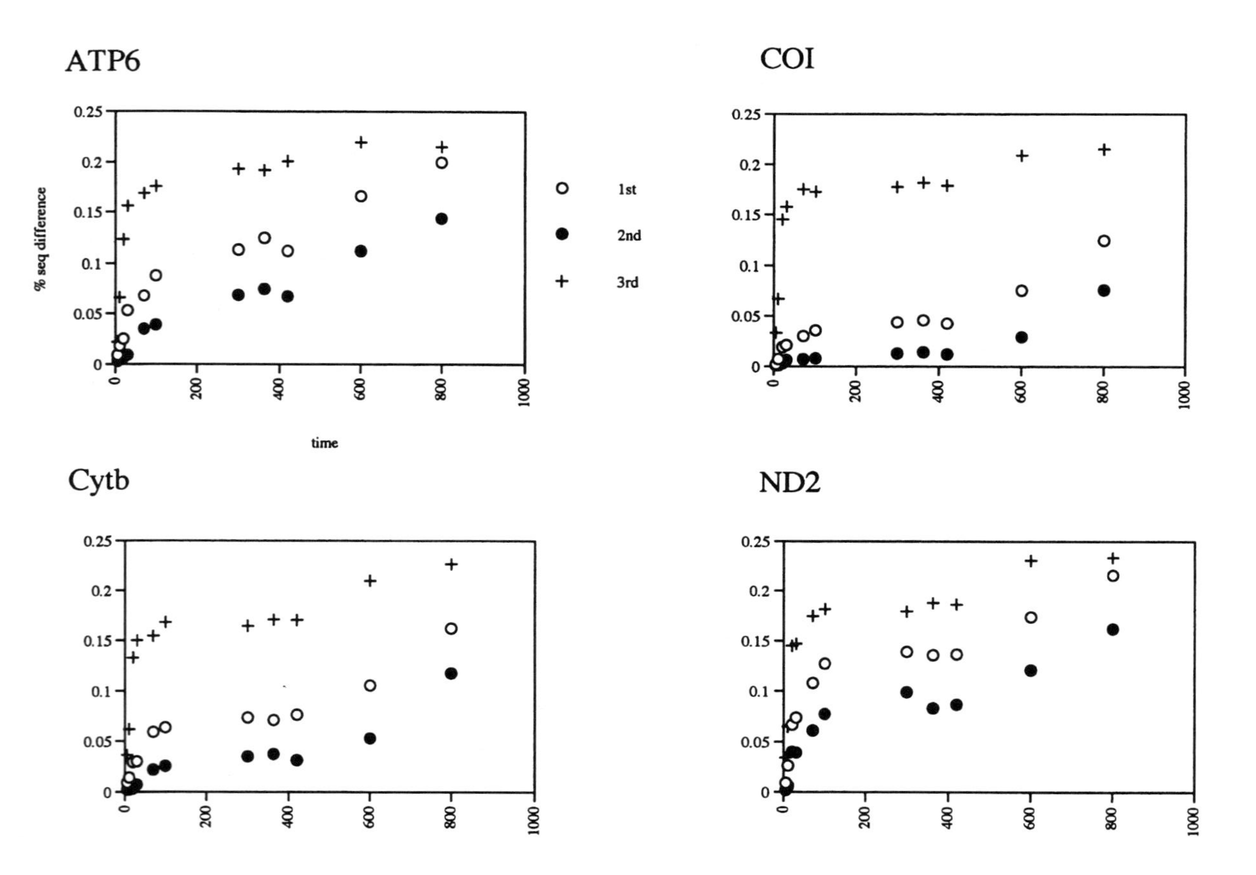
ATP6
COI
Cytb
ND2
% seq difference
time
0.25
0.2
0.15
0.1
0.05
0
200
400
600
800
1000
1st
2nd
3rd
(a)

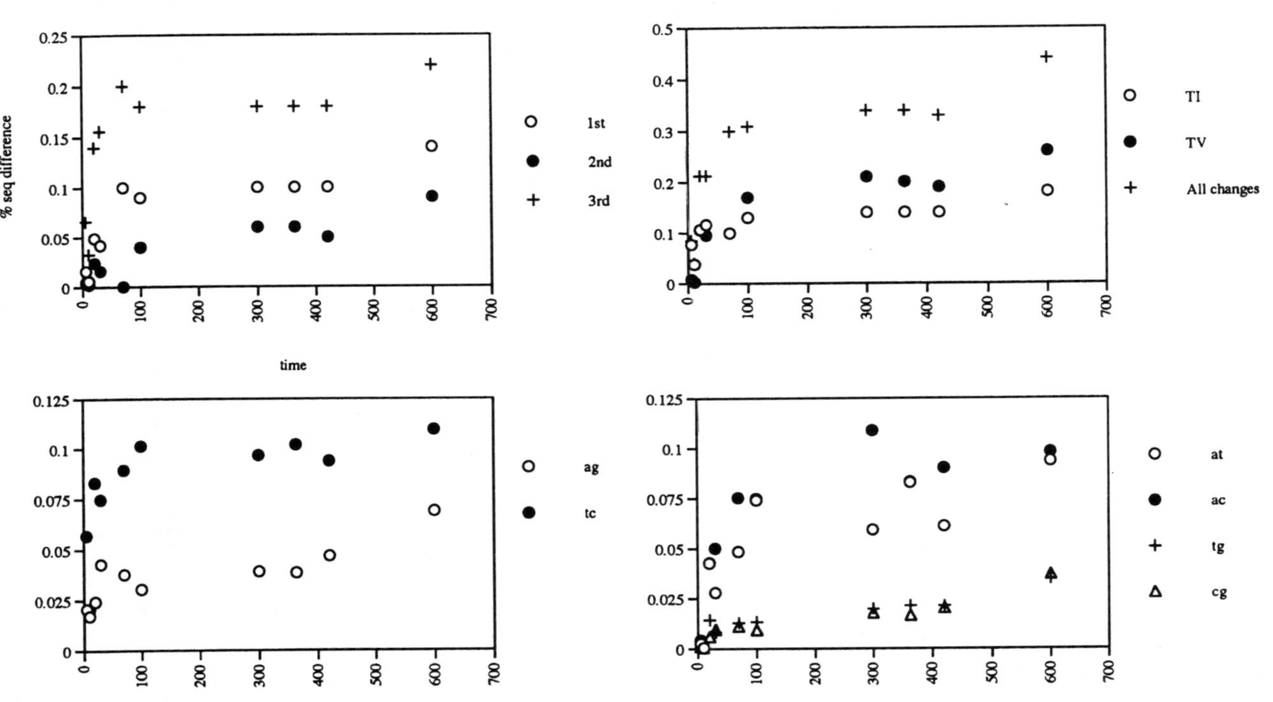

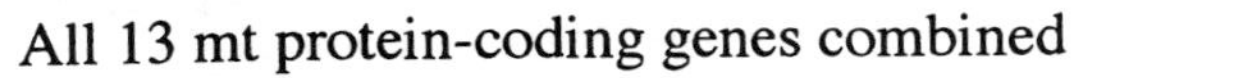

Figure 2 Comparative absolute rates of sequence change based on direct comparisons of species pairs. Plots show percentage DNA sequence difference versus estimated time (mya) since divergence for (*a*) four mitochondrial protein-coding genes, and for (*b*) 13 mitochondrial protein-coding genes combined (broken down by codon position and types of change; TI, transitions; TV, transversions). Comparisons involve the following 12 taxa (with database accessions numbers): *Cyprinius carpio,* carp, X61010; *Crossostoma lacustre,* loach, M91245; *Xenopus laevis,* frog, X02890, M10217, X01600, X01601; *Gallus gallus,* chicken, X52392; *Didelphus virginiana,* opossum, Z29573; *Mus musculus,* mouse, V00711; *Rattus norvegicus,* rat, X14848; *Homo sapiens,* human, V00662; *Balaenoptera physalus,* fin whale, X61145; *Balaenoptera musculus,* blue whale, X72204; *Phoca vitulina,* harbor seal, X63726, S37044; *Halichoerus grypus,* grey seal, X72004. Estimates of divergence times are based on fossil and biogeographic evidence as summarized in Benton (7).

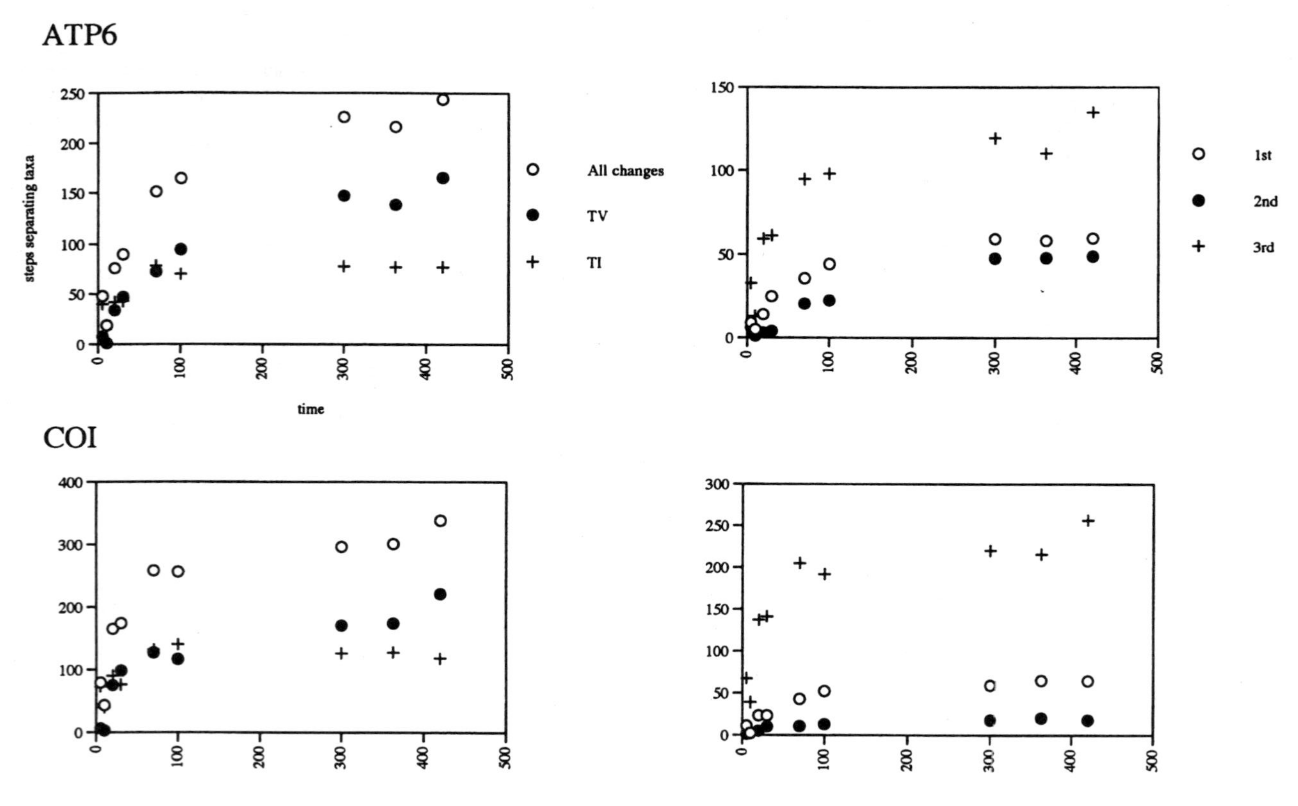

Figure 3 Comparative absolute rates of sequence change based on phylogenetic tree branch lengths. Plots show total numbers of character changes (counted from the two branch tips back to their first shared node) separating species pairs, versus estimated time since divergence for four individual mitochondrial protein-coding genes, and for 13 mitochondrial protein-coding genes combined. Comparisons involve the same 12 taxa as in Figure 2.

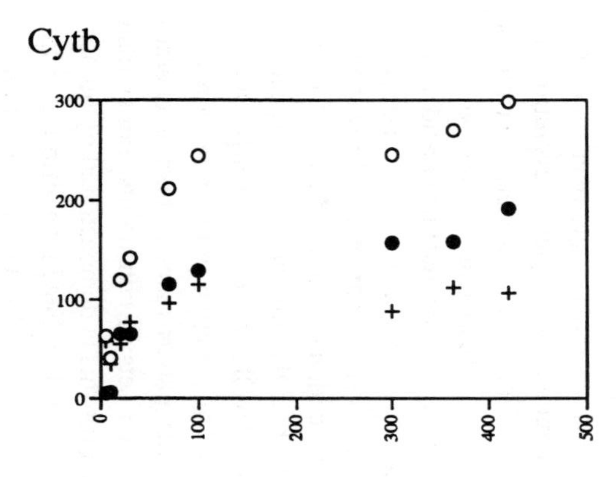

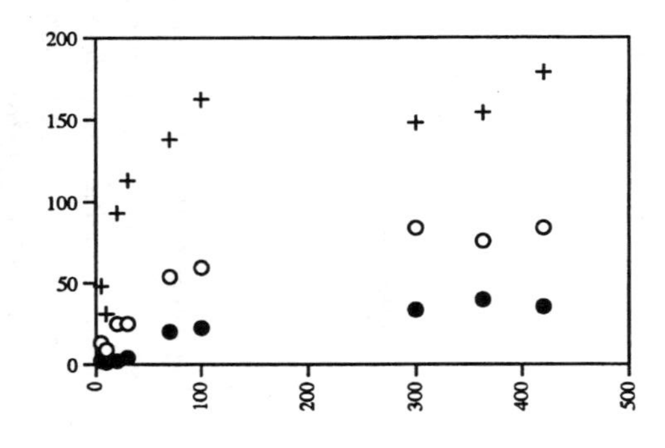

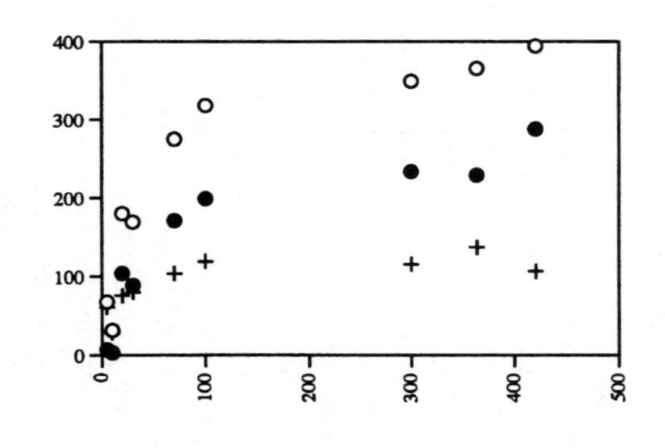

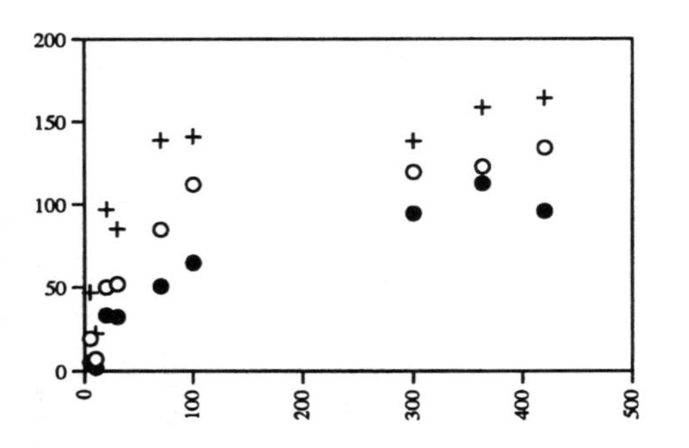

(b)

Figure 3 (*Continued*)

All 13 mt protein-coding genes combined

(c)

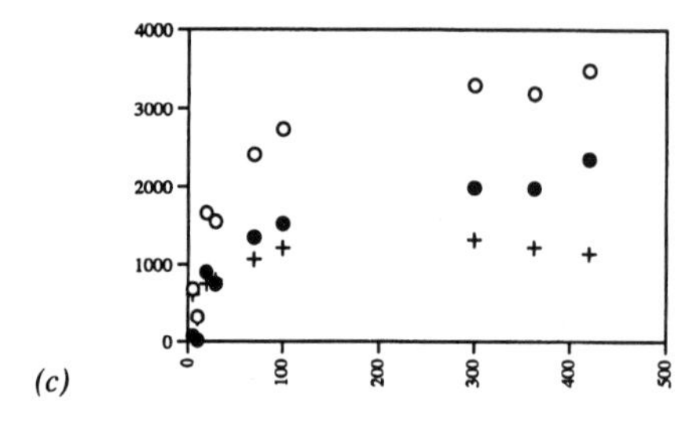

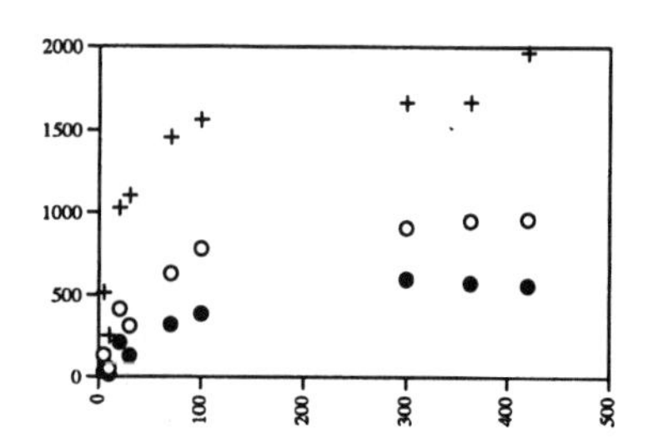

Figure 3 (*Continued*)

fidelity (different polymerase types), replication repair (different mechanisms and different enzymes), replication frequency, exposure to mutagens (especially DNA-damaging oxygen free-radicals), and the initial conditions of differential codon and nucleotide base composition for genes and taxa. These factors and others may influence variation in both the overall rate and the rate for particular kinds of change. The likelihood that one or more mutations become fixed is influenced by a set of continuous and overlapping constraints, including the genetic code, secondary or tertiary structure, gene function, population size, frequency of cladogenesis, and natural selection.

Variation in the nature of these constraints across taxa and over time contributes to patterns of variation in rate of molecular sequence evolution for different taxa and different kinds of character changes. However, determining which of the multiple overlapping constraints are responsible, and to what degree, may be difficult or impossible. So-called patterns must be considered as hypotheses to be judged on their individual merits for the set of taxa or genes in which they are proposed. In preliminary analyses, we characterized covariation among various types of character change (by codon position and by each of the six possible base substitution pairs) using principal component analysis, and we found that no single category of change served well as a proxy for a larger subset.

Recognizing that patterns in rate stem from variation in the constraints on mutation and fixation (Table 1), we consider patterns among taxa as characterizing groups whose members have similar sets of constraints, rather than characterizing individual clades. The need for this is demonstrated by numerous observations of rate differences among closely related taxa differing in the nature of their constraints (e.g. 8a, 9, 23, 37, 77, 126). Further illustrating a decoupling between patterns of rate variation and clades are studies showing that two or more sets of homologous genes need not show the same pattern of rate variability across the same set of organisms (76). That is, constraints vary

Table 1 Overlapping constraints and their hypothesized effects on evolutionary rates for molecular sequences among taxa.

Variable constraints	Hypothesized effect
A) ON MUTATION RATE	
Replication rate	Less frequent replication yields slower rates of change (11, 68, 89)
Replication and repair efficiency	Higher efficiency yields slower rates of change (49)
Polymerase type	Polymerases with lower error rates yield slower rates of change (e.g. low fidelity reverse transcriptase yields extremely fast rates for retroviruses compared to eukaryotes with higher fidelity nuclear-encoded DNA polymerases (24, 112))
Exposure to mutagens	Reduced exposure to mutagens, such as DNA-damaging oxygen free-radicals, yields slower rates of change (71, 105, reviewed in 95)
Base composition and codon usage	Unequal nucleotide base frequencies (8, 62, 94, 84, 58) and unequal codon usage (46, 84, 102) bias the frequency of alternative substitutions (e.g. initial low frequency of G yields relatively slow rate of G->T, A, or C)
B) ON FIXATION RATE	
Genetic code	Degeneracy of code contributes to slower rates for nonsynonymous than synonymous change, transversions than transitions, and progressively faster rates at second, first and third codon positions in many taxa.
Secondary and tertiary structure	Functional/structural importance and conservation across taxa is correlated with slower rates of primary sequence character change (123, 22, 48)
Gene or protein function	Functional importance (presence of a large proportion of sites of large effect) is correlated with slower rates of sequence character change in unchanging environments (reviewed in 67) and faster rates in changing environments (42a; see Directional selection below)
Purifying selection	Selection pressure against deleterious mutations yields slower rates of change. E.g. more stringent cellular environments such as extreme or fluctuating body temperatures are correlated with slower rates of change in some taxa (1, 6, 76, 114); relaxation of selection pressure is associated with faster rates in others (53)
Directional selection	Increased selection pressure for adaptive change yields faster rates of evolution, particularly for nonsynonymous substitutions (37, 57, 107)
Population size	Smaller effective population sizes can yield faster rates of change (23, 88, 82a)
Cladogenesis rate	Increased lineage splitting and isolation can preserve variation that might otherwise be lost, yielding faster rates of change (41, 77, 73, see 8a)

within and among genes as well as across taxa. Although the members of any particular clade, and the genes within any particular genome, may share similar constraints on mutation and fixation rates, this should not be assumed a priori.

With these caveats in mind, some general trends among clades, reflecting their variable constraints, are apparent. Mitochondrial and nuclear rates are reported as slower in primates, particularly humans, than in rodents (44, 68, 100, but see 26), slower in some birds than in some mammals (11, 59, 76), and slower in some fish, amphibians, and nonavian reptiles than in some mammals and birds (1, 4, 70, 117). Analogous rate differences have been observed within and among many other taxa, including plants (42, 124), bacteria (64), and viruses (24, 25). The broad trends noted among higher-level taxa entail slower rates in those taxa with one or more of the following: lower replication rates, decreased exposure to mutagenic oxygen free-radicals (decreased metabolic rate), higher replication fidelity, and greater purifying selection pressure. Variation in such constraints within and among taxa can yield exceptions in the trends (126). Effective population size and rate of cladogenesis are especially prone to within taxon variation and can influence rate variability. Patterns in rate heterogeneity for various kinds of molecular sequence character change also stem from multiple constraints (Table 1) and have recently been reviewed by Simon et al (108).

APPLICATION OF RATE VARIABILITY

Practice

We provide analyses to assess the ability of various weighting schemes to recover phylogenetic signal. Because it is not possible to know when any particular phylogenetic hypothesis is accurate, we use parsimony and corroboration among data from the largest available mt-DNA set to choose among competing hypotheses. We use the shortest tree based on the largest data set as a proxy for the most accurate tree and as a standard in comparing performance of alternative weighting schemes for various data sets. We analyze a set of 12 vertebrate animals (see Figure 2 for taxon names and GenBank accession numbers) representing taxa for which morphological and molecular analyses have been done and for which entire mitochondrial genomes are available (13, 19, 47, 54, 87).

We aligned amino acids for all 13 mitochondrial protein-coding genes using Clustal W (115) and imposed this alignment on the DNAs using DNA Stacks (27). Direct pairwise comparisons were done using MEGA (63), and branch lengths based on unambiguous changes only were determined for absolute rate comparisons using MacClade (69) and the most highly corroborated tree from our own and previous analyses (see references above). We conducted 100 replicate heuristic parsimony analyses, with randomized ordering of taxon addition, for five mitochondrial DNA data sets (all 13 protein-coding genes combined, ATP6, COI, Cytb, ND2) with each of four character weighting schemes (equal weight for all characters, transversions only, codon positions 1 and 2 only, codon position 1 and 2 transversions only) using PAUP (113). Loach (*Crossostoma lacustre*) and Carp (*Cyprinius carpio*) were designated as sister outgroup taxa. Character rescaled consistency indicies were used in iterative weighting bouts on analyses of all characters and of codon positions 1 and 2 only.

Using the 13 mitochondrial protein-coding genes combined, we found the same single most-parsimonious tree with all four weighting schemes (Figure 4), and this tree has the same topology as the most-parsimonious tree based on the entire mitochondrial genome (19). The 13 genes combined represent the largest data set in our analyses and is the least sensitive to alternative weighting schemes. This lack of sensitivity to alternative weighting schemes further indicates the robustness of this topology using congruence as an optimality criterion (119).

Does unequal weighting for individual genes, based on comparative absolute rates, improve congruence with our 13-gene-tree? The comparative absolute rate plots show both third codon position changes and TIs decreasing in rate after about 100 my (Figures 2 and 3). If these kinds of character change include proportionally more homoplasy, down-weighting them a priori should improve resolution for divergences that occurred 100 mya or more, involving the Frog, Chicken and Opossum. For the three individual genes (ATP6, COI, Cytb) in which equal weighting for all characters gives trees incongruent with the 13-gene tree, down-weighting does improve congruence for those older divergences (Figure 4), and the answer to the above question is "yes." The most stringent weighting scheme that we apply, using codon position 1 and 2 TVs only, brings Frog, Chicken, and Opossum into congruence with the 13-gene tree.

However, these unequal weighting schemes do not always bring placement of more recently diverged taxa, specifically Human, into congruence with the 13-gene tree. Divergence of Human from other mammalian orders is estimated to be about 70 mya, and the incongruence found suggests that unequal weighting yields a net loss of phylogenetic signal. Short internodes such as those thought to separate most mammalian orders are particularly susceptible to change under alternative weighting schemes, given the relatively small number of characters supporting them.

ND2 analyses differ from those for the other three individual genes in that both equal weighting for all characters and TVs only yield the 13-gene tree, but the other weighting schemes, involving codon positions 1 and 2 only, yield an incongruent placement of Chicken. The most stringent weighting, with codon positions 1 and 2 TVs only, does return Human to the position congruent with the

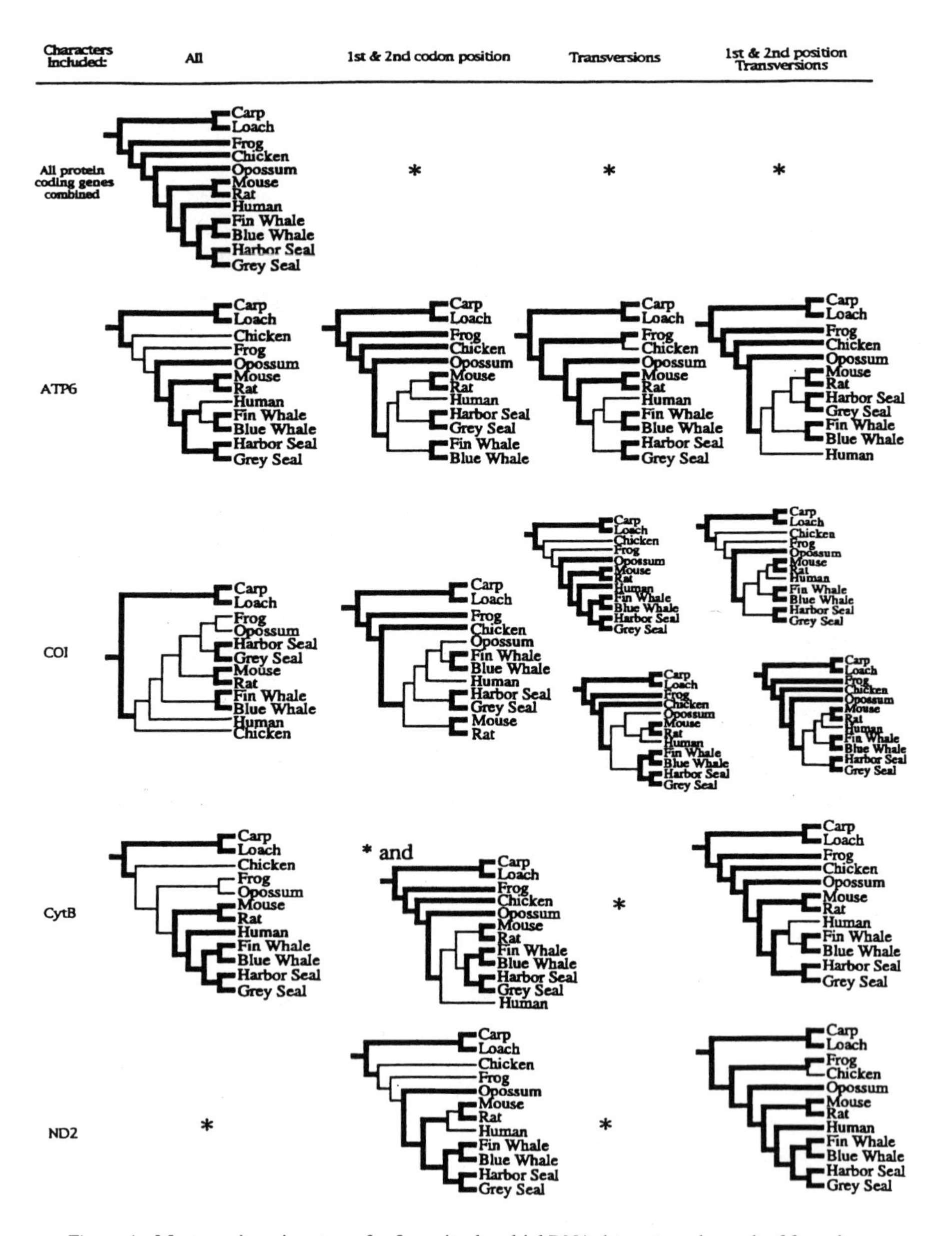

Figure 4 Most-parsimonious trees for five mitochondrial DNA data sets under each of four alternative weighting schemes. See Figure 2 for full taxon names. Asterisks (*) denote same topology as shown in the top left tree for all 13 protein-coding genes combined, all characters equally weighted (our most highly corroborated tree). Thin branches denote departure in phylogenetic placement relative to the most highly corroborated tree.

13-gene tree, though Chicken and Frog are sisters (Figure 4). Again, unequal weighting appears to yield a net loss of phylogenetic information. This is indicative of a general and persistent problem. Unequal weighting schemes may often yield a net increase in phylogenetic signal, though there are exceptions. We see the problem as involving lack of refinement in generalizations about relative rates of change for different kinds of characters that inform unequal weighting.

Theory

INDEPENDENCE OF SYSTEMATICS Consideration of weighting as something separate or additional reflects an operational view, with weighting perceived as an optional procedure to be done before, during, or after initial phylogenetic analysis. However, given that a decision not to weight characters is equivalent to deciding a priori that all characters deserve equal weight in phylogenetic analysis, it can be seen that weighting of characters, whether explicit or implicit, is an inextricable component of all phylogenetic analyses. Selection of taxa and characters for study from the set of all organisms and all potential characters is one of the first forms of weighting applied in all phylogenetic analysis, regardless of whether characters are collected de novo or compiled from previous analyses.

Equal weighting for all characters is consistent with the potential for all characters to be informative of phylogeny and the need to provide initial consideration of all characters. However, equal weighting as a stipulation for all studies conflicts with the view that not all characters are equally informative of phylogeny and also with the weight of evidence reviewed here on variable constraints and rates of change for molecular characters (Table 1), which supports that view. This support derives from the finite number of possible character states available and the inevitability of homoplasious similarity among taxa for rapidly evolving characters.

In an early recognition of the variable information content of different characters, Darwin (20) noted that the less any part of the organism is concerned with special habits, the more important it becomes for classification. He further suggested that characters found to be constant throughout large groups of species be given greater weight than more variable characters. In the same vein, Farris (30:587) states, "It is argued quite validly that a character which changes slowly is the best kind of character for discerning the evolutionary relationships of organisms." The notion of giving greater weight to characters changing less often, in order to reduce homoplasy, is widely accepted, though the means for determining weights is less clear. As noted by Eldredge & Cracraft (29:12), systematists have been able to avoid addressing the issue directly by recognizing "all nonconvergent characters [as] relevant to defining monophyletic groups at some level." However, this assumes an ability to identify nonconvergent characters and to apply them at the appropriate taxonomic level.

Character weights based on phylogenetically determined measures of homoplasy (31, 43) make no claims independent of phylogeny about the information content of characters. Other weighting schemes, whether based on measures of character compatibility (93, 101) or, as is more common, various estimates of comparative absolute rate (Figures 2 and 3; e.g. 54, 74) or frequency of co-occurrence of alternative states at homologous sites (62, 118), do make claims that are not derived from character distributions on a phylogenetic hypothesis. This apparent independence leads to the criticism of having to assume some knowledge of evolution not directly determined by phylogenetic analysis (43). This criticism is the same as that stated at the beginning of this review: putative knowledge about evolutionary process, not determined in a phylogenetic context, should not be used in estimating phylogeny because those assumptions weaken the empirical basis of analyses. If this criticism were applicable in the case of relative rate estimates, it would be a compelling reason not to use them as weights. However, we note four problems with the criticism that relative rate estimates are simply assumptions or hypotheses of evolutionary process, independent from, and inappropriate for, phylogenetic analysis:

1. The criticism entails misunderstanding of the source of physical-chemical constraints in dismissing them simply as evolutionary hypotheses. The effects of fundamental physical constraints, such as the genetic code, the initial base composition, and the replication rate (Table 1), do not rely on any particular evolutionary theory or phylogenetic hypothesis. They help to circumscribe physical limitations on sequence character change across all organisms, rather than describing relative frequencies for change that are unique to certain taxa, at certain times, and under certain historically variable conditions. As a consequence, not all character changes are equally probable.

2. The criticism asks systematists to impose a clean separation between evolutionary theory and phylogenetic practice (for justification of this separation see 10, 92) that, ultimately, cannot be achieved. Theory—our notions of how evolution has proceeded—impacts cladistic methods via choice of study taxa, choice of data sets, choice of outgroups, character polarity determinations (whether by outgroup or ontogenetic criteria) (see 60), character weighting (whether equal or unequal), and determination of which characters to compare across taxa (homology assessment). Absolute separation of theory and practice would seem to require random choice among all taxa and characters when initiating a study. Not only are theory and practice

integrated, they are mutually informing, and barring theory from informing practice would entail loss of evidence. Science is a cumulative enterprise. As Frost & Kluge (40:267) point out, "It is the reconciliation... of the results of several discovery operations that suggests global explanations," and we suggest that analyses of rates of character evolution represent a set of discovery operations. We agree that untested or weakly supported notions about evolution should not be used in phylogeny estimation. However, well-corroborated hypotheses about relative rate for the same characters and taxa whose relationships are being studied are useful and in some cases necessary. Several studies indicate the success of unequal character weighting in recovering "known" phylogenies, whether those phylogenies are well corroborated and "noncontroversial," simulated, or result from manipulation of populations in a laboratory (2, 3, 51, 56, 80 and see below).

3. Assessment of relative severity for other constraints does entail evolutionary hypotheses in many instances (e.g., the degree of directional selection for nonsynonymous change in some taxon, or the potential influence of speciation frequency on accumulation of character change within some clade); however, these hypotheses are amenable to testing, often in a phylogenetic context (e.g., 17, 77, 99).

4. Relative rate estimates based on discrete character branch lengths (Figure 3) are determined in the context of a phylogeny. Relative rate estimates based on pairwise comparisons cannot be justified in this way, although we have demonstrated their general similarity to rate estimates based on branch lengths in some instances (Figures 2 and 3).

A useful approach to character weighting is to assess relative rates of change for various character types as outlined above (though we consider only a small set of all the potential treatments), accord greater weight to types that are less common than expected in order to reduce homoplasy in the data set, and then apply a phylogeny-based iterative weighting method, focusing on homoplasy reduction in further analyses. Rate estimates should be based on the data set at hand and not on estimates of absolute or relative rates estimated for other taxa and other genes. The actual weight accorded to various characters is less important than their relative values. Iterative weighting methods (31, 121), while appealing, remain sensitive to the initial weights applied and can be misleading when data sets have high levels of homoplasy (50). This can be seen in our analyses for COI and Cytb including all characters. For these, we used individual character consistency indices as weights in successive bouts of parsimony analysis, and we found trees incongruent with the best corroborated tree based on whole mitochondrial genomes and our analyses of all 13 protein-coding genes combined (Figure 4).

Just as all phylogenetic analyses entail decisions of weighting, whether equal or unequal, all weighting schemes entail decisions of how to apply those weights across taxa and across characters. Application of weights based on comparative absolute rate curves assumes that general trends in relative rate inferred apply to all taxa and characters involved, and at some level this assumption will be violated. Violations may stem from a variety of sources, such as taxon-specific (or gene-specific) differences in base composition, sequence (and sub-sequence) function, and severity of selective effects. Such violations may underlie anomalous results from analyses such as ours for ND2 (Figure 4). That is, heterogeneity in evolutionary rate occurs across scales, and attempts to reduce homoplasy by shifting emphasis away from some characters can enhance the effects of homoplasious similarity in other characters. For example, down-weighting codon position 3 can enhance effects of any homoplasious similarity at positions 1 and 2.

TOTAL EVIDENCE, TOTAL DATA, AND THE LAW OF LARGE NUMBERS Considerable debate has focused on two related issues: 1) the relative merits of partitioning and combining data sets, and 2) taxonomic versus character congruence (12, 15, 16, 21, 28, 61, 79, 81). Partitioning of data sets for separate analyses is often associated with a preference for taxonomic over character congruence in estimating phylogeny. However, this is not required. Unequal weighting is a form of partitioning, giving differential weight to different sets of characters, yet character congruence is used rather than taxonomic congruence. Similarly, a preference for character congruence has become associated with a directive to use and equally weight all available data. This association is also not obligatory.

We distinguish between total data and total evidence. A total data approach accords equal weight to all characters regardless of their degree of homoplasy. A total evidence approach considers all characters as potentially informative but can use successive approximations (14, 31) and character analyses and rate estimates as discovery operations and sources of evidence in determining the relative information content of character sets, and it can apply weights accordingly. In this view, total evidence includes the characters themselves and our improving understanding of their history of change. This is in keeping with our view that theory and practice of systematics can be mutually informing and that informed unequal weighting can enhance phylogenetic accuracy.

The total data approach with equal weighting of all characters may be appropriate 1) in the absence of evolutionary rate heterogeneity (and the underlying variation in constraints,) (Table 1) or 2) if the law of large numbers can be shown to apply as a generality for phylogenetic analysis and if sufficient data are available. The law of large numbers states that in a series of independent trials, the frequency of a given event will tend toward the probability of the event in any one trial (39). Beyond such formal definitions, however, the law of large numbers is used to suggest that, as independent data accumulate, anomalies in the data will tend to cancel out or decrease in relative frequency. In the context of phylogenetic inference this might be paraphrased as supposing that as more and more data are included in analyses, the phylogenetic signal will dominate noise.

Numerous studies, including ours (Figure 4), demonstrate that larger data sets can perform better in recovery of a well-corroborated tree or a known tree based on simulation or laboratory manipulation. However, this observation does not necessarily mean the law of large numbers is well applied to phylogeny in all instances. The constraints outlined in Table 1, varying across taxa and characters sets, will have decidedly nonrandom effects on patterns of homoplasy accumulation, and it should not be assumed that homoplasies will always cancel out for any particular set of taxa and characters (see 65). The related assumption that homoplasy will always be randomly distributed on phylogenetic trees seems questionable, also due to the variability of constraints in Table 1. The addition of equally weighted character sets with high levels of homoplasy will not necessarily enhance phylogenetic resolution, and short internodes, due to similar divergence times, are particularly sensitive to additional homoplasy.

However, it appears that a threshold for number of DNA characters does exist at which phylogenetic accuracy is very likely with equal weighting of most characters. Cummings et al (19) found that parsimony trees from random samples of 7000 equally weighted mtDNAs (excluding the rapidly evolving D-loop region) yielded the same tree for 10 taxa as did the whole mitochondrial genome about 90% of the time. In a numerical simulation study of four taxa with equal rates of evolution Hillis et al (51) found that equally weighted parsimony analyses yielded the known tree 100% of the time with about 1500 bases. Parsimony analyses with unequal weighting required even fewer characters in recovering the known tree 100% of the time. Differences between these two studies suggest that larger numbers of taxa and (nonidealized) unequal rates of sequence change may require larger amounts of sequence data.

CONCLUSIONS

There have been recurrent calls to keep the practice of systematics separate and free from evolutionary theory in order to ensure its objectivity (10, 109). Presently, such prescriptions are mostly ignored, in part because the complete separation of practice and theory would leave systematists without the ability to choose study taxa and characters from the vast array of candidates. Systematists seek to avoid assumptions in their methods, yet the growing body of evidence in studies of organismal and molecular evolution indicate some patterns in rates of change that could be used to reduce confounding effects of homoplasy. Panchen (91:243) put the challenge succinctly, "[I]f it can be established that cladistics can yield a natural classification only in particular cases by the use of evolutionary hypotheses based on data extrinsic to the synapomorphy scheme, then phylogenetics would have to be reconsidered." We interpret his "evolutionary hypotheses" to include unequal character weights based on evolutionary rate comparisons and "phylogenetics" to include methods of unequal weighting. If our highly corroborated 13-gene tree in Figure 4 is acceptable as a standard for assessing a "natural classification," then we have demonstrated, as have many others, that "evolutionary hypotheses" can "yield a natural classification" where equal weighting does not.

Debate regarding the proper relationship between systematics practice and our understanding of evolution is continual. We see one particular issue that will be of increasing interest in the near future as data sets grow to include tens of thousands of characters for each taxon; that issue is the validity, however or not, of the law of large numbers in phylogenetic inference. If valid, as more and more data are included, phylogenetic signal will come to dominate noise, in all instances. Faith in this law of large numbers has been professed (implicitly or explicitly) by the pheneticists and systematists formerly known as pattern (or transformed) cladists, who share the ideal of theory-free systematics. We do not rule out the possibility that tens of thousands of bases may reduce or even eliminate the need for historical biological knowledge in applying unequal weighting schemes or other models in phylogenetic analysis for some taxa sets. Our analyses showing robustness of the largest data set (12,120 bases) to alternative weighting schemes (Figure 4) are consistent with predictions of the law of large numbers. However, it seems that biology will remain relevant in systematics for a number of reasons.

Minimally, biology (including our understanding of evolutionary rate heterogeneity) matters if we want to: 1) match sets of taxa with characters likely to be historically informative (neither invariant nor rife with homoplasy); 2) avoid comparing characters with different histories due to reticulations, horizontal transfer (21, 72), or lineage sorting (82); 3) be alerted to the possibility of long-branch attraction problems (33); 4) avoid the assumption inherent in the law of large numbers that homoplasy will tend to cancel itself out (or be randomly distributed in phylogenetic hypotheses); and 5) understand the cause of the hierarchy of taxa in nature as inheritance of DNA/RNA and descent with modification.

ACKNOWLEDGMENTS

We thank Arnold Kluge, Greg Gibson, and Mark Siddall for stimulating discussion and comments on an earlier draft of this chapter. This work was supported by grants from the National Science Foundation to DPM and by the Carl L and Laura Hubbs Fellowship Fund of the University of Michigan Museum of Zoology to CET.

Literature Cited

1. Adachi JY, Cao Y, Hasegawa M. 1993. Tempo and mode of mitochondrial DNA evolution in vertebrates at the amino acid sequence level: rapid evolution in warm-blooded vertebrates. *J. Mol. Evol.* 36:270–81
2. Allard MW, Miyamoto MM. 1992. Testing phylogenetic approaches with empirical data, as illustrated with the parsimony method. *Mol. Biol. Evol.* 9:778–86
3. Atchley WR, Fitch WM. 1991. Gene trees and the origins of inbred strains of mice. *Science* 254:554–58
4. Avise JC, Bowen BW, Lamb T, Meylan AB, Bermingham E. 1992. Mitochondrial DNA evolution at a turtle's pace: evidence for low genetic variability and reduced microevolutionary rate in the testudines. *Mol. Biol. Evol.* 9:457–73
5. Ball RM, Avise JC. 1992. Mitochondrial DNA phylogeographic differentiation among avian populations and the evolutionary significance of subspecies. *Auk* 109:626–36
6. Bargelloni L, Ritchie PA, Patarnello T, Battaglia B, Lambert DM, Meyer A. 1994. Molecular evolution at subzero temperatures: mitochondrial and nuclear phylogenies of fishes from Antarctica (suborder *Notothenioidei*), and the evolution of antifreeze glycopeptides. *Mol. Biol. Evol.* 11:854–63
7. Benton MJ. 1990. Phylogeny of the major tetrapod groups: morphological data and divergence dates. *J. Mol. Evol.* 30:409–24
8. Bernardi G, Bernardi G. 1986. Compositional constraints and genome evolution. *J. Mol. Evol.* 24:1–11

8a. Bousquet J, Strauss SH, Doerksen AH, Price RA. 1992. Extensive variation in evolutionary rate of *rbcL* gene sequences among seed plants. *Proc. Natl. Acad. Sci. USA* 89:7844–48

9. Bowen BW, Nelson WS, Avise JC. 1993. A molecular phylogeny for marine turtles: trait mapping, rate assessment, and conservation relevance. *Proc. Natl. Acad. Sci. USA* 90:5574–77
10. Brady RH. 1985. On the independence of systematics. *Cladistics* 1:114–26
11. Britten RJ. 1986. Rates of DNA sequence evolution differ between taxonomic groups. *Science* 231:1393–98
12. Bull JJ, Huelsenbeck JP, Cunningham CW, Swofford DL, Waddell PJ. 1993. Partitioning and combining data in phylogenetic analysis. *Syst. Biol.* 42:384–97
13. Cao Y, Adachi J, Janke A, Pääbo S, Hasegawa M. 1994. Phylogenetic relationships among eutherian orders estimated from inferred sequences of mitochondrial proteins: instability of a tree based on a single gene. *J. Mol. Evol.* 39:519–27
14. Carpenter J. 1988. Choosing among multiple equally parsimonious cladograms. *Cladistics* 4:291–96
15. Chippindale PT, Wiens JJ. 1994. Weighting, partitioning, and combining characters in phylogenetic analysis. *Syst. Biol.* 43:278–87
16. Cracraft J, Mindell DP. 1989. The early history of modern birds: a comparison of molecular and morphological evidence. In *The Hierarchy of Life*, ed. B Fernholm, K Bremer, H Jörnvall, pp. 389–403. Amsterdam: Elsevier Sci.

17. Coddington J. 1994. The roles of homology and convergence in studies of adaptation. In *Phylogenetics and Ecology*, ed. P Eggleton, RI Vane-Wright, pp. 53–78. London: Academic
18. Cooper A, Mourer-Chauviré C, Chambers GK, von Haeseler A, Wilson AC, Pääbo S. 1992. Independent origins of New Zealand moas and kiwis. *Proc. Natl. Acad. Sci. USA* 89:8741–44
19. Cummings MP, Otto SP, Wakeley J. 1995. Sampling properties of DNA sequence data in phylogenetic analysis. *Mol. Biol. Evol.* 12:814–22
20. Darwin C. 1859. *On the Origin of Species by Means of Natural Selection, or the Preservation of Favoured Races in the Struggle for Life.* London: Murray
21. deQueiroz A, Donoghue MJ, Kim J. 1995. Separate versus combined analysis of phylogenetic evidence. *Annu. Rev. Ecol. Syst.* 26:657–81
22. DeRijk P, Neefs JM, Van dePeer Y, DeWachter R. 1993. Compilation of small ribosomal subunit RNA sequences. *Nucleic Acids Res.* 20:S2075–89
23. DeSalle R, Templeton AR. 1988. Founder effects and the rate of mitochondrial DNA evolution in Hawaiian *Drosophila. Evolution* 42:1076–84
24. Domingo E, Holland JJ. 1994. Founder effects and the rate of mitochondrial DNA evolution in Hawaiian *Drosophila*. In *The Evolutionary Biology of Viruses,* ed. SS Morse, pp. 161–184. New York: Raven
25. Doolittle RF, Feng DF, Johnson MS, McClure MA. 1989. Origins and evolutionary relationships of retroviruses. *Q. Rev. Biol.* 64:1–30
26. Easteal S. 1991. The relative rate of DNA evolution in primates. *Mol. Biol. Evol.* 8:115–27
27. Eernisse DJ. 1992. DNA translator and aligner: hypercard utilities to aid phylogenetic analysis of molecules. *CABIOS* 8:177–84
28. Eernisse DJ, Kluge AG. 1993. Taxonomic congruence versus total evidence, and amniote phylogeny inferred from fossils, molecules, and morphology. *Mol. Biol. Evol.* 10:1170–95
29. Eldredge N, Cracraft J. 1980. *Phylogenetic Patterns and the Evolutionary Process.* New York: Columbia Univ. Press
30. Farris JS. 1966. Estimation of conservatism of characters by constancy within biological populations. *Evolution* 20:587–91
31. Farris JS. 1969. A successive approximations approach to character weighting. *Syst. Zool.* 18:374–85
32. Farris JS. 1983. The logical basis of phylogenetic analysis. In *Advances in Cladistics,* ed. NW Platnick, VA Funk, pp. 7–36. New York: Columbia Univ. Press
33. Felsenstein J. 1978. Cases in which parsimony or compatibility methods will be positively misleading. *Syst. Zool.* 27:401–10
34. Felsenstein J. 1984. Distance methods for inferring evolutionary trees: a justification. *Evolution* 38:16–24
35. Felsenstein J. 1988. Phylogenies from molecular sequences: inference and reliability. *Annu. Rev. Genet.* 22:521–65
36. Fitch WM, Bruschi M. 1987. The evolution of prokaryotic ferredoxins—with a general method correcting for unobserved substitutions in less branched lineages. *Mol. Biol. Evol.* 4:381–94
37. Fitch WM, Leiter JME, Li X, Palese P. 1991. Positive Darwinian evolution in human influenza A viruses. *Proc. Natl. Acad. Sci. USA* 88:4270–74
38. Fitch WM, Markowitz E. 1970. An improved method for determining codon variability in a gene and its application to the rate of fixation of mutations in evolution. *Biochem. Genet.* 4:579–93
39. Freund JE, Simon GA. 1992. *Modern Elementary Statistics.* Englewood Cliffs: Prentice Hall
40. Frost DR, Kluge AG. 1994. A consideration of epistemology in systematic biology, with special reference to species. *Cladistics* 10:259–94
41. Futuyma DJ. 1987. On the role of species in anagenesis. *Am. Nat.* 130:465–73
42. Gaut BS, Muse SV, Clegg MT. 1993. Relative rates of nucleotide substitution in the chloroplast genome. *Mol. Phylogenet. Evol.* 2:89–96
42a. Gillespie JH. 1991. *The Causes of Molecular Evolution.* New York: Oxford Univ. Press
43. Goloboff PA. 1993. Estimating character weights during tree search. *Cladistics* 9:83–91
44. Goodman M, Barnabas J, Matusda G, Moor GW. 1971. Molecular evolution in the descent of man. *Nature* 23:604–13
45. Goodman M, Moore GW, Barnabas J., Matsuda G. 1974. The phylogeny of human globin genes investigated by the maximum parsimony method. *J. Mol. Evol.* 3:1–48

46. Grantham R, Gautier C, Gouy M, Mercier R, Pave A. 1980. Codon *catalog usage* and the genome hypothesis. *Nucleic Acids Res.* 8:R49–62
47. Graur D. 1993. Towards a molecular resolution of the ordinal phylogeny of the eutherian mammals. *FEBS* 325:152–59
48. Gutell RR. 1993. Comparative studies of RNA: inferring higher-order structure from patterns of sequence variation. *Curr. Opin. Struct. Biol.* 3:313–22
49. Hanawalt PC. 1994. Transcription-coup- led repair and human disease. *Science* 266:1957–58
50. Hendy MD, Penny D. 1989. A framework for the quantitative study of evolutionary trees. *Syst. Zool.* 38:297–309
51. Hillis DM, Huelsenbeck JP, Cunningham CW. 1994. Application and accuracy of molecular phylogenies. *Science* 264:671–77
52. Hillis DM, Moritz C. 1990. An overview of applications of molecular systematics. In *Molecular Systematics,* ed. DM Hillis, C Moritz, pp. 502–15. Sunderland: Sinauer
53. Hoeh WR, Stewart DT, Sutherland BW, Zouros E. 1996. Cytochrome *c* oxidase sequence comparisons suggest an unusually high rate of mitochondrial DNA evolution in *Mytilus* (Mollusca: Bivalvia). *Mol. Biol. Evol.* 13:418–21
54. Honeycutt RL, Adkins RM. 1993. The higher level systematics of eutherian mammals: an assessment of molecular characters and phylogenetic hypotheses. *Annu. Rev. Ecol. Syst.* 24:279–305
55. Houde P. 1988. Paleognathous birds from the early Tertiary of the northern hemisphere. *Publs. of the Nuttall Ornithol. Club, No. 22.* Cambridge, Mass.
56. Huelsenbeck JP. 1995. The performance of phylogenetic methods in simulation. *Syst. Biol.* 44:17–18
57. Hughes AL, Nei M. 1989. Nucleotide substitution at major histocompatibility complex class II loci: evidence for overdominant selection. *Proc. Natl. Acad. Sci. USA* 86:958–62
58. Jermiin LS, Graur D, Crozier RH. 1995. Evidence from analyses of intergenic regions for strand-specific directional mutation pressure in metazoan mitochondrial DNA. *Mol. Biol. Evol.* 12:558–63
59. Kessler LG, Avise JC. 1985. A comparative description of mitochondrial DNA differentiation in selected avian and other vertebrate genera. *Mol. Biol. Evol.* 2:109–25
60. Kluge AG. 1985. Ontogeny and phylogenetic systematics. *Cladistics* 1:13–27
61. Kluge AG. 1989. A concern for evidence and a phylogenetic hypothesis for relationships among *Epicrates* (Boidae, Serpentes). *Syst. Zool.* 38:7–25
62. Knight R, Mindell DP. 1993. Substitution bias, a priori weighting of DNA sequence change, and the phylogenetic position of *Azemiops feae. Syst. Biol.* 42:18–31
63. Kumar S, Tamura K, Nei M. 1993. *MEGA: molecular evolutionary genetics analysis, version 1.01.* University Park: Penn. State Univ.
64. Lake JA. 1989. Origin of the eukaryotic nucleus determined by rate-invariant analyses of ribosomal RNA genes. In *The Hierarchy of Life,* ed. B Fernholm, K Bremer, H Jörnvall, pp. 87–101. New York: Elsevier
65. Lecointre G, Philippe H, Le HLV, Guyader HL. 1994. How many nucleotides are required to resolve a phylogenetic problem? The use of a new statistical method applicable to available sequences. *Mol. Phylogenet. Evol.* 3:292–309
66. Li P, Bousquet J. 1992. Relative-rate test for nucleotide substitutions between two lineages. *Mol. Biol. Evol.* 9:1185–89
67. Li W-H, Graur D. 1991. *Fundamentals of Molecular Evolution.* Sunderland: Sinauer
68. Li W-H, Tanimura M. 1987. The molecular clock runs more slowly in man than in apes. *Nature* 326:93–96
69. Maddison WP, Maddison DR. 1992. *MacClade: Analysis of Phylogeny and Character Evolution.* Sunderland: Sinauer
70. Martin AP, Naylor GJP, Palumbi SR. 1992. Rates of mitochondrial DNA evolution in sharks are slow compared with mammals. *Nature* 357:153–55
71. Martin AP, Palumbi SR. 1993. Body size, metabolic rate, generation time, and the molecular clock. *Proc. Natl. Acad. Sci. USA* 90:4087–91
72. Mindell DP. 1993. Merger of taxa and the definition of monophyly. *BioSystems* 31:130–33
73. Mindell DP. 1996. Positive selection and rates of evolution in immunodeficiency viruses from humans and chimpanzees. *Proc. Natl. Acad. Sci. USA* 90:4087–91
74. Mindell DP, Dick CW, Baker RJ. 1991. Phylogenetic relationships among megabats, microbats and primates. *Proc. Natl. Acad. Sci. USA* 88:10322–26

75. Mindell DP, Honeycutt RL. 1990. Ribosomal RNA in vertebrates: evolution and phylogenetic applications. *Annu. Rev. Ecol. Syst.* 21:541–66
76. Mindell DP, Knight A, Baer C, Huddleston CJ. 1996. Slow rates of molecular evolution in birds and the metabolic rate and body temperature hypotheses. *Mol. Biol. Evol.* 13:422–26
77. Mindell DP, Sites JW, Graur D. 1989. Speciational evolution: a phylogenetic test with allozymes in *Sceloporus* (REPTILIA). *Cladistics* 5:49–61
78. Mindell DP, Sites JW, Graur D. 1990. Assessing the relationship between speciation and evolutionary change. *Cladistics* 6:393–98
79. Miyamoto MM. 1985. Consensus cladograms and general classification. *Cladistics* 1:186–89
80. Miyamoto MM, Allard MW, Adkins RM, Janecek LL, Honeycutt RL. 1994. A congruence test of reliability using linked mitochondrial DNA sequences. *Syst. Biol.* 43:236–49
81. Miyamoto MM, Fitch WM 1995. Testing species phylogenies and phylogenetic methods with congruence. *Syst. Biol.* 44:64–76
82. Moore WS. 1995. Inferring phylogenies from mtDNA variation: mitochondrial-gene trees versus nuclear-gene trees. *Evolution* 49:718–26
82a. Moran NA. 1996. Accelerated evolution and Muller's rachet in endosymbiotic bacteria. *Proc. Natl. Acad. Sci. USA* 93:2873–78
83. Muse SV, Gaut BS. 1994. A likelihood approach for comparing synonymous and nonsynonymous nucleotide substitution rates, with application to the chloroplast genome. *Mol. Biol. Evol.* 11:715–24
84. Naylor GJP, Collins TM, Brown WM. 1995. Hydrophobicity and phylogeny. *Nature* 373:565–66
85. Nelson G. 1994. Homology and systematics. In *Homology: The Hierarchical Basis of Comparative Biology,* ed. B Hall, pp. 101–49. San Diego: Academic
86. Nelson G, Platnick N. 1981. *Systematics and Biogeography.* New York: Columbia Univ. Press
87. Novacek MJ. 1992. Mammalian phylogeny: shaking the tree. *Nature* 356: 121–25
88. Ohta T. 1976. Role of slightly deleterious mutations in molecular evolution and polymorphisms. *Theoret. Pop. Biol.* 10:254–75
89. Ohta T. 1993. An examination of the generation-time effect on molecular evolution. *Proc. Natl. Acad. Sci. USA* 90: 10676–80
90. Ovenden JR, Mackinlay AG, Crozier RH. 1987. Systematics and mitochondrial genome evolution of Australian rosellas (Aves: Platycercidae). *Mol. Biol. Evol.* 4:526–43
91. Panchen AL. 1992. *Classification Evolution and the Nature of Biology.* Cambridge: Cambridge Univ. Press
92. Patterson C. 1988. The impact of evolutionary theories on systematics. In *Prospects in Systematics,* ed. DL Hawksworth, pp. 59–91. Oxford: Clarendon
93. Penny D, Hendy MD. 1985. Testing methods of evolutionary tree construction. *Cladistics* 1:266–78
94. Perna NT, Kocher TD. 1995. Unequal base frequencies and the estimation of substitution rates. *Mol. Biol. Evol.* 12:359–61
95. Rand DM. 1994. Thermal habit, metabolic rate and the evolution of mitochondrial DNA. *TREE* 9:125–31
96. Rohlf FJ, Sokal RR. 1981. Comparing numerical taxonomic studies. *Syst. Zool.* 30:459–90
97. Sanderson MJ. 1990. Estimating rates of speciation and evolution: a bias due to homoplasy. *Cladistics* 6:387–91
98. Sarich VM, Wilson AC. 1967. Immunological time scale for hominid evolution. *Science* 158:1200–3
99. Seibert SA, Howell CY, Hughes MK, Hughes AL. 1995. Natural selection on the *gag, pol,* and *env* genes of human immunodeficiency virus 1 (HIV-1). *Mol. Biol. Evol.* 12:803–13
100. Seino S, Bell GI, Li W-H. 1992. Sequences of primate insulin genes support the hypothesis of a slower rate of molecular evolution in humans and apes than in monkeys. *Mol. Biol. Evol.* 9:193–203
101. Sharkey MJ. 1989. A hypothesis-independent method of character weighting for cladistic analysis. *Cladistics* 5:63–86
102. Sharp P, Li W-H. 1986. An evolutionary perspective on synonymous codon usage in unicellular organisms. *J. Mol. Evol.* 24:28–38
103. Sheldon FH. 1987. Rates of single-copy DNA evolution in herons. *Mol. Biol. Evol.* 4:56–69
104. Shields GF, Wilson AC. 1987. Calibration of mitochondrial DNA evolution in geese. *J. Mol. Evol.* 24:212–17

105. Shigenaga MK, Gimeno CJ, Ames BN. 1989. Urinary 8-hydroxy-2′-deoxyguanosine as a biological marker of in vivo oxidative DNA damage. *Proc. Natl. Acad. Sci. USA* 86:9697–701
106. Sibley CG, Ahlquist JE. 1990. *Phylogeny and Classification of Birds: A Study in Molecular Evolution.* New Haven: Yale Univ. Press
107. Simmonds P, Balfe P, Ludlam CA, Bishop JO, Leigh Brown, AJ. 1990. Discontinuous sequence change of human immunodeficiency virus (HIV) type 1 env sequences in plasma viral and lymphocyte-associated proviral populations in vivo: implications for models of HIV pathogenesis. *J. Virol.* 64:5840–50
108. Simon C, Frati F, Beckenbach A, Crespi B, Liu H, Flook P. 1994. Evolution, weighting, and phylogenetic utility of mitochondrial gene sequences and a compilation of conserved polymerase chain reaction primers. *Ann. Entomol. Soc. Am.* 87:651–701
109. Sneath PHA, Sokal RR. 1973. *Numerical Taxonomy*. San Francisco: Freeman
110. Springer MS, Kirsh JAW. 1989. Rates of single-copy DNA evolution in phalangeriform marsupials. *Mol. Biol. Evol.* 6:331–41
111. Sober E. 1988. *Reconstructing the Past.* Cambridge: MIT Press
112. Steinhauer DA, Holland JJ. 1987. Rapid evolution of RNA viruses. *Annu. Rev. Microbiol.* 41:409–33
113. Swofford D. 1993. *PAUP: Phylogenetic analysis using parsimony, version 3.1.1.* Champaign, IL: Ill. Nat. Hist. Surv.
114. Thomas WK, Beckenbach AT. 1989. Variation in salmonid mitochondrial DNA: evolutionary constraints and mechanisms of substitution. *J. Mol. Evol.* 29:233–45
115. Thompson JD, Higgins DG, Gibson TJ. 1994. CLUSTAL W: improving the sensitivity of progressive multiple sequence alignment through sequence weighting, positions-specific gap penalties and weight matrix choice. *Nucleic Acids Res.* 22:4673–80
116. Vawter L, Brown WM. 1993. Rates and patterns of base change in the small subunit ribosomal RNA gene. *Genetics* 134:597–608
116a. Wakeley J. 1994. Substitution-rate variation among sites and the estimation of transition bias. *Mol. Biol. Evol.* 11:436–42
117. Wallis GP, Arntzen JW. 1989. Mitochondrial-DNA variation in the crested newt superspecies: limited cytoplasmic gene flow among species. *Evolution* 43:88–104
118. Wheeler WC. 1990. Combinatorial weights in phylogenetic analysis: a statistical parsimony procedure. *Cladistics* 6:269–75
119. Wheeler WC. 1995. Sequence alignment, parameter sensitivity, and the phylogenetic analysis of molecular data. *Syst. Biol.* 44:321–31
120. Wiley EO. 1975. Karl R. Popper, systematics, and classification: a reply to Walter Bock and other evolutionary taxonomists. *Syst. Zool.* 24:233–42
121. Williams PL, Fitch WM. 1989. Finding the weighted minimal change in a given tree. In *The Hierarchy of Life,* ed. B Fernholm, K Bremer, H Jörnval, pp. 453–70. Cambridge: Elsevier
122. Wilson AC, Cann RL, Carr SM, George M, Gyllensten UB, et al. 1985. Mitochondrial DNA and two perspectives on evolutionary genetics. *Biol. J. Linn. Soc.* 26:375–400
123. Woese CR. 1987. Bacterial evolution. *Microbiol. Rev.* 51:221–71
124. Wolfe KH, Li WH, Sharp PM. 1987. Rates of nucleotide substitution vary greatly among plant mitochondrial, chloroplast, and nuclear DNAs. *Proc. Natl. Acad. Sci. USA* 84:9054–58
125. Wu CI, Li WH. 1985. Evidence for higher rates of nucleotide substitution in rodents than in man. *Proc. Natl. Acad. Sci. USA* 82:1741–45
126. Zhang Y, Ryder OA. 1995. Different rates of mitochondrial DNA sequence evolution in Kirk's Dik-dik (*Madoqua kirkii*) populations. *Mol. Phylogenet. Evol.* 4:291–97

Annu. Rev. Ecol. Syst. 1996. 27:305–35

HERBIVORY AND PLANT DEFENSES IN TROPICAL FORESTS

P. D. Coley and J. A. Barone
Department of Biology, University of Utah, Salt Lake City, Utah 84112

KEY WORDS: herbivory, tropical forests, plant defenses, pathogens, tree diversity, leaf phenology

ABSTRACT

In this review, we discuss the ecological and evolutionary consequences of plant-herbivore interactions in tropical forests. We note first that herbivory rates are higher in tropical forests than in temperate ones and that, in contrast to leaves in temperate forests, most of the damage to tropical leaves occurs when they are young and expanding. Leaves in dry tropical forests also suffer higher rates of damage than in wet forests, and damage is greater in the understory than in the canopy. Insect herbivores, which typically have a narrow host range in the tropics, cause most of the damage to leaves and have selected for a wide variety of chemical, developmental, and phenological defenses in plants. Pathogens are less studied but cause considerable damage and, along with insect herbivores, may contribute to the maintenance of tree diversity. Folivorous mammals do less damage than insects or pathogens but have evolved to cope with the high levels of plant defenses. Leaves in tropical forests are defended by having low nutritional quality, greater toughness, and a wide variety of secondary metabolites, many of which are more common in tropical than temperate forests. Tannins, toughness, and low nutritional quality lengthen insect developmental times, making them more vulnerable to predators and parasitoids. The widespread occurrence of these defenses suggests that natural enemies are key participants in plant defenses and may have influenced the evolution of these traits. To escape damage, leaves may expand rapidly, be flushed synchronously, or be produced during the dry season when herbivores are rare. One strategy virtually limited to tropical forests is for plants to flush leaves but delay "greening" them until the leaves are mature. Many of these defensive traits are correlated within species, due to physiological constraints and tradeoffs. In general, shade-tolerant species invest more in defenses than do gap-requiring ones, and species with long-lived leaves are better defended than those with short-lived leaves.

0066-4162/96/1120-0305$08.00

INTRODUCTION

In tropical forests the evolutionary relationships between herbivores and plants have resulted in an impressive variety of adaptations and interactions. Herbivore pressure has led to the evolution of chemical, mechanical, and phenological defenses in plants. Herbivores in turn have evolved to cope with food plants that are trying to starve or poison them. These relationships affect food webs, nutrient cycling, and community diversity, and thus every organism in tropical forests. In this review we examine the ecological and evolutionary outcomes of the interactions between herbivores and their host plants in lowland tropical rain forests. We begin by summarizing the patterns of herbivory in tropical forests, then turn to the herbivores, their diet breadth, and their strategies for dealing with plant defenses. Next we describe the ecological impact of herbivores and pathogens on the structure and diversity of tropical forests. We conclude with an examination of the wide array of plant defenses against herbivores. Throughout this review, we use the term "herbivory" to refer to leaf damage by insects, mammals, and pathogens. We have chosen this broad definition because all three groups have a profound effect on tropical plant ecology. While most research has focused on insect herbivores, we hope that this wider perspective will encourage more work on mammalian herbivores and, especially, pathogens. Our definition also reflects the practical problem that it is often difficult to distinguish the cause of damage to a leaf. This should be kept in mind in our discussion below of levels of herbivory in tropical forests.

RATES OF HERBIVORY

In this section we review the patterns of herbivory that have been documented in a number of tropical forests. Unfortunately, levels of leaf consumption have been measured in a variety of ways, making comparisons difficult. Nonetheless, we can identify several general results. Most importantly, herbivory in tropical forests is quantitatively and qualitatively different from that in the temperate zone. Moreover, within the tropics rainfall regimes, leaf age, and location in the canopy all influence damage rates.

Temperate vs Tropical Forests

Annual rates of leaf damage are higher in tropical forests than in temperate broad-leaved forests (Table 1). Herbivory averages 7.1% per yr in the temperate zone, and 11.1% for shade-tolerant species in the humid tropics ($p < 0.01$, based on 42 studies). Rates of damage to gap specialists are even higher (48.0%), but they comprise less than 15% of the individual trees in tropical forests (107). Although this latitudinal difference is statistically significant,

Table 1 Comparison of rates of herbivory in temperate and tropical forests.

	Annual		Mature Lves		Young Lves		Young/total	
	%	*N*	%/d	*n*	%/d	*n*	%	*n*
Temperate	7.1	13					27.0	
Tropical wet forest								
Shade-tolerant species	11.1	21	.03	105	.71	150	68.3	31
Gap specialists	48.0	4	.18	37	.65	37	47.3	30
Tropical dry forest	14.2	4	.07	78	.15	61	28.7	62

Annual is the average percentage damage per year, with N being the number of studies (each study included many species). Daily rates of herbivory are presented for young and mature leaves (%/d), young/total indicates the percentage of the total lifetime damage that occurs while leaves are expanding, *n* indicates the number of species. Data on young/total from the temperate zone are the average for an entire forest. (3, 5, 23, 30, 31, 47, 60, 62, 65, 78, 97, 118, 132, 137, 140, 144, 147, 160, 161, 163, 181, 182, 190)

it is not enormous, and given the paucity of accurate measures, it should be regarded as a working hypothesis. Moreover, when we discuss forest averages in herbivory, we are ignoring enormous temporal and spatial variation as well as consistent differences among species.

The tropics incur higher rates of damage despite the fact that tropical plants tend to be better defended (see below). This suggests that the high damage rates in the tropics must be due to greater pressure from herbivores, though few studies have attempted to measure the biomass of herbivores in different forests (131).

Assuming these latitudinal trends in herbivory are real, are they important ecologically? Although an annual leaf loss of 10% may not seem extreme, it is sufficient to reduce plant fitness (68). For example, Marquis found that 10% experimental defoliation of an understory shrub, *Piper arieianum*, reduced growth and seed production, delayed flowering, and decreased seed viability (144, 146). Annual survivorship is 85% for undamaged seedlings of *Dipteryx panamensis* and 0% for seedlings with 8% of their leaf area missing (38). Furthermore, most plants allocate only 10% of their resources to reproduction, an investment that obviously affects fitness (18). Hence, herbivory probably has a substantial impact on growth and survival of plants, and more so in the tropics than in the temperate zone.

Wet vs Dry Tropical Forests

Within the tropics, rainfall regimes influence the duration of the dry season. At sites with extended dry seasons, most or all of the tree species become deciduous for portions of each year, whereas in wetter sites, species are evergreen.

Herbivory differs significantly along this rainfall gradient. Dry forest species suffer higher rates of herbivory (14.2%/yr) than do shade-tolerant wet forest species (11.1%/yr) ($p < 0.05$; Table 1). These patterns result in part from the lower levels of defense in short-lived deciduous leaves (47), and in part from the fact that dry seasons reduce herbivore populations (45).

Young vs Mature Leaves

The most striking difference in patterns of herbivory within tropical forests is between mature leaves and young, expanding leaves. Due to the higher nutritional quality of young leaves, daily rates of damage are 5–25 times higher than on mature leaves (Table 1). Despite the fact that leaves are only expanding for a short 1–3 week period, the high rates of damage are significant over the lifetime of a leaf. For tropical shade-tolerant species, whose leaves last an average of 2–4 yr (43, 139), 68% of the lifetime damage occurs during the small window of leaf expansion (Table 1). This is in marked contrast to the temperate zone where only 27% of the lifetime damage accumulates while the leaf is expanding (Table 1). Young temperate leaves may partially escape damage by emerging in early spring when herbivore populations are reduced. Hence, for temperate species, most of the damage occurs on mature leaves, whereas for tropical species the majority of damage accrues while the leaf is young. Although the importance of young leaf herbivory is most pronounced for shade-tolerant species of wet forests, the pattern holds across the tropics and may be the most fundamental difference between temperate and tropical forests. The high absolute and relative rates of herbivory on young tropical leaves suggest that they have experienced stronger selection for defenses. Furthermore, because tropical herbivores depend on such an ephemeral food source, this may select for more elaborate host-finding mechanisms and tighter coupling between herbivore life-history traits and plant phenologies.

Canopy vs Understory Leaves

Recent innovations have led to easier and more reliable access to the tropical forest canopy and increased attention to herbivory on canopy leaves (See Lowman, this volume). In her pioneering work, Lowman examined herbivory on five tree species in subtropical and temperate rain forests in New South Wales, Australia, and found that leaves in the sun usually suffered significantly less damage than shade leaves and that herbivory was higher in the understory than in the canopy (138). Other work in Panama suggests a similar pattern, with lower damage from insect herbivores in the canopy, though pathogen damage may be greater there (15, 93).

Differences in plant chemistry, local microclimate, or predation rates on leaf herbivores have been suggested as possible causes for the decrease in herbivory with height in forests. Sun leaves are smaller, tougher, and have higher phenolic contents than do shade leaves (96, 138, 141). The canopy typically has a hotter, drier, and windier microclimate as well, which may severely affect many insect herbivores (138). Predation, such as by birds, may also reduce insect herbivore abundances in the canopy relative to the understory. Recent work in Puerto Rico showed that anolis lizards are important predators of insect herbivores in the canopy, thus reducing damage to canopy leaves, but whether their impact is greater in the canopy than in the understory is not known (67).

These three hypotheses concerning the effects of plant chemistry, microclimate, and predation are not mutually exclusive, and their impact may vary with the size and guild of the herbivores in question. In addition, the trend of greater herbivory in the understory suggests that many plant defenses, especially leaf phenology and greening (see below), may be the consequence of selection by herbivores in the understory and that the presence of these traits in adult trees may be holdovers from this earlier life stage.

Methods of Measuring Herbivory

Generalizations about herbivory are difficult, as many of the studies summarized above (Table 1) reported a one-time estimate of standing crop damage. Quantifying the amount of leaf tissue eaten with a single measure of missing leaf area is misleading. First, completely eaten leaves are not included in the sample, so the amount of damage is underestimated. In studies that have compared single measurements to rates derived from repeated measurements of marked leaves, underestimates averaged 50% (from 38–60% ; 39, 78, 137). Second, as leaf longevity differs among species, it is impossible to know the time scale over which damage has accumulated. Across tropical trees, leaves can live from 4 mo to over 14 yr (PD Coley, unpublished data; 43, 176). If these differences are not taken into account, it will be erroneously concluded that species with short-lived leaves have lower damage rates (161). Therefore, a single measure of standing crop damage cannot easily be translated into an annual rate of damage, nor can it be used meaningfully to compare herbivory on different species. Instead, we suggest that investigators measure marked leaves or expanding buds at two different times to calculate a rate of damage, and that they include both expanding and mature leaves in the study.

HERBIVORES

Leaves are subject to damage by an enormously diverse set of vertebrate and invertebrate herbivores as well as by pathogens. Studies document defoliation by all these enemies, but are they equally important? In this section we examine characteristics of each group and discuss how their seasonal and spatial distributions as well as the degree of diet specialization may influence patterns of herbivory.

Insects

TOTAL CONSUMPTION Folivorous insects are diverse taxonomically and physically, and they are the most important consumers in tropical forests. For example, there are at least 171 phytophagous insect families at La Selva, Costa Rica (147), and 95 different species feed on a single species of understory shrub (145). Although the biomass of vertebrate herbivores in a forest may be twice as high as for insects, insects cause most of the herbivory. On Barro Colorado Island (BCI) in Panama, 72% of the annual leaf consumption, or 575 kg/ha/yr, is eaten by chewing insects (131). The proportion of damage caused by insects may be even higher in other forests. For example, in Parque Nacional Manu, in Peru, arboreal vertebrate folivores are rarer, and together they consume about one sixth as much as their counterparts on BCI (131). So it is not unreasonable to suggest that chewing insects contribute 75% or more to the annual leaf consumption.

Not all herbivores leave obvious evidence of their consumption. The impact of leaf chewers has been best studied because the story can be partially read from the holes in leaves. Phloem feeders are numerically common, yet it is difficult to assess their impact on plant productivity. In a temperate hardwood forest, phloem feeders comprised 23% of the phytophage biomass (181). In Brunei, sucking insects in canopy crowns were as abundant numerically as chewing insects (191), and in a tropical eucalyptus forest, 79% of all herbivores were phloem feeders (77). Although phloem feeders tend to be smaller in body size, they appear to consume more per gram of body mass than chewers (181). Leigh (131) therefore suggested that phloem feeders may remove as much biomass as leaf chewers. Clearly this issue needs a great deal more attention.

SEASONALITY The extent of seasonality in insect numbers reflects the seasonality of rainfall in different forests (147, 205). In general, insect populations are depressed during the dry seasons, with a marked rebound at the beginning of the wet season followed by a gradual increase until the onset of the following dry season (116, 138, 147, 158, 205). Rates of herbivory mirror this pattern, being lowest in the dry season and highest in the rainy season.

DIET SPECIALIZATION OF INSECT HERBIVORES The ecological circumstances and evolutionary pressures that lead to narrow diet breadth in insect herbivores have received considerable attention (71, 88, 108). The topic is particularly important in tropical ecology because many explanations of the high species diversity in tropical forests assert that organisms in the tropics have smaller niche sizes, meaning that more can be packed together in a habitat (142, 143).

For insect herbivores, niche size is directly dependent on diet breadth, making information on host-range critical to understanding the processes that lead to high diversity. In addition, recent projections of both tropical and global species richness have relied on largely untested assumptions of the extent of host-specialization in tropical forests (74, 91). Finally, understanding the degree of host-specialization is critical to evaluating the Janzen-Connell model of tropical forest tree diversity, which assumes that herbivores are specialized (see below; 54, 109).

Two general factors are thought to favor specialization in herbivorous insects—plant defenses and natural enemies. Plant defenses, particularly chemical defenses, require energy to disarm or detoxify. Specialist herbivores should be more efficient in dealing with the defenses of their host plants and thus grow and reproduce more quickly than generalists (71, 95, 113, 128). Natural enemies may also select for a narrow diet breadth in herbivores, if they are better able to locate their prey on some plant species (or plant parts) than others. This differential predation pressure would favor those individual herbivores with a preference for host species where they suffered the lowest mortality, leading to a narrower diet breath (27, 172).

Both plant defenses and natural enemies may be stronger selective agents in tropical forests, leading to greater specialization than in temperate regions. Plants in tropical forests tend to be better defended chemically than their temperate counterparts (47), and the high diversity of plant species means that herbivores confront a greater array of defenses (113, 116). Conversely, the relative rarity of most plant species in tropical forests means that locating them is more difficult and costly for herbivores, both in terms of time and exposure to predators. Under these circumstances, selection would favor a more generalized habit (16, 20, 108, 113). This should be particularly true for smaller insects, which may be less efficient at locating hosts because of poorer dispersal abilities (144).

PATTERNS OF SPECIALIZATION The common assumption that insect herbivores are more specialized in tropical regions has been tested by relatively few studies (170, 171). Scriber (185) surveyed the global patterns of feeding specialization of the Papilionidae (Lepidoptera) and determined that generalists represent a higher proportion of species at temperate latitudes. Similar patterns hold for other insect groups. Butterflies from three families (Papilionidae, Pieridae, and Nymphalidae) typically have a narrower host range in the wet forest at La Selva, Costa Rica, than they do at five temperate sites (147). The percentage of specialist grasshoppers (Acrididae) is also higher at La Selva than at sites in Texas and Colorado (147). Basset (17) showed in feeding trials that tropical herbivores have a narrower diet than temperate ones, perhaps reflecting greater palatability of temperate trees. Within the tropics many groups show high levels of specialization (91, 145). *Ithomiine* and *Heliconius* butterflies (Nymphalidae) average between one and three host species (24, 32, 70). Most herbivorous bugs (Hemiptera) from Dumonga Bone National Park in Indonesia are restricted to a single host family (102).

Several studies have provided counterexamples to this level of specialization, however. Treehoppers (Membracidae) show greater host-specificity at higher latitudes (207), as do two families of wood-feeding beetles (Scolytidae and Platypodidae) (20). These beetles feed on fallen trees and branches, which are an unpredictable resource and decay rapidly in the warm, humid climate, making specialization difficult (20). In a study of the herbivores on a single species of tree, *Argyrodendron actinophyllum* (Sterculiaceae) in a subtropical forest in Australia, Basset (16) found that only 11% (of 156 folivorous species) were specialized, feeding on hosts from one or a few related host families. Despite the conclusions of these studies, most herbivorous insects in the tropics appear to be quite specialized, though this question requires more attention (145, 147). Since, in general, most insect herbivore species have narrow host ranges (26), it is difficult to assess whether tropical herbivores are more specialized than temperate ones. We suspect that, both for small herbivores that disperse poorly, such as treehoppers, and for those that feed on unpredictable resources, the problems of host location in diverse tropical forests may overwhelm the selective advantages of specialization, leading to broader diets in tropical forests. Mobile herbivores, such as many beetle species and most moths and butterflies, are probably more specialized than temperate species.

HERBIVORY AND SPECIALIZATION Little is known about the relative proportion of damage caused by specialist and generalist herbivores in tropical forests, yet such information is important to understanding the nature of the selection on plant defenses and how herbivores regulate plant populations. On BCI, about 60% of leaf damage in 9 tree species was due to specialist insect herbivores (feeding on plants in only one plant family), and 8% was from generalists, with the balance from fungal pathogens (JA Barone, unpublished data). The tree species showed considerable variation, but specialist herbivores always caused more damage than generalists. If this pattern of damage holds, it suggests that specialist insect herbivores are more important than generalists in ongoing selection for plant defenses. In addition, it supports the assumption of the Janzen-Connell model (54, 109) that host-specialists do most of the damage to plants in the tropics (see below).

Mammals

ABUNDANCE Although annual net production of vegetation is high in tropical forests and may exceed that of savannas, most of it is in the canopy, out of reach of terrestrial animals. Densities of terrestrial folivores such as deer and tapir vary significantly among lowland rain forest sites, but average 300 kg km^{-2} (28, 79, 131, 192), which is only 5% of the biomass typical for savannas (119). In rain forests, most of the folivores are arboreal, with a biomass 1.5 to 5 times as high as that of the terrestrial folivores (131). However, even considering both terrestrial and arboreal biomass of folivores, tropical forests have much lower mammalian densities than do savannas. This is because leaves are in the canopy, and there are difficulties associated with arboreality.

CONSTRAINTS ON ARBOREALITY The upper limit on body size for folivorous arboreal mammals is generally considered to be 13–15 kg (72), because size limits the ability to reach leaves at the ends of branches. However, small body size makes digestion of high-fiber and low-nutrient diets more difficult to the extent that animals less than 1 kg may be unable to survive on a strictly folivorous diet (58). Smaller animals need more energy per gram of body weight (122), but the digestive capacity and hence the ability to obtain energy is directly proportional to body size (63). Thus smaller animals have a higher mass-specific energetic demand that must be met by a proportionally smaller energetic input (58). Because of the low nutritional quality of leaves as compared to seeds or animal tissue, there is a limit for the body size of mammalian folivores below which they cannot obtain sufficient energy. The smallest body sizes of folivorous arboreal mammals are less than in terrestrial folivores, suggesting that arboreal mammals may be living closer to the metabolic limits imposed by digestion (58, 152). Because of the poor nutritional quality of mature leaves, many mammals supplement their diets with fruit or seeds, and even strict folivores consume the more nutritious young leaves when possible. On BCI, sloth mortality is highest in the late rainy season, when young leaves are rare, and when extended periods of cooler, rainy weather slow digestion (85, 86; PD Coley, personal observation). In years with late rainy season fruit failure, famine in herbivorous mammals is widespread (85, 86, 153). Terborgh & van Schaik hypothesized that in the Neotropics, seasonal shortages of fruit and new leaves coincide, and that this bottleneck in resource availability may explain the low biomass of primates in South America relative to Africa (194).

DIGESTIVE PHYSIOLOGY Folivores are confronted with a diet that is simultaneously poor in nutritional content and rich in defensive chemicals. A common solution has been to rely heavily on microbial symbionts to ferment the vegetation (200). Mammalian herbivores can be classified as hindgut (e.g. horses) or foregut (e.g. ruminants) fermenters, depending on whether the primary site of fermentation occurs before or after food passes through the stomach. Foregut fermentation is thought to be more efficient at digestion of high fiber/low nutrient diets, due to longer gut retention times and sieves that allow passage of digested material and retention of fiber in the foregut for continued fermentation (178). In addition, microbes in the foregut may aid detoxification of plant secondary metabolites (61). However, foregut fermentation of only leaves would not provide enough energy to small folivores because passage times are necessarily shorter in animals with small body size (58, 63). Instead, in arboreal herbivores, foregut fermentation is associated with a mixed diet of leaves and seeds or fruits. Nutritionally rich items can pass quickly through the digestive system, while more fibrous material is retained for further microbial fermentation. The one exception, sloths, have unusually low metabolic rates (159) that may allow them, with foregut fermentation, to survive on leaves.

Most arboreal folivores appear to be hindgut fermenters with an added ability for colonic separation (58). Separation of digesta in the cecum-colon allows retention of the nutritious parts of the digesta and rapid excretion of the larger, less digestible particles (80). Cork & Foley (58) suggested that selection for different digestive strategies results primarily from nutritional factors including high fiber and phenolic contents. The interplay between digestion and plant secondary metabolites is as yet unresolved.

Pathogens

Leaf pathogens, a taxonomically diverse and ecologically important group, have not received the research attention they deserve. Nonetheless, damage by these pathogens is common and widespread. In the lowland wet forest of Los Tuxlas, Mexico, Garcia-Guzman & Dirzo (90) found pathogen damage on 45% of the 67 understory species and 60% of the 30 canopy species surveyed. For 25 species on BCI, pathogens accounted for 29% of the damage for which culprits could be identified (PD Coley, TA Kursar, unpublished data). In the canopy of a seasonal dry forest in Panama, 5 kinds of pathogen damage were found on *Anacardium* leaves, and 75% of the *Luehea* leaves were diseased (93). For the tree *Quararibea* on BCI, pathogen damage accounted for 61% of the lost leaf area in the canopy, compared to 2% in the understory (15). In addition to leaf pathogens, fungi responsible for damping-off can kill large numbers of establishing seedlings (9), and stem cankers can attack and kill saplings (94).

ECOLOGICAL IMPACT OF HERBIVORES

Herbivory can have numerous negative effects on plant fitness by depressing growth and reproduction and by reducing competitive ability. In the following section, we examine the consequences of damage by different classes of herbivores to community composition.

Insects

Janzen (109) and Connell (54) proposed that host-specialized seed predators, herbivores, and pathogens can maintain the high diversity of tree species in tropical forests if they are more likely to damage and kill juvenile trees (seeds, seedlings, and saplings) growing at high densities or close to conspecific adults. Such a pattern could occur if the adult trees serve as reservoirs or cues for natural enemies. This higher rate of mortality near adults means that the chance of successful recruitment is likewise low near adults but increases with distance. This distance dependence results in turn in greater spacing between adults of competitively dominant species and permits more species to coexist (54, 105, 109). Most of the studies that have tested the prediction that levels of damage and mortality are higher near adults or at high densities have focused on seeds and seedlings. Of 36 studies[1], 28 have provided at least weak support for the prediction, with 63% (28 of 45) of the tree species across all the studies showing higher mortality or damage near conspecific adults. This distance-dependence in damage and mortality appears most likely when a single, host-specialized natural enemy is the main cause, though the responsible agent was not determined in most studies.

Although they have received little attention, leaf-chewing and leaf-sucking insects could generate the pattern of mortality predicted by the Janzen-Connell model, if damage rates are higher on young trees near conspecific adults (54). Indeed, unlike damage to seeds or seedlings, herbivore damage to saplings and older size classes could accumulate over many seasons, gradually killing off juvenile plants near conspecific adults. In addition, higher rates of herbivory near adults may reduce the growth of juveniles trees, making them more vulnerable to mortality from other causes such as falling branches or secondary pathogen infection.

Community-level studies have shown that these older size classes of young trees can suffer from distance-dependent mortality. In rain forests in Queensland, Australia, Connell and his coworkers frequently observed decreased growth and higher mortality when a tree's nearest neighbor was a conspecific, though such results were largely limited to small size-classes and to very short distances between individuals (55, 57). Condit, Hubbell & Foster have examined recruitment, growth, and mortality of woody stems on a 50-ha permanent plot on BCI (107). They have found that both proximity to conspecific adults and local conspecific density decreases growth and increases mortality of a few abundant tree species (53, 104). Recruitment into older size classes was also less likely near conspecific adults for 15 of the 80 (19%) woody species

[1]References: 6–8, 10, 22, 29, 35, 37, 38, 54, 56, 64, 66, 81–84, 87, 94, 103, 104, 110–112, 115, 117, 120, 121, 129, 130, 183, 193, 196, 198, 204, 208.

they examined, though for most of these species the negative effect disappeared beyond distance of 10 m (51, 52). While the strict critieria used in this analysis probably understate the importance of distance-dependence in this forest (131), their results do demonstrate conclusively that the distance and density dependence predicted by Janzen and Connell can continue past the seedling stage.

Insect herbivores likely play a role in generating these effects. Recent work on BCI showed that saplings of three abundant species near adults suffer higher levels of damage to young leaves than do those farther away, with most of this damage caused by specialist herbivores, as predicted by the Janzen-Connell model (J Barone unpublished data). This distance dependence was only seen when nearby conspecific adults were also flushing young leaves, suggesting that adults were either a source or an attractant of the herbivores. Thus, it appears likely that at least part of the distance and density dependence observed in older size classes of young trees is due to herbivores.

Mammals

The impact of arboreal mammals is difficult to assess experimentally. Mammalian folivory is much less than insect damage and is unlikely to account for more than 20% of the leaf area consumed in tropical forest canopies. Over evolutionary time, selection pressure from insects rather than arboreal mammals seems to have shaped leaf defenses (47, 58). In contrast to arboreal mammals, terrestrial mammals significantly depress survival of seeds and seedlings. Comparisons of neotropical areas with and without mammals have uniformly demonstrated increased seed and seedling survival where mammals are absent (66, 133, 187). Excluding vertebrates in Queensland, Australia, enhanced seedling survivorship and height growth (169). In experimental exclosures at Manu, Peru and BCI, Panama, sapling densities were approximately 20% higher than in open control plots after 2 yr (195), due to both increased recruitment and decreased mortality. In an on-going experimental exclosure on BCI, WP Carson (personal communication) has also found 2.5-fold increases in seedling densities, with particularly dense carpets under the parent tree. A compelling example of the long-term role of large mammals is seen in the comparison of two Mexican forests—Los Tuxlas, which has lost all of its browsing mammals, and Montes Azules, which retains most of the fauna (69). Los Tuxlas has 2.3 times the density of seedlings and saplings, but only one third the diversity, presumably because thinning of seedlings by mammalian herbivores offsets competitive dominance. Mammals also can damage seedlings in a distance-dependent fashion, with four out of five studies showing greater survivorship with increasing distance from the conspecific adult (66, 83, 84, 104, 183). Thus, while having a limited impact in terms of leaf area consumed, mammals may have dramatic effects on plant communities through their consumption of seeds and seedlings.

Pathogens

Pathogens are responsible for significant amounts of leaf damage and may also have impacts on the genetic and species diversity of host plants. Aylor (11) suggested that many pathogens have more restricted dispersal than insects or mammals. Thus we might expect pathogens to become locally abundant on adult trees or even genetically adapted to hosts (93). If adult trees serve as reservoirs of disease, they would be a source of infectious propagules for the seedlings below (54). Infection can be much more damaging to a small shaded plant with limited resources than to the adult. Two excellent examples consistent with the Janzen-Connell model show that juvenile mortality to damping-off disease (9, 121) and a stem canker (94) is greater close to the parent tree. A second consequence of adults serving as disease reservoirs is that local adaptation by the pathogen may occur. Limited dispersal and multiple generations could lead to more virulent pathogens better adapted to the parent genotype (93). As a consequence, offspring that are genetically different from the parent would be favored.

PLANT DEFENSES

Are the high rates of herbivory in tropical forests (Table 1) the result of poorly defended leaves? Apparently not. Leaves of tropical forests have both higher overall levels of defense and a greater diversity of defenses compared to their temperate counterparts (47). We suggest that, in part, this greater commitment to defense is an evolutionary response to elevated pressure from herbivores. In addition, mature leaves in evergreen tropical rain forests are extremely long-lived and must therefore be resistant to both abiotic and biotic damage. Average leaf lifetime for understory plants in tropical lowland rain forests is 3 yr with extremes of 14 yr (47, 139). So the combination of higher rates of herbivory and longer leaf lifetimes would select for higher defense in tropical leaves. Even more striking than the latitudinal patterns for mature leaf defenses are those for young expanding leaves. In the tropics almost 70% of a leaf's lifetime damage occurs while it is expanding (Table 1), suggesting that selection for young leaf defenses should be intense. Below we show that young tropical leaves have abundant and novel defenses, which in many cases, surpass levels seen in mature leaves. The opposite pattern occurs in the temperate zone, where young leaves tend to be less well defended than mature leaves.

Nutritional Quality

Nutritional content of leaves varies among species and across leaf ages. Protein, water, and fiber content may result from abiotic selection for different photosynthetic capabilities or protection from physical damage. However, nutritional content has consequences for herbivory and may also be partially shaped by selection from herbivores and pathogens (157). Low nitrogen and water contents have been repeatedly associated with reduced preference and performance of insects (189). Mature leaves of shade-tolerant tropical species have significantly lower nitrogen and water contents than do temperate leaves (47). Young leaves are almost uniformly higher in nitrogen and water than mature leaves—an apparently unavoidable consequence of cell growth. As this makes them more attractive to herbivores (189), we might expect selection to eliminate unnecessary nitrogen from the leaf. In a survey of more than 200 species from four lowland rain forests in Africa, SE Asia, and Central America, a significant positive relationship was found between young-leaf nitrogen and the rate of leaf growth during expansion (49). Apparently rapid leaf expansion requires high nitrogen, presumably in important metabolic enzymes. No physiological constraint prohibits high nitrogen in slowly expanding leaves, but since it is not required for slow expansion, selection by herbivores should favor reduced levels. Thus the nitrogen level of young leaves may reflect the balance between growth requirements and palatability to herbivores. Fiber poses digestive and mechanical problems to herbivores. Fiber content and leaf toughness, a frequently used composite measure of fiber, are both highly negatively correlated with herbivory (40, 178). In the tropics mature leaves are twice as tough as mature temperate leaves (47). Young tropical leaves are also significantly tougher than young temperate leaves, though both are less than half as tough as when they mature (47). Thus ontogenetic and latitudinal patterns of toughness are consistent with rates of herbivory. Perhaps because toughness is such an effective defense, young tropical leaves toughen rapidly as soon as they reach full size. Although the expansion period varies across species from about 6 to 60 days, all species toughen in only a few days immediately following cessation of leaf expansion (4, 123). In a study of daily herbivory on four species, rates of herbivory dropped fourfold during the 3–5 day period of leaf toughening (123). We therefore suggest that selection by herbivores may have caused toughening to occur as rapidly and as early as possible.

Rapid Leaf Expansion

Herbivory on young leaves comprises most of the lifetime damage for tropical species, so reducing the expansion period would lower overall damage (149, 168). Rapid leaf expansion should impose severe constraints on host-finding by specialist herbivores and shorten the period of exposure to generalists. Aide & Londoño (4) showed that the main herbivore specializing on young leaves of *Gustavia superba* (Lethycidaceae), a species with fast expansion, has only a 3-day window in which to successfully oviposit, even though the larvae have exceptionally quick developmental times. Ernest (73) found twice the damage on slow- as on fast-expanding *Pentagonia* (Rubiaceae) leaves.

Pathogens may be even more severely affected than insects by rapid expansion. To colonize the appropriate host species, a specialist pathogen must use an insect vector with similar host preferences, or it must produce sufficiently large numbers of spores that random dispersal by wind or rain will ensure arrival at the target species. Yet in a study of 25 understory species on BCI (50), no correlation was shown between expansion rate and pathogen damage in the field, even though extracts from rapidly expanding leaves were less toxic in laboratory assays. This suggests that pathogens have limited dispersal ability that makes them less capable of colonizing fast-expanders. Thus pathogens may be a key selective factor favoring rapid expansion of young leaves. Expansion rates vary by an order of magnitude among species, with some leaves doubling in size in less than a day, and others needing more than 15 days (49). So, although rapid expansion appears to reduce damage by both herbivores and pathogens, many species have slow expansion and must rely on alternative defenses.

Secondary Metabolites

The diversity and abundance of plant secondary metabolites appear to be greater in tropical than in temperate forests. For example, a survey of the distribution and activity of alkaloids shows that they are more common and more toxic in the tropics (134, 135). About 16% of the temperate species surveyed contain alkaloids as compared to more than 35% of the tropical species (47). Phenolic compounds, as measured by the Folin Denis assay, do not show obvious latitudinal trends. In a literature review of mature leaves of 282 species in temperate, tropical dry, and tropical wet forests, concentrations of phenols average 6.9% dry weight [range, 6.5%–7.4% dry weight (dw)] and do not significantly differ between forest types (47). However, condensed tannins in mature leaves, measured by the BuOH/HCl method, are almost three times higher in the tropical forests ($n = 268$ species, temperate mean = 1.9% dw, tropical mean = 5.5% dw) (21, 47, 197). Tropical leaves contain many other classes of secondary metabolites, but we know of no comparative studies across tropical or temperate communities. Although mature leaves in the tropics appear to invest heavily in secondary metabolites, the young expanding leaves show the most dramatic commitment to chemical defense (49).

Toughness, the most effective defense (4, 40, 127, 141), is not compatible with leaf expansion, so young leaves must rely on other defenses. Although it was originally suggested that the problems of sequestering secondary metabolites during cell division and expansion would pose insurmountable problems for young leaves (148, 168), that appears not to be the case. In tropical trees, mono-, sesqui-, and diterpenes reach higher concentrations in young as compared to mature leaves (59, 127), and levels of simple phenolics and condensed tannins are almost twice as high ($n = 125$ species) (21, 47, 197). In contrast, young temperate leaves have twice the level of total phenols, but only half the level of condensed tannins as mature leaves ($n = 7$ species; 47).

Young tropical leaves also have high concentrations of anthocyanins, which cause the dramatic red coloration that has captured scientific and casual interest for decades (34, 100, 167). Several investigators have argued that the selective advantage of anthocyanins is to screen harmful UV (129) or to protect against photoinhibition (98). However, since anthocyanins are associated primarily with shade-tolerant plants and are only present during leaf expansion, adaptive explanations relating anthocyanin to light have been questioned (46). Instead, it has been demonstrated that anthocyanins have antifungal properties (46) that may be particularly important during leaf expansion when the cuticle is poorly developed and risk from pathogen attack is high (PD Coley, TA Kursar, unpublished data; 46, 90). Data on other secondary metabolites are spotty, although we suggest they may also be more common in young expanding leaves than in mature leaves. Coley & Kursar (49) suggest that at full expansion, toughness may play a more important role, and investment in chemical defenses can be relaxed. It would therefore be advantageous for expanding leaves to invest in compounds that could easily be reclaimed (149, 150). The most likely candidates for this are low molecular weight compounds such as monoterpenes, toxic proteins and amino acids, cyanogenic compounds, alkaloids, and saponins. Furthermore, if costs are associated with turnover of these compounds (92), it may be too expensive to use them as defense in long-lived mature leaves, but reasonable for defense during expansion (48). For these reasons we expect the diversity and quantity of low molecular weight compounds to be extremely high in young leaves.

Investment in secondary metabolites may be lower in rapidly expanding young leaves because of a greater risk of autotoxicity (149, 177) or because resource input into the leaf may simply not be sufficient for both rapid expansion and synthesis of secondary metabolites (156, 168). This hypothesis is supported by studies showing that shoot tips of chemically well-defended plants elongate more slowly than those of less-protected relatives (168, 174). In general, rapidly expanding young leaves have significantly higher damage rates in the field than do slow expanders (49), apparently due to differences in chemical defense. Extracts from fast-expanding young leaves were preferred by insects in feeding trials and supported greater fungal growth than extracts from slow-expanders ($n = 25$ species; PD Coley, TA Kursar, unpublished).

Delayed Greening

In many tropical species, young leaves have reduced chlorophyll contents and appear white, pink, or red. Because of the visually dramatic impact of red young leaves, most investigators have focused on understanding the role of anthocyanins (see above). However, more remarkable is the fact that these leaves have altered development such that the normal process of greening is delayed until after full leaf expansion (12, 13, 49). This developmental pattern is extremely common and has apparently arisen independently many times. In a survey of 250 tropical tree species in 44 families, 33% of the species and 61% of the families had delayed greening (49). TA Kursar & PD Coley (195) argue that delayed greening has evolved because it reduces the amount of resources lost for a given amount of herbivory. In delayed greening, chloroplast development is postponed until after the leaf has reached full size, toughened, and is better protected from herbivores (125). As a consequence, young leaves with delayed greening have approximately 10–20% lower levels of light harvesting proteins, photosynthetic enzymes, chlorophyll, and lipid-rich membranes than do normally greening leaves (13, 123, 124). Although ultimately the mature leaves of species with normal and delayed greening have similar photosynthetic characteristics and construction costs, the timing of investment differs (49).

The benefits of delayed greening occur because lower protein and energy contents during expansion translate to a lower loss of resources for a given amount of herbivory (49). The cost is reduced photosynthesis (124, 125). Kursar & Coley compared costs and benefits in habitats with different light regimes to determine if delayed greening was ever cost effective (49, 123–125). Their analysis shows that at the light and herbivory levels typical of tropical forest understories, leaves with delayed greening cost less. However, at the higher light levels of gaps or even temperate forest understories, rates of herbivory would have to be near 100% to balance the increased cost of forfeited photosynthesis. Thus, the analysis suggests that the understory of tropical forests is the only habitat where light is sufficiently low and herbivory sufficiently high to favor delayed greening. And, in fact, delayed greening is restricted to shade-tolerant tropical species (123). Not all shade-tolerant species delay greening. There is a significant negative correlation between expansion rate and chlorophyll content (49). Delayed greening is therefore most common in species with rapid leaf expansion and low investments in chemical defense. As we argued for secondary metabolites, resource limitation in rapid expanders may make simultaneous investment in growth and greening impossible. Furthermore, delayed greening would reduce the impact of intense herbivory. Thus both physiological constraints and selection would favor delayed greening in rapid expanders. Slow expanders, which are also better defended chemically, would gain little benefit from delayed greening but would pay the cost over a long period of time. So although delayed greening is physiologically possible in slow expanders, selection would favor normal development.

Leaf Phenology

Another strategy plants in tropical forests may use to avoid herbivory on young leaves is to alter the phenology of leaf production (1, 75, 136, 149). This can be done in two ways. First, leaf production may be shifted to peak during the time of year when herbivore abundance is lowest, which is the dry season in most forests. Second, leaves can be flushed synchronously, saturating herbivores with an abundance of leaves to ensure that some escape damage, an idea analogous to mast fruiting as a way to avoid seed predators (3).

LEAF PRODUCTION WHEN HERBIVORES ARE RARE During times of the year when herbivores are rare in a forest, rates of damage to young leaves are lower. In the Accra dry forest in southeastern Ghana, herbivore damage to young leaves was lowest at the start of the wet season (136). Likewise, in two dry forests in south India, trees that flushed new leaves during the dry season suffered significantly less damage than those that produced new leaves during the wet season (158). This pattern was also found in the moist, semideciduous forest on BCI (1, 3). Aide (2) experimentally demonstrated a seasonal escape from herbivory on BCI using the shrub *Hybanthus prunifolius*, which normally flushes in the dry season when herbivore numbers are low (2). Plants forced to produce new leaves in the wet season suffered significantly more herbivore damage than those that naturally produced leaves in the dry season (2). Nevertheless, it is not clear that a seasonal shift in leaf production to avoid herbivory is a viable strategy for plants in forests with weak or short dry seasons, since the abundances of insect herbivores do not decline dramatically under such conditions (206). Have low rates of herbivory during the dry or early wet season been the determining factor in the timing of leaf production? Three factors, water availability, solar radiation, and herbivory, potentially are critical to the evolution of the timing of leaf production (199). In dry, deciduous forests, present evidence suggests that water stress in the dry season limits leaf production (175, 199, 210). In wetter forests where dry seasons are shorter and water stress is less severe, trees may concentrate leaf production in the sunniest times of the year to avoid light limitation during the rainy season (199, 210), though this hypothesis has been disputed (175). Alternatively, trees in wetter forests may produce leaves in response to individual and "endogenous" factors, with individual trees within a species behaving independently (175).

This current emphasis on abiotic factors in the evolution of the timing of leaf production does not rule out a role for herbivores, but assessing their influence is difficult, largely because of the coincidence of low herbivore numbers and peak irradiance at the end of the dry season (199, 209). Wright & van Schaik (210) noted that one way to disentangle these factors is to look at leaf production in forests where herbivore abundances are high during the dry season, as occurs in Gabon (101). They found that leaf production peaks when both irradiance and insect abundances are greatest, suggesting that for this forest, herbivores are less important than light levels in determining when leaves are produced (210). More of such comparative studies are needed, however, before the role of herbivores in influencing leaf production can be fairly assessed.

SYNCHRONY OF LEAF FLUSHING Increased synchrony in leaf production may also be an adaptation to avoid herbivory. In Ghana, tree species that were more synchronous were less likely to suffer insect damage to young leaves (136). On BCI, Aide found that for the 10 most self-synchronous species he studied, herbivore damage was significantly higher on leaves produced outside of the peak months of leaf production, suggesting that herbivore pressure is maintaining synchrony in these species (3). He also showed, however, that species producing leaves more or less continuously also suffered lower rates of herbivore damage, presumably by using chemical defenses (3, 49).

If herbivores do play an important role in selecting for synchrony in at least some plant species in tropical forests, then the degree of diet-specialization by herbivores should have an impact on the level of synchrony in the forest. If the most damaging herbivores in a forest are generalists, then selection would favor synchrony at the community level. On the other hand, if most damage is done by herbivores that are specialized to a single species, as seems to be the case, then selection would favor individuals that were synchronous with conspecifics, but there would be no particular advantage (or disadvantage) to flushing simultaneously with any other species.

Compared to herbivores, pathogens have probably had a negligible role in selecting for synchronous leaf production in their host plants. Because insect herbivores actively seek young host leaves, the chance that any particular young host leaf will be discovered decreases with higher numbers of young leaves. This is why leaves produced synchronously have lower damage rates from insects than those produced at other times. But for pathogens dispersed haphazardly through the forest by wind and rain, the chance that any young host leaf will be colonized is independent of the abundance of young host leaves. In other words, insect herbivores can be satiated, but pathogens cannot be. If anything, synchrony of leaf production may result in higher rates of pathogen damage, if spore release by pathogens coincides with leaf production. Unfortunately, no data are available to test these hypotheses.

Third Trophic Level

Thus far, we have focused largely on the interactions between plants and their herbivores. Yet, as has been recognized for some time, the predators, parasitoids, and pathogens of herbivores greatly influence these interactions, and this is reflected in the defenses employed by plants (see 172 for a review). Here we focus briefly on a few issues as they relate to tropical forests.

The antiherbivore defenses of tropical plants have evolved within the context of a community that includes the natural enemies of their herbivores. For this reason, "quantitative defenses" such as tannins or toughness are effective against herbivores even though they do not present an absolute barrier to herbivores. Instead these defenses slow herbivore growth and lengthen the time that herbivores are exposed to predators and parasitoids (75, 172, 177). As the majority of feeding occurs in the final instars, predation will reduce leaf damage. The high levels of these defenses in mature leaves of tropical plants and the relative rarity of insects that feed on mature leaves (JA Barone, unpublished data) suggest that the plants have consistently relied on the enemies of herbivores throughout their evolution. We believe that tropical forests are "green" in large part because the natural enemies of herbivores make quantitative defenses effective.

Plants have also evolved adaptations to attract ants and use them as a defense (25). The production of ant attractants has been predicted to be more common in plants with short-lived leaves, because a continuous investment is needed to feed the ants, and as a leaf ages this cost eventually exceeds that of investing in quantitative defenses (150). Gap species typically have high rates of leaf turnover (48) and, with readily available light, have an abundance of carbon, which makes sugar and lipid awards relatively cheap (19, 165). In a survey of 243 plant species on BCI, gap species were indeed more likely to have attractants than were understory species (184).

Unlike ants, the defensive role of mites on plants has received little attention. Plants may use mites as "bodyguards" against fungal and bacterial pathogens (201, 202). Domatia, specialized chambers in leaf axils, presumably function to house mites, and these occur in 28% of the world's dicot families. In a survey of North Queensland rain forests, 15% of trees had domatia, and 50% of the domatia contained mites (166). Over 80% of the mites were scavengers or fungivores, with only a few being plant parasites (202). Although this suggests mites may benefit the plant by feeding on fungal spores and thereby reducing infection rates, no studies have tested this idea.

Interactions Among Defensive Characteristics of Young Leaves

The high rates of herbivory on young tropical leaves might suggest that selection should favor investment in a large fraction of the defenses described above. However, each species invests in only a small subset of possible defenses. Furthermore, we consistently see the same suites of co-occurring traits across unrelated species. Convergence on similar combinations of traits suggests tradeoffs or physiological constraints that limit the defensive possibilities. We discussed specifics of these relationships in previous sections, so here we describe only the general patterns.

Common defensive patterns were identified primarily from a survey of more than 200 tropical woody species from four forests in Africa, SE Asia, and Central America (49). Each forest had the same emergent associations of rapid leaf expansion, high nitrogen, delayed greening, low toughness, low secondary metabolites, and synchronous leaf flushing. We suggest that by examining the relationships among traits, we can explain why particular sets of traits co-occur. For example, although high nitrogen makes leaves more palatable, it is required for rapid expansion. Because of resource limitation, it is physiologically impossible simultaneously to expand rapidly, green normally, and synthesize secondary metabolites. Because rapid expanders suffer high rates of herbivory, the added protection of synchrony and delayed greening would be favored by selection. And finally, ant defense is most effective on species with continuous leaf production. Thus, the various defensive traits are connected by physiology or selective advantages such that an individual trait is predictably and somewhat inflexibly tied to the entire suite of defenses.

INTERSPECIFIC PATTERNS OF DEFENSE

We have presented general patterns of defense and have highlighted differences between tropical and temperate systems. However, within the tropics, the variation in both herbivory and defense is enormous, dwarfing the latitudinal differences. In the following section, we discuss several clear interspecific trends in defense.

Leaf Lifetime and Defense

A positive correlation exists between leaf lifetime and the commitment to defenses, presumably because the value to the plant and the risk of discovery both increase with leaf lifetime (75, 149, 177, 188). Most gap-demanding species have leaves that last less than six months and are relatively palatable to herbivores (Table 1; 47, 163). Leaf lifetimes for shade-tolerant species are longer, from 1 to 14 yr, and leaves are better defended (40). Furthermore, even within shade-tolerant species growing in the same habitat, leaf lifetime is negatively correlated with herbivory and positively correlated with defenses such as tannins and fiber (43, 47). Comparisons of vertebrate herbivory in deciduous and evergreen forests are consistent with increased defense in long-lived leaves. Two deciduous forests in India support 10 times the biomass of large herbivores than do three evergreen forests in Africa and the Neotropics (119). In Madagascar, rainfall is significantly negatively correlated with the biomass of folivorous lemurs; lemur biomass in drier, deciduous forests is eight times higher (89). Similar patterns are seen with colobine monkeys in Africa and Asia (164, 203). Ganzhorn (89) attributes the lower biomass of herbivores in evergreen forests to a relatively higher fiber content in the long-lived leaves.

Light and Nutrients

A common evolutionary response to habitats where light or nutrients are limiting is slow growth and lower rates of leaf turnover (36, 99). For species with slow growth, it is hypothesized that opportunity costs of defense will be lower, and the relative impact of herbivory will be higher, than in faster growing species (48, 99). Furthermore, because more resources have been invested in long-lived leaves and replacement is costly, leaf lifetime should be positively correlated with defense (75, 99, 114, 149, 150, 177). Consistent with these hypotheses is the common observation that species from nutrient poor forests have well-defended leaves (42, 48, 114, 151, but see 197).

Light gaps made by treefalls create a mosaic of high light habitats within a rain forest. Presumably because of their rapid growth and short leaf lifetimes, species that specialize on gaps are poorly defended and as a consequence suffer higher rates of herbivory (Table 1). Gap-specialists have similar concentrations of simple phenols but significantly lower condensed tannin, toughness, and fiber contents as compared to shade-tolerant species (21, 47, 197).

INTRASPECIFIC PHENOTYPIC VARIATION IN DEFENSE

The environment can also strongly modify the phenotypic expression of defenses in a given individual. The carbon/nitrogen balance hypothesis (33) suggests that resources in excess of growth demands are shunted to defenses. Thus, high light should lead to elevated photosynthesis and carbohydrate accumulation, which would cause an increase in carbon-based defenses such as tannins and terpenes. Much evidence from the tropics supports this (41, 65, 76, 126, 154, 162, 179, 180). Because the carbon/nitrogen ratio of tissue is high in the light, nitrogen-based defenses should decrease (33), though adequate tests of this prediction are lacking. The phenotypic response to light is opposite to the pattern seen across species (18). Selection has favored high levels of carbon-based defenses in shade-tolerant species, yet moving a plant from sun to shade reduces these same defenses. Although this seems to cause confusion, there is no reason to expect phenotypic responses to imbalances in source/sink relationships to follow the same trends as evolved differences among species (44). It is therefore misleading and inappropriate to apply the carbon/nitrogen balance hypothesis to explain interspecific defensive patterns. Furthermore, comparing defense levels of gap specialists in the sun with shade-tolerant species in the shade confounds these opposing phenotypic and genetic trends (14). Comparisons between closely related species with different habitat requirements frequently show patterns more typical of phenotypic responses, e.g. similar or higher levels of phenolic compounds in the sun species (40, 65). We suggest that this results from phylogenetic constraints, as species in the same genus have pathways for secondary metabolism that shared a recent common ancestor. Therefore, although selection may be favoring a downregulation of carbon-based metabolites in the gap species, this may be masked by a phenotypic increase because plants exist in a high light habitat. Thus, defensive patterns displayed by congeners may not be "optimal" for their current habitats, especially when they do not mirror common patterns seen in ecologically similar but unrelated species.

CONCLUSIONS

Overall, tropical forests have been shaped by strong ecological and evolutionary interactions between plants and herbivores. The elevated rates of leaf damage, compared to those of the temperate zone, have apparently selected for a greater investment in a diversity of defenses. Particularly distinctive is the fact that the majority of leaf damage in tropical forests occurs on the young, expanding leaves. Mature leaves commonly have defenses, such as tannins and toughness, that function by slowing the growth of herbivores, making them more vulnerable to predators and parasitoids. Insect herbivores display a high degree of diet specialization in the tropics and are responsible for a majority of the leaf damage.

These conclusions suggest several areas for productive research in the future. First, the high degree of diet specialization by insect herbivores should lead to tight linkages between the population dynamics of herbivores and their hosts. Moreover, as the majority of herbivores depend on young leaves, a well-defended and ephemeral resource, host plants may exert particularly strong selection on herbivore life histories and detoxification abilities. Second, although pathogens may cause a third of the leaf damage in tropical forests and may be as important as insects in determining the success of different host genotypes or species, they have received very little attention. Third, insect herbivores and pathogens are strongly influenced by differences in the length of the dry season, and thus the strength of their impact may vary with precipitation. Furthermore, because of the generally lower levels of defense in deciduous leaves, mammals may be more common and may make a relatively larger contribution to herbivory in dry forests. Comparative studies between forests with different rainfall regimes would no doubt prove very informative. Finally, what ultimately distinguishes tropical forests from other ecosystems is that their community structure results from long-term and intricate biotic interactions involving plants, their consumers and natural enemies. With the rapid destruction of tropical forests and the threat of global climate change, a greater understanding of the importance of these interactions, and how they are altered by fragmentation, is essential to the preservation of tropical forests.

ACKNOWLEDGMENTS

We are extremely grateful to Thomas A Kursar who helped develop most of the ideas presented here. The manuscript benefited from comments by D Dearing, B Howlett, T Kursar, R Lee, E Leigh, V Sork, S Talley, and S Torti. We appreciate financial support from the National Science Foundation (DEB-9420031 to PDC) and the Smithsonian Tropical Research Institute (to PDC and JAB).

Literature Cited

1. Aide TM. 1988. Herbivory as a selective agent on the timing of leaf production in a tropical understory community. *Nature* 336:574–75
2. Aide TM. 1992. Dry season leaf production: an escape from herbivory. *Biotropica* 24:532–37
3. Aide TM. 1993. Patterns of leaf development and herbivory in a tropical understory community. *Ecology* 74:455–66
4. Aide TM, Londoño EC. 1989. The effects of rapid leaf expansion on the growth and survivorship of a lepidopteran herbivore. *Oikos* 55:66–70
5. Aide TM, Zimmerman JK. 1989. Patterns of insect herbivory, growth, and survivorship in juveniles of a neotropical liana, *Connarus turczaninowii* (Connaraceae). *Ecology* 71:1412–21
6. Augspurger CK. 1983. Offspring recruitment around tropical trees: changes in cohort distance with time. *Oikos* 40:198–96
7. Augspurger CK. 1983. Seed dispersal of the tropical tree, *Platypodium elegans,* and the escape of its seedlings from fungal pathogens. *J. Ecol.* 71:759–71
8. Augspurger CK. 1984. Seedling survival of tropical tree species: interactions of dispersal distance, light-gaps, and pathogens. *Ecology* 65:1705–12
9. Augspurger CK. 1990. The potential impact of fungal pathogens on tropical plant reproductive biology. In *Reproductive Ecology of Tropical Forest Plants,* ed. KS Bawa, M Hadley, pp. 237–45. Paris: Parthenon

10. Augspurger CK, Kitajima K. 1992. Experimental studies of seedling recruitment from contrasting seed distributions. *Ecology* 73:1270–84
11. Aylor DE. 1978. Dispersal in time and space: aerial pathogens. In *Plant Disease: An Advanced Treatise,* ed. JG Horsfall, EB Cowling, pp. 159–80. New York: Academic
12. Baker NR, Hardwick K. 1973. Biochemical and physiological aspects of leaf development in cacao (*Theobroma cacao*). I. Development of chlorophyll and photosynthetic activity. *New Phytol.* 72:1315–24
13. Baker NR, Hardwick K, Jones P. 1975. Biochemical and physiological aspects of leaf development in cacao (*Theobroma cacao*). II. Development of chloroplast ultrastructure and carotenoids. *New Phytol.* 75:513–18
14. Baldwin IT, Schultz JC. 1988. Phylogeny and the patterns of leaf phenolics in gap- and forest-adapted *Piper* and *Miconia* understory shrubs. *Oecologia* 75:105–9
15. Barone JA. 1994. Herbivores and herbivory in the canopy and understory on Barro Colorado Island, Panama. *1st Int. Canopy Conf.,* Selby Bot. Gard., Sarasota, FL. (Abstr.)
16. Basset Y. 1992. Host specificity of arboreal and free-living insect herbivores in rain forests. *Biol. J. Linn. Soc.* 47:115–33
17. Basset Y. 1994. Palatability of tree foliage to chewing insects: a comparison between a temperate and a tropical site. *Acta Oecol.* 15:181–91
18. Bazzaz FA, Chiariello NR, Coley PD, Pitelka LF. 1987. Allocating resources to reproduction and defense. *Bioscience* 37:58–67
19. Beattie AJ. 1985. *The Evolutionary Ecology of Ant-Plant Mutualisms.* Cambridge: Cambridge Univ. Press. 182 pp.
20. Beaver RA. 1979. Host-specificity of temperate and tropical animals. *Science* 281:139–41
21. Becker P. 1981. Potential physical and chemical defenses of *Shorea* seedling leaves against insects. *Malay. For.* 2/3:346–56
22. Becker P, Wong M. 1985. Seed dispersal, seed predation, and juvenile mortality of *Aglaia* sp. (Meliaceae) in lowland Dipterocarp Rainforest. *Biotropica* 17:230–37
23. Benedict F. 1976. *Herbivory rates and leaf properties in four forests in Puerto Rico and Florida.* PhD thesis. Univ. Fla., Gainesville
24. Benson WW. 1978. Resource partitioning in passion vine butterflies. *Evolution* 32:493–518
25. Bentley B. 1977. Extra-floral nectaries and protection by pugnacious bodyguards. *Annu. Rev. Ecol. Syst.* 8:407–27
26. Bernays EA, Chapman RF. 1994. *Host-Plant Selection by Phytophagous Insects.* New York/London: Chapman & Hall. 312 pp.
27. Bernays EA, Graham M. 1988. On the evolution of host specificity in phytophagous arthropods. *Ecology* 69:886–92
28. Bodmer RE. 1989. Ungulate biomass in relation to feeding strategy within Amazonian forests. *Oecologia* 81:547–50
29. Boucher DH. 1981. Seed predation by mammals and forest dominance by *Quercus oleoides*, a tropical lowland oak. *Oecologia* 49:409–14
30. Bray JR. 1961. Measurement of leaf utilization as an index of minimum level of primary consumption. *Oikos* 12:70–74
31. Brown BJ, Ewel JJ. 1987. Herbivory in complex and simple tropical successional ecosystems. *Ecology* 68:108–16
32. Brown KS Jr. 1987. Chemistry at the Solanaceae/Ithomiinae interface. *Ann. Mo. Bot. Gard.* 74:359–97
33. Bryant JP, Chapin FS III, Klein DR. 1983. Carbon/nutrient balance of boreal plants in relation to vertebrate herbivory. *Oikos* 40:357–68
34. Burgess PF. 1969. Colour changes in the forest. *Malay. Nat. J.* 22:171–73
35. Burkey TV. 1994. Tropical tree species diversity: a test of the Janzen-Connell Model. *Oecologia* 97:533–40
36. Chapin FS III. 1980. The mineral nutrition of wild plants. *Annu. Rev. Ecol. Syst.* 11:261–85
37. Clark DA, Clark DB. 1984. Spacing dynamics of a tropical rain forest tree: evaluation of the Janzen-Connell model. *Am. Nat.* 124:769–88
38. Clark DB, Clark DA. 1985. Seedling dynamics of a tropical tree: impacts of herbivory and meristem damage. *Ecology* 66:1884–92
39. Coley PD. 1982. Rates of herbivory on different tropical trees. See Ref. 132, pp. 123–32
40. Coley PD. 1983. Herbivory and defensive characteristics of tree species in a lowland tropical forest. *Ecol. Monogr.* 53:209–33
41. Coley PD. 1986. Costs and benefits of defense by tannins in a neotropical tree. *Oecologia* 70:238–41
42. Coley PD. 1987. Patrones en las defensas de las plantas: ¿ porque los herbivo-

ros prefrieren ciertas especies? *Rev. Biol. Trop.* 35:251–63
43. Coley PD. 1988. Effects of plant growth rate and leaf lifetime on the amount and type of anti-herbivore defense. *Oecologia* 74:531–36
44. Coley PD. 1993. Gap size and plant defenses. *Trends Ecol. Evol.* 8:1–2
45. Coley PD. 1996. Effects of climate change on plant-herbivore interactions in moist tropical forests. *Clim. Change*. In press
46. Coley PD, Aide TM. 1989. Red coloration of tropical young leaves: a possible antifungal defence? *J. Trop. Ecol.* 5:293–300
47. Coley PD, Aide TM. 1991. A comparison of herbivory and plant defenses in temperate and tropical broad-leaved forests. See Ref. 173, pp. 25–49
48. Coley PD, Bryant JP, Chapin FS III. 1985. Resource availability and plant antiherbivore defense. *Science* 230:895–99
49. Coley PD, Kursar TA. 1996. Antiherbivore defenses of young tropical leaves: physiological constraints and ecological tradeoffs. See Ref. 186, pp. 305–36
50. Coley PD, Kursar TA, Fikstad T, Rosengreen L. 1995. Chemical defense against pathogens and herbivores is related to expansion rates of young leaves. *Ecol. Soc. Am.* (Abstr.) 76:
51. Condit R, Hubbell SP, Foster RB. 1992. Recruitment near conspecific adults and the maintenance of tree and shrub diversity in a neotropical forest. *Am. Nat.* 140:261–86
52. Condit R, Hubbell SP, Foster RB. 1992. Short-term dynamics of a neotropical forest. *BioScience* 42:822–28
53. Condit R, Hubbell SP, Foster RB. 1994. Density dependence in two understory tree species in a neotropical forest. *Ecology* 75:671–80
54. Connell JH. 1971. On the role of natural enemies in preventing competitive exclusion in some marine animals and in rain forest trees. In *Dynamics of Populations, Proceedings of the Advanced Study Institute on Dynamics of Numbers in Populations, Oosterbeek, 1970*, ed. PJ den Boer, GR Gradwell, pp. 298–312. Wageningen, Netherlands: Cent. Agric. Publ. Doc.
55. Connell JH. 1978. Diversity in tropical rain forests and coral reefs. *Science* 199:1302–10
56. Connell JH. 1979. Tropical rain forests and coral reefs as open nonequilibrium systems. In *Population Dynamics*, ed. RM Anderson, BD Turner, LR Taylor, pp. 141–61. Oxford/London: Blackwell Sci.
57. Connell JH, Tracey JG, Webb LJ. 1984. Compensatory recruitment, growth and mortality as factors maintaining rain forest tree diversity. *Ecol. Monogr.* 54:141–64
58. Cork SJ, Foley WJ. 1991. Digestive and metabolic strategies of arboreal mammalian folivores in relation to chemical defenses in temperate and tropical forests. In *Plant Defenses Against Mammalian Herbivory*, ed. RT Palo, CT Robbins, pp. 133–66. Boca Raton: CRC
59. Crankshaw DR, Langenheim JH. 1981. Variation in terpenes and phenolics through leaf development in *Hymenaea* and its possible significance to herbivory. *Biochem. Syst. Ecol.* 9:115–24
60. Crossley DA Jr, Gist CS, Hargrove WW, Ridley LS, Schowalter TD, Seatedt TR. 1988. Foliage consumption and nutrient dynamics in canopy insects. In *Forest Hydrology and Ecology at Coweeta*, ed. WT Swank, DA Crossley Jr, pp. 193–205. New York: Springer-Verlag
61. Dasilva GL. 1992. The western black and white colobus as a low-energy strategist: activity budgets, energy expenditure and energy intake. *J. Anim. Ecol.* 61:79–91
62. de la Cruz M, Dirzo R. 1987. A survey of the standing levels of herbivory in seedlings from a Mexican rain forest. *Biotropica* 19:98–106
63. Demment MW, Van Soest PJ. 1985. A nutritional explanation for body-size patterns of ruminant and nonruminant herbivores. *Am. Nat.* 125:641–72
64. Denslow JS. 1980. Notes on the seedling ecology of a large-seeded species of Bombacaceae. *Biotropica* 12:220–21
65. Denslow JS, Schultz JC, Vitousek PM, Strain BR. 1990. Growth responses of tropical shrubs to treefall gap environments. *Ecology* 71:165–79
66. DeSteven D, Putz FE. 1984. Impact of mammals on early recruitment of a tropical canopy tree, *Dipteryx panamensis*, in Panama. *Oikos* 43:207–16
67. Dial R, Roughgarden J. 1995. Experimental removal of insectivores from rain forest canopy: direct and indirect effects. *Ecology* 76:1821–34
68. Dirzo R. 1984. Herbivory, a phytocentric overview. In *Perspectives in Plant Population Biology*, ed. R Dirzo, J Sarukhan, pp. 141–65. Sunderland, MA: Sinauer
69. Dirzo R, Miranda A. 1991. Altered patterns of herbivory and diversity in the forest understory: a case study of possible consequences of contemporary defauna-

tion. See Ref. 173, pp. 273–87

70. Drummond BA III. 1986. Coevolution of Ithomiine butterflies and Solanaceous plants. In *Solanaceae: Biology and Systematics,* ed. WG D'Arcy, pp. 307–27. New York: Columbia Univ. Press
71. Ehrlich PR, Raven PH. 1964. Butterflies and plants: a study in coevolution. *Evolution* 18:586–608
72. Eisenberg JF. 1978. The evolution of arboreal folivores in the class Mammalia. In *The Ecology of Arboreal Folivores,* ed. GG Montgomery, pp. 135–52. Washington, DC: Smithsonian Inst.
73. Ernest KA. 1989. Insect herbivory on a tropical understory tree: effects of leaf age and habitat. *Biotropica* 21:194–99
74. Erwin TL. 1982. Tropical forests: their richness in Coleoptera and other arthropod species. *Coleopt. Bull.* 36:74–75
75. Feeny PP. 1976. Plant apparency and chemical defense. In *Biochemical Interactions Between Plants and Insects. Recent Advances in Phytochemistry,* ed. J Wallace, RL Mansell, 10:1–40. New York: Plenum
76. Feibert EB, Langenheim JH. 1988. Leaf resin variation in *Copaifera langsdorfii*: relation to irradiance and herbivory. *Phytochemistry* 27:2527–32
77. Fensham RJ. 1994. Phytophagous insect-woody sprout interactions in tropical eucalypt forest. II. Insect community structure. *Aust. J. Ecol.* 19:189–96
78. Filip V, Dirzo R, Maass JM, Sarukhan J. 1995. Within- and among-year variation in the levels of herbivory on the foliage of trees from a Mexican tropical deciduous forest. *Biotropica* 27:78–86
79. Fittkau EJ, Klinge H. 1973. On biomass and trophic structure of the Central Amazonian rainforest ecosystem. *Biotropica* 5:2–14
80. Foley WJ, Hume ID. 1987. Passage of digesta markers in two species of arboreal folivorous marsupials: the greater glider (*Petauroides volans*) and the brushtail possum (*Trichosurus vulpecula*). *Physiol. Zool.* 60:103–13
81. Forget PM. 1989. La regeneration naturelle d'une espece autochore de la foret guyanaise: *Eperua falcata* Aublet (Caesalpiniaceae). *Biotropica* 21:115–25
82. Forget PM. 1991. Comparative recruitment patterns of two non-pioneer canopy tree species in French Guiana. *Oecologia* 85:434–39
83. Forget PM. 1992. Regeneration ecology of *Eperua grandiflora* (Caesalpiniaceae), a large-seeded tree in French Guiana. *Biotropica* 24:146–56
84. Forget PM. 1994. Recruitment pattern of *Vouacapoua americana* (Caesalpiniaceae), a rodent-dispersed tree species in French Guiana. *Biotropica* 26:408–19
85. Foster RB. 1982. Famine on Barro Colorado Island. See Ref. 132, pp. 201–12
86. Foster RB. 1982. The seasonal rhythm of fruitfall on Barro Colorado Island. See Ref. 132, pp. 151–72
87. Fowler HG. 1979. Seed predator responses. *Oecologia* 41:361–63
88. Futuyma DJ. 1983. Evolutionary interactions among herbivorous insects and plants. In *Coevolution,* ed. DJ Futuyma, M Slatkin, pp. 207–31. Sunderland, MA: Sinauer
89. Ganzhorn JU. 1992. Leaf chemistry and the biomass of folivorous primates in tropical forests. *Oecologia* 91:540–47
90. Garcia-Guzman G, Dirzo R. 1991. Plant-pathogen-animal interactions in a tropical rain forest. *Am. Inst. Biol. Sci.* (Abstr.)
91. Gaston KJ. 1993. Herbivory at the limits. *Trends Ecol. Evol.* 8:193–94
92. Gershenzon J. 1994. The cost of plant chemical defense against herbivory. In *Insect-Plant Interactions,* ed. EA Bernays, 5:105–73. Boca Raton, FL: CRC
93. Gilbert GS. 1995. Rainforest plant diseases: the canopy understory connection. *Selbyana* 16:75–77
94. Gilbert GS, Hubbell SP, Foster RB. 1994. Density and distance-to-adult effects of a canker disease of trees in a moist tropical forest. *Oecologia* 98:100–8
95. Gilbert LE, Singer MC. 1975. Butterfly ecology. *Annu. Rev. Ecol. Syst.* 6:365–97
96. Givnish TJ. 1988. Adaptation to sun and shade: a whole-plant perspective. In *Ecology of Photosynthesis in Sun and Shade,* ed. JR Evans, S von Caemmerer, WW Adams III, pp. 63–92. CSIRO: Melbourne
97. Gosz FR, Likens GE, Bormann FH. 1972. Nutrient content of litter fall on Hubbard Brook experimental forest, New Hampshire. *Ecology* 53:769–84
98. Gould KS, Kuhn DN, Lee DW, Oberbauer SF. 1995. Why leaves are sometimes red. *Nature* 378:241–42
99. Grime JP. 1979. *Plant Strategies and Vegetation Processes.* Chichester, UK: Wiley & Sons
100. Harborne JB. 1979. Function of flavonoids in plants. In *Chemistry and Biochemistry of Plant Pigments,* ed. TW Goodwin, pp. 736–88. New York: Academic

101. Hladik A. 1978. Phenology of leaf production in rain forest of Gabon: distribution and composition of food for folivores. See Ref. 155, pp. 1–72
102. Hodkinson ID, Casson D. 1991. A lesser predilection for bugs: Hemiptera (Insecta) diversity in tropical rain forests. *Biol. J. Linn. Soc.* 43:101–9
103. Howe HF, Primack RB. 1975. Differential seed dispersal by birds of the tree *Casearia nitida* (Flacourtiaceae). *Biotropica* 7:278–83
104. Howe HF, Schupp EW. 1985. Early consequences of seed dispersal for a neotropical tree (*Virola surinamensis.*) *Ecology* 66:781–91
105. Howe HF, Smallwood J. 1982. Ecology of seed dispersal. *Annu. Rev. Ecol. Syst.* 13:201–28
106. Hubbell SP, Condit R, Foster RB. 1990. Presence and absence of density dependence in a neotropical tree community. *Philos. Trans. R. Soc. London Ser. B* 330:269–81
107. Hubbell SP, Foster RB. 1986. Commonness and rarity in a neotropical forest: implications for tropical tree conservation. In *Conservation Biology: The Science of Scarcity and Diversity,* ed. ME Soule, pp. 205–31. Sunderland, MA: Sinauer
108. Jaenike J. 1990. Host specialization in phytophagous insects. *Annu. Rev. Ecol. Syst.* 21:243–73
109. Janzen DH. 1970. Herbivores and the number of tree species in tropical forests. *Am. Nat.* 104:501–28
110. Janzen DH. 1971. Escape of juvenile *Dioclea megacarpa* (Leguminosae) vines from predators in a deciduous tropical forest. *Am. Nat.* 105:97–112
111. Janzen DH. 1972. Association of a rainforest palm and seed-eating beetles in Puerto Rico. *Ecology* 53:258–61
112. Janzen DH. 1972. Escape in space by *Sterculia apetala* seeds from the bug *Dysdercus fasciatus* in a Costa Rican deciduous forest. *Ecology* 53:350–56
113. Janzen DH. 1973. Comments on host-specificity of tropical herbivores and its relevance to species richness. In *Taxonomy and Ecology,* ed. VH Heywood, pp. 201–11. New York: Academic
114. Janzen DH. 1974. Tropical blackwater rivers, animals and mast fruiting by the Dipterocarpaceae. *Biotropica* 6:69–103
115. Janzen DH. 1975. Interactions of seeds and their insect predators/parasitoids in a tropical deciduous forest. In *Evolutionary Strategies of Parasitic Insects and Mites,* ed. PW Price, pp. 154–86. New York: Plenum
116. Janzen DH. 1985. Plant defences against animals in the Amazonian rainforest. In *Amazonia,* ed. GT Prance, TE Lovejoy, pp. 207–17. Oxford/New York: Persimmon
117. Janzen DH, Miller GA, Hackforth-Jones J, Pond CM, Hooper K, Janos DP. 1976. Two Costa Rican bat-generated seed shadows of *Andira inermis* (Leguminosae). *Ecology* 57:1068–75
118. Kaczmarek W. 1967. Elements of organization in the energy flow of forest ecosystems (preliminary notes). In *Secondary Productivity in Terrestrial Ecosystems,* ed. K Petrusewica, pp. 683–85. Warsaw: Panstwowe Wydawnictwo Naukowe
119. Karanth KU, Sunquist ME. 1992. Population structure, density and biomass of large herbivores in the tropical forests of Nagarohole, India. *J. Trop. Ecol.* 8:21–35
120. Kiltie RA. 1981. Distribution of palm fruits on a rain forest floor: Why white-lipped peccaries forage near objects. *Biotropica* 13:141–45
121. Kitajima K, Augspurger CK. 1989. Seed and seedling ecology of a monocarpic tropical tree, *Tachigalia versicolor. Ecology* 70:1102–14
122. Kleiber M. 1975. *The Fire of Life.* New York: Wiley & Sons. 453 pp.
123. Kursar TA, Coley PD. 1992. The consequences of delayed greening during leaf development for light absorption and light use efficiency. *Plant Cell Environ.* 15:901–9
124. Kursar TA, Coley PD. 1992. Delayed development of the photosynthetic apparatus in tropical rain forest species. *Funct. Ecol.* 6:411–22
125. Kursar TA, Coley PD. 1992. Delayed greening in tropical leaves: an antiherbivore defense? *Biotropica* 24:256–62
126. Langenheim JH, Arrhenius SP, Nascimento JC. 1981. Relationship of light intensity to leaf resin composition and yield in the tropical leguminous genera *Hymenaea* and *Copaifera. Biochem. Syst. Ecol.* 9:27–37
127. Langenheim JH, Macedo CA, Ross MK, Stubblebine WH. 1986. Leaf development in the tropical leguminous tree *Copaifera* in relation to microlepidopteran herbivory. *Biochem. Syst. Ecol.* 14:51–59
128. Lawton JH. 1978. Host-plant influences on insect diversity: the effects of space and time. In *Diversity of Insect Faunas,* ed. LA Mound, N. Waloff, pp. 105–25.

Oxford/London: Blackwell Sci.
129. Lee DW, Brammeier S, Smith AP. 1987. The selective advantages of anthocyanins in developing leaves of mango and cacao. *Biotropica* 19:40–49
130. Lee MAB. 1985. Dispersal of *Panadanus tectorius* by the land crab *Cardisoma carnifex. Oikos* 45:169–73
131. Leigh EG Jr. 1997. *Ecology of Tropical Forests: The View from Barro Colorado.* New York: Oxford Univ. Press
132. Leigh EG Jr, Rand AS, Windsor DM, eds. 1982. *Ecology of a Tropical Forest: Seasonal Rhythms and Long-term Changes.* Washington, DC: Smithsonian Inst. Press. 468 pp.
133. Leigh EG Jr, Wright SJ, Herre EA. 1993. The decline of tree diversity on newly isolated tropical islands: a test of a null hypothesis and some implications. *Evol. Ecol.* 7:76–102
134. Levin DA. 1976. Alkaloid-bearing plants: an ecogeographic perspective. *Am. Nat.* 110:261–84
135. Levin DA, York BM. 1978. The toxicity of plant alkaloids: an ecogeographic perspective. *Biochem. Syst. Ecol.* 6:61–76
136. Lieberman D, Lieberman M. 1984. Causes and consequences of synchronous flushing in a dry tropical forest. *Biotropica* 16:193–201
137. Lowman MD. 1984. An assessment of techniques for measuring herbivory: Is rainforest defoliation more intense than we thought? *Biotropica* 16:264–68
138. Lowman MD. 1985. Temporal and spatial variability in insect grazing of the canopies of five Australian rainforest tree species. *Aust. J. Ecol.* 10:7–24
139. Lowman MD. 1992. Herbivory in Australian rain forests, with particular reference to the canopies of *Doryphora sassafras* (Monimiaceae). *Biotropica* 24:263–72
140. Lowman MD. 1993. Forest canopy research: Old World, New World comparisons. *Selbyana* 14:1–2
141. Lowman MD, Box JD. 1983. Variation in leaf toughness and phenolic content among five species of Australian rain forest trees. *Aust. J. Ecol.* 8:17–25
142. MacArthur R. 1967. Limiting similarity, convergence and divergence of coexisting species. *Am. Nat.* 101:377–85
143. MacArthur RH. 1969. Patterns of communities in the tropics. *Biol. J. Linn. Soc.* 1:19–30
144. Marquis RJ. 1984. Leaf herbivores decrease fitness of a tropical plant. *Science* 226:537–39
145. Marquis RJ. 1991. Herbivore fauna of Piper (Piperaceae) in a Costa Rican wet forest: diversity, specificity, and impact. See Ref. 173, pp. 179–99
146. Marquis RJ. 1992. A bite is a bite is a bite? Constraints on response to folivory in Piperarieianum (Piperaceae). *Ecology* 73:143–52
147. Marquis RJ, Braker HE. 1994. Plant-herbivore interactions: diversity, specificity, and impact. In *La Selva: Ecology and Natural History of a Neotropical Rainforest,* ed. LA McDade, KS Bawa, HA Hespenheide, GS Hartshorn, pp. 261–81. Chicago/London: Univ. Chicago Press
148. McKey DD. 1974. Adaptive patterns in alkaloids physiology. *Am. Nat.* 108:305–20
149. McKey DD. 1979. The distribution of secondary compounds within plants. In *Herbivores: Their Interactions with Secondary Plant Metabolites,* ed. GA Rosenthal, DH Janzen, pp. 55–133. New York: Academic
150. McKey DD. 1984. Interaction of the ant-plant *Leonardoxa africana* (Caesalpiniaceae) with its obligate inhabitants in a rainforest in Cameroon. *Biotropica* 16:81–99
151. McKey DD, Waterman PG, Mbi CN, Gartlan SJ, Struhsaker TT. 1978. Phenolic content of vegetation in two African rain forests: ecological implications. *Science* 202:61–64
152. McNab BK. 1978. Energetics of arboreal folivores: physiological problems and ecological consequences of feeding on an ubiquitous food supply. See Ref. 155, pp. 153–62
153. Milton K. 1990. Annual mortality patterns of a mammal community in central Panama. *J. Trop. Ecol.* 6:493–99
154. Mole S, Ross JAM, Waterman PG. 1988. Light-induced variation in phenolic levels in foliage of rain-forest plants. I. Chemical changes. *J. Chem. Ecol.* 14:1–21
155. Montgomery GG, ed. 1978. *Ecology of Arboreal Folivores.* Washington, DC: Smithsonian Inst. 574 pp.
156. Mooney HA, Chu C. 1974 Seasonal carbon allocation in *Heteromeles arbutifolia,* a California shrub. *Oecologia* 14:295–306
157. Moran N, Hamilton WD. 1980. Low nutritive quality as a defense against herbivores. *J. Theor. Biol.* 86:247–54
158. Murali KS, Sukumar R. 1993. Leaf flushing phenology and herbivory in a tropical dry deciduous forest, southern India. *Oe-*

cologia 94:114–19
159. Nagy KA, Montgomery GG. 1980. Field metabolic rate, water flux, and food consumption in three-toed sloths (*Bradypus variegatus*). *J. Mamm.* 61:465–72
160. Nascimento MT, Hay JD. 1993. Intraspecific variation in herbivory on *Metrodorea pubescens* (Rutaceae) in two forest types in central Brazil. *Rev. Brasil. Biol.* 53:143–53
161. Newberry DM, de Foresta H. 1985. Herbivory and defense in pioneer gap and understory trees in tropical rain forests in French Guiana. *Biotropica* 17:238–44
162. Nichols-Orians CM. 1991. The effects of light on foliar chemistry, growth and susceptibility of seedlings of a canopy tree to an attine ant. *Oecologia* 86:552–60
163. Nuñez-Farfan J, Dirzo R. 1989. Leaf survival in relation to herbivory in two tropical pioneer species. *Oikos* 55:71–74
164. Oates JF, Whitesides GH, Davies AG, Waterman PG, Green SM, et al. 1990. Determinants of variation in tropical forest primate biomass: new evidence from West Africa. *Ecology* 71:328–43
165. O'Dowd DJ. 1979. Foliar nectar production and ant protection on a neotropical tree, *Ochroma pyramidale. Oecologia* 43:233–48
166. O'Dowd DJ, Willson MF. 1989. Leaf domatia and mites on Australasian plants: ecological and evolutionary implications. *Biol. J. Linn. Soc.* 37:191–236
167. Opler PA, Frankie GW, Baker HG. 1980. Comparative phenological studies of treelet and shrub species in tropical wet and dry forest in the lowlands of Costa Rica. *J. Ecol.* 68:167–88
168. Orians GH, Janzen DH. 1974. Why are embryos so tasty? *Am. Nat.* 108:581–92
169. Osunkoya OO, Ash JE, Graham AW, Hopkins MS. 1993. Growth of tree seedlings in tropical rain forests of North Queensland. *J. Trop. Ecol.* 9:1–18
170. Price PW. 1980. *Evolutionary Biology of Parasites.* Princeton, NJ: Princeton Univ. Press. pp. 105–33. 237 pp.
171. Price PW. 1991. Patterns in communities along latitudinal gradients. See Ref. 173, pp. 51–69
172. Price PW, Bouton CE, Gross P, McPheron BA, Thompson JN, Weis AE. 1980. Interactions among three trophic levels: influence of plants on interactions between insect herbivores and natural enemies. *Annu. Rev. Ecol. Syst.* 11:41–65
173. Price PW, Lewinsohn TM, Fernandes GW, Benson WW, eds. 1991. *Plant-Animal Interactions: Evolutionary Ecology in Tropical and Temperate Regions.* New York: Wiley & Sons. 639 pp.
174. Rehr SS, Feeny PP, Janzen DH. 1973. Chemical defense in Central American non-ant acacias. *Biochem. Syst.* 1:63–67
175. Reich PB. 1995. Phenology of tropical forests: patterns, causes and consequences. *Can. J. Bot.* 73:164–65
176. Reich PB, Uhl C, Walters MB, Ellsworth DS. 1991. Leaf lifespan as a determinant of leaf structure and function among 23 Amazonian tree species. *Oecologia* 86:16–24
177. Rhoades DF, Cates RG. 1976. Toward a general theory of plant antiherbivore chemistry. In *Biochemical Interactions Between Plants and Insects. Recent Adv. Phytochem.,* ed. J Wallace, RL Mansell, 10:168–213. New York: Plenum
178. Robbins CT. 1993. *Wildlife Feeding and Nutrition.* New York: Academic. 2nd ed.
179. Sagers CL. 1992. Manipulation of host plant quality: herbivores keep leaves in the dark. *Funct. Ecol.* 6:741–43
180. Sagers CL, Coley PD. 1995. Benefits and costs of defense in a neotropical shrub. *Ecology* 76:1835–43
181. Schowalter TS, Crossley DA Jr. 1988. Canopy arthropods and their response to forest disturbance. In *Forest Hydrology and Ecology at Coweeta,* ed. TW Swank, DA Crossley Jr, pp. 207–18. New York: Springer-Verlag
182. Schowalter TS, Webb JW, Crossley DA Jr. 1981. Community structure and nutrient content of canopy arthropods in clearcut and uncut forest ecosystems. *Ecology* 62:1010–19
183. Schupp EW. 1988. Seed and early seedling predation in the forest understory and in treefall gaps. *Oikos* 51:71–78
184. Schupp EW, Feener DH Jr. 1991. Phylogeny, lifeform, and habitat dependence of ant-defended plants in a Panamanian forest. In *Ant-Plant Interactions,* ed. CR Huxley, DF Cutler, pp. 175–97. Oxford: Oxford Univ. Press
185. Scriber JM. 1973. Latitudinal gradients in larval feeding specialization of the world Papilionidae (Lepidoptera). *Psyche* 80:355–73
186. Smith AP, Mulkey SS, Chazdon RL, eds. 1996. *Tropical Forest Plant Ecophysiology.* New York: Chapman & Hall. 675 pp.
187. Sork VL. 1987. Effect of predation and light on seedling establishment in *Gustavia superba. Ecology* 68:1341–50
188. Southwood TRE, Brown VK, Reader

PM. 1986. Leaf palatability, life expectancy and herbivore damage. *Oecologia* 70:544–48

189. Stamp NE, Casey TM, eds. 1993. *Caterpillars: Ecological and Evolutionary Constraints on Foraging*. New York: Chapman & Hall. 587 pp.
190. Sterck F, van der Meer P, Bongers F. 1992. Herbivory in two rain forest canopies in French Guyana. *Biotropica* 24:97–99
191. Stork NE. 1987. Guild structure of arthropods from Bornean rain forest trees. *Ecol. Entomol.* 12:69–80
192. Terborgh J, Emmons LH, Freese C. 1986. La fauna silvestre de la Amazonia: el dispilfarro de un recurso de un renovable. *Bol. Lima* 46:77–85
193. Terborgh J, Losos E, Riley MP, Bolanos Riley M. 1993. Predation by vertebrates and invertebrates on the seeds of five canopy tree species of an Amazonian forest. *Vegetatio* 107/108:375–86
194. Terborgh J, van Schaik CP. 1987. Convergence vs. nonconvergence in primate communities. In *Organization of Communities*, ed. JHR Gee, PS Giller, pp. 205–26. Oxford: Blackwell Sci.
195. Terborgh J, Wright SJ. 1994. Effects of mammalian herbivores on plant recruitment in two neotropical forests. *Ecology* 75:1829–33
196. Traveset A. 1990. Post-dispersal predation of *Acacia farnesiana* seeds by *Stator vachelliae* (Bruchidae) in Central America. *Oecologia* 84:506–12
197. Turner IM. 1995. Foliar defences and habitat adversity of three woody plant communities in Singapore. *Funct. Ecol.* 9:279–84
198. Vandermeer JH. 1977. Notes on density-dependence in *Welfia georgii* Wendl. ex. Burret (Palmae) a lowland rainforest species in Costa Rica. *Brenesia* 10/11:9–15
199. van Schaik CP, Terborgh JW, Wright SJ. 1993. Phenology of tropical forest: adaptive significance and consequences for primary consumers. *Annu. Rev. Ecol. Syst.* 24:353–77
200. van Soest PJ. 1982. *Nutritional Ecology of the Ruminant*. Portland: Durham & Downey
201. Walter DE, O'Dowd DJ. 1992. Leaves with domatia have more mites. *Ecology* 73:1514–18
202. Walter DE, O'Dowd DJ. 1995. Beneath biodiversity: factors influencing the diversity and abundance of canopy mites. *Selbyana* 16:12–20
203. Waterman PG, Ross JAM, Bennett EL, Davis AG. 1988. A comparison of the floristics and leaf chemistry of the tree flora in two Malaysian rainforests and the influence of leaf chemistry on populations of colobine monkeys in the Old World. *Biol. J. Linn. Soc.* 34:1–32
204. Wilson DE, Janzen DH. 1972. Predation on *Scheelea* palm seeds by bruchid beetles: seed density and distance from the parent palm. *Ecology* 53:954–59
205. Wolda H. 1978. Fluctuations in abundance of tropical insects. *Am. Nat.* 112:1017–45
206. Wolda H. 1988. Insect seasonality: Why? *Annu. Rev. Ecol. Syst.* 19:1–18
207. Wood TK, Olmstead KL. 1984. Latitudinal effects on treehopper species richness (Homoptera: Membracidae). *Ecol. Entomol.* 9:109–15
208. Wright SJ. 1983. Dispersion of eggs by a bruchid beetle among *Scheelea* palm seeds and the effect of distance to the parent palm. *Ecology* 64:1016–21
209. Wright SJ. 1996. Phenological responses to seasonality in tropical forest plants. See Ref. 186, pp. 440–60
210. Wright SJ, van Schaik CP. 1994. Light and phenology of tropical trees. *Am. Nat.* 143:192–99

Annu. Rev. Ecol. Syst. 1996. 27:337–63

MECHANISMS CREATING COMMUNITY STRUCTURE ACROSS A FRESHWATER HABITAT GRADIENT

Gary A. Wellborn[1], *David K. Skelly*[2], *and Earl E. Werner*[3]
[1]Department of Biology, Yale University, New Haven, Connecticut 06520, [2]School of Forestry and Environmental Studies, Yale University, 370 Prospect Street, New Haven, Connecticut 06511, [3]Department of Biology, University of Michigan, Ann Arbor, Michigan 48109-1048

KEY WORDS: community ecology, disturbance, evolution, permanence, predation

ABSTRACT

Lentic freshwater habitats in temperate regions exist along a gradient from small ephemeral ponds to large permanent lakes. This environmental continuum is a useful axis for understanding how attributes of individuals ultimately generate structure at the level of the community. Community structure across the gradient is determined by both (*a*) physical factors, such as pond drying and winter anoxia, that limit the potential breadth of species distributions, and (*b*) biotic effects mediated by ecological interactions, principally predation, that determine the realized success of species. Fitness tradeoffs associated with a few critical traits of individuals often form the basis for species turnover along the gradient. Among species that inhabit temporary ponds, distributions are often constrained because traits that enhance developmental rate and competitive ability also increase susceptibility to predators. In permanent ponds, changes in the composition of major predators over the gradient limit distributions of prey species because traits that reduce mortality risk in one region of the gradient cause increased risk in other regions of the gradient. Integrated across the gradient, these patterns in species success generate distinct patterns in community structure. Additionally, spatial heterogeneity among habitats along the gradient and the fitness tradeoffs created by this heterogeneity may hold important evolutionary implications for habitat specialization and lineage diversification in aquatic taxa.

0066-4162/96/1120-0337$08.00

INTRODUCTION

Mechanistic approaches in ecology seek to functionally link traits of individuals to higher level processes such as the dynamics of populations and multispecies interactions, determinants of species distributions, and development and maintenance of community structure (175, 191). Studies conducted across environmental gradients can greatly enhance our understanding of the ways in which individual traits act to shape these higher level processes because they can reveal patterns of concordance in species traits and species assemblages across the changing ecological conditions of the gradients (24).

In this review we are concerned with a well-known gradient in lentic freshwater habitats (e.g. pools, marshes, ponds, and lakes) in temperate regions. These habitats can be placed on an axis ranging from small, highly ephemeral habitats to extremely large habitats that have been present for millennia (202). Ecologists have long recognized this environmental gradient as a critical axis along which aquatic communities are organized (17, 74, 196). Virtually every type of animal known to inhabit freshwater is also known to have a restricted distribution across this habitat gradient. Representatives from nearly every class of free-living freshwater animals sort among habitats according to their permanence or in relation to the distribution of predators whose own distributions are related to permanence (Table 1). With few exceptions, however, sorting occurs at the family level and below; most higher taxa (phyla, classes, orders) are not restricted to particular regions of the gradient. Among the groups in which these patterns have been quantified, species are often restricted to a subset of habitats, while the distribution of even a single genus can encompass a large portion of the entire range of available habitats (Figure 1). These patterns imply that overall body plan differences associated with higher order taxonomic classification do not usually represent constraints to use of different habitat types, but if constraints do exist, they occur among species, genera, and families.

The integration of restricted species distributions across the gradient leads to highly characteristic shifts in community structure (17, 23, 32, 67, 89, 149, 154, 195). Our thesis is that restricted species distributions and turnover in community composition along the gradient result largely from a relatively few important constraints on the life-styles of aquatic animals. As developed below, certain attributes of organisms such as body size, developmental rate, activity, and life history form axes for fitness tradeoffs across the habitat gradient. In each case, substantial evidence suggests that success at one point on the gradient entails having a phenotype that will hinder performance at other points along the gradient. We argue that changes in community structure along the gradient are best understood in terms of the critical fitness tradeoffs that determine a species' pattern of performance among habitats. Elucidating these mechanistic

Table 1 Free-living animals of lentic freshwater habitats whose distributions are known to vary with frequency of habitat drying or predator distribution. Representative references are given for each taxon.

Taxon	Common name	Permanence	Predators	Reference
Proifera	Sponges	x		200
Turbellaria	Flatworms	x		196
Nematoda	Roundworms	x		7
Rotifera	Rotifers	x		139
Bivalvia	Clams	x	x	90, 133
Gastropoda	Snails	x	x	181, 18, 19, 86
Hirudinea	Leeches	x	x	196, 94
Oligochaeta	Worms	x		196
Arachnida	Mites	x	x	196, 75
Anostraca	Fairy Shrimp	x	x	196, 89
Cladocera	Water Fleas	x	x	17, 196, 139, 89
Conchostraca	Clam Shrimp	x		9, 196, 139, 89
Notostraca	Tadpole Shrimp	x		51, 139
Copepoda	Copepods	x		22, 6, 89
Amphipoda	Amphipods	x	x	196, 188
Decapoda	Crayfish, Shrimp	x		196
Isopoda	Isopods	x		196
Ostracoda	Seed Shrimp	x		196
Anisoptera	Dragonflies	x	x	196, 95
Coleoptera	Beetles	x	x	80, 41
Diptera	True Flies	x	x	36, 91, 179, 182
Ephemeroptera	Mayflies	x		196
Hemiptera	True Bugs	x	x	27, 157
Megaloptera	Alderflies	x		200
Trichoptera	Caddisflies	x		114
Zygoptera	Damselflies	x	x	196, 95
Osteichthyes	Fishes	x		127
Anura	Frogs	x	x	23, 151
Caudata	Salamanders	x	x	23

links between individual- and community-level processes provides a fundamental understanding of how constituent species shape community structure and, conversely, how community structure may influence the distribution and evolution of species.

A Schematic Model

Changes in community structure along this freshwater habitat gradient are determined by the coupled effects of (*a*) physical environmental factors that limit the potential breadth of species distributions and (*b*) biotic interactions that determine the realized success of species. In Figure 2, we present a schematic model of community structure across the gradient. The model depicts key negative effects, both biotic and abiotic, that underlie significant shifts in community structure across the gradient. Bold arrows indicate strong effects and thinner arrows represent weaker effects.

The physical environment forms the template along which communities develop on the gradient. At the broadest scale, physical factors constrain the potential pool of community members by eliminating species unable to cope with the physical stress. In temporary pond habitats, the need to cope with periodic drying imposes severe constraints on a species' behavior, development, and life history, and only those species able to deal with drying are successful

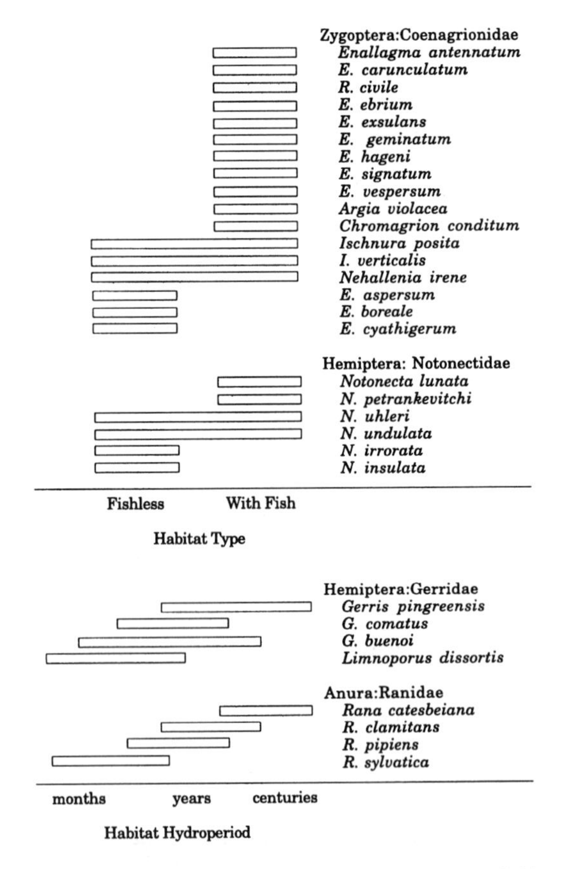

Figure 1 Distribution ranges of Zygoptera (95) and Notonecta (27) across habitats with and without predatory fish, and Hemiptera (157) and Anura (23) across habitats that differ in hydroperiod. Genera and families are distributed across a broad region of the gradient, but individual species are often restricted to a narrow region.

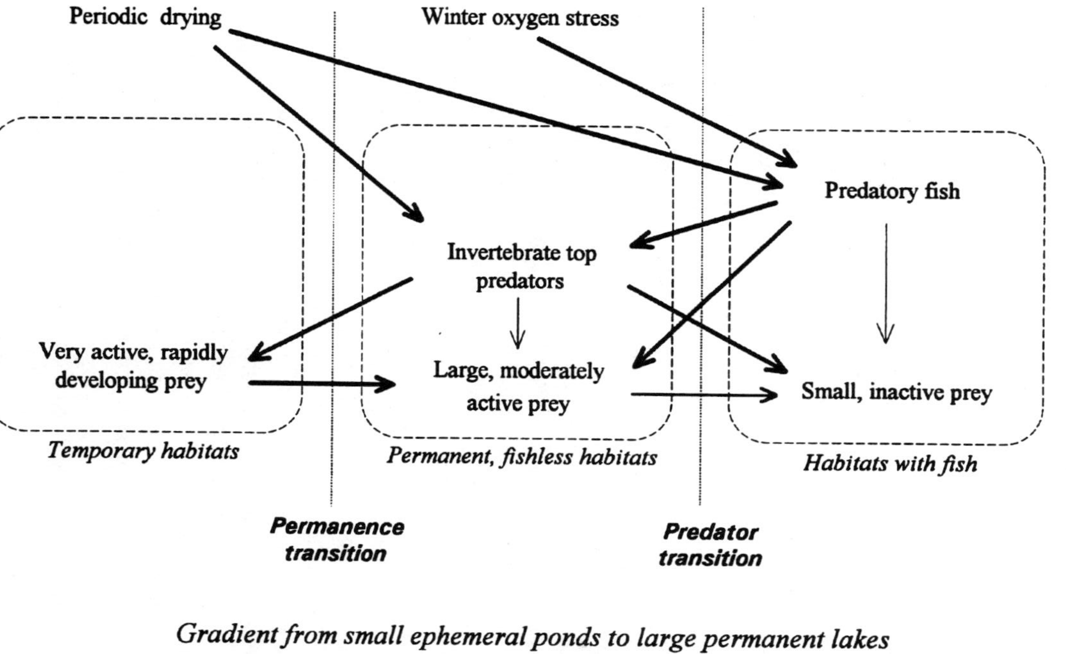

Figure 2 A schematic model of mechanisms generating community structure along the freshwater habitat gradient. Arrows indicate direction of negative effects. Bold arrows indicate very strong effects that act to constrain the distribution of affected species. Thinner arrows indicate weaker trophic interactions that do not prevent coexistence of interacting species. Strong interactions resulting primarily from predation cause distinct transitions in community structure along the gradient. One major transition, the permanence transition, occurs between temporary habitats that contain few predators and permanent habitats that contain significant invertebrate predators. A second major transition in community structure on the gradient, the predator transition, occurs between permanent fishless and fish-containing habitats. Mechanisms causing differential interaction strengths and the critical phenotypic tradeoffs of individuals that mediate species and community transitions along the gradient are described in the text.

in these habitats (196). Importantly, this constraint prevents colonization by many key predators found elsewhere on the gradient because fish and important predatory invertebrates such as dragonfly larvae are highly susceptible to pond drying. Temporary pond habitats thus often contain fewer predators than do more permanent habitats (149, 208). In shallow permanent ponds, physical stresses such as low oxygen levels during periods of ice cover can impose heavy mortality on fish (28, 127, 177). Thus well-developed fish communities are often restricted to relatively deep permanent habitats (177).

While species-specific tolerances to physical factors determine the pool of species potentially able to exist in any habitat, biotic interactions determine the actual species composition of each community. Strong negative interactions, through either predation or competition, may prevent survival of affected populations, and thus restrict the distribution of these species to a narrow range of the gradient (10, 17, 95, 184). Weaker negative interactions may limit a population's density but do not cause the population to be eliminated (172, 189). It is the relative strengths of biotic interactions, under the backdrop of physical constraints, that shape community structure on the gradient.

This interplay of physical and biotic factors along the gradient tends to produce distinct, alternative community types (delineated by boxes in Figure 2), determined largely by hydroperiod and the key predators (if any) able to survive in the habitat. Temporary habitats are characterized by very active and rapidly developing species and few predators (149, 208). Permanent but fishless habitats have communities composed of large predatory invertebrates, such as larval dragonflies and pelagic Diptera, and large, moderately active prey species (95, 186, 212). In permanent habitats that contain fish, fish are often key predators, and animal species at lower trophic levels are small-bodied (186, 212) and inactive (96, 165). Between these community types are distinct transitions in community structure. We refer to the transition between temporary habitats and permanent fishless habitats as the permanence transition and the transition between permanent fishless habitats and habitats with fish as the predator transition. These transitions and the alternative community types they form arise on the gradient because strong interactions across transitions lead to selective elimination of species, while weaker interactions within community types do not.

To be sure, the model presents an idealized and highly simplified view of the habitat gradient, and numerous additional complexities and mitigating factors are certainly important. Nonetheless, we believe the model captures essential features of an important and prominent pattern among aquatic systems. This perspective on the structure of freshwater habitats is not new (145, 208, 212). Our goal here is to emphasize the insights gained by evaluating species distributions and community structure in light of the effects of individual traits and key fitness tradeoffs expressed along the habitat gradient.

THE PERMANENCE GRADIENT

In this section we consider ponds that range from extremely temporary through those that may be permanent for years at a time, but which do not contain fish. Within these habitats, what factors determine the shape and limits of a species' distribution? We argue that both pond hydroperiod and predation are prevalent influences with important consequences for a species' distributional pattern that may be understood through a mechanistic analysis of key phenotypic traits.

Permanence and Developmental Rate

Temporary waters have often been considered distinct from permanent ponds (196, 200). Much of the early literature on temporary pond fauna was concerned with the question of survival in an ephemeral aquatic environment (7, 9, 51, 74, 138, 196). This focus on the adaptations to life in temporary ponds is understandable. The existence of a dry phase presents an obvious challenge to any "aquatic" animal. The persistence of these organisms requires that they survive the dry period in situ, migrate to other aquatic habitats, undergo a transition to a terrestrial phase, or recolonize following extinction. All of these strategies are found among temporary pond fauna (196, 200).

While there is a great variety of particular adaptations to ephemeral ponds, most resident species persist by reaching some critical developmental stage prior to pond drying (51, 200). Examples include holometabolous and hemimetabolous insects and amphibians that undergo metamorphosis into a terrestrial form. In addition, many invertebrates (e.g. cladocera, clam shrimp, copepods, fairy shrimp, mayflies, seed shrimp, tadpole shrimp) must reach adulthood and then produce encysted eggs that can endure the dry phase within the pond basin (200). For these groups, hydroperiod represents a maximum limit on development time. That this limit is frequently an active constraint on the performance and distribution of freshwater organisms is suggested by numerous reports of catastrophic mortality (34, 67, 108, 116, 131, 142, 170) and even local extinction (23, 67) associated with pond drying events. Experiments directed at the role of hydroperiod have been of two primary types. A number of artificial pond experiments have been conducted in which hydroperiod is imposed as a treatment (e.g. 136, 141, 197). In addition, several experiments have transplanted species among natural ponds that vary in permanence (92, 149, 153, 154). Together, these studies provide evidence that drying-associated mortality is often high and is correlated with developmental rate of the species.

In order to succeed in the most temporary aquatic environments, developmental rates must be phenomenal. A seed shrimp (Conchostroca) of the Saharan Desert is able to achieve maturity within just five days after pond filling (138). Spadefoot toads (*Scaphiopus couchii*) of the southwestern deserts of the United States can reach metamorphosis in just 10 days after hatching (176). In more permanent environments, the restriction on developmental rates is relaxed. North American frogs, for instance, have larval periods that range up to three years in length (209). Studies of tadpole distributions show that species with longer larval periods tend to be found in more permanent ponds (23, 32, 149, 154). As hydroperiod increases, so does the potential species pool. Correspondingly, a number of studies have noted correlations between permanence and species richness (36, 89, 111, 116, 211). While the total number of species might increase with permanence, there is also significant turnover in species composition. Most species that live in temporary ponds are absent from permanent ponds (23, 196). This pattern suggests that some factor, or suite of factors, can often prevent species from occupying habitats that are apparently suitable given developmental constraints.

Predation and Permanence

While there are many possible reasons why a species may be restricted from living in more permanent habitats than those it occupies (see caveats below), predation may be the most widespread. Predatory species are often entirely absent within the most temporary habitats (91, 139, 153) and are typically numerous and diverse within permanent fishless ponds (151). Along the intervening portion of the gradient, several studies have shown that the identity, density, and size distribution of predators changes with even small differences in permanence (41, 59, 67, 74, 80, 151, 153, 154, 208). For example, Woodward (208) reported an approximately fourfold increase in aquatic predator density in permanent versus temporary desert ponds in New Mexico. Similar trends have been found among Michigan ponds where predator density triples between annual temporary ponds and permanent ponds (149, 151). In these permanent ponds, average body lengths of predatory salamander, beetle larvae, and dragonfly larvae were longer by 100%, 36%, and 46%, respectively, than those of their counterparts in temporary ponds (151).

The association between permanence and predator distribution suggests that predation could be an important means by which prey distribution becomes limited in more permanent environments. Species may experience a refuge within more temporary ponds where they are relatively invulnerable to resident predators (68, 156). Indeed, many nonfish predators have strong effects on prey abundance (15, 44, 76, 199, 204). However, other studies from a range of habitats also suggest that aquatic predators can have little or no effect on prey abundance (71, 166). Because the influence of a given predator is known to vary substantially among prey species (e.g. 198), evaluation of predator limitation requires comparing prey susceptibility to predation at multiple points along the permanence gradient. While these more directed analyses are relatively few, they uniformly suggest that predation has greater effects on prey survivorship in larger and more permanent habitats (115, 135, 149, 153, 154) and can lead to elimination of prey species (153).

Permanence, Predation, and Mechanistic Tradeoffs

Thus far we have argued that the negative effects of pond hydroperiod and predation are important and widespread. Because predator distributions are also limited by hydroperiod, these factors covary inversely. A generalist strategy would require mastering the very different situations imposed by predator-free temporary environments as well as permanent environments where numerous and diverse predators are found. The rarity of true generalists in these habitats (see below) suggests that restricted distributions may have some mechanistic basis.

Behavior may form the foundation of an important tradeoff for many aquatic taxa. Put simply, many animals must move to forage, but moving can also increase the likelihood of being detected by movement-oriented predators (82, 85, 145, 193, 208). Experiments substantiate that movement (or activity) is correlated with acquisition of food as well as both growth and developmental rates (147, 149, 152). Both interspecific comparisons and intraspecific manipulation of activity show that more active individuals are also at greater risk of predation (4, 21, 72, 148). Additionally, studies of diet preference show that activity is a strong predictor of relative risk of potential prey in aquatic environments (12, 84, 146).

If activity is a critical trait for assortment among aquatic habitats, then species from different regions of the habitat gradient should differ in activity. As expected, larval amphibians (82, 149, 154, 195, 208) and insects (27) from more temporary ponds tend to have higher activity levels than do their counterparts from permanent habitats. Where these comparisons have been coupled with field transplants across the permanence gradient, studies have demonstrated a strong association between phenotype and performance (149, 154).

Other Mechanisms and Some Caveats

Mechanistic approaches to the study of temporary pond communities are relatively new (154). More substantial evidence exists for the roles of activity and other traits in contributing to patterns of distribution in fish-containing versus fishless habitats (see below). In addition, information on temporary pond communities has a strong taxonomic bias. Most studies of distributional mechanisms in temporary ponds have been conducted using larval amphibians. We know far less about the relationships among phenotype, performance, and distribution for the invertebrates in these habitats. Even if research on these other taxa also finds that activity is an important trait, there are other important mechanisms that can lead to restriction of a species distribution. Some species require drying or water level fluctuation as a physiological trigger for the initiation of embryonic development (51, 118). Many other species are sessile and not subject to activity tradeoffs. For these and other species for which movement is not related to risk, different traits may underlie patterns of distribution. For example, allocation of resources among morphological defense, development, and reproduction may provide critical axes for the sorting of groups such as snails among ponds that vary in permanence (19, 86). Specialization on food resources available in ponds of different permanence could be important for some taxa (1, 11).

Finally, interspecific competition has often been advanced to explain distributional patterns along the permanence gradient (e.g. 18, 37). There is strong evidence that interspecific competition can have severe effects, including competitive exclusion, in artificial pond experiments (e.g. 102, 104, 197, 198). While competition may have similar effects on distributional patterns in the field, relatively few experimental analyses have been conducted in natural environments. From these it appears that although competition can lead to species exclusion in some cases (91, 113, 205), it often has relatively small effects on survivorship, growth, and reproduction in other cases (86, 149, 150, 154, 195).

THE PREDATOR TRANSITION

That strong predator-prey interactions across the predator transition act as a sieve for community organization (Figure 2) is apparent when one compares studies that manipulate predator density within a single community type, where relatively weak interactions are expected, to studies that introduce a predator into a community type that does not normally contain that predator, a manipulation expected to produce pronounced changes in community composition (172, 190). When introduced into a previously fishless habitat, fish precipitate major changes in the community by substantially reducing the density of some prey species, sometimes to the point of local extinction (17, 30, 47, 57, 101, 124, 195). Similar conclusions are drawn from studies comparing the fauna of fishless and fish-containing habitats (27, 55, 95, 123, 163, 184). In contrast, manipulations of fish density in habitats normally containing fish often produce only minor changes in prey density (14, 43, 58, 173, 174, 190). Some studies of this sort have indicated more substantial effects of predatory fish (100, 103); however, only prey abundance, not species composition, is affected in these studies. Similarly, when predatory invertebrates normally found only in fishless habitats are introduced into fish habitats (with fish excluded), these predators cause substantial reductions in prey density (95, 132, 179) and elimination of some taxa (179). Predatory invertebrates may have weaker effects, however, in their native fishless habitats (95).

To briefly illustrate the degree of change in community structure across the predator transition, we used published surveys to calculate average community similarity [using species relative abundance and an index of overlap (140)] between the two community types relative to the average similarity within community types. For Brooks & Dodson's (17) classic study of zooplankton in New England lakes with and without zooplanktivorous fish, community similarity of habitats across the predator transition was only 16% of that observed among habitats of the same lake type. Furthermore, 47% of the species sampled exhibited complete segregation by lake type. Similar patterns are seen in littoral communities. McPeek (95) sampled the larval odonate fauna of three fishless and three fish-containing lakes in Michigan. For *Enallagma* damselflies, a diverse group that constituted the majority of damselflies in the lakes, all seven species exhibited complete segregation by lake type (Figure 1). For larval dragonflies, community similarity of habitats across the predator transition was 57% of that observed among habitats of the same lake type. This comparatively high value is due to the presence of a single common species that is ubiquitous among lakes and does not reflect a general lack of habitat segregation among dragonflies in these lakes. In fact, of the eight species that occurred in more than one lake, four occurred in only one lake type.

The substantial changes in species composition that are observed across the predator transition arise because interactions between species from similar community types are weaker than those between species from different community types. For predator-prey interactions, interaction strengths are determined by the coupled effects of predator traits that determine foraging characteristics and prey traits that determine susceptibility. Thus, elucidation of mechanisms underlying changes in community structure across the transition requires an understanding of these predator and prey qualities.

Foraging Characteristics of Predators

Changes in community composition observed across the transition occur primarily because foraging characteristics of predatory fish and invertebrates are qualitatively different, and thus favorable prey defense attributes in one community type are generally unfavorable in another community type.

SIZE-BIASED PREDATION Predatory fish and invertebrates often differ in the size of prey items consumed. Most fish disproportionately consume larger individuals across the range of prey sizes that they are physically able to ingest (33, 49, 95, 99, 103, 107, 186, 212). In part, size selectivity may be a behavioral adaptation to maximize foraging gain (99, 194). Additionally, because fish are generally much larger than their invertebrate prey, only a few large prey species are able to grow to an invulnerable size (161). Thus the coupled effects of fish selectivity and their comparatively large size lead to the general trend of greatest predatory effect on larger prey species. In contrast to fish, predatory invertebrates exhibit less size bias in prey consumption, generally consuming all prey within the range of prey sizes that they can capture and handle (2, 117, 126, 167, 171, 186). Predatory invertebrates typically have a disproportionate impact on small-bodied prey, however, because functional constraints often limit the ability of these predators to consume larger prey. Additionally, because of the similarity in size to their invertebrate predators, many prey species may grow large enough to become invulnerable to invertebrate predation (178, 186, 203, 207).

MECHANISMS OF PREY DETECTION Most fish species that have strong impacts in these habitats rely on vision for prey detection (54). Predatory invertebrates, in contrast, primarily use tactile cues to detect prey (117), although some may use a combination of visual and tactile cues (112). An important result of this difference in detection mechanisms is that fish can detect prey from a greater distance than can invertebrates (210, and see 95). Additionally, although both fish and invertebrates detect moving prey more easily than stationary prey (40, 54, 65, 125, 210), the acute vision of fish may enable them to detect less conspicuous prey motion than do predatory invertebrates.

SEARCH MODES Predatory fish and invertebrates also tend to differ in search mode. Fish actively search for prey over a comparatively broad spatial area (39, 54, 65). Predatory invertebrates, in contrast, often employ a passive ambush, or "sit-and-wait," foraging mode in which they remain motionless and capture prey that come near them (117). The ambush foraging mode of predatory invertebrates greatly limits the area over which each predator forages and produces relatively low encounter rates with prey. Some predatory invertebrates do actively stalk prey, but these are usually limited to slow movements along a plant stem or other substrate (96).

PURSUIT SPEED While some predatory invertebrates are able to swim (e.g. dragonfly larvae and some beetle larvae), the pursuit speed of fish is greater than any typical predatory invertebrate. This disparity implies that while invertebrate prey may sometimes be able to escape from predatory invertebrates by swimming, the prey would have little chance of swimming away from a fish (96).

Traits of Prey that Influence Susceptibility to Predation

The contrasting foraging modes of predators imply that prey with traits that allow them to coexist with predators in one community type will fare poorly with predators in the alternative community type. Some key traits of prey that seem to underlie turnover on the gradient have been identified in several taxa.

BEHAVIOR Prey activity may be the most important trait shaping changes in species composition across the predator transition. Higher activity levels increase encounter rates with predators and enhance a predator's ability to detect prey. Furthermore, because prey movement is usually for foraging, locating mates, or other essential activities, prey can not remain inactive, and thus are forced to strike an appropriate balance between the benefits and risks associated with activity (63, 144, 193). For most freshwater invertebrate species, the resolution of this tradeoff will involve lower activity levels in species that coexist with fish than in species coexisting with invertebrate top predators in fishless habitats (13, 26, 29, 70, 77, 96, 111, 165) because of the greater mortality risk associated with activity in the presence of fish (13, 26, 27, 56, 70, 77, 96).

The importance of the tradeoff between activity and risk in shaping the distribution of species across the predator transition seems especially strong for predatory invertebrates. For example, larvae of the dragonfly *Anax junius*, important top predators in many fishless habitats (95, 77, 132), are very active when foraging, a trait probably contributing to their success in fishless habitats (77). *Anax*, however, are thereby highly susceptible to fish predation (30, 77, 195), sufficiently so that they are excluded from habitats with predatory fish (30, 77, 95, 195). Similarly, the activity patterns of *Enallagma* damselfly larvae, intermediate level predators in freshwater communities, contribute substantially to their habitat distribution over the gradient (13, 96, 121). Species that coexist with fish are less active than those that face only predatory invertebrates (13, 96).

Besides general activity level, other behaviors contribute to changes in species composition over the predator transition. Both the encounter response (56, 96) and the use of protective microhabitats (8, 77, 120, 121, 137, 162, 189) are important in some taxa. For example, Pierce (120) examined microhabitat use and vulnerability to fish predation of larval dragonflies from fish-containing and fishless habitats. A species found commonly in fishless habitats often used exposed microhabitats, even in the presence of fish, while two related species that are abundant in habitats with fish overwhelmingly used habitats offering cover when fish were present. Accordingly, the species from the fishless habitat was more vulnerable to fish predation than the species native to the fish-containing habitat (120).

LIFE HISTORIES Tradeoffs in prey life histories across the predator transition may contribute to patterns of species turnover, especially for species that complete their entire life cycle in the aquatic environment (46, 78, 88, 186, 206). A fundamental life history trait that shapes species' success over the gradient is the pattern of reproductive allocation over ontogeny (186, 188). For size-structured populations, the optimal resolution of the tradeoff between allocation of resources to growth versus reproduction over ontogeny will be strongly influenced by the ontogenetic pattern of mortality experienced by the population (81, 186–188). Thus the differences between fish-containing and fishless habitats in the form of size-biased predation may restrict prey species to habitats for which their life histories are appropriate. For example, differences in reproductive allocation patterns between species of a littoral amphipod, *Hyalella,* appear to be the primary determinant of species segregation between fish-containing and fishless habitats (186). Predators in these two habitats impose qualitatively different regimes of size-specific predation mortality on *Hyalella,* a difference manifest in the overall mortality schedules of the amphipods in the two habitats (186). Whereas mortality increases with body size in the fish-containing habitat, it declines with size in the fishless habitat. *Hyalella* in lakes with fish initiate reproduction at a smaller body size and have a greater size-specific reproductive effort than do *Hyalella* from habitats with large dragonfly predators. Thus, in the habitat where larger individuals are more susceptible to predators, *Hyalella* shifts reproduction to smaller sizes, but in the habitat where smaller individuals are more susceptible, *Hyalella* shifts reproductive allocation to larger size classes, thus allowing rapid growth through the most vulnerable size range (186).

MORPHOLOGY For many species, especially those occupying open water, the most important morphological feature affecting their relative success across the predator transition is body size (212), although other traits may also play a role in some groups (52, 98). In habitats with fish, where larger individuals are most vulnerable to predation, small-bodied prey species are most successful (64, 83, 122, 128, 160, 183, 186, 188). In fishless habitats, where smaller individuals are disproportionately vulnerable to predation, large-bodied prey species are most successful (50, 131, 183, 184, 186).

In summary, the contrasting modes of predation by major predators in fishless and fish-containing habitats appear to be the primary cause underlying the development of alternative community types across the predator transition. Prey traits effective at mitigating risk of predation in one community type unavoidably produce inimical effects in the alternative community type. Specifically, the relatively large body size, high activity levels, and conspicuous microhabitat use of prey that allow them to be successful in fishless habitats also make them highly conspicuous to predatory fish. Conversely, the relatively small body size, reduced activity, and cryptic behavior of those prey species that are successful in fish-containing habitats make them vulnerable to the invertebrate top predators found in fishless habitats.

Competition

Competitive effects may play some role in the segregation of species across the transition (105, 155, 180), but these effects are less important than predation (3, 95, 121, 179). Although early work in this system proposed that interspecific competition with larger-bodied zooplankton was the primary vehicle by which small-bodied zooplankton species were excluded from fishless habitats (17, 48), subsequent empirical studies demonstrated that predation by size-selective predatory invertebrates was often the principal cause of exclusion of these species (35, 179, 212). Although this dichotomous view of the effects of predation and competition is overly simplistic, since the two may work interactively to shape patterns of species success over the gradient (192), predation is clearly the major factor driving community changes over the predator transition.

SPECIALIZATION AND PHENOTYPIC PLASTICITY

Many aquatic species are habitat specialists, occurring in only one region of the gradient. That specialization is common on the gradient is perhaps to be expected since circumstances favoring the evolution of specialization (42, 60, 61) are common across much of the gradient. Specialization is especially likely to evolve (*a*) in spatially heterogeneous but temporally predictable environments, (*b*) when individuals face fitness tradeoffs associated with occupying different habitat types, and (*c*) when hard selection (e.g. strong mortality selection from predators and physical environmental factors) predominates. These qualities, common for permanent ponds and lakes, may foster the evolution of specialization in species occupying these habitats. Temporary ponds, however, often exhibit temporal changes in species composition caused by high variation in hydroperiod. Alternate wetter or drier years can alter species composition of habitats as "permanent pond taxa" spread out across the landscape during wet periods, only to "retreat" with the next dry period (67). Specialization is less likely to develop in species occupying these temporally heterogeneous habitats (60).

Holt (60, 61) has stressed the potential importance of habitat selection in the evolution of specialization. Habitat selection can limit the scope of natural selection on a population because only agents of selection operating within the preferred habitat can guide evolutionary change in the population. This effect can lead to a self-reinforcing evolutionary process when individuals face fitness tradeoffs between habitat types (60). Habitat selection causes ever more precise adaptation to one habitat type (and consequently reduced performance in alternative habitat types), and this habitat specialization in turn favors ever greater proficiency for selecting the most appropriate habitat type. Many aquatic taxa have highly mobile terrestrial (amphibians) or aerial (insects) adult stages, and thus they have potential for selecting appropriate habitats along the gradient (73, 129, 130). Natal philopatry may also reinforce specialization, even in species unable to discriminate between habitats containing different community types (93).

Many freshwater species alter activity (26, 33, 53, 66, 69, 96, 143, 162, 201), microhabitat use (120, 121, 137, 163), morphology (52, 119), and life history (31, 62, 87, 109, 164, 168, 184, 185) in response to predators. For many aquatic prey species, such phenotypic plasticity appears to function as a mechanism for dealing with spatial or temporal variation in predator density within a single habitat type rather than as a mechanism allowing broader distribution across the habitat gradient. Thus, many species exhibiting plasticity in behavioral or other traits in response to predators are specialist species that occur in only one habitat type. For example, *Enallagma* damselfly species that occur only in fish-containing habitats and those species that occur only in fishless habitats respond to the predators with which they coexist by altering their behavior in ways that reduce predation risk (13, 96, 121). They do not respond appropriately, however, to predators with which they do not coexist (96). Thus, predator-mediated plasticity in *Enallagma* does not allow a broader distribution on the gradient.

Although phenotypic plasticity may often be of little importance in allowing species to maintain a broad distribution on the habitat gradient, it is probably of substantial importance in the success of species within a habitat type. Because the ability to deal with predation risk entails substantial fitness tradeoffs (193), species that facultatively alter their phenotype in response to changes in the immediate threat of predation can partially mitigate detrimental effects of these tradeoffs (145). Since this topic has been reviewed elsewhere (85, 145, 193), we will simply stress here that the prevalence of plasticity in response to predators among taxonomically diverse freshwater organisms underscores the considerable importance of these tradeoffs in this system.

For some species phenotypic plasticity does appear to facilitate a broader distribution on the gradient under some environmental conditions, but restricted breadth of reaction norms seems to limit the conditions under which a broader distribution might occur. For example, although their distributions may overlap, larval chorus frogs (*Pseudacris triseriata*) are most successful in more temporary habitats with few predators, and larval spring peepers (*P. crucifer*) are most successful in more permanent habitats with predatory invertebrates; differential susceptibility to predation contributes importantly to this pattern (149). Both species facultatively respond to the presence of an invertebrate predator by reducing activity, but chorus frogs do not reduce activity to levels observed for spring peepers (149). This limit to the breadth of their reaction norm probably explains the low survival of chorus frogs in habitats with predatory invertebrates (149).

ADAPTATION AND SPECIATION ON THE HABITAT GRADIENT

The lentic freshwater habitat gradient presents a potentially potent template for evolutionary change and diversification in freshwater taxa (79, 97). Both the substantial heterogeneity among lentic habitats in critical selective agents and the inherent insularity of these habitats may drive adaptation and foster speciation. Relatively few studies, however, have explicitly examined evolutionary processes mediated by the key ecological agents shaping the habitat gradient, though many have interpreted species or population comparisons in adaptive terms.

Size-biased predation is one potentially important evolutionary agent that has been studied in some detail. Size-biased mortality will produce an evolutionary response in populations if heritable phenotypic variation in body size or associated life history traits exists in populations. At least moderate levels of heritability are common for a broad range of traits and taxa (134), and for morphological and life history traits of aquatic taxa specifically (169). These observations imply that an evolutionary response to size-biased predation may be common in aquatic prey populations, a conclusion supported by several studies. For example, natural (158) or simulated (38) size-biased predation on cladoceran zooplankton species cultured in the laboratory does produce adaptive change in body size and life history traits. Similarly, a natural population of *Daphnia galeata mendotae* (169) that experiences strong predation by fish displayed a genetically based reduction in mean body size and size at maturity over time, a response consistent with the form of size-biased predation experienced by the population.

For species distributed across major transitions on the habitat gradient, local adaptation may be an important source of differentiation among populations (20, 45). For example, Spitze (159) found genetic differentiation in body size among *Daphnia obtusa* populations. He suggested that variation among populations in predation regime is the likely cause of the divergence. Also, Neill (106) found genetic differentiation in vertical migration patterns between two neighboring populations of a calanoid copepod, each of which appears to be locally adapted to deal with the form of size-specific predation it experiences.

Macroevolutionary diversification in aquatic taxa may occur in the context of the freshwater habitat gradient through the mechanism of evolutionary habitat shifts (98). This mechanism is a form of peripheral isolates speciation (16) in which a small founder population of a species normally found in one habitat type may sometimes disperse to an alternative habitat type. Rarely, these dispersing individuals might establish a population that persists long enough to adapt to local conditions. Speciation may be promoted by both the potential for rapid genetic change in small founder populations (5) and factors impeding backcrossing with ancestral populations, such as assortitive mating, natal habitat philopatry, and low dispersal rates due to the inherent insularity of ponds and lakes. Although historical evidence for this mechanism is sketchy, many closely related species do segregate on the gradient (Table 1, Figure 1), indicating that phylogenetic inertia need not constrain a lineage to one habitat type and that habitat shifts may occur across a wide range of freshwater taxa. The potential importance of evolutionary habitat shifts in fostering diversification can best be evaluated through phylogenetic techniques (16, 97, 98). Phylogenetic analysis of *Enallagma* damselflies suggests that coexistence with fish is the ancestral condition in the genus, and that species have independently invaded fishless habitats containing large dragonfly predators at least twice (97), indicating that habitat shifts have occurred in this group and may contribute to the high species diversity of the genus. More studies using phylogenetics are needed before we can draw general conclusions concerning the role of the habitat gradient in shaping lineage diversification.

Besides shedding light on historical patterns of diversification and speciation mechanisms, phylogenetic methods can also serve as a valuable tool for identifying causal ecological agents driving adaptive evolution and for identifying those phenotypic traits responsible for species success (16). Again, *Enallagma* damselflies are illustrative on this point. Through convergence, species invading fishless habitats have evolved robust abdomens and large caudal gills (97), characteristics that allow these species to evade predatory dragonflies (98).

CONCLUSION

As pointed out by Connell (25), if an assemblage of species is to be regarded as possessing organization or structure, it is the form and strengths of interactions among the species that must produce that structure. We would further suggest that it is the traits of individuals that determine the form and strengths of species interactions and thus ultimately produce structure at the level of the community.

Gradient studies greatly facilitate our ability to detect the ways in which individual traits give rise to community structure. On the freshwater habitat gradient, transitions in community structure are best understood in terms of critical fitness tradeoffs that limit species distributions along the gradient. This perspective would not emerge, however, if each habitat type were examined in isolation. For example, examining the influence of fish predation on littoral invertebrate communities of fish-containing lakes might lead to the conclusion that predation often has little effect on these communities. While this conclusion would often be correct within the context of this single habitat type, the broader perspective gained by the gradient approach would reveal the fundamental role of predation in organizing the community by limiting community membership to only a subset of potential species. Indeed, the permanent removal of fish from such habitats would likely result in wholesale changes in community structure (10). Moreover, evaluating the functional basis of differential species success across community types elucidates the mechanisms by which individual traits create community structure. Finally, these same ecological mechanisms may serve as important evolutionary agents of selection for aquatic taxa, driving adaptive evolution and forming a template for lineage diversification (97, 98).

ACKNOWLEDGMENTS

We thank LK Freidenburg, MA McPeek, DG Fautin, and SA McCollum for insightful comments on the manuscript.

Literature Cited

1. Alexander DG. 1965. *An Ecological Study of the Swamp Cricket Frog,* Pseudacris nigrita feriarum *(Baird), with Compparartive Notes on Two Other Hylids of the Chapel Hill, North Carolina Region.* PhD thesis. Univ. NC, Chapel Hill
2. Allan JD. 1982. Feeding habits and prey consumption of three setipalpian stoneflies (Plecoptera) in a mountain stream. *Ecology* 63:26–34
3. Arnott SE, Vanni MJ. 1993. Zooplankton assemblages in fishless bog lakes: influence of biotic and abiotic factors. *Ecology* 74:2361–80
4. Azevedo-Ramos C, Van Sluys M, Hero JM, Magnusson WE. 1992. Influence of tadpole movement on predation by odonate naiads. *J. Herpetol.* 26:335–38
5. Barton NH. 1989. Founder effect speciation. In *Speciation and Its Consequences,* ed. D Otte, JA Endler, pp. 229–56. Sunderland, MA: Sinauer
6. Bayly IAE. 1978. Variation in sexual dimorphism in nonmarine calanoid copepods and its ecological significance. *Limnol. Oceanogr.* 23:1224–28
7. Bayly IAE. 1982. Invertebrate fauna and ecology of temporary pools on granite outcrops in southwestern Australia. *Aust. J. Mar. Freshw. Res.* 33:599–606

8. Bennett DV, Streams FA. 1986. Effects of vegetation on Notonecta (Hemiptera) distribution in ponds with and without fish. *Oikos* 46:62–69
9. Bishop JA. 1967. Some adaptations of *Limnadia stanleyana* King (Crustacea: Branchiopoda: Conchostraca) to a temporary freshwater environment. *J. Anim. Ecol.* 36:599–609
10. Black RW II, Hairston NG Jr. 1988. Predator driven changes in community structure. *Oecologia* 77:468–79
11. Blandenier P, Perrin N. 1989. A comparison of the energy budgets of two freshwater pulmonates: *Lymnaea peregra* (Muller) and *Physa acuta* (Drap.). *Rev. Suisse Zool.* 96:325–33
12. Blinn DW, Runck C, Davies RW. 1993. The impact of prey behaviour and prey density on the foraging ecology of *Ranatra montezuma* (Heteroptera): a serological examination. *Can. J. Zool.* 71:387–91
13. Blois-Heulin CP, Crowley H, Arrington M, Johnson DM. 1990. Direct and indirect effects of predators on the dominant invertebrates of two freshwater littoral communities. *Oecologia* 84:295–306
14. Bohanan RE, Johnson DM. 1983. Response of littoral invertebrate populations to a spring fish exclusion experiment. *Freshw. Invertebr. Biol.* 2:28–40
15. Brockelman WY. 1969. An analysis of density effects and predation in *Bufo americanus* tadpoles. *Ecology* 50:632–44
16. Brooks DR, McLennan DA. 1991. *Phylogeny, Ecology and Behavior: A Research Program in Comparative Biology.* Chicago: Univ. Chicago Press
17. Brooks JL, Dodson SI. 1965. Predation, body size and the composition of the plankton. *Science* 150:28–35
18. Brown KM. 1982. Resource overlap and competition in pond snails: an experimental analysis. *Ecology* 63:412–22
19. Brown KM, DeVries DR. 1985. Predation and the distribution and abundance of a pulmonate pond snail. *Oecologia* 66:93–99
20. Brown KM, DeVries DR, Leathers BK. 1985. Causes of life history variation in the freshwater snail *Lymnaea elodes. Malacologia* 26:191–200
21. Chovanec A. 1992. The influence of tadpole swimming behaviour on predation by dragonfly nymphs. *Amphibia-Reptilia* 13:341–49
22. Cole GA. 1966. Contrasts among calanoid copepods from permanent and temporary ponds in Arizona. *Am. Midl. Nat.* 76:351–68
23. Collins JP, Wilbur HM. 1979. Breeding habits and habitats of the amphibians of the Edwin S. George Reserve, Michigan, with notes on the local distribution of fishes. *Occ. Pap. Mus. Zool. Univ. Mich.* 686:1–34
24. Connell JH. 1961. The influence of interspecific competition and other factors on the distribution of the barnacle *Chthamalus stellatus. Ecology* 42:710–23
25. Connell JH. 1975. Some mechanisms producing structure in natural communities: a model and evidence from field experiments. In *Ecology and Evolution of Communities,* ed. ML Cody, JM Diamond, pp. 460–91. Cambridge, MA: Belknap
26. Convey P. 1988. Competition for perches between larval damselflies: the influence of perch use on feeding efficiency, growth rate and predator avoidance. *Freshw. Biol.* 19:15–28
27. Cook WL, Streams FA. 1984. Fish predation on *Notonecta* (Hemiptera): relationship between prey risk and habitat utilization. *Oecologia* 64:177–83
28. Cooper GP, Washburn GN. 1946. Relation of dissolved oxygen to winter mortality of fish in Michigan lakes. *Trans. Am. Fish. Soc.* 76:23–33
29. Cooper SD, Smith DW, Bence JR. 1985. Prey selection by freshwater predators with different foraging strategies. *Can. J. Fish. Aquat. Sci.* 42:1720–32
30. Crowder LB, Cooper WE. 1982. Habitat structural complexity and the interaction between bluegills and their prey. *Ecology* 63:1802–13
31. Crowl TA, Covich AP. 1990. Predator-induced life-history shifts in a freshwater snail. *Science* 247:949–51
32. Dale JM, Freedman B, Kerekes J. 1985. Acidity and associated water chemistry of amphibian habitats in Nova Scotia. *Can. J. Zool.* 63:97–105
33. Dixon SM, Baker RL. 1988. Effects of size on predation risk, behavioural response to fish, and cost of reduced feeding in larval *Ischnura verticalis* (Coenagrionidae: Odonata). *Oecologia* 76:200–5
34. Dodd CK. 1995. The ecology of a sandhills population of the eastern narrow mouthed toad, *Gastrophryne carolinensis,* during a drought. *Bull. Fla. Mus. Nat. Hist.* 38:11–41
35. Dodson SI. 1974. Zooplankton competition and predation: an experimental test of the size-efficiency hypothesis. *Ecology* 55:605–13
36. Driver EA. 1977. Chironomid communities in small prairie ponds: some char-

acteristics and controls. *Freshw. Biol.* 7:121–33
37. Dumas PC. 1964. Species-pair allopatry in the genera Rana and Phrynosoma. *Ecology* 45:178–81
38. Edley MT, Law R. 1988. Evolution of life histories and yields in experimental populations of *Daphnia magna*. *Biol. J. Linn. Soc.* 34:309–26
39. Ehlinger TJ. 1989. Learning and individual variation in bluegill foraging: habitat-specific techniques. *Anim. Behav.* 38:643–58
40. Etienne AS. 1972. The behavior of the dragonfly larva *Aeschna cyanea* M. after a short presentation of prey. *Anim. Behav.* 20:724–31
41. Eyre MD, Carr R, McBlane RP, Foster GN. 1992. The effects of varying site-water duration on the distribution of water beetle assemblages, adults and larvae (Coleoptera: Haliplidae, Dytiscidae, Hydrophilidae). *Arch. Hydrobiol.* 124:281–91
42. Futuyma DJ, Moreno G. 1988. The evolution of ecological specialization. *Annu. Rev. Ecol. Syst.* 19:207–33
43. Gilinsky E. 1984. The role of fish predation and spatial heterogeneity in determining benthic community structure. *Ecology* 65:455–68
44. Griffiths RA, de Wijer P, May RT. 1994. Predation and competition within an assemblage of larval newts (Triturus). *Ecography* 17:176–81
45. Hairston NG Jr, Olds EJ. 1984. Population differences in the timing of diapause: adaptation in a spatially heterogeneous environment. *Oecologia* 61:42–48
46. Hairston NG Jr, Walton WE, Li KT. 1983. The causes and consequences of sex specific mortality in a freshwater copepod. *Limnol. Oceanogr.* 28:935–47
47. Hall DJ, Cooper WE, Werner EE. 1970. An experimental approach to the production dynamics and structure of freshwater animal communities. *Limnol. Oceanogr.* 15:829–28
48. Hall DJ, Threlkeld ST, Burns CW, Crowley PH. 1976. The size-efficiency hypothesis and the size structure of zooplankton communities. *Annu. Rev. Ecol. Syst.* 7:177–208
49. Hambright KD, Trebatoski RJ, Drenner RW, Kettle D. 1986. Experimental studies of the impacts of bluegill (*Lepomis macrochirus*) and largemouth bass (*Micropterus salmoides*) on pond community structure. *Can. J. Fish. Aquat. Sci.* 43:1171–76
50. Hanazato T, Yasuno M. 1989. Zooplankton community structure driven by vertebrate and invertebrate predators. *Oecologia* 81:450–58
51. Hartland-Rowe R. 1972. The limnology of temporary waters and the ecology of Euphyllopoda. See Ref. Clark & Wootton, pp. 15–32
52. Havel JE. 1987. Predator-induced defenses: a review. In *Predation: Direct and Indirect Impacts on Aquatic Communities*, ed. WC Kerfoot, A Sih, pp. 263–78. Hanover, NH: Univ. Press New Engl.
53. Heads PA. 1985. The effect of invertebrate and vertebrate predators on the foraging movements of *Ischnura elegans* larvae (Odonata: Zygoptera). *Freshw. Biol.* 15:559–71
54. Healey M. 1984. Fish predation on aquatic insects. In *The Ecology of Aquatic Insects*, ed. VH Resh, DM Rosenberg, pp. 255–88. New York: Praeger
55. Hemphill N, Cooper SD. 1984. Differences in the community structure of stream pools containing or lacking trout. *Verh. Int. Ver. Theor. Angew. Limnol.* 22:1858–61
56. Henrikson B-I. 1988. The absence of antipredator behavior in the larvae of *Leucorrhinia dubia* (Odonata) and the consequences for their distribution. *Oikos* 51:197–83
57. Henrikson L, Oscarson HG. 1978. Fish predation limiting the abundance and distribution of *Glaenocorisa p. propinqua*. *Oikos* 31:102–5
58. Hershey AE. 1985. Effects of predatory sculpin on the chironomid communities in an arctic lake. *Ecology* 66:1131–38
59. Heyer WR, McDiarmid RW, Weigmann DL. 1975. Tadpoles, predation and pond habitats in the tropics. *Biotropica* 7:100–11
60. Holt RD. 1985. Population dynamics in two-patch environments: some anomalous consequences of an optimal habitat distribution. *Theor. Popul. Biol.* 28:181–208
61. Holt RD. 1987. Population dynamics and evolutionary processes: the manifold roles of habitat selection. *Evol. Ecol.* 1:331–47
62. Hornbach DJ, Deneka T, Dado R. 1991. Life-cycle variation of *Musculium partumeium* (Bivalvia: Sphaerlidae) from a temporary and a permanent pond in Minnesota. *Can. J. Zool.* 69:2738–44
63. Houston AI, McNamara JM, Hutchinson JMC. 1993. General results concerning the trade-off between gaining energy and

avoiding predation. *Philos. Trans. R. Soc. London Ser. B* 341:375–97
64. Hrbácek J, Dvorakova M, Korínek V, Procházkóva L. 1961. Demonstration of the effect of the fish stock on the species composition of zooplankton and the intensity of metabolism of the whole plankton association. *Verh. Int. Ver. Limnol.* 14:192–95
65. Janssen J. 1982. Comparison of the searching behavior for zooplankton in an obligate planktovore, blueback herring (*Alosa aestivalis*) and a facultative planktivore, bluegill (*Lepomis macrochirus*). *Can. J. Fish. Aquat. Sci.* 39:1649–54
66. Jeffries M. 1990. Interspecific differences in movement and hunting success in damselfly larvae (Zygoptera: Insecta): responses to prey availability and predation threat. *Freshw. Biol.* 23:191–96
67. Jeffries M. 1994. Invertebrate communities and turnover in wetland ponds affected by drought. *Freshw. Biol.* 32:603–12
68. Jeffries MJ, Lawton JH. 1984. Enemy free space and the structure of ecological communities. *Biol. J. Linn. Soc.* 23:269–86
69. Johansson F. 1993. Effects of prey types, prey density and predator presence on behavior and predation risk in a larval damselfly. *Oikos* 68:481–89
70. Johnson DM. 1991. Behavioral ecology of larval dragonflies and damselflies. *TREE* 6:8–13
71. Johnson DM, Pierce CL, Martin TH, Watson CN, Bohanan RE, Crowley PH. 1987. Prey depletion by odonate larvae: combining evidence from multiple field experiments. *Ecology* 68:1459–65
72. Kanou M, Shimozawa T. 1983. The elicitation of the predatory labial strike of dragonfly larva in response to a purely mechanical stimulus. *J. Exp. Biol.* 107:391–404
73. Kats LB, Sih A. 1992. Oviposition site selection and avoidance of fish by streamside salamanders (*Ambystoma barbouri*). *Copeia* 1992:468–73
74. Kenk R. 1949. The animal life of temporary and permanent ponds in southern Michigan. *Misc. Publ. Mus. Zool. Univ. Mich.* 71:1–66
75. Kerfoot WC. 1982. A question of taste: crypsis and warning coloration in freshwater zooplankton communities. *Ecology* 63:538–54
76. Kesler DH, Munns WR Jr. 1989. Predation by *Belostoma flumineum* (Hemiptera): an important cause of mortality in freshwater snails. *J. North Am. Benth. Soc.* 8:342–50
77. Kime JB. 1974. *Ecological Relationships Among Three Species of Aeshnid Dragonfly Larvae.* PhD thesis. Univ. Wash., Seattle. 142 pp.
78. Koufopanou V, Bell G. 1984. Measuring the cost of reproduction. IV. Predation experiments with *Daphnia pulex. Oecologia* 64:81–86
79. Kraus F, Petranka JW. 1989. A new sibling species of *Ambystoma* from the Ohio River drainage, USA. *Copeia* 1989:94–110
80. Larson DJ. 1985. Structure in temperate predaceous diving beetle communities (Coleoptera: Dytiscidae). *Holarc. Ecol.* 8:18–32
81. Law R. 1979. Optimal life histories under age-specific predation. *Am. Nat.* 114:399–417
82. Lawler SP. 1989. Behavioural responses to predators and predation risk in four species of larval anurans. *Anim. Behav.* 38:1039–47
83. Lazzaro X. 1987. A review of planktivorous fishes: their evolution, feeding behaviors, selectivity and impacts. *Hydrobiologia* 146:97–167
84. Leff LG, Bachmann MD. 1988. Basis of selective feeding by the aquatic larvae of the salamander, *Ambystoma tigrinum. Freshw. Biol.* 19:87–94
85. Lima SL, Dill LM. 1990. Behavioral decisions made under risk of predation: a review and prospectus. *Can. J. Zool.* 68:619–40
86. Lodge DM, Brown KM, Klosiewski SP, Stein RA, Covich AP, et al. 1987. Distribution of freshwater snails: spatial scale and the relative importance of physicochemical and biotic factors. *Am. Malacol. Bull.* 5:73–84
87. Lünig J. 1992. Phenotypic plasticity of *Daphnia pulex* in the presence of invertebrate predators: morphological and life history responses. *Oecologia* 92:383–90
88. Lynch M. 1980. The evolution of cladoceran life histories. *Q. Rev. Biol.* 55:23–42
89. Mahoney DL, Mort MA, Taylor BE. 1990. Species richness of calanoid copepods, cladocerans and other branchiopods in Carolina bay temporary ponds. *Am. Midl. Nat.* 123:244–58
90. McKee PM, Mackie GL. 1980. Desiccation resistance in *Sphaerium occidentale* and *Musculium securis* (Bivalvia:Sphaeriidae) from a temporary pond. *Can. J. Zool.* 58:1693–96
91. McLachlan AJ. 1985. What determines the species present in a rain-pool? *Oikos*

45:1–7

92. McLachlan AJ, Cantrell MA. 1980. Survival strategies in tropical rain pools. *Oecologia* 47:344–51
93. McPeek MA. 1989. Differential dispersal tendencies among *Enallagma* damselflies (Odonata) inhabiting different habitats. *Oikos* 56:187–95
94. McPeek MA. 1989. *The Determination of Species Composition in the* Enallagma *damselfly Assemblages (Odonata: Coenagrionidae) of Permanent Lakes.* PhD thesis. Mich. State. Univ., East Lansing
95. McPeek MA. 1990. Determination of species composition in the *Enallagma* damselfly assemblages of permanent lakes. *Ecology* 71:83–98
96. McPeek MA. 1990. Behavioral differences between *Enallagma* species (Odonata) influencing differential vulnerability to predators. *Ecology* 71:1714–26
97. McPeek MA. 1995. Morphological evolution mediated by behavior in the damselflies of two communities. *Evolution* 49:749–69
98. McPeek MA, Schrot AK, Brown JM. 1996. Adaptation to predators in a new community: swimming performance and predator avoidance in damselflies. *Eco logy.* 77:617–29
99. Mittelbach GG. 1981. Foraging efficiency and body size: a study of optimal diet and habitat use by bluegills. *Ecology* 62:1370–86
100. Mittelbach GG. 1988. Competition among refuging sunfishes and effects of fish density on littoral zone invertebrates. *Ecology* 69:614–23
101. Modenutti BE, Balseiro EG. 1994. Zooplankton size spectrum in four lakes of the Patagonian Plateau. *Limnologica* 24:51–56
102. Morin PJ. 1983. Predation, competition, and the composition of larval anuran guilds. *Ecol. Monogr.* 53:119–38
103. Morin PJ. 1984. The impact of fish exclusion on the abundance and species composition of larval odonates: results of short-term experiments in a North Carolina farm pond. *Ecology* 65:53–60
104. Morin PJ, Johnson EA. 1988. Experimental studies of asymmetric competition among anurans. *Oikos* 53:398–407
105. Neill WE. 1984. Regulation of rotifer densities by crustacean zooplankton in an oligotrophic montane lake in British Columbia. *Oecologia* 61:175–81
106. Neill WE. 1992. Population variation in the ontogeny of predator-induced vertical migration of copepods. *Nature* 356:54–57
107. Nemjo J. 1990. The impact of colonization history and fish predation on larval odonates (Odonata: Anisoptera) in a central New Jersey farm pond. *J. Freshw. Ecol.* 5:297–305
108. Newman RA. 1987. Effects of density and predation on *Scaphiopus couchi* tadpoles in desert ponds. *Oecologia* 71:301–7
109. Newman RA. 1988. Adaptive plasticity in development of *Scaphiopus couchi* tadpoles in desert ponds. *Evolution* 42:774–83
110. Nilsson AN, Svensson BW. 1994. Dytiscid predators and culicid prey in two boreal snowmelt ponds differing in temperature and duration. *Annu. Zool. Fenn.* 31:365–76
111. Nilsson B-I. 1981. Susceptibility of some Odonata larvae to fish predation. *Verh. Int. Ver. Limnol.* 21:1612–15
112. Oakley B, Palka JM. 1967. Prey capture by dragonfly larvae. *Am. Zool.* 7:727–28
113. Odendaal FJ, Bull CM. 1983. Water movements, tadpole competition and limits to the distribution of the frogs *Ranidella riparia* and *R. signifera. Oecologia* 57:361–67
114. Otto C. 1976. Habitat relationships in the larvae of three Trichoptera species. *Arch. Hydrobiol.* 77:505–17
115. Pearman PB. 1995. Effects of pond size and consequent predator density on two species of tadpoles. *Oecologia* 102:1–8
116. Pechmann JHK, Scott DE, Gibbons JW, Semlitsch RD. 1989. Influence of wetland hydroperiod on diversity and abundance of metamorphosing juvenile amphibians. *Wetlands Ecol. Manage.* 1:1–11
117. Peckarsky BL. 1984. Predator-prey interactions among aquatic insects. In *The Ecology of Aquatic Insects,* ed. VH Resh, DM Rosenberg, pp. 196–254. New York: Praeger
118. Petranka JW. 1990. Observations on nest site selection, nest desertion, and embryonic survival in marbled salamanders. *J. Herpetol.* 24:229–34
119. Pfennig D. 1990. The adaptive significance of an environmentally-cued developmental switch in an anuran tadpole. *Oecologia* 85:101–7
120. Pierce CL. 1988. Predator avoidance, microhabitat shift, and risk-sensitive foraging in larval dragonflies. *Oecologia* 77:81–90
121. Pierce CL, Crowley PH, Johnson DM. 1985. Behavior and ecological interactions of larval odonata. *Ecology* 66:1504–12
122. Pont D, Crivalli AJ, Gillot F. 1991.

The impact of three-spined sticklebacks on zooplankton of a previously fish-free pool. *Freshw. Biol.* 26:149–63

123. Pope GF, Carter JCH, Power G. 1973. The influence of fish on the distribution of *Chaoborus* spp. (Diptera) and density of larvae in the Matamek river system, Quebec. *Trans. Am. Fish. Soc.* 102:707–14
124. Post JR, Cucin D. 1984. Changes in the benthic community of a small precambrian lake following the introduction of yellow perch, *Perca flavescens*. *Can. J. Fish. Aquat. Sci.* 41:1496–501
125. Pritchard G. 1965. Prey capture by dragonfly larvae (Odonata: Anisoptera). *Can. J. Zool.* 43:271–89
126. Pritchard G, Leischner TG. 1973. The life history and feeding habits of *Sialis cornuta* Ross in a series of abandoned beaver ponds (Insecta: Megaloptera). *Can. J. Zool.* 51:121–31
127. Rahel FJ. 1984. Factors structuring fish assemblages along a bog successional gradient. *Ecology* 65:1276–89
128. Reinertsen H, Jensen A, Koksvik JI, Langeland A, Olsen Y. 1990. Effects of fish removal on the limnetic ecosystem of a eutrophic lake. *Can. J. Fish. Aquat. Sci.* 47:166–73
129. Resetarits WJ, Wilbur HM. 1989. Choice of oviposition site by *Hyla chrysoscelis:* role of predators and competitors. *Ecology* 70:220–28
130. Resetarits WJ, Wilbur HM. 1991. Calling site choice by *Hyla chrysoscelis*: effect of predators, competitors, and oviposition sites. *Ecology* 72:778–86
131. Riessen HP, Sommerville JW, Chiappari C, Gustafson D. 1988. *Chaoborus* predation, prey vulnerability, and their effect on zooplankton communities. *Can. J. Fish. Aquat. Sci.* 45:1912–20
132. Robinson JV, Wellborn GA. 1987. Mutual predation in assembled communities of odonate species. *Ecology* 68:921–27
133. Robinson JV, Wellborn GA. 1988. Ecological resistance to the invasion of a freshwater clam, *Corbicula fluminea:* fish predation effects. *Oecologia* 77:445–52
134. Roff DA. 1992. *The Evolution of Life Histories: Theory and Analysis.* New York: Chapman & Hall
135. Roth AH, Jackson JF. 1987. The effect of pool size on recruitment of predatory insects and on mortality in a larval anuran. *Herpetologica* 43:224–32
136. Rowe CL, Dunson WA. 1995. Impacts of hydroperiod on growth and survival of larval amphibians in temporary ponds of Central Pennsylvania, USA. *Oecologia* 102:397–403
137. Ryazanova GI, Mazokhin-Porshnyakov GA. 1993. Effects of presence of fish on the spatial distribution of dragonfly larvae, *Calopteryx splendens* (Odonata). *Entomol. Rev.* 72:90–96
138. Rzoska J. 1961. Observations on tropical rainpools and general remarks on temporary waters. *Hydrobiologia* 17:265–86
139. Rzoska J. 1984. Temporary and other waters. In *Sahara Desert,* ed. JL Cloudsley-Thompson, pp. 105–14. Oxford: Pergamon. 348 pp.
140. Schoener TW. 1970. Non-synchronous spatial overlap of lizards in patchy habitats. *Ecology* 51:408–18
141. Semlitsch RD, Reyer H-U. 1992. Performance of tadpoles from the hybridogenetic *Rana esculenta* complex: interactions with pond drying and interspecific competition. *Evolution* 46:665–76
142. Shoop CR. 1974. Yearly variation in larval survival of *Ambystoma maculatum. Ecology* 55:440–44
143. Short TM, Holomuzki JR. 1992. Indirect effects of fish on foraging behaviour and leaf processing by the isopod *Lirceus fontinalis. Freshw. Biol.* 27:91–97
144. Sih A. 1980. Optimal behavior: can foragers balance two conflicting demands? *Science* 210:1041–43
145. Sih A. 1987. Predators and prey lifestyles: an evolutionary and ecological overview. In *Predation, Direct and Indirect Impacts on Aquatic Communities,* ed. WC Kerfoot, A Sih, pp. 203–24. Hanover, NH: Univ. Press New Engl.
146. Sih A, Moore RD. 1990. Interacting effects of predator and prey behavior in determining diets. In *Behavioural Mechanisms of Food Selection,* ed. RN Huges, pp. 771–95. Berlin: Springer-Verlag
147. Skelly DK. 1992. Field evidence for a behavioral antipredator response in a larval amphibian. *Ecology* 73:704–8.
148. Skelly DK. 1994. Activity level and the susceptibility of anuran larve to predation. *Anim. Behav.* 47:465–68
149. Skelly DK. 1995. A behavioral trade-off and its consequences for the distribution of *Pseudacris* treefrog larvae. *Ecology* 76:150–64
150. Skelly DK. 1995. Competition and the distribution of spring peeper larvae. *Oecologia* 103:203–7
151. Skelly DK. 1996. Pond drying, predators, and the distribution of *Pseudacris* tad-

poles. *Copeia.* In press
152. Skelly DK, Werner EE. 1990. Behavioral and life historical responses of larval American toads to an odonate predator. *Ecology* 71:2313–22
153. Smith DC. 1983. Factors controlling tadpole populations of the chorus frog (*Pseudacris triseriata*) on Isle Royale, Michigan. *Ecology* 64:501–10
154. Smith DC, Van Buskirk J. 1995. Phenotypic design, plasticity, and ecological performance in two tadpole species. *Am. Nat.* 145:211
155. Smith DW, Cooper SD. 1982. Competition among cladocera. *Ecology* 63:1004–15
156. Soderstrom O, Nilsson AN. 1987. Do nymphs of *Parameletus chelifer* and *P. minor* (Ephemeroptera) reduce mortality from predation by occupying temporary habitats? *Oecologia* 74:39–46
157. Spence JR. 1989. The habitat templet and life history strategies of pond skaters (Heteroptera: Gerridae): reproductive potential, phenology, and wing dimorphism. *Can. J. Zool.* 67:2432–47
158. Spitze K. 1991. *Chaoborus* predation and life-history evolution in *Daphnia pulex:* temporal patterns of population diversity, fitness, and mean life history. *Evolution* 45:82–92
159. Spitze K. 1993. Population structure in *Daphnia obtusa*: quantitative genetic and allozyme variation. *Genetics* 135:367–74
160. Starkweather PL. 1990. Zooplankton community structure of high elevation lakes: biogeographic and predator-prey interactions. *Verh. Int. Ver. Limnol.* 24:513–17
161. Stein RA. 1977. Selective predation, optimal foraging, and the predator-prey interaction between fish and crayfish. *Ecology* 58:1237–53
162. Stein RA, Magnuson JJ. 1976. Behavioral response of crayfish to a fish predator. *Ecology* 57:751–61
163. Stenson JAE. 1978. Differential predation by fish on two species of *Chaoborus* (Diptera, Chaoboridae). *Oikos* 31:98–101
164. Stibor H, Lünig J. 1994. Predator-induced phenotypic variation in the pattern of growth and reproduction in *Daphnia hyalina* (Crustacea: Cladocera). *Funct. Ecol.* 8:97–101
165. Streams FA. 1986. Foraging behavior in a Notonectid assemblage. *Am. Midl. Nat.* 117:353–61
166. Strohmeier KL, Crowley PH, Johnson DM. 1989. Effects of red-spotted newts (*Notophthalmus viridescens*) on the densities of invertebrates in a permanent fish free pond: a one month enclosure experiment. *J. Freshw. Ecol.* 5:53–66
167. Swift MC. 1992. Prey capture by the four larval instars of *Chaoborus crystallinus*. *Limnol. Oceanogr.* 37:14–24
168. Tejedo M, Reques R. 1994. Plasticity in metamorphic traits of natterjack tadpoles: the interactive effects of density and pond duration. *Oikos* 71:295–304
169. Tessier AJ, Young A, Leibold M. 1992. Population dynamics and body size selection in *Daphnia*. *Limnol. Oceanogr.* 37:1–13
170. Tevis L. 1966. Unsuccessful breeding by desert toads at the limit of their ecological tolerance. *Ecology* 47:766–75
171. Thompson DJ. 1978. Prey size selection by larvae of the damselfly *Ischnura elegans* (Odonata). *J. Anim. Ecol.* 47:769–85
172. Thorp JH. 1986. Two distinct roles for predators in freshwater assemblages. *Oikos* 47:75–82
173. Thorp JH, Bergey EA. 1981. Field experiments on responses of a freshwater, benthic macroinvertebrate community to vertebrate predators. *Ecology* 62:365–75
174. Thorp JH, Bergey EA. 1981. Field experiments on interactions between vertebrate predators and larval midges (Diptera: Chironomidae) in the littoral zone of a reservoir. *Oecologia* 50:285–90
175. Tilman D. 1987. The importance of the mechanisms of interspecific competition. *Am. Nat.* 129:769–74
176. Tinsley RC, Tocque K. 1995. The population dynamics of a desert anuran, *Scaphiopus couchii*. *Aust. J. Ecol.* 20:376–84
177. Tonn WM, Magnuson JJ. 1982. Patterns in the species composition and richness in fish assemblages in northern Wisconsin lakes. *Ecology* 63:1149–66
178. Travis J, Keen WH, Juilianna J. 1985. The role of relative body size in a predator-prey relationship between dragonfly naiads and larval anurans. *Oikos* 45:59–65
179. Vanni MJ. 1988. Freshwater zooplankton community structure: introduction of large invertebrate predators and large herbivores to a small-species community. *Can. J. Fish. Aquat. Sci.* 45:1758–70
180. Vanni MJ. 1986. Competition in zooplankton communities: suppression of small species by *Daphnia pulex*. *Limnol. Oceanogr.* 31:1039–56

181. Vermeij GJ, Covich AP. 1978. Coevolution of freshwater gastropods and their predators. *Am. Nat.* 112:833–43
182. Vonder Brink RH, Vanni MJ. 1993. Demographic and life history response of the cladoceran *Bosmina longirostris* to variation in predator abundance. *Oecologia* 95:70–80
183. von Ende CN. 1979. Fish predation, interspecific predation, and the distribution of two *Chaoborus* species. *Ecology* 60:119–28
184. von Ende CN, Dempsey DO. 1981. Apparent exclusion of the cladoceran *Bosmina longirostris* by invertebrate predator *Chaoborus americanas*. *Am. Midl. Nat.* 105:240–48
185. Weider LJ, Pijanowska J. 1993. Plasticity of *Daphnia* life histories in response to chemical cues from predators. *Oikos* 67:385–92
186. Wellborn GA. 1994. Size-biased predation and the evolution of prey life histories: a comparative study of freshwater amphipod populations. *Ecology* 75:2104–17
187. Wellborn GA. 1994. The mechanistic basis of body size differences between two *Hyalella* (Amphipoda) species. *J. Freshw. Biol.* 9:159–67
188. Wellborn GA. 1995. Predator community composition and patterns of variation in life history and morphology among *Hyalella* (Amphipoda) populations in southeast Michigan. *Am. Midl. Nat.* 133:322–32
189. Wellborn GA, Robinson JV. 1987. Microhabitat selection as an antipredator strategy in the aquatic insect *Pachydiplax longipennis* Burmeister (Odonata: Libellulidae). *Oecologia* 71:185–89
190. Wellborn GA, Robinson JV. 1991. The impact of fish predation on an experienced macroarthropod community. *Can. J. Zool.* 69:2515–22
191. Werner EE. 1977. Species packing and niche complementarity in three sunfishes. *Am. Nat.* 111:553–78
192. Werner EE. 1991. Nonlethal effects of a predator on competitive interactions between two anuran larvae. *Ecology* 72:1709–20
193. Werner EE, Anholt BR. 1993. Ecological consequences of the trade-off between growth and mortality rates mediated by foraging activity. *Am. Nat.* 142:242–72
194. Werner EE, Hall DJ. 1974. Optimal foraging and the size selection of prey by the bluegill sunfish (*Lepomis macrochirus*). *Ecology* 55:1042–52
195. Werner EE, McPeek MA. 1994. Direct and indirect effects of predators on two anuran species along an environmental gradient. *Ecology* 75:1368–82
196. Wiggins GB, Mackay RJ, Smith IM. 1980. Evolutionary and ecological strategies of animals in annual temporary ponds. *Arch. Hydrobiol. Suppl.* 58:97–206
197. Wilbur HM. 1987. Regulation of structure in complex systems: experimental temporary pond communities. *Ecology* 68:1437–52
198. Wilbur HM, Fauth JE. 1990. Experimental aquatic food webs: interactions between two predators and two prey. *Am. Nat.* 135:176–204
199. Wilbur HM, Morin PJ, Harris RN. 1983. Salamander predation and the structure of experimental communities: anuran responses. *Ecology* 64:1423–29
200. Williams DD. 1987. *The Ecology of Temporary Waters*. London: Croom Helm
201. Williams DD, Moore KA. 1985. The role of semiochemicals in benthic community relationships of the lotic amphipod *Gammarus pseudolimnaeus*: a laboratory analysis. *Oikos* 44:280–86
202. Williams WD. 1983. *Life in Inland Waters*. Melbourne: Blackwell Sci.
203. Williamson CE. 1987. Predator-prey interactions between omnivorous diaptomid copepods and rotifers: the role of prey morphology and behavior. *Limnol. Oceanogr.* 32:167–77
204. Wilson CC, Hebart PDN. 1993. Impact of copepod predation on distribution patterns of *Daphnia pulex* clones. *Limnol. Oceanogr.* 38:1304–10
205. Wiltshire DJ, Bull CM. 1977. Potential competitive interactions between larvae of *Pseudophryne bibroni* and *P. semimarmorata* (Anura: Leptodactylidae). *Aust. J. Zool.* 25:449–54
206. Winfield IJ, Townsend CR. 1983. The cost of copepod reproduction: increased susceptibility to fish predation. *Oecologia* 60:406–11
207. Wissinger SA. 1988. Effects of food availability on larval development and interinstar predation among larvae of *Libellula lydia* and *Libellula luctuosa* (Odonata: Anisoptera). *Can. J. Zool.* 66:543–49
208. Woodward BD. 1983. Predator-prey interactions and breeding-pond use of temporary-pond species in a desert anuran community. *Ecology* 64:1549–55
209. Wright AH, Wright AA. 1949. *Handbook of Frogs and Toads of the United States and Canada*. Ithaca, NY: Comstock

210. Wright DJ, O'Brien WJ. 1982. Differential location of *Chaoborus* larvae and *Daphnia* by fish: the importance of motion and visible size. *Am. Midl. Nat.* 108:68–73
211. Wyngaard GA, Taylor BE, Mahoney DL. 1991. Emergence and dynamics of cyclopoid copepods in an unpredictable environment. *Freshw. Biol.* 25:219–32
212. Zaret TM. 1980. *Predation and Freshwater Communities.* New Haven, CT: Yale Univ. Press

Annu. Rev. Ecol. Syst. 1996. 27:365–

NATURAL FREEZING SURVIVAL IN ANIMALS

Kenneth B. Storey and Janet M. Storey
Institute of Biochemistry and Department of Biology, Carleton University, Ottawa, Ontario, Canada K1S 5B6

KEY WORDS: freeze-tolerance, cold hardiness, cryobiology, cryoprotectants, cell volume regulation

ABSTRACT

Natural freeze-tolerance supports the winter survival of many animals including numerous terrestrial insects, many intertidal marine invertebrates, and selected species of terrestrially hibernating amphibians and reptiles. Freeze-tolerant animals typically endure the conversion of 50% or more of total body water into extracellular ice and employ a suite of adaptations that counter the negative consequences of freezing. Specific adaptations control the sites and rate of ice formation to prevent physical damage by ice. Other adaptations regulate cell-volume change: Colligative cryoprotectants minimize cell shrinkage during extracellular ice formation; other protectants stabilize membrane structure; and a high density of membrane transporter proteins ensure rapid cryoprotectant distribution. Cell survival during freezing is also potentiated by anoxia tolerance, mechanisms of metabolic rate depression, and antioxidant defenses. The net result of these protective mechanisms is the ability to reactivate vital functions after days or weeks of continuous freezing. Magnetic resonance imaging has allowed visual examinations of the mode of ice penetration through the body of freeze-tolerant frogs and turtles, and cryomicroscopy has illustrated the effects of freezing on the cellular and microvasculature structure of tissues. Various metabolic adaptations for freezing survival appear to have evolved out of pre-existing physiological capacities of animals, including desiccation-resistance and anoxia-tolerance.

INTRODUCTION

Seasonally cold temperatures are a reality over much of our planet, and both animals and plants have evolved many ways of coping with low temperatures and with the restriction of feeding, growth, and reproduction that typically

0066-4162/96/1120-0365$08.00

accompany prolonged cold exposure (20, 49). Most organisms seek some refuge from the lowest extremes of air temperature by spending the winter underwater, underground, or under the snowpack, but others winter in sites that lack thermal buffering. Strategies of underwater or deep underground hibernation generally eliminate the probability of encountering freezing temperatures (53). The insulation of the snowpack typically holds temperatures in the subnivean (under the snow) environment close to 0°C, but lows of −6 to −8°C can occur (49, 60, 75), so animals wintering near the soil surface require some capacity to deal with subzero temperatures. Species wintering above the snowpack, however, may have to deal with prolonged periods of very low subzero temperatures.

To cope with exposure to temperatures below the freezing point (FP) of their body fluids (FP ranges from about −0.5°C for terrestrial species to −1.86°C for osmoconforming marine invertebrates in full strength seawater), ectothermic animals use one of two general strategies: freeze-avoidance or freeze-tolerance. Both include adaptations at behavioral, physiological, and biochemical levels, and indeed, some elements are shared between the two strategies (for review see 1–3, 22, 45, 53, 62, 70, 71, 74, 81). Their fundamental difference is that, whereas freeze-avoiding animals preserve the liquid state of body fluids even at very low temperatures (e.g. some Arctic insects can supercool to −50°C), freeze-tolerant animals defend only the liquid state of the cytoplasm, allowing ice to form in extracellular and extraorgan spaces of their bodies. The combination of adaptations used and the magnitude of their expression varies among species but deals with the specific winter habitat conditions of each.

In phylogenetic terms, both freeze-avoidance and freeze-tolerance have appeared many times in unrelated animal groups. The actual "choice" of strategy was probably a compromise that arose in each species due to the interplay of several factors including (*a*) the need to retain mobility and an active lifestyle at low temperatures, (*b*) the environmental realities of the species habitat or hibernation site, (*c*) pre-existing physiological capacities of the species, and (*d*) the ability to develop specific metabolic and physiological adaptations to perfect one form of cold hardiness. A few examples illustrate these points.

Some teleosts live year round in polar seas at a virtually constant water temperature of −1.86°C. If they were to freeze, they would never again thaw, for the melting point (MP) of their body fluids is above −1°C. However, contrary to expectations for most solutions, the FP of the blood of these fish is not equal to the MP but is pushed below −2°C by the action of specific antifreeze peptides or proteins (21). Various terrestrial invertebrates also benefit from a freeze-avoidance strategy to remain active under the snowpack in winter, and they also employ antifreeze proteins as their primary defense (22). The freeze-avoidance strategy is very common among terrestrial insects and other arthropods (e.g. spiders, mites) (3, 62), and its development may have taken advantage of various pre-existing factors such as a waterproof cuticle, a protective silk cocoon, autumn elimination of gut contents that contain nonspecific ice nucleators (e.g. bacteria, food particles), and various metabolic changes associated with the cessation of feeding and entry into winter dormancy. Indeed, in summer many insects can supercool to at least −8°C in the absence of any apparent antifreeze or cryoprotectant, and winter hardening further extends this ability (22, 81).

Supercooling (that is, remaining liquid below the FP) is a metastable state in which the probability of spontaneous freezing increases with decreasing temperature until it reaches 100% at the crystallization temperature (T_c; also known as the supercooling point). Freeze-avoiding species have developed adaptations (including antifreeze proteins, high concentrations of colligative cryoprotectants, masking or eliminating nucleators, partial dehydration) that effectively lower whole animal T_c to a value well below the anticipated environmental minima (3, 22, 71, 81). For example, in midwinter the T_c of goldenrod gall moth caterpillars (*Epiblema scudderiana*) was −38°C in a locale where the lowest ambient air temperature was about −25°C (71). However, should freezing occur in such deeply supercooled animals, it is lethal because the instantaneous conversion of a very high percentage of total body water into ice allows no time for compensatory and protective responses by cells. Only a few exceptions to this are known, most notably an Arctic beetle that survived freezing after spontaneous nucleation at about −54°C (55). For other types of animals, the combination of winter environment with pre-existing physiology could make freezing virtually unavoidable; hence, to populate seasonally cold environments, they had to develop freeze-tolerance.

Many species are highly susceptible to inoculative freezing if they come in contact with environmental ice at or below the FP of their body fluids. This was probably the reason for the development of freeze-tolerance among intertidal marine invertebrates and terrestrially hibernating frogs. For example, aerial exposure at low tide can rapidly lower body temperatures of marine molluscs and barnacles to below 0°C (1). Although the animals close themselves off within strong shells, seawater is also trapped within the shell, and when this freezes, the animal comes into direct contact with ice that will seed freezing in tissues. However, an interesting advantage of freezing has been reported for intertidal species: Freezing actually minimizes the net subzero temperature stress on the animals, because the latent heat released during crystallization stabilizes body temperature at a high value (theoretically at the FP) until equilibrium ice content is approached. For animals that normally experience only a few hours of low-tide exposure per day, this can represent a significant percentage of the total aerial exposure time. For example, the body temperature of mussels (*Mytilus edulis*) freezing in −10°C air remained at about −2°C for nearly 3 h before continuing a descent toward ambient (78). This same effect can apply on a larger scale to all animals in a tide pool that gain thermal buffering at low tide from the layer of surface ice growing on the pool (20).

Frogs that hibernate terrestrially also live in habitats where freezing may be unavoidable. Unlike salamanders and toads that retreat underground to hibernate, frogs remain at the soil surface in sites with a good cover of damp leaf litter to prevent desiccation (8). When ice penetrates these sites, frogs cannot avoid freezing because their highly water-permeable skin presents no barrier to the propagation of ice. Indeed, wood frogs supercooled to −1.5 to −2°C began freezing within about 30 sec when an ice crystal was dropped onto the damp paper on which they were resting (38); most frogs cooled in contact with damp paper or moss show no supercooling but begin to freeze at the FP of body fluids (K Storey, J Storey, unpublished observations).

FREEZE-TOLERANT ANIMALS

Although it is the most challenging method of winter survival, freeze-tolerance has, nonetheless, developed independently in many species (70). Freeze-tolerance is characteristic of hundreds of species of terrestrial insects (especially among Hymenoptera, Diptera, Coleoptera, and Lepidoptera) (22, 43, 55, 81), various intertidal marine invertebrates (including barnacles, bivalves, gastropods) (1, 45), and selected terrestrially hibernating amphibians and reptiles (74). Freeze-tolerance has also been reported for centipedes (76), one species of woodland slug (54), and nematodes (77). As first reported by Schmid (60), several species of terrestrially hibernating frogs are freeze-tolerant (*Rana sylvatica, Pseudacris crucifer, P. triseriata, Hyla versicolor, H. chrysoscelis*), but of urodeles that have been tested, only the Siberian newt, *Salamandrella* (formerly *Hynobius*) *keyserlingii,* with a range that extends onto the tundra, appears to tolerate freezing (74). Among reptiles, box turtles (*Terrepene carolina*), which hibernate in shallow burrows, and hatchling painted turtles (*Chrysemys picta*), which winter in shallow nests on exposed banks, show well-developed freeze-tolerance, surviving several days with more than 50% of body water frozen (5, 13, 16, 67, 75). Other reptiles, including garter snakes, hatchling *Pseudemys scripta*, and various lizards, endure some freezing (6, 7, 12, 14, 15, 41).

Overall, however, the capacity for freeze-tolerance is quite weakly developed in the Reptilia, and the widely variable limits of time, temperature, and ice content endured by reptiles have generated considerable debate about the relevance of freeze-tolerance to the winter survival of these species. Geographic variation in the freeze-tolerance of painted turtles seems to be considerable; both

C. p. marginata and *C. p. bellii* from Canadian populations (in Ontario and Manitoba, respectively) show good freeze-tolerance, whereas *C. p. bellii* from more southern populations (Nebraska) appear to rely on freeze-avoidance for winter survival (5, 52, 75). Furthermore, lizards, garter snakes, and red-eared sliders can endure only a few hours of freezing at relatively mild temperatures; often they cannot recover after longer times when ice content has reached its maximum.

Because temperature change is often slow in the protected hibernacula used by these animals, it becomes difficult to imagine situations in which freezing exposures in the hibernacula could be brief enough to permit recovery after thawing (remembering also that animals might supercool to −2 or −3°C before beginning to freeze but will not melt until temperature rises to the MP of about −0.5°C). It has been proposed that the ability to endure brief freezing stresses is adaptive in dealing with occasional low temperatures when these reptiles are active in the spring and fall, but for long-term hibernation, protected sites are chosen to avoid freezing temperatures (53, 74). At high latitudes, for example, garter snakes migrate large distances to hibernate by the hundreds in underground dens where temperature does not fall below 0°C (47).

Indeed, recent studies with vertebrates have indicated the need to differentiate between "ecologically relevant freeze-tolerance" as an integral component of the winter hardiness strategy of a species and the ability of an animal to endure brief freezing stress. The former probably grew out of the latter, and indeed, survival during and recovery after brief periods of freezing affecting the body extremities including skin and underlying musculature is possible in many species. Ecologically relevant or true freeze-tolerance, however, should be reserved to describe situations in which animals endure long periods (days, weeks) of continuous freezing at temperatures routinely encountered in the hibernaculum, with ice content rising to its maximum, and with ice penetration throughout the core of the body such that vital functions (movement, breathing, heart beat) are interrupted.

Given the physiological challenges of freeze-tolerance, as well as the fact that an animal is totally helpless while frozen, one wonders what advantages are to be gained from wintering in freezing sites. There are probably at least three. The first is early spring emergence. Animals in less protected hibernation sites can detect and respond to the warming temperatures of spring sooner than those in underground or underwater sites. Wood frogs and spring peepers, for example, are active at breeding ponds very early in the spring, weeks before aquatic-hibernating frogs. From this, they gain a long growing season for tadpoles and can make good use of temporary ponds. The second advantage is predator avoidance. Throughout their range, painted turtles that hatch late in the season remain in their subterranean nests over the first winter, living off stored internal yolk. Delayed emergence offers protection from predation until such time as conditions are favorable for rapid juvenile growth (24), but in the north, this behavior requires mechanisms for enduring subzero temperature. The third advantage is range extension—the ability to penetrate into environments that are not compatible with a freeze-avoidance strategy. Thus, the diversity of invertebrate species in the intertidal zone falls dramatically approaching the polar seas, but freeze-tolerant macrofauna such as mussels, littorines, and barnacles abound. Freeze tolerance has also allowed insects to penetrate some very harsh environments such as the high Arctic or to winter in exposed sites above the snowpack. The woolly bear caterpillar of Elsemere Island, for example, is freeze-tolerant throughout the year, actively feeds for only about one month in the summer, requires as much as 14 years to reach adulthood, and can withstand temperatures as low as −70°C in winter (33).

FREEZING INJURY

To understand the complexity of natural freeze-tolerance and the adaptations that are needed for survival, a brief examination of the damage done by freezing to nontolerant organisms is required. Intracellular freezing is apparently lethal for all organisms in nature due to the physical damage done to subcellular architecture by growing ice crystals. Under laboratory conditions, using isolated cells and extremely high rates of freeze/thaw, some cases of survivable intracellular ice formation have been documented (50). Although some evidence of natural intracellular freezing has been reported in insect fat body cells and nematodes (41, 77), the evidence remains unsatisfying. Ice can also do physical damage in extracellular spaces because the expansion of water when it crystallizes can break delicate capillaries; indeed, the loss of vascular integrity after thawing is a critical problem in cryomedical attempts at freezing organs (57). Rapid rates of ice formation (as occur when there is extensive pre-freeze supercooling) are also highly injurious because of the extreme osmotic stress placed on cells and the very limited time available to make metabolic adjustments before organ functions are shut down by advancing ice. For example, although their overall ability to endure freezing is marginal, European wall lizards uniformly died if the instantaneous ice surge upon nucleation was greater than 5% of total body water (12).

As extracellular ice forms, the osmolality of remaining extracellular fluids rises, and this in turn causes an efflux of water from cells and a reduction of cell volume. Ice continues to accumulate and cells continue to dehydrate and shrink until the osmolality of the remaining fluids rises to a level at which its melting point is equivalent to the subzero temperature of the tissues. The osmotic shock and cell-volume collapse that this causes are probably the most devastating effects of freezing on unprotected cells. Membranes are highly vulnerable, for these can withstand only so much compression before the lipid bilayer collapses irreversibly into a gel state. Cellular proteins and metabolic functions can also be adversely affected by the increase in intracellular ion concentrations that are the consequence of sequestering a high percentage of water as ice. Other injuries during freezing can result from ischemia, since freezing halts the circulation of blood or hemolymph. Furthermore, the rapid reintroduction of oxygen during thawing may initiate a burst of damage by oxygen free-radicals, similar to the well-documented injuries associated with recovery from ischemia in other systems (28). Finally, freezing halts vital functions including skeletal and smooth muscle movements, breathing, and heart beat.

ADAPTATIONS SUPPORTING FREEZE TOLERANCE

From this list of freezing injuries, we can identify the types of adaptations needed for freezing survival (22, 50, 70, 81). These include: 1. ice control—mechanisms to induce extracellular ice formation, modulate its rate of accumulation, and minimize the physical damage that it can cause (17, 22, 38, 56, 58); 2. cell-volume regulation—including colligative mechanisms that prevent shrinkage below the critical minimum cell-volume (CMCV), transporters for the movement of cryoprotectants and water across membranes, and stabilizers of membrane structure (30, 50, 59, 66); 3. mechanisms of anoxia/ischemia tolerance and of metabolic arrest to sustain cellular viability over long-term freezing (28, 29, 70); and 4. mechanisms for the spontaneous reactivation of vital signs after thawing (34, 35, 67, 74).

Successful freeze-tolerant animals can typically endure days or weeks of continuous freezing with at least 50%, and very often about 65%, of total body water frozen (70, 74); up to 80% ice has been reported in barnacles (1). However, the temperature at which this amount of ice accumulates can vary widely as can the lower lethal temperature (LLT) endured. For example, gall fly larvae reached 64% ice when frozen at −23°C and had an LLT of −27.5°C (42), whereas wood frogs accumulated 65% ice at only −2 to −3°C and did not survive freezing at −5.5°C (36). The LLT seems to be determined primarily by the temperature at which the maximal amount of tolerable extracellular ice is formed, or more correctly, by the CMCV that can be endured. The freezing temperature at which the CMCV is reached is inversely proportional to the osmolality of body fluids, and the CMCV may also be lower in freeze-tolerant than in nontolerant species due to adaptations that stabilize membrane bilayer structure. LLT decreased progressively when marine bivalves were acclimated to progressively higher seawater salinities (51). Indeed, marine molluscs seem to need only the naturally high osmolality of their body fluids (which are isosmotic with seawater) to defend cell volume during freezing as no specific cryoprotectant is produced.

Freeze-tolerant frogs as well as most insects elevate cellular osmolality by the synthesis of low molecular weight cryoprotectants. Wood frogs and spring peepers produce glucose in rapid response to ice forming in body extremities (11, 63), whereas insects slowly accumulate polyhydric alcohols (glycerol is the most common) over several weeks of autumn cold hardening and sustain polyol pools throughout the winter (for review see 71, 74). Some amphibians (*H. versicolor, S. keyserlingii*) also accumulate glycerol (74), but the pattern of accumulation and clearance of the polyol has not been examined. Perhaps, like insects, these sustain glycerol pools throughout the winter, whereas other frogs, because they use glucose as the cryoprotectant, clear the sugar after each thaw (64). Glucose clearance may be necessary because of the many negative effects of sustained high glucose on metabolism (as occur in diabetes) (23). Apart from glycerol, some insects use other polyols including sorbitol, mannitol, myoinositol, ribitol, erythritol, threitol, and ethylene glycol; some also employ sugars like trehalose as cryoprotectants (70, 73).

No studies have yet determined whether a phylogenetic pattern to cryoprotectant choice can be discerned, but glycerol clearly has metabolic advantages over the others, the most important being that glycerol production maximizes the number of osmotically active particles produced (two C3 glycerol molecules per one C6 hexose-phosphate unit cleaved off glycogen) without any loss from the total carbon pool (syntheses of C2, C4, or C5 polyols all involve CO_2 release) (73). Various freeze-tolerant insects accumulate both glycerol and sorbitol, each synthesized and catabolized with different seasonal patterns. Glycerol, whose synthesis is ATP-dependent, is accumulated early in the fall, well before freezing could impede aerobic energy metabolism, whereas sorbitol can be produced under anaerobic conditions (73). However, the cryoprotective advantage of the dual polyol system, compared with glycerol alone, is unknown. The dual system may benefit the repartitioning of carbohydrate reserves in the spring, since sorbitol carbon is quantitatively reconverted into glycogen, but glycerol carbon has other fates (oxidation, lipid biosynthesis) (70, 73).

Given that freeze-tolerance has arisen in numerous species from diverse groups, it is reasonable to suggest that the capacity arose, at least in part, by potentiating one or more pre-existing physiological capacities. Three of these are immediately obvious. The first is the capacity to deal with wide variations in cell volume and in the osmolality and ionic strength of body fluids. It is not surprising, then, that two groups that are highly tolerant of these variables (amphibians and intertidal invertebrates) (25, 61) also include within their ranks numerous freeze-tolerant species. From the point of view of the cell, extracellular freezing is simply a form of water stress; whether water is lost to the external environment (as during desiccation or exposure to hypersaline conditions) or temporarily sequestered in extracellular ice masses makes no difference to cells. The second capacity that aids freeze-tolerance is good ischemia-resistance. Again, the ability to survive for extended periods of time without oxygen is well developed in various lower vertebrates (particularly freshwater turtles) and many invertebrates (46). Gill-breathing intertidal invertebrates are particularly good facultative anaerobes for they must deal with oxygen deprivation during each low tide aerial exposure (65). The third factor is metabolic rate depression, the ability to lower basal metabolic rate manyfold and so gain a comparable extension of the time that a fixed reserve of endogenous body fuels can support metabolism. Metabolic depression is always a component of facultative anaerobiosis but is also a widespread response to stresses including heat, cold, and dryness, and indeed, winter dormancy is common for many animals (72). Diapausing insects, air-exposed mussels, and submerged turtles typically have metabolic rates that are only 10% or less of their nondiapausing or aerobic resting rates at the same temperature. A state of metabolic arrest, whether pre-existing or induced during freezing, would both increase the potential survival time while the organism is frozen and minimize the accumulation of deleterious metabolic end products.

In several previous studies we have shown that well-developed mechanisms of anoxia tolerance are important for sustaining cellular energy metabolism during freezing (69, 70). Thus, freeze-tolerant wood frogs and insects show slow declines in tissue ATP content and energy charge over time during freezing, along with accumulation of lactate and alanine as glycolytic end products, but these are readily reversed upon thawing (69, 70). Recently, we have also analyzed the ischemia/reperfusion event of freeze/thaw from a different perspective. Studies with mammalian ischemia/reperfusion models have identified injuries to cellular macromolecules due to a burst of reactive oxygen species (ROS) generation when oxygen is reintroduced at the end of an ischemic episode (26). All animals maintain antioxidant defenses in the form of enzymes such as superoxide dismutase, catalase, and glutathione peroxidase, and metabolites such as glutathione and vitamin E, but these can apparently be overwhelmed by sudden bursts of ROS generation. To determine how freeze-tolerant animals deal with this problem, we compared antioxidant enzyme activities, glutathione levels, and the accumulation of lipid peroxidation damage products in freeze-tolerant (*R. sylvatica*) and freeze-intolerant (*R. pipiens*) frogs (29). Using two different methods, we found no evidence for accumulated lipid peroxidation damage products in four tissues of wood frogs after 24 h of freezing or up to 4 h of thawing. Furthermore, freeze/thaw had little effect on the glutathione status of wood-frog organs, and together, these results indicate that wood frogs experience little or no oxidative stress over this ischemia-reperfusion event.

The metabolic basis for the lack of oxidative damage during freeze/thaw can be traced to high constitutive activities of antioxidant enzymes in wood-frog tissues; activities of superoxide dismutase, catalase, and glutathione peroxidase in wood-frog liver and skeletal muscle were two- to threefold higher than in the same organs of leopard frogs (29) or weakly freeze-tolerant garter snakes (28). Freezing-induced modification of enzyme activities also occurred in some wood-frog organs (29). Thus, it appears that mechanisms to deal with potential damage due to ROS formation during thawing are another of the important metabolic adaptations supporting natural freeze-tolerance.

In the remainder of this review, we deal with some new advances in the understanding of how freeze-tolerant animals control ice formation and regulate cell volume, and the influence of cell-volume changes on the expression of metabolic adaptations for freeze-tolerance.

ICE CONTROL

Ice growth can be initiated in two ways. Body fluids can be "seeded" across the epidermis when an animal comes in contact with environmental ice at or below the freezing point of body fluids. Freezing can also occur spontaneously in supercooled body fluids. For many species, freezing generally begins by seeding. The advantages of initiating freezing at a high subzero temperature are time (the slower the rate of ice formation, the longer the time available to make metabolic adjustments) and minimal osmotic shock, for the equilibrium content of ice will be low. Frogs frozen at −2.5°C, for example, take about 24 h to reach maximum ice content (36). The lower the temperature at which freezing begins (compared with the FP), the faster the rate of freezing, the greater the percentage of water that will freeze in the initial ice surge, and the less the time that will be available to implement cryoprotective measures. For example, when wood frogs were frozen at −2.5°C, there was plenty of time for a wide distribution of glucose from the liver to all core organs, but on a subsequent freeze, when frogs were held at −4°C, the rate of ice formation was too fast for extensive glucose distribution, and most cryoprotectant remained locked in the liver where it was made (64).

To minimize the freezing stress and maximize the ability to implement protective responses, freeze-tolerant animals, if not seeded by contact with environmental ice, stimulate crystallization themselves by employing ice nucleators. These initiate freezing usually only a few degrees Celsius below the FP of body fluids. Sometimes nucleators are nonspecific, and in different species they have been linked to bacteria on the skin surface or in the gut, other gut contents, or frass (22, 39). In other species, specific plasma or hemolymph proteins are synthesized seasonally, and these reproducibly nucleate at a precise temperature. Hemolymph ice-nucleating proteins occur widely in insects and have been reported in some marine snails, wood frogs, and painted turtle hatchlings (22, 48, 68, 79, 80).

The process of freezing and the mode of ice propagation through the body of a freeze-tolerant animal have been examined in detail in wood frogs and painted turtles. It is apparent that ice formation is carefully controlled to ensure survival. Using proton magnetic resonance imaging (MRI), ice formation can be monitored in real time in intact, living animals. Figure 1 shows selected images taken over the course of freezing and thawing an individual frog (58). The freezing front moves directionally through the body of the frog. The striated appearance of skeletal muscle (Figure 1A, B) shows that extracellular freezing is constrained by the morphology of the tissue, with crystals growing along the length of the muscle fibers. Within the abdominal cavity, the dark outlines around organs (Figure 1B, C) show that freezing occurs first in the extraorgan fluid. Close examination of the liver shows that the lobes shrink in size as freezing progresses and that the liver is the last organ to freeze fully (Figure 1C–E). Freezing monitored at another cross section that highlighted the brain and spinal cord of the frog revealed similar phenomena, with freezing occurring first in the spinal fluid and within the ventricles of the brain before moving into the tissue itself (58). A similar pattern appeared when the freezing of painted turtle hatchlings was monitored by MRI (56).

In contrast to the directional mode of freezing, the pattern of thawing revealed by MRI was quite different. In both wood frogs and painted turtles, thawing began uniformly throughout the entire body with images from all organs lightening in concert (56, 58). Organs clearly melted while still surrounded by extraorgan ice, a phenomenon that was particularly striking for the core organs of frogs. The same phenomenon was seen in the brain and spinal cord; tissues thawed before ice melted in brain ventricles, spinal fluid, or the vitreous humor of eyes (in turtles) (56, 58). The reason for this pattern of thawing can be traced to the higher osmolality of fluids in contact with ice within the organ vasculature (due to the presence of cryoprotectants as well as normal plasma solutes) compared with the large mass of nearly pure ice in extraorgan spaces. While a frog freezes, glucose is rapidly produced by the liver and distributed by the blood to other organs (70). As the freezing front moves inward, circulation (and cryoprotectant distribution) is progressively cut off first to peripheral, and then to core, sites. Final cryoprotectant levels are highest, therefore, in liver

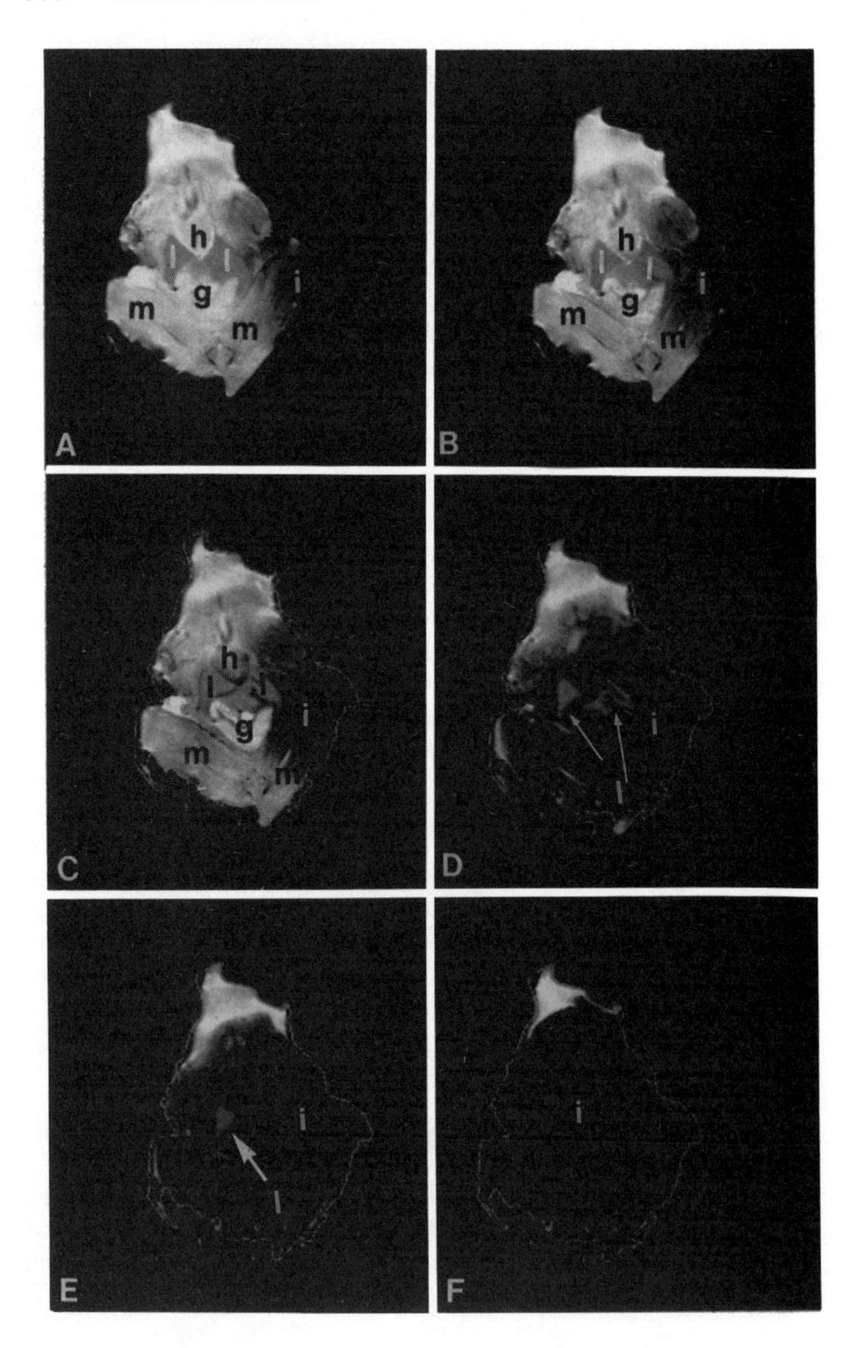
h
l
l
g
i
m
m
A
h
l
l
g
i
m
m
B
h
l
l
g
i
m
m
C
i
l
D
i
l
E
i
F

and heart, somewhat lower in brain and other abdominal organs, and lowest in skeletal muscle and skin (64). During thawing, the high glucose in core organs causes these to melt first. Melting from the inside out seems odd, but it has the physiological benefit of allowing the vital functions of the heart and lungs to reactivate as soon as possible. Not surprisingly, then, cardiac function is the last vital sign to cease during freezing (arrest occurs 11–21 h after freezing starts) (37) and is the earliest vital sign detected during thawing, occurring within 1 h at 3–5°C (32, 67). Following the resumption of heart beat, blood flow to the skin is detected soon thereafter, followed by spontaneous breathing, and finally skeletal muscle reflexes recover (35). Sciatic nerves regained excitability with 5 h of thawing, and recovery times for hind limb retraction and righting responses were 8 and 14 h, respectively (32). Differences in the physiology of nerve function between cold-sensitive, cold-resistant, and freeze-tolerant anurans appear to be important in the ability to recover after freezing exposure (19).

The MRI images suggested that liver and other organs shrink in size during freezing, and this can also be observed when dissecting frozen frogs. Huge masses of ice fill the abdominal cavity, and organs are visibly shrunken and encased in ice. Large flat crystals are also sandwiched between the skin and skeletal muscles of body and limbs. This extraorgan sequestration of ice has been quantified for frogs frozen slowly at −2.5°C (17, 40); organ water contents decreased by 2.8, 8.7, 12.7, 19.5, and 24.2% for eye, brain, skeletal muscle, liver, and heart, respectively, compared with organs from unfrozen animals. This appears to be an important method of avoiding physical damage by ice. By evacuating large amounts of water from organs and sequestering it innocuously as ice in extraorgan sites, the potential for damage due to excessive ice expansion within microvasculature of organs is greatly reduced. Such damage is a recognized problem in cryomedical organ preservation (57). Cryomicroscopy of tissue slices from both mammals and freeze-tolerant frogs shows that freezing

←

Figure 1 Proton magnetic resonance imaging showing the progress of ice formation during simulated natural freezing in a wood frog *Rana sylvatica*. Shown is a dorsal cross section through the frog during freezing at −7°C. The frog was placed in a cylinder around which the radio-frequency coil was wound, and then the cylinder was cooled from the bottom through contact with circulating chilled fluid. Frozen areas of the frog are darker because the protons in ice are invisible to standard proton MRI. In this individual, at this cross section highlighting the abdominal core, the freezing front can be seen moving inward from the right side (*A–C*). The liver is the last organ to freeze (*D*, *E*). Timed from the initiation of freezing, images were taken at 1 h 16 min (*A*), 1 h 28 min (*B*), 1 h 48 min (*C*), 2 h 10 min (*D*), 2 h 23 min (*E*), and 4 h 24 min (*F*). Labels are: (*h*) heart, (*l*) liver, (*m*) skeletal muscle, (*g*) gut, (*i*) ice; photographs are full size. Edge detection was added to some of the images during data processing. Taken from Rubinsky et al (58).

results in shrunken cells with much ice accumulated in an expanded vasculature space (66). Good freeze-tolerant animals, then, appear to move water out of their tissues, minimize the amount of ice forming within organ vasculature, and pack their cells with cryoprotectant to maintain the CMCV.

CELL-VOLUME REGULATION

The formation of ice in extracellular fluid spaces places an osmotic stress on cells that results in the net outflow of water from cells, a net influx of low molecular weight solutes, and a net decrease in cell volume. To be survivable, the reduction in cell volume cannot exceed the CMCV, which is usually associated with about 65% of total body water frozen as extracellular ice. This amount of cellular dehydration still leaves considerable intracellular free water, as illustrated by cryomicroscopy of tissue slices from freeze-tolerant frogs and turtles (56, 66). Indeed, image analysis of cryomicrographs of turtle organs indicated that extracellular ice constituted 36% of the total tissue volume in liver at −4°C, and 61% in skeletal muscle and heart (56); at these levels of dehydration, intracellular structure was clearly maintained. However, when tissue slices were frozen at −20°C (not survivable in nature), total tissue ice values rose to 65% , 79% , and 72% in the three tissues, respectively, and the severe cellular dehydration that ensued disrupted subcellular organization (56).

At least three types of adaptations appear to be required for cell-volume regulation in freeze-tolerant animals: 1. mechanisms that stabilize membrane bilayer structure under the compression stress of cell-volume reduction, 2. mechanisms that limit cell-volume reduction and prevent shrinking below the CMCV within the range of naturally encountered freezing temperatures, and 3. adaptations of membrane transport systems to allow rapid redistribution of water and solutes between intra- and extracellular compartments. Membrane stabilization is achieved through the action of specific low molecular weight cryoprotectants, such as trehalose and proline, that interact directly with the polar head groups of membrane lipids to stabilize the bilayer structure. The actions of these compounds have been well studied in species that endure extreme low water stress (anhydrobiosis) (18) and confirmed for freezing preservation of isolated membranes (59). Notably, both trehalose and proline levels are elevated in freeze-tolerant insects during the winter, and proline is often one of the major intracellular free amino acids in euryhaline marine invertebrates, one whose concentration can change rapidly in response to osmotic stress (25, 70).

The second component of volume regulation is to minimize cell-volume decrease during freezing via the colligative actions of low molecular weight solutes, generally specifically synthesized cryoprotectants (sugars, polyols).

Regulation of cryoprotectant biosynthesis (glucose in frogs; glycerol, sorbitol, or other polyols in insects) and the actions of these compounds in regulating cell volume and stabilizing macromolecules have been extensively reviewed (18, 70, 71, 73, 80).

The third facet of volume regulation, which has only recently received attention, is the regulation of water and solute fluxes across the plasma membrane during freezing and thawing. The lipid bilayer of the plasma membrane allows few compounds to cross by simple diffusion. For most compounds, entry into or exit from cells is gated by transport proteins that span the membrane and provide facilitated transport for compounds moving in the direction of an osmotic gradient and active transport to move compounds against their concentration gradient. During extracellular freezing, the osmotic and ionic imbalance set up by the exclusion of solutes from rapidly growing ice crystals requires a redistribution of water, ions, and cryoprotectants across cell membranes; reverse movements accompany thawing. Glucose movement across cell membranes is carrier-mediated by transporter proteins, and glucose has proven to be a poor cryoprotectant in cryomedical trials with mammalian tissues because it can not enter cells quickly. Similarly, most mammalian cells are impermeable to sorbitol, yet this is one of the major polyols accumulated by freeze-tolerant insects (73). Hence, adaptations of membrane sugar and polyol transporters must have accompanied the use of these compounds as cryoprotectants in nature.

Recent studies have targeted glucose transporters in freeze-tolerant wood frogs. Within just a few hours after freezing begins, wood frogs catabolize a huge liver glycogen reserve (as much as 600–700 μmol/g wet weight (gww) in glucose equivalents) and export glucose to other organs to raise their sugar content as high as 200 μmol/gww before full freezing halts circulation (64, 70). Not surprisingly, the number of glucose transporters in *R. sylvatica* plasma membranes is specifically increased to deal with the demands of rapid cryoprotectant movement (30, 31). Membrane vesicles prepared from liver of autumn-collected wood frogs had an 8.2-fold greater rate of carrier-mediated glucose transport and a 4.7-fold higher number of glucose transporters (quantified by cytochalasin B binding) than did liver membrane vesicles from the freezing-intolerant, aquatic-hibernating leopard frog. Glucose transport rate by wood-frog skeletal muscle vesicles was also 8-fold higher than in leopard-frog vesicles, showing that transporter systems of both the cryoprotectant-producing organ and a receiving organ are modified in concert (30). Furthermore, the rate of carrier-mediated glucose transport by liver vesicles was 6-fold higher and the number of transporters 8.5-fold higher in liver vesicles from wood frogs collected in September, compared with June animals (31). The importance of elevated glucose transport capacity for freeze-tolerance in frogs raises the question of whether other specific transporters are involved in cell-volume regulation during freezing. For example, water moves across cell membranes both by simple diffusion and by channel-mediated facilitated transport by aquaporins (4). An important next step in studying natural freeze-tolerance will be to determine whether adaptations of aquaporin numbers or activity aid water fluxes during freezing and thawing.

RELATIONSHIP BETWEEN FREEZE TOLERANCE AND DEHYDRATION TOLERANCE

The probable development of freeze-tolerance as an extension of pre-existing mechanisms of dealing with water stress in animals is further supported by some recent experiments on the metabolic responses to dehydration in frogs. Because ice forms extracellularly, the major stress perceived by cells during freezing is a sharp volume reduction. Hence, it seemed logical to predict that protective metabolic responses to freezing might be triggered or regulated by dehydration or changes in cell volume. Indeed, this seems to be the case for cryoprotectant synthesis in frogs. Two freeze-tolerant species, the wood frog *R. sylvatica* and the spring peeper *P. crucifer*, were subjected to controlled, whole-body dehydration stress at 5°C at a rate of 0.5–1% of total body water lost per hour (achieved by holding frogs in closed containers with desiccant in the bottom). Both species tolerated the loss of 50–60% of total body water, putting them among the best of the desiccation-tolerant anurans (61). Both species responded to dehydration with rapid glycogenolysis in liver and glucose export to other organs. In all six organs tested of autumn-collected wood frogs, glucose rose progressively as animals were dehydrated (Figure 2 shows liver, heart, and brain); the maximal increase ranged from 9-fold in gut to 313-fold in liver of frogs that had lost 50% of total body water. Final liver glucose was 127 μmol/gww, a value not much less than the 200–300 μmol/gww typically stimulated by freezing exposure (8). A similar response was seen with autumn-collected *P. crucifer*; glucose rose by 120-fold to 2690 ± 400 nmol/mg protein

Figure 2 Effect of dehydration and rehydration on glucose levels in liver, heart, and brain of the freeze-tolerant frogs *R. sylvatica* and *P. crucifer* and the freeze-intolerant *R. pipiens*. All frogs were autumn-collected, acclimated at 5°C, and then dehydrated at 5°C at a rate of 0.5–1% of total body water lost per hour in closed containers over a layer of silica gel desiccant. For rehydration, 50% dehydrated frogs were placed in containers with 1–2 cm of distilled water and sampled after 24 h at 5°C. Bars are: open, controls at 5°C; rising right, dehydrated to 25% of total body water lost; crosshatched, dehydrated to 50% of total body water lost; solid, 50% dehydrated then fully rehydrated. Data compiled from Churchill & Storey (8–10). Storey & Storey (30).

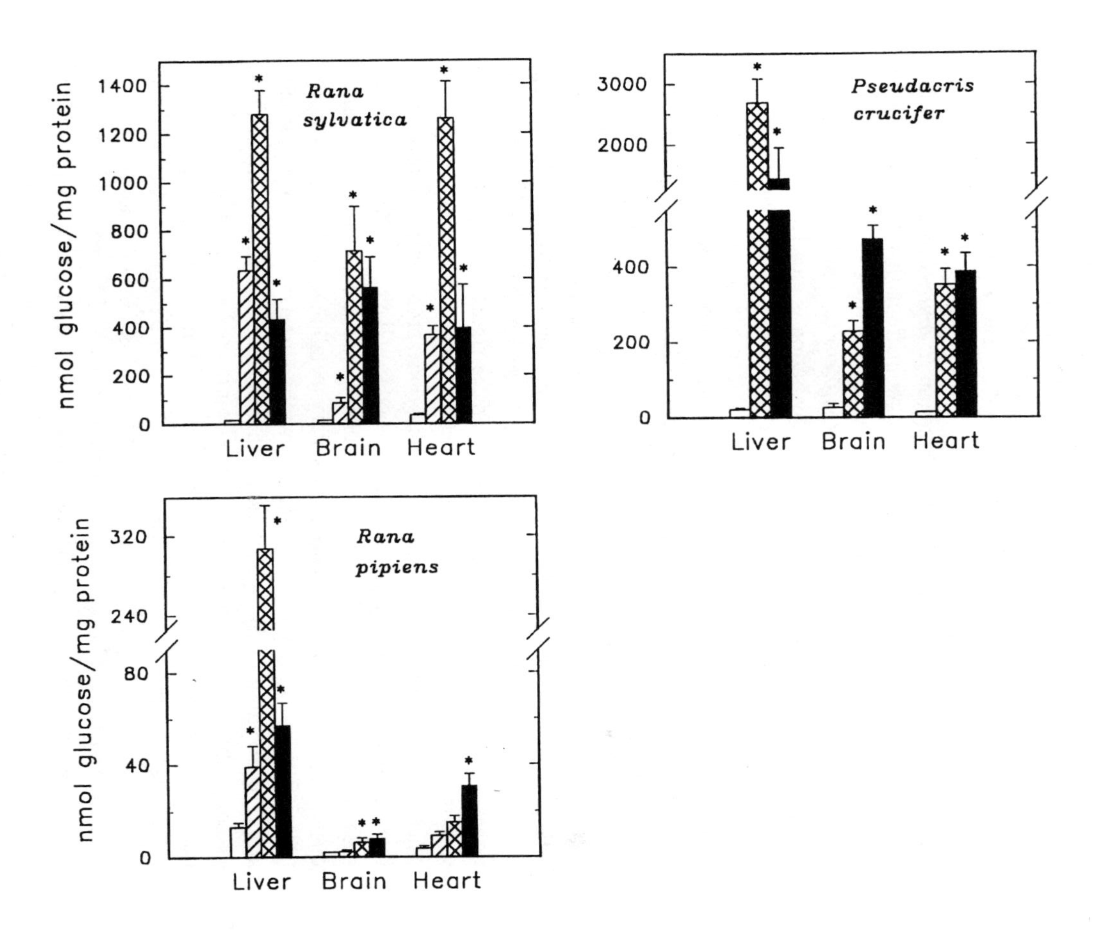
Rana sylvatica
nmol glucose/mg protein
1400 1200 1000 800 600 400 200 0
Liver Brain Heart
Pseudacris crucifer
3000 2000 400 200 0
Liver Brain Heart
Rana pipiens
nmol glucose/mg protein
320 240 80 40 0
Liver Brain Heart

or 220 μmol/gww in liver of 50% dehydrated frogs (Figure 2) (9). Glucose in other organs of *P. crucifer* rose by 3- to 60-fold. Glucose levels in both species fell when animals were rehydrated (Figure 2), which also occurs when frogs are thawed, and the hyperglycemic response to dehydration was much greater in autumn- versus spring-collected frogs, which again occurs with freezing-induced cryoprotectant synthesis (8, 9). Parallel experiments with the aquatic-hibernating leopard frogs showed that liver glycogenolysis and glucose output responses to dehydration were also shared by a freeze-intolerant species (10). Glucose rose progressively with dehydration in *R. pipiens* liver, rising by 24-fold overall to a final value of 20 μmol/gww in frogs that had lost 50% of total body water (Figure 2). Thus, although the magnitude of the dehydration-induced hyperglycemia is much lower in *R. pipiens* than in the freeze-tolerant species, the glycogenolytic response to dehydration is clearly in place in the freeze-intolerant species, and this suggests that the cryoprotectant response to freezing grew out of a more primitive hyperglycemic response to dehydration.

Also intriguing about this hyperglycemic response is that it is not a direct response to water loss by liver cells themselves. In all three species undergoing dehydration, water was lost first from extraorgan pools, and water content of core organs was defended until a high percentage of total body water was lost (8–10). Indeed, *R. sylvatica* excelled at this, and even when 50–60% of total body water was lost, liver water content remained unchanged (8). Rather, it appears that water loss (due to dehydration or freezing) is detected by peripheral target cells, perhaps in the skin, and the signal is transmitted by nervous or hormonal stimuli to the liver. The signal may be catecholamine-based since administration of the β-adrenergic antagonist, propranolol, suppresses the freezing-induced synthesis of glucose by wood frog liver (74). Recent research with mammalian systems has shown that cell-volume change can trigger numerous effects, including changes to intermediary metabolism and gene expression (27); insulin, for example, stimulates cell swelling in rat liver, whereas glucagon and catecholamines (glycogenolytic hormones) induce shrinkage (27). Thus, glycogenolysis has an ancient link to cell-volume reduction, and this may underlie the hyperglycemic response to freezing by frog liver. A key area for future research in freeze-tolerance will be to explore the range of cryoprotective responses stimulated and coordinated by changes in cell volume.

In summary, then, freeze-tolerance is one of the most fascinating and complex adaptations that has evolved among animals. Although much is known about some aspects of freeze-tolerance, notably cryoprotectant biosynthesis, many other areas are largely unexplored, and completely new elements of freeze-tolerance still remain to be identified. For example, in the newest studies by our lab, we have constructed and screened a cDNA library prepared from liver of freezing-exposed wood frogs to identify genes that are upregulated during freezing. Expecting to identify genes related to one of the known areas of adaptation that have been the subject of this article, we found instead that the first two clones we identified are genes for the α and γ subunits of fibrinogen (Q Cai, KB Storey, submitted for publication). Fibrinogen is a plasma protein produced by liver that is key to the clotting process, and these results suggest that we have stumbled into a previously unrecognized area of adaptation in freeze-tolerance—the repair mechanisms that may be needed to deal with mechanical injuries to tissues by ice crystals. These new results suggest that the body's clotting defenses are potentiated while the animal is still frozen in order to deal swiftly and effectively with any damage to the vasculature that is detected upon thawing.

ACKNOWLEDGMENTS

We are indebted to numerous colleagues and graduate students, past and present, for their many valuable contributions to research on animal freeze-tolerance. Research in our laboratory is supported by grants from the National Institutes of Health, GM 43796, for vertebrate studies and from the NSERC Canada for invertebrate studies.

Literature Cited

1. Aarset AV. 1982. Freezing tolerance in intertidal invertebrates (a review). *Comp. Biochem. Physiol. A* 73:571–80
2. Bale JS. 1987. Insect cold hardiness: freezing and supercooling—an ecophysiological perspective. *J. Insect Physiol.* 33:899–908
3. Cannon RJC, Block W. 1988. Cold tolerance of microarthropods. *Biol. Rev.* 63:23–77
4. Chrispeels MJ, Agre P. 1994. Aquaporins: water channel proteins of plant and animal cells. *Trends Biochem. Sci.* 19:421–25
5. Churchill TA, Storey KB. 1992. Natural freezing survival by painted turtles *Chrysemys picta marginata* and *C. p. bellii*. *Am. J. Physiol.* 262:R530–37
6. Churchill TA, Storey KB. 1992. Responses to freezing exposure by hatchling turtles *Trachemys scripta elegans*: factors influencing the development of freeze tolerance by reptiles. *J. Exp. Biol.* 167:221–33
7. Churchill TA, Storey KB. 1992. Freezing survival of the garter snake *Thamnophis sirtalis*. *Can. J. Zool.* 70:99–105
8. Churchill TA, Storey KB. 1993. Dehydration tolerance in wood frogs: a new perspective on the development of amphibian freeze tolerance. *Am. J. Physiol.* 265:R1324–32
9. Churchill TA, Storey KB. 1994. Effects of dehydration on organ metabolism in the frog *Pseudacris crucifer*: hyperglycemic responses to dehydration mimic freezing-induced cryoprotectant production. *J. Comp. Physiol. B* 164:492–98
10. Churchill TA, Storey KB. 1995. Metabolic effects of dehydration on an aquatic frog

Rana pipiens. J. Exp. Biol. 198:147–54

11. Churchill TA, Storey KB. 1996. Organ metabolism and cryoprotectant synthesis during freezing in spring peepers *Pseudacris crucifer. Copeia.* In press
12. Claussen DL, Townsley MD, Bausch RG. 1990. Supercooling and freeze tolerance in the European wall lizard, *Podarcis muralis. J. Comp. Physiol. B* 160:137–43
13. Costanzo JP, Claussen DL. 1990. Natural freeze tolerance in the terrestrial turtle, *Terrapene carolina. J. Exp. Zool.* 254:228–32
14. Costanzo JP, Claussen DL, Lee RE. 1988. Natural freeze tolerance in a reptile. *Cryo-Lett.* 9:380–85
15. Costanzo JP, Grenot C, Lee RE. 1995. Supercooling, ice inoculation and freeze tolerance in the European common lizard, *Lacerta vivipara. J. Comp. Physiol. B* 165:238–44
16. Costanzo JP, Iverson JB, Wright MF, Lee RE. 1995. Cold hardiness and overwintering strategies of hatchlings of an assemblage of northern turtles. *Ecology* 76:1772–85
17. Costanzo JP, Lee RE, Wright MR. 1992. Cooling rate influences cryoprotectant distribution and organ dehydration in freezing wood frogs. *J. Exp. Zool.* 261:373–78
18. Crowe JH, Crowe LM, Carpenter JF, Wistrom CA. 1987. Stabilization of dry phospholipid bilayers and proteins by sugar. *Biochem. J.* 242:1–10
19. Dalo NL, Hackman JC, Storey KB, Davidoff RA. 1995. Changes in motoneuron membrane potential and reflex activity induced by sudden cooling of isolated spinal cords: differences among cold-sensitive, cold-resistant, and freeze-tolerant amphibian species. *J. Exp. Biol.* 198:1765–74
20. Davenport J. 1992. *Animal Life at Low Temperature.* London: Chapman & Hall. 245 pp.
21. DeVries AL. 1982. Biological antifreeze agents in coldwater fishes. *Comp. Biochem. Physiol. A* 73:627–40
22. Duman JG, Wu DW, Xu L, Tursman D, Olsen TM. 1991. Adaptations of insects to subzero temperatures. *Q. Rev. Biol.* 66:387–410
23. Furth A, Harding J. 1989. Why sugar is bad for you. *New Sci.* 123:44–47
24. Gibbons JW, Nelson DH. 1978. The evolutionary significance of delayed emergence from the nest by hatchling turtles. *Evolution* 32:297–303
25. Gilles R. 1979. *Mechanisms of Osmoregulation in Animals.* New York: Wiley-Intersci.
26. Halliwell B, Gutteridge JMC. 1989. *Free Radicals in Biology and Medicine.* London: Clarendon. 2nd ed.
27. Haussinger D, Lang F, Gerok W. 1994. Regulation of cell function by the cellular hydration state. *Am. J. Physiol.* 267:E343–55
28. Hermes-Lima M, Storey KB. 1993. Antioxidant defenses in the tolerance of freezing and anoxia by garter snakes. *Am. J. Physiol.* 265:R646–52
29. Joanisse DR, Storey KB. 1996. Oxidative damage and antioxidants in *Rana sylvatica,* the freeze tolerant wood frog. *Am. J. Physiol.* In press
30. King PA, Rosholt MN, Storey KB. 1993. Adaptations of plasma membrane glucose transport facilitate cryoprotectant distribution in freeze-tolerant frogs. *Am. J. Physiol.* 265:R1036–42
31. King PA, Rosholt MN, Storey KB. 1995. Seasonal changes in plasma membrane glucose transport in freeze-tolerant wood frogs. *Can. J. Zool.* 73:1–9
32. Kling KB, Costanzo JP, Lee RE. 1994. Post-freeze recovery of peripheral nerve function in the freeze-tolerant wood frog, *Rana sylvatica. J. Comp. Physiol. B* 164:316–20
33. Kukal O, Serianni AS, Duman JG. 1988. Glycerol metabolism in a freeze-tolerant Arctic insect: an in vivo ^{13}C NMR study. *J. Comp. Physiol. B* 158:175–83
34. Layne JR. 1992. Postfreeze survival and muscle function in the leopard frog (*Rana pipiens*) and the wood frog (*Rana sylvatica*). *J. Therm. Biol.* 17:121–24
35. Layne JR, First MC. 1991. Resumption of physiological functions in the wood frog (*Rana sylvatica*) after freezing. *Am. J. Physiol.* 261:R1324–37
36. Layne JR, Lee RE. 1987. Freeze tolerance and the dynamics of ice formation in wood frogs (*Rana sylvatica*) from southern Ohio. *Can. J. Zool.* 65:2062–65
37. Layne JR, Lee RE, Heil TL. 1989. Freezing-induced changes in the heart rate of wood frogs (*Rana sylvatica*), after freezing. *Am. J. Physiol.* 257:R1046–49
38. Layne JR, Lee RE, Huang JL. 1990. Inoculation triggers freezing at high subzero temperatures in a freeze-tolerant frog (*Rana sylvatica*) and insect (*Eurosta solidaginis*). *Can. J. Zool.* 68:506–10
39. Lee MR, Lee RE, Strong-Gunderson JM, Minges SR. 1995. Isolation of ice-nucleating active bacteria from the freeze-tolerant frog, *Rana sylvatica. Cryobiology* 32:358–65

40. Lee RE, Costanzo JP, Davidson EC, Layne JR. 1992. Dynamics of body water during freezing and thawing in a freeze-tolerant frog (*Rana sylvatica*). *J. Therm. Biol.* 17:263–66
41. Lee RE, Denlinger DL. 1991. *Insects at Low Temperature.* New York: Chapman & Hall. 513 pp.
42. Lee RE, Lewis EA. 1985. Effect of temperature and duration of exposure on tissue ice formation in the gall fly, *Eurosta solidaginis* (Diptera, Tephritidae). *Cryo-Lett.* 6:25–34
43. Lee RE, McGrath JJ, Morason RT, Taddeo RM. 1993. Survival of intracellular freezing, lipid coalescence and osmotic fragility in fat body cells of the freeze-tolerant gall fly *Eurosta solidaginis. J. Insect Physiol.* 39:445–50
44. Lemos-Espinal JA, Ballinger RE. 1992. Observations on the tolerance to freezing by the lizard, *Sceloporus grammicus,* from Iztaccihuatl volcano, Mexico. *Herpetol. Rev.* 23:8–9
45. Loomis SH. 1987. Freezing in intertidal invertebrates: an update. *Cryo-Lett.* 8:186–95
46. Lutz PL, Storey KB. 1996. Adaptations to variations in oxygen tension by vertebrates and invertebrates. In *Handbook of Comparative Physiology.* Boca Raton: CRC Press. In press
47. Macartney JM, Larsen KW, Gregory PT. 1989. Body temperatures and movements of hibernating snakes (*Crotalus* and *Thamnophis*) and thermal gradients of natural hibernacula. *Can. J. Zool.* 67:108–14
48. Madison DL, Scrofano MM, Ireland RC, Loomis SH. 1991. Purification and partial characterization of an ice nucleator protein from the intertidal gastropod *Melampus bidentatus. Cryobiology* 28:483–90
49. Marchand PJ. 1991. *Life in the Cold.* Hanover: Univ. New Engl. Press. 239 pp.
50. Mazur P. 1984. Freezing of living cells: mechanisms and implications. *Am. J. Physiol.* 247:C125–42
51. Murphy DJ, Pierce SK. 1975. The physiological basis for changes in the freezing tolerance of intertidal molluscs. *J. Exp. Zool.* 193:313–22
52. Packard GC, Packard MJ. 1993. Hatchling painted turtles (*Chrysemys picta*) survive exposure to subzero temperatures during hibernation by avoiding freezing. *J. Comp. Physiol. B* 163:147–52
53. Pinder AW, Storey KB, Ultsch GR. 1992. Estivation and hibernation. In *Environmental Biology of the Amphibia,* ed. ME Feder, WW Burggren, pp. 250–74. Chicago: Univ. Chicago Press
54. Riddle WA. 1983. Physiological ecology of land snails and slugs. In *The Mollusca,* ed. WD Russell-Hunter, 6:431–61. New York: Academic
55. Ring RA. 1981. The physiology and biochemistry of cold tolerance in Arctic insects. *J. Therm. Biol.* 6:219–29
56. Rubinsky B, Hong JS, Storey KB. 1994. Freeze tolerance in turtles: visual analysis by microscopy and magnetic resonance imaging. *Am. J. Physiol.* 267:R1078–88
57. Rubinsky B, Lee CY, Bastacky J, Onik J. 1987. The process of freezing and the mechanism of damage during hepatic cryosurgery. *Cryobiology* 27:85–97
58. Rubinsky B, Wong STS, Hong JS, Gilbert J, Roos M, Storey KB. 1994. 1H magnetic resonance imaging of freezing and thawing in freeze-tolerant frogs. *Am. J. Physiol.* 266:R1771–77
59. Rudolph AS, Crowe JH. 1985. Membrane stabilization during freezing: the role of two natural cryoprotectants, trehalose and proline. *Cryobiology* 22:367–77
60. Schmid WD. 1982. Survival of frogs in low temperature. *Science* 215:697–98
61. Shoemaker VH. 1992. Exchange of water, ions, and respiratory gases in terrestrial amphibians. In *Environmental Physiology of the Amphibians,* ed. ME Feder, WW Burggren, pp. 125–59. Chicago: Univ. Chicago Press
62. Somme L. 1989. Adaptations of terrestrial arthropods to the alpine environment. *Biol. Rev.* 64:367–407
63. Storey JM, Storey KB. 1985. Triggering of cryoprotectant synthesis by the initiation of ice nucleation in the freeze tolerant frog, *Rana sylvatica. J. Comp. Physiol. B* 156:191–95
64. Storey KB. 1987. Organ-specific metabolism during freezing and thawing in a freeze-tolerant frog. *Am. J. Physiol.* 253:R292–97
65. Storey KB. 1993. Molecular mechanisms of metabolic arrest in mollusks. In *Surviving Hypoxia: Mechanisms of Control and Adaptation,* ed. PW Hochachka, PL Lutz, TJ Sick, M Rosenthal, G van den Thillart, pp. 253–69. Boca Raton: CRC Press
66. Storey KB, Bischof J, Rubinsky B. 1992. Cryomicroscopic analysis of freezing in liver of the freeze-tolerant wood frog. *Am. J. Physiol.* 263:R185–94
67. Storey KB, Layne JR, Cutwa MM,

Churchill TA, Storey JM. 1993. Freezing survival and metabolism of box turtles, *Terrapene carolina. Copeia* 1993(3):628–34
68. Storey KB, McDonald DG, Duman JG, Storey JM. 1991. Blood chemistry and ice nucleating activity in hatchling painted turtles. *Cryo-Lett.* 12:351–58
69. Storey KB, Storey JM. 1986. Freeze tolerant frogs: cryoprotectants and tissue metabolism during freeze/thaw cycles. *Can. J. Zool.* 64:49–56
70. Storey KB, Storey JM. 1988. Freeze tolerance in animals. *Physiol. Rev.* 68:27–84
71. Storey KB, Storey JM. 1989. Freeze tolerance and freeze-avoidance in ectotherms. In *Advances in Comparative & Environmental Physiology,* ed. LCH Wang, 4:51–82. Heidelberg: Springer-Verlag. 441 pp.
72. Storey KB, Storey JM. 1990. Facultative metabolic rate depression: molecular regulation and biochemical adaptation in anaerobiosis, hibernation and estivation. *Q. Rev. Biol.* 65:145–74
73. Storey KB, Storey JM. 1991. Biochemistry of cryoprotectants. In *Insects at Low Temperature,* ed. DL Denlinger, RE Lee, pp. 64–93. New York: Chapman & Hall. 513 pp.
74. Storey KB, Storey JM. 1992. Natural freeze tolerance in ectothermic vertebrates. *Annu. Rev. Physiol.* 54:619–37
75. Storey KB, Storey JM, Brooks SPJ, Churchill TA, Brooks RJ. 1988. Hatchling turtles survive freezing during winter hibernation. *Proc. Natl. Acad. USA* 85:8350–54
76. Tursman D, Duman JG, Knight CA. 1994. Freeze tolerance adaptations in the centipede, *Lithobius forficatus. J. Exp. Zool.* 268:347–53
77. Wharton DA. 1995. Cold tolerance strategies in nematodes. *Biol. Rev.* 70:161–85
78. Williams RJ. 1970. Freezing tolerance in *Mytilus edulis. Comp. Biochem. Physiol.* 35:145–61
79. Wolanczyk JP, Storey KB, Baust JG. 1990. Nucleating activity in the blood of the freeze-tolerant frog, *Rana sylvatica. Cryobiology* 27:328–35
80. Zachariassen KE. 1980. The role of polyols and nucleating agents in cold-hardy beetles. *J. Comp. Physiol. B* 140:227–34
81. Zachariassen KE. 1985. Physiology of cold tolerance in insects. *Physiol. Rev.* 65:799–832

Annu. Rev. Ecol. Syst. 1996. 27:387–421

DEMOGRAPHIC AND GENETIC MODELS IN CONSERVATION BIOLOGY: Applications and Perspectives for Tropical Rain Forest Tree Species

E. R. Alvarez-Buylla, R. García-Barrios, C. Lara-Moreno, and M. Martínez-Ramos
Centro de Ecología, Ap. Postal 70-275, México D.F. 04510, México

KEY WORDS: tropical rain forest trees, population models, inbreeding, conservation, effective population size

ABSTRACT

We review deterministic and stochastic demographic models as well as classical population genetic models that have been applied to tropical rain forest tree species. We discuss their implications for conservation. The main conclusions of deterministic demographic models are the key importance of species' longevity in determining susceptibility of population growth rate to harvesting of individuals at different life-stages, the critical effect of patch dynamics, and the importance of density-dependent mechanisms at least for abundant species. Population viability analysis to predict extinction times of tropical rain forest tree species has only been performed for four tropical rain forest tree species using the simplest Lefkovitch matrix linear model. Results obtained are in accordance with results of simple stochastic models for nonstructured populations that have been solved analytically. Population genetic models have shown that tropical rain forest trees: (*a*) possess high levels of genetic diversity, (*b*) maintain greater proportions of genetic variation within than among populations, (*c*) are predominantly outcrossed, and (*d*) have high levels of gene flow. These results suggest that tropical tree species may not be in immediate danger of extinction from genetic factors if actual conditions are maintained. However, the impact of forest fragmentation is expected to be particularly strong for most tropical rain forest tree species due mainly to the high genetic load kept by their present population genetic structures. Recent theoretical demographic-genetic models for simple systems suggest that the fixation of new mildly detrimental mutations may be comparable in importance to environmental stochasticity, implying minimum viable populations as large as a few thousands. It is urgent to develop a model that integrates genetic

0066-4162/96/1120-0387$08.00

and demographic factors, that enables evaluations of their relative importance in long-term persistence of tropical rain forest tree species.

HIGH DIVERSITY AND RARENESS OF TROPICAL RAIN FOREST TREE SPECIES: CHALLENGE AND URGENCY

Conservation of tropical forests is one of the most urgent and challenging tasks of biologists, social scientists, environmental managers, entrepreneurs, nonprofit social organizations, and public administrators (166, 169). Much has been said on the great importance of tropical forests and the many species they harbor (see 12, 17, 121–124, 135, 170, for a few examples). Alarming projections of the rates of deforestation of tropical forests and the consequent threat of a mass extinction have also been put forward in several occasions (e.g. 12, 17, 121–124, 148, 149, 169). Of all tropical forests, those referred to as lowland tropical rain forests (TRF) harbor the greatest number of species, and their destruction would have the worst and most extensive global effects (121–124, 132). In this review we concentrate on this type of forest.

Conservation analyses of natural populations consider demographic and genetic criteria (52, 87, 141a, 149). The former refer to the size of populations and their temporal change, and the latter to the genetic variation and its distribution within populations. Both criteria have been used to estimate minimum viable populations (MVP), defined according to extinction probabilities and persistence times (24, 101, 149, 157, 158). It is impossible to obtain MVP estimates applicable to all species, but they are useful reference values to analyze the relative effects of different demographic stochasticity and genetic factors; both are likely to cause detrimental population growth and fitness effects, respectively, as population size is reduced (see reviews in 87, 149, 158). Demographic models incorporate two types of stochasticity: environmental and demographic. Genetic and evolutionary models incorporate three types of factors: loss of genetic variability for adaptive evolution, random fixation of deleterious mutations or alleles by genetic drift, and inbreeding depression (e.g. 47, 70, 90).

Tropical rain forest long-term persistence depends upon conservation of trees that are the community's dominant physiognomic elements and among which there are several key species for the subsistence of many other plant and animal species (see for example 51, 168). The challenge of TRF conservation relies mainly on the high diversity and low population densities or rarity of most tree species even in protected forests (see Table 1). For example, in Los Tuxtlas forest, more than half of the tree species had less than one individual > 10 cm diameter per ha; another 20% of the species had two or less individuals; and

Table 1 Number of tree species found in a 5 ha plot with different densities per ha (individuals > 10 cm stem diameter) at Los Tuxtlas tropical rain forest (México). Data from Martínez-Romas & Alvarez-Buylla 1995.

Density range (number of individuals > 10 cm stem diam.)	Canopy trees	Pioneer trees	Medium & understory trees & palms	Total	%
< 1	29	7	18	54	55.1
≤ 2	12	3	4	19	19.4
2–10	9	1	5	15	15.3
10–22	3	1	6	10	10.2
TOTAL NUMBER OF SPP	53	12	33	98	100

only 10% had more than 10 individuals per ha. This is the general situation of TRFs (75, 78, 81). These low densities may imply that to encompass MVP (see Table 3) even all available untouched TRF areas (169) may not be enough for the long-term conservation of many TRF tree species. Furthermore, different types of rarity among TRF tree species imply different genetic and demographic constraints for their conservation (see 17 for a review).

Extinction probability increases as population size decreases (see 56, 88, 158); however, the form of the function that relates extinction risk to population size is not clear (99). For example, the relative importance of demographic and genetic factors in determining extinction probabilities in natural populations is still unclear. Lande (87) stressed the priority of demographic factors, and his view has permeated the recent literature of conservation biology (24, 113). However, Lande's own research and other studies have recently shown that fixation of deleterious mutations might be particularly important even in sexual outbreeding populations. These models suggest that this genetic factor could be as important as environmental stochasticity in populations of considerable size, determining MVP in the order of several thousands and census population sizes of more than 10,000 individuals (89, 98–100).

Although such simple models are valuable rigorous analytical tools for benchmark analyses, habitat loss or degradation and ecological factors constitute the ultimate threats of species extinction. Hence population viability analyses (PVA), traditionally based mainly on stochastic demographic models and genetic considerations, should include as much ecological reality as possible for particular analyses (24, 149, 158).

In this paper we review the demographic and genetic models that have been developed or applied to TRF tree species. These are only starting points of rigorous population viability analyses of TRF tree species. At the beginning of each section we summarize the general mathematical models or theory relevant in each case, and then we describe their applications, developments, or potentialities for TRF tree species. We pay special attention to studies of two model systems for which abundant demographic and genetic data have been accumulated: *Cecropia obtusifolia*, a light-demanding, short-lived, pioneer tree, and *Astrocaryum mexicanum*, a shade-tolerant, long-lived palm species. These are also typical species of the extremes of a continuum of life-history types of TRF tree species (8, 105). We contend that further theoretical developments that help identify critical parameters and rigorous empirical estimates of these should be urgently pursued for TRF tree species or for model experimental plant systems. PVA should consider alternative scenarios of the dynamics of the whole community's species diversity (37, 77).

DEMOGRAPHIC-ECOLOGICAL MODELS

Simple demographic models of exponential growth rate can be used as analytical benchmarks to compare predictions of more complicated models that include ecological factors such as habitat heterogeneity, succession, density-dependence, and environmental and demographic stochasticity. In this section we sequentially analyze results of models incorporating these factors for TRF tree species. We then review their conservation implications.

Lefkovitch Linear Matrix Models

GENERAL THEORY Population models applied to TRF tree species should explicitly consider demographic differences among individuals belonging to different age, size, or stage categories. Age or size structure per se may have significant effects on population trajectories (92). Caswell (29) provides a complete review of Leslie and Lefkovitch matrix models for individuals distinguished by age or growth stage, respectively; the latter is the most appropriate for plants. The general matrix model is $\mathbf{N}_{t+1} = \mathbf{P} \cdot \mathbf{N}_t$, where $\mathbf{N}$ represents the population state vector, whose elements are the numbers of individuals in a particular stage category, and $\mathbf{P}$ is a square non-negative matrix of vital rates. For large t, the proportions of individuals in different stages will become constant, and the population will grow exponentially at a constant finite rate, λ, which is the largest positive eigenvalue of $\mathbf{P}$. Therefore, $\lambda < 1$ implies the population's exponential decrease to extinction. From these models it is possible to derive sensitivities and elasticities of λ with respect to any of the matrix entries. Elasticities are recommended because they measure relative sensitivity (i.e. they avoid the problem of comparing measurements on different scales) (39) and can be compared across populations and species.

APPLICATIONS TO TRF TREE SPECIES A Lefkovitch model was first applied to demographic estimates of *Pentaclethra macroloba* (67, 68) and *Stryphnodendron excelsum* (67), both primary long-lived canopy species found at La Selva, Costa Rica. Later, the model was applied to other TRF tree species (2, 3, 28, 43, 48, 67, 68, 109a, 128x, 131, 133, 134, 137; Table 1): seven understory long-lived palms, six canopy long-lived trees, and one pioneer short-lived tree. Although confidence limits of λ have been obtained only in three cases (3, 109a, 128, 131), previous simulations (10, 11) suggest that all λ's are not significantly different from 1, and hence, according to this model, most TRF tree populations would be close to a numerical equilibrium.

For long-lived TRF tree species, the largest elasticities are found for probabilities of remaining in the same life-stage (i.e. survival) during preadult and adult stages for long-lived palms and canopy tree species (see Table 2). Elasticities of λ with respect to survivorship of younger life-stages (seeds, seedlings, and juveniles), fecundities, and probabilities of transition from one stage to another (growth) are much smaller. In contrast, the short-lived pioneer tree *Cecropia obtusifolia* had greatest elasticities of λ for the transitions from seedlings to juveniles and for fecundity. These trends have been observed in other types of organisms (24).

Table 2 Estimates of population finite growth rates (λ) obtained for TRF tree species using a matrix Lefkovitch model.

Species	Life-history	λ	Largest elasticity	Longevity (yr)	Reference
TREES:					
+*Araucaria hunsteinii*	canopy-slow	0.9889–1.0884	Pij-a	100	(48)
+*Araucaria cunninghamii*	canopy-slow	1.0115–1.0202	Pij-a	100	(48)
Cecropia obtusifolia$^{\psi}$	pioneer	0.9928–1.0346*	Pij-s, Gij-j, Fij	35	(11)
Brosimum alicastrum$^{\psi}$	canopy-fast	1.0635	Pij-a	120	(134)
+*Pentaclethra macroloba*$^{\psi}$	canopy-slow	1.0021	Pij-pa	150	(67, 68)
Omphalea oleifera	canopy-medium	1.0085	Pij-a	140	(133)
Stryphnodendron excelsum	canopy-slow	1.0471	Pij-pa	150	(67)
PALMS:					
+*Astrocaryum mexicanum*$^{\psi}$	understory-slow	0.9890–1.0120	Pij-a	125	(137)
Chamaedorea tepejilote	understory-slow	0.9699–1.1232	Pij-pa	60	(131)
Podococus barterii	understory-slow	1.0125	Pij-pa	75	(28)
Pseudophoenix sargentii	understory-slow	1.0080–1.1995	Pij-pa	80	(43)
Thrinax radiata	understory-slow	0.9890–1.0120	Pij-pa	120	(128)
Coccothrinax readii	understory-slow	1.0129–1.0969*	Pij-pa	> 145	(128)

+Species included in Menges MVP study (111).

$^{\psi}$Species with genetic and demographic data, see references in text.

*These ranges correspond to 95% confidence limits estimated according to Alvarez-Buylla and Slatkin (10, 11).

Largest elasticities found for: Pij-probabilities of remaining in the same stage (survival); Gij-probabilities of advancing to following stages (growth) and Fij-fecundities or seedling recruitment. Life-stages: s-seeds, sd-seedlings, j-juveniles, pa-preadults, a-adults.

Longevities are only approximations.

The use of the relationship between species, life-histories, and aggregate elasticities of vital rates has been proposed as an attractive and suggestive shortcut for management and life-history evolution analysis (146, 147). Important mathematical constraints exist on the elasticities of linear transition matrices, and these must be considered when making biological interpretations from elasticity analysis (52a). Structural errors in the specification of the matrix model may distort the elasticities by artificially altering the mathematical restrictions imposed on them. The sign and magnitude of these biases for such structural errors are yet to be formally analyzed.

Abundant ecological data suggest that linear matrix models are structurally incorrect for most organisms under natural conditions (24). For example, it is difficult to match the assumption of constant, density-independent vital rates with the fact that all TRF tree species have λ's very close to 1 (Table 2). Rather, density-dependent mechanisms may be affecting vital rates and limiting population growth. It is also unlikely that λ is not sensitive in the long term to environmental changes. Most TRF tree species are affected, for example, by the dynamics of gap formation and closure (see below) that varies in time and space (102–104). Detailed field work on plant vital rates and the factors that make them vary are, therefore, indispensable for an adequate specification of demographic-ecological models and for estimation of the parameters included in them.

Metapopulation Models

GENERAL THEORY A metapopulation may be defined as an assemblage of local populations that grow, become extinct, or evolve more or less independently, but are connected by dispersion. Metapopulation models, first formalized by Levins (95), have generated increasing interest among theoretical and field ecologists and managers. Most plant (82) and animal (64) populations grow and evolve either in natural heterogeneous environments or in fragmented habitats due to human perturbation. Therefore, local populations are segregated in patches of different sizes and qualities; landscape level processes as well as local conditions affect their demography and genetics. Metapopulation models may be used also to draw recommendations about how landscapes should be modified and managed to maintain or enhance biodiversity (64, 166).

Metapopulation analysis focuses on the balance between extinction and recolonization in patchy environments (69, 95). We distinguish two types of metapopulation models. The first includes simple stochastic models incorporating either constant recolonization and extinction rates or isolation-dependent recolonization rates and patch-area–dependent extinction rates (64). In these models, recolonization and extinction rates do not depend on patch state, which is usually defined by presence or absence of the species under analysis (occupancy). In this case, it is assumed that within-patch change and equilibrium occur instantaneously, in a static environment subdivided in many similar and uniformly distributed patches. At equilibrium, the metapopulation is described as a shifting mosaic of occupied and unoccupied patches, although below a threshold number of available patches, the metapopulation goes extinct because the rate of colonization is lower than the rate of extinction (64).

More recent models assume rates of colonization and extinction to depend on patch state (69, 112). The environment is assumed to be either static or dynamic (64). In the latter, the environment is a shifting mosaic of different successional stages, where in any one place the right kind of habitat is present only for a limited period of time. Thus, local extinctions are inevitable, but a species may survive in the environment if it is able to establish new local populations elsewhere, where the right kind of habitat has appeared. This metapopulation approach generally considers local population dynamics and dispersal among patches in an environmental shifting mosaic (5, 6, 72, review in 82). An interesting case developed for TRF tree populations occurs when processes at different scales have coupled dynamics. For example, patch successional stages and demographic conditions of particular local populations may be correlated (3, 5).

APPLICATIONS TO TRF TREE SPECIES Tropical forests are shifting mosaics of patches at different successional stages (25, 26, 76, 168). The forest canopy is recurrently disrupted by branches snapping or trees falling (25, 36, 106). Canopy gaps enable more direct light to reach the understory, and the humidity, soil, nutrient conditions, and biotic interactions in gaps contrast with those in closed forest (13, 25, 33, 42, 108, 140, 142–144, 164). Microclimatic conditions of gaps depend on their size and origin (33). Gaps are closed by lateral crown expansion of neighboring trees, sprouting of surviving falling trees, growth of seedlings and saplings established before the gap aperture, and by growth of newly established seedlings (36). Four main patch states (environmentally different forest patches) are distinguished (168): 1. recently formed gaps (< 1 yr since opened) that may be small (ca. < 100 m^2) or 2. large (> 100 m^2), 3. building or successional patches (2–35 yr since opened) dominated by pioneer tree species, and 4. mature patches (e.g. > 35 yr since formed) where mainly mature long-lived trees grow (106, 168).

TRF tree populations are segregated among the different forest regeneration patches. Recruitment of some species is highly dependent on gap occurrence and cohorts of different life-cycle stages are found in patches of different successional stages (3, 5). Many species recruit individuals and grow across all the regeneration mosaic; cohorts of different ages are mixed within each patch-type (107). The role of forest patch dynamics in the ecology and evolution of TRF plants has been reviewed extensively elsewhere (40, 102, 107).

Metapopulation models in TRF mosaics consider population dynamics within each patch-type, transition probabilities among patch-types determined by rate of gap formation and closure, and dispersal among patch-types (5). Patch-specific population dynamics is modeled with matrix models as described above. Within-patch dynamics has been coupled to forest dynamics by assuming that the latter is independent of the dynamics of any particular species. This is reasonable in species-rich TRF (3). A finite linear Markovian process has been assumed to model forest dynamics as: $f(t+1) = \mathrm{D} \cdot f(t)$, where D is the forest matrix with entries d^{kl} representing the constant probabilities that a patch-type l becomes type k from t to $t+1$ and f is a vector of the number of patches, if all patches are assumed to be of equal size (72), or a proportion of area of each patch-type, if patches are allowed to vary in size (3, 5).

Overall metapopulation dynamics is modeled as $n(t+1) = \mathrm{G}\ n(t)$, where G is a matrix of all possible transitions ($g_{ij}^{kl} = p_{ij}\ d^{kl}$) among life-stages ($ij$) and patch-types ($kl$); and the vectors $n(t+1)$ and $n(t)$ contain number of individuals (n_i^k) in each life-stage (i) and patch-type (k) at time $t+1$ and t, respectively. Therefore, the dimension of a metapopulation matrix would be equal to the number of patch-types times the number of life-stages. If it incorporates regeneration dynamics, it has entries in all submatrices for which valid transitions among patch-types were defined in the Markov matrix D. Along the main diagonal of matrix G are patch-specific, population-dynamics submatrices. A metapopulation matrix with zeros in all but these submatrices would model environmental heterogeneity but no regeneration or successional dynamics. Finally, seed dispersal among patch-types has been explicitly considered (3, 5).

A two-patch model to simulate *A. mexicanum* metapopulation dynamics, at Los Tuxtlas, showed that the chance of λ being smaller than 1 only increased with significant increases in gap size and fast forest turnover rates (108). This is explained by the fact that only in large gaps are the adults' survival rates reduced, and in this species, λ is most sensitive to changes in this transition (see Table 2). In contrast, for the pioneer tree *C. obtusifolia*, colonization but not extinction was affected by large-gap formation in the same forest (3). The λ estimated with this model was significantly larger than that obtained when the environment was assumed to be homogeneous and constant (see Table 2), because in stable-stage conditions, the area occupied by building patches (where mature *C. obtusifolia* grow) would be greater than that found in present conditions. Analytical and simulation analyses of this model have shown that the response of λ to changes in different entries of the patch-dynamics matrix (d^{kl}'s above) are not intuitively obvious even in a species with such a tight dependence on gaps for regeneration as this one (3). Such nonintuitive results highlight the importance of formal models to guide analyses of the effects of different disturbance regimes on TRF tree species dynamics.

Similar generalized Lefkovitch models with immigration were used to predict seed densities in different patch-types and to quantify the relative contribution of seeds from different sources to seedling recruitment (5). For constant seed rains, an analytical solution of the equilibrium number of seeds per patch-type was provided. For seed rain modeled as a function of forest patch structure, simulation results were presented. These models were validated because the equilibrium seed densities per patch-type they predicted were very similar to field estimates independently obtained from soil seed samples (5, 7). More than 90% of the yearly recruited seedlings in a 5-ha plot originated from seeds < 1 year-old (seed-rain). Analytical solutions of both models (ER Alvarez-Buylla, R García-Barrios, in preparation) yield general results and these models may be used to estimate seed densities per patch-type and number of years to attain equilibrium densities for any species.

Besides being prone to patch dynamics within natural forests, TRF tree species are also expected to have metapopulation structures at the regional level that result from the fragmentation of the forest, caused, for example, by human deforestation. Metapopulation models developed for TRF tree species have only considered gap dynamics (5), corresponding to metapopulation models of the second type described above. This is a shortcoming if gap dynamics in a particular fragment depends on regional metapopulation dynamics.

Nonlinear Models: Density-Dependence

GENERAL THEORY Density-independent factors may cause fluctuations in population size, but they are not related to the population's density or growth rate, which determine the pattern of temporal fluctuations of population size in density-dependent models. These models consider demographic rates to be functions of density, instead of constant, and in them an equilibrium population density (K) is attained (21). Accurate empirical estimates of these models' parameters are particularly important because, depending on the values of these parameters, complex dynamics may be predicted (109). Despite this fact, and that density-dependent resource limitation is expected even in some rare species (24), rather few examples of empirically parameterized density-dependent models are available for plants (145). Density-dependence is also age and/or size-dependent (156), and including these factors in matrix models makes analytical solutions difficult to derive.

APPLICATIONS TO TRF TREE SPECIES Several empirical studies have evaluated the negative effect of density on seed and seedling survival (7, 35, 142, 143), on juveniles's growth and survival (37), and on adults' growth, survival, and fecundity (49, 55, 74) for several TRF tree species. However, overall population dynamics models may be developed for only a few of these species for which complete life-cycle information is available (3, 37, 49, 74).

Some authors have suggested that density-dependent effects are significant only among high-density TRF tree species (37, 74). Data collected for tree species at Barro Colorado Island tropical rain forest suggest that very few tree species have density values at or near carrying capacities, where density-dependent regulation could be important. Density-dependent effects have been detected only in the abundant canopy tree *Trichilia tuberculata* (37). These results suggest that most tree populations fluctuate due to density-independent factors. Detailed field and model analyses for *C. obtusifolia* and *A. mexicanum*, at Los Tuxtlas, suggest that density-dependent factors may be important in the population dynamics of these species (3; M Martínez-Ramos, in preparation). Both are among the most abundant of their type (see Table 1), but the first has a very patchy distribution, and density effects were evaluated locally (37, 74).

Field evidence of density effects on survivorship and fecundity of females > 20 cm diameter were incorporated into *C. obtusifolia*'s matrix models with and without patch dynamics. A negative exponential function and an increasing linear function were adjusted for fecundity and mortality, respectively. This model was analyzed with simulations. Without patch dynamics, a stable population was reached, and numbers of all life-stages leveled off at a total population size (K) very similar to that observed in the field. This model predicted a population structure that was not statistically different to that in the field. With patch dynamics and seed dispersal, the number of individuals per life-stage and per patch-type attained equilibrium values more than twice those in the field. This suggests that gap availability varies with time (i.e. in some years no adequate gaps are available for this species, and thus population size fluctuates below K) (3).

The inability of seed predators to detect seeds when these are not clumped determines that in *A. mexicanum* the probability of seed removal increases with seed density and the transition probability from seed to seedling stage decreases (139; J. Sarukhán et al unpublished). Also, juvenile survivorship decreases as juvenile density increases; no other density-dependent effects have been detected at other life-stages (49, 108). These effects were introduced into a Lefkovitch matrix by making relevant matrix entries linear functions of density. The model predicted a value of K six times larger than the observed density in the forest (M Martínez Ramos, unpublished data). According to these results, the present population at Los Tuxtlas should be increasing and long-term data have actually shown population increase (J Sarukhán and collaborators; 49).

Perhaps severe adult mortality events caused by climatological events (hurricanes) keep the population below its carrying capacity and growing (i.e. $\lambda > 1$).

Relative Effects of Different Factors

When one projects the population dynamics of a species with a particular model that has been parameterized with field data, it is difficult to know if it incorporates the key factors affecting the population. This is the approach that most tropical and nontropical applications of population models have followed (but see 22 and 156). An alternative approach is to map the contrasting results of population growth rate (in density-independent models), total size (carrying capacity population size in density-dependent models), and population structure, obtained when applying different models, to the assumptions of the models (3). This approach was used to quantify the isolated and combined effect of two factors affecting population change and structure in TRF: density-dependent demographic rates and density-independent environmental change caused by patch dynamics (3).

This approach was used for analyses of four models discussed in the preceding sections for *C. obtusifolia*. Model 1 was a simple Lefkovitch model that did not incorporate density-dependence or patch dynamics. Model 2 added only density dependence; model 3 added only patch dynamics. Finally, model 4 combined both. The dependence that this species has on gaps, as documented by extensive field data (7, 8), may be evaluated theoretically by comparing finite population growth rates and structures predicted by the linear models 1 and 3, or by comparing the population growth rates at particular times and the carrying capacities predicted by density-dependent models 2 and 4. We proceeded in a similar way to evaluate the role of density-dependence (i.e. we compared model 1 vs model 2, and model 3 vs model 4). While patch-dynamics significantly affects population growth rate and size, density-dependence regulates growth and significantly affects population structure (3).

This approach yielded general theoretical considerations as well. For example, for species that actually grow in heterogeneous environments, assuming a homogeneous environment in the population dynamics model may obscure or exaggerate the regulating effects of density-dependent factors (see also 65). For example, horizontal field studies may detect prolonged positive growth rates (i.e. those obtained in model 4 when $N < K$), and these may be erroneously interpreted as slow density-independent growth in a static environment. This might be the case of the observations made, for example, for the *A. mexicanum* populations (see above). This erroneous interpretation is due to a misspecification of the model (structural errors of the model) (cf 24), and it may yield an erroneous evaluation of the effect of density-dependent mechanisms with respect to density-independent ones on population dynamics (3).

When density-dependent mechanisms are detected (see e.g. in 74), their effect may be evaluated erroneously if patch dynamics are not considered. If patch dynamics affect population dynamics, the effect of density-dependence should be evaluated by comparing results of a model 4 with those yielded by a model 3. However, when patch dynamics are not considered, the comparison is actually between a type 2 model and a type 1 model (see e.g. 74). Erroneous evaluations of density-dependent effects due to misspecifications of the true model underlying the data may be quantified by comparing the areas (A_{12} and A_{34}) between the curves of the constant functions with respect to N (λ vs N/K) of growth rates of models 1 or 3, and the power growth rate functions with respect to N ([dN/t]/N vs N/K) of models 2 and 4, respectively. For *C. obtusifolia*, the effect of density-dependence would have been underestimated 13 times if models 1 and 2 had been compared, instead of models 3 and 4 ($A_{12}/A_{34} = 1/13$). In this case, the environmental change that results when applying model 3 (i.e. increase in gap area) in comparison to model 1, causes an increase in *C. obtusifolia*'s growth rate. This explains why, if environmental heterogeneity is not considered, the effect of density-dependence is underestimated. In contrast, if we modeled the dynamics of a shade-tolerant species (*Astrocaryum mexicanum*), the same environmental change predicted by our model 3 would cause a decrease in the species population growth rate. Likewise, if models 1 and 2 (no environmental heterogeneity) were compared instead of models 3 and 4, the effect of density-dependence would be overestimated. If environmental heterogeneity and density-dependence are suspected to affect a species population dynamics under natural conditions, a model that incorporates both of these factors should be used to project evolutionary dynamics and harvesting regimes (3).

Stochastic Models: Demographic and Environmental Factors

GENERAL THEORY All population models reviewed up to here assume deterministic factors and yield deterministic predictions of constant population growth rates and stable-stage distributions (linear models), or of population size at equilibrium (K) and population structure (nonlinear, density-dependent models). However, processes affecting population dynamics operate in a stochastic fashion and yield stochastic behavior of population growth rate, population size, and carrying capacity. Therefore, both demographic parameters and outputs of population dynamics should be treated as random variables, and deterministic models should be used only as heuristic tools. Furthermore, stochastic models may yield unexpected results because of large variances, nonlinearities, or skewed probability distributions, and only these types of models may be used to generate probability distributions of times to extinction (24).

There are three sources of stochasticity in demographic-ecological dynamics and the estimation of its parameters: sampling, demographic, and environmental. Estimates of population growth rates are subject to sampling and experimental errors because they are obtained from experimental and census data on the survival, growth, and fecundity rates (i.e. vital rates or life-table parameters) of a sample of individuals. For example, if the demographic rates are estimated repeatedly from different random samples of only 10 individuals, the survivorship and fecundity estimates will vary from sample to sample in a random fashion even if the population is of infinite size and has constant vital rates. Assignment of confidence limits to deterministic and probabilistic outputs of population models is rarely performed but are indispensable (9–11, 24, 29, 160). For example, in two of three studies that we reviewed, conclusions contrasting to those reached by the original papers were suggested when confidence limits were assigned to estimates of population growth rates (11).

Uncertainty due to sampling should be clearly distinguished from the effects of random variability of life-table parameters arising from the chance realization of individual probabilities of death and reproduction. In this case, vital rates will be random variables that will fluctuate independently of sampling errors and will affect the probability of extinction of the whole population (88). Uncorrelated fluctuations across individuals give rise to demographic stochasticity, whereas perfectly positively correlated fluctuations (within each age or stage class) generate environmental stochasticity. Hence, the variance of the mean growth rate under demographic stochasticity will decrease with population size but will remain constant under environmental stochasticity. Extinction may also be due to random catastrophies, defined as large environmental perturbations that produce sudden major reductions in population size without affecting vital rates other than mortalities (review in 24).

Environmental and demographic stochasticity may be thought of as extremes along a continuum of correlations among individual vital rates. This perspective has not received attention in the theoretical literature, perhaps due to the analytical difficulties it implies, but it may help clarify the underlying causes of different forms of stochasticity (149). Environmental stochasticity may be assumed to represent a limited number of small or moderate perturbations that similarly affect the vital rates of all individuals (88), while demographic stochasticity may correspond to the effect of a relatively large number of simultaneous, uncorrelated environmental perturbations, each one affecting a very small proportion of the population ("demographic accidents"). More interesting is that demographic stochasticity may also correspond to the statistical effect of the uncorrelated responses to a single environmental perturbation of a relatively large number of different phenotypes in the population. Hence, individuals' responses to fluctuating external factors (i.e. environmental stochasticity) are expected to be more correlated in populations with simpler age-structure, lower genetic variability, or more homogeneous microenvironments than in populations with greater among-individual heterogeneity.

Single-species stochastic models have been extensively reviewed elsewhere (see for example, 24, 163). Ecologically meaningful stochastic models are generally analytically intractable and are explored through simulation (24). However, Lande (88) has recently proposed a useful analytical approach to evaluate the relative effect of demographic and environmental stochasticity and random catastrophes on the expected time to extinction of a population without age structure and a constant per capita growth rate (r), except at the carrying capacity, K, where growth ceases. This approach also enables the evaluation of the relative role of genetic and demographic factors in population risks to extinction.

In Lande's model (88), with demographic stochasticity, average time to extinction increases in proportion to exp $[aK]/K$, where $a = 2r^*/V_i$, and r^* and V_i are, respectively, the mean Malthusian fitness and its variance among individuals. Under environmental stochasticity, the average extinction time is asymptotically proportional to K^c, where $c = 2r^*/V_e - 1$, and r^* and V_e are respectively the mean and environmental variance of r. Contrary to Goodman (56), expected time to extinction may scale greater than linearly with carrying capacity if $c > 1$, or equivalently, if $r^* > V_e$. Therefore, expected persistence time under environmental stochasticity may be extremely long, even for populations of modest size. However, it will increase as K increases more slowly than under demographic stochasticity, regardless of the constants of proportionality or the values of $\boldsymbol{a}$ and $\boldsymbol{c}$. Finally, under random catastrophes, average time to extinction is also a potential function of K that may scale greater than linearly if r greatly exceeds the catastrophe rate multiplied by the catastrophe size. The similarity of scaling laws for extinction risks under environmental stochasticity and random catastrophes makes intuitive sense because these constitute extreme manifestations of a fluctuating environment.

APPLICATIONS TO TRF TREE SPECIES Only one example of stochastic population modeling is available for four tropical rain forest tree species (see Table 2). Menges (111) based his simulations on Lefkovitch matrices to analyze impacts of environmental and demographic stochasticity on the average time to extinction. Environmental stochasticity was modeled by allowing population model matrix entries to vary independently from year to year. To model demographic stochasticity, the same matrix was used each year but appropriate transition probabilities were applied to each individual in the population. The results were similar to those predicted by Lande for non-structured populations. Environmental stochasticity had comparatively larger effects than demographic stochasticity on the extinction risks (i.e. larger MVP were necessary if environmental stochasticity was included). Also, larger environmental stochasticity slowed average population growth, produced greater fluctuations, reduced average time to extinction, and increased extinction probability. With highest λ, extinction never occurred unless environmentally induced variation in demographic parameters was high.

In the analysis of the effects of environmental stochasticity, some demographic stochasticity was allowed due to the uncorrelated variation in the vital rates of different plant-stages. Demographic stochasticity had minor effects relative to the deterministic case, and these effects were only felt at very low population sizes. However, it represented a threat to extinction for species with low population density and λ near 1, which includes the four TRF tree species considered. One important limitation of this study is that no correlations of environmentally produced fluctuations of matrix entries were allowed (i.e. they were assumed to be independent from each other). Menges found that increasing the number of independent vital rates buffered populations against random mortality or reproduction failure. To the extent that vital rates may be relatively uncorrelated in structured populations, as Menge's study assumed, this may represent a legitimate biological phenomenon. However, correlations are likely to occur in natural conditions to different degrees depending on the biology of the species (111). Effects of such correlations on the variability of λ may be investigated by means of simulations and their comparison to an analytical approximation, originally developed to study the effect of different correlation matrices of error estimates of matrix entries for a wide range of population life-history structures (9–11). The main result is that the variance of λ is a monotonically increasing function of correlation. The role of stochasticity in TRF tree species with different life-histories and including density-dependence and patch dynamics remains to be explored.

Applications of Demographic Models to Conservation

Some of the demographic models developed for TRF have made explicit management and conservation considerations (3, 5, 128, 134, 136, 165). Most are straightforward extensions of the linear Lefkovitch model (94, 167). In a deterministic demographic model of sustainable management, the objective would be to find the harvesting conditions that maintained λ equal to or greater than one. The sensitivities and elasticities of λ may be used directly to make inferences about the consequences of altering different matrix entries (145). But harvesting regimes may also be simulated by explicitly testing the effect of removing different numbers of individuals of each stage-category (see review in 104, 134, 136), or more elaborate harvesting regimes (130). Analyses of these simple models suggest that long-lived species are very sensitive to adult harvesting, but seeds, seedlings, and juveniles or parts of adults may be extracted in considerable numbers without altering their populations. In contrast, short-lived species (ca. <50 yr) could be heavily logged if recruitment of young individuals could be ensured. These models may provide useful guidelines, but we must keep in mind that they have unrealistic assumptions.

Environmental heterogeneity, density-dependence, and stochasticity likely mediate the effect of different harvesting regimes. In the case of linear patch dynamics models, sensitivity and elasticity analyses could also be used to infer effects of harvesting regimes. Such analyses suggest that forest patches that have not been disturbed during the last 35 yr should be set aside, and management should be performed in building patches (less than 35 years since the last disturbance) (5); contrasting results are obtained for long-lived species (104). Harvesting regimes assuming nonlinearities have been analyzed only with simulations (3). Results vary depending on whether harvesting and gap formation are coupled. In the former case, *C. obtusifolia* attains maximum carrying capacity under moderate levels of adult removal. Stochastic patch dynamics has been considered also, and by linking it to replacement probabilities of tree species with different light requirements (T Vázquez, ER Alvarez-Buylla, M Martínez-Ramos, in preparation), the effect of different disturbance regimes on the relative abundance of species has been investigated. Finally, explicit consideration of the effect of varying spatial fragmentation showed that deforestation of a wide area would decrease the potential for regeneration significantly more than if the same deforestation rate (total area opened per year) was performed by opening areas encapsuled in a matrix of undisturbed forest (5).

GENETIC MODELS

Conservation genetics estimates minimum viable populations by combining models of population genetics with estimates of rates at which different mutations arise (Table 3 and references therein). Until the early 1990s, estimates emphasized only two types of genetic effects on extinction risks of small populations: inbreeding depression (1-MVP in Table 3), caused mainly by segregation of partially recessive lethal alleles, and the loss of potentially adaptive variation in quantitative characters due to genetic drift (1-MVP and 3-MVP in Table 3) (reviews in 14, 47, 53, 70, 86). More recent models incorporate the effect of new mildly detrimental mutations that accumulate and might become fixed by random genetic drift and gradually decreasing fitness (4-MVP) (89, 90, 98–100). These models suggest that this factor may pose a serious risk of extinction in small populations.

MVP estimates summarized in Table 3, have been obtained only as reference values, rather than as definite recommendations (91), and they guide our discussion on the conservation implications of the estimators that have been published for TRF tree species. The first three types of MVP estimates

Table 3 Estimates of Minimum Viable Population (MVP) considering only one or various combinations of three types of genetic factors: segregation of recessive (partially) deleterious alleles, loss of adaptive variation and fixation of deleterious mutations by genetic drift.

MVP (Ne)	Genetic effect	Evolutionary forces	Type of genetic variation	Reference
1–50	Segregation	inbreeding	deleterious, recessive alleles (mainly dominance)	(159)
2–500	Loss of variation	mutation, drift	quantitative; all mutations	(91, 159)
2–5000	Loss of variation	mutation, drift	quantitative; quasineutral mutations only; ca. 10% of all mutations	(90)
3–1414–2000*	Loss of variation	mutation, drift, stabilizing selection	quantitative; all mutations	(89)
3–4472–10,000*	Loss of variation	mutation, drift, stabilizing selection	quantitative; quasineutral mutations only; ca. 10% of all mutations	(89)
4–100–1000's**	Accumulation & fixation of mutations	mutation, drift, selection§	mildly detrimental mutations with additive effects	(89, 90, 98)
4–100***	Accumulation & fixation of mutations & segregation	mutation, drift, selection, inbreeding§	mildly detrimental mutations with additive effects plus deleterious mutations with any degrree of dominance	(99)

Arabic number (1–4) preceding MVP indicates criteria to determine MVP as follows: 1-the effective population size (Ne) with a 1% increase per generation in inbreeding coefficeint, 2-Ne in which generation of genetic variation by mutation equals loss of variation by genetic drift, 3-Ne that enables maintenance of 80% or more of the variation with respect to an infinite population, or 4-Ne for which mean time to extinction is in the order of 100 generations. Note that these type-4 etimates are the only ones generated from stochastic population dynamic models that yield probability distributions of times to extinction and actually enable strict population viability analysis. See text for more details.

*Low values for house-of-cards model and high value for gaussian model; these are appropriate for loci with strong effects and low mutation rates (few alleles/locus) and for loci with small effects and high mutation rate (many alleles), respectively (see Ref. 90)

§ Constant growth rate until maximum $N = K$, discrete and non-overlapping generations.

**Low values for constant selection coefficients (s) against mildly deleterious mutations (standard deviation/mean = c = 0); high limit for c = 1, that indicate the existence of variance in s.

***Constant selection coefficients (c = 0), see** for futher explanations.

are based only on population genetic considerations assuming populations at equilibrium. Estimates of effective population size, inbreeding coefficients, genetic variation and structure are useful to evaluate the role of inbreeding depression and loss of potentially adaptive variation in natural populations stressed in 1-3 MVP estimates. We review data on these for TRF trees. All studies have used electrophoretic isozyme loci. DNA-level studies have recently started to appear (66, 162), and these and the analytical tools being developed are promising (113, 151, 153, 154). Studies on genetic variation of quantitative traits are important, but practically absent from the literature of TRF trees (see preliminary results in 84).

Effective Population Size and Genetic Drift

GENERAL THEORY Genetic drift refers to chance fluctuations in allele frequencies due to random sampling among gametes (172). The relative importance of genetic drift compared to other evolutionary forces can be assessed by means of the effective population size (N_e). N_e is defined as the size of an idealized population that would have the same amount of inbreeding (inbreeding effective size) or of random gene frequency drift (variance effective size) as the population under consideration; these two quantities are not necessarily equal, and they differ especially when population size varies (85). The rate of loss of genetic diversity due to genetic drift will be higher in populations with smaller effective sizes, whereas in large populations its effects are negligible. If population numbers are rapidly recovered after a reduction in size, the effects of drift will be smaller than if populations were constantly kept small (14).

In most cases, population size (N = number of reproductive adults) are different from N_e due mainly to fluctuations in population size, high variance in reproductive success, sex ratios different from one, and overlapping generations (38). The N_e/N ratio has been found to vary between 0.25 and 1 for animal species (127). Plants are likely to have low N_e/N ratios because of sessile condition, restricted pollen and seed dispersal, and partial or complete selfing (110). Estimates of N_e assume that a discrete population can be identified. However, it is sometimes difficult to delimit objectively a discrete group of individuals in natural conditions; in this case, information on the size of populations is provided by estimates of neighborhood size (N_b), defined as a group equivalent to a panmictic unit within a continuous distribution of individuals (38). Neighborhood size is related to deme population density by the relationship: $N_b = Ad$, where A is the neighborhood area and d the density of breeding individuals (38).

APPLICATIONS TO TRF TREE SPECIES Unfortunately, few estimates of N_e or N_b exist for plant species in general, and they have been obtained for only five

TRF tree species. N_e estimates vary greatly depending on method as shown in data for only one species (45). Estimates of N_b assume a stepping-stone population structure model and are based on the effective number of immigrants per generation (Nm), obtained using Wright's F_{ST} statistic (152). N_b values are: for *A. mexicanum*, 27.84 (many times smaller than direct estimates, 560 and 187) (45); for the canopy tree *Cordia alliodora*, 15–76 (23); for *C. obtusifolia*, 19.4–87.4 (C Lara-Moreno, A Garay, ER Alvarez-Buylla, in preparation); for the lower canopy trees *Psychotria faxlucen*, 70 (45); and for *Combretum fruticosum*, 6 (45). We obtained N_b estimates for another 14 TRF tree species (data from 60): values varied between 15.9 for *Acalypha diversifolia* and 69.8 for *Gustavia superba* and *Quararibea asterolepis*, with a mode around 25.

In an attempt to relate N_e to loss of genetic variation, Hamrick & Murawski (62, also see 117) compared 16 common tropical tree species with 13 uncommon species and showed that less genetic diversity is maintained in less dense populations. They argued that this was due to low effective population sizes of the less common species. However, estimates of N_e were not obtained, and it is not clear what the relationship of the density estimates would be to the species' census and effective sizes. If the population distribution is uniform, then population densities will be partial indicators of neighborhood size.

Breeding and Mating Systems, Inbreeding Depression, and Heterosis

GENERAL THEORY Mating systems determine the mode of transmission of genes from one generation to the next and are thus important factors affecting levels of genotypic variation and its distribution within and among populations. Genetic markers can be used to obtain quantitative estimations of the mating system and of biparental inbreeding of tree species by means of the single (t_s) and multiple (t_m) outcrossing rates (see review in 27, 138). The outcrossing rate can affect the degree of isolation between demes in continuous populations through its effect on the neighborhood area: Low outcrossing rates reduce neighborhood areas, and a maximum neighborhood area is achieved at $t = 1$ (38).

Theoretical and empirical reviews of the relationship between mating systems and inbreeding depression (the reduced fitness of inbred offspring compared with outcrossed offspring) are numerous (32, 83, 93, 141). This is an important issue to consider in conservation genetics because small populations generally undergo inbreeding. High inbreeding levels, however, do not necessarily result in inbreeding depression; levels of inbreeding depression depend mainly on the underlying genetic mechanism and the previous breeding history of the species (14, 32). If the detrimental effects of inbreeding are caused by lethal or deleterious recessive alleles, after prolonged and intense inbreeding most of the genetic load will be purged and inbreeding depression will disappear. Purging, however, is not always feasible (14, see also 90 for other causes that impede purging). If inbreeding depression is caused by overdominant loci, it may increase with the selfing rate. Only in the case of asymmetrical overdominance, inbreeding depression is eventually eliminated due to the loss of less fit alleles (14, 32).

Mechanisms underlying the relationship between heterozygosity and fitness have been the subject of intense research and are still a matter of debate (31, 71, 114). Many animal and plant species exhibit heterosis, in which more heterozygous individuals perform better than more homozygous ones. Two main hypotheses exist to explain this phenomenon: (*a*) overdominance, where heterozygosity per se confers an advantage, and (*b*) dominance, that proposes that more homozygous inbreds are merely expressing a higher proportion of deleterious recessive alleles (57). Smouse (155) proposed a model (the adaptive distance model) to distinguish between these two hypotheses, but similar results can be obtained with either genetic mechanism of heterosis under nonrandom mating (71).

APPLICATIONS TO TRF TREE SPECIES TRFs have high levels of dioecious and self-incompatible, hermaphroditic species, which led to the idea that tropical trees are predominantly outcrossed (15, 16, 17a, 18, 19, 23). Quantitative estimates of the amount of outcrossing further showed that most tropical tree species [studied by means of Ritland & Jain's (138) multilocus mixed-mating method] have high outcrossing rates (t_m) (23, 46, 61, 129, 130; reviews in 58, 61, 97). In some species, however, inbreeding was not negligible (129, 130). Genus-wide (*Acacia*) surveys have revealed variation in outcrossing rates (0.62–0.97) among species (116).

Lowest levels of outcrossing were found for pioneers *Cavanillesia platanifolia*, a canopy tree with hermaphroditic flowers, ($t_m = 0.57$ and 0.35) and *Ceiba pentandra*, ($t_m = 0.689$). These values suggested that early successional, colonizing tropical trees have highly plastic mating systems that ensure seed production when a single individual colonizes a large gap (58). However, the dioecious pioneer *C. obtusifolia* did not show evidence of biparental inbreeding (4) with $t_m = 0.974$ (SE = 0.024). The outcrossing rate in TRF trees was also found to be strongly influenced by nongenetic factors. Trees with lower population densities had lower t_m values than did more abundant species (120), and population reductions in density due to logging were associated with a significantly higher proportion of seeds produced through selfing (118).

Few and scattered data exist on the harmful consequences of inbreeding for TRF trees. In *Acacia mearnsii* and *A. decurrens*, the average height of two experimental groups of selfed progeny was 26% and 15%, respectively, less than open-pollinated controls (115, cited in 57). Inbreeding depression was also evidenced in *A. mangium,* introduced from Australia into Malaysia from a single tree, and in *Hevea brasiliensis* (57).

Heterosis has been documented for only a few tropical species, although it seems to be a widespread phenomenon in forest trees (114). Decreasing fixation indices exist along life-cycles in *A. mexicanum, C. obtusifolia, Shorea megistophylla,* and *Cavanillesia platanifolia* (3a, 46, 118). In *A. mexicanum,* a positive significant correlation was found between the adults' heterozygosity and trunk growth (46). Smouse's (155) adaptive distance model applied to *C. obtusifolia* suggested that overdominance was the most likely cause of heterosis; however, associative overdominance cannot be completely ruled out as the underlying mechanism (E Alvarez-Buylla, C Lara-Moreno, AA Garay, unpublished information).

Genetic Variation, Population Genetic Structure, and Gene Flow

GENERAL THEORY The most widely used parameters to quantify genetic variation are the proportion of polymorphic loci within species (P_s) and within populations (P_p), and the overall genetic diversity within species (H_s) and within populations (H_p) (59). The distribution of this variation within and between populations can be found by means of Wright's F statistics or Nei's G_{ST}. However, the scale at which the genetic structure of populations is considered is important; patterns may differ depending on the relative importance of factors (mating systems, selection, population size, and pollen and seed dispersal distances) that affect the genetic structure of populations at each spatial scale. Spatial autocorrelation models can be used to assess the fine-scale genetic structure of populations (50, 150). Another method, proposed by Hamrick et al (63), determines the mean number of alleles in common (NAC) for individual pairs at increasing distances.

Theory predicts that in subdivided populations, random genetic drift will result in genetic differentiation among subpopulations (172). Gene flow between these subpopulations, however, will set the limit to how much genetic divergence can occur; reduced gene flow between populations will increase the effects of genetic drift. Several models take into account the effects of gene flow under several assumptions of population structure (152; review in 150). For example, under an island model and at drift-mutation equilibrium, a value of Nm greater than 1 will be required to prevent genetic divergence of subpopulations resulting from genetic drift (171). Allendorf (1) further suggests for management purposes that an average exchange rate of exactly one reproductively successful migrant among demes per generation ($Nm = 1$) should be maintained to avoid genetic drift and still enable local adaptation. In two-dimensional stepping-stone models, drift predominates over other forces if Nm

is much smaller than 1; if it is greater than 4, subpopulations behave as a single panmictic unit (150).

Several indirect methods yield estimates of gene flow (Nm or N_b) (150, 152). In a subdivided population at demographic equilibrium, methods based on Wright's F_{ST} statistic [$Nm = (1 - F_{ST})/(4F_{ST})$] for an infinite population number and an island model provide the most accurate and practical estimates of Nm or N_b (152). Direct methods for estimating gene flow depend on observations of dispersing individuals or gametes; these measure current gene flow rather than average levels of gene flow obtained with indirect methods (see review 150).

APPLICATIONS TO TRF TREE SPECIES The proportion of polymorphic loci within species (P_s) for tropical woody species was found to be high (mean $P_s = 50.6\%$), as well as levels of genetic diversity within species (mean H_s = 0.160). Greater proportions of variation were found to be maintained within rather than among populations for tropical species (mean G_{ST}-0.135) (58, 59). The most important determinants of genetic diversity among tropical tree species are the density of populations (62) and geographic distribution (97). The proportions of variation maintained among populations were very low for 10 TRF tree species (average $G_{ST} = 0.05$) (59). *C. obtusifolia* and *A. mexicanum* had also very low levels of among population variation ($F_{ST} = 0.029$ and 0.040 respectively) (4, 46).

Microspatial genetic structure has been documented for a few species. For several TRF trees, Hamrick et al (63) found that near neighbors have more alleles in common than more distant ones; however, family structure disappeared in older life-stages. These patterns were related to the seed dispersal syndromes because NAC values were greater for species with limited seed dispersal. Spatial autocorrelation statistics showed that *C. obtusifolia* has a marked genetic substructure among seedlings within canopy gaps, perhaps caused by limited or correlated seed dispersal; this structure was less evident for older life-stages but was maintained up to the adult stage (B Epperson, ER Alvarez-Buylla, in preparation). Patterns of local genetic structure for this species were further supported by F_{ST} estimates which showed that between-patch (local scale) genetic differentiation was high, suggesting that gap dynamics has significant effects on this species' genetic microspatial structure (3a). The local scale pattern for this species strongly contrasts with macrogeographical patterns (up to 130 km) that did not show significant genetic structuring ($F_{ST} = 0.029$) (4).

TRF trees are pollinated predominantly by animals rather than by wind. However, animal and pollen vectors do involve a wide range of sizes, foraging strategies, and potential for flying long distances. Seed dispersal by animals is also common, although wind or gravity dispersal occur more often (61). Hamrick & Loveless (60) found that the potential for gene movement and

G_{ST} values were highly correlated. These same authors provide Nm estimates based on F_{ST} that ranged from 2.53 for *A. diversifolia* to 11.11 for *Quararibea asterolepis* and *Gustavia superha*. Another method to study gene flow that has been applied to tropical trees is paternity analysis (41). General results show that a great potential exists for long-distance pollen dispersal (review in 126). This and other similar results, obtained by other methods (see above), led Hamrick & Murawski (61) to suggest that effective breeding units for common tropical tree species may be on the order of 25 to 50 ha (126). It is important to note from these results that pollen dispersal may sometimes be restricted. Using paternity analyses Boshier et al (23) found evidence of an increase in localized matings among genetically correlated individuals for *Cordia alliodora*, and S Kaufman, P Smouse, and ER Alvarez-Buylla (in preparation) found that *C. obtusifolia* exhibited a pattern of isolation by distance at a local scale; a greater percentage of matings occurred between near neighbors.

Genetic Effects of Fragmentation and Applications to Conservation

It is a subject of debate in conservation biology whether to keep a single large population or many small populations of total equal size (SLOSS) (149). Which of these strategies is more appropriate for conservation of a particular species will depend on the species biology (breeding structure, mating system, genetic structure, etc).

If populations are subdivided and gene flow restricted, genetic drift may cause genetic differentiation. Small population size (a result of habitat fragmentation) may also increase the number of matings between relatives, thus increasing levels of inbreeding. This may be particularly important in obligate and predominantly outcrossing species, due to the possibility of inbreeding depression. Habitat fragmentation may also cause disruption of pollen and seed vectors, making isolated populations more vulnerable to drift and inbreeding effects. The effects of population subdivision need not result from habitat fragmentation. Isolation will occur if seed or pollen are dispersed over short distances, causing neighborhood areas to be small (e.g. 38). Other researchers have emphasized the beneficial consequences of population subdivision. Under certain circumstances, population subdivision may enhance population survival through the maintenance of genetic variation. Conservation strategies that incorporate this effect for animal species have been suggested (30, 34); however, these must take into consideration the particular biological characteristics of the managed species so as to avoid other possibly detrimental consequences (170a).

To decide what strategy is adequate for a given species, it is relevant to consider genetic data from continuous populations. The effects of fragmentation on population viability will depend on conditions prior to fragmentation (126). Outcrossing rates of tropical trees decrease with decreasing density. Outcrossing species should have high genetic loads, and if a sudden reduction in population size or density occurs, increased inbreeding may lead to inbreeding depression. Negative effects of inbreeding have been reported for some tropical tree species. Tropical tree species should then be particularly vulnerable to changes in density and population size.

DEMOGRAPHIC-GENETIC MODELS

General Theory

Recent theoretical developments that integrate explicit genetic factors to stochastic demographic models are providing new estimates of MVP and important insights concerning the relative role of genetic and demographic factors in determining risk to extinction of populations of different sizes (89, 90, 98, 99, see Table 3). Minimum viable population estimates that stress the accumulation of mutations (4-MVP estimates in Table 3) depend critically on the rate of spontaneous mutation, fitness effects of different mutations, and coefficients of variation of selection coefficients of mildly deleterious mutations (90, 98 and Table 3). All MVP estimates in Table 3 except the first one have used data from *Drosophila melanogaster*. There is much need of this type of estimates for other organisms, particularly for plants (see review in 99). Incorporation of inbreeding depression due to segregation of preexisting mutations does not yield a significant increase in estimates of MVP (see Table 3) (99). Both initial N_e and demographic parameters r and K affect the rate of fixation of deleterious mutations and the population decline to extinction after the population rate of increase has become negative (see also 88 discussed above). Epistatic interactions among mildly detrimental mutations and compensatory mutations at different loci, not incorporated in models, could increase mean times to extinction. However, variance in selection coefficients guarantees that the mean time to extinction is asymptotically proportional to a low power of N_e (90).

These demographic-genetic models (see review in 90) have enabled analyses of how the risk of eventual extinction from fixation of new mutations scales with population size, and how this compares with the risks from stochastic demographic factors. In contrast to previous expectations (87), the fixation of new mildly detrimental mutations may be comparable in importance to environmental stochasticity. Therefore, the effect of this genetic factor implies MVP as large as a few thousands. Both factors imply a power relationship between mean time to extinction and population size. In contrast, demographic stochasticity, inbreeding depression, and fixation of new mutations, assuming constant selection coefficients, determine a nearly exponential relationship between mean time to extinction and population size (see also 98, 99).

Perspectives of Applications to TRF Tree Species

It is clear that the effect of demographic and environmental factors will depend on the details of the life-history of the species being studied, while the effects of genetic factors are more general because of universal genetic laws. Traits of TRF tree species (overlapping generations, response to patch dynamics, population genetic structure at different scales, dioecy, and varying sex ratios) are likely to affect the results of demographic-genetic models. These models should be taken as a basis to analyze the sequential incorporation of these traits. Gene flow among subpopulations as another source of genetic variation is likely to affect the models' outcomes. Analyses will probably have to rely almost entirely on simulations or on transition matrix approaches such as those proposed by Lynch et al (99). This type of stochastic model will enable rigorous PVA for TRF tree species. Experimental plant systems should also be implemented to estimate some of these key parameters. The obvious candidate is *Arabidopsis thaliana*.

The few TRF tree species for which both demographic and genetic data have been obtained help us to illustrate the value of integrating both types of information. For example, the available demographic and genetic results obtained for seed banks of *C. obtusifolia*, show that soil–seed storage in this species has limited demographic relevance, but it is potentially important from an evolutionary standpoint (3, 4, 9). Soon after dispersal, most seeds of this species are eaten by ants; predation rates are density-dependent, however, and the few scattered seeds that remain may survive for long time periods (3, 4). Isozyme data suggest that rare alleles may arise or accumulate among these stored seeds and could constitute a pool of genetic variation not available in any of the other life-stages (9). The other example is that of *A. mexicanum*, for which the availability of demographic and genetic data enabled the first direct estimation of neighborhood size and effective population size in plant species (45).

PERSPECTIVES AND CONCLUSIONS

The deterministic demographic models reviewed here yield generalizations with relevant implications for designing conservation strategies for TRF tree species. Simple Lefkovitch matrix models that have been applied to TRF tree species and their extensions to simulate harvesting regimes show that the species' longevity is a key parameter. In long-lived species later life-cycle stages (preadult and adult) are the most important, while in short-lived species the regenerative phases (seeds, seedlings, and juveniles) are the most critical. For conservation purposes, therefore, the study of the critical life-stages of the species being considered should be emphasized. Metapopulation models suggest that long- and short-lived species will also have contrasting responses to different temporal and spatial regimes of perturbation. However, future developments of metapopulation models for TRF tree species should consider dynamics that result from regional fragmentation regimes caused by deforestation by humans and stochastic fluctuations of the patch-dynamics regime (24, 88). The possibility of aging and of locating past treefalls at Los Tuxtlas make this forest an ideal model study site to address questions on the effects of long-term gap dynamics on persistence of species with contrasting life histories (106). Finally, density-dependent factors seem to be important in regulating populations of at least the most abundant species. However, sporadic catastrophic events seem to maintain populations fluctuating in a density-independent fashion with densities below their carrying capacities.

Such catastrophic events are of a stochastic nature, and their impact on population long-term viability should be considered in more detail in future models. Only one demographic study has performed population viability analysis for four TRF tree species using the simplest linear Lefkovitch matrix model. Its results agree with those derived analytically for nonstructured populations. Hence, more elaborate stochastic analytical theory might guide further analyses of the effect of demographic and environmental stochasticity on determining extinction times of TRF tree species. We have emphasized that quantitative evaluations of the relative effects of the key factors affecting TRF tree populations should be based on comparisons of model outputs with contrasting assumptions, rather than on comparison of field data with outputs of single models. The obvious next step is to extend this approach by incorporating the effect of different types of stochasticity to perform more realistic viability analyses of TRF tree species.

The use of classical population genetic models has shown that (*a*) TRF tree species possess high levels of genetic variation; (*b*) most of their variation is found within rather than among populations; (*c*) such species are predominantly outcrossed; and (*d*) they have high levels of gene flow. These results and neighborhood size-estimates suggest that TRF tree species have large effective population sizes. Hence, genetic drift is not expected to play an important role in the evolutionary dynamics of the type of TRF tree species studied up to now (59). Therefore, under present conditions, genetic factors do not seem to be critical for the long-term persistence of TRF tree species. However, drastic population reductions due to fragmentation are likely to have strong impact because of the high genetic load kept by most TRF tree species. Also, the correlations between the genetic structure and the type of dispersers and pollinators support the idea that conservation plans of TRF trees should keep in mind the effects of fragmentation regimes on animal vectors (96, 126). Regional analyses of population structure also suggest that preserving a few populations would ensure the conservation of most genetic variation existing in these species (58). However, recent studies are revealing significant genetic structuring at microspatial scales that may result from the action of important ecological factors such as seed dispersal modes and gap dynamics. The latter results imply that several of such local populations should be preserved.

The specifications of the models and the values of the parameters used determine the MVP estimates reached, which vary from 50 to more than 1000 individuals. Therefore, details of the natural history of species and accurate estimates of parameters may be crucial for establishing realistic demographic-genetic models that provide useful MVPs for guiding conservation decisions (24, 149). However, time and resource limitations hinder long-term exhaustive studies of many species. Therefore, we suggest that detailed, rigorous, and standardized studies of model systems should be pursued. These will provide guidelines of key demographic-genetic factors and conservation priorities for different types of TRF tree species. TRF tree species may be arranged along a continuum of life history (i.e. demographic and genetic traits) types (8, 105). Species of similar types along such a continuum should share traits that may determine their population and evolutionary dynamics and their response to disturbance. We have chosen to study the extremes of the continuum: the pioneer *C. obtusifolia* and the long-lived palm *A. mexicanum*. The results reviewed here for these two species clearly show their contrasting demographic and genetic dynamics.

Present models do not enable evaluations of the relative importance of demographic and genetic factors in the long-term persistence of TRF species with contrasting life histories. Rigorous demographic-genetic models (89, 90, 99, 100) that incorporate population substructuring and gene flow should aid in resolving the effect of fragmentation on the long-term subsistence of TRF tree populations. A general conclusion from demographic-genetic models and empirical evidence that should apply to TRF tree species is that, in large populations that are suddenly reduced in size, inbreeding depression based on segregating detrimental mutations carried by the founders constitutes an important risk of rapid extinction (90, 159). More gradual population reduction will make fixation of new detrimental mutations a more serious risk of eventual extinction (90).

Most remaining natural TRF areas are very reduced in size and many of them are highly fragmented (168). This situation is likely to worsen due to the pressure that human populations are exerting on natural resources because of the unequal distribution of wealth and technologies among and within countries and social groups, because of market and institutional failures and the consequent inefficiencies in resource management (20, 54, 138a, 161). In this scenario it is not conceivable to think of long-term conservation of untouched, large, and continuous forested areas. Present conservation plans should be concerned with the demographic and genetic consequences of forest fragmentation and the design of plans of sustainable management for ecologically or economically key species.

Most TRF tree species are rare and large areas seem to be required to meet even the most modest MVP estimates (see Tables 1 and 3). Most demographic and genetic studies, however, have been performed for abundant species. It is urgent to have model systems of the group to which most TRF tree species belong: long-lived and rare, and/or long-lived and fragmented. From the demographic perspective, for example, it is important to establish density-dependent factors that may be important at a local scale and that may regulate these species' populations, of which many have clumped distributions. From the genetic perspective, for example, it would be important to document effective population sizes of this type of species with different degrees and patterns of fragmentation. Finally, however powerful the population approach may be, conservation and management strategies of natural complex systems should keep in mind the question of how independent the dynamics of single species populations are. The "equilibrium" and "non-equilibrium" hypotheses of community structure and dynamics constitute contrasting ecological scenarios (37, 79, 80) and suggest two different strategies and challenges concerning the conservation of TRF. In the equilibrium scenario, conservation programs directed to preserve particular species without consideration of other species may not be appropriate because we expect a stable system in equilibrium, in which complex biotic interactions have evolved for long periods of time and yielded an equilibrium in a coevolved complex of species. However, these same conditions could determine lower MVP, and species with fewer individuals could be preserved for long periods. In the non-equilibrium scenario, we may assume more independent dynamics and population models may be particularly appropriate. For such systems, the challenge would be to conserve dynamic systems that maintain, at any particular time, a high tree diversity; but the specific composition would change randomly as a result of stochastic loss of species and origin of new ones.

ACKNOWLEDGMENTS

Rigorous demographic work at Los Tuxtlas TRF Station (UNAM) was initiated by J Sarukhán, our teacher and friend. Because he has guided, inspired, and supported much of the work developed at Los Tuxtlas, we would like to dedicate this paper to him. Many thanks to M Franco, D Piñero, and F Bongers, who carefully read previous versions and made useful comments, to J Hamrick and M Lynch for sending hard-to-access or still unpublished papers, and to K Oyama, R Dirzo, and L Eguiarte for providing useful references. The help of A Cortez and R Tapia-López in many aspects, and particularly while preparing the final version of the manuscript, was indispensable; many thanks to both of them. EAB and MMR were supported by grants from the Mexican Council of Science and Technology (CONACYT) and the National Autonomous University of Mexico (UNAM).

Literature Cited

1. Allendorf FW. 1983. Isolation, gene flow, and genetic differentiation among populations. See Ref. 141a, pp. 51–65
2. Alvarez-Buylla ER. 1986. *Demografía v dinámica poblacional de Cecropia obtusifolia (Moraceae) en la selva de los Tuxtlas, México.* MSc thesis. UNAM, Mexico
3. Alvarez-Buylla ER. 1994. Density dependence and patch dynamics in tropical rain forests: matrix models and applications to a tree species. *Am. Nat.* 143:155–91

3a. Alvarez-Buylla ER, Chaos A, Piñero D, Garay AA. 1996. Evolutionary consequences of patch dynamics, seed dispersal, and seed banks: demographic genetics of *Cecropia obtusifolia*, a pioneer tropical tree species. *Evolution.* 50:1155–66

4. Alvarez-Buylla ER, Garay AA. 1994. Population genetic structure of *Cecropia obtusifolia* a tropical pioneer tree species. *Evolution* 48:437–53
5. Alvarez-Buylla ER, García-Barrios R. 1991. Seed and forest dynamics: a theoretical framework and an example from the neotropics. *Am. Nat.* 137:133–54
6. Alvarez-Buylla ER, García-Barrios R. 1993. Models of patch dynamics in tropical forests. *Trends Ecol. Evol.* 8(6):201–4
7. Alvarez-Buylla ER, Martínez-Ramos M. 1990. Seed bank versus seed rain in the regeneration of a tropical pioneer tree. *Oecologia* 84:314–25
8. Alvarez-Buylla ER, Martínez-Ramos M. 1992. Demography and allometry of *Cecropia obtusifolia*, a neotropical pioneer tree—evaluation of the climax-pioneer paradigm for tropical rain forests. *J. Ecol.* 80:275–90
9. Alvarez-Buylla ER, Slatkin M. 1991. Finding confidence limits on population growth rates. *Trends Ecol. Evol.* 6:221–24
10. Alvarez-Buylla ER, Slatkin M. 1993. Finding confidence limits on population growth rates: montecarlo test of a simple analytic method. *Oikos* 68:273–82
11. Alvarez-Buylla ER, Slatkin M. 1994. Finding confidence limits on population growth rates: three real examples revised. *Ecology* 72:852–63
12. Ashton PS. 1984. Biosystematics of tropical forest plants: a problem of rare species. In *Plant Biosystematics*, ed. WF Grant, pp. 495–518. New York: Academic
13. Augspurger CK, Kelly CK. 1984. Pathogen mortality of tropical tree seedlings: experimental studies of effects of dispersal distance, seedling density and light conditions. *Oecologia* 61:211–17
14. Barrett SCH, Kohn JR. 1991. Genetic and evolutionary consequences of small population size in plants: implications for conservation. See Ref. 51a, pp. 3–30
15. Bawa KS. 1974. Breeding systems of

three species of a lowland tropical community. *Evolution* 28:85–92
16. Bawa KS. 1979. Breeding systems of trees in a tropical wet forest. *NZ J. Bot.* 17:521–24
17. Bawa KS, Ashton PS. 1991. Conservation of rare trees in tropical rain forests: a genetic perspective. See Ref. 51a, pp. 62–71
17a. Bawa KS, Hadley M, eds. 1990. *Reproductive Ecology of Tropical Forest Plants*. Paris: UNESCO
18. Bawa KS, Krugman SL. 1991. Reproductive biology and genetics of tropical trees in relation to conservation and management. In *Rain Forest Regeneration and Management*, ed. A Gómez-Pompa, TC Whitmore, M Hadley, pp. 119–36. Paris: UNESCO
19. Bawa KS, Opler PA. 1975. Dioecism in tropical forest trees. *Evolution* 29:167–79
20. Bawmol WJ, Oates WE. 1988. *The Theory of Environmental Policy*. New York: Cambridge Univ. Press. 2nd ed.
21. Begon M, Harper JL, Townsend CR. 1986. *Ecology: Individuals, Populations and Communities*. Oxford: Blackwell Sci.
22. Bierzychudek P. 1982. The demography of jack-in-the-pulpit, a forest perennial that changes sex. *Ecol. Monogr.* 52:335–51
23. Boshier DH, Chase MR, Bawa KS. 1995. Population genetics of *Cordia alliodora* (Boraginaceae), a neotropical tree. 2. Mating system. *Am. J. Bot.* 82:476–83
24. Boyce MS. 1992. Population viability analysis. *Annu. Rev. Ecol. Syst.* 23:481–506
25. Brokaw NVL. 1985. Treefalls, regrowth and community structure in tropical forests. In *Natural Disturbance: The Patch Dynamic Perspective,* ed. STA Pickett, PS White, pp. 53–59. New York: Academic
26. Brokaw NVL. 1987. Gap phase regeneration of three pioneer tree species in a tropical forest. *J. Ecol.* 75:9–19
27. Brown AHD. 1990. Genetic characterization of plant mating systems. In *Plant Population Genetics, Breeding, and Genetic Resources,* ed. AHD Brown, MT Clegg, AL Kahler, BS Weir, pp. 145–63. Sunderland: Sinauer
28. Bullock SH. 1980. Demography of an under-growth palm in littoral Cameroon. *Biotropica* 12:247–55
29. Caswell H. 1989. *Matrix Population Models*. Sunderland, MA: Sinauer. 328 pp.
30. Chambers SM. 1995. Spatial structure, genetic variation, and the neighborhood adjustment to effective population size. *Conserv. Biol.* 9:1312–15
31. Charlesworth D. 1991. The apparent selection on neutral marker loci in partially inbreeding populations. *Genet. Res.* 57:159–75
32. Charlesworth D, Charlesworth B. 1987. Inbreeding depression and its evolutionary consequences. *Annu. Rev. Ecol. Syst.* 18:237–68
33. Chazdon R. 1988. Sunflecks and their importance to forest understory plants. *Adv. Ecol. Res.* 18:1–63
34. Chesser RK. 1983. Isolation by distance: relationship to the management of genetic resources. See Ref. 141a, pp. 66–77
35. Clark DA, Clark DB. 1984. Spacing dynamics of a tropical rain forest tree: evaluation of the Janzen-Connell model. *Am. Nat.* 142:769–88
36. Clark DB, Clark DA. 1991. The impact of physical damage on the seedling mortality of a neotropical rain forest. *Oikos* 55:225–30
37. Condit R, Hubbell SP, Foster RB. 1992. Recruitment near conspecific adults and the maintenance of tree and shrub diversity in a neotropical forest. *Am. Nat.* 140:261–86
38. Crawford TJ. 1984. What is a population? In *Evolutionary Ecology,* ed. B Shorrocks, pp. 135–74. Oxford: Blackwell Sci.
39. De Kroon H, Plaiser A, van Groenendael J, Caswell H. 1986. Elasticity: the relative contribution of demographic parameters to population growth rate. *Ecology* 67:1427–31
40. Denslow JS. 1987. Tropical rainforest gaps and tree species diversity. *Annu. Rev. Ecol. Syst.* 18:431–51
41. Devlin B, Ellstrand NC. 1988. Fractional paternity assignment: theoretical development and comparison to other methods. *Theor. Appl. Gen.* 76:369–80
42. Dirzo R, Domínguez C. 1986. Seed shadows, seed predation, and the advantages of dispersal. See Ref. 51, pp. 237–50
43. Durán R. 1992. *Variabilidad intraespecífica y dinámica poblacional de* Pseudophoenix sargentii. PhD thesis. UNAM, México
44. Eguiarte LE. 1990. *Genética de poblaciones de Astrocaryum mexicanum*

Liebm. en Los Tuxtlas, Veracruz. PhD thesis. UNAM, México
45. Eguiarte LE, Búrquez A, Rodríguez J, Martínez-Ramos M, Sarukhán J, et al. 1993. Direct and indirect estimates of neighborhood and effective population size in a tropical palm *Astrocaryum mexicanum. Evolution* 47:75–87
46. Eguiarte LE, Pérez-Nasser N, Piñero D. 1992. Genetic structure, outcrossing rate, and heterosis in *Astrocaryum mexicanum* (tropical palm): implications for evolution and conservation. *Heredity* 69:217–28
47. Ellstrand NC, Elam DR. 1993. Population genetic consequences of small population size: implications for plant conservation. *Annu. Rev. Ecol. Syst.* 24:217–42
48. Enrigt N, Ogden J. 1979. Applications of transition matrix models in forest dynamics: *Araucaria* in Papua New Guinea and *Notophagus* in New Zealand. *Aust. J. Ecol.* 4:3–23
49. Enriquez A. 1991. *Variación especial en los patrones demográficos de una palma tropical.* BSc thesis. UNAM, México
50. Epperson BK. 1995. Fine-scale spatial structure: correlations for individual genotypes differ from those for local gene frequencies. *Evolution* 45:1022–26
51. Estrada A, Fleming TH, eds. 1986. *Frugivores and Seed Dispersal.* Dordrecht: Dr W Junk
51a. Falk DA, Holsinger KE, eds. 1991 *Genetics and Conservation of Rare Plants.* New York: Oxford Univ. Press. 283 pp.
52. Fiedler PL, Jain SK. 1992. *Conservation Biology: The Theory and Practice of Nature Conservation and Management.* New York/London: Chapman & Hall
52a. Franco M, Silvertown J. 1994. On tradeoffs, elasticities and the comparative method: a reply to Shea, Rees & Wood. *J. Ecol.* 82:958
53. Frankham R. 1995. Conservation genetics. *Annu. Rev. Genet.* 29:305–27
54. García-Barrios R. 1994. Biodiversity and market failure in Mexico. *Late Drummers* September Issue
55. Gilberts GS, Hubell SP, Foster RB. 1994. Density and distance-to-adult effects of a canker disease of trees in a moist tropical forest. *Oecologia* 98:100–10
56. Goodman D. 1987. Consideration of stochastic demography in the design and management of biological reserves. *Nat. Res. Model.* 1:205–34
57. Griffin AR. 1990. Effects of inbreeding on growth of forest trees and implications for management of seed supplies for plantation programmes. See Ref. 17a, pp. 355–74
58. Hamrick JL. 1994. Distribution of genetic diversity in tropical tree populations: implications for the conservation of genetic resources. In *Resolving Tropical Forest Resource Concerns Through Tree Improvement, Gene Conservation and Domestication of New Species,* ed. CC Lambeth, W Dvorak, pp. 74–82. Raleigh: NC State Univ. Press
59. Hamrick JL. 1994. Genetic diversity and conservation in tropical forests. *Proc. Int. Symp. Genetic Conservation Production of Tropical Forest Tree Seed,* ed. M Drysdale, SET John, AC Yapa, pp. 1–9. Asean-Canada Forest Tree Seed Center
60. Hamrick JL, Loveless MD. 1989. The genetic structure of tropical tree populations: associations with reproductive biology. In *Evolutionary Ecology of Plants,* ed. J Bock, YB Linhart, pp. 129–46. Boulder: Westview
61. Hamrick JL, Murawski DA. 1990. The breeding structure of tropical tree forest populations. *Plant Species Biol.* 5:157–65
62. Hamrick JL, Murawski DA. 1991. Levels of allozyme diversity in populations of uncommon neotropical tree species. *J. Trop. Ecol.* 7:395–99
63. Hamrick JL, Murawski DA, Nason JD. 1993. The influence of seed dispersal mechanisms on the genetic structure of tropical tree populations. *Vegetatio* 107/108:281–97
64. Hanski I. 1994. Patch-occupancy dynamics in fragmented landscapes. *Trends Ecol. Evol.* 9:131–35
65. Hanski I, Gilpin M. 1991. Metapopulation dynamics: brief history and conceptual domain. *Biol. J. Linn. Soc.* 42:3–16
66. Harris SA. 1995. Systematics and randomly amplified polymorphic DNA in the genus *Laucaena. Plant Syst. Evol.* 197:195–208
67. Hartshorn G. 1972. *The ecological life history and population dynamics of* Pentaclethra macroloba, *a tropical wet forest dominant and* Stryphnodendron excelsum, *an occasional associate.* PhD thesis. George Washington Univ., Washington DC
68. Hartshorn G. 1975. A matrix model of tree population dynamics. In *Tropical Ecological Systems, Trends in Terrestrial and Aquatic Research,* ed. FB Golley, E Medina, pp. 41–51. Berlin:

Springer-Verlag
69. Hastings A, Harrison S. 1994. Metapopulation dynamics and genetics. *Annu. Rev. Ecol. Syst.* 25:167–88
70. Hedrick PW, Miller PS. 1992. Conservation genetics: techniques and fundamentals. *Ecol. Appl.* 2:30–46
71. Houle D. 1994. Adaptative distance and the genetic basis of heterosis. *Evolution* 48:1410–17
72. Horvitz CC, Schemske DW. 1986. Seed dispersal and environmental heterogeneity in a neotropical herb: a model of population and patch dynamics. See Ref. 51, pp. 169–86
73. Deleted in proof
74. Hubbell SP, Condit R, Foster RB. 1990. Presence and absence of density dependence in a neotropical tree community. *Philos. Trans. R. Soc. London Ser. B* 330:269–82
75. Hubbell SP, Foster RB. 1983. Diversity of canopy trees in a neotropical forest and implications for conservation. In *Tropical Rain Forest Ecology and Management,* ed. TC Whitmore, AC Chadwick, pp. 25–41. Oxford: Blackwell Sci.
76. Hubbell SP, Foster RB. 1986. Canopy gaps and the dynamics of a neotropical forest. In *Plant Ecology,* ed. M Crawley, pp. 77–95. Oxford: Blackwell Sci.
77. Hubbell SP, Foster RB. 1986. Biology, chance and history and structure of tropical rain forest tree communities. In *Community Ecology,* ed. J Diamond, TJ Case, pp. 314–29. New York: Harper & Row
78. Hubbell SP, Foster RB. 1986. Commonness and rarity in a neotropical forest: implications for tropical tree conservations. See Ref. 157, pp. 205–23
79. Hubbell SP, Foster RB. 1987. The spatial context of regeneration in a neotropical forest. In *Colonization, Succession and Stability,* ed. PJ Edwards, pp. 395–412. Oxford: Blackwell Sci.
80. Hubbell SP, Foster RB. 1990. The fate of juvenile trees in a neotropical forest: implications for the natural maintenance of tropical tree diversity. See Ref. 17a, pp. 522–41
81. Hubbell SP, Foster RB. 1992. Short-term dynamics of a neotropical forest: Why ecological research matters to tropical conservation and management. *Oikos* 63:48–61
82. Husband BC, Barrett SCH. 1996. A metapopulation perspective in plant population biology. *J. Ecol.* In press
83. Jarne P, Charlesworth D. 1993. The evolution of the selfing rate in functionally hermaphrodite plants and animals. *Annu. Rev. Ecol. Syst.* 24:441–66
84. Kageyama PY. 1990. Genetic structure of tropical tree species of Brazil. See Ref. 17a, pp. 375–87
85. Kimura M, Crow JF. 1963. The measurement of effective population number. *Evolution* 17:279–88
86. Lacy RC. 1987. Loss of genetic diversity from managed populations: interacting effects of drift, mutation, immigration, selection and population subdivision. *Conserv. Biol.* 1:143–58
87. Lande R. 1988. Genetics and demography in biological conservation. *Science* 241:1455–60
88. Lande R. 1993. Risks of population extinction from demographic and environmental stochasticity and random catastrophes. *Am. Nat.* 142:911–27
89. Lande R. 1994. Risk of population extinction from fixation of new deleterious mutations. *Evolution* 48:1460–69
90. Lande R. 1995. Mutation and conservation. *Conserv. Biol.* 9:782–91
91. Lande R, Barrowclough GF. 1987. Effective population size, genetic variation, and their use in population management. See Ref. 158, pp. 87–123
92. Lande R, Orzack SH. 1988. Extinction dynamics of age-structured populations in a fluctuating environment. *Proc. Natl. Acad. Sci. USA* 85:7418–21
93. Lande R, Schemske DW. 1985. The evolution of self fertilization and inbreeding depression in plants. I. Genetic models. *Evolution* 39:24–40
94. Lefkovitch LP. 1967. A theoretical evaluation of population growth after removing individuals from some age groups. *Bull. Entomol. Res.* 57:437–45
95. Levins R. 1970. Extinction. In *Some Mathematical Questions in Biology. Lecture Notes on Mathematics in the Life Sciences,* ed. M Gerstenhaber, pp. 75–107. Providence, RI: Am. Math. Soc.
96. Loiselle BA, Sork VL, Nason JD, Graham C. 1996. Spatial genetic structure of a tropical understory shrub, *Psychotria officinalis* (Rubiaceae). *Am. J. Bot.* In press
97. Loveless MD. 1992. Isozyme variation in tropical trees: patterns of genetic organization. *New For.* 6:67–94
98. Lynch M, Conery J, Bürger R. 1995. Mutational meltdowns in sexual populations. *Evolution* 49:1067–80
99. Lynch M, Conery J, Bürger R. 1995. Mutation accumulation and the extinction of small populations. *Am. Nat.* 146:489–

518
100. Lynch M, Gabriel W. 1990. Mutation load and the survival of small populations. *Evolution* 44:1725–37
101. Mace GM, Lande R. 1991. Assessing extinction threats: toward a reevaluation of IUCN threatened species categories. *Conserv. Biol.* 5:148–57
102. Martínez-Ramos M. 1985. Claros, ciclos vitales de los árboles tropicales y regeneración natural de las selvas altas perennifolias. In *Investiaciones sobre la regeneracíon natural de las selvas altas perennifolias*, ed. A Gómez-Pompa, S del Amo, pp. 191–240. México: Alhambra
103. Martínez-Ramos M, Alvarez-Buylla ER. 1986. Gap dynamics, seed dispersal and tree recruitment at Los Tuxtlas, México. See Ref. 51, pp. 323–46
104. Martínez-Ramos M, Alvarez-Buylla ER. 1995. Ecología de Poblaciones de Plantas en una Selva Húmeda de México. *Bol. Soc. Bot. Méx.* 56:121–53
105. Martínez-Ramos M, Alvarez-Buylla ER, Sarukhán J. 1989. Tree demography and gap dynamics in a tropical rain forest. *Ecology* 70:555–58
106. Martínez-Ramos M, Alvarez-Buylla ER, Sarukhán J, Piñero D. 1988. Treefall age determination and gap dynamics in a tropical forest. *J. Ecol.* 76:700–16
107. Martínez-Ramos M, Samper C. 1996. Tree life history patterns and forest dynamics: a conceptual model for the study of plant demography in patchy environments. *J. Sustainable Forestry.*
108. Martínez-Ramos M, Sarukhán J, Piñero D. 1988. The demography of tropical trees in the context of gap dynamics: the case of *Astrocaryum mexicanum* at Los Tuxtlas tropical rain forest. In *Plant Population Ecology*, ed. AJ Davy, MJ Hutchings, AR Watkinson, pp. 293–313. Oxford: Blackwell Sci.
109. May RM. 1976. Simple mathematical models with very complicated dynamics. *Nature* 261:459–67
109a. Mendoza AE. 1994. *Demografía e integración clonal en Reinhardtia gracilis, una palma tropical.* PhD thesis. UNAM, Mexico
110. Menges ES. 1991. The application of minimum viable population theory to plants. See Ref. 51a, pp. 45–61
111. Menges E. 1992. Stochastic modeling of extinction in plant populations. In *Conservation Biology: the Theory and Practice of Nature Conservation and Management,* ed. PL Fiedler, SK Jain, pp. 253–76. New York: Chapman & Hall
112. Metz JAJ, Diekmann O. 1986. *The Dynamics of Physiologically Structured Populations*, New York: Springer-Verlag
113. Milligan BG, Leebens-Mack J, Strand E. 1994. Conservation genetics: beyond the maintenance of marker diversity. *Mol. Ecol.* 3:423–35
114. Mitton JB, Grant MC. 1984. Associations among protein heterozygosity, growth rate, and developmental homeostasis. *Annu. Rev. Ecol. Syst.* 15:479–99
115. Moffett AA, Nixon KM. 1974. The effects of self-fertilization on green wattle (*Acacia decurrens*) and black wattle (*Acacia mearnsii*). *Wattle Res. Inst. Rep.* 1973–1974:66–84
116. Moran GF, Muona O, Bell JC. 1989. Breeding systems and genetic diversity in *Acacia auriculiformis* and *Acacia crassicarpa. Biotropica* 21:250–56
117. Moran GF, Muona O, Bell JC. 1989. *Acacia mangium*: a tropical forest tree of the coastal lowlands with low genetic diversity. *Evolution* 43:231–35
118. Murawski DA, Gunatilleke IAUN, Bawa KS. 1994. The effects of selective logging on inbreeding in *Shorea megistophylla* (Dipterocarpaceae) from Sri Lanka. *Conserv. Biol.* 8:997–1002
119. Deleted in proof
120. Murawski DA, Hamrick JL. 1991. The effect of the density of flowering individuals on the mating systems of nine tropical tree species. *Heredity* 67:167–74
121. Myers N. 1980. *Conversion of Tropical Moist Forest.* Washington, DC: Natl. Acad. Sci.
122. Myers N. 1988. Tropical forests and their species. Going, going . . . ? In *Biodiversity,* ed. EO Wilson, FM Peter, pp. 28–35. Washington, DC: Natl. Acad. Sci.
123. Myers N. 1989. *Deforestation Rates in Tropical Countries and Their Climatic Implications.* London: Friends of the Earth
124. Myers N. 1990. The biological challenge: extended hot-spots analysis. *Environmentalist* 10:243–56
125. Deleted in proof
126. Nason JD, Aldrich PR, Hamrick JL. 1996. Dispersal and the dynamics of genetic structure in fragmented tropical tree populations. In *Tropical Forest Remnants,* ed. WF Laurance. In press
127. Nunney L, Campbell KA. 1993. As-

sessing minimum viable population size: demography meets population genetics. *Trends Ecol. Evol.* 8:234–43

128. Olmsted I, Alvarez-Buylla E. 1995. Sustainable harvesting of tropical trees: demography and matrix models of two palm species in Mexico. *Ecol. Appl.* 5:484–500
129. O'Malley DM, Bawa KS. 1987. Mating system of a tropical rain forest tree species. *Am. J. Bot.* 74:1143–49
130. O'Malley DM, Buckley DP, Prance GT, Bawa KS. 1988. Genetics of Brazil nut (*Bertholletia excelsa*: Lecythidaccae). 2. Mating system. *Theor. Appl. Genet.* 76:929–32
131. Oyama K. 1987. *Demografía y Dinámica poblacional en Chamaedorea tepejilote de Liebm. (Palmae) en la selva de los Tuxtlas, Ver, (México).* MSc thesis. UNAM, Mexico
132. Oyama K. 1993. Conservation biology of tropical trees: demographic and genetic considerations. *Environ. Update* 1:17–32
133. Palomeque R. 1996. *Demografía y herbivoría en* Omphalea oleifera (Euphorbiaceae). BSc thesis. UNAM, Mexico
134. Peters CM. 1991. Plant demography and the management of tropical forest resources: a case study from *Brosimum alicastrum* in Mexico. In *Rain Forest Regeneration and Management,* ed. A Gómez-Pompa, TC Whitmore, M Hadley, pp. 91–118. Paris: UNESCO
135. Peters CM, Gentry A, Mendelsohn R. 1989. Valuation of an Amazonian rainforest. *Nature* 339:655–56
136. Pinard M. 1993. Impacts of stem harvesting on populations of *Iriartea deltoide* (Palmae) in an extractive reserve in Acre, Brazil. *Biotropica* 25:2–14
137. Piñero D, Martínez-Ramos M, Sarukhán J. 1984. A population model of *Astrocaryum mexicanum* and a sensitivity analysis of its rate of increase. *J. Ecol.* 72:977–91
138. Ritland K, Jain SK. 1981. A model for the estimation of outcrossing rate and gene frequencies based on independent loci. *Heredity* 47:37–54

138a. Robles H, García-Barrios R. 1994. Fallas estructurales del mercado de maiz y la logica de la produccioń campesina. *Econ. Mexicana (Nueva Epoca)* 3:225–85

139. Rodríguez-Vlázquez J. 1994. *Efecto del mosaico de regeneracion y la densidad sobre la remoción post-dispersion de las diásporas de una palma tropical.* BSc thesis. UNAM, Mexico
140. Samper C. 1992. *Natural disturbance and plant establishment in an Andean cloud forest.* PhD thesis. Harvard Univ., MA
141. Schemske DW, Lande R. 1985. The evolution of self-fertilization and inbreeding depression in plants. II. Empirical observations. *Evolution* 39:41–52

141a. Schonewald-Cox CM, Chambers SM, MacBryde B, Thomas L, eds. 1983. *Genetics and Conservation: a Reference for Managing Wild Animal and Plant Populations.* Menlo Park, CA: Benjamin/Cummings. 722 pp.

142. Schupp EW. 1988. Seed and early seedling predation in the forest understory and in treefall gaps. *Oikos* 51:525–30
143. Schupp EW. 1988. Factors affecting post-disersal seed survival in a tropical forest. *Oecologia* 76:525–30
144. Schupp EW, Frost EJ. 1989. Differential predation of *Welfia georgii* seeds in treefall gaps and in the forest understory. *Biotropica* 21:200–3

144a. Shea K, Rees M, Wood SN. 1994. Trade-offs, elasticities and the comparative method. *J. Ecol.* 82:951–57

145. Silvertown J. 1987. *Introduction to Plant Population Ecology.* Harlow: Longman
146. Silvertown J, Franco M, Menges E. 1996. Interpretation of elasticity matrices as an aid to the management of plant populations for conservation. *Conserv. Biol.* In press
147. Silvertown J, Franco M, Pisanty I, Mendoza A. 1993. Comparative plant demography—relative importance of life-cycle components to the finite rate of increase in woody and herbaceous perennials. *J. Ecol.* 81:465–76
148. Simberloff D. 1986. Are we on the verge of mass extinction in tropical rain forests? In *Dynamics of Extinction,* pp. 165–80. New York: Wiley
149. Simberloff D. 1988. The contribution of population and community biology to conservation science. *Annu. Rev. Ecol. Syst.* 19:473–511
150. Slatkin M. 1985. Gene flow in natural populations. *Annu. Rev. Ecol. Syst.* 16:393–430
151. Slatkin M. 1991. Inbreeding coefficients and coalescence times. *Genet. Res.* 58:167–75
152. Slatkin M, Barton NH. 1989. A comparison of three indirect methods for estimating average levels of gene flow. *Evolution* 43:1349–68

153. Slatkin M, Maddison WP. 1989. A cladistic measure of gene flow inferred from the phylogenies of alleles. *Genetics* 123:603–13
154. Slatkin M, Maddison WP. 1990. Detecting isolation by distance using phylogenies of genes. *Genetics* 126:249–60
155. Smouse P. 1986. The fitness consequences of multiple-locus heterozygosity under the multiplicative overdominance and inbreeding depression models. *Evolution* 40:946–57
156. Solbrig O, Sarandon R, Bossert W. 1988. A density-dependent growth model of a perennial herb: *Viola frimbriatula*. *Am. Nat.* 131:385–400
157. Soulé ME, ed. 1986. *Conservation Biology: the Science of Scarcity and Diversity.* Sunderland, MA: Sinauer. 584 pp.
158. Soulé ME, ed. 1987. *Viable Populations for Conservation.* Cambridge: Cambridge Univ. Press. 189 pp.
159. Soulé ME, Wilcox BA, eds. 1980. *Conservation Biology: An Evolutionary-Ecological Perspective.* Sunderland, MA: Sinauer
160. Taylor BL. 1995. The reliability of using population viability analysis for risks classification of species. *Conserv. Biol.* 9:551–58
161. Taylor P, García-Barrios R. 1996. The social analysis of ecological change. *Soc. Sci. Info.* In press
162. Terauchi R. 1994. DNA analysis of *Dryobalanops lanceolata* (Dipterocarpaceae). In *Plant Reproductive Systems and Animal Seasonal Dynamics: Long Term Study of Dipterocarp Forest in Sarawak,* ed. T Inoue, AA Hamid, pp. 114–17. Kyoto: Kyoto Univ.
163. Tuljapurkar SD. 1990. *Population Dynamics in Variable Environments.* New York: Springer-Verlag
164. Uhl CK, Dezzeo CN, Maquino P. 1988. Vegetation dynamics in Amazonian treefall gaps. *Ecology* 69:751–63
165. Uhl CK, Guimarães Vieira IC. 1989. Ecological impacts of selective logging in the Brazilian Amazon: a case study from the Paragontinas region of the state of Pará. *Biotropica* 21:98–106
166. United Nations Envirronmental Program. 1995. *Global Biodiversity.* Cambridge: Cambridge Univ. Press
167. Usher MB. 1976. Extensions to models, used in renewable resource management, which incorporate an arbitrary structure. *J. Environ. Manage.* 4:123–40
168. Whitmore TC. 1982. *Tropical Rain Forests of the Far East.* Oxford: Clarendon. 2nd ed.
169. Whitmore TC, Sayer JA. 1992. *Tropical Deforestation and Species Extinction.* London/New York: Chapman & Hall. 153 pp.
170. Wilson EO. 1988. *Biodiversity.* Washington, DC: Natl. Acad. Sci.
170a. Wilson MH, Kepler CB, Snyder NFR, Derrickson SR, et al. 1994. Puerto Rican parrots and potential limitations of the metapopulation approach to species conservation. *Conserv. Biol.* 8:114–23
171. Wright S. 1931. Evolution in Mendelian populations. *Genetics* 16:97–159
172. Wright S. 1969. *Evolution and Genetics of Populations: The Theory of Gene Frequencies.* Chicago: Univ. Chicago Press

Annu. Rev. Ecol. Syst. 1996. 27:423–50

GENE TREES, SPECIES TREES, AND SYSTEMATICS: A Cladistic Perspective

A. V. Z. Brower[1], *R. DeSalle*[1], *and A. Vogler*[2]

[1]Department of Entomology, American Museum of Natural History, Central Park West at 79th Street, New York, NY 10024; [2]Department of Entomology, The Natural History Museum, Cromwell Road, London SW7 5BD, United Kingdom; and Department of Biology, Imperial College at Silwood Park, Ascot, Berkshire, SL5 7PY, United Kingdom

KEY WORDS: mtDNA, phylogeny, coalescence, cladistics, empiricism

ABSTRACT

The proliferation of molecular data in systematics has opened a Pandora's box of alternate approaches to inferring hierarchical patterns of relationship among taxa. In this review, we examine practical and theoretical reasons for employing some methods and avoiding others. We offer a philosophical overview of the relationship between systematics patterns and evolutionary processes, and we discuss the differential emphasis given to each of these areas by opposing methodological camps. We review the sources and types of incongruence between data partitions from different sources and recommend a specific procedure for contending with incongruence. We then focus on inference of relationships among closely related taxa, with particular emphasis on mtDNA as a source of characters, its advantages and potential pitfalls. We conclude with a review of several widely cited empirical studies and suggest that the gene tree–species tree problem may be less severe than its prevalence in the literature would suggest.

INTRODUCTION

The architects of the New Synthesis (e.g. 54, 114, 164) unified disparate disciplines under the explanatory umbrella of evolutionary theory, which they argued provided an empirically sound philosophical framework for all biology (165). Nevertheless, the day-to-day work of microevolutionists studying changes of allele frequencies in populations, and of macroevolutionists studying historical

0066-4162/96/1120-0423$08.00

patterns of diversity, was carried on in largely separate empirical and intellectual realms. Systematists recognized a logical lower bound for their methods (85): Below the point where relationships between taxa become nonhierarchical (i.e. when dichotomously branching trees dissolve into anastomosing networks connected by interbreeding), systematic methods become philosophically and practically inappropriate for describing character relations and inferring relationships among organisms (46, 56, 131). By contrast, the upper bound of population genetics was limited by the physical constraint of the disjunction of gene pools between species. In recent years, technological advances have made it increasingly possible to apply genetic techniques and data to evolutionary questions beyond the traditional species boundary, and the success of these efforts has led many researchers to view larger-scale patterns of diversity as interpretable with the same methodological tools, or at least within the same philosophical framework, as population-level patterns (e.g. 4, 12, 111, 189).

In this review, we compare population genetic and cladistic approaches to interpreting patterns of relationship among closely related taxa. As some of the most interesting issues in evolution are accessible only by study of such groups, our main objective is to investigate the problem of discovering hierarchical patterns around the species level, with respect to the complications of ancestral polymorphism, lineage sorting, and introgression. Discovery of hierarchical patterns of relationship among taxa is widely regarded to be the province of systematics (20, 139). In principle, the cladistic method permits inference of relationships as soon as characters are discovered that allow diagnosis of discrete groups of populations (taxa). However, theoretical studies have argued that the probability of error of cladistic methods remains high for long periods after gene flow has ceased between divergent populations, if the cladograms are inferred from molecular data from single genes (38, 89, 127, 138, 178, 181). To counter and correct for these potential difficulties, numerous alternative approaches have been devised.

Our essay begins with a brief review of philosophical issues that underlie the differences between population genetic and systematic approaches to the gene tree–species tree problem. The fundamental role of philosophy as arbiter between competing interpretive paradigms is acknowledged only rarely in the methodologically or empirically focused literature, but it is in this realm that we believe methods must compete. We hope this discussion provides a starting point for philosophical debate among the various methodological schools. We then examine ideas of species and speciation, showing how the cladistic approach circumvents many of the ontological difficulties besetting these complex theoretical issues. Because congruent hierarchical character distributions provide the evidence for inferring patterns of phylogenetic relationship, character incongruence among closely related taxa is theoretically problematical (47). We review theories, models, and observed patterns of character incongruence in some detail. We draw empirical examples from animal mitochondrial data sets, as their interpretation has been paradigmatic in the gene tree–species tree theories of systematists and population geneticists alike (6, 49, 56, 82, 86, 120, 141). Many readers will disagree with our philosophical stance, but we hope our exploration of ideas and of empirical studies provides a framework for better understanding the advantages and limits of alternate approaches.

EPISTEMOLOGY AND "TRUTH"

Scientific hypotheses are framed in the context of plausible prior theories that are accepted as background knowledge (assumptions not tested by the data testing the hypothesis in question, but that are in principle testable). Accepting a particular body of background knowledge (an ontology) is justifiable only by recourse to an infinite regress of prior, more general assumptions of background knowledge (144). Epistemology is the questioning of the adequacy of these assumptions: How do we construct a rational framework to interpret and explain our observations? Epistemological questioning diminishes when a discipline reaches a consensus about an ontological paradigm; at that point the discipline becomes a "normal science" (105, 139).

Both systematics and population genetics depend on generally accepted, underlying ontological tenets; they are thus Kuhnian normal sciences. However, these tenets differ greatly between the two fields. Population genetics is a deterministic, regular science that draws deductive inferences about evolutionary process from established laws of heredity (109, 147, 193), while systematics is a historical science that attempts to discover and describe the intricacies of the evidently hierarchical pattern of nature (70, 139, 149). Population genetics seeks to document and explain the origin and maintenance of diversity in systems where continuity of process is explicitly postulated (evolution happens only because of deviations from Hardy-Weinberg equilibrium conditions). Systematics seeks to document hierarchical patterns among disjunct entities and needs to postulate little except that a tree-like hierarchy exists and is recoverable by studying attributes of individual organisms.

Reviewing tree-building algorithms and philosophies lies beyond our scope here (see 162, 176 for references), but some comments about their assumptions may highlight the pattern-process dichotomy discussed above. Distance-based methods such as UPGMA (166) depend on prior knowledge about rates of evolution. These methods were popular in the 1970s, when optimism about the existence of a molecular clock ran high. More recent distance methods, such as neighbor joining (155), add additional model parameters to buffer against manifest violations of clock-like regularity in DNA evolution (e.g. 73).

Like distance methods, each of the myriad variations of phylogeny estimation via maximum-likelihood (e.g. 64, 66, 74) also relies on an explicit underlying model of character transformation. These models range from simple but unrealistic one- or two-parameter approximations to complex models that incorporate relative frequencies of amino acid changes (103). Extra parameters increase "realism" by improving the fit of the model to some "true" topology (see below), but adding them sacrifices both explanatory power and computational speed.

Further, because methods that rely on explicit, a priori models of evolution are acknowledged to be poor estimators of hierarchical pattern when the assumptions of the models are violated (65, 200), the underpinnings of a chosen model must be defensible on empirical grounds. Although statistical computer simulations of model-fit and robustness-to-error (e.g. 60, 92, 153) may provide quantitative rigor, the plausibility of a particular model of evolution, and therefore the verisimilitude of results from its employment in systematic inference, has no ultimate extrinsic appeal except to congruence with "known" topologies, derived from real data and real organisms by an independent method (121, 122).

It is frequently stated that all phylogenetic inferences depend on underlying models of the evolutionary process (e.g. 27, 74). This does not mean that all models are equally general or entail results with equal explanatory potential. When population geneticists, evolutionary systematists, and "phylogenetic" systematists[1] examine hierarchical patterns of relationship, they operate implicitly or explicitly with the presupposition that evolutionary processes (such as phylogeny) are responsible for the observed patterns of biotic diversity. The more explicit the process model, the more precise the estimate of relationships may be. But precision is not a substitute for accuracy, and the accuracy of the method (an ontological claim) is only measurable indirectly, by reference

[1]Both Hennig (85) and Wiley (199) made the term "phylogenetic" synonymous with "genealogical." As recognized by early evolutionists (e.g. 45, 94), propinquity of descent is not the cause of, but a theory that explains, our continuing successful discovery of a hierarchical Natural System of classification that was well established long before Darwin's or anyone else's materialist causal hypothesis was proposed to explain it (see e.g. 17, 173, 195, and discussion in 130, 135, 172). As pointed out by Brady (20), systematics must exist independent from process theories if evolution is to avoid tautology. The term "phylogeny" is so vague and metaphysical as to be embraced by researchers with antithetical systematic philosophies (e.g. 35 contra 161; 129 contra 49). For clarity, we avoid "phylogeny" and "phylogenetic" when referring to cladograms or trees, but rather use the term only to describe the process theoretically responsible for the patterns we observe.

to the plausibility of the background knowledge underlying the model (167, 194).

To a cladist, phylogeny is a compelling and well-corroborated process theory that provides a causal explanation for the inferred patterns. It is not part of the framework of background knowledge within which results are interpreted (except in the sense that some single hierarchical pattern is postulated to exist; 142). Cladistic methods rely on general ontological claims that lie deeper in the infinite regress of background knowledge than the claims underlying methods based on more explicit process models, instilling them with generally greater epistemological credibility (62). More importantly, cladistic hypotheses are free from specific evolutionary assumptions (143) and therefore represent independent explananda for adaptive or phylogenetic process theories (20). Process theories that do not fit the observed pattern may be rejected empirically, rather than dictating the pattern discovered.

At the bottom of the ontological problem of understanding phylogeny is the epistemological question, "How much about the evolutionary process are we willing to accept as background knowledge when inferring particular patterns from empirical observations, and constructing causal hypotheses to explain them?" In our view, a strong justification of the plausibility of underlying assumptions is the sine qua non of a particular method. It is not difficult to imagine myriad possible explanations for patterns we observe in nature, once we abandon the position that logical consistency is a necessary underlying feature of the chosen methodology.

THE PROCESS OF SPECIATION

To proceed with our discussion of pattern inference at the species boundary, we need to ask, "What are species, how do they come to be, and how do we discover them?" Understanding the gradation of differences between taxa has remained perhaps the most elusive and intractable challenge to natural historians (and later, evolutionary biologists) since the beginnings of modern systematics in the eighteenth century (19). The vexedness of the problem continues to inspire philosophical tracts, definitions, and discussions of the "reality" of species (e.g. 12, 13, 48, 49, 70, 77, 110, 111, 196), but we are unconvinced that the conception of the process has advanced beyond Hennig's (85) oft-reproduced cartoon (Figure 1). This diagram represents a simple ontological model of the speciation process, as the bifurcation of a braided lineage of interbreeding individuals into two such lineages. Each daughter lineage retains its own genetic connectedness, yet gains a discrete genetic identity with respect to its sister. As Hennig recognized, however realistic such a model might seem, our ability to assess its validity springs from, and is constrained by our access to, empirical

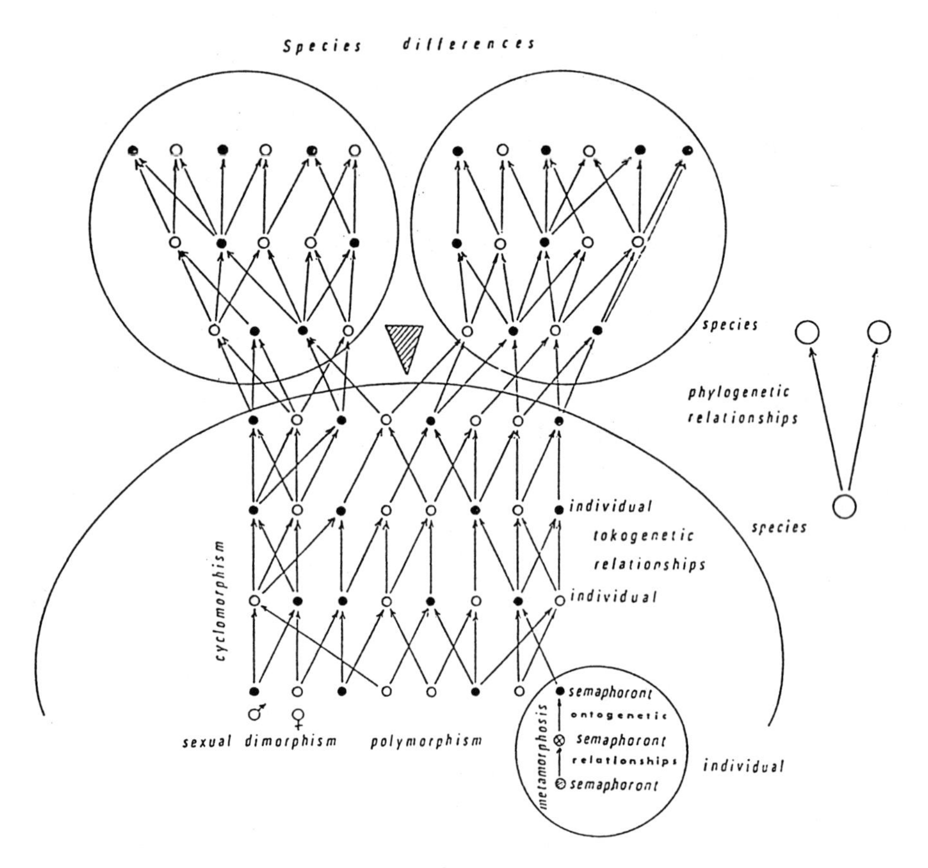

Figure 1 Hennig's (Figure 6 in 85) diagram of "the total structure of hologenetic relationships and the differences in form associated with its individual parts." Development of single individuals (ontogenetic relationships), interbreeding between individuals (tokogenetic relationships), and bifurcation of species (phylogenetic relationships) are depicted as coexisting at nested levels of organization. Note that (coincidentally) the mtDNA of individuals in each descendant species would be fixed after one generation for alternate haplotypes from the ancestral species. Copyright 1979 by the Board of Trustees of the University of Illinois. Used with permission of the University of Illinois Press.

information. The process is an explanation for the pattern inferred from comparative data, which almost always is limited to observed differences in holomorphology ("the multidimensional gestalt of the semaphoronts," p. 66).

Rieppel (149) argued that species have a necessarily dual nature: Depending on the question being asked, species must be seen as either individuals or classes. He characterized these alternate views as nominalist and essentialist, and he argued compellingly that they are both potentially valid, although epistemologically incompatible. We suggest that the contrast between population genetics and systematics developed above reflects these two viewpoints. The nominalist population-genetic view of species is an inclusive one, of species made contemporaneously coherent by gene flow and historically connected by the coalescence of alleles. There are no breakpoints in the continua of lineages, and divisions between species are not recognizable, except after the fact and by arbitrary criteria, or "statistical distributions" (111). By contrast, many cladists identify species as exclusive classes composed of individuals united by sharing unique combinations of characters (34, 130, 131). Metaphysical questions concerning the true nature of species are not important to the operation of identifying groups by this criterion. Trying to shoehorn additional information from the nominalist perspective, such as cohesion (185) and exclusivity (13, 49, 88a), into cladistic discovery of patterns serves only to obfuscate the process-independent, empirical basis of the method, which is its chief strength.

GENE TREES AND INCONGRUENCE

The cladistic method implies the most parsimonious cladograms for the taxa examined, given a particular data set. Whether these cladograms correspond to the "true" phylogenetic history of the taxa is not subject to empirical proof or disproof. Instead, they should be considered as theories, subject to corroboration or refutation by comparison with additional data (144, 198). We do not address theoretical issues of data partitioning versus simultaneous analysis in systematic inference in depth, as the topic has been reviewed recently (47, 122, 132). If the simultaneous analysis approach is rigorously applied, there are no separate data sets or topologies to conflict with one another. However, many systematists are not satisfied with this view, and all recognize that some variable traits of organisms have poor potential as characters. A particular set of characters[2] may not reflect "organismal history" (or be corroborated by the bulk of other evidence that we believe does reflect organismal history). Concern

[2]The idea that logically separable data sets can be drawn from the same taxa is problematical (47, 104, 122, 132). For convenience, we refer to data drawn from different sources (e.g. mtDNA sequence versus morphology, or one gene versus another) as different "data sets," although they might more appropriately be called subsets or hypothetical "process partitions" (27, 122).

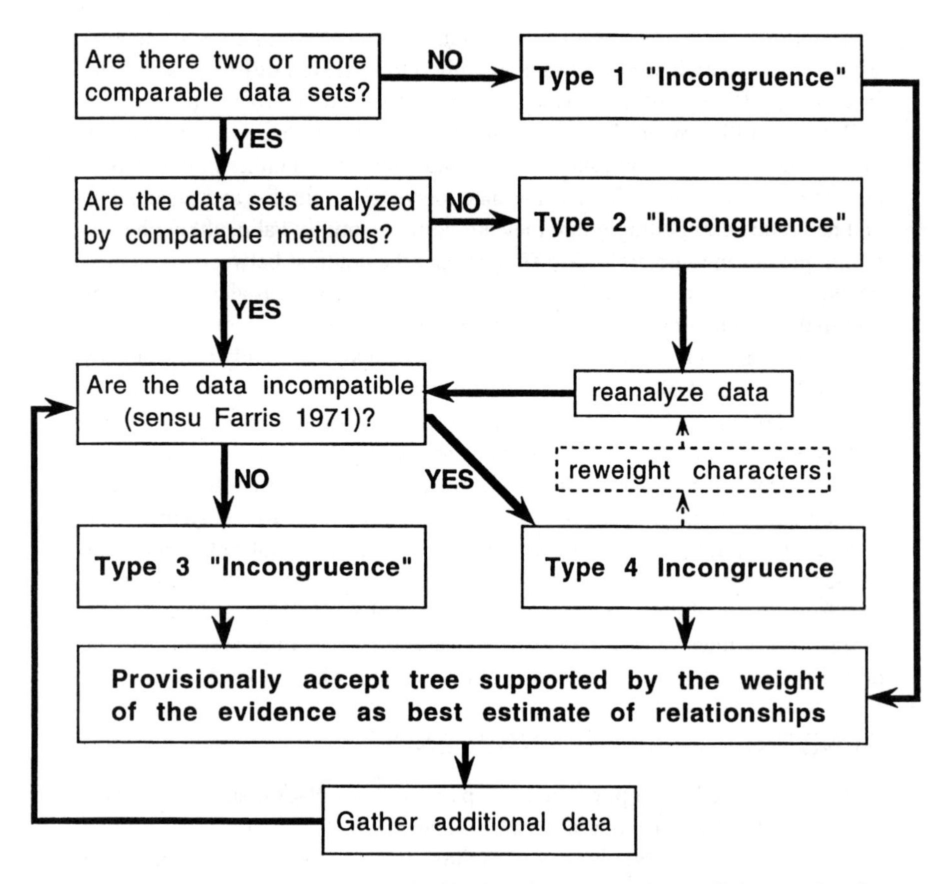

Figure 2 Flow chart illustrating the classification of congruence types and the procedure for distinguishing them, described in the text. Differential character weighting of incongruent data sets is shown as an alternate pathway, with dashed lines signifying our reservations about such techniques (see also 24).

over data incongruence is increasing as multiple molecular data sets become available for comparison with one another and with morphology for the same taxa. The remainder of this paper addresses the sources, hypothesized causes, and prevalence of this pernicious difficulty. First, we enumerate a hierarchy of proximate explanations for conflict between alternate data sets (Figure 2).

The idea of assessing congruence or incongruence of a gene tree with the species tree presupposes the existence of at least two data sets with features amenable to formal comparison (117). Frequently in the literature (e.g. 23, 53), however, gene trees are compared to traditional classifications based on non-explicit interpretations of intuitively analyzed data, or concern about the gene tree's relation to the "true history" of the taxa is expressed in the absence of any conflicting data at all. A hypothesis of relationships with limited or absent empirical foundations is easily refuted by the first data that contradict it. Thus, the first type of "incongruence" we identify is incongruence with non-empirical expectations, which we argue represent nothing more than arbitrary, conjectural starting points for empirical investigation. Under such circumstances, the gene tree should be provisionally accepted as the best available hypothesis of taxonomic relationship.

Traditional, informal ideas about relationships are not without value in our methodological scheme. They can (and should) guide the discovery of characters for explicit systematic analysis. Some types of comparative data are not amenable to discovery of hierarchical patterns among taxa, however, and should be ignored as sources of systematic evidence, in our opinion. These include morphometric data (18), DNA-DNA hybridization data (35), and differential frequencies of traits (such as allozymes) among polymorphic populations (46).

A second type of incongruence between trees can be a product of incomparability of analytical methods. The same data set can yield different topologies under alternate optimization criteria (117, 118). If two data sets are analyzed by different methods, it is difficult to distinguish incongruence as an artifact of the analyses from incongruence intrinsic to the data. For example, Omland (134) showed that the topological incongruence between molecular and morphological analyses of dabbling duck relationships was largely due to the analysis of the mitochondrial DNA restriction fragment data with the UPGMA method. When both mtDNA and morphology were analyzed using the parsimony method, the data were found to be almost entirely congruent.

One reason topologies from noncladistic analytical methods should not be compared is mosaic incongruence (61), which results from unequal amounts of state change in different characters, yielding either differences in the degree of resolution or unequal branch lengths between topologies from alternate data sets. Farris contrasted this to incompatibility incongruence due to homoplasy, or active disagreement in hierarchical groupings implied by alternate characters (or data sets). Of course, one of the main criticisms against tree-building algorithms based on distances, such as the UPGMA method criticized by Omland (134), is that they conflate mosaic and incompatibility incongruence. In cladistic methodology, only incompatibility incongruence is relevant, because lack of information or symplesiomorphic similarity due to lack of change in one character has no bearing on the informativeness of another.

Type 3 incongruence is based on minor character incompatibility (homoplasy) discovered in separate analyses of partitioned data sets, rather than on strong conflict between data sets. Farris et al (63) described the incongruence length difference (D), which measures the extra homoplasy entailed by combining data sets, and a computer algorithm that tests this value against random partitions of the same data. It is quite unlikely that two data partitions drawn at random from an identical distribution will imply completely congruent topologies, and thus some basic level of incongruence is expected between any real data partitions. Empirical cases have been described from the Solanaceae (133), primates (201), and josiine moths (119). Incongruence between the separate data sets in these cases was ignored, and the cladogram resulting from analysis of all data was presented as the best current hypothesis of relationships. It is notable in these cases that data partitions implying a different topology in separate analysis add additional branch support (22) to the topology from simultaneous analysis. By contrast, some recently proposed topological congruence methods (11, 107, 175) might interpret these results as conflicting because they conflate character incompatibility with mosaic incongruence, lack of information, or weak character support for conflicting nodes in alternate data partitions. Because trees are summaries of character data, comparing them instead of the data themselves can add no additional knowledge to the analysis and can result in erroneous conclusions (104, 197).

As the amount of disagreement between data sets increases, attributing incompatibility incongruence to homoplasy (Type 3) becomes more and more challenging to the assumption that the data reflect a single underlying hierarchical pattern (27). Incongruence due to inferred separate histories (Type 4) is the only kind of incongruence that challenges the validity of the single hierarchical pattern postulate. Most of the current debate over combined versus separate analysis (see 47, 132) revolves around determining how much homoplasy (Type 3) we are willing to allow before invoking separate histories (Type 4). Beyond this threshold, we must decide if and by which criteria the data should be discounted (e.g. by alternative weighting), and we must invoke ad hoc explanations to discount homoplasy in one or another of the data sets to meet the underlying assumption of a single hierarchy.

One possibility is to develop a general model of expected data behavior, and statistical tests to estimate when the data are misbehaving (47, 108, 174). How to explain, rationalize, and factor out "significant" differences discovered by such methods remains problematical. A first step is to identify potential processes that could systematically bias certain types of data. The next section describes some possible reasons why different fragments of DNA sequence from a sample of taxa might strongly imply incongruent hierarchical patterns of relationship.

SOURCES OF CONFOUNDING INCONGRUENCE IN MOLECULAR DATA

In the 1980s, there was a sense of optimism that DNA would be the Rosetta Stone of systematics, allowing us empirical access to objective phylogenetic patterns obscured in morphologies warped by adaptation (e.g. 76, 158). More recently, Patterson et al (141) presented a more pessimistic view, that sequence data were not contributing much to our knowledge of hierarchical relations, except in areas where morphological analysis is not applicable, such as the "tree of life" and patterns of relationship in human mitochondrial DNA. One of the worst problems with DNA data is that they are noisier than we would like, yielding cladograms with as much or more homoplasy than morphological data sets of comparable size (156). This homoplasy is thought to be caused by drastically different rates of evolution in characters (nucleotide sites) within and among genes (100, 137). Thus, some sites are invariant, some sites are too variable, and some sites contain the information we desire—the hierarchical pattern.

As argued by Farris (62; see above), if homoplasy is random with respect to the "phylogenetic signal," then it doesn't make any difference how homoplasious the data are: The hierarchical pattern will emerge from the noise. However, evidence exists for differential mechanistic constraints on DNA sequences that may produce patterns in the data that do not reflect the "true phylogeny." Nucleotide bias or codon bias in protein-encoding genes varies greatly from gene to gene and from taxon to taxon (14, 95, 159); it may result in grouping due to convergent similarity rather than putative ancestry (32, 83, 162, 171). While nucleotide bias may lead astray evolutionary models based on neutral assumptions, it is not a problem for cladistic analysis unless there are independent, convergent origins of differential bias among the ingroup taxa. To our knowledge, such a pattern has not yet been demonstrated among closely related taxa (e.g. 112, but see 1). Of course, saturation of variable sites in a particular sequence may result in poor resolution among taxa, but poor resolution results in Type 3 incongruence between data sets, which should be ignored in light of other data bearing a strong signal. The latter will predominate in simultaneous analysis.

Of more serious concern are other hypothetical evolutionary processes that could result in a particular character (or set of characters, such as a gene) being actively incongruent with the preponderance of other evidence (e.g. morphology or other genes) (27). Three such processes are horizontal transfer, introgression, and ancestral polymorphism. Again, we emphasize that detection of incongruence depends fundamentally on the prior recognition of a dominant pattern of relationships: If, for example, horizontal transfer of DNA were rampant, the hierarchical pattern of diversity we observe would not exist, and the inference of horizontally transferred DNA fragments would be impossible. Our ability to discover well-corroborated cladograms (and instances of homoplasy) implies that these phenomena are the exception rather than the norm, and the attention paid them in the literature may reflect more the unease they inspire in our phylogenetic paradigm than their prevalence in nature.

Horizontal Transfer

In eukaryotes, DNA generally passes only from the parents' genome(s) to the offspring's genome, and it is this unidirectional pattern of inheritance that provides the basis for the inference of descent with modification. However, certain small bits of DNA (transposable elements) are capable of transfer between lineages that are not connected by interbreeding (e.g. P-elements; 44), or even, it appears, between insect orders (e.g. mariner; 150). Transposable elements (TEs) tend to occur in multiple copies, and they appear capable of inserting and excising themselves from particular sites throughout the genome. The mechanism for transfer of TEs both within and between genomes is unknown but may be mediated by retroviruses or other parasitic vectors (reviewed in 99). Fortunately for hierarchical pattern inference, the 90 or so classes of TEs recognized to date appear to play an insignificant role in the disruption of taxon boundaries. However, if horizontal transfer turns out to be a more prevalent phenomenon than we currently believe it to be, we may need to retool our entire conceptual and methodological battery for inferring patterns of relationship (e.g. 128, 192).

Introgression

Unlike horizontal transfer, introgression is *anastomosis* of clades due to intrinsic hybridization and differential gene flow between the taxa themselves. Evidence for introgression consists of discovery of particular "foreign" characters (e.g. electrophoretic alleles, mitochondrial haplotypes) against a mostly "normal" genetic or morphological background. The classic situation for discovery of introgression is in hybrid zones between formerly allopatric, closely related taxa that are capable of interbreeding (114). For many years, it was not considered possible to distinguish introgression between historically separated populations from primary intergradation of divergent traits across environmental gradients (59), but cladistic approaches (81, 190) allow distinction between the two, at least under certain biogeographical circumstances. Introgression may be balanced by selection in tension zones, resulting in a long-term, stable equilibrium through which unlinked, neutral alleles may flow extensively (10, 183). It may also lead to fusion of taxa that had been differentiated to some degree in the past. The latter seems especially prevalent in taxa that have been brought into contact by human intervention (e.g. 5, 58, 148).

Ancestral Polymorphism

Closely related taxa may share more than one allele at a locus, due to stabilizing selection or simply to failure of neutral alleles present at the time of lineage splitting to drift to fixation. Analysis of hierarchical patterns among such alleles amounts to an orthology-paralogy problem (68, 75, 136, 140). If we are interested in inferring relationships among taxa, it is necessary to distinguish the splitting event that gave rise to the alternate alleles (which may represent the deepest node in the topology), from the splitting event that gave rise to the taxa. Of course, these alternate alleles can become fixed in the taxa via lineage sorting (127), yielding hierarchical patterns that do not reflect the course of taxon splitting. Such patterns may be further complicated at autosomal loci by recombination, which can result in a mosaic pattern of relative degree of identity by descent among alleles (89). A practical recommendation for molecular systematics is to avoid genes that are suspected of being subject to balancing selection, such as genes of the major histocompatibility complex (MHC) (e.g. DQB1; 71).

Doyle (57) elegantly discussed the related problem of heterozygosity in diploid loci: Two alleles occurring in a single individual need not be closely related in the hierarchy of relationships for alleles at that locus. The nonsensical outcome of drawing systematic inferences from such data is that, depending on which allele is sequenced, the same organism will occupy different positions in the gene tree. Doyle recognized that systematic methods are inappropriate under such conditions, and proposed a method akin to population aggregation analysis (46) that delimits the boundaries of the gene pool by discovering all alleles cohabiting with other alleles in heterozygote combinations among individuals. This procedure is tantamount to an empirical investigation of the boundaries of biological species (114, 115), and although rigorous, it is not likely to provide a feasible solution to most systematic problems at or around the species level.

This section has briefly described some genetic processes that could cause gene trees to be misleading relative to species trees. However, any parsimonious explanation of data could be wrong, and speculating in the realm of the possible does not advance our knowledge very far. In order to know how much we should trust our molecular trees, we would like to know how often incongruence occurs between a particular gene tree and trees based on other data from the same taxa.[3]

[3]Recall that we are not concerned about the correspondence of a particular result with "truth," but only with other empirical results. Suggesting that one data set is false because it is incongruent with another that is true both opens the door for Ockham's *genius malignus* (how do we know that all the data aren't misleading?) and begs the point of further inquiry (149; see "Epistemology and Truth," above).

In the next two sections, we examine theoretical and empirical results based on hierarchical pattern analysis of mitochondrial DNA in order to address two questions, "Are gene trees theoretically likely to be misleading?" and "Are these theoretical concerns reflected by empirical results?"

THEORETICAL APPROACHES TO GENE TREES AND SPECIES TREES: THE MTDNA PARADIGM

MtDNA has well-known advantages as a source of characters for systematics and biogeography (6, 15, 80, 124). It is usually maternally inherited (effectively haploid), lacks recombination, evolves relatively rapidly, and segregates independently from the nuclear genome (but see 202). These features have made mtDNA a tractable subject for genetic models as well. Birky et al (16) pointed out that the effective population size (Ne) of mtDNA is one fourth that of an autosomal locus in a random-mating diploid population. This means that fixation due to genetic drift occurs four times more rapidly among selectively neutral mitochondrial haplotypes than among neutral nuclear alleles, and that gene flow sufficient to maintain nuclear panmixia may allow differentiation of mitochondrial lineages in different local demes. These effects are amplified if dispersal is male-biased. Under ideal circumstances, therefore, mtDNA would seem to be a superior marker for studying relations among closely related taxa (6, 123, but see 42, 116).

However, because mtDNA does not undergo the anastomosis of recombining nuclear loci, unambiguous yet spurious hierarchical relationships exist among mtDNA haplotypes of individual organisms within interbreeding populations. If patterns of haplotype relationship are used to infer patterns of taxonomic relationship, it is critical that these patterns agree: Hierarchical structure must exist at both levels (46). As we noted at the outset, if there is no hierarchical structure among terminal taxa being compared, systematic methods are inappropriate for inferring their relations.

Even when hierarchical structure exists among the taxa, an extensive body of theory suggests that hierarchical patterns from mtDNA (or any other loci) may not reflect the historical order of divergence events among the examined taxa. Introgression of unlinked, neutral alleles across hybrid zones is both likely and potentially far-reaching (183), even if the zone is maintained by quite strong selective forces acting on nuclear loci. Ancestral polymorphism of mtDNA in populations is also theoretically likely (but not, of course, heterozygosity of individuals). Tajima (178) showed that, even for very simple situations such as two neutral alleles drawn from each of two populations, the probability of inferring the correct tree is small if the time since the branching event is less than 2Ne generations. This is because the mean time to fixation (monophyly) of neutral alleles in each of the populations is 4Ne (Ne for mtDNA) under neutral expectations (101), and the daughter populations are therefore likely to contain alleles older than the population-splitting event. Avise and colleagues (7, 127) extended this theory to predict patterns of mtDNA evolution under various demographic scenarios. Pamilo & Nei (138) used genetic distances to show that increasing the number of alleles sampled did not help solve this problem.

A neutral model of the genealogical process called coalescent theory has been increasingly important to incorporating hierarchical information into the interpretation of genetic diversity (55, 89, 91, 102). Coalescent theory relates the age of a gene clade to its degree of genetic diversity, given a stochastic mutation process and temporal constancy of a series of factors that can affect the rate of diversification (or, looking backwards, coalescence). These factors include population size, selection, linkage and recombination, geographic structure, and migration between populations. Takahata (181) used coalescent theory to argue that larger sample sizes of alleles could increase the probability of inferring the correct historical pattern, contrary to Pamilo & Nei's (138) view. Recently, Moore (123) used a coalescent argument to suggest that mtDNA is a substantially more reliable marker, based on its smaller Ne, for estimating phylogenetic patterns than are nuclear genes.

Templeton et al (188) and Crandall (37) used elements of coalescent theory to develop a novel cladogram estimation procedure. The Templeton et al algorithm (188) uses a statistical approach based on Hudson's (89) neutral coalescent model to establish the probability of obtaining a "non-parsimonious" inference from RFLP data. This "statistical parsimony" procedure determines the probability of genealogical connections in haplotype networks being parsimonious (sensu 37, 89) by providing a measure of the probability that connections between two haplotypes differing by some number of steps are nonhomoplasious. We stress that coalescent methods so far developed are for single genes (see for instance 89, 186–188) or for linked clusters of genes such as mtDNA or small phage genomes (see 123 and 37, respectively). Crandall & Templeton (38) tested several predictions of coalescent theory using many empirical data sets from both mtDNA and nuclear DNA. Their tests suggest that Templeton's approach to coalescent theory represents a valuable innovation for study of gene genealogies and may offer insights to the various causal factors responsible for the pattern observed.

Coalescent models have also been used to argue that certain markers are more likely than others to recover the species genealogy because of their population genetics (e.g. 123) and to infer causes of organismic geographical distribution

(189). Crandall (37) examined the performance of the algorithm, using a subsampling technique on T7 bacteriophage data, and showed that it performs more efficiently than maximum parsimony at recovering the "true" phylogeny when very few restriction sites are available for analysis. This led him to suggest that the choice of discovery method should depend on the level of diversity at which a gene genealogy is being constructed and the amount of character information available.

Such extrapolations of coalescent theory are only as plausible as the assumptions underlying the particular model applied. Most common coalescent models depend upon constant historical population size, lack of selection, panmixia, and several other assumptions (90, 163). The chance that these assumptions are met by any real data set is difficult to assess but is probably quite small. Simulations of the effects of altering single parameters show that current or historical selection on the gene or a gene it is linked to (98, 113, 182), recombination (84, 90), and fluctuations in population size (126, 151, 179) can each have major impacts on allelic diversity and times to coalescence. If two or more of these "constants" are not constant, then it becomes difficult to infer which process is responsible for observed patterns (180). Unfortunately, actual data from natural populations often appear to violate at least some of the underlying assumptions of coalescent theory (9, 31, 123, 126), throwing in doubt the explanatory relevance of the models in those systems.

It is arguable, in fact, that neutral models of allele genealogy must be unrealistic. As noted above, if taxa diverge over periods less than 4Ne generations, then the probability of any particular gene tree reflecting the population history is low (138, 154). Under such circumstances, most genes will have allele genealogies that are incongruent with one another and do not reflect the "species phylogeny." If this is what actually happens, it is difficult for us to imagine how divergent evolution can occur, unless it is driven by a special subset of "speciation genes" that drive phylogenesis in the face of homoplasious evolution of the majority of the genome. If this were the case, studying those genes alone might provide the key to discovering the true course of phylogenetic history. However, evidence supporting this proposition is decidedly sparse (but see 33). A more plausible hypothesis is that Ne is often small at speciation (due to selection or any number of population subdivision scenarios), and thus all genes are likely to track population history to some extent, unless times between splitting events have been very short (in which case, we do not expect to see much character support for any topology; 24).

In any event, determining regularities of population splitting should start from an empirical framework: Theoretical speculation, even in the guise of elegant mathematics, does not advance our knowledge unless it is tested with actual data. The T7 studies (37) offer a beginning, but it is notable that while parsimony recovered the "true" tree with strong resolution from all the data, neither method performed well on the tiny subsets of data employed in this comparative test. Further, generalizations drawn from contrived experimental phylogenies are bound to be oversimplified, and whether they tell us anything at all about gene trees from more complex taxa under "real" circumstances is subject to debate (87, 168).

EMPIRICAL APPROACHES TO GENE TREES AND SPECIES TREES: EVIDENCE FROM MTDNA

To exhaustively review empirical evidence for and against congruence between mtDNA trees and trees from other data would fill a book (see 4). There are many cases showing that molecules are congruent with morphology in closely related taxa (e.g. 3, 8, 23, 25, 119, 125, 184, 191). Nevertheless, the caveats in quite a few papers list the same, rather small, group of publications as evidence that gene trees are not congruent with species trees, so we examine some of these here.[4]

In a widely cited paper, Powell (145) examined mtDNA RFLP data from sympatric and allopatric populations of *Drosophila pseudoobscura* and *D. persimilis* in the western United States, Mexico, and Colombia. His title, "Interspecific cytoplasmic gene flow in the absence of nuclear gene flow: evidence from *Drosophila*," suggested that he had observed introgression of mtDNA across the species boundary. However, this interpretation is based on mosaic (Type 2) incongruence (61), rather than compatibility (Type 4) incongruence (indeed, no cladogram or network was presented). Later (146), Powell suggested that the mtDNA paraphyly due to lineage sorting is an equally parsimonious explanation of the observed pattern. Another plausible alternative is that populations of *D. pseudoobscura* have no apomorphies with respect to *D. persimilis*, a conclusion that is unproblematic for the cladistic view of species (46, 85, 131). A more extensive study of mtDNA (79), as well as evidence from RFLP analyses of the ADH region of the fourth chromosome (157) and the AMY region of the third chromosome (2), support the species-level paraphyly of *D. pseudoobscura* populations with respect to *D. persimilis*.

A strongly analogous situation was described in desert tortoises (106). In this case, lineage sorting was invoked to explain the apparent paraphyly of mtDNA among individuals of *Xerobates agassizi* with respect to *X. berlandieri* mtDNA,

[4]Ironically, Avise & Saunders (8) is frequently cited as a case exhibiting mtDNA introgression, in spite of the authors' unambiguous statement (p. 252): "in the present study we have no mtDNA (or allozyme) evidence for introgression between species of *Lepomis*."

yet the morphological authority cited (21) viewed the latter as a peripheral isolate of the former, implying that other characters also support the pattern observed in the mtDNA. This is another example of Type 2 incongruence.

Another case of apparent introgression of mtDNA was reported between white-tailed deer (*Odocoileus virginianus*) and mule deer (*0. hemionus*) (28). Mule deer are traditionally held to be conspecific with black-tailed deer, but the mule deer mtDNA in this study differed from that of black-tailed deer by 17 restriction sites, while sharing haplotypes with white-tailed deer. Further study (39, 41, 43) corroborated the affinity of mtDNA between the two species. It is interesting that mtDNA data (39) suggest that the black-tailed deer is more closely related to the neotropical red brocket (*Mazama americana*) than to the other *Odocoileus* deer. By contrast, allozymes (40, 50, 72) show mule and black-tailed deer to be phenetically more similar to one another than either is to white-tailed deer. All these papers attributed the discrepancy between the markers to historical mtDNA introgression from white-tailed deer to mule deer. If this is the case, it is surprising that hybrid zones between the putative conspecific taxa (mule and black-tailed deer) reveal no introgression of mtDNA (41), while interspecific hybridization has evidently resulted in complete replacement of black-tail–like mtDNA with white-tail–like mtDNA in the mule deer. Cronin (42) criticized the reliability of mtDNA for study of closely related taxa, in general, based on these results. However, granting the validity of phenetically analyzed allozyme frequency trees, the reported patterns can again be equally parsimoniously explained by mosaic incongruence (Type 2), because none of the analyses are rooted, except by the clock-dependent long-branch criterion. In fact, three-taxon statements do not allow inference of relationships at all, unless a root is provided. Thus, the white-tailed deer and mule deer may be sister taxa relative to black-tailed deer, as implied by mtDNA.

Of course, some cases of character incongruence between mtDNA and other data do seem most easily explained by mtDNA lineage sorting, introgression, or some other process. For example, directional introgression of mtDNA from *Mus domesticus* to *M. musculus* appears to have occurred across a hybrid zone in Denmark (67). Gyllensten & Wilson (78) suggested that the observed pattern may reflect the colonization of Sweden by a small number of hybrid mice from another hybrid zone in north Germany, in conjunction with the spread of agriculture around 4000 years ago. Although the amount of genetic divergence between the taxa suggests that *M. domesticus* and *M. musculus* diverged at least a million years ago (67), Danish and Swedish house mouse populations must have immigrated relatively recently, since the region was under ice during the Pleistocene. As noted above, observed patterns of genetic introgression are frequently associated with recent, human-mediated dispersal events. Peripatetic commensal organisms like house mice and *Drosophila melanogaster* may be uniquely poor study taxa for phylogeographic analysis, in spite of their popularity in genetic research.

Spolsky & Uzzell (169, 170) documented the presence of *Rana lessonae* mtDNA in *Rana ridibunda* populations from eastern Europe, suggesting introgressive hybridization. They hypothesized that the mtDNA had leapfrogged the species boundary via *R. esculenta,* a hybridogenetic species that eliminates the *R. lessonae* component of its nuclear genome during gametogenesis. The unusual genetics of these species may account for the dramatic discordance of mtDNA with other markers: Other hybridizing European amphibians, such as *Bombina bombina* and *B. variegate* (177), and *Triturus cristatus* and *T. mannoratus* (3), exhibit largely congruent patterns of introgression between nuclear and mtDNA.

Another interesting case occurs in Sri Lankan macaques (*Macaca sinica*). Based on RFLP analysis, Hoelzer et al (88) found mtDNA sequence divergence of over 3% between adjacent social groups in a local population that, based on allozyme samples, is otherwise apparently genetically panmictic. The strong differential between these character systems may be maintained by sex-biased gene flow between troops. Female macaques are mostly philopatric, while males typically disperse more readily. This behavioral difference seems unlikely to account for the high degree of differentiation, which suggests the lineages have been separate for 1.5 million years, if the standard clock calibration applies (26). Based on morphology, Fooden (69) recognized two subspecies of *M. sinica* in Sri Lanka, and the contact zone between them evidently falls quite near the mtDNA study site. The sampled population could thus lie on a secondary contact zone between two formerly separate groups. Allozyme data (160), however, suggest that the nuclear alleles are not differentiated strongly between the subspecies.

Perhaps the most extensively documented example of mtDNA incongruence is drawn from the closely related Hawaiian species quartet of *Drosophila silvestris, D. heteroneura, D. planitibia* and *D. differens.* These four taxa have been the target of intense morphological, behavioral, and genetic research over the past 30 years (reviewed in 30, 52, 97). There are no fixed chromosomal or allozyme differences among the four species, and morphological differences among them are mostly autapomorphies (29, 36). Using RFLP, DeSalle & Giddings (51) found the mtDNA pattern of relationship for these taxa (d(p(s, h))) to be incongruent with the topology implied by dendrograms from DNA-DNA hybridization (93) and allozymes (36) ((d,p)(s,h)).

DeSalle & Giddings (51; p. 6906) argued that, "the mtDNA phylogeny represents the more accurate evolutionary history of these species, and the nuclear

DNA phylogeny is indicative of a lack of differentiation of nuclear genetic components of *D. differens* and *D. planitibia*." Under the criteria laid out in Figure 2, this was an appropriate conclusion, since the alternative morphological, karyotypic, allozyme, and DNA-DNA hybridization data sets each fail at Stage 1 or 2, in that they contain no cladistic information or are intrinsically incompatible with the hierarchical pattern discovery operation. However, subsequent sequence data from nuclear genes (ADH, 152; vitellogenin yp-1, 96) support the ((d,p)(s,h)) topology. Since these data are comparable by cladistic methods, the data from the various genes may be formally compared for active (Type 4) incongruence. Using the incongruence length difference test (63), we find that the two nuclear genes are completely congruent (D = 0.0). However, the mtDNA data are strongly incongruent with the nuclear sequences (ILD with ADH = 0.256; ILD with yp-1 = 0.186).

How are these incongruent data sets best interpreted? The weight of the evidence supports the nuclear gene topology ((d,p)(s,h)), so the position that the mtDNA pattern is superior must be abandoned, regardless of population genetic arguments that can be made concerning the higher relevance of the maternal marker (51, 123). This conclusion does not mean that the current answer is correct in the ontological sense. Neither does it imply that it is not interesting to investigate causes for the observed pattern of incongruence. It simply means that ((d,p)(s,h)) is the best current estimate of the relationships among these taxa.

CONCLUSION

Should we expect gene trees to conflict with species trees? In our view, any expectation that comparative biologists bring to their data must be based on relevant empirical results. First, there must be some reason to hypothesize the existence of hierarchical relationships among the study taxa. If these do not exist, using a hierarchy-based discovery technique is simply inappropriate. If we believe there is a hierarchical pattern, then we should expect the data to reflect that pattern alone. The relevant characters should be evaluated parsimoniously and interpreted without prejudice derived from false notions of incongruence (Types 1, 2, and 3) based on process theories. If the data do not exhibit incongruence in the first place, ad hoc theories explaining incongruence are unwarranted. Topological comparison of gene trees to traditional classifications or trees based on ill-conceived or inappropriately analyzed data is likewise an inappropriate measure of "incongruence." Only character incongruence (Type 4) is problematical for the hierarchy paradigm. Even when such incongruence is discovered, the best estimate of hierarchical relationships is still derived from parsimonious interpretation of all the data, analyzed simultaneously, unless compelling rationalizations are provided for weighting or otherwise discrediting some of the evidence in favor of the remainder of the evidence. Explaining incongruence is not a task of systematics.

ACKNOWLEDGMENTS

We thank J Gatesy, P Goldstein, H Rosenbaum, V Schawaroch, BI Vane-Wright, D yeats, A. Burt, and our AMNH discussion group for helpful comments on our manuscript, and A de Queiroz for providing a preprint of his ARES manuscript. A Brower was supported by NSF DEB-9303251 and BSR-9106517.

Literature Cited

1. Anderson CL, Carew EA, Powell JR. 1993. Evolution of the Adh locus in the *Drosophila willistoni* group: the loss of an intron, and shift in codon usage. *Mol. Biol. Evol.* 10:605–18
2. Aquadro CF, Weaver AL, Schaeffer SW, Anderson WW. 1991. Molecular evolution of inversions in *Drosophila pseudoobscura*: the amylase gene region. *Proc. Natl. Acad. Sci. USA* 88:305–9
3. Arntzen JW, Wallis GP. 1991. Restricted gene flow in a moving hybrid zone of the newts *Triturus cristatus* and *T. marmoratus* in western France. *Evolution* 45:805–26
4. Avise JC. 1994. *Molecular Markers, Natural History and Evolution.* New York: Chapman & Hall
5. Avise JC, Ankney CD, Nelson WS. 1990. Mitochondrial gene trees and the evolutionary relationship of mallard and black ducks. *Evolution* 44:1109–19
6. Avise JC, Arnold J, Ball RM, Bermingham E, Lamb T, et al. 1987. Intraspecific phylogeography: the mitochondrial DNA bridge between population genetics and systematics. *Annu. Rev. Ecol. Syst.* 18:489–522
7. Avise JC, Neigel JE, Arnold J. 1984. Demographic influences on mitochondrial DNA lineage survivorship in animal populations. *J. Mol. Evol.* 20:99–105
8. Avise JC, Saunders NC. 1984. Hybridization and introgression among species of sunfish (*Lepomis*): analysis by mitochondrial DNA and allozyme markers. *Genetics* 108:237–55
9. Ballard JWO, Kreitman M. 1995. Is mitochondrial DNA a strictly neutral marker? *Trends Ecol. Evol.* 10:485–88
10. Barton NH, Hewitt GM. 1985. Analysis of hybrid zones. *Annu. Rev. Ecol. Syst.* 16:113–48
11. Baum BR. 1992. Combining trees as a way of combining data sets for phylogenetic inference, and the desirability of combining gene trees. *Taxon* 41:3–10
12. Baum DA, Donoghue MJ. 1995. Choosing among alternative "phylogenetic" species concepts. *Syst. Bot.* 20:560–73
13. Baum DA, Shaw KL. 1995. Genealogical perspectives on the species problem. In *Experimental and Molecular Approaches to Plant Biosystematics,* ed. PC Hoch, AG Stephenson, pp. 289–303. St. Louis: Mo. Bot. Gard.
14. Bernardi G, Bernardi G. 1986. Compositional constraints and genome evolution. *J. Mol. Evol.* 24:1–11
15. Birky CW Jr. 1991. Evolution and population genetics of organelle genes: mechanisms and models. In *Evolution at the Molecular Level,* ed. RK Selander, AG Clark, TS Whittam, pp. 112–34. Sunderland, MA: Sinauer
16. Birky CW Jr, Maruyama T, Fuerst P.

1983. An approach to population and evolutionary genetic theory for genes in mitochondria and chloroplasts, and some results. *Genetics* 103:513–27
17. Blyth E. 1836. Observations on the various seasonal and other external changes which regularly take place in birds, more particularly in those which occur in Britain; with remarks on their great importance in indicating the true affinities of species; and upon the natural system of arrangement. *Mag. Nat. Hist.* 9:393–409, 505–14
18. Bookstein FL. 1994. Can biometrical shape be a homologous character? In *Homology: The Hierarchical Basis of Comparative Biology,* ed. BK Hall, pp. 197–227. San Diego: Academic
19. Bowler PJ. 1989. *Evolution, the History of an Idea.* Berkeley: Univ. Calif. Press. 2nd ed.
20. Brady RH. 1985. On the independence of systematics. *Cladistics* 1:113–26
21. Bramble DM. 1971. *Functional morphology, evolution and paleoecology of gopher tortoises.* PhD. Diss., Univ. Calif., Berkeley
22. Bremer K. 1994. Branch support and tree stability. *Cladistics* 10:295–304
23. Brower AVZ. 1994. Phylogeny of *Heliconius* butterflies inferred from mitochondrial DNA sequences (Lepidoptera: Nymphalidae). *Mol. Phylogenet. Evol.* 3:159–74
24. Brower AVZ, DeSalle R. 1994. Practical and theoretical considerations for choice of a DNA sequence region in insect molecular systematics, with a short review of published studies using nuclear gene regions. *Ann. Entomol. Soc. Am.* 87:702–16
25. Brown JM, Pellmyr O, Thompson JN, Harrison RG. 1994. Phylogeny of *Greya* (Lepidoptera: Prodoxidae), based on nucleotide sequence variation in mitochondrial cytochrome oxidase I and II: congruence with morphological data. *Mol. Biol. Evol.* 11:128–41
26. Brown WM, George M Jr, Wilson AC. 1979. Rapid evolution of animal mitochondrial DNA. *Proc. Natl. Acad. Sci. USA* 76:1967–71
27. Bull JJ, Huelsenbeck JP, Cunningham CW, Swofford DL, Waddell PJ. 1993. Partitioning and combining data in phylogenetic analysis. *Syst. Biol.* 42:384–97
28. Carr SM, Ballinger SW, Derr JN, Blankenship LH, Bickham JW. 1986. Mitochondrial DNA analysis of hybridization between sympatric white-tailed deer and mule deer in west Texas. *Proc. Natl. Acad. Sci. USA* 83:9576–80
29. Carson HL. 1983. Evolution of *Drosophila* on the newer Hawaiian volcanoes. *Heredity* 48:3–25
30. Carson HL. 1987. Tracing ancestry with chromosomal sequences. *Trends Ecol. Evol.* 2:203–7
31. Choudhary M, Singh RS. 1987. Historical effective size and the level of genetic diversity in *Drosophila melanogaster* and *Drosophila pseudoobscura. Biochem. Genet.* 25:41–51
32. Collins TM, Kraus F, Estabrook G. 1994. Compositional effects and weighting of nucleotide sequences for phylogenetic analysis. *Syst. Biol.* 43:449–59
33. Coyne JA. 1992. Genetics and speciation. *Nature* 355:511–15
34. Cracraft J. 1983. Species concepts and speciation analysis. In *Current Ornithology,* ed. RF Johnston, 1:159–87. New York: Plenum
35. Cracraft J. 1987. DNA hybridization in avian systematics. In *Evolutionary Biology,* ed. MK Hecht, B Wallace, GT Prance, 21:47–96. New York: Plenum
36. Craddock EM, Johnson WE. 1979. Genetic variation in Hawaiian *Drosophila.* 5. Chromosomal and allozymic diversity in *Drosophila silvestris* and its homosequential species. *Evolution* 33:137–55
37. Crandall KA. 1994. Intraspecific cladogram estimation: accuracy at higher levels of divergence. *Syst. Biol.* 43:222–35
38. Crandall KA, Templeton AR. 1993. Empirical tests of some predictions from coalescent theory with applications to intraspecific phylogeny reconstruction. *Genetics* 134:959–69
39. Cronin MA. 1991. Mitochondrial DNA phylogeny of deer (Cervidae). *J. Mammal.* 72:553–66
40. Cronin MA. 1991. Mitochondrial DNA and nuclear genetic relationships in deer (*Odocoileus* spp.) in western North America. *Can. J. Zool.* 69:1270–79
41. Cronin MA. 1992. Intraspecific variation in mitochondrial DNA of North American cervids. *J. Mammal.* 73:70–82
42. Cronin MA. 1993. Mitochondrial DNA in wildlife taxonomy and conservation biology: cautionary notes. *Wildl. Soc. Bull.*. 21:339–48
43. Cronin MA, Vyse ER, Cameron DG. 1988. Genetic relationships between mule deer and white-tailed deer in Montana. *J. Wildl. Manage.* 52:320–28
44. Daniels SB, Peterson KR, Strausbaugh

LD, Kidwell MG, Chovnick A. 1990. Evidence for horizontal transmission of the *P* transposable element between *Drosophila* species. *Genetics* 124:339–55

45. Darwin C. 1859. *On the Origin of Species.* London: John Murray. (Facsimile of 1st ed., 1964. Harvard Univ. Press)
46. Davis JI, Nixon KC. 1992. Populations, genetic variation, and the delimitation of phylogenetic species. *Syst. Biol.* 41:421–35
47. de Queiroz A, Donoghue MJ, Kim J. 1995. Separate versus combined analysis of phylogenetic evidence. *Annu. Rev. Ecol. Syst.* 26:657–81
48. de Queiroz K, Donoghue MJ. 1988. Phylogenetic systematics and the species problem. *Cladistics* 4:317–38
49. de Queiroz K, Donoghue MJ. 1990. Phylogenetic systematics or Nelson's version of cladistics? *Cladistics* 6:61–75
50. Derr JN. 1991. Genetic interactions between white-tailed deer and mule deer in the southwestern United States. *J. Wildl. Manage.* 55:228–37
51. DeSalle R, Gaddings LV. 1986. Discordance of nuclear and mitochondrial DNA phylogenies in Hawaiian *Drosophila. Proc. Natl. Acad. Sci. USA* 83:6902–6
52. DeSalle R, Hunt JA. 1987. Molecular evolution in Hawaiian drosophilids. *Trends Ecol. Evol.* 2:212–16
53. Disotell TR. 1994. Generic level relationships of the Papionini (Cercopithecoidea). *Am. J. Phys. Anthropol.* 94:47–57
54. Dobzhansky T. 1937. *Genetics and the Origin of Species.* New York: Columbia Univ. Press
55. Donnelly P, Tavaré S. 1995. Coalescents and genealogical structure under neutrality. *Annu. Rev. Genet.* 29:401–21
56. Doyle JJ. 1992. Gene trees, and species trees: molecular systematics as a 1-character taxonomy. *Syst. Bot.* 17:144–63
57. Doyle JJ. 1995. The irrelevance of allele tree topologies for species delimitation, and a non-topological alternative. *Syst. Bot.* 20:574–88
58. Echelle AA, Connor PJ. 1989. Rapid, geographically extensive genetic introgression after secondary contact between two pupfish species (*Cyprinodon,* Cyprinodontidae). *Evolution* 43:717–27
59. Endler JA. 1977. *Geographic Variation, Speciation, and Clines.* Princeton, NJ: Princeton Univ. Press
60. Excoffier L, Smouse PE. 1994. Using allele frequencies and geographic subdivision to reconstruct gene trees within a species: molecular variance parsimony. *Genetics* 136:343–59
61. Farris JS. 1971. The hypothesis of nonspecificity and taxonomic congruence. *Annu. Rev. Ecol. Syst.* 2:277–302
62. Farris JS. 1983. The logical basis of phylogenetic analysis. In *Advances in Cladistics,* ed. NI Platnick, VA Funk, 2:7–36. New York: Columbia Univ. Press
63. Farris JS, Källersjö M, Kluge AG, Bult C. 1994. Testing significance of congruence. *Cladistics* 10:315–19
64. Felsenstein J. 1982. Numerical methods for inferring evolutionary trees. *Q. Rev. Biol.* 57:379–404
65. Felsenstein J. 1988. Phylogenies from molecular sequences: inferences and reliability. *Annu. Rev. Genet.* 22:521–65
66. Felsenstein J, Churchill GA. 1996. A hidden Markov model approach to variation among sites in rate of evolution. *Mol. Biol. Evol.* 13:93–104
67. Ferris SD, Sage RD, Huang C-M, Nielsen JT, Ritte U, Wilson AC. 1983. Flow of mitochondrial DNA across a species boundary. *Proc. Natl. Acad. Sci. USA* 80:2290–94
68. Fitch WM. 1970. Distinguishing homologous from analogous proteins. *Syst. Zool.* 19:99–113
69. Fooden J. 1979. Taxonomy and evolution of *sinica* group of macaques: I. Species and subspecies accounts of *Macaca sinica. Primates* 20:109–40
70. Frost DR, Kluge AG. 1994. A consideration of epistemology in systematic biology, with special reference to species. *Cladistics* 10:259–94
71. Gaur LK, Hughes AL, Heise ER, Gutknecht J. 1992. Maintenance of *DQB1* polymorphisms in primates. *Mol. Biol. Evol.* 9:599–609
72. Gavin TA, May B. 1991. Taxonomic status and genetic purity of Columbian white-tailed deer. *J. Wildl. Manage.* 52:1–10
73. Gillespie JH. 1991. *The Causes of Molecular Evolution.* Oxford: Oxford Univ. Press
74. Goldman N. 1993. Statistical tests of models of DNA substitution. *J. Mol. Evol.* 36:182–98
75. Goodman M, Czelusniak J, Moore GW, Romero-Herrera AE, Matsuda G. 1979. Fitting the gene lineage into its species

lineage, a parsimony strategy illustrated by cladograms constructed from globin sequences. *Syst. Zool.* 28:132–63
76. Gould SJ. 1985. A clock of evolution: we finally have a method for sorting out homologies from "subtle as subtle can be" analogies. *Nat. Hist.* 94:12–25
77. Graybeal A. 1995. Naming species. *Syst. Biol.* 44:237–50
78. Gyllensten U, Wilson AC. 1987. Interspecific mitochondrial DNA transfer and the colonization of Scandinavia by mice. *Genet. Res.* 49:25–29
79. Hale LR, Beckenbach AT. 1985. Mitochondrial DNA variation in *Drosophila pseudoobscura* and related species in Pacific Northwest populations. *Can. J. Genet. Cytol.* 27:357–64
80. Harrison RG. 1989. Animal mitochondrial DNA as a genetic marker in population and evolutionary biology. *Trends Ecol. Evol.* 4:6–11
81. Harrison RG. 1990. Hybrid zones: windows on evolutionary process. In *Oxford Surveys in Evolutions Biology,* ed. D Futuyma, J Antonovics, 7:69–128. Oxford: Oxford Univ. Press
82. Harrison RG. 1991. Molecular changes at speciation. *Annu. Rev. Ecol. Syst.* 22:281–308
83. Hasegawa M, Hashimoto T. 1993. Ribosomal RNA trees misleading? *Nature* 361:23
84. Hein J. 1990. Reconstructing evolution of sequences subject to recombination using parsimony. *J. Math. Biosci.* 98:185–200
85. Hennig W. 1966. *Phylogenetic Systematics.* Urbana: Univ. Ill. Press. 263 pp.
86. Hillis DM. 1987. Molecules versus morphological approaches to systematics. *Annu. Rev. Ecol. Syst.* 18:23–42
87. Hillis DM, Bull JJ, White ME, Badgett MR, Molineux IJ. 1993. Experimental approaches to phylogenetic analysis. *Syst. Biol.* 42:90–92
88. Hoelzer GA, Dittus WPJ, Ashley MV, Melnick DJ. 1994. The local distribution of highly divergent mitochondrial DNA haplotypes in toque macaques *Macaca sinica* at Poponnaruwa, Sri Lanka. *Mol. Ecol.* 3:451–58
88a. Hoelzer GA, Melnick DJ. 1994. Patterns of speciation and limits to phylogenetic resolution. *Trends Ecol. Evol.* 9:104–7
89. Hudson RR. 1990. Gene genealogies and the coalescent process. In *Oxford Surveys in Evolutionary Biology,* ed. D Futuyma, J Antonovics, 7:1–44. Oxford: Oxford Univ. Press
90. Hudson RR. 1993. The how and why of generating gene genealogies. In *Mechanisms of Molecular Evolution,* ed. N Takahata, AG Clark, pp. 23–36. Sunderland, MA: Sinauer
91. Hudson RR, Kreitman M, Aguadé M. 1987. A test of neutral molecular evolution based on nucleotide data. *Genetics* 116:153–59
92. Huelsenbeck JP. 1995. Performance of phylogenetic methods in simulation. *Syst. Biol.* 44:17–48
93. Hunt JA, Carson HL. 1982. Evolutionary relationships of four species of Hawaiian *Drosophila* as measured by DNA reassociation. *Genetics* 104:353–64
94. Huxley TH. 1874. On the classification of the animal kingdom. *Nature* 11:101–2
95. Ikemura T. 1985. Codon usage and tRNA content in unicellular and multicellular organisms. *Mol. Biol. Evol.* 2:13–34
96. Kambysellis MP, Ho K-F, Craddock EM, Piano F, Parisi M, Cohen J. 1995. Pattern of ecological shifts in the diversification of Hawaiian *Drosophila* inferred from a molecular phylogeny. *Curr. Biol.* 5:1129–39
97. Kaneshiro KY, Boake CRB. 1987. Sexual selection and speciation: issues raised by Hawaiian *Drosophila. Trends Ecol. Evol.* 2:207–12
98. Kaplan NL, Hudson RR, Langley CH. 1989. The "hitchhiking effect" revisited. *Genetics* 123:887–99
99. Kidwell MG. 1993. Lateral transfer in natural populations of eukaryotes. *Annu. Rev. Genet.* 27:235–56
100. Kimura M. 1983. *The Neutral Theory of Molecular Evolution.* Cambridge: Cambridge Univ. Press
101. Kimura M, Ohta T. 1969. The average number of generations until fixation of a mutant gene in a finite population. *Genetics* 61:763–71
102. Kingman JFC. 1982. On the genealogy of large populations. *J. Appl. Probab. A* 19:27–43
103. Kishino H, Miyata T, Hasegawa M. 1990. Maximum likelihood inference in protein phylogeny and the origin of chloroplasts. *J. Mol. Evol.* 21:151–60
104. Kluge AJ. 1989. A concern for evidence and a phylogenetic hypothesis of relationships among *Epicrates* (Boidae, Serpentes). *Syst. Zool.* 38:7–25
105. Kuhn TS. 1970. *The Structure of Scientific Revolutions.* Chicago: Univ. Chicago Press. 2nd ed.
106. Lamb T, Avise JC, Gibbons JW. 1989.

Phylogeographic patterns in mitochondrial DNA of the desert tortoise (*Xerobates agassizi*), and evolutionary relationships among the North American gopher tortoises. *Evolution* 43:76–87
107. Lanyon SM. 1993. Phylogenetic frameworks: towards a firmer foundation for the comparative approach. *Biol. J. Linn. Soc.* 49:45–61
108. Larson A. 1994. The comparison of morphological and molecular data in phylogenetic systematics. In *Molecular Ecology and Evolution: Approaches and Applications,* ed. B Schierwater, B Streit, GP Wagner, R DeSalle, pp. 371–90. Basel: Birkhauser
109. Lewontin RC. 1974. *The Genetic Basis of Evolutionary Change.* New York: Columbia Univ. Press
110. Luckow M. 1995. Species concepts: assumptions, methods, and applications. *Syst. Bot.* 20:589–605
111. Maddison W. 1995. Phylogenetic histories within and among species. In *Experimental and Molecular Approaches to Plant Biosystematics,* ed. PC Hoch, AG Stephenson, pp. 273–87. St. Louis: Mo. Bot. Gard.
112. Martin AP. 1995. Mitochondrial DNA sequence evolution in sharks: rates, patterns, and phylogenetic inferences. *Mol. Biol. Evol.* 12:1114–23
113. Maruyama T, Birky CW Jr. 1991. Effects of periodic selection on gene diversity in organelle genomes and other systems without recombination. *Genetics* 127:449–51
114. Mayr E. 1942. *Systematics and the Origin of Species.* New York: Columbia Univ. Press (1982 ed.)
115. Mayr E. 1963. *Animal Species and Evolution.* Cambridge, MA: Belknap
116. Melnick DJ, Hoelzer GA. 1992. Differences in male and female macaque dispersal lead to contrasting distributions of nuclear and mitochondrial DNA variation. *Int. J. Primatol.* 13:379–93
117. Mickevich MF. 1978. Taxonomic congruence. *Syst. Zool.* 27:143–58
118. Mickevich MF, Johnson MS. 1976. Congruence between morphological and allozyme data in evolutionary inference and character evolution. *Syst. Zool.* 25:260–70
119. Miller JS, Brower AVZ, DeSalle R. 1996. Phylogeny of the neotropical moth tribe Josiini (Notodontidae: Dioptinae): evidence from DNA sequences and morphology. *Biol. J. Linn. Soc.* In press
120. Mitton JB. 1994. Molecular approaches to population biology. *Annu. Rev. Ecol. Syst.* 25:45–69
121. Miyamoto MM, Cracraft J. 1991. Phylogenetic inference, DNA sequence analysis, and the future of molecular systematics. See Ref. 121a, pp. 3–17
121a. Miyamoto MM, Cracraft J, eds. 1991. *Phylogenetic Analysis of DNA Sequences.* Oxford, UK: Oxford Univ. Press
122. Miyamoto MM, Fitch WM. 1995. Testing species phylogenies and phylogenetic methods with congruence. *Syst. Biol.* 44:64–76
123. Moore WS. 1995. Inferring phylogenies from mtDNA variation: mitochondrial-gene trees versus nuclear-gene trees. *Evolution* 49:718–26
124. Moritz C, Dowling TE, Brown WM. 1987. Evolution of animal mitochondrial DNA: relevance for population biology and systematics. *Annu. Rev. Ecol. Syst.* 18:269–92
125. Moritz C, Schneider CJ, Wake DB. 1992. Evolutionary relationships within the *Ensatina eschscholtzii* complex confirm the ring species interpretation. *Syst. Biol.* 41:273–91
126. Nei M, Graur D. 1984. Extent of protein polymorphism and the neutral mutation theory. *Evol. Biol.* 17:73–118
127. Neigel JE, Avise JC. 1986. Phylogenetic relationships of mitochondrial DNA under various demographic models of speciation. In *Evolutionary Processes and Theory,* ed. E Nevo, S Karlin, pp. 515–34. San Diego: Academic
128. Nelson G. 1983. Reticulation in cladograms. In *Advances in Cladistics,* ed. NI Platnick, VA Funk, 2:105–11. New York: Columbia Univ. Press
129. Nelson G. 1989. Cladistics and evolutionary models. *Cladistics* 5:275–89
130. Nelson G, Platnick N. 1981. *Systematics and Biogeography.* New York: Columbia Univ. Press
131. Nixon KC, Wheeler QD. 1990. An amplification of the phylogenetic species concept. *Cladistics* 6:211–23
132. Nixon KC, Carpenter JM. 1996. On simultaneous analysis. *Cladistics.* In press
133. Olmstead RG, Sweere JA. 1994. Combining data in phylogenetic systematics: an empirical approach using three molecular data sets in the Solanaceae. *Syst. Biol.* 43:467–81
134. Omland KC. 1994. Character congruence between a molecular and morphological phylogeny for dabbling ducks (*Anas*). *Syst. Biol.* 43:369–86

135. Ospovat D. 1981. *The Development of Darwin's Theory.* Cambridge: Cambridge Univ. Press
136. Page RDM. 1993. Genes, organisms, and areas: the problem of multiple lineages. *Syst. Biol.* 42:77–84
137. Palumbi SR. 1989. Rates of molecular evolution and the fraction of nucleotide positions free to vary. *J. Mol. Evol.* 29:180–87
138. Pamilo P, Nei M. 1988. Relationships between gene trees and species trees. *Mol. Biol. Evol.* 5:568–83
139. Panchen AL. 1992. *Classification, Evolution and the Nature of Biology.* Cambridge: Cambridge Univ. Press
140. Patterson C. 1988. Homology in classical and molecular biology. *Mol. Biol. Evol.* 5:603–25
141. Patterson C, Williams DM, Humphries CJ. 1993. Congruence between molecular and morphological phylogenies. *Annu. Rev. Ecol. Syst.* 24:153–88
142. Platnick NI. 1977. Cladogams, phylogenetic trees, and hypothesis testing. *Syst. Zool.* 26:438–42
143. Platnick NI. 1979. Philosophy and the transformation of cladistics. *Syst. Zool.* 28:537–46
144. Popper KR. 1965. *The Logic of Scientific Discovery.* New York: Harper & Rowe
145. Powell JR. 1983. Interspecific cytoplasmic gene flow in the absence of nuclear gene flow: evidence from *Drosophila. Proc. Natl. Acad. Sci. USA* 80:492–95
146. Powell JR. 1991. Monophyly/paraphyly/polyphyly and gene/species trees: an example from *Drosophila. Mol. Biol. Evol.* 8:892–96
147. Provine WB. 1971. *The Origins of Theoretical Population Genetics.* Chicago: Univ. Chicago Press
148. Rhymer JM, Williams MJ, Braun MJ. 1994. Mitochondrial analysis of gene flow between New Zealand mallards *(Anas platyrhynchos)* and grey ducks *(A. superciliosa). Auk* 111:970–78
149. Rieppel OC. 1988. *Fundamentals of Comparative Biology.* Basel: Birkhauser
150. Robertson HM. 1995. The Tcl-mariner superfamily of transposons in animals. *J. Insect Physiol.* 41:99–105
151. Rogers AR, Harpending H. 1992. Population growth makes waves in the distribution of pairwise genetic markers. *Mol. Biol. Evol.* 9:552–69
152. Rowan RG, Hunt JA. 1991. Rates of DNA change and phylogeny from the DNA sequences of the alcohol dehydrogenase gene for five closely related species of Hawaiian *Drosophila. Mol. Biol. Evol.* 8:49–70
153. Rzhetsky A, Nei M. 1995. Tests of applicability of several substitution models for DNA sequence data. *Mol. Biol. Evol.* 12:131–51
154. Saitou N, Nei M. 1986. The number of nucleotides required to determine the branching order of three species with special reference to the human-chimpanzee-gorilla divergence. *J. Mol. Evol.* 24:189–204
155. Saitou N, Nei M. 1987. The neighbor-joining method: a new method for reconstructing phylogenetic trees. *Mol. Biol. Ecol.* 4:406–25
156. Sanderson MJ, Donoghue MJ. 1989. Patterns of variation in levels of homoplasy. *Evolution* 43:1781–95
157. Schaeffer SW, Aquadro CF, Anderson WW. 1987. Restriction-map variation in the alcohol dehydrogenase region of *Drosophila pseudoobscura. Mol. Biol. Evol.* 4:254–65
158. Selander RK. 1982. Phylogeny. In *Perspectives on Evolution,* ed. R Milkman, pp. 32–59. Sunderland, MA: Sinauer
159. Sharp PM, Li W-H. 1987. The rate of synonymous substitutions in enterobacterial genes is inversely related to codon usage bias. *Mol. Biol. Evol.* 4:222–30
160. Shotake T, Nozawa K, Santiapilai C. 1991. Genetic diversity within and between the troops of toque macaque, *Macaca sinica,* in Sri Lanka. *Primates* 32:283–99
161. Sibley CG, Ahlquist JA, Sheldon FH. 1987. DNA hybridization and avian phylogenetics. In *Evolutionary Biology,* ed. MK Hecht, B Wallace, GT Prance, 21:97–125. New York: Plenum
162. Simon C, Frati F, Beckenbach A, Crespi B, Liu H, Flook P. 1994. Evolution, weighting and phylogenetic utility of mitochondrial gene sequences and a compilation of conserved polymerase chain reaction primers. *Ann. Entomol. Soc. Am.* 87:651–701
163. Simonsen KL, Churchill GA, Aquadro CF. 1995. Properties of statistical tests of neutrality for DNA polymorphism data. *Genetics* 141:413–29
164. Simpson GG. 1944. *Tempo and Mode in Evolution.* New York: Columbia Univ. Press
165. Smocovitis VB. 1992. Unifying biology: the evolutionary synthesis and evolutionary biology. *J. Hist. Biol.* 25:1–65

166. Sneath P, Sokal R. 1973. *Numerical Taxonomy.* San Francisco: Freeman
167. Sober E. 1988. *Reconstructing the Past.* Cambridge, MA: MIT Press
168. Sober E. 1993. Experimental tests of phylogenetic inference methods. *Syst. Biol.* 42:85–89
169. Spolsky C, Uzzell T. 1984. Natural interspecies transfer of mitochondrial DNA in amphibians. *Proc. Natl. Acad. Sci. USA* 81:5802–5
170. Spolsky C, Uzzell T. 1986. Evolutionary history of the hybridogenetic hybrid frog *Rana esculenta* as deduced from mtDNA analysis. *Mol. Biol. Evol.* 3:44–56
171. Steel MA, Lockhart PJ, Penny D. 1993. Confidence in evolutionary trees from biological sequence data. *Nature* 364:440–42
172. Stevens PF. 1983. Augustin Angier's "Arbre botanique" (1801), a remarkable early botanical representation of the natural system. *Taxon* 32:203–11
173. Strickland HE. 1841. On the true method of discovering the natural system in zoology and botany. *Ann. Mag. Nat. His.* 6:184–94
174. Sullivan J, Holzinger KE, Simon C. 1995. Among-site variation and phylogenetic analysis of 12S RRNA in sigmodontine rodents. *Mol. Biol. Evol.* 12:988–1001
175. Swofford DL. 1991. When are phylogeny estimates from molecular and morphological data incongruent? See Ref. 121a, pp. 295–333
176. Swofford DL, Olsen GJ, Waddell PJ, Hillis DM. 1996. Phylogenetic inference. In *Molecular Systematics,* ed. DM Hillis, C Moritz, BK Mable, pp. 407–14. Sunderland, MA: Sinauer. 2nd ed.
177. Szymura JM, Spolsky C, Uzzell T. 1985. Concordant change in mitochondrial and nuclear genes in a hybrid zone between two frog species (genus *Bombina*). *Experientia* 41:1469–70
178. Tajima F. 1983. Evolutionary relationship of DNA sequences in finite populations. *Genetics* 105:437–60
179. Tajima F. 1989. DNA polymorphism in a subdivided population: the expected number of segregating sites in the two-subpopulation model. *Genetics* 123:229–40
180. Tajima F. 1993. Measurement of DNA polymorphism. In *Mechanisms of Molecular Evolution,* ed. N Takahata, AG Clark, pp. 37–59. Sunderland, MA: Sinauer
181. Takahata N. 1989. Gene genealogy in three related populations: consistency probability between gene and population trees. *Genetics* 122:957–66
182. Takahata N. 1990. A simple genealogical structure of strongly balanced allelic lines and trans-species evolution of polymorphism. *Proc. Natl. Acad. Sci. USA* 87:2419–23
183. Takahata N, Slatkin M. 1984. Mitochondrial gene flow. *Proc. Natl. Acad. Sci. USA* 81:1764–67
184. Tegelström H, Gelter HP. 1990. Haldane's rule and sex-biased gene flow between two hybridizing flycatcher species (*Fidecula albicollis* and *F. hypoleuca,* Aves: Muscicapidae). *Evolution* 44:2012–21
185. Templeton AR. 1989. The meaning of species and speciation: a genetic perspective. In *Speciation and Its Consequences,* ed. D Otte, JA Endler, pp. 3–27. Sunderland, MA: Sinauer
186. Templeton AR. 1994. Biodiversity at the molecular genetic level: experiences from disparate macroorganisms. *Philos. Trans. R. Soc. London Ser. B* 345:59–64
187. Templeton AR, Boerwinkle E, Sing CF. 1987. A cladistic analysis of phenotypic associations with haplotypes inferred from restriction endonuclease mapping. I. Basic theory and an analysis of alcohol dehydrogenase activity in *Drosophila. Genetics* 117:343–51
188. Templeton AR, Crandall KA, Sing CF. 1992. A cladistic analysis of phenotypic associations with haplotypes inferred from restriction endonuclease mapping and DNA sequence data. III. Cladogram estimation. *Genetics* 132:619–33
189. Templeton AR, Routman E, Phillips CA. 1995. Separating population structure from population history: a cladistic analysis of the geographical distribution of mitochondrial DNA haplotypes in the tiger salamander, *Ambystoma tigrinum. Genetics* 140:767–82
190. Thorpe RS. 1984. Primary and secondary transition zones in speciation and population differentiation: a phylogenetic analysis of range expansion. *Evolution* 38:233–43
191. Vogler AP, Knisley CB, Glueck SB, Hill JM, DeSalle R. 1993. Using molecular and ecological data to diagnose endangered populations of the puritan tiger beetle *Cicindela puritana. Mol. Ecol.* 2:375–83
192. von Haeseler A, Churchill GA. 1993. Network models for sequence evolution.

J. Mol. Evol. 37:77–85
193. Weir BS. 1990. Intraspecific differentiation. In *Molecular Systematics,* ed. DM Hillis, C Moritz, pp. 373–410. Sunderland, MA: Sinauer
194. Wenzel JW, Carpenter JM. 1994. Comparing methods: adaptive traits and tests of adaptation. In *Phylogenetics and Ecology,* ed. P Eggleton, RI Vane-Wright, pp. 79-101. London: Academic
195. Westwood JO. 1840. Observations upon the relationships existing amongst natural objects, resulting from more or less perfect resemblance, usually termed affinity and analogy. *Mag. Nat. Inst.* (NS) 6:141–44
196. Wheeler QD, Nixon KC. 1990. Another way of looking at the species problem: a reply to de Queiroz and Donoghue. *Cladistics* 6:77–81
197. Wheeler WC. 1991. Congruence among data sets: a Bayesian approach. See Ref. 121a, pp. 334–46
198. Wiley EO. 1975. Karl R. Popper, systematics, and classification: a reply to Walter Bock and other evolutionary systematists. *Syst. Zool.* 24:233–43
199. Wiley EO. 1981. *Phylogenetics.* New York: Wiley & Sons
200. Yang Z, Goldman N, Friday A. 1994. Comparison of models for nucleotide substitution used in maximum-likelihood phylogenetic estimation. *Mol. Biol. Evol.* 11:316–24
201. Yoder AD. 1994. Relative position of the Cheirogaleidae in strepsirhine phylogeny: a comparison of morphological and molecular methods and results. *Am. J. Phys. Anthropol.* 94:25–46
202. Zouros E, Ball AO, Saavedra C, Freeman KR. 1994. An unusual type of mitochondrial DNA inheritance in the blue mussel *Mytilus. Proc. Natl. Acad. Sci. USA* 91:7463–67

Annu. Rev. Ecol. Syst. 1996. 27:451–76

INCIDENCE AND CONSEQUENCES OF INHERITED ENVIRONMENTAL EFFECTS

MaryCarol Rossiter
Institute of Ecology, University of Georgia, Athens, Georgia 30602

KEY WORDS: maternal effects, nongenetic cross-generational transmission, non-Mendelian inheritance, paternal effects, time-lagged components of phenotypic variance

ABSTRACT

Inherited environmental effects are those components of the phenotype that are derived from either parent, apart from nuclear genes. Inherited environmental effects arise as the product of parental genes and the parental environment, or their interation, and can include contributions that reflect the abiotic, nutritional, and other ecological features of a parental environment. Separating the impact of inherited environmental effects from inherited genetic effects on offspring phenotype variation has been and continues to be a challenge. This complexity is represented in the presentation of a qualitative model that distinguishes the possible paths of nongenetic cross-generational transmission. This model serves as the framework for considering the nature, in published works, of what was actually measured. Empirical evidence of inherited environmental effects arising from these pathways is documented for a diversity of plant and animal taxa. From these results one can conclude that the impact of inherited environmental effects on offspring can be positive or negative depending on the nature of the contribution and the ecological context in which the offspring exists. Finally, there is a description of theoretical and experimental efforts to understand the consequences of parental effects relative to their impact on population dynamics, the expression of adaptive phenotype plasticity, and character evolution.

INTRODUCTION AND TERMINOLOGY

Inherited environmental effects include those components of an offspring's phenotype that are derived from the parent, apart from the nuclear genes. This definition serves as the broad designation for the outcome of multiple cross-

0066-4162/96/1120-0451$08.00

generational processes described in the following section. In this review, you will see that inherited environmental effects are documented for species from a broad taxonomic range, and that the external parental environment often plays a significant role in the magnitude and nature of the expression of these effects. Moreover, most of the parental environmental variables known to contribute to inherited environmental effects are permanent components of species' environments (e.g. seasonal features, density, or food and habitat quality); this fact suggests their sustained importance in ecological and evolutionary processes. Much of the work that addresses the complexity of transmission of inherited environmental effects uses quantitative genetic analysis. I hope to distill, but not to minimize, this complexity by including a general framework that provides a way to think about how previous environmental experience interacts with genes to alter individual and population responses in unexpected ways.

Since an array of terms are used for the processes involved in inherited environmental effects, I first define those used frequently. The most common and most variably defined are "maternal effects" and "paternal effects." These designations include the inheritance of nuclear genes from the mother or father (84, 134, 211), the inheritance of cytoplasmic genes (mitochondria or plastids) from the mother or father (95, 134), or the transmission of information derived from parental quality or parental environment (11, 40, 116, 161). Sometimes these terms are used without distinguishing the source of impact on offspring phenotype. For example, the presence of a maternal or paternal effect is inferred when offspring from reciprocal crosses differ in mean phenotype (68, 150, 157, 184, 190).

The term "maternal genetic inheritance" or "maternal additive inheritance" is used when a component of the maternal effect is of genetic origin, that is, due to maternal performance genes expressed in the mother and received as an environmental component of offspring phenotype. After the maternal genetic effects are accounted for, the remainder of the inherited environmental effect is often called a "maternal environmental effect" (40, 50, 56, 160, 217, 218). When confirmation of nongenetic cross-generational transmission arises (e.g. from variation in the parental environment due to photoperiod or nutritional quality), the terms "environmentally based" maternal, paternal, or parental effects (69, 168) and "general environmental" and "specific environmental" maternal or paternal effects (180) are used, depending on the experimental design. Durrant (51) found that flax varieties grown under several nutrient regimes responded with changes in their own growth parameters and appeared to transmit these changes to their offspring in the next and later generations, a phenomenon he called maternal "conditioning."

Inherited environmental effects that extend beyond one offspring generation are called "permanent" (218) or "persistent" (160) environmental effects. For example, an environmental effect on a grandmother's maternal performance influences the maternal performance of her daughter and thus affects phenotypic values in grandchildren, and so on (56). Multiple generation reverberation is also called "environmental inheritance" because the maternal performance phenotype of the mother can directly modify the maternal performance phenotype of the daughter, even in the absence of genetic heritability (160).

To facilitate discussion of the evolutionary impact of inherited environmental effects, Kirkpatrick & Lande (91) used the term "maternal inheritance" for the transmission of non-Mendelian contributions from parent to offspring, and "maternal selection" for the effect of this transmission on offspring fitness. Similarly, Lombardi (109) suggested that "maternal influence," which is the equivalent of "maternal inheritance," should refer to the source rather than to the phenotypic impact on offspring expressing inherited environmental effects.

PROCESSES THAT PRODUCE INHERITED ENVIRONMENTAL EFFECTS

The multiplicity of meanings and underlying assumptions encompassed by the term "maternal effects" likely arises from difficulties in (*a*) distinguishing inherited environmental effects from direct genetic effects (i.e. transmission of nuclear genes to offspring), and (*b*) determining the extent to which inherited environmental effects are mediated by the parental genotype. Some of this complexity is represented in Figure 1, which is based on a quantitative genetics model with inherited environmental effects and a variable offspring environment of Eisen & Saxton (53). This figure differs from their model by the addition of a variable parental environment, a condition often measured or manipulated in empirical studies of inherited environmental effects. To describe the processes involved in the production of inherited environmental effects, it is best to start with the offspring phenotype, then consider what sources may contribute to it.

In Figure 1, offspring phenotype P_o may derive from any or all of eight sources shown. Since this figure represents the involvement of only one (either) parent, interactions between maternal and paternal sources (e.g. as occur in plant endosperm) are not addressed. The source of a contribution to offspring phenotype is indicated as genetic (G) or environmental (E), originating in the parental (m) or offspring (o) generation. The parental performance phenotype, $P_{m(t)}$, represents those traits related to the quality of parenting. P_m arises from parental genes for performance traits, $G_{m(t-1)}$, whose expression can be modified by the parental environment, $E_{m(t-1)}$ or offspring environment, $E_{o(t)}$.

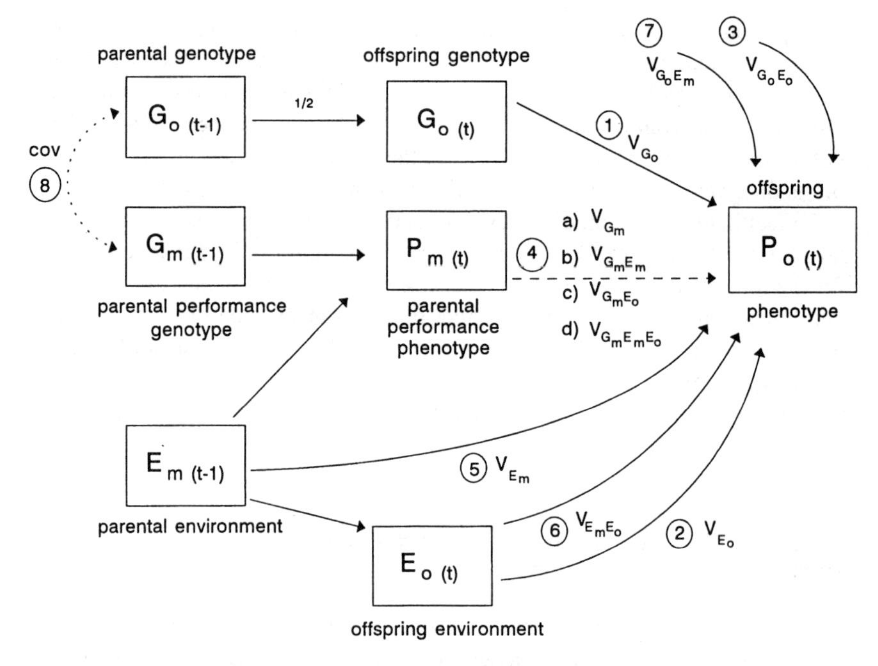

Figure 1 The components of offspring phenotype (P_o) expressed in time *t*, deriving from the direct contribution of nuclear genes by one parent (G_o), a time-lagged presentation of the parental environment (E_m), a time-lagged expression of parental performance genes (G_m) and their interactions with the parental environment to produce the parental performance phenotype (P_m), plus the offspring's own environment (E_o). For simplicity of presentation, G indicates additive genetic effects with dominance and epistasis assumed to be negligible. The numbered sources indicate possible routes of contribution to the offspring phenotype; Source 8 is any genetic covariance (cov) between genes expressed in two generations such as $covG_mG_o$ or $cov(G_mE_o)(G_oE_o)$; see text for full description of variables.

The contributions to offspring phenotype typically considered come from: source 1, the offspring genotype, V_{Go}, which arrives as some fraction (usually, but not always 1/2) of the parental genotype; source 2, the offspring environment, V_{Eo}; and source 3, interaction between offspring genotype and offspring environment, V_{GoEo}. Modification of offspring phenotype due to inherited environmental effects comes from sources 4 through 8: source 4, contribution of the parental performance phenotype to offspring phenotype due to parental performance genotype, V_{Gm} (4a); interaction between the parental performance genotype and parental environment, V_{GmEm} (4b); interaction between parental performance genotype and offspring environment, V_{GmEo} (4c); or interaction between parental performance genotype, parental environment, and offspring environment, V_{GmEmEo} (4d); source 5, contribution of the parental environment,

V_{Em}; source 6, the interaction between parental and offspring environment, V_{EmEo}; source 7, interaction between parental environment and offspring genotype, V_{GoEm}; and source 8, covariance (correlation) between parental performance genes expressed in the parental $(t-1)$ and in subsequent $(t, t+1\ldots)$ generations. Here, covariance can be simple, $covG_mG_o$, or associated with interaction effects, e.g. $cov(G_mE_o)(G_oE_o)$. The contribution of these covariance components to offspring phenotype can be positive or negative, a feature of great consequence in interpreting experimental results and predicting the consequences of inherited environmental effects (6, 91, 160, 218).

There is empirical evidence for each of these sources, although a single experimental design seldom assesses all of them: Source 4a: References 98, 142, 180 (plants), 35 (rotifers), 153 (guppies), 6, 43 (mammals); source 4b: 38, 98, 142, 145, 180 (plants), 89 (insects), 35 (rotifers), 64 (daphnia), 9, 165 (rodents), 107 (fish); source 4d, specifically: 180 (inferred, plants), 107 (fish); sources 4c, 4d, or 6: 1, 60, 124, 125, 180, 182, 198, 220 (plants), 58, 59, 65, 66, 67a, 89, 92, 168, 212 (insects), 24 (daphnia), 20 (voles), 87 (frogs), 74, 107 (fish), 29 (cattle); source 5: 180 (plants), 78 (insects), 9 (rats); source 7: 98 (plants); source 8: 19 (insects), 6, 160, 192, 217 (mammals).

While this model can act as a guide to considering the processes involved in the expression of inherited environmental effects, an additional complexity must be recognized. It is often difficult to assess the presence, source, or magnitude of inherited environmental effects (*a*) when selection acts on the gametophyte, which means that selection acts on the parental genotype, a situation that can give the appearance of inherited environmental effects, or (*b*) when selection acts during the embryonic stage, a situation often difficult to measure (59, 194). Experiments can be designed to distinguish (60, 119a) or minimize (102, 103, 147, 173, 208) the confounding effects of selection on the measurement of inherited environmental effects.

HOW THE PARENTAL ENVIRONMENT PARTICIPATES

In addition to direct nuclear and cytoplasmic genetic contributions, parents can provide offspring with nutrients. This aspect of inherited environmental effects is the most extensively studied. Data on egg and seed quality often come from studies of parental resource allocation patterns and tradeoffs between propagule size and number, long-standing topics in the study of life-history evolution (12, 37, 76, 153, 183, 188, 191, 201, 210a). Tools and techniques for quantification and manipulation (e.g. 10) provide a tractable approach to studying this aspect of inherited environmental effects. In addition to nutrients, parents can provide offspring with a preregulated genome, defensive agents, symbionts, pathogens, toxins, hormones, enzymes, and cultural conditioning. Depending on the species, parental environmental contributions can be limited to gamete packaging (egg or seed production) or extended through prenatal and postnatal periods of care. What follows is a sampling of references that illustrate each type of environmental participation.

Parental Imprinting

The notion that the environment might determine regulation in a gametic genome was suggested by Giesel from work on *Drosophila* (61). He offered that a photoperiod-induced paternal effect on offspring development time might be attributed to "memory effects" in the male germline arising from changes in some aspect of gene regulation. Since that time, considerable work has been done on parental imprinting wherein the expression of a gene (or an entire parental genome) in the offspring varies according to its maternal or paternal origin (210b). It has been argued that DNA methylation (which determines the expression or repression of a gene) is the mechanism of regulatory modification (8, 115, 199). To support this argument, Barlow (8) developed a model that assumes that regulatory modifications occur during gametogenesis because that is the only period when maternal and paternal gametic genomes are separate and can be subjected to different influences. Since methylation patterns are heritable, it is reasonable to consider that any environmental influences on DNA methylation patterns in gametes could result in inherited environmental effects. In the absence of any empirical work, Jablonka & Lamb (80) developed a model wherein an environmental stimulus induces heritable chromatin modifications that are specific and predictable, and which might result in an adaptive response to the environmental stimulus.

Nutritional Factors in the Parental Environment

The availability of nutrients in the parental generation is known to influence offspring phenotype. For example, minerals such as iron and zinc are taken up from the parental diet and transmitted to eggs (insects) or developing embryos (rodents) (108, 202). In some insect species, the male provides the female with nuptial gifts such as prey items or products of his accessory glands (17) which provide energy sources (104, 114) and minerals (54, 141). While these gifts may be used by the female for her own needs, there have been numerous confirmations that the paternally derived materials end up in the egg itself (17).

In species with extended maternal care, the nutritional status of the mother during prenatal and postnatal care can influence offspring phenotype. In laboratory rats, the impact of alcohol intake by mothers on fetal development was modified by the quality of her nutrition during pregnancy, such that the negative effects of alcohol ingestion on offspring were greater when the mother had a low-protein diet (209). In wild squirrels, juveniles born to mothers who received supplemental food (sunflower seeds left daily at burrow entrances) were

28% heavier compared to offspring of untreated mothers (owing to faster growth versus longer development time), and this mass difference was maintained after juveniles left the natal burrow (205).

Abiotic and Ecological Factors in the Parental Environment

The parental thermal experience can influence cold acclimation of offspring in plants (145) and cold tolerance of offspring in *Drosophila* (213). Density of the parental population can influence yolk quality in quail (7) and early behavior of offspring in tent caterpillars (216). From correlations between temporal or spatial environmental heterogeneity and variable provisioning of eggs or seeds by a mother, it has been argued that variable provisions increase the probability of maternal fitness when habitat quality for emerging young is unpredictable (e.g. 47, 153).

Defensive Agents

Plant secondary compounds can be sequestered by either parent and passed on to offspring where they serve to defend offspring against natural enemies. Paternal contributions of plant-derived alkaloids are made to offspring in moths (52, 72), butterflies (28), and grasshoppers (22), and terpenoid contributions are made in moths (30).

Symbionts

Whether the transmission of symbionts across generations is a form of genetic or environmental inheritance depends on point of view (111). Beyond the familiar cases of mitochondria and chloroplast and other plastid transmission, other examples may be more pertinent to inherited environmental effects. For example, many marine and terrestrial animal species harbor vertically transmitted bacterial or algal symbionts. The influence of these symbionts on offspring vigor can range from mild to complete (31, 204). Cytoplasmically inherited bacteria can be transmitted to offspring from either parent (e.g. 27, 195). In some species, cytoplasmically inherited bacteria interfere with paternal chromosome incorporation into fertilized eggs and cause reproductive incompatibility (an example of selection on the parental genotype), or the bacteria prevent segregation of chromosomes in unfertilized eggs and cause parthenogenesis (196, 197). In these cases, normal sexual reproduction can be reinstated by treatment with antibiotics (23b, 32, 48, 156, 195). Given this potential for "curing," the transmission of these symbionts through a lineage might be changed in response to the parental environment, for example, by a diet that includes secondary compounds from plants or fungi with molecules active against microbes.

Parental Gene Products

The parental performance phenotype is responsible for the passage of parental gene products such as enzymes or proteins (3, 4, 90, 177, 221), hormones

(77, 130, 136, 163), resistance factors (132), self-made defensive compounds (110), or parental mRNA that is essential for offspring development (usually during the early embryonic period) (13, 88, 155). Whether this transmission is subject to modification by the parental environment is less studied. For insects, the parental photoperiod modifies the hormones that are passed on to offspring in locusts and aphids (130). Kerver & van Delden (90) suggested that the parental alcohol experience in *Drosophila* (i.e. parental environment) may influence the quantity or quality of ADH (or its precursor) passed on. Based on comparisons of life history and physiology in viviparous fish, Lombardi (109) hypothesized that the maternal environmental experience can influence offspring phenotype through transfer of disease immunity factors and physiological conditioning of endocrine, osmoregulatory, and thermoregulatory capacity.

Cultural Conditioning

Cultural conditioning ranges from the choice of appropriate propagule release sites in limited care species to teaching/learning phenomena in species with extended parental care (41). For many species with relatively immobile early stages and no obvious parental care, choice of oviposition site is a component of the parental performance phenotype, a critical component that determines the first external environment encountered by offspring (148). Choice of oviposition site, an expression of the parental performance phenotype, can be influenced by the quality of the parental environment, for example, faunal composition of ponds (frogs, 149) and floral composition of fields (insects, 189). Postnatal conditioning results from parental behavior which can have both genetic and environmental components. For example, sound reception capacity in infant bats is at least partly determined by the sound reception phenotype of mothers, and the maternal reception phenotype is dictated, in part, by her own environmental experience (83). In birds, evidence from quantitative genetic analysis indicates that socially learned foraging-site fidelity is a component of the correlation between mother and offspring body size (101). Similarly, lambs show socially induced preference for the host plants eaten by their mothers (126), as do mice, even when the mother's food is less palatable than other available food (207).

Toxins and Pathogens

When parental toxicity or infection alters the parental performance phenotype, inherited environmental effects can be expressed. This may or may not involve transmission of the toxin or pathogen itself. In mice, reciprocal crosses between strains characterized by high and low susceptibility to an encephalitis virus indicated that infection was under the influence of the maternal environment (i.e. parental performance phenotype), but not that of the offspring genotype (133).

In molluscs, infectious bacteria from parents were found associated with gametes (158). In the Indian meal moth, sublethal viral infection in either mother or father resulted in reduced egg production and hatchability (178). Heavy metals modify grasshoppers' parental performance phenotype; heavy metals are also transmitted into eggs, the net effect of which is to modify offspring metabolism and growth (181). For vertebrates, Lewis (106) provided a comprehensive medical/occupational hazard reference on the adverse effects on the fitness of offspring of thousands of substances ingested by parents; substances include drugs, food additives, pesticides, metals, and natural plant products. The evidence of inherited environmental effects is enormous, and depending on the half-life of the ingested substance, reverberations across multiple generations are possible. For example, Swain (200) used data on the half-life of PCBs and the success of their transmission to human offspring through transplacental passage and postpartum nursing to predict that PCB transmission would continue for at least five generations based on the intake of just the original mother.

EMPIRICAL EVIDENCE OF INHERITED ENVIRONMENTAL EFFECTS

Most citations in this review come from relatively recent work; citations for earlier work can be found in existing reviews on maternal effects. These reviews include examples of maternal and paternal genetic effects that contribute to the parental performance phenotype, non-nuclear genetic effects, and contributions from the parental environment: 67b, 97, 161, 176 (plants); 17, 96, 130, 131 (insects); 109, 151 (fish); 86 (amphibians); 186 (reptiles); 42 (mammals), 11 (viviparous species). Empirical evidence of inherited environmental effects comes chiefly from studies that determined whether there is a relationship between (*a*) a specific aspect of the external parental environment and offspring quality at the egg or seed stage, (*b*) a specific aspect of the external parental environment and postnatal offspring quality, or (*c*) parental performance phenotype and offspring quality. In addition, the presence of maternal or paternal effects has been (*d*) inferred from breeding experiments, although these lack information about sources of environmental variation in the parental generation, and (*e*) directly measured after premature separation of mothers and offspring by embryo transplant and cross-fostering.

Parental Environment and Offspring Quality at Egg or Seed Stage

Early fitness traits, particularly propagule quality, are those most likely to express inherited environmental effects (118, 130, 161, 162, 215). Propagule size, often correlated with the level of nutritional provisions supplied to offspring

(18, 119), can contribute to offspring fitness throughout development (119, 167, 169, 198). Egg or seed qualities are desirable traits for experimental work because it is often easiest to quantify an inherited environmental effect, particularly the parental environmental component, before the influence of the offspring environment dominates; some examples are given in Table 1.

Parental Environment and Postnatal Offspring Quality

For plants, invertebrates, and vertebrates, excellent demonstrations exist of the degree to which the environment in one generation can influence individual and population quality of the next generation. The examples in Table 2 come from experiments that confirm the presence and magnitude of inherited environmental effects on postnatal traits when the parental environment was manipulated or its natural variation accounted for. In most cases, the experimental designs prevented (or minimized) the possibility that offspring phenotypic patterns were due to selection during the parental generation. The abbreviation of experimental results seen in Table 2 bypasses the creativity and complexity of these experimental designs and analyses. (Original references contain full details; particularly recommended are 60, 79, 107, 124, 142, 180, 213.)

Parental Performance Phenotype and Offspring Quality

In contrast to the previous studies in which variation in the external parental environment was measured or manipulated, some studies demonstrate inherited environmental effects without accounting for the external parental environment. Some demonstrations are based on a correlation between the parental performance phenotype and offspring phenotype, after genetic contributions are taken into account. For example, a plant's maternal provisions to the seed can

Table 1 Incidence of inherited environmental effects related to parents' environmental experience, expressed in propagule features

Reference	Organism	Parental environmental variable[a]	Offspring trait with inherited environmental effects
39	Plants	temperature	seed weight, nutrient constitution, germinability
175	Insects	foliage phenolic levels	egg weight
174	Insects	host species	yolk provisions (mean & variation around)
64	Daphnia	food rations	egg weight
152	Fish	food rations	size, fat reserves (allocation differed by species)
153	Fish	food rations	number, size, fat reserves
113	Fish	temperature	egg weight
137	Chickens	food rations	yolk quality
7	Quail	density	quantity, quality of yolk provisions
85	Frogs	temperature, food availability	egg size

[a] Variation in the parental environmental variable was accomplished through manipulation or measurement of natural environmental variation.

provide an environment so favorable for early growth that inbreeding effects are temporarily masked (127, 219). Radiolabelling techniques have been used to document the incorporation of nongenetic ejaculate materials into insect eggs (112, 139, 140). In polychaetes, the quality of the offspring changes in terms of survival and juvenile size as a function of the mother's age (25, 105). In echinoderms, experimental alteration of yolk provisions influences early developmental traits, the most interesting of which is the degree of dependence on exogenous food during the pelagic period, an aspect of a species ecology with implications for life-history evolution (121). In turtles, maternal ability to select a thermally appropriate nest site, relative to egg size, influences the time to reproduction in daughters (166). In birds, parental performance phenotype, in terms of foraging behavior, has a significant influence on offspring growth (146).

Other studies that examine the relationship between parental performance phenotype and offspring quality are based on inference from population genetics principles using reciprocal cross designs, sibling analysis, and parent-offspring regression: (215—plants); (143, 214—insects); (203b—frogs); (144, 179—birds); (21, 71—wild rodents).

Embryo transplantation or cross-fostering experiments have been used to assess the relative contributions of temporally disjunct aspects of the parental performance phenotype on offspring phenotype. For example, Cowley et al (44) showed that regardless of offspring genotype, mice transferred as embryos to the uterus of mothers from an inbred line with large body size had greater body weights, longer tails, and higher growth rates compared to those transferred to mothers from an inbred line with small body size. Bolton (18) showed that although egg size was a component of the maternal performance phenotype that made significant contributions to chick fledgling success, parental care during nesting was of greater importance. Further documentation of inherited environmental effects stemming from prenatal and postnatal care can be found in (5, 34, 120—rodents) and (167—turtles).

ECOLOGICAL AND EVOLUTIONARY CONSEQUENCES

Inherited environmental effects influence survival and fecundity, particularly at the earliest stages of offspring life, a time when mortality and selection are often greatest (154, 215). By virtue of cross-generational transmission, inherited environmental effects may delay the impact of environmental factors on population growth and delay the opportunity for selection to act on the parental performance phenotype and the offspring genotype. These considerations have led to the development of hypotheses and models to predict the consequences

Table 2 Incidence of inherited environmental effects related to parents' environmental experience, expressed through postnatal stages.

Reference	Organism	Parental environmental variable[a]	Offspring trait with inherited environmental effects[b,c]
2	Plants	altitude	frost hardiness
142		competition intensity	seed weight, time to germination, dormancy
69		field vs lab	resin production
60		habitat productivity	flower production, plasticity (#)
193		microhizal infection or not	germination, vegetative & reproductive yield
94		microrhizal infection or not	development, nutrient content, reproductive allocation
124		nutrients, pulse intensity, timing	spike biomass for both species tested (#)
67b		photo period	seed germination
206		seasonality	salt tolerance variables
220		soil nutrients	seed weight, growth (#)
198		soil nutrients	seed and plant size (#)
1		soil nutrients	seed weight, survival & size under nutrient stress
180		soil quality	seed weight, germination, harvest size (*) (#)
38		temperature	life history traits through adult stages (#)
145		temperature	seed weight, nutrient content, cold acclimation
98		temperature (pre- post zygotic)	seed weight, growth, onset of reproduction(*) (#)
78, 79	Insects	density at 3 developmental stages	degree of gregarization, coloration (*) (#)
123		density: pre- & post-natal	wingedness
117		dietary cadmium	hatch success, enzyme activity
89		dietary iron availability	survival, development time & stability
65		deiary quercitin or not	development time and weight (#)
36		field vs lab	development time, diapause, growth variables
82		field vs lab	heat resistance

(*continued*)

of inherited environmental effects on ecological and evolutionary processes. In short, it has been predicted that inherited environmental effects can influence population dynamics and character evolution (see below). Clearly, any significant changes in population dynamics and character evolution owing to inherited environmental effects will be dependent on ecological circumstance and the population's evolutionary history (172).

The consequences of inherited environmental effects will depend on the transience or sustainability of both environmental input and organism response. Inherited environmental effects that are due to transient environmental input come from experiences that are unique or intermittent (e.g. toxin spill or drought). Sustained environmental input comes from experiences that are predictable such as photoperiod (61), average temperature (222), successive seasonal change (187, 206), or average competition (198). In some cases, the length of a species' generation time and population structure will determine whether the environmental source of an inherited environmental effect is sustained or transient. For example, the experience of soil contamination by a toxin with a 50-year half-life will

Table 2 *continued*

Reference	Organism	Parental environmental variable[a]	Offspring trait with inherited environmental effects[b,c]
58		host species	survival, development time, body size (*) (#)
59		host species	survival (#)
168		host species & quality	pre-dispersal period, development time, pupal weight
181		mercury, cadmium in soil	hatch, development time, adult weight
57		nutritional quality	survival, development time
164		photoperiod	diapause propensity
99		photoperiod	body weight, temperature sensitivity
66		photoperiod	age of first reproduction, initial fecundity (#)
92		photperiod, temperature	diapause propensity
23a		photoperiod, temperature	diapause propensity
26		temp, daylength, maternal age	diapause propensity
67		temperature	age of first reproduction, body size (#)
223		temeprature	body size, territorial success in sons
212		temperature	cold resistance for both species tested (*) (#)
213		temperature	cold resistance, development time, fecundity
45		temperature	heat tolerance
73		temperature	fecundity
35	Rotifers	density	proportion of mictic offspring
107	Fish	salinity, food ration	time to hatch, hatch success (#)
187	Lizards	seasonal variation	relationship between egg size & survival
20	Rodents	field density	growth rate, age of sexual maturity
9		toxin dose	survival, development, stress response
138		PCB exposure	behavior
205		food availability	initial & subsequent body mass
55		pre-reproduc.nutritional quality	physiological variables
29	Cattle	host species	pre-weaning size

[a]See Footnote a in Table 1.
[b](*)indicates paternal effect on at least one trait listed
[c](#)Indicates that offspring response was measured in more than 1 offspring environment

likely be different for a hardwood species than for a soil invertebrate species. Which particular environmental experiences are involved in inherited environmental effects will be species-specific or population-specific. Further, the more variable the magnitude of a sustained environmental input, the more likely will be the adaptation for buffering through inherited environmental effects (33, 36, 46, 75, 107, 122, 129, 169).

Relative to an organism's response, expression of an inherited environmental effect is transient if it is confined to one offspring generation (161). It is more difficult to decide when transient expression grades into sustained expression. However, multiple-generation reverberation from a single environmental input is known for plants and animals (38, 94, 125, 171, 181, 200, 212, 218), although the intensity usually wanes over time (56, 100). It is conceivable that an inherited environmental effect is initiated by a transient environmental input and sustained indefinitely. From a model of parental imprinting that includes the environmentally induced regulation of germline genes by DNA methylation, Jablonka & Lamb (80) inferred that inherited environmental effects could be responsible for apparent reversible mutations, thus drawing a fine line between genetic and inherited environmental effects. Along similar lines, Koch (93) argues that environmentally induced changes in bacterial gene regulation, including the activation of previously silent genes, are part of an adaptive plasticity response in the face of extreme environmental fluctuation.

Population Dynamics

Inherited environmental effects can generate a delay or acceleration in the response of a population to its environment. (And the environment can include the organism's own phenotype e.g. parental body size or behavior.) It has been hypothesized that inherited environmental effects could act as a proximal source of time-lagged effects on population dynamics, with the potential to cause cycles or destabilization (either noncyclic outbreak or extinction), depending on other environmental parameters (21, 62, 81, 170, 171, 216). Relative to population ecology, a time lag occurs when the per capita rate of increase of a population is adjusted by the environment experienced in a previous generation(s). If the time-lagged component of population growth arises from a $t-2$ effect (environmental experience two generations prior) or greater, population cycles can occur (14). Inherited environmental effects can produce a time lag of $t-2$ or greater when, for example, they influence fecundity in adult offspring (e.g. 73, 168, 193). The capacity of inherited environmental effects to influence population dynamics is supported by a theoretical model that includes the influence of population quality expressed on a time delay (62), although the model is equally applicable to other proximal sources of time-lagged effects (15, 63). Empirical work, based on extrapolation of laboratory results to the field, supports the hypothesis that inherited environmental effects can influence population dynamics: (142—plants), (25—polychaetes), (21—lemmings), (20, 135—voles), (224—wild mice), (78, 79, 168, 171, 178—outbreak insect species).

Cross-Generational Phenotypic Plasticity and Life History Evolution

It has been hypothesized that as the environmental quality becomes less predictable or more heterogeneous, parents will adopt (i.e. selection will favor) a strategy of producing offspring of variable phenotype to increase the probability of reproductive success. For example, it has been predicted that inherited environmental effects will be important to the initial colonization ability of plants at infertile sites (124), and for early survival in insects with life-history features that preclude or reduce the ability of the parents to place eggs in locations that will predictably maximize the probability of future success (172).

Consequently, inherited environmental effects have been characterized as a form of bet-hedging wherein the production of greater phenotypic variation in offspring phenotype increases the likelihood of survival and reproduction in the face of environmental uncertainty (33, 36, 122, 185). With bet-hedging, a mother (or father) produces a range of offspring phenotypes, presumably to increase the probability that some subset of phenotypes will be appropriate for the environment encountered. The presence of such bet-hedging is interpreted to be a response to unpredictable components of the environment (such as the degree of annual divergence from seasonal means for temperature and rainfall). Inherited environmental effects have also been characterized as a form of plasticity if the parental environmental experience (e.g. photoperiod) produces an average adjustment of offspring phenotype which is directional and repeatable (e.g. 36, 203a), or if it produces a change in the magnitude of offspring phenotypic variation (e.g. 36, 169). Whether inherited environmental effects alter the magnitude of variance around the offspring mean phenotype or shift the offspring mean phenotype in a "programmed" direction, there is no certainty that the effect will be adaptive (i.e. increase the long-term fitness of a lineage), although a positive outcome is often seen (130). Bernardo (11) supplies some examples of nonadaptive maternal effects in viviparous species and cautions researchers about denoting maternal effects as a form of plasticity because the term contains an adaptive connotation for many.

Boggs (16) is working toward the integration of paternally based inherited environmental effects and life-history evolution. She developed a model to predict the consequences of paternal nuptial gifts on life-history evolution as it relates to female reproductive allocation of nutrients and the likelihood of sexual selection, mediated by paternal environmental contributions. Data from the literature corroborated the prediction that male nuptial contributions influenced female fecundity only when her feeding was restricted.

Character Evolution

Scientists have long been aware of the potential for inherited environmental effects to influence evolution. For example, a correlation between the genetic component of the parental performance phenotype and direct genetic effects (source 8 in Figure 1) can accelerate, retard, or change the direction of evolution depending on its magnitude and sign (49, 56, 70, 217, 218). Inherited environmental effects can also alter the correlation between offspring genotype and offspring phenotype (e.g. 36), and thus the response to selection. Evolution will be accelerated or retarded depending on whether sustained inherited environmental effects improve or diminish the relationship between offspring genotype and phenotype (159). Inherited environmental effects can also produce a delay of one or more generations in the response to selection on the parental phenotype. When this time lag occurs, the trajectory of character evolution cannot be predicted as it would normally be, by defined aspects of selection pressure and inheritance (91, 100, 160). Kirkpatrick & Lande (91) developed a quantitative genetic model for the evolution of multiple traits under maternal inheritance and found that populations could continue to evolve for an indefinite number of generations after selection was relaxed, and that the consequences could be of greater magnitude (+ or −) or in a different direction than expected from simple Mendelian inheritance. Under frequency-dependence and maternal selection, wherein response to selection depends on a fitness function and the degree of resemblance between parents and offspring, evolution away from fitness optima is possible.

The results of empirical work lend support to the general prediction that inherited environmental effects can influence character evolution (128, 198, 213). In particular, from a study to estimate the genetic component of body size, development time, and propensity to diapause in an herbivorous insect, Carriére (36) discovered that inherited environmental effects can adjust both among- and within-family components of variance. He presents data supporting the contention that a change in the relative magnitude of these variance components (and thus a change in the heritability value) would diminish the potential for response to selection. Much the same conclusion was drawn by Galloway (60), who did an experiment on an annual herbaceous plant species from an environment with an unpredictable moisture level. The experiment was designed to determine how the parental environment influenced offspring phenotype—through selection in the parental generation or through transmission of inherited environmental effects. She found that both processes were involved and concluded from the data that the ability of environmentally based maternal effects to reduce phenotypic variation in offspring flower production would slow the rate of evolutionary change.

Groeters & Dingle (66) found that response to selection on reproductive traits in the milkweed bug was modified by a maternal effect; this resulted in a lack of opportunity to select strongly for long delays in the onset of reproduction for two successive generations. From this result they concluded that inherited environmental effects could constrain adaptation toward a genetic optimum but, by providing an override mechanism to the genetically based response of offspring to some environmental condition, could allow greater realization of a phenotypic optimum.

Lacey (98) subjected the parental generation of an herbaceous perennial plant species to several temperature regimes during the pre- and post-zygotic phases of seed production in order to quantify the relative contributions of genetic and inherited environmental effects at sequential stages of development. In the light of the results, she describes the potential for evolutionary change in flowering time based on the direction of the parental effect (parental temperature hastens or slows offspring flower development) relative to the profile of a population's genetic variation (frequency distribution of genes for flowering time).

Plantenkamp & Shaw (142) studied an annual plant species where the parental generation was reared under different competition regimes. They found, among other things, the presence of a $G_m \times E_m$ interaction (source 4b on Figure 1) for seed weight. From this and other results they discuss the role of inherited environmental effects on the evolution of reaction norms for seed weight and suggest that such reaction norms may contribute to the maintenance of maternally based genetic variation.

A reading of the papers cited in this section reveals that empirical demonstration of the impact of inherited environmental effects on evolution is challenging, particularly when there are temporal changes in the direction and magnitude of their expression (e.g. 179, 198), and the underlying biology is unknown (91).

ACKNOWLEDGMENTS

A sincere thank you is extended to all colleagues (nearly 40) who responded to a letter of inquiry with bibliographic information, copies of work in press, encouragement, and the occasional joke. I also thank Mark Hunter for discussion and enthusiastic support.

Literature Cited

1. Aarssen LW, Burton SM. 1990. Maternal effects at four levels in *Senecio vulgaris* (Asteraceae) grown in a soil nutrient gradient. *Am. J. Bot.* 77:1231–40
2. Andersson B. 1994. After effects of maternal environment on autumn frost hardiness in *Pinus sylvestris* seedlings in relation to cultivation techniques. *Tree Phys.* 14:313–22
3. Aratake H, Deng LR, Fujii H, Kawaguchi Y, Koga K. 1990. Chymotrypsin inhibitors in haemolymph and eggs of the silkworm, *Bombyx mori*: developmental changes in inhibitory activity. *Comp. Biochem. Physiol. A* 97:205–9
4. Aratake H, Deng LR, Fujii H, Kawaguchi Y, Koga K. 1990. Incorporation of a haemolymph α-chymotrypsin inhibitor into eggs of *Bombyx mori* (Lepidoptera: Bombycidae). *Appl. Entomol. Zool.* 25:148–50
5. Atchley WR, Logsdon T, Cowley DE, Eisen EJ. 1991. Uterine effects epigenetics and postnatal skeletal development in the mouse. *Evolution* 45:891–909
6. Atchley WR, Newman S. 1989. A quantitative-genetics perspective on mammalian development. *Am. Nat.* 134:486–512
7. Bandyopadhyay UK, Ahuja SD. 1990. Effect of cage density on some of the egg quality traits in Japanese quail. *Indian J. Poult. Sci.* 25:159–62
8. Barlow DP. 1993. Methylation and imprinting: from host defense to gene regulation. *Science* 260:309–10

9. Barton SJ, Bode G, Sterz HG, Fukunishi K, Kobayashi Y. 1994. Reproductive and developmental toxicity study: effect of naftopidil on fertility and general reproductive performance in rats. *Pharmacometrics* 48:17–30 (In Japanese)
10. Bernardo J. 1991. Manipulating egg size to study maternal effects on offspring traits. *TREE* 6:1–2
11. Bernardo J. 1996. Maternal effects in animal ecology. *Am. Zool.* 36:83–105
12. Bernardo J. 1996. The particular maternal effect of propagule size, especially egg size: patterns, models, quality of evidence and interpretations. *Am. Zool.* 36:216–36
13. Berry SJ. 1982. Maternal direction of oogenesis and early embryogenesis in insects. *Annu. Rev. Entomol.* 27:205–27
14. Berryman AA. 1981. *Population Systems: A General Introduction.* New York: Plenum
15. Berryman AA. 1995. Population cycles: a critique of the maternal and allometric hypotheses. *J. Anim. Ecol.* 64:290–93
16. Boggs CL. 1990. A general model of the role of male-donated nutrients in female insects' reproduction. *Am. Nat.* 136:598–617
17. Boggs CL. 1995. Male nuptial gifts: phenotypic consequences and evolutionary implications. In *Insect Reproduction,* ed. SR Leather, J Hardie, pp. 215–42. Boca Raton, FL: CRC
18. Bolton M. 1991. Determinants of chick survival in the lesser black-backed gull relative contributions of egg size and parental quality. *J. Anim. Ecol.* 60:949–60
19. Bondari KR, Willham L, Freeman AE. 1978. Estimates of direct and maternal genetics correlations for pupa weight and family size of *Tribolium. J. Anim. Sci.* 47:358–65
20. Boonstra R, Boag PT. 1987. A test of the Chitty hypothesis: inheritance of life-history traits in meadow voles *Microtus pennsylvanicus. Evolution* 41:929–47
21. Boonstra R, Hochachka WM. 1996. Maternal effects and additive genetic inheritance in the collared lemming (*Dicrostonyx groenlandicus.*) *Evol. Ecol.* In press
22. Boppre M, Fischer OW. 1994. Zonocerus and Chromolaena in West Africa. In *New Trends in Locust Control,* ed. S Krall, H Wilps, pp. 108–26. D-Eschborn: Gesell. Tech. Zusammenarbeit
23a. Bradford MJ, Roff DA. 1995. Genetic and phenotypic sources of life history variation along a cline in voltinism in the cricket *Allonemobius socius. Oecologia* 103:319–26
23b. Breeuwer JAJ, Werren JH. 1990. Microorganisms associated with chromosome destruction and reproductive isolation between two insect species. *Nature* 346:558–60
24. Brett MT. 1993. Resource quality effects on *Daphnia longispina* offspring fitness. *J. Plankton Res.* 15:403–12
25. Bridges TS, Heppell S. 1996. Fitness consequences of maternal effects in *Streblospio benedicti* (Annelida: Polychaeta). *Am. Zool.* 36:1342–46
26. Brodeur J, McNeil JN. 1989. Biotic and abiotic factors involved in diapause induction of the parasitoid *Aphidius nigripes* (Hymenoptera:Aphidiidae). *J. Insect Physiol.* 35:969–74
27. Brough CN, Dixon AFG. 1990. Ultrastructural features of egg development in oviparae of the vetch aphid *Megoura viciae* Buckton. *Tissue Cell* 22:51–64
28. Brown KS. 1984. Adult-obtained pyrrolizidine alkaloids defend ithomiine butterflies against a spider predator. *Nature* 309:707–8
29. Brown MA, Tharel LM, Brown AH Jr, Jackson WG, Miesner JR. 1993. Genotype χ environment interactions in preweaning traits of purebred and reciprocal cross angus and Brahman calves on common Bermudagrass and endophyte-infected tall fescue pastures. *J. Anim. Sci.* 71:326–33
30. Brust GE, Barbercheck ME. 1992. Effect of dietary cucurbitacin C on southern corn rootworm (Coleoptera: Chrysomelidae) egg survival. *Environ. Entomol.* 21:1466–71
31. Buchner P. 1965. *Endosymbiosis of Animals With Plant Microorganisms.* New York: Interscience
32. Campbell BC, Bragg TS, Turner CE. 1992. Phylogeny of symbiotic bacteria of four weevil species (Coleoptera: Curculionidae) based on analysis of 16S ribosomal DNA. *Insect Biochem. Mol. Biol.* 22:415–21
33. Capinera JL. 1979. Qualitative variation in plants and insects: effects of propagule size on ecological plasticity. *Am. Nat.* 114:350–61
34. Carlier M, Roubertoux PL, Pastoret C. 1991. The Y-chromosome effect on intermale aggression in mice depends on the maternal environment. *Genetics* 129:231–36

35. Carmona MJ, Serra M, Miracle MR. 1994. Effect of population density and genotype on life-history traits in the rotifer *Brachionus plicatilis* O.F. Mueller. *J. Exp. Marine Biol. Ecol.* 182:223–35
36. Carrière Y. 1994. Evolution of phenotypic variance: non-Mendelian parental influences on phenotypic and genotypic components of life-history traits in a generalist herbivore. *Heredity* 72:420–30
37. Carrière Y, Roff DA. 1995. The evolution of offspring size and number: a test of the Smith-Fretwell model in three species of crickets. *Oecologia* 102:389–96
38. Case AL, Lacey EP, Hopkins RG. 1996. Parental effects in *Plantago lanceolata* L. II. Manipulation of grandparental temperature and parental flowering time. *Heredity.* 76:287–95
39. Charest C, Potvin C. 1993. Maternally-induced modification of progeny phenotypes in the C-4 weed *Echinochloa crus galli*: an analysis of seed constituents and performance. *Oecologia* 93:383–88
40. Cheverud JM. 1984. Evolution by kin selection: a quantitative genetic model illustrated by maternal performance in mice. *Evolution* 38:766–77
41. Cheverud JM, Moore AJ. 1994. Quantitative genetics and the role of the environment provided by relatives in behavioral evolution. In *Quantitative Genetic Studies of Behavioral Evolution,* ed. CRB Boake, pp. 67–100. Chicago, IL: Univ. Chicago Press
42. Cowley DE. 1991. Prenatal effects on mammalian growth: embryo transfer results. See Ref. 50, pp. 762–79
43. Cowley DE, Atchley WR. 1992. Quantitative genetic models for development: epigenetic selection and phenotypic evolution. *Evolution* 46:495–518
44. Cowley DE, Pomp D, Atchley WR, Eisen EJ, Hawkins-Brown D. 1989. The impact of maternal uterine genotype on postnatal growth and adult body size in mice. *Genetics* 122:193–204
45. Crill WD. 1991. High parental and developmental temperatures increase heat resistance of adult *Drosophila melanogaster*. *Am. Zool.* 31:58 (Abstr.)
46. Crump ML. 1981. Variation in propagule size as a function of environmental uncertainty for tree frogs. *Am. Nat.* 117:724–37
47. Cummins CP. 1986. Temporal and spatial variation in egg size and fecundity in *Rana temporaria*. *J. Anim. Ecol.* 55:303–16
48. Degrugillier ME. 1994. Testicular microorganisms in *Heliothis:* ultrastructure and distribution in *H. virescens, H. subflexa,* F-1 hybrid and backcross males. *J. Invert. Pathol.* 64:77–88
49. Dickerson GE. 1947. Composition of hog carcass as influenced by heritable differences in rate and economy of gain. *Iowa Agric. Exp. Stn. Res. Bull.* 354:489–524
50. Dudley EC, ed. 1991. *The Unity of Evolutionary Biology.* Portland, OR: Dioscorides
51. Durrant A. 1958. Environmental conditioning of flax. *Nature* 181:928–29
52. Dussourd DE, Ubik K, Harvis C, Resch J, Meinwald J, Eisner T. 1988. Biparental defensive endowment of eggs with acquired plant alkaloid in the moth *Utetheisa ornatrix*. *Proc. Natl. Acad. Sci. USA* 85:5992–96
53. Eisen EJ, Saxton AM. 1983. Genotype by environment interactions and genetic correlations involving two environmental factors. *Theor. Appl. Genet.* 67:75–86
54. Engebreston JA, Mason WH. 1980. Transfer of ^{65}Zn at mating in *Heliothis virescens*. *Environ. Entomol.* 9:119–21
55. Eriksson UJ, Swenne I. 1993. Diabetes in pregnancy: fetal macrosomia, hyperinsulinism, and islet hyperplasia in the offspring of rats subjected to temporary protein-energy malnutrition early in life. *Pediatr. Res.* 34:791–95
56. Falconer DS. 1965. Maternal effects and selection response. In *Genetics Today, Proc. 11th International Congress Genetics,* ed. SJ Geerts, 3:763–74. Oxford: Pergamon
57. Fox CW, Dingle H. 1994. Dietary mediation of maternal age effects on offspring performance in a seed beetle (Coleoptera: Bruchidae). *Funct. Ecol.* 8:600–6
58. Fox CW, Waddell KJ, Mousseau TA. 1995. Parental host plant affects offspring life histories in a seed beetle. *Ecology* 76:402–11
59. Futuyma DJ, Herrmann C, Milstein S, Keese MC. 1993. Apparent transgenerational effects of host plant in the leaf beetle *Ophraella notulata* (Coleoptera: Chrysomelidae). *Oecologia* 96:365–72
60. Galloway LF. 1995. Response to natural environmental heterogeneity: maternal effects and selection on life-history characters and plasticities in *Mimulus guttatus*. *Evolution* 49:1095–1107
61. Giesel JT. 1986. Effects of parental pho-

toperiod regime on progeny development time in *Drosophila simulans*. *Evolution* 40:649–51
62. Ginzburg LR, Taneyhill DE. 1994. Population cycles of forest Lepidoptera: a maternal effect hypothesis. *J. Anim. Ecol.* 63:79–92
63. Ginzburg LR, Taneyhill DE. 1995. Higher growth rate implies shorter cycle, whatever the cause: a reply to Berryman. *J. Anim. Ecol.* 64:294–95
64. Glazier DS. 1992. Effects of food genotype and maternal size and age on offspring investment in *Daphnia magna*. *Ecology* 73:910–26
65. Gould F. 1988. Stress specificity of maternal effects in *Heliothis virescens* (Lepidoptera: Noctuidae) larvae. In *Paths from a Viewpoint: The Wellington Festschrift on Insect Ecology,* ed. TS Sahota, CS Holling, pp. 191–97. *Mem. Entomol. Soc. Can., No. 146.* Ottawa: Entomol. Soc. Can.
66. Groeters FR, Dingle H. 1987. Genetic and maternal influences on life history plasticity in response to photoperiod by milkweed bugs (*Oncopeltus fasciatus*). *Am. Nat.* 129:332–46
67a. Groeters FR, Dingle H. 1988. Genetic and maternal influences on life history plasticity in milkweed bugs (Oncopeltus): response to temperature. *J. Evol. Biol.* 1:317–33
67b. Gutterman Y. 1992. Maternal effects on seeds during development. In *Seeds: The Ecology of Regeneration in Plant Communities,* ed. M Fenner, pp, 27–59. Wallingford, Engl: CAB Int.
68. Han K, Lincoln DE. 1994. The evolution of carbon allocation to plant secondary metabolites: a genetic analysis of cost in *Diplacus aurantiacus*. *Evolution* 48:1550–63
69. Han K, Lincoln DE. 1996. The impact of plasticity and maternal effects on the evolution of leaf resin production in *Diplacus aurantiacus*. *Evol. Ecol.* In press
70. Hanrahan JP. 1976. Maternal effects and selection response with an application to sheep data. *Anim. Prod.* 22:359–69
71. Hansson L. 1988. Parent-offspring correlations for growth and reproduction in the vole *Clethrionomys glareolus* in relation to the Chitty hypothesis. *Z. Saeugetierkd.* 53:7–10
72. Hare JF, Eisner T. 1993. Pyrrolizidine alkaloid deters ant predators of *Utetheisa ornatrix* eggs: effects of alkaloid concentration, oxidation state, and prior exposure of ants to alkaloid-laden prey. *Oecologia* 96:9–18
73. Huey RB, Wakefield T, Crill WD, Gilchrist GW. 1995. Within- and between-generation effects of temperature on early fecundity of *Drosophila melanogaster*. *Heredity* 74:216–23
74. Hutchings JA. 1991. Fitness consequences of variation in egg size and food abundance in brook trout *Salvelinus fontinalis*. *Evolution* 45:1162–68
75. Hutchings JA. 1996. Adaptive phenotypic plasticity in brook trout, *Salcelinus fontinalis*, life histories. *Ecoscience.* 3:25–32
76. Hutchings JA. 1996. Life history responses to environmental variability in early life. In *Early Life History and Recruitment in Fish Populations,* ed. RC Chambers, EA Trippel. New York: Chapman & Hall. In press
77. Isaac RE, Rees HH. 1985. Metabolism of maternal ecdysteriod-22-phosphates in developing embryos of the desert locust, *Schistocerca gregaria*. *Insect Biochem.* 15:65–72
78. Islam MS, Roessingh P, Simpson SJ, McCaffery AR. 1994. Effects of population density experienced by parents during mating and oviposition on the phase of hatchling desert locusts, *Schistocerca gregaria*. *Proc. R. Soc. London Ser. B* 257:93–98
79. Islam MS, Roessingh P, Simpson SJ, McCaffery AR. 1994. Parental effects on the behaviour and colouration of nymphs of the desert locust *Schistocerca gregaria*. *J. Insect Physiol.* 40:173–81
80. Jablonka E, Lamb MJ. 1989. The inheritance of acquired epigenetic variations. *J. Theor. Biol.* 139:69–84
81. Janssen GM, DeJong G, Joose ENG, Scharloo W. 1988. A negative maternal effect in springtails. *Evolution* 42:828–34
82. Jenkins NL, Hoffmann AA. 1994. Genetic and maternal variation for heat resistance in *Drosophila* from the field. *Genetics* 137:783–89
83. Jones G, Ransome RD. 1993. Echolocation calls of bats are influenced by maternal effects and change over a lifetime. *Proc. R. Soc. London. Ser. B* 252:125–28
84. Kahn TL, Adams CJ, Arpaia ML. 1994. Paternal and maternal effects on fruit and seed characteristics in Cherimoya (*Annona cherimola* Mill). *Sci. Hortic.* 59:11–25
85. Kaplan RH. 1987. Developmental plasticity and maternal effects of reproduc-

tive characteristics in the frog *Bombina orientalis*. *Oecologia* 71:273–79
86. Kaplan RH. 1991. Developmental plasticity and maternal effects in amphibian life histories. See Ref. 50, pp. 794–99
87. Kaplan RH. 1992. Greater maternal investment can decrease offspring survival in the frog *Bombina orientalis*. *Ecology* 73:280–88
88. Kastern WH, Watson CA, Berry SJ. 1990. Maternal messenger RNA distribution in silkmoth eggs. I. Clone EC4B is associated with the cortical cytoskeleton. *Development* 108:497–506
89. Keena MA, ODell TM, Tanner JA. 1995. Phenotypic response of two successive gypsy moth (Lepidoptera: Lymantriidae) generations to environment and diet in the laboratory. *Ann. Entomol. Soc. Am.* 88:680–89
90. Kerver JWM, van Delden W. 1985. Development of tolerance to ethanol in relation to the alcohol dehydrogenase locus in *Drosophila melanogaster.* 1. Adult and egg-to-adult survival in relation to ADH activity. *Heredity* 55:355–67
91. Kirkpatrick M, Lande R. 1989. The evolution of maternal effects. *Evolution* 43:485–503
92. Kobayashi J. 1990. Effects of photoperiod on the induction of egg diapause of tropical races of the domestic silkworm, *Bombyx mori,* and the wild silkworm, *B. mandarina. Jpn. Agric. Res. Q.* 23:202–5
93. Koch AL. 1993. Genetic response of microbes to extreme challenges. *J. Theor. Biol.* 160:1–21
94. Koide RT, Lu X. 1992. Mycorrhizal infection of wild oats: maternal effects on offspring growth and reproduction. *Oecologia* 90:218–26
95. Kondo R, Satta Y, Matsuura ET, Ishiwa H, Takahata N, Chigusa SI. 1990. Incomplete maternal transmission of mitochondrial DNA in *Drosophila. Genetics* 126:657–63
96. Labeyrie V. 1988. Maternal effects and biology of insect populations. *Mem. Entomol. Soc. Can.* 146:153–70 (In French)
97. Lacey EP. 1991. Parental effects on life-history traits in plants. See Ref. 50, pp. 735–44
98. Lacey EP. 1996. Parental effects in *Plantago lanceolata* L. I. A growth chamber experiment to examine pre-and post-zygotic temperature effects. *Evolution* 50:865–78
99. Lanciani CA, Giesel JT, Anderson JF. 1990. Seasonal change in metabolic rate of *Drosophila simulans*. *Comp. Biochem. Phys. A* 97:501–4
100. Lande R, Kirkpatrick M. 1990. Selection response in traits with maternal inheritance. *Genet. Res.* 55:189–97
101. Larsson K, Forslund P. 1992. Genetic and social inheritance of body and egg size in the barnacle goose *Branta leucopsis. Evolution* 46:235–44
102. Lau TC, Stephenson AG. 1993. Effects of soil nitrogen on pollen production, pollen grain size, and pollen performance in *Cucurbita pepo* (Cucurbitaceae). *Am. J. Bot.* 80:763–68
103. Lau TC, Stephenson AG. 1994. Effects of soil phosphorous on pollen production, pollen size, pollen phosphorus content, and the ability to sire seeds in *Cucurbita pepo* (Cucurbitaceae). *Sex Plant Reprod.* 7:215–20
104. Leopold RA. 1976. The role of male accessory glands in insect reproduction. *Annu. Rev. Entomol.* 21:199–221
105. Levin LA, Zhu J, Creed EL. 1991. The genetic basis of life-history characters in a polychaete exhibiting planktotrophy and lechithotrophy. *Evolution* 45:380–97
106. Lewis RJ Sr. 1991. *Reproductively Active Chemicals: A Reference Guide.* New York: Van Nostrand Reinhold
107. Lin H-C, Dunson WA. 1995. An explanation of the high strain diversity of a self-fertilizing hermaphroditic fish. *Ecology* 76:593–605
108. Locke M, Nichol H. 1992. Iron economy in insects: transport metabolism and storage. *Annu. Rev. Entomol.* 37:195–215
109. Lombardi J. 1996. Postzygotic maternal influences and the maternal-embryonic relationship of viviparous fishes. *Am. Zool.* 36:106–15
110. Magrath R, Mithen R. 1993. Maternal effects on the expression of individual aliphatic glucosinolates in seeds and seedlings of *Brassica napus. Plant Breed.* 111:249–52
111. Margulis L. 1992. Symbiosis theory: Cells as microbial communities. In *Environmental Evolution: Effects of the Origin and Evolution of Life on Planet Earth,* ed. L Margulis, L Olendzenski, pp. 149–72. Cambridge, MA: MIT Press
112. Markow TA, Gallagher PD, Krebs RA. 1990. Ejaculate-derived nutritional contribution and female reproductive success in *Drosophila mojavensis* Patterson

and Crow. *Funct. Ecol.* 4:67–74
113. Marsh E. 1984. Egg size variation in central Texas populations of *Etheostoma spectabile* (Pisces: Percidae). *Copeia* 2:291–301
114. Marshall LD, McNeil JN. 1989. Spermatophore mass as an estimate of male nutrient investment: a closer look in *Pseudaletia unipuncta* (Haworth) (Lepidoptera: Noctuidae). *Funct. Ecol.* 3:605–12
115. Marx JL. 1988. A parent's sex may affect gene expression. *Science* 239:352–53
116. Mather K, Kinks JL. 1971. *Biometrical Genetics*. Ithaca, NY: Cornell Univ. Press
117. Mathova A. 1990. Biological effects and biochemical alterations after long-term exposure of *Galleria mellonella* (Lepidoptera: Pyralidae) larvae to cadmium containing diet. *Acta Entomol. Bohemoslov.* 87:241–48
118. Mazer SJ. 1987. Parental effects on seed development and seed yield in *Raphanus raphanistrum*: implications for natural selection and sexual evolution. *Evolution* 41:355–71
119. Mazer SJ. 1989. Family mean correlations among fitness components in wild radish controlling for maternal effects on seed weight. *Can. J. Bot.* 67:1890–97
119a. Mazer SJ, Gorchov DL. 1996. Paternal effects on progeny phenotype in plants: distinguishing genetic and environmental causes. *Evolution* 50:44–53
120. McCarty R, Fields-Okotcha C. 1994. Timing of preweanling maternal effects on development of hypertension in SHR rats. *Physiol. Behav.* 55:839–44
121. McEdward LR. 1996. Experimental manipulation of parental investment in *Echinoid echinoderms*. *Am. Zool.* 36:169–79
122. McGinley MA, Temme DH, Geber MA. 1987. Parental investment in offspring in variable environments: theoretical and empirical considerations. *Am. Nat.* 130:370–98
123. Messina FJ. 1993. Effect of initial colony size on the per capita growth rate and alate production of the Russian wheat aphid (Homoptera: Aphididae). *J. Kansas Entomol. Soc.* 66:365–71
124. Miao SL, Bazzaz FA, Primack RB. 1991. Effects of maternal nutrient pulse on reproduction of two colonizing *Plantago* spp. *Ecology* 72:586–96
125. Miao SL, Bazzaz FA, Primack RB. 1991. Persistence of maternal nutrient effects in *Plantago major* the third generation. *Ecology* 72:1634–42
126. Mirza SN, Provenza FD. 1994. Socially induced food avoidance in lambs: direct or indirect maternal influence? *J. Anim. Sci.* 72:899–902
127. Montalvo AM. 1994. Inbreeding depression and maternal effects in *Aquilegia caerulea,* a partially selfing plant. *Ecology* 75:2395–409
128. Montalvo AM, Shaw RG. 1994. Quantitative genetics of sequential life-history and juvenile traits in the partially selfing perennial, *Aquilegia caerulea. Evolution* 48:828–41
129. Mousseau TA. 1991. Geographic variation in maternal-age effects on diapause in a cricket. *Evolution* 45:1053–59
130. Mousseau TA, Dingle H. 1991. Maternal effects in insect life histories. *Annu. Rev. Entomol.* 36:511–34
131. Mousseau TA, Dingle H. 1991. Maternal effects in insects: examples, constraints, and geographic variation. See Ref. 50, pp. 745–61
132. Munkittrick KR, Dixon DG. 1988. Evidence for a maternal yolk factor associated with increased tolerance and resistance of feral white sucker (*Catostomus commersoni*) to waterborne copper. *Ecotoxicol. Environ. Saf.* 15:7–20
133. Murakami Y, Miura K, Fujisaki Y. 1989. Maternal effect on the vertical infection in inbred mouse with Japanese encephalitis virus. *Bull. Natl. Inst. Anim. Health* 94:1–6 (In Japanese)
134. Nakamura RR, Stanston ML. 1989. Embryo growth and seed size in *Raphanus sativus* maternal and paternal effects in-vivo and in-vitro. *Evolution* 43:1435–43
135. Nelson RL. 1991. Maternal diet influences reproductive development in male prairie vole offspring. *Physiol. Behav.* 50:1063–66
136. Nijhout HF, Wheeler DE. 1982. Juvenile hormone and the physiological basis on insect polymorphisms. *Q. Rev. Biol.* 57:100–33
137. Olawuni KA, Ubosi CO, Alaku SO. 1992. Effects of feed restriction on egg production and egg quality of exotic chickens during their second year of production in a Sudano-Sahelian environment. *Anim. Feed Sci. Technol.* 38:1–9
138. Pantaleoni G, Fanini D, Sponta AM, Palumbo G, Giorgi R, Adams PM. 1988. Effects of maternal exposure to polychlorobiphenyls PCBs on F1 generation behavior in the rat. *Fund. Appl. Toxicol.* 11:440–49
139. Pardo MC, Camacho JPM, Hewitt GM.

1994. Dynamics of ejaculate nutrient transfer in *Locusta migratoria. Heredity* 73:190–97
140. Pitnick S, Markow TA, Riedy MF. 1991. Transfer of ejaculate and incorporation of male-derived substances by females in the nannoptera species group (Diptera: Drosophilidae). *Evolution* 45:774–80
141. Pivnick KA, McNeil JN. 1987. Puddling in butterflies: sodium affects reproductive success in *Thymelicus lineola. Physiol. Entomol.* 12:461–72
142. Plantenkamp GAJ, Shaw RG. 1993. Environmental and genetic maternal effects on seed characters in *Neomophila menziesii. Evolution* 47:540–55
143. Posthuma L, Hogervorst RF, Joosse ENG, Van Straalen NM. 1993. Genetic variation and covariation for characteristics associated with cadmium tolerance in natural populations of the springtail *Orchesella cincta* (L.). *Evolution* 47:619–31
144. Potti J, Merino S. 1994. Heritability estimates and maternal effects on tarsus length in pied flycatchers, *Ficedula hypoleuca. Oecologia* 100:331–38
145. Potvin C, Charest C. 1991. Maternal effects of temperature on metabolism in the C-4 weed *Echinochloa crus galli. Ecology* 72:1973–79
146. Pugesek BH. 1995. Offspring growth in the California gull: reproductive effort and parental experience hypotheses. *Anim. Behav.* 49:641–47
147. Quesada MR, Bollman K, Stephenson AG. 1995. Leaf damage decreases pollen production and hinders pollen performance in *Cucurbita texana. Ecology* 76:437–43
148. Resetarits WJ. 1996. Oviposition site choice and life history evolution. *Am. Zool.* 36:In press
149. Resetarits WJ, Wilbur HM. 1989. Choice of oviposition site by *Hyla chrysoscelis*: role of predators and competitors. *Ecology* 70:220–28
150. Reznick DN. 1981. "Grandfather effects": the genetics of interpopulation differences in offspring size in the mosquitofish. *Evolution* 35:941–53
151. Reznick DN. 1991. Maternal effects in fish life histories. See Ref. 50, pp. 780–93
152. Reznick D, Callahan H, Llauredo R. 1996. Maternal effects on offspring quality in Ploeciliid fishes. *Am. Zool.* 36:147–56
153. Reznick D, Yang AP. 1993. The influence of fluctuating resources on life history: patterns of allocation and plasticity in female guppies. *Ecology* 74:2011–19
154. Rice JA, Miller TJ, Rose KA, Crowder LB, Marschall EA, 1993. Growth rate variation and larval survival: inferences from an individual-based size-dependent predation model. *Can. J. Fish. Aquat. Sci.* 50:133–42
155. Richards WG, Carroll PM, Kinloch RA, Wassarman PM, Strickland S. 1993. Creating maternal effect mutations in transgenic mice: Antisense inhibition of an oocyte gene product. *Dev. Biol.* 160:543–53
156. Richardson PM, Holmes WP, Saul GB. 1987. The effect of tetracycline on nonreciprocal cross incompatibility in Mormoniella [= Nasonia] vitripennis.] *J. Invert. Pathol.* 50:176–83
157. Richardson TE, Stephenson AG. 1992. Effects of parentage and size of the pollen load on progeny performance in *Campanula americana. Evolution* 46:1731–39
158. Riquelme CE, Chavez P, Morales Y, Hayashida G. 1994. Evidence of parental bacterial transfer to larvae in *Argopecten purpuratus* (Lamarck, 1819). *Biol. Res.* 27:129–34
159. Riska B. 1989. Composite traits, selection response, and evolution. *Evolution* 43:1172–91
160. Riska B, Rutledge JJ, Atchley WR. 1985. Covariance between direct and maternal genetic effects in mice, with a model of persistent environmental influences. *Genet. Res.* 45:287–97
161. Roach DA, Wulff RD. 1987. Maternal effects in plants. *Annu. Rev. Ecol. Syst.* 18:209–35
162. Rocha OJ, Stephenson AG. 1990. Effect of ovule position on seed production, seed weight, and progeny performance in *Phaseolus coccineus* L. (Leguminosae). *Am. J. Bot.* 77:1320–29
163. Rockey SJ, Miller BB, Denlinger DL. 1989. A diapause maternal effect in the flesh fly (*Sarcophaga bullata*): transfer of information from mother to progeny. *J. Insect Physiol.* 35:553–58
164. Rockey SJ, Yoder JA, Denlinger DL. 1991. Reproductive and developmental consequences of a diapause maternal effect in the flesh fly, *Sarcophaga bullata. Physiol. Entomol.* 16:477–83
165. Rogowitz GL. 1996. Trade-offs in energy allocation during lactation. *Am. Zool.* 36:197–204
166. Roosenburg WM. 1996. Maternal condi-

tion and nest site choice: an alternative for the maintenance of environmental sex determination? *Am. Zool.* 36:157–68

167. Roosenburg WM, Kelly KC. 1996. The effect of egg size and incubation temperature on growth in the turtle, *Malaclemys terrapin. J. Herpetol.* In press
168. Rossiter MC. 1991. Environmentally-based maternal effects: a hidden force in insect population dynamics. *Oecologia* 87:288–94
169. Rossiter MC. 1991. Maternal effects generate variation in life history: consequences of egg weight plasticity in the gypsy moth. *Funct. Ecol.* 5:386–93
170. Rossiter MC. 1992. The impact of resource variation on population quality in herbivorous insects: a critical component of population dynamics. In *Resource Distribution and Animal-Plant Interactions,* ed. MD Hunter, T Ohgushi, PW Price, pp. 13–42. San Diego: Academic
171. Rossiter MC. 1994. Maternal effects hypothesis of herbivore outbreak. *BioScience* 44:752–63
172. Rossiter MC. 1995. Impact of life history evolution on population dynamics: predicting the presence of maternal effects. In *Population Dynamics: New Approaches and Synthesis,* ed. N Cappuccino, PW Price, pp. 251–75. San Diego: Academic
173. Rossiter MC. 1997. Assessment of genetic variation in the presence of maternal or paternal effects in herbivorous insects. In *Genetic Structure in Natural Insect Populations: Effects of Host Plants and Life History,* ed. S Mopper, S Strauss. New York: Chapman & Hall. In press
174. Rossiter MC, Cox-Foster DL, Briggs MA. 1993. Initiation of maternal effects in *Lymantria dispar*: genetic and ecological components of egg provisioning. *J. Evol. Biol.* 6:577–89
175. Rossiter MC, Schultz JC, Baldwin IT. 1988. Relationships among defoliation, red oak phenolics, and gypsy moth growth and reproduction. *Ecology* 69:267–77
176. Rowe JS. 1964. Environmental preconditioning, with special reference to forestry. *Ecology* 45:399–403
177. Ruder FJ, Fischer R, Busen W. 1990. A sequence-specific protease cleaves a maternal cortical protein during early embryogenesis in *Sciara coprophila* (Diptera). *Dev. Biol.* 140:231–40
178. Sait SM, Begon M, Thompson DJ. 1994. The effects of a sublethal baculovirus infection in the Indian meal moth, *Plodia interpunctella. J. Anim. Ecol.* 63:541–50
179. Schluter D, Gustafsson L. 1993. Maternal inheritance of condition and clutch size in the collard flycatcher. *Evolution* 47:658–67
180. Schmid B, Dolt C. 1994. Effects of maternal and paternal environment and genotype on offspring phenotype in *Solidago altissima* L. *Evolution* 48:1525–49
181. Schmidt GH, Ibrahim NMM, Abdallah MD. 1991. Toxicological studies on the long-term effects of heavy metals (Hg, Cd, Pb) in soil on the development of *Aiolopus thalassinus* Fabr. (Saltatoria: Acrididae). *Sci. Total Environ.* 107:109–34
182. Schmitt J, Niles J, Wulff RD. 1992. Norms of reaction of seed traits to maternal environments in *Plantago lanceolata. Am. Nat.* 139:451–66
183. Schultz DL. 1991. Parental investment in temporally varying environments. *Evol. Ecol.* 5:415–27
184. Schwaegerle KE, Levin DA. 1990. Quantitative genetics of seed size variation in *Phlox. Evol. Ecol.* 4:143–14
185. Seger J, Brockmann H. 1987. What is bet-hedging? In *Oxford Surveys in Evolutionary Biology,* ed. PH Harvey, L. Partridge, pp. 182–211. Oxford: Oxford Univ. Press
186. Sinervo B. 1991. Experimental and comparative analyses of egg size in lizards constraints on the adaptive evolution of maternal investment per offspring. See Ref. 50, pp. 725–34
187. Sinervo B, Doughty P. 1996. Interactive effects of offspring size and timing of reproduction on offspring reproduction: experimental, maternal, and quantitative genetic aspects. *Evolution.* 50:1314–27
188. Sinervo B, McEdward LR. 1988. Developmental consequences of an evolutionary change in egg size an experimental test. *Evolution* 42:885–99
189. Singer MC, Thomas CD, Billington HL, Parmesan C. 1989. Variation among conspecific insect populations in the mechanistic basis of diet breadth. *Anim. Behav.* 37:751–59
190. Singh BN, Pandey MB. 1993. Evidence for additive polygenic control of pupation height in *Drosophila ananassae. Hereditas* 119:111–16
191. Smith CC, Fretwell SD. 1974. The optimal balance between size and number

of offspring. *Am. Nat.* 108:499–506
192. Southwood OI, Kennedy BW. 1990. Estimation of direct and maternal genetic variance for litter size in Canadian Yorkshire and Landrace swine using an animal model. *J. Anim. Sci.* 68:1841–47
193. Srivastava D, Mukerji KG. 1995. Field response of mycorrhizal and nonmycorrhizal *Medicago sativa* var. local in the F1 generation. *Mycorrhiza* 5:219–21
194. Stephenson AG, Erickson CW, Lau TC, Quesada MR, Winsor JA. 1994. Effects of growing conditions on the male gametophyte. In *Pollen-Pistil Interactions and Pollen Tube Growth,* ed. AG Stephenson, TC Kao, 12:220–29. Am. Soc. Plant Physiol. Ser., Rockville, MD
195. Stevens L. 1993. Cytoplasmically inherited parasites and reproductive success in *Tribolium* flour beetles. *Anim. Behav.* 46:305–10
196. Stevens L, Wade MJ. 1988. Effect of antibiotics on the productivity of genetic strains of *Tribolium confusum* and *Tribolium castaneum* (Coleoptera: Tenebrionidae). *Environ. Entomol.* 17:115–19
197. Stouthamer R, Breeuwer JAJ, Luck RF, Werren JH. 1993. Molecular identification of microorganisms associated with parthenogenesis. *Nature* 361:66–68
198. Stratton DA. 1989. Competition prolongs expression of maternal effects in seedlings of *Erigeron annuus* (Asteraceae). *Am. J. Bot.* 76:1646–53
199. Swain JL, Stewart TA, Leder P. 1987. Parental legacy determines methylation and expression of an autosomal transgene a molecular mechanism for parental imprinting. *Cell* 50:719–28
200. Swain WR. 1988. Human health consequences of consumption of fish contaminated with organochlorine compounds. *Aquat. Toxicol.* 11:357–77
201. Tallamy DW. 1994. Nourishment and the evolution of paternal investment in subsocial arthropods. In *Nourishment and Evolution in Insect Societies,* ed. JH Hunt, CA Nalepa, pp. 21–55. Boulder, CO: Westview
202. Taubeneck MW, Daston GP, Rogers JM, Keen CL. 1994. Altered maternal zinc metabolism following exposure to diverse developmental toxicants. *Reprod. Toxicol.* 8:25–40
203a. Travis J. 1994. Evaluating the adaptive role of morphological plasticity. In *Ecological Morphology: Integrative Organismal Biology,* ed. PC Wainwright, SM Reilly, pp. 99–122. Chicago: Univ. Chicago Press
203b. Travis J, Emerson SB, Blouin M. 1987. A quantitative-genetic analysis of larval life-history traits in *Hyla crucifer. Evolution* 41:145–56
204. Trench RK. 1992. Microalgal invertebrate symbiosis: current trends. In *Encyclopedia of Microbiology,* ed. J Lederberg, pp. 129–42. San Diego, CA: Academic
205. Trombulak SC. 1991. Maternal influence on juvenile growth rates in Belding's ground squirrel *Spermophilus beldingi. Can. J. Zool.* 69:2140–45
206. Ungar IA. 1988. Effects of the parental environment on the temperature requirements and salinity tolerance of *Spergularia marina* seeds. *Bot. Gaz.* 149:432–36
207. Valsecchi P, Mainardi M, Sgoifo A, Taticchi A. 1989. Maternal influences on food preferences in weanling mice *Mus domesticus. Behav. Proc.* 19:155–66
208. Van Herpen MMA. 1986. Biochemical alterations in the sexual partners resulting from environmental conditions before pollination regulate processes after pollination. In *Biotechnology and Ecology,* ed. DL Mulcahy, GB Mulcahy, E Ottaviano, pp. 131–37. New York: Springer-Verlag
209. Vavrousek-Jakuba EM, Baker RA, Shoemaker WJ. 1991. Effect of ethanol on maternal and offspring characteristics comparison of three liquid diet formulations fed during gestation. *Alcohol Clin. Exp. Res.* 15:129–35
210a. Venable DL. 1992. Size-number tradeoffs and the variation of seed size with plant resource status. *Am. Nat.* 140:287–304
210b. Villar AJ, Pedersen RA. 1994. Parental imprinting of the Mas protooncogene in mouse. *Nat. Genet.* 8:373–79
211. Wade MJ, Beeman RW. 1994. The population dynamics of maternal-effects selfish genes. *Genetics* 138:1309–14
212. Watson MJO, Hoffmann AA. 1995. Cross-generation effects for cold resistance in tropical populations of *Drosophila melanogaster* and *D. simulans. Aust. J. Zool.* 43:51–58
213. Watson MJO, Hoffmann AA. 1996. Acclimation, cross-generation effects, and the response to selection for increased cold resistance in *Drosophila. Evolution.* 50:1182–92
214. Webb KL, Roff DA. 1992. The quantitative genetics of sound production in

Gryllus firmus. *Anim. Behav.* 44:823–32
215. Weis AE, Hollenbach HG, Abrahamson WG. 1987. Genetic and maternal effects on seedling characters of *Solidago altissima* compositae. *Am. J. Bot.* 74:1476–86
216. Wellington WG. 1965. Some maternal influences on progeny quality in the western tent caterpillar, *Malacosoma pluviale* (Dyar). *Can. Entomol.* 97:1–14
217. Willham RL. 1963. The covariance between relatives for characters composed of components contributed by related individuals. *Biometrics* 19:18–27
218. Willham RL. 1972. The role of maternal effects in animal breeding. III. Biometrical aspects of maternal effects in animals. *J. Anim. Sci.* 35:1288–93
219. Wolfe LM. 1993. Inbreeding depression in *Hydrophyllum appendiculatum:* role of maternal effects, crowding, and parental mating history. *Evolution* 47:374–86
220. Wulff RD, Bazzaz FA. 1992. Effect of the parental nutrient regime on growth of the progeny in *Abutilon theophrasti malvaceae*. *Am. J. Bot.* 79:1102–7
221. Yamamoto Y, Zhao XF, Suzuki AC, Takahashi SY. 1994. Cysteine proteinase from the eggs of the silkmoth, *Bombyx mori*: site of synthesis and a suggested role in yolk protein degradation. *J. Insect Physiol.* 40:447–54
222. Yanega D. 1993. Environmental influences on male production and social structure in Halictus rubicundus (Hymenoptera: Halictidae). *Insect Soc.* 40: 169–80
223. Zamudio KR, Huey RB, Crill WD. 1995. Bigger isn't always better; body size, developmental and parental temperatures, and male territorial success in *Drosophila melanogaster. Anim. Behav.* 49:671–77
224. Zielinski WJ, Vandenbergh JG, Montano MM. 1991. Effects of social stress and intrauterine position on sexual phenotype in wild-type house mice *Mus musculus*. *Physiol. Behav.* 49:117–24

Annu. Rev. Ecol. Syst. 1996. 27:477–500

RECRUITMENT AND THE LOCAL DYNAMICS OF OPEN MARINE POPULATIONS

M. J. Caley[1], *M. H. Carr*[2], *M. A. Hixon*[3], *T. P. Hughes*[1], *G. P. Jones*[1], *and B. A. Menge*[3]

[1]Department of Marine Biology, James Cook University, Townsville, Queensland 4811, Australia; [2]Marine Science Institute, University of California, Santa Barbara, California 93106, USA; [3]Department of Zoology, Oregon State University, Corvallis, Oregon 97331-2914, USA; (Authorship is alphabetical.)

KEY WORDS: benthic invertebrates, density dependence, limitation, population dynamics, reef fishes, regulation, settlement

ABSTRACT

The majority of marine populations are demographically open; their replenishment is largely or exclusively dependent on a supply of juveniles from the plankton. In spite of much recent research, no consensus has yet been reached regarding the importance of recruitment relative to other demographic processes in determining local population densities. We argue 1. that demographic theory suggests that, except under restrictive and unlikely conditions, recruitment must influence local population density to some extent. Therefore, 2. the question as to whether the size of a particular population is limited by recruitment is misguided. Finally, 3. the effect of recruitment on population size can be difficult to detect but is nonetheless real. A major weakness of most existing studies is a lack of attention to the survival of recruits over appropriate scales of time and space. Acknowledgment of the multifactorial determination of population density should guide the design of future experimental studies of the demography of open populations.

INTRODUCTION

Recruitment, in the broadest sense, is the addition of new individuals to populations or to successive life-cycle stages within populations. This process is clearly important for understanding a range of ecological phenomena, from the

0066-4162/96/1120-0477$08.00

genetic structure of populations (1, 46, 64, 81, 140), to population dynamics (the subject of this review), to community structure (10, 75, 104, 110, 125, 128, 165). While interest in recruitment of marine populations has a long history (58, 161, 175), the last decade has seen renewed interest and substantial progress in this area. This resurgence stems in part from the recognition 1. that most local populations of marine organisms are demographically open, where local recruitment is uncoupled from local reproduction by a dispersive larval phase, and 2. that marine species are often organized into non-equilibrial communities whose structure and dynamics depend on the interactions of a suite of biotic and abiotic processes that affect both recruitment and postrecruitment survival (29, 47, 95, 103, 104, 126, 128, 154, 160). Thus, for open marine populations, recruitment is generally defined more restrictively as the addition of individuals to local populations following settlement from the pelagic larval phase to the benthic or demersal early juvenile phase. Recruitment, in this case, is analogous to births in closed populations.

Seven recent reviews of recruitment in marine species (10, 42, 47, 50, 95, 110, 127) attest to the rapid growth of interest in this topic. However, no consensus has yet been reached regarding the relative importance of recruitment and other processes in determining the dynamics of local, open populations of marine species (42, 65, 87, 110, 172). Therefore, our goals are fourfold: 1. to review demographic theory as it applies to recruitment and subsequent population density in open populations; 2. to review methods for investigating the importance of recruitment relative to other processes in driving the dynamics of these populations; 3. to examine how issues of temporal and spatial scale affect the interpretation of the importance of recruitment in setting local population size; and 4. to suggest a protocol for future studies of the dynamics of open populations of marine species. We limit our review mainly to the substantial literature published since 1980 for temperate and tropical reef fishes and for intertidal and subtidal invertebrates living on hard substrata.

GENERAL CONCEPTS

Population Dynamics in Open Systems

Understanding the processes that underlie local population dynamics requires knowledge of the rates of birth (here recruitment), death, immigration, and emigration. In open marine populations, various combinations of these demographic rates could be examined for both the pelagic phase and the benthic or demersal phase; usually they are studied only after settlement into the adult habitat. For most benthic invertebrates and demersal fishes, the local production of offspring has little or no direct role in setting local population size because larval recruitment from elsewhere provides the only substantial input of new individuals. If recruitment fails, the local population will decline to extinction, regardless of local fecundity. Conversely, the local population will persist as long as recruitment continues, even if these adults produce no viable offspring. As a result, a local population cannot be regulated by its own fecundity, even if density-dependent effects on local reproductive output are evident (e.g. 54, 74). Postsettlement movement of juveniles and adults is potentially important in local population dynamics only for highly mobile species; they can safely be ignored for most sessile species. However, localized dispersal of benthic stages does occur among marine invertebrates that are capable of detachment or that undergo fragmentation (76), and also among some nonterritorial reef fishes (118, 119, JL Fredrick, unpublished data).

Historically, fisheries studies have provided the major focus on the dynamics of marine species (3, 5, 35, 36, 50, 68, 117, 121, 138, 144–146). Documenting strong year classes in various fish stocks, Hjort (67) was the first to articulate the idea that variation in larval survival may drive adult population dynamics. However, much of this research is not directly relevant to understanding local dynamics because fisheries biologists consider entire stocks (of fish, molluscs, crustaceans, etc), which, by definition, are reproductively closed. Recruitment in fisheries also typically refers to individuals entering the exploitable stock, which usually occurs at the subadult or adult stage. Consequently, a linear relationship is often postulated between recruitment and subsequent stock size (35, 36). In contrast, most population studies of marine invertebrates and reef fishes occur at localized sites, where fecundity and recruitment, estimated at the time of settlement, are decoupled.

The theory required to understand population dynamics is fundamentally different for open versus closed systems, and the appropriate choice of models is clearly scale-dependent. For example, at a sufficiently large spatial scale groups of open local populations become closed at the level of the "metapopulation" [i.e. an interconnected group of subpopulations linked by dispersal (59)]. Metapopulations, however, are not the subject of this review for three reasons. First, and most importantly, field studies of benthic invertebrates and demersal fishes generally have been conducted at the level of local populations. Thus, linking empirical studies with population dynamics theory can only be done presently at a relatively small spatial scale. Although an immense potential for dispersal exists between subpopulations during the larval stage (46, 91, 132, 133, 140), much more work on rates of larval exchange between local populations is required before metapopulation models can be tested for marine species (54). Second, many terrestrial-based metapopulation models are not directly relevant to marine populations because they assume the subpopulations are mostly closed, with only infrequent dispersal between patches (38, 60). In contrast, for most sessile invertebrates and fishes, the length of larval life ensures that virtually all local recruitment is from elsewhere. Third, most existing metapopulation theory addresses systems in which subpopulations are subject to high probabilities of extinction followed by rapid recolonization (54, 60). Local marine populations appear to be much more persistent; typically they do not go extinct and are reestablished on a regular basis, except at very small spatial scales (e.g. 26, 28). Only a handful of metapopulation models incorporate these features, which are therefore directly applicable to marine systems (11, 22, 27, 80, 111, 122).

Limitation vs Regulation

Confusion is evident in the marine literature regarding definitions associated with population limitation and regulation. Limitation occurs when any process adds individuals to or subtracts individuals from a population (142). In contrast, regulation occurs when at least one demographic rate is density-dependent—i.e. per capita rates of birth (recruitment) or immigration decrease, or the rates of death or emigration increase, as population density increases (62). Regulation leads to bounded fluctuations in population size, such that a population neither increases to infinity nor goes extinct (16, 107). A single process such as recruitment may limit a population if it is density-independent in one instance but regulate the population if it is density-dependent in another.

Recruitment to open populations typically varies spatially and temporally by several orders of magnitude (e.g. reviews in 42, 47 for reef fishes, and 31, 160 for benthic invertebrates). For a large peak of recruitment to result in not even one more adult, exact compensation is required, which could only occur if the strength of density-dependent mortality increased with density (143). The empirical evidence indicates that density-dependent postrecruitment mortality is unlikely to increase to such an extent that it completely eliminates a recruitment signal, i.e. fluctuations in recruitment invariably account for some of the variation in local density. Therefore, given that the dynamics of all open populations are driven to varying extents by both recruitment and mortality, a multifactorial approach is required to evaluate the contribution of recruitment relative to other limiting and regulating processes in setting population density (e.g. 65, 66, 74, 87, 88, 99, 166).

The importance of recruitment in determining local population size cannot be resolved by measuring rates of recruitment without additional knowledge of subsequent mortality patterns (89). However, as noted by Underwood & Denley (160), it is often assumed that the amount of recruitment in many marine organisms is so large that it could not possibly be limiting. This approach is unconvincing unless the postrecruitment survival rate is measured, and it can be demonstrated that even more recruits would not result in a further increase in local population size. Of course, patterns of mortality are much more difficult to measure in the field than is recruitment, particularly for mobile organisms such as fish. This logistical difficulty probably accounts for the greater emphasis on recruitment in studies of reef fishes compared to sessile invertebrates (where mortality of marked individuals is relatively easy to record). However, studying reef fishes has the advantage that newly settled individuals are relatively easy to identify to species compared to many sessile invertebrates.

THEORY OF OPEN POPULATION DYNAMICS

Conceptual Models

Although the importance of recruitment has been recognized repeatedly through this century (reviewed in 58, 161, 175), it was not until 1981 that Doherty (39) formalized the recruitment-limitation hypothesis, asserting that "the planktonic supply of larvae, far from being endless, may often be the limiting factor that forces future population size" (p. 470). At the time, many researchers believed that larval supply was generally sufficient to saturate habitats with juveniles, so that postsettlement competition was inevitable unless mediated by postsettlement predation or physical disturbance. As originally presented, the recruitment-limitation hypothesis stated simply that low rates of settlement could limit densities below levels where substantial competition occurred. Subsequently, Doherty & Williams (47) provided explicit predictions of the recruitment-limitation hypothesis, including that 1. postrecruitment mortality should be density-independent, 2. differences in the sizes of consecutive cohorts resulting from annual pulses of recruits should be preserved in the age-structure of populations without distortion, and 3. local population size should be highly correlated with variations in recruitment (i.e. recruitment determination sensu 51). As demonstrated below, however, the first and third predictions are not necessary conditions for the limitation of population size by recruitment, and the second may be true even if postrecruitment mortality is density-dependent. Clarifying these concepts is a major goal of this review.

Larval supply and subsequent settlement are difficult to measure due to problems of defining the number of larvae capable of settling at a particular place and time, and of counting small and/or cryptic individuals. Consequently, recruitment has come to be operationally defined as the initial sighting of a recently settled juvenile in the adult habitat (30, 89, 116). By this definition, recruitment necessarily incorporates early postsettlement losses (31, 89). Early mortality is often very high (reviews by 65 for reef fishes; 31, 169, 176 for invertebrates). To account for this pattern, Victor (163) proposed the dichotomy between primary and secondary recruitment limitation. Primary recruitment limitation is identical to Doherty's (39) original formulation, i.e. due to limited larval supply, the density of new recruits at the time of settlement is below levels where competition for limiting resources occurs. Secondary recruitment limitation is said to occur when the initial density of new recruits is sufficient for competition to occur if they were all to survive, but subsequent mortality reduces their density below any competitive threshold before they reach the adult stage. In this context, however, "recruitment" refers strictly to the number of older juveniles (up to several years postsettlement) entering the adult population, rather than larval input to the benthic population.

Mathematical Models

Hughes (73) and Roughgarden et al (123) were the first to model local population dynamics of corals and barnacles, respectively, using an external input of recruits rather than an intrinsic birth rate. More recent modeling studies have used the same approach for coral-reef fishes (166), bryozoans (74), and kelp (4, 108). These models demonstrate that local populations may be regulated, not just limited, by recruitment. A form of regulation occurs because, for a given absolute recruitment rate, the per capita recruitment rate into the adult population is higher when the density of a local population is low. In order for the per capita rate to remain steady across a range of adult densities, the number of recruits per unit area would have to increase over time in direct proportion to the increase in adult density. This is unlikely, except where recruitment is strongly aggregative (7, 69, 84, 112, 113, 152, 153). Thus, in many circumstances, the per capita recruitment rate is likely to decline as population size increases, i.e. recruitment is effectively density-dependent (4, 71, 73, 74, 166). This mechanism of regulation is not due to density dependence in the conventional sense, because no biological interaction is necessary for it to occur, but it is nonetheless real. In addition, density-dependent interactions may occur among recruits or between recruits and adults, which will cause a further depression of per capita recruitment rates at higher density. Few field biologists are aware of the implications of changes in the rate of per capita recruitment because of the convention of measuring inputs on an areal basis (e.g. per hectare or per m^2). In contrast, mortality is rarely expressed as the number of deaths per unit area; instead it is routinely quantified as a percentage (i.e. a per capita rate). This historical mixture of currencies has hampered studies of the relative roles of recruitment and mortality in limiting and regulating local, open populations.

A density-related decline in per capita recruitment into local populations cannot regulate an entire metapopulation. For regulation to occur at this larger spatial scale, "true" density dependence (in recruitment, fecundity, or mortality) must operate among the local populations or in the plankton. Most existing metapopulation models include density dependence at some level (72, 107,

PL Chesson, personal communication). Indeed, model metapopulations with density-independent dynamics take the same random walk to extinction as do local populations (23).

To date, models of open population dynamics have demonstrated that 1. a local population may fluctuate around an equilibrium level due to regulation by recruitment even in the absence of density-dependent mortality (4, 73, 74), and 2. when recruitment limitation occurs, by definition variation in recruitment will cause fluctuations in total population size (166). This correlation is the most common empirical test of the influence of recruitment on population size. However, computer simulations show that 3. even minor variation in rates of mortality will decouple the relationship between recruitment and population size (or year-class strength) (166). Finally, 4. where recruitment is inhibited by adults, cyclical variations in local population size can result from the time lag between recruitment and adulthood (4, 123). Thus, competition can result in wide fluctuations in numbers that might easily be mistaken as the result of peaks and troughs of recruitment. Clearly, many of the criteria used in field studies to characterize the effect of recruitment have been inadequate.

A Graphical Model of Open Population Regulation

To clarify the above issues, consider a simple graphical model of the dynamics of a local, open population (Figure 1). As noted earlier, the obligatory density-related decline in per capita recruitment will regulate a local population, even where postrecruitment mortality is density-independent (73, 166). The lack of density-dependent mortality will allow peaks of recruitment to translate directly into peaks in adult numbers, with no damping of fluctuations (Figure 1A). If mortality is high, large fluctuations in recruitment will obviously translate into relatively smaller fluctuations in the number of adults, compared to longer-lived organisms receiving the same variation in recruitment. Conversely, if mortality is low, the total number of adults will be the result of many recruitment events that remain "stored" in the population, resulting in a larger population size (24, 165). Consequently, fluctuations in total adult numbers in response to recruitment will be most evident in short-lived taxa that have one or a few dominant age-classes generated by recruitment pulses. Where density-independent mortality is variable, the correlation between recruitment peaks and the resultant number of adults will be scrambled and difficult to detect, especially in longer lived organisms. Thus, it is no accident that many existing empirical demonstrations of recruitment limitation are based on studies of relatively short-lived taxa whose populations are comprised of only a few cohorts (e.g. 74, 151, 162, 163), or those in which mortality rates are remarkably uniform (44).

Where density-dependent mortality occurs, it will tend to damp out large peaks in recruitment. However, the population size is still likely to rise and fall in response to peaks and troughs of recruitment, even if the correspondence

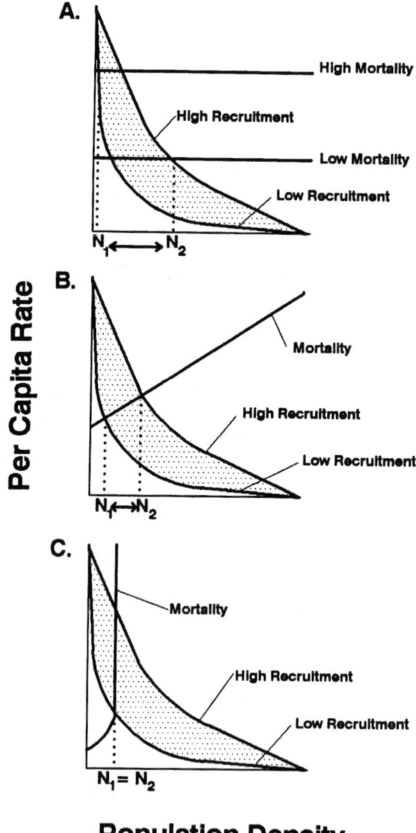

Figure 1 A graphical model of local population regulation in an open system. Recruits are generated elsewhere and arrive in numbers independent of local adult density. Unless the number of recruits increases in direct proportion to adult population density, the per capita recruitment rate must decline as the number of established individuals increases, i.e. per capita recruitment is effectively density-dependent. At very low population density, the per capita recruitment rate increases sharply, converging toward infinity when there are no established adults. Conversely, as adult density increases, the per capita rate of recruitment declines asymptotically toward zero. The stippled area bounded by the high and low recruitment curves represents variable recruitment levels. Three situations are illustrated: In (A), mortality is density-independent, and the local population is regulated solely by recruitment. Population size will fluctuate in response to changes in mortality or recruitment, between N_1 (low recruitment, high mortality) and N_2 (high recruitment, low mortality). In case (B), mortality is weakly density-dependent, which will partially damp out large peaks in recruitment, but the population size will still reflect recruitment fluctuations. Finally, in case (C), mortality is strongly density-dependent and increases in strength with increasing population density. This is the only case where population size will not reflect recruitment fluctuations.

between recruitment and population size is not perfectly linear (Figure 1B). For example, density-dependent mortality may be sufficiently strong to ensure that a doubling of the per capita recruitment rate results in only a 10% increase in population size. In this case, the local population is clearly regulated both by recruitment and by mortality. This is probably the most common case for marine invertebrates and fishes, in which density-dependent mortality is not uncommon, but in which its strength varies insufficiently to completely eliminate large pulses of recruitment. Complete elimination of large recruitment pulses would occur only if every recruit died above a certain threshold density that was exceeded even in years of poor recruitment (Figure 1C). Thus, the often-stated dichotomy that local population size is determined either by recruitment limitation or by density-dependent mortality is misguided. Certainly, an empirical demonstration of density-dependent mortality does not rule out the likelihood that recruitment is also important, nor does an empirical demonstration of recruitment effects on population size rule out the possibility that density-dependent mortality or emigration affects population size.

EMPIRICAL STUDIES OF OPEN MARINE POPULATIONS

Various interacting processes have the potential to influence the dynamics of open marine populations. Their relative importance will depend in part on where and when the study is conducted, and the spatial scale and temporal duration of the study. Much disagreement in the ecological literature stems from the use of spatial or temporal scales that differ among studies (21, 37, 109, 170, references in 56 and 135). For example, processes affecting larval supply and recruitment can operate at scales much larger than postsettlement processes such as competition and predation (47, 102). Hence, large-scale studies may overemphasize the former processes, and small-scale studies the latter. The relative importance of different processes may also depend on whether researchers are interested in total population size or the densities of reproductively mature adults (see 87), and on their operational definition of recruitment. Collectively, these differences constrain the way we define populations and which parameters we measure. Here we consider how operational decisions made prior to commencing a study can affect our interpretation of what drives the dynamics of local, open populations.

Selecting Study Sites

Ideally, several sites should be chosen that span a representative range in recruitment levels (e.g. 31, 55, 70, 86, 99, 115, 151). However, recruitment limitation has frequently been tested in habitats where rates of replenishment are unusually low. For example, rates of recruitment by the coral-reef fish *Pomacentrus moluccensis* onto lagoonal patch reefs of the southern Great Barrier Reef (44) are less than those onto contiguous reef slopes (BD Mapstone, unpublished data). The relative impact of recruitment fluctuations on population size is likely to be greater in the former habitat than in the latter. Conversely, Connell (31) argued that many experimental results on barnacles may have underestimated the importance of recruitment because workers frequently select high density sites where recruitment rates are relatively low and levels of competition and predation are high. Since sites are likely to vary in their rates of larval supply (98, 105) and/or the availability and quality of settlement habitat (e.g. benthic invertebrates: 34, 96; reef fishes: 17–19, 21, 92–94, 129, 155, 167), generalizations based on a single location will be inadequate.

Selecting Spatial and Temporal Scales

The minimum spatial scale for studying local population dynamics is constrained by the need for reliable estimates of recruitment and mortality. For mobile organisms, the area examined must be large enough to assume postsettlement movement is negligible, or at least measurable, throughout the life of the cohort. Where population densities are low, larger areas will be required to obtain reliable estimates of recruitment and mortality.

The relative importance of various processes may change with scale, so that a multiscale approach is required for a full understanding of population dynamics. In marine populations, density-dependent effects on mortality rates are often detected in small-scale experiments (e.g. for invertebrates: 31, 33, 148, 158, 159; for reef fishes: 8, 14, 51, 52, 78, 82, 85, 119, 120, 134, 147, 156, 157). It is quite possible that these effects may be less important in regulating numbers at larger scales, particularly in mobile animals where density-dependent emigration can ameliorate competition and where competition may occur only at a limited number of sites (see also 63). However, investigating the relative importance of limiting and regulating processes on very large spatial scales is problematic because all parameters must be estimated by scaling up from small-scale observations. For example, the dynamics of marine populations on a single habitat-type within a coral reef (e.g. 44, 86, 131), cannot be assumed to be representative of patterns at the scale of entire reefs.

The optimal temporal scale for determining the relative importance of recruitment and mortality in population dynamics is set by the longevity of a cohort (61, 75). The biggest effect of a new cohort on local population size occurs immediately after settlement. Because this impact diminishes with time, an inappropriately short time-scale may lead to an overestimate of the importance of recruitment. The time-scales of most marine population studies have not been justified, and few workers have provided estimates of longevity for the species under investigation (but see for example, 28, 48, 75).

Measuring Population Densities and Recruitment

The local densities of juveniles and adults are crucial variables in population studies. Since a cohort is always most abundant at settlement, it should not be surprising if the total local population size (i.e. all age classes combined) fluctuated in response to recent pulses of recruitment, particularly in short-lived species. Therefore, knowledge of the size and density of the resultant adult population is required to adequately assess the extent of local population regulation. For organisms with patchy spatial distributions, standard sampling techniques based on areal counts may underestimate the importance of density-dependent processes. This underestimation occurs because individuals at the center of aggregations experience higher densities compared to those at the edge. For juveniles in particular, it may be more appropriate to estimate the average density experienced by randomly chosen individuals rather than average densities based on quadrats or transects (83).

How recruitment is defined and measured is critical to testing and understanding the role of both pre- and postsettlement processes and the extent to which these processes are dependent or independent of population density. Measures of larval supply, settlement, and subsequent survival through to the adult population are all relevant response variables in experimental studies. However, methods of quantifying rates of delivery of larvae to benthic populations have only recently been developed (25, 41, 53, 105, 136, 178), and the manipulation of larval supply is extremely difficult. Furthermore, because settlement is typically sporadic, often sparse, nocturnal and/or cryptic, difficulties in measuring rates of settlement have forced most ecologists to adopt the standard operational definition of recruitment: the number of individuals that have settled and survived between census intervals (30, 89, 116).

Reef-fish ecologists have measured recruitment across time-scales ranging from hours to years, often using recruitment as a proxy for settlement. The median duration between censuses was 6 days in 105 published studies surveyed for this review. For some sessile invertebrates, quantifying recruitment is relatively easy, because settling stages can be detected shortly after metamorphosis, and settlers attach to both natural and artificial substrata (e.g. 12, 31, 49, 55, 57, 100, 101, 174). For others, however, recruits cannot be identified as to species until they attain diagnostic morphological characteristics months or years after settlement (e.g. 28, 75, 77).

The length of time between settlement and when recruitment is estimated can affect in a variety of ways interpretations of how population density is determined. First, the greater the time lag, the more likely patterns of abundance established at settlement will be modified by postsettlement mortality. For both reef fishes and sessile invertebrates, most studies of postsettlement mortality have reported highest mortality rates within the first few days after settlement (e.g. fish: reviewed in 65; see also 6, 9, 20, 66, 94, 156; invertebrates: 31, 169, 176). During this brief period, mortality may be sufficiently strong or variable to modify patterns of abundance established at settlement (e.g. 13, 15, 32, 66, 93, 94, 156). Furthermore, species differ markedly in early postsettlement mortality rates (45, 48, 130), which may alter patterns of relative abundance established at settlement (13, 20). The greater the lag between the occurrence of settlement and the measurement of recruitment, the more likely juvenile and subsequent adult densities will be positively correlated. Thus, long delays between settlement and measured recruitment can overestimate the importance of recruitment in determining the number of adults.

The frequency with which recruitment is sampled can also substantially influence estimates of the mean and variance of recruitment (e.g. 6, 90, 106) and can make comparisons of studies with different intensities of sampling virtually impossible. Of course, the ideal sampling frequency is a continuous record of arrival of settlers. Examples are provided by Yund et al (178) and Gaines & Bertness (53), who adapted sediment traps for use as passive, continuous samplers of barnacle larvae.

Another method of estimating recruitment in studies of reef fishes is the back calculation of daily settlement rates from circuli of otoliths (inner ear stones), which has become a common technique (reviewed by 164). No comparable method is yet commonly used in recruitment studies of invertebrates, although high-frequency banding (e.g. daily growth bands) could potentially be used in some taxa. Of the 28 published studies of reef fishes that estimated daily recruitment, 30% were based on otolith back-calculations. Such estimates assume constant mortality through time (i.e. across settlement episodes) and little measurement error (114, 162). These assumptions become increasingly tenuous with increased duration between settlement and the collection of fish for otolith samples (97). Therefore, this method of estimating recruitment and/or settlement rates should be avoided unless these assumptions can be met.

Relative Importance of Recruitment and Postrecruitment Processes

Field tests of the relative importance of recruitment and patterns of mortality can be observational or experimental. Neither approach is superior. They provide complementary information, and both are useful for testing theory or parameterizing models. Observational studies provide information on natural rates of recruitment and mortality, and changes in adult population size. However, experimental manipulations are necessary to examine the relative contribution of these factors to determining population size. The relative importance of factors in experiments (judged from statistical effect sizes) must be evaluated in the context of long-term observational data at larger spatial scales. For example, competition may show strong effects in experiments, but recruitment may not always reach a level that leads to competition in the field (see 86).

EFFECTS OF RESIDENT POPULATIONS ON RECRUITMENT In most published studies of open marine populations, a tacit yet untested assumption is that recruitment is determined solely by the incoming supply of larvae competent to settle, i.e. that recruitment is not a product of competent larvae interacting with resident populations. Correlations between recruitment and adult densities, both negative (e.g. 149, 177) and positive (e.g. 99, 151, 163), can provide evidence that such interactions are important. Experimental manipulation of resident densities has demonstrated that residents may facilitate (7, 69, 84, 112, 113, 134, 152, 153), inhibit (52, 113, 124, 134, 139, 153), or have no effect on recruitment (40, 84, 93, 155, 171). So far, experimental tests have been restricted to small patches of habitat. We do not know whether the significant effects that have been detected would be detected at larger spatial scales. Likewise, it is not clear whether nonsignificant results genuinely indicate no effect, or merely stem from high variability among small patches, inadequate sample sizes, or both.

NATURAL PATTERNS OF RECRUITMENT, JUVENILE MORTALITY, AND SUBSEQUENT ADULT DENSITIES The extent to which local populations are influenced by patterns of recruitment and mortality may be partially revealed by detailed demographic observation. Numerous studies of marine organisms provide estimates of population size, recruitment, and mortality monitored over multiple sites and/or years (e.g. 2, 20, 28, 74, 75, 77, 79, 141, 173). Data on mortality rates, however, are often limited or absent, especially from studies of mobile organisms such as fish, in which monitoring of individuals is difficult. The relationship between the abundance of recruits and adults over time (with an appropriate lag phase based on the time to maturity) has been used to distinguish between recruitment limitation and density dependence. If adult population size tracks recruitment levels, variable recruitment can be inferred to be the major determinant of population variation (e.g. 43, 44, 55, 151, 171). Alternatively, if adult population size is stable despite fluctuating recruitment, density dependence is likely to predominate (e.g. 86). The absence of a relationship between recruit and adult densities, however, does not necessarily imply that recruitment does not affect adult numbers. Instead, variable mortality, emigration, and/or postsettlement immigration may obscure the relationship (99, 118, 119, 166). In addition, the form of the relationship observed in a particular population may change through time. For instance, in a study of the goby *Sagamia geneionema*, Sano (personal communication) detected significant relationships between recruitment and adult densities in some years but not others.

To explore the consequences of various combinations of demographic variables, consider the graphical model in Figure 2. As noted previously, if mortality is density-independent, fluctuating recruitment will be translated directly into fluctuating numbers of adults (Figure 2, Case 1 or 2). Thus, the absence of a relationship between recruitment and subsequent mortality is usually attributed to density-independent mortality (163). However, it is only in situations in which mortality is constant from site to site (and/or generation to generation) that a strong linear relationship between recruitment levels and the size of the resulting adult population will be observed (Case 1C). If mortality is density-independent, but highly variable, patterns of recruitment will be modified by mortality, and there will be a relatively poor relationship between recruitment and subsequent adult population size (Case 2C). In this case, knowledge of mortality rates and factors affecting mortality could provide a better predictor of population size than recruitment, even though the population is recruitment-limited in the sense that, if an individual cohort of recruits was larger, on average more individuals would subsequently be added to the adult population.

If mortality is density-dependent but highly variable (Figure 2, Case 3A), there will be an indistinct or "density-vague" (150) relationship between recruitment and adult numbers (Case 3C), which might be difficult to distinguish from the situation in which mortality is primarily density-independent and variable (Case 2C). With constant density-dependent mortality (Case 4A), cohorts of different strength will tend to converge in size, and the number of adults derived from increasing recruitment will tend toward an asymptote (Case 4C), as observed in some species (e.g. 8, 52, 82, 86, 156), or even generate a unimodal relationship (117).

PATTERNS OF RECRUITMENT REFLECTED IN THE AGE STRUCTURE OF ADULT POPULATIONS If larval supply of a species is variable and early mortality is not strongly density-dependent, then age-class strength should be variable among years, with the size of each cohort reflecting the recent history of recruitment. This observation has been the basis of much research into the role of recruitment in marine populations (e.g. 43, 44, 162, 163, 168). Variable and persistent age-class strength, however, can also occur in populations subjected to density-dependent processes (70). Therefore, persistent age-class structure is not diagnostic of density-independent mortality. Furthermore, variable postrecruitment mortality may be sufficient to hide the relationship between recruitment and year-class strength, even in the absence of density-dependent mortality (Figure 2, Case 2C). Future experiments which manipulate the magnitude of successive cohorts may enable us to better evaluate the effect of fluctuating

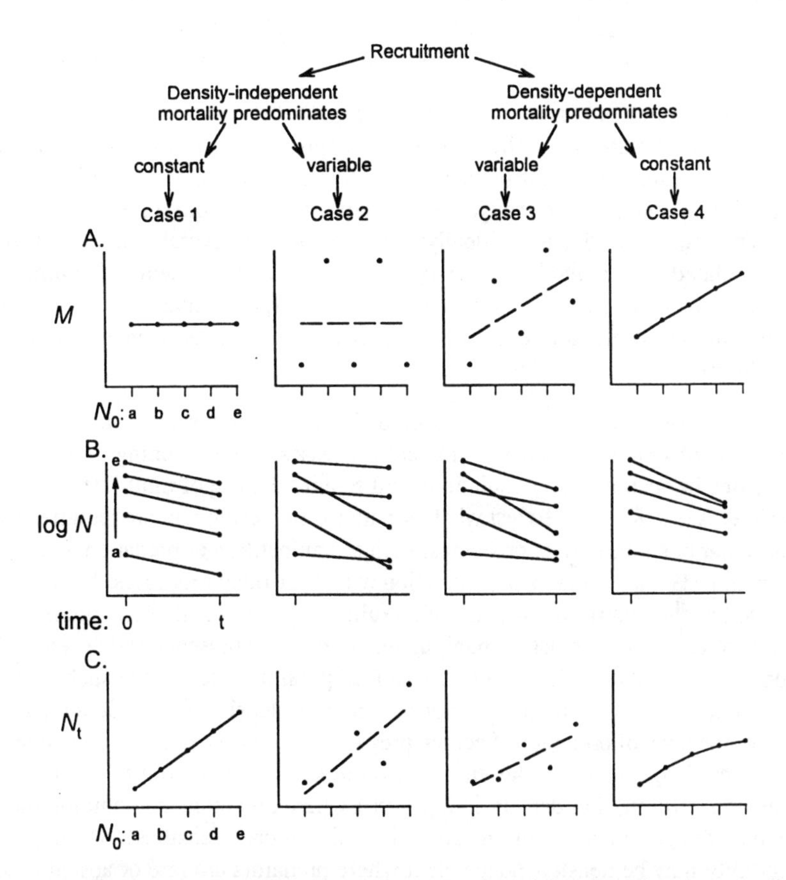

Figure 2 The effect of postrecruitment mortality on local population dynamics when mortality is constant and density-independent (Case 1), variable and density-independent (Case 2), variable and density-dependent (Case 3), or constant and density-dependent (Case 4). Following a fixed pattern of recruitment at time 0 (five cohorts of density *a* through *e*), we illustrate the patterns of, *A*, per capita mortality rate (*M*) as a function of recruitment density (N_0), *B*, the resulting survivorship curve of log *N* as a function of time from initial density at time 0 to later densities of juveniles or adults at time *t*, and *C*, densities of adults and juveniles (N_t) as a function of N_0 (the recruit-stock function). Between-cohort variation in mortality can cause unclear relationships (dashed curves) between recruitment and mortality both in terms of slope and curvature (Cases 2A and 3A), as well as unclear recruit-stock relationships (Cases 2C and 3C). Note that all recruit-stock functions must pass through the origin, so the illustrated curves are only portions of entire functions. Note also that all these plots are for cohorts and, therefore, do not consider the cumulative affects of recruitment and mortality of sequential cohorts on total population size.

recruitment on age-distributions as well as the overall population size. Differences in the initial strength of cohorts will tend to disappear over time if density-dependent mortality is strong or to persist if it is weak or absent.

RECRUITMENT MANIPULATIONS: EFFECTS ON JUVENILE MORTALITY AND SUBSEQUENT ADULT DENSITY The most direct approach to testing the effects of varying recruitment on subsequent population size is to manipulate levels of recruitment and monitor changes in the size of cohorts with different starting densities (e.g. 40, 51, 83). Ideally, recruitment into several cohorts will be manipulated over a number of years (86). In these experiments, recruitment must be manipulated in a realistic way, and the experimental units should be large enough to provide realistic measures of mortality rates under different conditions.

MULTIFACTORIAL EXPERIMENTS Ultimately, assessment of the importance of recruitment relative to other biological processes in determining population densities in open marine populations will require the use of multifactorial experiments (65, 87, 99). To test whether patterns of recruitment are modified by particular postrecruitment processes, such as competition or predation, recruitment must be manipulated in conjunction with these other processes. For example, to test the relative importance of recruitment and postsettlement predation, levels of recruitment must be manipulated, in both the presence and absence of appropriate predators. There are at least four possible outcomes of such an experiment. First, postrecruitment mortality may be density-independent regardless of the level of predation. Second, prey mortality may be density-dependent where predators are rare or absent (due to competition or a mortality agent other than competition), but density-independent where predators are common (due to the reduction in competition caused by reduced prey densities). Third, prey mortality may be density-independent where predators are rare or absent (due to low rates of recruitment), but density-dependent where predators are common. Finally, prey mortality may be density-dependent regardless of the level of predation, indicating either that competition (or a mortality agent other than predation) operates independently of predation, or that density-dependent mortality is the result of, for example, competition in the absence of predation but is the result of predation when predators are present. To date, this multifactorial experiment has been conducted on two species of damselfishes (*Chromis cyanea* in the Bahamas, Hixon and Carr unpublished data; and *Pomacentrus amboinensis* on the northern Great Barrier Reef, Jones & Hixon, unpublished data). For both species, mortality was density-independent in the absence of predators, but was weakly density-dependent in their presence.

FUTURE DIRECTIONS

Key issues remain unresolved in each of three main areas regarding recruitment. First, understanding of why recruitment varies in space and time is still limited. In particular, we need to understand better the causes of variation in production and survival of larvae, the processes that transport them, and mechanisms of habitat selection (see 91, 137). Second, the extent to which temporal and spatial variation in recruitment underlie variation in adult populations remains poorly understood. Third, and perhaps the most difficult and important question to answer for benthic and demersal populations, concerns the relative importance of recruitment vs postrecruitment processes in determining population size and structure. That is, to what extent does abundance in natural populations depend on recruitment, the process that establishes initial patterns, in comparison to factors such as competition, predation, facilitation, or disturbance—the processes that can modify these patterns? As noted above, the relative importance should be examined by multifactorial experiments in which recruitment and postrecruitment processes are manipulated simultaneously. A major problem in executing such experimental designs will be the manipulation of recruitment in realistic ways. Increasingly, investigators are finding ways to study and manipulate the early life history stages of both fishes and invertebrates, and we are optimistic that this trend will continue.

ACKNOWLEDGMENTS

Many of the ideas in this review were developed during the authors' participation in an international workshop, entitled "Recruitment and Population Dynamics of Coral Reef Fishes," funded by the Australian Department of Industry, Science, and Technology's Bilateral Science and Technology Collaboration Program grant 94/2599 to GP Jones and the US National Science Foundation (NSF) grant INT-94-18018 to MA Hixon. MJ Caley, TP Hughes and GP Jones were supported by research grants and an Australian Postdoctoral Research Fellowship to MJ Caley from the Australian Research Council. MA Hixon and MH Carr were supported by NSF grant OCE-92-17163, the National Undersea Research Program, and U.S. Minerals Management Service grant 14-35-0001-30758. BA Menge was supported by NSF grant OCE-92-17459 and a grant from the Andrew W. Mellon Foundation. We thank B Black, B Bryne, D Fautin, P Sale, and J Tanner for helpful comments on an earlier draft, E Dinsdale for help with the bibliography and figures, and L Schwarzkopf for help throughout this project. This is contribution number 144 of the Coral Group at James Cook University.

Literature Cited

1. Ayre DJ, Duffy S. 1994. Evidence for restricted gene flow in the viviparous coral *Seriatopora hystrix. Evolution* 48:1183–201
2. Babcock RC. 1985. Growth and mortality in juvenile corals (*Goniastrea, Platygyra* and *Acropora*): the first year. *Proc. 5th Int. Coral Reef Congress, Tahiti* 4:355–60
3. Bailey KM, Houde ED. 1989. Predation on eggs and larvae of marine fishes and the recruitment problem. *Adv. Mar. Biol.* 25:1–83
4. Bence JR, Nisbet RM. 1989. Space-limited recruitment in open systems: the importance of time delays. *Ecology* 70:1434–41
5. Beverton RJH, Holt SJ. 1957. On the dynamics of exploited fish populations. *Fisheries Investment.* Vol 19, Ser. 2. London: UK Ministry Agric. Fish
6. Booth DJ. 1991. The effects of sampling frequency on estimates of recruitment of the domino damselfish *Dascyllus albisella. J. Exp. Mar. Biol. Ecol.* 145:149–59
7. Booth DJ. 1992. Larval settlement patterns and preferences by domino damselfish *Dascyllus albisella* Gill. *J. Exp. Mar. Biol. Ecol.* 155:85–104
8. Booth DJ. 1995. Juvenile groups in a coral-reef damselfish: density-dependent effects on individual fitness and demography. *Ecology* 76:91–106
9. Booth DJ, Beretta GA. 1994. Seasonal recruitment, habitat associations and survival of pomacentrid reef fish in the US Virgin Islands. *Coral Reefs* 13:81–89
10. Booth DJ, Brosnan DM. 1995. The role of recruitment dynamics in rocky shore and coral reef fish communities. *Adv. Ecol. Res.* 26:309–85
11. Botsford LW, Moloney CL, Hastings A, Largier JL, Powell TM, et al. 1994. The influence of spatially and temporally varying oceanographic conditions on meroplanktonic metapopulations. *Deep-Sea Res. II* 41:107–45
12. Caffey HJ. 1985. Spatial and temporal variation in settlement and recruitment of intertidal barnacles. *Ecol. Monogr.* 55:313–32
13. Caley MJ. 1993. Predation, recruitment and the dynamics of communities of coral-reef fishes. *Mar. Biol.* 117:33–43
14. Caley MJ. 1995. Community dynamics of tropical reef fishes: local patterns between latitudes. *Mar. Ecol. Progr. Ser.* 129:7–18
15. Caley MJ, St. John J. 1996. Refuge availability structures assemblages of tropical reef fishes. *J. Anim. Ecol.* 65:414–28
16. Cappuccino N, Price PW. 1995. *Population Dynamics: New Approaches and Synthesis.* San Diego: Academic. 429 pp.
17. Carr MH. 1991. Habitat selection and recruitment of an assemblage of temperate zone reef fishes. *J. Exp. Mar. Biol. Ecol.* 146:113–37
18. Carr MH. 1994. Effects of macroalgal dynamics on recruitment of a temperate reef fish. *Ecology* 75:1320–33
19. Carr MH. 1994. Predicting recruitment of temperate fishes in response to changes in macrophyte density caused by disturbance. In *Theory and Application in Fish Feeding Ecology,* ed. DJ Stouder, KL Fresh, RJ Feller, pp. 255–69. Columbia, SC: Univ. S. Carolina Press
20. Carr MH, Hixon MA. 1995. Predation effects on early post-settlement survivorship of coral-reef fishes. *Mar. Ecol. Progr. Ser.* 124:31–42
21. Caselle JE, Warner RR. 1996. Variability in recruitment of coral reef fishes: the importance of habitat at two spatial scales. *Ecology.* In press
22. Chesson PL. 1985. Coexistence of competitors in spatially and temporally varying environments: a look at the combined effects of different sorts of variability. *Theor. Pop. Biol.* 28:263–87
23. Chesson PL. 1996. Matters of scale in the dynamics of populations and communities. In *Frontiers of Population Ecology,* ed. RB Floyd, AW Sheppard. CSIRO. In press
24. Chesson PL, Warner RR, 1981. Environ-

mental variability promotes coexistence in lottery competitive systems. *Am. Nat.* 117:923–43
25. Choat JH, Doherty PJ, Kerrigan BA, Leis JM. 1993. A comparison of towed nets, purse seine, and light-aggregation devices for sampling larvae and pelagic juveniles of coral reef fishes. *Fish. Bull.* 91:195–209
26. Coe WR. 1957. Fluctuations in littoral populations. *Geol. Soc. Am. Mem.* 67 1:935–40
27. Comins HN, Noble IR. 1985. Dispersal, variability and transient niches: species coexistence in a uniformly variable environment. *Am. Nat.* 126:706–23
28. Connell JH. 1973. Population biology of reef-building corals. In *Biology and Geology of Coral Reefs*, Vol. II: *Biology 1*, ed. OA Jones, R Endean, pp. 205–43. New York: Academic
29. Connell JH. 1978. Diversity in tropical rain forests and coral reefs. *Science* 199:1302–10
30. Connell JH. 1983. On the prevalence and relative importance of interspecific competition: evidence from field experiments. *Am. Nat.* 122:661–96
31. Connell JH. 1985. The consequences of variation in initial settlement vs. post-settlement mortality in rocky intertidal communities. *J. Exp. Mar. Biol. Ecol* 93:11–45
32. Connell SD, Jones GP. 1991. The influence of habitat complexity on postrecruitment processes in a temperate reef fish population. *J. Exp. Mar. Biol. Ecol* 151:271–94
33. Creese RG. 1980. An analysis of distribution and abundance of populations of the high-shore limpet, *Notoacmea petterdi* (Tenison-Woods). *Oecologia* 45:252–60
34. Crisp DJ. 1974. Factors influencing the settlement of marine invertebrate larvae. In *Chemoreception in Marine Organisms*, ed. PT Grant, AN Mackie, pp. 177–265. London: Academic
35. Cushing DH. 1975. *Marine Ecology and Fisheries.* London: Cambridge Univ. Press 278. pp.
36. Cushing DH. 1995. *Population Production and Regulation in the Sea.* Cambridge: Cambridge Univ. Press. 354 pp.
37. Dayton PK, Tegner MJ. 1984. The importance of scale in community ecology: a kelp forest example with terrestrial analogs. In *A New Ecology: Novel Approaches to Interactive Systems,* ed. PW Price, CM Slobodchikoff WS Gaud, pp. 457–81. New York: Plenum
38. Doak DF, Mills LS. 1994. A useful role for theory in conservation. *Ecology* 75:615–29
39. Doherty PJ. 1981. Coral reef fishes: recruitment-limited assemblages? *Proc. 4th Int. Coral Reef Symp.* 2:465–70
40. Doherty PJ. 1983. Tropical territorial damselfishes: Is density limited by aggression or recruitment? *Ecology* 64:176–90
41. Doherty PJ. 1987. Light traps: selective but useful devices for quantifying the distribution and abundances of larval fishes. *Bull. Mar Sci.* 41:423–31
42. Doherty PJ. 1991. Spatial and temporal patterns in recruitment. See Ref. 128a, pp. 261–93
43. Doherty PJ, Fowler A. 1994. Demographic consequences of variable recruitment to coral reef fish populations: a congeneric comparison of two damselfishes. *Bull. Mar. Sci.* 54:297–313
44. Doherty PJ, Fowler T. 1994. An empirical test of recruitment limitation in a coral reef fish. *Science* 263:935–39
45. Doherty PJ, Sale PF. 1985. Predation on juvenile coral reef fishes: an exclusion experiment. *Coral Reefs* 4:225–34
46. Doherty PJ, Planes S, Mather P. 1995. Gene flow and larval duration in seven species of fish from the Great Barrier Reef. *Ecology* 76:2373–91
47. Doherty PJ, Williams DMcB. 1988. The replenishment of coral reef fish populations. *Oceanogr. Mar. Biol. Annu. Rev.* 26:487–551
48. Eckert GJ. 1987. Estimates of adult and juvenile mortality for labrid fishes at One Tree Reef, Great Barrier Reef. *Mar. Biol.* 95:167–71
49. Farrell TM, Bracher D, Roughgarden J. 1991. Cross-shelf transport causes recruitment to intertidal populations in central California. *Limnol. Oceanogr.* 36:279–88
50. Fogarty MJ, Sissenwine MP, Cohen EB. 1991. Recruitment variability and the dynamics of exploited marine populations. *Trends Ecol. Evol.* 6:241–46
51. Forrester GE. 1990. Factors influencing the juvenile demography of a coral reef fish. *Ecology* 71:1666–81
52. Forrester GE. 1995. Strong density-dependent survival and recruitment regulate the abundance of a coral reef fish. *Oecologia* 103:275–82
53. Gaines SD, Bertness M. 1993. The dynamics of juvenile dispersal: why

field ecologists must integrate. *Ecology* 74:2430–35

54. Gaines SD, Lafferty KD. 1995. Modeling the dynamics of marine species: the importance of incorporating larval dispersal. In *Ecology of Marine Invertebrate Larvae,* ed. L McEdward, pp. 389–412. New York: CRC
55. Gaines SD, Roughgarden J. 1985. Larval settlement rate: a leading determinant of structure in an ecological community of the marine intertidal zone. *Proc. Natl. Acad. Sci. USA* 82:3707–11
56. Giller PS, Hildrew AG, Raffaelli DG. 1994. *Aquatic Ecology: Scale, Pattern and Process*. Oxford: Blackwell Sci. 649 pp.
57. Gotelli NJ. 1988. Determinants of recruitment, juvenile growth, and spatial distribution of a shallow-water gorgonian. *Ecology* 69:157–66
58. Grosberg RK, Levitan DR. 1992. For adults only? Supply-side ecology and the history of larval biology. *Trends Ecol. Evol.* 7:130–33
59. Hanski I, Gilpin ME. 1991. Metapopulation dynamics: brief history and conceptual domain. In *Metapopulation Dynamics,* ed. ME Gilpin, I Hanski, pp. 3–16. London: Academic
60. Harrison S. 1991. Local extinction in a metapopulation context: an empirical evaluation. *Biol. J. Linnean Soc.* 42:73–88
61. Harrison S, Cappuccino N. 1995. Using density-manipulation experiments to study population regulation. In *Population Dynamics: New Approaches and Synthesis,* ed. N Cappuccino, PW Price, pp. 131–47. San Diego: Academic
62. Hassell MP. 1986. Detecting density dependence. *Trends Ecol. Evol.* 1:90–93
63. Hassell MP, Southwood TRE, Reader PM. 1987. The dynamics of the viburnum whitefly (*Aleurotrachelus jelinekii*): a case study of population regulation. *J. Anim. Ecol.* 56:283–300
64. Hedgecock D. 1986. Is gene flow from pelagic larval dispersal important in the adaptation and evolution of marine invertebrates? *Bull. Mar. Sci.* 39:550–64
65. Hixon MA. 1991. Predation as a process structuring coral-reef fish communities. See Ref. 128a, pp. 475–508
66. Hixon MA, Beets JP. 1993. Predation, prey refuges, and the structure of coral-reef fish assemblages. *Ecol. Monogr.* 63:77–101
67. Hjort J. 1914. Fluctuations in the great fisheries of northern Europe viewed in the light of biological research. Rapp. P.-V. *Cons. Int. Explor. Mer* 20:1–228
68. Houde ED. 1987. Fish early life dynamics and recruitment variability. *Am. Fish. Soc. Symp. Ser* 2:17–29
69. Hoffman DL. 1989. Settlement and recruitment patterns of a pedunculate barnacle, *Pollicipes polymerus* Sowerby, off La Jolla, California. *J. Exp. Mar. Biol. Ecol.* 125:83–98
70. Holm ER. 1990. Effects of density-dependent mortality on the relationship between recruitment and larval settlement. *Mar. Ecol. Progr. Ser.* 60:141–46
71. Holt RD. 1985. Population dynamics in two patch environments: some anomalous consequences of an optimal habitat distribution. *Theor. Pop. Biol.* 28:181–208
72. Holt RD. 1993. Ecology at the mesoscale: the influence of regional processes on local communities. In *Species Diversity in Ecological Communities,* ed. RE Ricklefs, D Schluter, pp. 77–88. Chicago: Univ. Chicago Press
73. Hughes TP. 1984. Population dynamics based on individual size rather than age: a general model with a reef coral example. *Am. Nat.* 123:778–95
74. Hughes TP. 1990. Recruitment limitation, mortality, and population regulation in open systems: a case study. *Ecology* 71:12–20
75. Hughes TP. 1996. Demographic approaches to community dynamics: a coral reef example. *Ecology*. In press
76. Hughes TP, Ayre D, Connell JH. 1992. The evolutionary ecology of corals. *Trends Ecol. Evol.* 7:292–95
77. Hughes TP, Jackson JBC. 1985. Population dynamics and life histories of foliaceous corals. *Ecol. Monogr.* 55:141–66
78. Hunt von Herbing I, Hunte W. 1991. Spawning and recruitment of the bluehead wrasse *Thallassoma bifasciatum* in Barbados, West Indies. *Mar. Ecol. Progr. Ser.* 72:49–58
79. Hurlbut CJ. 1991. Community recruitment: settlement and juvenile survival of seven co-occurring species of sessile marine invertebrates. *Mar. Biol.* 109:507–15
80. Iwasa Y, Roughgarden J. 1986. Interspecific competition among metapopulations with space-limited subpopulations. *Theor. Pop. Biol.* 30:194–214
81. Johnson MS, Black R. 1984. Pattern beneath the chaos: the effect of recruitment on genetic patchiness in an inter-

tidal limpet. *Evolution* 38:1371–83
82. Jones GP. 1984. Population ecology of the temperate reef fish *Pseudolabrus celidotus* Bloch & Schneider (Pisces: Labridae). II. Factors influencing adult density. *Mar. Ecol. Progr. Ser.* 75:277–303
83. Jones GP. 1987. Competitive interactions among adults and juveniles in a coral reef fish. *Ecology* 68:1534–47
84. Jones GP. 1987. Some interactions between residents and recruits in two coral reef fishes. *J. Exp. Mar. Biol. Ecol.* 114:169–82
85. Jones GP. 1988. Experimental evaluation of the effects of habitat structure and competitive interactions on the juveniles of two coral reef fishes. *J. Exp. Mar. Biol. Ecol.* 123:115–26
86. Jones GP. 1990. The importance of recruitment to the dynamics of a coral reef fish population. *Ecology* 71:1691–98
87. Jones GP. 1991. Post-recruitment processes in the ecology of coral reef fish populations: a multifactorial perspective. See Ref. 128a, pp. 294–328
88. Keough MJ. 1988. Benthic populations: Is recruitment limiting or just fashionable? *Proc. 6th Int. Coral Reef Symp., Townsville,* 1:141–48
89. Keough MJ, Downes BJ. 1982. Recruitment of marine invertebrates: the role of active larval choices and early mortality. *Oecologia* 54:348–52
90. Keough MJ, Riamondi PT. 1992. Robustness of estimates of recruitment rates for sessile marine invertebrates. In *Recruitment Processes*, ed. DA Hancock, 16:33–39. Canberra: Bur. Rur. Resour. Proc
91. Leis JM. 1991. The pelagic stage of reef fishes: the larval biology of coral reef fishes. See Ref. 128a, pp. 183–230
92. Levin PS. 1991. Effects of microhabitat on recruitment variation in a Gulf of Maine reef fish. *Mar. Ecol. Progr. Ser.* 75:183–89
93. Levin PS. 1993. Habitat structure, conspecific presence and spatial variation in the recruitment of a temperate reef fish. *Oecologia* 94:176–85
94. Levin PS. 1994. Fine-scale temporal variation in recruitment of a demersal fish: the importance of settlement versus post-settlement loss. *Oecologia* 97:124–33
95. Mapstone BD, Fowler AJ. 1988. Recruitment and the structure of assemblages of fish on coral reefs. *Trends Ecol. Evol.* 3:72–76
96. Meadows PS, Campbell JI. 1972. Habitat selection by aquatic invertebrates. *Adv. Mar. Biol.* 10:271–382
97. Meekan MG. 1992. Limitations to the back-calculation of recruitment patterns from otoliths. *Proc. 7th Int. Coral Reef Symp. Guam* 1:624–28
98. Meekan MG, Milicich MJ, Doherty PJ. 1993. Larval production drives temporal patterns of larval supply and recruitment of a coral reef damselfish. *Mar. Ecol. Progr. Ser.* 93:217–25
99. Menge BA. 1991. Relative importance of recruitment and other causes of variation in rocky intertidal community structure. *J. Exp. Mar. Biol. Ecol.* 146:69–100
100. Menge BA, Berlow EL, Blanchette CA, Navarrete SA, Yamada SB. 1994. The keystone species concept: variation in interaction strength in a rocky intertidal habitat. *Ecol. Monogr.* 64:249–86
101. Menge BA, Farrell TM. 1989. Community structure and interaction webs in shallow marine hard-bottom communities: tests of an environmental stress model. *Adv. Ecol. Res.* 18:189–262
102. Menge BA, Olson AM. 1990. Role of scale and environmental factors in regulation of community structure. *Trends Ecol. Evol.* 5:52–57
103. Menge BA, Sutherland JP. 1976. Species diversity gradients: synthesis of the role of predation, competition and temporal heterogeneity. *Am. Nat.* 110:351–69
104. Menge BA, Sutherland JP. 1987. Community regulation: variation in disturbance, competition, and predation in relation to environmental stress and recruitment. *Am. Nat.* 130:730–57
105. Milicich MJ, Meekan MG, Doherty PJ. 1992. Larval supply: a good predictor of recruitment of three species of reef fish (Pomacentridae). *Mar. Ecol. Progr. Ser.* 86:153–66
106. Minchinton TW, Scheibling RW. 1993. Variations in sampling procedure and frequency affect estimates of recruitment of barnacles. *Mar. Ecol. Prog. Ser.* 99:83–88
107. Murdoch WW. 1994. Population regulation in theory and practice. *Ecology* 75:271–87
108. Nisbet RM, Bence JR. 1989. Alternative dynamic regimes for canopy-forming kelp: a variant on density-vague population regulation. *Am. Nat.* 134:377–408
109. Ogden JC, Ebersole JP. 1981. Scale and community structure of coral reef fishes: a long-term study of a large artificial reef. *Mar. Ecol. Progr. Ser.* 4:97–103

110. Olafsson EB, Peterson CH, Ambrose WG Jr. 1994. Does recruitment limitation structure populations and communities of macro-invertebrates in marine soft sediments: the relative significance of pre- and post-settlement processes. *Oceangr. Mar. Biol. Annu. Rev.* 32:65–109
111. Pacala SW. 1987. Neighborhood models of plant population dynamics. 3. Models with spatial heterogeneity in the physical environment. *Theor. Pop. Biol.* 31:359–92
112. Paine RT. 1974. Intertidal community structure: experimental studies on the relationship between a dominant competitor and its principle predator. *Oecologia* 15:93–120
113. Petersen JH. 1984. Larval settlement behavior in competing species: *Mytilus californianus* Conrad and *M. edulis* L. *J. Exp. Mar. Biol. Ecol.* 82:147–59
114. Pitcher CR. 1988. Validation of a technique for reconstructing daily patterns in the recruitment of coral reef damselfish. *Coral Reefs* 7:105–11
115. Raimondi PT. 1990. Patterns, mechanisms, consequences of variability in settlement and recruitment of an intertidal barnacle. *Ecol. Monogr.* 60:283–309
116. Richards WJ, Lindeman KC. 1987. Recruitment dynamics of reef fishes: planktonic processes, settlement and demersal ecologies, and fisheries analysis. *Bull. Mar. Sci.* 41:392–410
117. Ricker WE. 1954. Stock and recruitment. *J. Fish. Res. Bd. Can.* 11:559–623
118. Robertson DR. 1988. Settlement and population dynamics of *Abudefduf saxitalis* on patch reefs in Caribbean Panama. *Proc. 6th Int. Coral Reef Symp., Townsville* 2:839–44
119. Robertson DR. 1988. Abundances of surgeonfishes on patch reefs in Caribbean Panama: due to settlement, or post-settlement events? *Mar. Biol.* 97:495–501
120. Robertson DR. 1996. Interspecific competition controls abundance and habitat use of territorial Caribbean damselfishes. *Ecology* 77:885–99
121. Rothschild BJ. 1986. *Dynamics of Marine Fish Populations.* Cambridge: Harvard Univ. Press. 277 pp.
122. Roughgarden J, Iwasa Y. 1986. Dynamics of a metapopulation with space-limited subpopulations. *Theor. Pop. Biol.* 29:235–61
123. Roughgarden J, Iwasa Y, Baxter C. 1985. Demographic theory for an open marine population with space-limited recruitment. *Ecology* 66:54–67
124. Sale PF. 1976. The effect of territorial adult pomacentrid fishes on the recruitment and survival of juveniles on patches of coral rubble. *J. Exp. Mar. Biol. Ecol.* 24:297–306
125. Sale PF. 1977. Maintenance of high diversity in coral reef fish communities. *Am. Nat.* 111:337–59
126. Sale PF. 1980. Assemblages of fish on patch reefs—predictable or unpredictable? *Environ. Biol. Fish.* 5:243–49
127. Sale PF. 1990. Recruitment of marine species: Is the bandwagon rolling in the right direction? *Trends Ecol. Evol.* 5:25–27
128. Sale PF. 1991. Reef fish communities: open nonequilibrial systems. See Ref. 128a, pp. 564–98
128a. Sale PF, ed. 1991. *The Ecology of Fishes on Coral Reefs*. San Diego: Academic
129. Sale PF, Douglas WA, Doherty PJ. 1984. Choice of microhabitats by coral reef fishes at settlement. *Coral Reefs* 3:91–99
130. Sale PF, Ferrell DJ. 1988. Early survivorship of juvenile coral reef fishes. *Coral Reefs* 7:117–24
131. Sale, PF, Guy, JA., Steel, WJ. 1994. Ecological structure of assemblages of coral reef fishes on isolated patch reefs. *Oecologia* 98:83–99
132. Scheltema RS. 1971. Larval dispersal as a means of genetic exchange between geographically separated populations of shallow-water benthic marine gastropods. *Bull. Mar. Biol. Lab., Woods Hole* 140:284–322
133. Scheltema RS. 1986. On dispersal and planktonic larvae of benthic invertebrates: an eclectic overview and summary of problems. *Bull. Mar. Sci.* 39:290–322
134. Schmitt RJ, Holbrook SJ. 1996. Local-scale patterns of settlement: do they predict recruitment? *Mar. Freshw. Res.* In press
135. Schneider DC. 1994. *Quantitative Ecology: Spatial and Temporal Scaling*. San Diego: Academic. 395 pp.
136. Setran AC. 1992. A new plankton trap for use in the collection of rocky intertidal zooplankton. *Limnol. Oceanogr.* 37:669–74
137. Shanks AL. 1995. Mechanisms of cross-shelf dispersal of larval invertebrates and fish. In *Ecology of Marine Invertebrate Larvae,* ed. L McEdward, pp. 232–67.

Boca Raton: CRC
138. Sheperd JG, Cushing DH. 1980. A mechanism for density dependent survival of larval fish as the basis for a stock-recruitment relationship. *J. Cons. Int. Explor. Mer.* 39:160–67
139. Shulman MJ. 1984. Resource limitation and recruitment patterns in a coral reef fish assemblage. *J. Exp. Mar. Biol. Ecol.* 74:85–109
140. Shulman MJ, Bermingham EL. 1995. Early life histories, ocean currents, and the population genetics of Caribbean reef fishes. *Evolution* 49:897–910
141. Shulman MJ, Ogden JC. 1987. What controls tropical reef fish populations: recruitment or benthic mortality? An example in the Caribbean reef fish *Haemulon flavolineatum. Mar. Ecol. Progr. Ser* 39:233–42
142. Sinclair ARE. 1989. Population regulation in animals. In *Ecological Concepts: The Contribution of Ecology to an Understanding of the Natural World,* ed. JM Cherrett, pp. 197–241. Oxford: Blackwell Sci.
143. Sinclair ARE, Pech RP. 1996. Density dependence, stochasticity, compensation and predator regulation. *Oikos.* 75:164–73
144. Sinclair M. 1988. *Marine Populations: an Essay on Population Regulation and Speciation.* Washington Sea Grant Program; Seattle, Washington
145. Sissenwine P. 1984. Why do fish populations vary? In *Exploitation of Marine Communities,* ed. RM May, pp. 59–94, Berlin: Springer-Verlag
146. Smith TD. 1994. *Scaling Fisheries: The Science of Measuring the Effects of Fishing, 1855–1955.* Cambridge: Cambridge Univ. Press. 392 pp.
147. Steele MA 1996. The relative importance of processes affecting recruitment of two temperate reef fishes. *Ecology.* In press
148. Stimson J, Black R. 1975. Field experiments on population regulation in intertidal limpets of the genus *Acmaea. Oecologia* 18:111–20
149. Stimson JS. 1990. Density dependent recruitment in the reef fish *Chaetodon miliaris. Environ. Biol. Fish.* 29:1–13
150. Strong DR Jr. 1986. Density-vague population change. *Trends Ecol. Evol.* 1:39–42
151. Sutherland JP. 1990. Recruitment regulates demographic variation in a tropical intertidal barnacle. *Ecology* 71:955–72
152. Sweatman HPA. 1983. Influence of conspecifics on choice of settlement sites by larvae of two pomacentrid fishes (*Dascyllus aruanus* and *D. reticularus*). *Mar. Biol.* 75:225–29
153. Sweatman HPA. 1985. The influence of adults of some coral reef fishes on larval recruitment. *Ecol. Monogr.* 55:496–85
154. Talbot FH, Russell BC, Anderson GRV. 1978. Coral reef fish communities: unstable, high diversity systems? *Ecol. Monogr.* 48:425–40
155. Tolimieri N. 1995. Effects of microhabitat characteristics on the settlement and recruitment of a coral reef fish at two spatial scales. *Oecologia* 102:52–63
156. Tupper M, Boutilier RG. 1995. Effects of conspecific density on settlement, growth and post-settlement survival of a temperate reef fish. *J. Exp. Mar. Biol. Ecol.* 191:209–22
157. Tupper M, Hunte W. 1994. Recruitment dynamics of coral reef fishes in Barbados. *Mar. Ecol. Prog. Ser.* 108:225–35
158. Underwood AJ. 1978. An experimental evaluation of competition between three species of intertidal gastropods. *Oecologia* 33:185–208
159. Underwood AJ. 1984. Vertical and seasonal patterns in competition for microalgae between intertidal gastropods. *Oecologia* 64:211–22
160. Underwood AJ, Denley EJ. 1984. Paradigms, explanations and generalizations in models for the structure of intertidal communities on rocky shores. In *Ecological Communities: Conceptual Issues and the Evidence,* ed. DR Strong Jr, D Simberloff, LG Abele, AB Thistle, pp. 151–80. Princeton: Princeton Univ. Press
161. Underwood AJ, Fairweather PG. 1989. Supply-side ecology and benthic marine ecology. *Trends Ecol. Evol.* 4:16–20
162. Victor BC. 1983. Recruitment and population dynamics of a coral reef fish. *Science* 219:419–20
163. Victor BC. 1986. Larval settlement and juvenile mortality in a recruitment-limited coral reef fish population. *Ecol. Monogr.* 56:145–60
164. Victor BC. 1991. Settlement strategies and biogeography of reef fishes. See Ref 128a, pp. 231–60
165. Warner RR, Chesson PL. 1985. Coexistence mediated by recruitment fluctuations: a field guide to the storage effect. *Am. Nat.* 125:769–87
166. Warner RR, Hughes TP. 1988. The population dynamics of reef fishes. *Proc. 6th Int. Coral Reef Symp. Townsville,* 1:149–

55
167. Wellington GM. 1992. Habitat selection and juvenile persistence control the distribution of two closely related Caribbean damselfishes. *Oecologia* 90:500–8
168. Wellington GM, Victor BC. 1985. El Niño mass coral mortality: a test of resource limitation in a coral reef damselfish population. *Oecologia* 68:15–19
169. Wethey DS. 1986. Local and regional variation in settlement and survival in the littoral barnacle *Semibalanus balanoides* (L.): patterns and consequences. In *The Ecology of Rocky Coasts*, ed PG Moore, R Seed, pp. 194–202. Sevenoaks, CA: Hodder & Stoughton
170. Wiens JA. 1989. Spatial scaling in ecology. *Funct. Ecol.* 3:385–97
171. Williams DMcB. 1980. Dynamics of the pomacentrid community on small patch reefs in One Tree lagoon (Great Barrier Reef). *Bull. Mar. Sci.* 30:159–70
172. Williams DMcB. 1991. Patterns and processes in the distribution of coral reef fishes. See Ref. 128a, pp. 437–74
173. Williams DMcB, Sale PF. 1981. Spatial and temporal patterns of recruitment of juvenile coral reef fishes to coral habitats within "One Tree Lagoon", Great Barrier Reef. *Mar. Biol.* 65:245–53
174. Wing SR, Largier JL, Botsford LW, Quinn JF. 1995. Settlement and transport of benthic invertebrates in an intermittent upwelling region. *Limnol. Oceanogr.* 40:316–29
175. Young CM. 1987. Novelty of "supply-side" ecology. *Science* 235:415–16
176. Young CM, Chai FS. 1984. Microhabitat-associated variability in survival and growth of subtidal solitary ascidians during the first twenty-one days after settlement. *Mar. Biol.* 81:61–68
177. Young CM, Gotelli NJ. 1988. Larval predation by barnacles: effects on patch colonization in a shallow subtidal community. *Ecology* 69:624–34
178. Yund PO, Gaines SD, Bertness MD. 1991. Cylindrical tube traps for larval sampling. *Limnol. Oceanogr.* 36:1167–77

Annu. Rev. Ecol. Syst. 1996. 27:501–42

WHEN DOES MORPHOLOGY MATTER?

M. A. R. Koehl
Department of Integrative Biology, University of California, Berkeley, California 94720-3140

KEY WORDS: performance, ecomorphology, novelty, constraint, Reynolds number

ABSTRACT

The performance of an organism is the crucial link between its phenotype and its ecological success. When does an organism's morphology affect its performance? Quantitative mechanistic analyses of how function depends on biological form have shown that the relationship between morphology and performance can be nonlinear, context-dependent, and sometimes surprising. In some cases, small changes in morphology or simple changes in size can lead to novel functions, while in other cases changes in form can occur without performance consequences. Furthermore, the effect of a specific change in morphology can depend on the size, shape, stiffness, or habitat of an organism. Likewise, a particular change in posture or behavior can produce opposite effects when performed by bodies with different morphologies. These mechanistic studies not only reveal potential misconceptions that can arise from the descriptive statistical analyses often used in ecological and evolutionary research, but they also show how new functions, and novel consequences of changes in morphology, can arise simply as the result of changes in size or habitat. Such organismal-level mechanistic research can be used in concert with other tools to gain insights about issues in ecology (e.g. foraging, competition, disturbance, keystone species, functional groups) and evolution (e.g. adaptation, interpretation of fossils, and origin of novelty).

INTRODUCTION

The biological literature abounds with qualitative arguments about the selective advantages of particular morphological traits; more recently such qualitative arguments have been replaced by quantitative correlations between structural or performance characteristics of organisms and their fitness or ecological role.

0066-4162/96/1120-0501$08.00

Such qualitative or statistical statements are often made without a mechanistic understanding of how the morphological traits affect performance. Nonetheless, the performance of an organism is recognized as the crucial link between its phenotype and its ecological success (e.g. 7, 8, 20, 45, 91, 104, 198, 228, 229).

The purpose of this article is to draw together for ecologists and evolutionary biologists examples of the nonlinear, context-dependent, and sometimes surprising relationships between the morphology and performance of organisms. These nonintuitive effects, which have been revealed by mechanistic organismal-level investigations, are often missed in descriptive statistical or phylogenetic studies that use morphological or performance data. I have two goals in reviewing this information: One is to warn about the misconceptions that can arise from descriptive statistical studies that are blind to mechanism, and the other is to point out ways in which such organismal-level mechanistic information can be used to gain insights about issues in ecology (e.g. foraging, competition, disturbance, keystone species, functional groups) and evolution (e.g. adaptation, interpretation of fossils, and origin of novelty).

Some Definitions

I define the *morphology* of an organism as its structure on any level of organization from molecular to organismal, and I define *performance* as a measure of ability to carry out a specific function. Although some authors (8, 45) consider behavioral and physiological traits as morphology, I view them here as functions (although this distinction can sometimes be blurred—75). Furthermore, while some authors (8, 45) define performance as a measure of whole-organism capacity, I also consider performance of parts of organisms (e.g. appendages, enzymes). A *function* of a structure is simply a function the structure is capable of doing [i.e. *fundamental niche* sensu, (198); *performance* sensu, (63)], whereas a *role* of a structure is a use to which the structure is put by an organism in a given environment [i.e. *realized niche* sensu, (98); *behavior* sensu, (63)] (20, 60, 134). How well a structure performs a role (such as food-gathering) is often assumed to affect the fitness of the organism (e.g. 60, 134), although fitness may depend most on the performance of rare life-or-death roles (such as escape maneuvers) (198). *Fitness* is the number of zygotes or surviving offspring, corrected for rate of population growth, produced by an individual during its lifetime (45).

The Biomechanical Approach to Studying Effects of Morphology on Performance

There is a long history of research on the relationship between biological structure and function (reviewed by 137, 158, 229, 232, 234). One approach to functional morphology is biomechanics, the application of quantitative engineering techniques to study how organisms perform mechanical functions and interact with their physical environments. Biomechanists are concerned with elucidating the basic physical rules governing how biological structures operate, identifying physical constraints on what organisms can do, evaluating which structural characteristics affect performance, and analyzing the mechanisms responsible for the effects of morphological differences on performance (e.g. 3, 4, 35, 40, 44, 45, 54, 60, 104, 131, 134, 153, 166, 185, 198, 224, 226, 228, 230, 234). Although some biomechanists have been accused of assuming that natural selection has led to the morphologies being studied (e.g. 232), many of us simply focus on the mechanisms by which form affects function without making inferences about evolutionary origin. In addition to being a legitimate field on its own, biomechanics has also served as the handmaiden of other disciplines (232), providing useful tools for studying questions in ecology as well as in evolutionary biology and paleontology.

EXAMPLES OF SURPRISES THAT ORGANISMAL-LEVEL MECHANISTIC STUDIES REVEAL ABOUT HOW MORPHOLOGY AFFECTS PERFORMANCE

Many quantitative studies of the effects of morphology on performance are reviewed in biomechanics books (e.g. 3, 35, 40, 153, 166, 185, 224, 226, 230). My purpose here is not to summarize the field, but rather to focus on examples of the nonlinear and context-dependent ways in which performance depends on structure. After introducing basic types of nonlinear relationships between structure and function, I describe two examples of how the relationship between morphology and performance can be surprising (fluid dynamics of little hairs, and effects of body shape and texture on drag). I then discuss in more general terms the categories of nonintuitive effects of morphology on performance that we should keep in mind when using morphological data to address ecological or evolutionary questions.

Overview of Nonlinear Effects of Morphology on Performance

If the quantitative relationship between a measure of performance and a measure of morphology is nonlinear, then there are ranges of the morphological parameter where modifications of structure make little difference, and other ranges where small morphological changes can have large consequences. For example, an asymptotic curve is shown in Figure 1a: increasing the number of receptor sites on a cell increases the rate at which it adsorbs molecules when receptor numbers are low, but offers little improvement when receptor numbers are high (17).

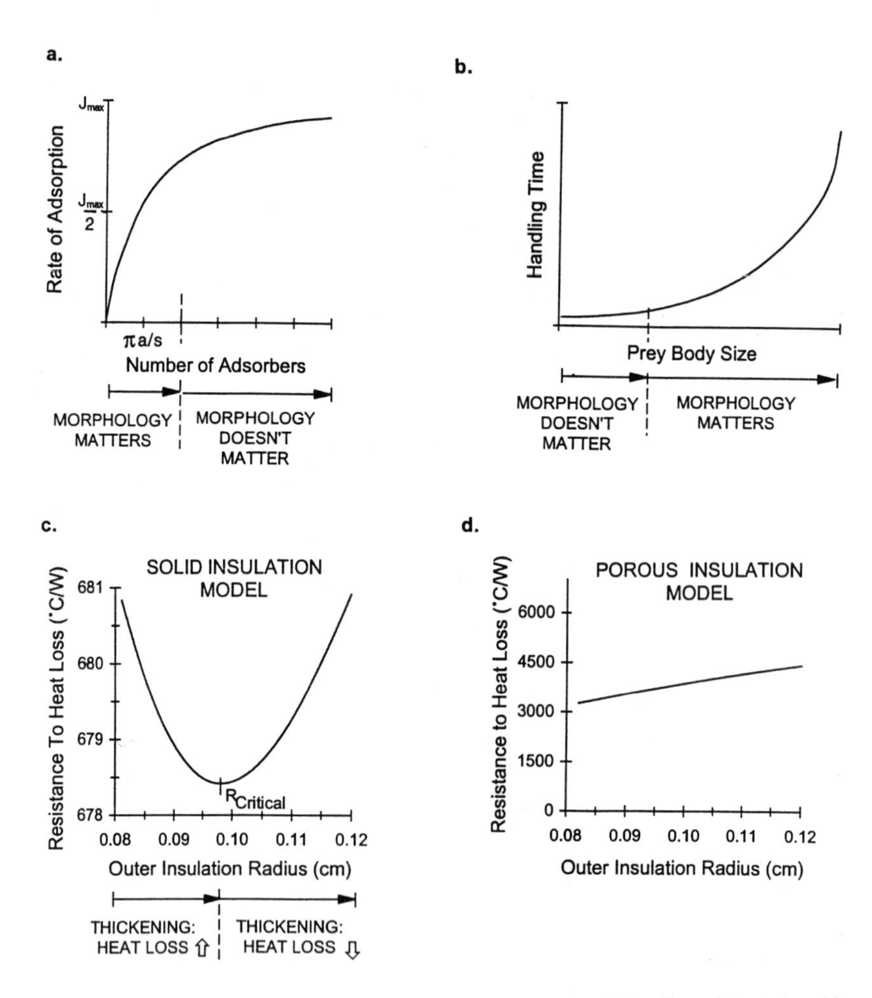

Figure 1 Examples of the relationship between performance (*y* axis) and morphology (*x* axis). (*a*) Rate of molecule adsorption (molecules per time, where J_{max} = number of molecules per time captured by a sphere whose entire surface is adsorbing receptor sites) of a spherical cell, plotted as a function of the number of adsorbing receptor sites on the cell (where a = radius of cell, s = radius of a single receptor site). [Curve calculated using equation 8 in (17)]. (*b*) Handling time for a predator to eat a prey organism (the higher the value, the better the performance of the prey), plotted as a function of prey body size. [Redrawn by digitizing one of the curves in Figure 6.3 of (46)]. (c) Resistance to heat loss by a heated cylindrical body (diameter = 0.16 cm) surrounded by a layer of solid insulation, plotted as a function of the outer radius of the body plus insulation. Resistance was calculated for heat loss by conduction through the insulation, and by free convection and radiation from the outer surface of the insulation. [Redrawn by digitizing the total resistance curve in Figure 2 of (192)]. (*d*) Resistance to heat loss by the same heated cylindrical body, but with porous insulation. [Redrawn by digitizing the free-convection curve in Figure 6 of (192)].

An example of an exponential curve is shown in Figure 1b: Differences in the size of small prey have little effect on predator handling time and hence on the prey's likelihood of being eaten, whereas differences in body size between larger prey can have a big effect on the danger of becoming a meal (46). [Of course, once prey become large enough that they escape in size from predation (179), differences in size once again become unimportant to the risk of being eaten.] Many aspects of mechanical performance also have exponential relationships to morphological features [e.g. deflection of a bending beam bearing a given load $\propto$ length^3; weight borne by a skeleton $\propto$ body volume $\propto$ length^3; volume flow rate through a pipe $\propto$ diameter^4; and many others described in e.g. (2, 3, 154, 166, 226, 230)]. Thus, performance of functions like skeletal support should be insensitive to structural variation at small size but very sensitive to morphological changes at large size.

If the relationship between performance and a morphological variable goes through a maximum or a minimum (Figure 1c), then the effect of increasing the morphological variable reverses once it passes a critical value. We are used to trying to relate such maxima and minima to the peaks and troughs in adaptive landscapes (e.g. 58, 104). In addition, we might also consider that passing through such an inflection point represents the acquisition of a novel consequence for a particlar type of morphological change. For example, if a heated body is surrounded by a non-heat-producing layer (e.g. extracellular cuticle, mucus, or fur), thickening that layer enhances the rate of heat loss from the body until a critical outer radius is reached, above which further thickening of the layer reduces heat loss (Figure 1c) (192). This critical radius concept from heat transfer physics was used to argue that naked baby mammals and birds would lose heat faster if they had feathers or fur (12), but calculations by Porter et al (192) showed the critical radius to be too small to be relevant (Figure 1c). Furthermore, when the non-heat-producing layer surrounding the body was assumed to be porous (like feathers or fur containing air spaces), the calculated resistance to heat loss was much greater than when the insulating layer was assumed to be solid (Figure 1d) (192). This example illustrates the importance of doing quantitative assessments of how morphology affects performance and of using biologically relevant assumptions in calculations.

Now, armed with the idea that the effect of morphology on performance is sometimes nonlinear, I provide some examples of various types of surprising relationships between morphology and performance.

Performance of Hairy Little Legs

Many animals from different phyla use appendages bearing arrays of hairs to perform important biological functions such as suspension-feeding, gas exchange, olfaction, mechanoreception, and swimming or flying (Figure 2a-e).

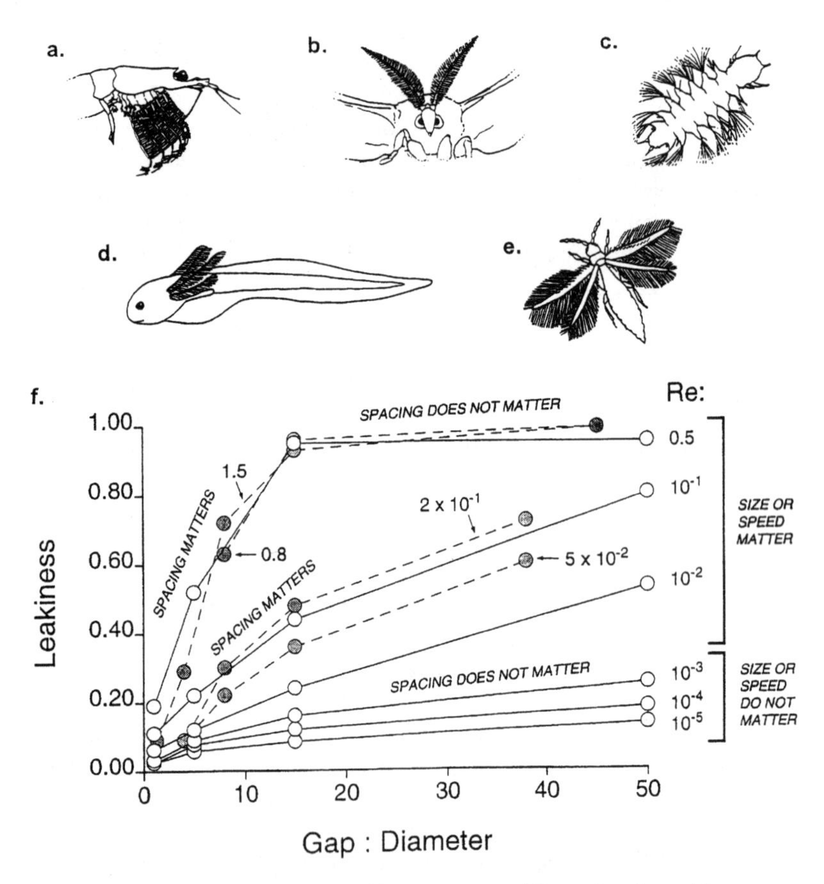

Figure 2 Examples of hair-bearing appendages that serve different functions: (*a*) suspension-feeding legs of a euphausid, Phylum Arthropoda; (*b*) olfactory antenna of a male moth, Phylum Arthropoda; (*c*) swimming parapodia of a nereid larva, Phylum Annelida; (*d*) external gills of a larval African lungfish, Phylum Chordata; wings of a thrips, Phylum Arthropoda. (*e*) Plot of leakiness (volume of fluid flowing through the gap between adjacent hairs divided by the volume of fluid that would flow through a space of that width if the hairs were not there) as a function of gap:diameter ratio of neighboring hairs. Open circles and solid lines represent leakiness calculated using the model of Cheer & Koehl (30). Grey circles and dashed lines represent leakiness measured during towing experiments with comb-like physical models of Hansen & Tiselius (78). Each line represents a different Re, as indicated by the numbers near the lines. [Redrawn from Figures 1 and 3 in (119)].

To carry out any of these functions, an array of hairs must interact with the water or air around it; thus, to understand how appendage morphology affects performance we must analyze the fluid dynamics of arrays of hairs.

Reynolds number (Re) represents the relative importance of inertial to viscous forces for a particular flow situation ($Re = LU/\nu$) where L is a linear dimension such as hair diameter, U is fluid velocity relative to the hair, and ν is kinematic viscosity of the fluid (226). At high Re's (e.g. large, rapidly moving structures), inertial forces predominate, so flow is messy and turbulent, whereas at low Re's (e.g. small, slowly moving structures), viscosity damps out disturbances in the fluid, hence flow is smooth and orderly. When fluid flows past a solid surface, the fluid in contact with the surface does not slip relative to the surface, and a velocity gradient (boundary layer) develops between the surface and the freestream flow. At low Re's, boundary layers are thick relative to the dimensions of the structure.

Most of the types of hairs listed above operate at Re's of order 10^{-5} to 10 (119). If the layers of fluid stuck to and moving with the hairs in an array are thick relative to the gaps between hairs, little fluid leaks through the array. Since performance of the functions listed above depends on the leakiness of hair-bearing appendages (reviewed in 119, 120), the effects of hair spacing and Re (size or speed) on leakiness have been explored using mathematical and physical models (Figure 2f) (29, 30, 78, 118–120). Although hairy appendages look like sieves, they are not always leaky: at $Re < 10^{-3}$, so little fluid leaks through the gaps between neighboring hairs that arrays of hairs function like paddles; in contrast, at Re's $> 10^{-2}$, fluid flows readily between the hairs and arrays behave like leaky filters. Another surprising discovery is that at Re's $< 10^{-3}$, changes in morphology (hair diameter or spacing) or behavior (speed) have little effect on leakiness (i.e. there is permission for morphological and behavioral diversity without performance consequences), whereas at Re's of 10^{-2} to 1, changes in size or speed can have a big effect on leakiness. Moreover, at Re's of 10^{-2} to 10^{-1}, decreasing gap width reduces leakiness, whereas at $Re = 1$, changes in hair spacing affect leakiness only when hairs are quite close together. The effect of a morphological change can also reverse at a critical Re: adding more hairs to an array reduces leakiness if $Re < 1$, but has the opposite effect if $Re > 1$ (D Abdullah, personal communication; 119). The leakiness of an array is increased when it moves near a wall (such as the body surface) if $Re < 10^{-2}$ (146)—thus the behavior that can alter leakiness changes as an animal grows (i.e. altering appendage distance from the body when $Re < 10^{-2}$, versus changing appendage speed when $Re > 10^{-2}$).

The hairy feeding appendages (second maxillae, M2's) of calanoid copepods (Figure 3) provide a biological example of the consequences of these physical

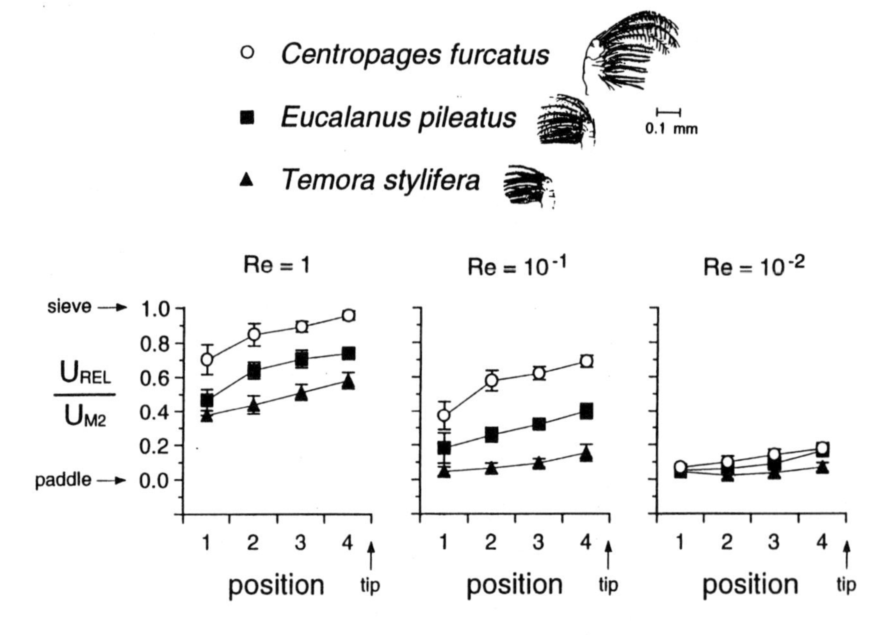

Figure 3 Fluid movement through dynamically scaled physical models of the M2's (pictured at the top) of *C. furcatus* (open circles), *E. pileatus* (grey squares), and *T. stylifera* (black triangles). The mean values of U_{REL}/U_{M2} (where U_{REL} is the absolute value of the fluid velocity relative to the M2, and U_{M2} is the velocity of the M2) for each section along the length of the model (section 1 near base, 4 near tip) are plotted for models run at a Re of 1 (*left graph*), 10^{-1} (*middle graph*), and 10^{-2} (*right graph*). Error bars indicate one standard deviation ($N = 3$ to 12). The lower the value of U_{REL}/U_{M2}, the less leaky (i.e. the more paddle-like) the M2. Note that the coarseness of the mesh of the M2's affects leakiness at Re's of 1 and 10^{-1}, but not at 10^{-2}. *C. furcatus* operate their M2's at Re $\simeq$ 1 (at which the M2's are sieve-like), *T. stylifera* at Re $\simeq 10^{-2}$ (at which the M2's are paddle-like), and *E. pileatus* at Re $\simeq 10^{-2}$ to 10^{-1} (a range in which the M2 leakiness varies). (Redrawn from Figure 14 in 119).

rules (118, 119). Copepods capture single-celled algae by flinging apart their pair of M2's and then squeezing them back together (114). Some species (e.g. *Centropage typicus*) that have coarsely meshed M2's, whose setae (hairs) operate at Re = 1, have leaky M2's and filter their food from the water during the squeeze; in contrast, other species (e.g. *Temora stylifera*) that have finely meshed, slowly moving M2's, whose setae operate at Re = 10^{-2}, have paddle-like M2's that capture food by drawing a parcel of water containing an algal cell toward the mouth during the fling. Thus, even though their M2 feeding motions look qualitatively similar, the physical mechanisms by which these two copepods capture food are different because they operate at Re's above and below the transition from paddle to sieve. Some copepods (e.g. *Eucalanus pileatus*) are plastic in their behavior and can switch their M2 speed, and thus leakiness, for different functions; note that only organisms operating in this transitional Re range can alter their leakiness by this means.

Thus, quantitative study of mechanism has revealed the conditions under which permission exists for morphological diversity of hairy appendages with little consequence to performance, versus conditions under which simple changes in hair speed, size, or spacing can lead to novel physical mechanisms of operation.

Effects of Body Shape and Texture on Fluid Dynamic Drag

Drag is the hydrodynamic force tending to push a body in the direction of fluid movement relative to the body (explained in e.g. 25, 40, 113, 226), hence drag tends to dislodge sessile organisms and to resist the motion of swimming, flying, and sinking creatures. At low Re's, drag is due to skin friction (the viscous resistance of the fluid in the boundary layer around the body to being sheared as the fluid moves past the body), so greater wetted area leads to higher skin friction. At high Re's drag is due to skin friction plus form drag (the pressure difference across the body due to the formation of a wake on the downstream side of the body). The bigger the wake, the higher the form drag; hence any morphological feature that moves the flow separation point (i.e. the place the wake starts to form) rearward along a body reduces drag at high Re. The drag coefficient (C_D) is a dimensionless measure of the drag-inducing effect of body shape.

Streamlining (putting a long, tapered end on the downstream side of a body) is one familiar way to reduce form drag, although the increased area raises skin friction. For large, fast organisms operating at high Re, streamlining reduces the net drag, but for small, slow organisms at low Re, streamlining increases drag. For example, C_D's of globose ammonoid shells are lower than C_D's of flat, streamlined shells at $Re < 100$, but the reverse is true for larger shells at higher Re (95). Similarly, drag on small ($Re = 1$ to 10) benthic stream invertebrates is lowered if their shape becomes more hemispherical, but is lowered on larger animals ($Re = 1000$) if they become more flattened (216). Nonetheless, most lotic invertebrates do not change shape as they grow, having streamlined profiles even when small (215). However, even though streamlining doesn't work when small stream insects are exposed to slow currents, flat body shapes do reduce hydrodynamic-resistance to their higher-Re escape maneuvers (34). For animals like these insects that can cross a Re transition by changing their speed, the Re of the activity that has the greatest impact on fitness (e.g. escape) appears to be the Re for which the body shape is drag-reducing.

Another morphological feature that has different effects on drag at different Re's is surface roughness (25, 109, 223, 226) (Figure 4). As the Re of a bluff body increases (i.e. as a nonstreamlined organism grows or moves more rapidly), C_D drops when flow in the boundary layer along the body's surface suddenly becomes turbulent and carries the separation point rearward, producing a smaller wake and lower form drag. At Re's below point A in Figure 4, surface texture is buried in the boundary layer and has no effect on drag, whereas at very high Re's surface bumps can protrude through the boundary layer and increase skin friction drag. However, surface roughness can trip the boundary layer to go turbulent at a lower Re than for a smooth body. Thus, there is a range of Re's (between A and B, Figure 4) in which a bumpy surface reduces drag on a bluff body. The shape of an organism's body affects whether or not this drag-reducing effect of bumpy skin occurs: Net drag on streamlined bodies is simply increased by surface texture once the critical Re is reached (A, Figure 4). The verrucae on sea anemones do not affect drag because the animals' Re's are below the transition Re (109). In contrast, tubercles increase

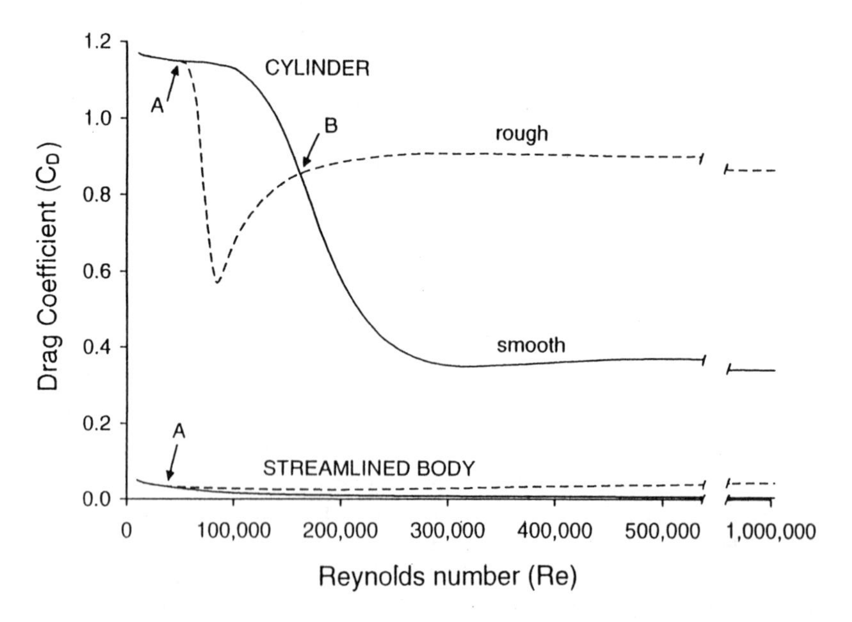

Figure 4 Plot of drag coefficient ($C_D = 2D/\{\rho SU^2\}$, where D = drag, ρ = fluid density, S = plan area of body, U = fluid velocity relative to body) as a function of Reynolds number ($\text{Re} = \rho UL/\mu$, where L = linear dimension of body, μ = dynamic viscosity of fluid) for a cylinder at right angles to the flow, and for a streamlined body (note that both axes of graph are log scales). The solid lines indicate bodies with smooth surfaces, and the dashed lines indicate bodies with rough surfaces. *A* indicates the onset of turbulence in the boundary layer, and *B* indicates the point beyond which roughness increases drag on the cylinder. (Redrawn using data digitized from Figure 5.8 in 226).

the drag on a swordfish's high-Re sword, but induce a turbulent boundary layer over the fish's body such that net drag on the whole fish is reduced (25, 223).

Both the examples described above (leakiness of hair-bearing appendages and drag on bodies) involve structures moving relative to the surrounding air or water, and both illustrate that the performance consequences of particular morphological characters depend on Re. I now discuss in more general terms the categories of nonintuitive effects of morphology on performance that we should keep in mind when using morphological data to address ecological or evolutionary questions.

Morphology Does Not Affect Performance

When the morphology of a structure does not affect the performance of some function, there is permission for diversity of form and for assumption of other functions.

MORPHOLOGICAL FEATURES THAT HAVE LITTLE EFFECT ON PERFORMANCE

Denny (42) found that lift is more important than drag in removing limpets from the substratum; thus features affecting only drag do not influence limpet performance at resisting ambient flow. Limpets show high diversity in the shell characters that affect drag.

Organisms that swim by flapping appendages at high Re can generate thrust to propel the body by using either drag or lift on the appendages. Vogel (226) noted that appendage shape has a big influence on lift-based swimming performance but makes little difference to drag-based propulsion, and he thus predicted that multifunctional appendages should use drag to generate thrust. Indeed, the walking appendages of polychaetes, ducks, muskrats, and freshwater turtles all use drag-based propulsion during swimming, whereas the lift-based flippers of sea turtles serve poorly as walking legs.

In some cases only part of a structure is critical to performance, so there is permission for diversity of form of the noncritical regions of the structure. For example, the morphology of the petiole and basal lobes of a tree leaf determine how easily it rolls up in the wind, but the diversity of form of the rest of the leaf does not affect performance of this drag-reducing rolling (225, 226). Butterflies bask in the sun to warm up, using their wings as solar panels. Dark wings absorb more heat, but since most of the heat transferred to the body comes from the basal region of a wing, there is permission for the rest of the wing to sport defensive or cryptic color patterns without interfering with thermoregulatory ability (235). At Re's where surface roughness affects drag, bumps on the anterior and widest regions of a body are very important to drag, whereas texture on the posterior region of a body makes little difference (25).

PERMISSION FOR DIVERSITY OF MORPHOLOGY AND KINEMATICS AT SMALL SIZE As mentioned above, skeletal structures should be insensitive to structural variation at small size. Indeed, there is variability in the ossification of bones (i.e. in their material stiffness—35; in very small salamanders—77). Similarly, the tiny stalks of the fruiting bodies of cellular slime molds show simple geometric scaling, in contrast to large biological columns (e.g. tree trunks, leg bones) that as they grow must become disproportionately wide relative to their length to support body weight (22, 154).

There are also biofluiddynamic functions whose performance is insensitive to morphology or kinematics at small size, such as the hair-spacing and surface roughness examples described above. Many small free-swimming organisms create feeding currents past themselves by flapping appendages. Calculation of the scanning currents produced by different types of appendage motions shows that for each technique, the energy cost per volume of water scanned changes very little if animals depart from optimal appendage kinematics (although which scanning technique is most efficient depends on the size of an animal's target zone—the distance at which it can perceive and capture prey) (31).

Another example of permission for kinematic diversity at small size is provided by basilisk lizards, which run on the surface of water (66, 67). The force to support the lizard's body during this sort of locomotion is provided by an upward impulse as the foot slaps onto the water surface, followed by an upward impulse as the foot strokes down into the water. Comparison of water-surface running by basilisks of different sizes revealed that small animals, which have the capacity to generate a large force surplus relative to their body weight, varied their kinematics considerably without performance consequences, whereas larger animals, which can generate barely enough force to support their weight, were constrained to a narrow range of leg and foot motions to run successfully on water. Indeed, in the field juveniles often run on water simply to move to another sunning spot, whereas adults venture onto the water only under duress.

Small Changes in Morphology or Simple Changes in Size Lead to Novel Functions

We should expect transitions in hydrodynamic or aerodynamic function as organisms grow or clades evolve through different Re ranges. Examples of such transitions were described above for the leakiness of hairy appendages and the drag on streamlined or rough bodies. Other examples can be found in ontogenetic studies of swimming. For instance, as brine shrimp larvae get bigger, even though the flapping motion of their appendages does not change, their propulsive mechanism switches from drag-based rowing at low Re to inertial swimming at higher Re (241, 242). Similarly, larval fish switch from drag-based swimming at low Re to inertial propulsion when they grow to higher Re (14, 176), and intermittent swimming becomes more energetically advantageous as the importance of viscous force declines at higher Re (237). Another example is provided by scallops, which swim by jet propulsion by squirting water out of the mantle cavity while clapping their shells together. Very small juvenile scallops cannot use this inertial mode of locomotion effectively and are sedentary; larger scallops can jet, and once at Re > 3000, they can also use lift to get up off the substratum; however, when very large they become poor swimmers again, as their shells grow too heavy relative to the thrust they can generate (36, 147).

Functional transitions accompanying size changes can also be found for organisms moving through air. For example, wing shapes that optimize gliding performance of plant seeds or animals depend on Re: short, wide wings are better at small size, whereas long, narrow wings enhance gliding at large size (51). An example of how isometric size changes in the absence of shape changes have the potential to generate novel functions is provided by the experiments of Kingsolver & Koehl (105, 107) that tested the aerodynamic and thermoregulatory consequences of changes in the length of protowings on models of fossil insects. At small body size, short thoracic protowings can improve thermoregulatory performance, although they have negligible effect on aerodynamic gliding, parachuting, or turning performance; in contrast, protowings of the same relative length on a larger insect can improve aerodynamic performance. This illustrates that it is physically possible for a simple increase in body size to cause a novel function (i.e. a solar panel can become a wing) without requiring the invention of a novel structure. (However, whether protowings served these aerodynamic or thermoregulatory roles in early insects is just as speculative as other feasible hypotheses, like sexual signaling, gas exchange, or skimming along the surface of a body of water.)

Another example of a functional switch accompanying a simple continuous change in morphology is provided by the chitinous exoskeleton (perisarc) of hydroid colonies (92) (Figure 5a). If bent too far, perisarc kinks like a beer can, damaging the tissue inside (Figure 5b). Perisarc, which has annulated regions and internodes, is thickened with time. Tissue damage from kinking is worse in annulated regions than internodes when perisarc is thin near the growing tips of colonies, but as the perisarc is thickened, these roles reverse and the annulated regions provide protection from damage when the colony is subjected to large bends (Figure 5c,d).

Dimensionless numbers, such as Re, that express the relative importance of various physical factors affecting a process, can provide us with hints of other places to look for functional shifts. For example, Froude number (gravity

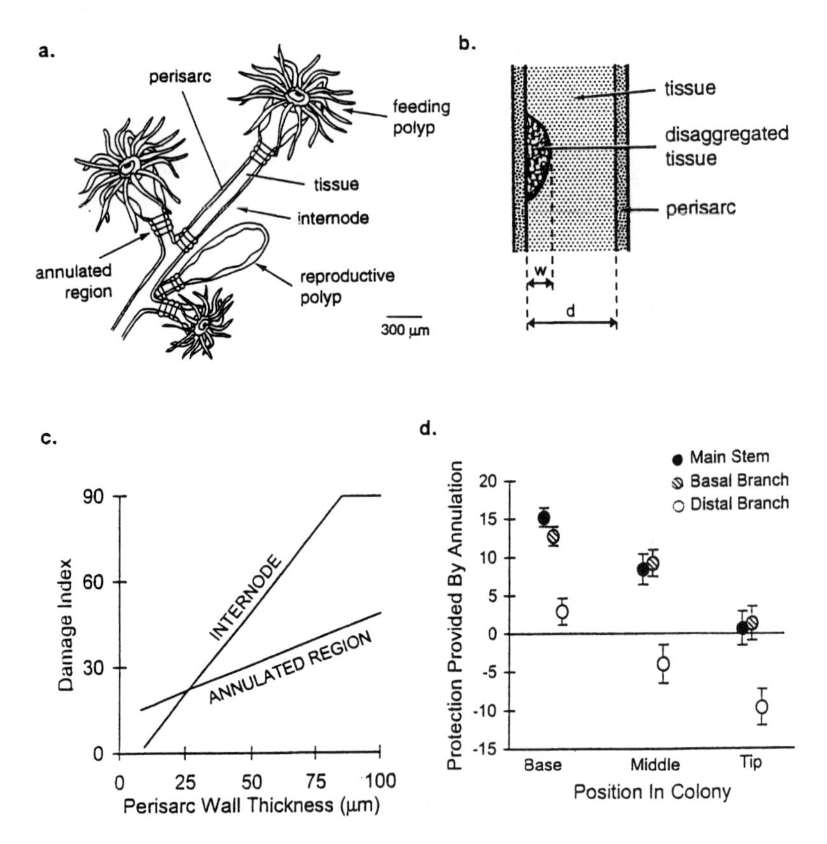

Figure 5 Perisarc of the hydroid *Obelia longissima*: (*a*) Diagram of the tip of a branch of a colony, showing the annulated and internode regions of the perisarc. (*b*) Diagram of the tissue damage (disaggregated tissue) caused by perisarc kinking when bent (w = width of damaged tissue, d = diameter of tissue inside the perisarc). (*c*) Damage index (arcsin w/d), plotted as a function of perisarc wall thickness for internode and annulated regions. The internode line levels off at a damage index of 90, indicating complete tissue damage. [Redrawn by digitizing the regression lines in Figure 7, Chapter 1, of (92).] (*d*) Index of protection from damage (a measure of the protection of soft tissue afforded by the presence of annulations in the perisarc, given by the difference between tissue damage in the internode region and tissue damage in the annulated region) for different regions of hydroid colonies. Error bars = 95% confidence intervals (n = 93 per group). Positive values indicate that annulations protect from damage, whereas negative values indicate that annulations make damage worse. [Redrawn using data digitized from Figure 8, Chapter 1, of (92).]

relative to inertia) is a good predictor of gait changes in pedestrian locomotion (e.g. 5), while reduced frequency (accelerational relative to steady-state flow) indicates the importance of nonsteady-state mechanisms of generating lift and thrust in swimming or flying (e.g. 37), and Péclet number (fluid convection relative to molecular diffusion) indicates the importance of bulk air or water movement in getting molecules to the surface of a collecting device such as a gill or olfactory antenna (120).

Effects of a Morphological Trait Depend on Other Characteristics of an Organism's Body

Single traits should not be studied in isolation (68), not only because multiple traits can affect a particular aspect of performance (e.g. 4), but also because both the magnitude and direction of the performance consequences of a particular morphological change can depend on other aspects of an organism's structure.

An example of the interactive effect of several traits on performance is provided by flying frogs, tree frogs that glide through the forest canopy and that have a unique suite of derived morphological characters, including enlarged hands and feet. An aerodynamic study using physical models of flying and nonflying frogs on which such characters could be modified one at a time revealed that the effects of the flyer traits on aerodynamic performance were nonadditive (48). For example, all the flyer traits occurring together improved turning performance significantly more than expected from the sum of their individual effects. However, for certain aspects of aerodynamic performance, the effect of the co-occurrence of flyer traits depended on body size: Gliding performance was improved more than expected only for small frogs, whereas parachuting performance was improved less than expected only for large frogs.

Performance of a structure at one level of organization can depend on morphology at another level of organization. In the following examples, the deformability of a structure (which depends on tissue microarchitecture and molecular composition—e.g. 224, 230) can affect the consequences of variation in gross morphology. While Lauder (132) has proposed a phylogenetic method to examine the independence of different levels of organization during evolution, mechanistic studies like those cited below reveal the physical reasons that performance depends on the interaction of different levels of structure.

Flexible sessile organisms experience lower drag forces than do rigid ones of the same shape because the deformable organisms are passively blown into more streamlined shapes (e.g. 109, 115, 225, 226). Flexibility also determines whether or not body shape even affects flow forces. Because of passive streamlining, the drag coefficients of various species of floppy intertidal algae are similar when water velocities are high enough to cause damage, even though they have very different shapes (28). In contrast, blade shape does affect drag for the less flexible blades of bull kelp (121). Denny (42) has suggested that once a lineage has become sufficiently flexible, shape may be removed from further selection by drag.

Flexibility can also determine the consequences of growth for organisms of a given shape, as illustrated by model studies of planar sessile organisms (122). If a planar rigid organism (e.g. a plating hydrocoral) lengthens in a wave-swept habitat, the hydrodynamic force it bears rises, whereas if a very flexible organism lengthens (e.g. a floppy alga that can move back and forth with the flow), the force on its holdfast remains low (Figure 6). However, a flexible organism in waves must grow to a critical size before it can benefit from "going with the flow": A floppy creature can move with the flow only until it reaches the end of its tether, at which point the water moves past it and it must bear the hydrodynamic force (115). Thus, algae that are short relative to the distance the water travels in a wave before it reverses direction do experience an increase in force as they grow (64). Furthermore, an organism of intermediate flexural stiffness can deflect enough to move with the flow only after the organism has become sufficiently long (deflection of a cantilever $\propto$ length^3), so as it grows, the force rises, then plateaus, and then decreases (Figure 6). Flexibility also determines which sort of flow habitat is most mechanically stressful: For rigid organisms, waves produce larger forces than do unidirectional currents of the

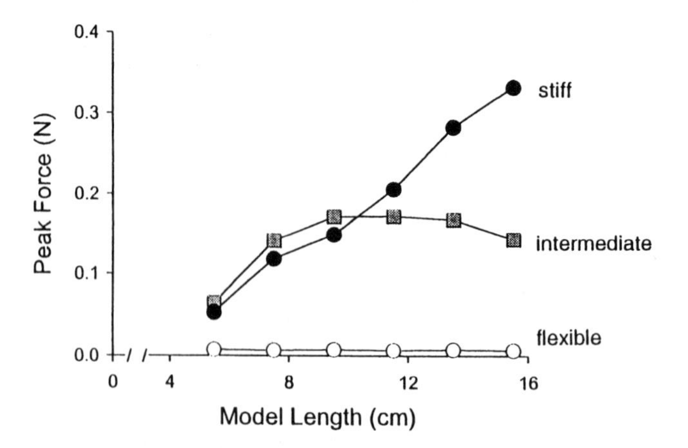

Figure 6 Peak hydrodynamic force measured on models of planar benthic organisms of different stiffnesses exposed to oscillatory flow in a wave tank, plotted as a function of model length. Error bars, which represent 95% confidence intervals ($n = 5$ per point), are smaller than the symbols used on the graph. The models maintained constant width and thickness as they "grew" (M Koehl, T Hunter, and J Jed, unpublished data).

same peak velocity, whereas the opposite is true for flexible organisms (122). Another example of the consequences of flexibility is provided by protozoans that locomote by undulating flagella that bear hairs (mastigonemes). Both fluid dynamical analysis (52) and observations of swimming protozoans (86) show that when mastigonemes are stiff, protozoans swim in the same direction as the direction of wave propagation along the flagellum, whereas when mastigonemes are flexible, the organisms swim in the opposite direction (as do protozoans without mastigonemes).

Effects of a Morphological Trait Depend on Habitat

The performance of an organism is meaningful only in the context of the environment in which the organism lives (e.g. 63, 74, 109). The following examples illustrate not only that changes in habitat can have profound effects on performance, but also that the consequences of a morphological modification can depend on the environment.

One environmental variable whose effect on performance has received much attention from biologists is temperature. Temperature is perhaps the most commonly used abscissa on graphs in physiology books, and the effects of temperature on important biological and ecological processes such as locomotor performance, predator-prey interactions and foraging strategies, development rate and life history patterns, and habitat use are well documented (e.g. 32, 63, 81, 82, 87, 90, 189–191). Other more subtle effects of temperature on mechanical performance include changes in the mechanical properties (such as stiffness, resilience, strength, and toughness) of biological tissues (e.g. 110, 112, 224, 230), and a shift in the Reynolds number of aquatic organisms [due to changes in kinematic viscosity, which nearly doubles between 0° and 20°C, as well as to changes in the rate of movement of some species (e.g. 187, 188, 215, 226, 245)]. Therefore, the body speeds and sizes at which Re-dependent functional shifts occur depends on habitat temperature. Obviously the temperature changes that accompany diurnal and seasonal cycles, climate shifts, microhabitat changes, and species range extensions can have profound effects on the performance of organisms with given morphologies.

Another obvious feature of the physical environment that can have enormous mechanical and physiological consequences is whether an organism is surrounded by air or water. The transition between aquatic and terrestrial habitats has been made in the evolution of many lineages. This transition between water and air is also made during the ontogeny of some species (e.g. with aquatic larvae and terrestrial adults) and is made daily by intertidal organisms and by animals that locomote between the two habitats (e.g. littoral crabs, diving birds). The consequences to performance of moving between these two media are reviewed by Denny (41) and Vogel (226).

Benthic marine organisms are exposed to unidirectional currents at some sites, but to waves (oscillatory flow with high accelerations) at others. The morphology of a spawning animal in waves does not affect gamete dispersal (123), although it does in gentler unidirectional flow (221). Similarly, streamlining works only if flow direction can be predicted, and hence it is ineffective at wavy sites (42, 109). Furthermore, the consequences of growth to a stiff organism's risk of dislodgment by hydrodynamic forces are different in the two types of habitats. The attachment strength of a sessile organism depends on holdfast or foot area ($\propto L^2$), and the drag and lift it must withstand in unidirectional flow depends on plan area ($\propto L^2$), whereas the acceleration reaction force in waves depends on body volume ($\propto L^3$) (e.g. 40, 109). Therefore, wave-swept organisms have a physical constraint on body size not experienced by creatures in steady currents (43).

Flow conditions in the environment affect the feeding performance of many aquatic animals that make their living by filtering small organisms and particles from the surrounding water. For example, ambient flow velocity affects not only the volume of water per time that a passive benthic suspension feeder can process for food and the amount of particulate material stirred up in the water (e.g. 171), but the velocity also determines the efficiency (proportion of encountered particles that are captured) and the size-selectivity of a filter of a given morphology (e.g. 204, 209). In addition, the turbulence (random fluctuations in velocity) of the flow can alter feeding performance of both planktonic (e.g. 151, 203, 210) and benthic (e.g. 183) suspension feeders. However, whether turbulence enhances or inhibits feeding rate depends on the morphology, swimming mode, and physical capture mechanism(s) used by an organism (209, 210).

An organism's performance can depend on the morphology of its neighbors. For example, the structure of canopies of terrestrial plants, or of aquatic sessile animals or macrophytes, affects the microclimate that they and the other organisms living among them encounter, often buffering them from fluctuations and extremes in environmental conditions (e.g. 56, 109, 115, 121, 157, 244). Similarly, the arrangement of individuals in aggregations of sessile benthic invertebrates, such as tubiculous phoronids and polychaetes, can affect the feeding and spawning performance of individuals within the aggregation (100, 101, 220, 221). Whether neighbors improve or harm the performance of a benthic animal depends on morphology. For example, the feeding performance of encrusting bryozoans is enhanced by upstream neighbors (but only if ambient flow is fast), whereas that of arborescent forms is reduced (172–174). Furthermore, physical constraints on organisms in aggregations can sometimes be different from those on solitary individuals: e.g. sea palm kelp in dense stands can support their crowns of blades on slimmer, taller stipes without buckling because they lean on their neighbors (85).

Another example of how habitat can alter the effect of a morphological feature on performance is provided by arboreal lizards. The effect of leg length on sprint speed depends on the diameter of the branch on which an animal runs (145).

Consequences of Behavior Depend on Morphology

I define behavior as an action taken by an organism, ranging from simple kinematics or assumption of a posture to complex interactions with other organisms. The consequences of a particular behavior can depend on the morphology of the organism. For example, the flying frogs described above take on a characteristic posture when airborne, with knees pulled forward and feet spread out laterally. Emerson & Koehl (47) discovered via model experiments that when frogs assume the flying posture, parachuting performance of the flyer morph improves whereas that of the nonflyer morph worsens; gliding performance worsens much more for the nonflyer than the flyer; and only turning performance improves for both morphs. Another example is provided by copepods (119): Although *E. pileatus* can alter the leakiness of their M2's by changing speed (from $Re \simeq 10^{-2}$ to $Re \simeq 10^{-1}$), if *T. stylifera* changed the speed of their more finely meshed M2's over the same Re range, leakiness would not change (Figure 2).

Functional Equivalency: "There's More Than One Way to Skin A Cat"

Evolutionary biologists (e.g. 58, 152) and ecologists (e.g. 219) have recognized that organisms can play the same role in a variety of ways. Studies of natural history and behavior reveal how different organisms perform the same task. For example, several taxa of lizards have the ability to dive into sand but utilize different techniques that involve distinct exaptations in each case (6); three species of snakes eat whiptail lizards, but use different methods to capture them (73); and a variety of mechanisms can play the same role in predator defense (50).

Biomechanical studies elucidate the physical principles underlying how different structures can serve the same mechanical function. For example, many animals in different phyla reinforce their soft tissues with mineral inclusions (e.g. spicules, sclerites, ossicles) whose bizarre shapes are so specific that they are used as taxonomic characters. Analysis of the mechanics of spicule-reinforced tissues as filled-polymer composite materials revealed that the stiffening effect of spicules depends on the surface area of the spicule-tissue interaction, but that the particular combination of spicule sizes, shapes, and numbers used to produce that surface area in a volume of tissue matters little to performance (112). Many other examples of functional equivalency can be found by considering the flexural stiffness (resistance to bending, EI) or torsional stiffness (resistance to twisting, τJ) of organisms. Both depend on the product of the elastic modulus (resistance to deformation, E in tension or τ in shear) of the tissues from which a structure is made, and to I or J, shape properties of the structure (proportional to radius^4) (e.g. 230). Thus, organisms can produce a flexible structure via the microarchitecture of their tissues or via the gross morphology of cross-sectional shape. Simple examples of both can be found among cnidarians: Flexible joints in some sea fans occur at regions of lower E (due to sclerite microarchitecture of the tissue) (162), whereas the bending joints in sea anemones (e.g. 111) and the torsional joints in sea pens (18) are due solely to local reductions in I or J.

Another example of functional equivalency is provided by the phenotypic plasticity of giant bull kelp in different water-flow habitats. As they grow, the kelp maintain the same ratio of stress (force per cross-sectional area) required to break the stipe (stem) to stress imposed on the stipe by hydrodynamic forces; they can do so by altering a variety of morphological traits: blade shape (affecting drag), stipe diameter (affecting stress), or stipe material properties (affecting strength) (102).

Summary: Nonlinear Context-Dependent Effects of Morphology on Performance

When does the morphology of an organism affect its performance? For structures that perform mechanical functions (e.g. skeletal support, locomotion, food capture), the relationships between morphological dimensions and measures of performance can be quantified using physical principles. Although many biomechanical studies have shown how particular aspects of performance are affected by defined changes in morphology, others have revealed cases in which changes in form can occur without performance consequences. Quantitative mechanistic studies of how function depends on form have also produced some intriguing surprises. For example, in some cases small changes in morphology or simple changes in size can lead to novel functions. Furthermore, the effect of a specific change in morphology can depend on the size, shape, stiffness, or habitat of an organism. Likewise, a particular change in posture or behavior can produce opposite effects when performed by bodies with different morphologies.

What implications do these findings have for ecologists and evolutionary biologists? I devote the rest of this review to pointing out ways in which quantitative mechanistic organismal-level research can be a useful tool in the arsenal of approaches for attacking ecological and evolutionary questions. I mention misconceptions that can arise by ignoring mechanism, but I also point out limitations of the mechanistic approach.

WHY SHOULD ECOLOGISTS CARE ABOUT THE MECHANISMS BY WHICH MORPHOLOGY AFFECTS PERFORMANCE?

Mechanistic Versus Phenomenological Approaches

Why should ecologists worry about how individual organisms work when they are studying populations, communities, or ecosystems? Both quantitative empirical studies and mathematical models of ecological processes can use either a phenomenological or a mechanistic approach (see historical review in 180). If we focus on phenomenological analysis of a population, community, or ecosystem, we are concerned *that* organisms perform certain processes (e.g. consume certain prey, overgrow neighbors, migrate, produce offspring) at defined rates, rather than worrying about the details of *how* they perform these activities. In contrast, mechanistic studies assume that particular processes at the organismal level are important in governing the behavior of a system at a larger level of organization, such as a population, community, or ecosystem. The pros and cons of phenomenological versus mechanistic approaches are reviewed in (116, 138, 194, 200, 206). Phenomenological models can be powerful tools for making short-term predictions about systems for which descriptive data are available. Although mechanistic models generally do not fit the data as well as phenomenological models and may be complicated and slow to provide answers, the development of mechanistic models can lead to an understanding of how a system works. A number of examples of how mechanistic studies have provided ecological insights are reviewed in (194, 206).

Organismal-level mechanistic information about how performance depends on morphology not only can reveal limitations to the interpretation of phenomenological data, they can also provide insights about the mechanisms underlying ecological processes.

Morphology as a Tool to Infer Function or Ecological Role: Usefulness and Problems of this Phenomenological Approach

BACKGROUND One common type of ecomorphological study is the statistical description of patterns of distribution of morphologies with aspects of the environment, community structure, or ecological roles organisms play (reviewed by 198, 199, 207). Such studies do not directly assess the functional meaning of morphological variables, but rather they assume that the ecological characteristics of a species can be inferred from its morphology (198). These descriptive studies are an effective way to reveal patterns that can guide further mechanistic research and that can aid in interpretation of fossil communities or poorly studied recent communities, but they are limited in their ability to establish cause and effect (88). The dangers of making spurious conclusions about causes using statistical tests based on descriptive models have been reviewed in 96.

IMPORTANCE OF MECHANISTIC INFORMATION The examples described above of the many ways in which the relationship between morphology and function can be surprising and complex should caution us against expecting simple correlations between structure and function to yield reliable predictions of performance. However, mechanistic studies can yield quantitative expressions of the basic physical rules governing how a type of biological structure operates. Such mechanistic equations can be powerful tools for predicting the effects of specific morphological parameters on defined aspects of function, even in cases where the effects are nonlinear and context-dependent.

Although there are certainly instances when the function of an organism has been inferred successfully from its structure alone (reviewed in 133, 228), many other cases exemplify the problems of trying to read function from morphology without the aid of mechanistic information (discussed by 59, 68). Anyone trying to infer function from morphology should be aware of the following potential problems when descriptive statistical studies are done without mechanistic analysis:

1. Statistical analyses may not reveal a connection between structure and function in cases for which the effects of morphology on performance are nonlinear, or for which different mechanisms can play similar roles, as illustrated by the examples described above.

2. Statistical studies can also fail to reveal a mechanistic relationship between a structural feature and performance if the feature studied is only one of several that affects the performance (4). For example, the adhesive force holding a tree frog to a surface is proportional to toepad area, but measures of angles of surfaces at which frogs slipped off (sticking performance) did not correlate with toepad area because sticking performance is also inversely related to a frog's weight (44). Similarly, frog leg length did not correlate with jumping distance (45), and fish streamlining did not correlate with swimming speed, because both aspects of performance also depend on muscle mass, arrangement, and power output (4).

3. Statistical studies can find correlations between morphological features and performance or fitness that are not causally related when other correlated

but unmeasured morphological variables are responsible for the performance differences assessed (7, 88, 96, 133, 134).

4. Greene (72), who found that morphology was a poor predictor of lizard diets, stressed the importance of the ecological context in which an organism operates. For example, function is difficult to infer from morphology when information is lacking about the trade-offs to which a structure is subjected if it serves more than one role (133). Furthermore, we may waste time correlating unimportant aspects of performance to morphological characters if we do not base our studies on natural history observations of what organisms actually do in the field (e.g. 73) and on quantification of physical conditions actually encountered by organisms in nature (e.g. 40, 42, 102, 109, 113, 190).

Organismal Mechanistic Studies Shed Light on Ecological Questions

BACKGROUND Processes acting at the level of individual organisms can determine the properties of populations, communities, and ecosystems (reviewed by 116, 198), hence the effects of morphology on performance can have important ecological consequences (e.g. 7, 16, 44, 53, 88, 206, 232). For example, biomechanical analyses reveal the mechanisms responsible for differences in susceptibility of intertidal organisms of various morphologies to removal by waves (e.g. 40, 42, 43, 109, 110, 113, 115, 124); such wave-induced disturbance is important in determining the structure of intertidal communities (e.g. 180, 181, 213). Biophysical analyses of heat and water exchange between animals and the environment reveal where and when particular species can be active, and hence such analyses point out morphological constraints on habitat use, on ecological interactions such as competition or predation (e.g. 71, 81, 82, 189–191), and on reproductive strategies (e.g. 103). Similarly, flight aerodynamics provides a mechanistic explanation for the patterns of foraging and competition by hummingbirds living at different altitudes (55), and of foraging and habitat use by bats of different morphologies (170), while swimming hydrodynamics and head biomechanics do so for fish (e.g. reviewed by 228, 236). Likewise, biophysical studies reveal physical constraints on the distribution and ecological interactions of plants of different morphologies (e.g. 166, 169). Mechanistic studies such as these also enhance theoretical ecology by elucidating factors that can be ignored versus those that must be included in mechanistic ecological models, by testing the assumptions of such models, and by providing realistic values for parameters used in model calculations (88, 116).

INSIGHTS PROVIDED BY RECOGNITION OF NONLINEAR, CONTEXT-DEPENDENT EFFECTS OF MORPHOLOGY ON PERFORMANCE Although the literature abounds with examples of how ecological studies are enriched by information about how organisms function, the nonlinear context-dependent effects of morphology on performance reviewed here may be especially useful in providing insights in the developing area of "context-dependent ecology." Evidence is accumulating (193, 219) that the ecological role played by a particular species, as well as its impact on community structure and ecosystem dynamics, depends on the ecological context (e.g. physical conditions, time since disturbance, ecosystem productivity).

Keystone species (e.g. 155, 156, 178), now defined as those species whose impact on a community or ecosystem is disproportionately large relative to their abundance or biomass (193), may not be dominant controlling agents in all parts of their range or at all times in the succession of a community (examples tabulated in 193). By weaving together organismal-level studies of how habitat affects performance with data about the ecological patterns characterizing situations in which a sometimes-keystone species does play a significant role, we may reveal the mechanisms responsible for the context-dependency of its importance. Although such studies have not, to my knowledge, been conducted yet, an example can be pieced together using information in the literature. Biomechanical analyses reveal that kelp with weak, deformable tissues can resist breakage by stretching like extensible shock absorbers when hit by waves. A context-dependent performance consequence of this mechanism (which depends on the microarchitecture of the tissue) is that such kelp are generally quite tough but can break easily if the long-duration waves that accompany storms stretch them beyond their limit (113, 114). A storm can clear an area of kelp when broken plants become entangled with their neighbors, which then also break (114, 124). An ecological study of the role of sea urchins in benthic communities showed that these animals are keystone grazers that control community composition in areas where kelp are absent, but not where kelp are present and they have plenty of drift algae to eat; storms can cause a community to convert from a kelp bed to a "barrens" controlled by urchin grazing (79). Thus, information about the organismal-level mechanical performance of kelp can shed light on the issue of when sea urchins are keystone species.

Functional groups (e.g. 218, 219) are suites of species that play equivalent roles in an ecosystem. Understanding the mechanisms responsible for functional equivalency at the organismal level may help us identify the circumstances under which one species can play the same ecological role as another. Again, an example can be pieced together from published studies about the convergence of ecological roles played by mussels (*Mytilus californianus*) on wave-swept rocky shores in Washington state and tunicates (*Pyura praeputialis*) in similar habitats in Chile (182). Both species are competitive dominants that can form mat-like monocultures of individuals attached to each other; interstices in these mats provide protected habitats for an assemblage of small organisms. The formation of holes ("patches") in these mats of competitive dominants is an important process affecting the diversity of the rocky shore community by providing space on the substratum to sessile species that would otherwise be out-competed (180, 181). A biomechanical analysis of the physical mechanisms by which patches are produced in mussel beds revealed that the same morphological features that lead to the ecological convergence of these mussels and tunicates also are responsible for patch initiation. The pressure difference between the slowly moving water in the interstices below the mats and the rapidly moving water in a breaking wave above the mats cause lift forces high enough to rip chunks of the mat away (39). Analysis of forces on individual mussels indicated that waves do not exert forces large enough to wash them away (39). Thus, evidence that the performance consequences of a given morphology are very different when in an aggregation than when isolated leads to this insight about why competitively dominant mat-forming intertidal species are also subject to patch formation.

Conclusions

Ecologists should care about the mechanisms by which morphology affects performance for two reasons. Knowledge of these mechanisms can reveal the limitations of interpretation of descriptive phenomenological information. Mechanistic information also can provide insights about processes affecting the structure of populations, communities, and ecosystems.

WHY SHOULD EVOLUTIONARY BIOLOGISTS CARE ABOUT THE MECHANISMS BY WHICH MORPHOLOGY AFFECTS PERFORMANCE?

Observations about the nonlinear context-dependent relationship between morphology and performance can provide insights about the evolution of biological structure to researchers using a variety of approaches: the externalists, who emphasize natural selection and the performance or fitness of different phenotypes in the environment (reviewed by 10, 229, 233); the paleontologists, who interpret fossil evidence about the history of evolution; and the internalists, who focus on the generation of form and on the ontogenetic mechanisms that might constrain phenotypic variation or produce novelty (reviewed by 10, 23, 65, 83, 161, 233).

Externalists: The Study of Adaptations

BACKGROUND Traditionally, when biologists noted correlations between particular morphological features and certain habitats or lifestyles of organisms, they referred to such features as adaptations (discussed by 20, 63). However, since Gould & Lewontin (68) harpooned this plausible-argument approach to identifying which traits are adaptations, the topic of adaptation has been contentious (144, 152). Today a morphological feature can strictly be called an adaptation only if it promotes the fitness of the organism and if it arose via natural selection for its present role (69). Although these requirements are difficult to satisfy, various research methodologies for identifying adaptations have been proposed (6–8, 16, 60, 62, 63, 73, 130, 131, 134, 136, 197, 198, 239). Many of these schemes incorporate the "morphology⟶performance⟶fitness" paradigm.

Arnold (7) formalized an emerging conceptual framework for studying the selective advantage of morphological features: The morphology of an organism can determine its performance, which in turn can affect fitness. This approach, which has become the "central paradigm in ecomorphology" (198), uses natural variation in populations to seek correlations between morphology and performance, and between performance and fitness (e.g. 7, 8, 63, 228). When this paradigm is followed, the primary goal of studying performance is to identify how morphological features interact with each other and the environment to affect fitness (45). This popular quantitative approach is a powerful tool for demonstrating natural selection in the field and for revealing patterns that suggest which morphological features might be adaptive in which ecological contexts.

IMPORTANCE OF MECHANISTIC INFORMATION The examples described above of the nonlinear ways in which morphology can affect performance illustrate that the "morphology⟶performance" connection can be complex and surprising. Nonetheless, many studies using the "morphology⟶performance⟶fitness" methodology have relied on statistical correlations between morphological features and performance or fitness but have not included mechanistic analyses of how the features cause the correlated effects (e.g. 11, 16, 44, 49, 61, 63, 70, 97, 98, 127, 142, 143, 198, 199, 228). Even the classic studies correlating garter snake performance with morphology (11, 97), morphology with fitness (9), and performance with fitness (98) have not been complemented by experimental studies investigating the mechanisms by which vertebral number or tail length produce differences in burst speed, or by which burst speed improves survivorship.

Because the morphology⟶performance⟶fitness methodology is descriptive rather than mechanistic, a major limitation of this approach, discussed by

Arnold (7) and others (e.g. 88, 96, 133, 134), is that unmeasured morphological variables (that correlate with those that are measured) may be responsible for the performance differences assessed, and that unmeasured aspects of performance (that correlate with those assessed) may be the actual focus of selection. In addition, we must remember the other warnings (listed in the Morphology as a Tool to Infer Function or Ecological Role section above) about misconceptions that can arise when mechanism-blind correlations are made between morphology and performance. Hence, one means by which organismal-level mechanistic studies can enhance research in evolutionary biology is by providing the information necessary to prevent such misinterpretations of correlational data.

ADAPTATION CANNOT BE INFERRED FROM EFFECTS OF MORPHOLOGY ON PERFORMANCE Both mechanistic and correlational studies that focus only on the relationship between structure and performance can be misleading when used to infer adaptation. An untested assumption underlying many such studies is that a performance advantage translates into increased fitness (discussed by 7, 15, 45). There are a number of limitations of performance testing that call this assumption into question:

1. The aspect of performance measured may not be important to the biology of the organism in nature (74, 135, 228), or may play a different role in the life of the organism than we assumed. For example, tall, slim benthic organisms made of stiff, brittle tissues are susceptible to breakage in waves (seemingly "poor" performance), but breakage can be an important mechanism of asexual reproduction and dispersal by corals with such morphologies, which can therefore thrive on wave-swept reef crests (reviewed in 113). Similarly, rapidly growing seaweeds with weak stipes and holdfasts ("poor" performance) may be as successful in habitats where they can reproduce before seasonal storms hit as are stronger kelp ("good" performance) that grow more slowly, but that survive the storms (115).

2. Most performance studies are done on adults, even though organisms change properties as they grow and environments vary with time (diurnally, seasonally, and from year to year). The examples of the size-dependent and context-dependent effects of morphology on performance cited above should make us realize the importance of assessing performance at different stages in an organism's ontogeny. One way to deal with this problem is to devise performance measures, such as the environmental stress factor described in (102), that relate the performance of an organism at each stage of its ontogeny to the environmental conditions it encounters at that stage.

3. Lack of information on the genetic basis of the morphological or performance differences studied limits the evolutionary conclusions that can be drawn from such experiments (108).

4. Morphological features that improve performance do not necessarily arise via natural selection (discussed by e.g. 69, 83, 131). Some features may be epiphenomena of how a structure is produced (e.g. 68, 83), such as the ridges on clam shells that may improve burrowing (208), or the shapes of sea urchin skeletons that correlate with their water-flow habitats (13). Sometimes wear and tear in the environment can improve the performance of a structure. For example, pruning of kelp by limpet foraging can reduce their chances of being ripped away by storm waves (19), chipping of barnacle shells by wave-borne debris can produce more breakage-resistant shapes (186), wear of radular teeth in snails can sharpen their cutting edges (83), and passive orientation of gorgonian sea fans by hydrodynamic forces (231) can increase their suspension-feeding rates (139). Of course, the growth rules and breakage patterns described above could themselves be the result of natural selection.

Paleontologists: The Interpretation of Fossils

BACKGROUND The ways in which morphological data are used to infer the function of fossil organisms are reviewed by Hickman (83, 84), Lauder (133), and Van Valkenburgh (222). Perhaps the most commonly used approach is analogy with living species of similar morphology. Analogy arguments are most convincing if the living organisms that possess a particular structure all use it in the same way, and if the structure does not appear in the fossil record before its hypothesized function was possible (e.g. features for arboreality should not precede the origin of vascular plants—222). Homology among living species can also be used to infer the functions of extinct organisms: Ancestral character states of functions are determined by mapping functions of living organisms onto a phylogeny; then the functions of extinct taxa are inferred by their position within particular clades (75). Another approach to inferring the function of extinct organisms is the paradigm method in which morphological features are compared with theoretical optimal designs for particular functions. If a fossil structure is close to the ideal design for accomplishing some function, it is inferred that the fossil structure probably served that function (84, 133, 222, 229). This approach has limited usefulness since there are many reasons that a structure might not be optimal for a function that it serves (e.g. 42, 57, 58, 68, 72, 152, 177, 217, 232). Both the analogy and paradigm methods suffer from the problems (discussed above) of assuming that morphology is a reliable predictor of function, while the homology method is only as reliable as the phylogenetic hypothesis on which it is based.

USEFULNESS AND LIMITATIONS OF MECHANISTIC STUDIES One way to avoid these problems is to conduct performance tests using physical or mathematical models of fossil organisms (e.g. 105, 107, 133, 196). Obviously this approach is limited to testing hypotheses about physical functions. Furthermore, even if such biomechanical studies show that a fossil structure could have carried out some task or improved the performance of some function, that does not reveal the role that morphological feature served in the life of the organism; the best we can hope to accomplish with such quantitative studies is to reject functional hypotheses that are physically impossible (105–107, 133).

Several potential pitfalls of mechanistic analyses of fossil function are illustrated by the study of Marden & Kramer (149, 150), who presented an intriguing argument by analogy with living stoneflies that the protowings of early insects served in skimming or sailing locomotion on the surface of water. They showed by wing-trimming experiments that skimming and sailing performance are improved by increasing wing length. However, in interpreting these results they fell prey to a flaw in logic and they ignored available evidence on the phylogenetic relationships of the organisms involved. The flaw in logic was the assertion (148) that evidence supporting one functional hypothesis (surface skimming) implies rejection of alternate hypotheses (e.g. parachuting, gliding, thermoregulation), even though these alternative functions may not be mutually exclusive (107). The phylogenetic faux pas was the proposition that surface skimming represents an intermediate stage in the evolution of insect wings in Pterygotes, and this ancestral function has been retained by primitive stoneflies. This interpretation ignores the fact that stoneflies are members of the Neoptera, whose wing characteristics are considered to represent a derived condition (240). Without phylogenetic support, all the feasible scenarios proposed for the evolution of insect wings remain speculative (240).

Internalists: Study of the Origin of Evolutionary Novelty

The mechanisms by which novel phenotypes arise during evolution and the mechanisms responsible for the rapid morphological transformations that are recorded in the fossil record are challenging and contentious issues in evolutionary biology (history reviewed by 58, 65). Evidence emerging from mechanistic studies about the nonlinear size- and context-dependent effects of morphology on performance suggest another simple mechanism by which evolutionary novelty might arise.

BACKGROUND Evolutionary novelty or innovation has been defined in various ways (233): Some investigators require that it be a qualitative deviation in morphology (10, 160, 161), whereas others refer to a morphological, physiological, or behavioral change that permits the assumption of a new function (15, 94, 168). A key innovation is a novel feature that characterizes a clade and allows a subsequent diversification of the lineage (20, 134, 136, 140, 175, 202). The concept of key innovation has been criticized (e.g. 33, 94, 136) for a variety of reasons, including the difficulty of choosing which feature is the novelty and of demonstrating the causal link between that feature and a subsequent increased speciation rate. Nonetheless, various methodologies have been proposed to identify key innovations (e.g. 136, 212), and a number of examples of key innovations have been proposed (e.g. 134). A key adaptation is a novelty that reduces the costs of tradeoffs between various functions a species performs, thereby permitting that species to invade a niche when the incumbent species in that niche becomes extinct (201). A preadaptation is a feature that acquires a new biological role when organisms interact with their environment in a different way (e.g. 20, 57, 227). A preadaptation becomes an exaptation, a trait whose origins in a clade were due to selective pressures different from those that currently maintain it (69).

The idea of uncoupling (or decoupling) has provided a conceptual framework for much of the discussion of the origin of novelty (e.g. 10, 57, 128, 134, 202, 233). The basic argument is that coupling (e.g. one structure serving several functions, some function depending on several interrelated structures, or a change in one structure necessitating changes in others via pleiotropic effects or via their interconnection during morphogenesis) leads to evolutionary stasis because of the difficulty of changing one trait without negative effects on other features coupled to it (132). Examples of decoupling permitting evolutionary change (reviewed in 128, 202) include duplication of structural elements (if one set takes on a new function, the original function is not compromised), and loss of an old function (the structures that once performed it are free to be involved in new functions). However, some authors have argued that phenotypic plasticity permits suites of coupled characters to change in a coordinated way such that a complex organism's phenotype can shift rapidly with little genetic change (160, 238).

SOURCES OF NOVELTY: NEW BEHAVIORS AND CHANGES IN DEVELOPMENTAL PROGRAM

There are different views about the origin of novelty. While some investigators argue that behavioral shifts precede structural changes, others focus on the origins of new morphologies during development.

Behavioral shifts may precede morphological or physiological changes because behavior is more labile than morphology, and because natural selection should favor individuals showing compensatory behavior if the environment changes (e.g. 89, 161, 168, 227, 238; and others reviewed by 47, 72, 234).

Range expansion into a novel habitat can also provide a new set of selective forces on a population (examples discussed in 134). Furthermore, changes in the motor pattern controlling the kinematics of existing structures can produce novel functions (133). Once new behaviors or functions are acquired, selection should favor morphological variations that facilitate the new activity (161).

Small modifications in developmental program can lead to large changes in morphology (i.e. novelties) (e.g. 1, 23, 65, 94, 160, 161, 195). While the basic conceptual framework for this view has been formalized in terms of heterochrony (changes in the relative rates of different developmental processes) (e.g. 1), the nuts-and-bolts evidence for how changes in development can occur is coming from mechanistic studies, such as those of homeobox genes (reviewed by 65) and of the biomechanics of morphogenesis (reviewed by 38, 117).

MECHANISTIC STUDIES REVEAL ANOTHER POTENTIAL SOURCE OF NOVELTY New functions and novel consequences of changes in morphology can arise simply as the result of physics. As the examples described above illustrate, a simple change in environmental physical conditions or in body size can sometimes suffice to alter function. Although I think that both behavioral changes and alterations in developmental programs are important sources of novelty, it is possible for innovation to occur without either.

Mechanistic studies also illustrate that a common form of decoupling can be simply the lack of dependence of performance on morphology. Such permission for diversity of form without performance consequences may free structures to vary randomly or to respond to selection on other functions.

Although many ecomorphologists view size as a confounding factor in their analyses and propose various statistical techniques to eliminate size effects (e.g. 63, 228), I think it is important to consider size effects if one is addressing evolutionary questions. Most studies of size in biology have focused on the allometric changes required to maintain function as organisms grow or lineages evolve (e.g. 2, 27, 46, 102, 125, 126, 154, 167, 184, 205). If size changes over evolutionary time, such allometric growth of different parts of organisms might lead to new arrangements of these components and hence to innovations (e.g. 1, 136, 165). Of course, another way to think about allometry is to consider that if organisms do not change their form as they enlarge, their function does change, and such functional changes might be a source of evolutionary novelty.

There is ample evidence for selection on body size (reviewed by e.g. 24), and there are many examples in the fossil record of size changes within lineages over evolutionary time (reviewed by e.g. 93, 125, 126). The evolutionary trend in many, but certainly not all (93), lineages is that size increased with time (Cope's rule). This may be an artifact of better preservation and bias in observation of large organisms in the fossil record (125), or it may be that the founders of lineages tended to be small, and as size diversified, descendants on average got larger (93, 214). Fossil evidence indicates that many higher taxa arose from small ancestors (214). Stanley (214) suggested that small organisms were more likely to be founders of lineages than were large ones because little organisms are less subject to allometric constraints and therefore are more likely to give rise to novel types, while LaBarbera (126) pointed out that, even if the probability of breakthrough is the same at all body sizes, there are more small species.

The species diversity of small organisms is greater than that of large ones (e.g. 125, 163, 211), but the causes remain the subject of speculation (e.g. 33). One view is that there are fewer physical constraints on body form at small size (e.g. 21, 214). Another view is that ecosystems have more niches at small size (125) due to the fractal nature of habitats (159, 243).

The observations compiled above lead me to speculate about another potential mechanism to add to the list of ways of generating evolutionary innovation. Morphological and kinematic diversity may accumulate at small size without functional consequence, but such novelties may not assume new functional roles until there is a size increase and morphology matters. The structural diversity that did not affect performance at small size might gain functional significance at larger size; not only might features that were selectively neutral at small size become subject to selection at larger size, but novel functions might also become physically possible. Müller (160) has also suggested that evolutionary innovation should be associated with changes in size, basing his argument on evidence that size changes in developing embryos can affect pattern formation, thereby producing novelties in adult morphology.

If size changes tend to lead to evolutionary innovation, then I might speculate that the rate of evolutionary change would correlate with the rate of size change in a lineage. If we turn to the fossil record for evidence, and if we assume that short taxon longevity is a rough indication of rapid evolutionary change (i.e. high rates of modification or extinction), then the data from Hallam (76) for Jurassic ammonites and bivalves (replotted in Figure 7) is consistent with my speculation, but this obviously bears further investigation.

The Phylogenetic Approach and the Usefulness of Mechanistic Morphological Research

Modern studies of adaptation stress the importance of integrating analyses of structure, function, and fitness with phylogenetic history (16, 26, 45, 63, 68, 73, 80, 88, 89, 99, 129, 131, 136, 141, 144, 198, 222, 239). Unfortunately, enthusiasm for this approach has produced a climate in which mechanistic research can be dismissed when done without a phylogeny in hand. This dismissal ignores

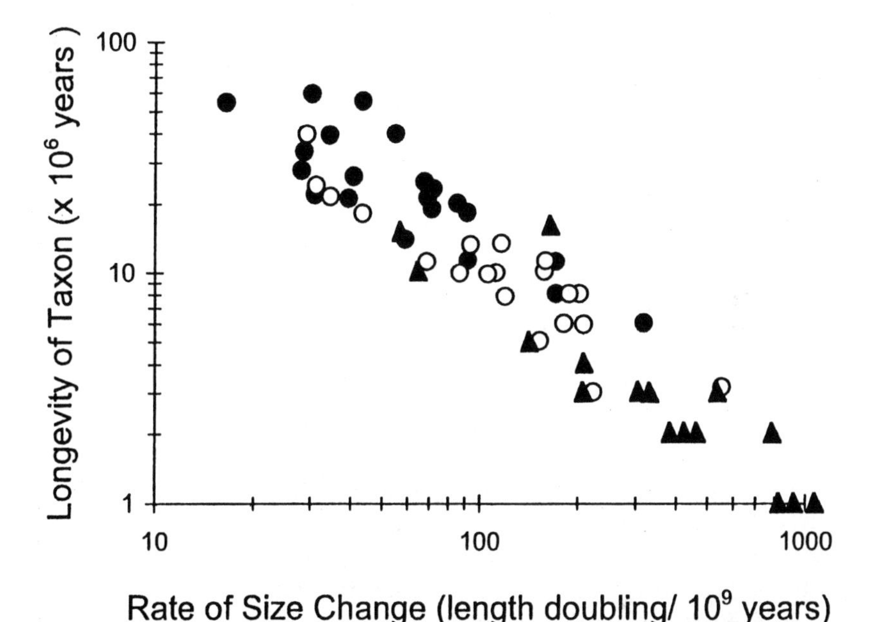

Figure 7 Longevity of each taxon in the fossil record, plotted as a function of the rate of change of body size, for Jurassic bivalved mollusks (*circles*) and ammonites (*triangles*). Solid symbols represent genera, and open symbols represent species. [Replotted using data digitized from Figure 1 of (76.)]

the fact many mechanistic studies do not have the identification of adaptations as their goal, but rather simply seek to unravel how (i.e. mechanism of operation), not why (i.e. evolutionary history) performance depends on morphology. I hope that the examples cited in this review will serve as a reminder that such mechanistic "how" information can also provide insights about evolutionary questions.

CONCLUSIONS

Quantitative mechanistic analyses of how function depends on biological form, and on the ecological context in which an organism operates, should complement descriptive statistical and phylogenetic studies to provide insights about ecological and evolutionary questions. Such quantitative studies have shown that the relationship between morphology and performance is often nonlinear and sometimes surprising. These mechanistic studies not only reveal potential misconceptions that can arise from the descriptive statistical analyses often used in ecological and evolutionary research, but they also show how new functions, and novel consequences of changes in morphology, can arise simply as the result of changes in size or habitat.

> "... You ask me whom the Macrocystis alga hugs in its arms?
> Study, study it, at a certain hour, in a certain sea I know.
> ... Or the crystal architecture of the sea anemone?
> ... I will tell you the ocean knows this, that life in its jewel boxes
> is endless as the sand,...
> ... I walked around as you do, investigating..."
>
> Pablo Neruda, *Enigmas* (164)

ACKNOWLEDGMENTS

I am grateful to the following people for leads into the literature and/or discussions that helped shape my thinking about the issues addressed above: S Arnold, S Emerson, R Full, H Greene, R Huey, T Hunter, D Jablonsky, J Kingsolver, A Kohn, G Lauder, G Oster, R Paine, M Power, M Slatkin, S Stearns, J Valentine, S Vogel, D Wake, M Wake, S Wainwright, and the participants in Integrative Biology 231 and 290 at University of California, Berkeley. I thank H Greene, K Quillin, and P Wainwright for helpful comments on the manuscript. My data presented here were gathered with support from National Science Foundation Grant #OCE-9217338 and Office of Naval Research Grant #N00014-90-J-1357. M O'Donnell helped prepare the manuscript; T Cooper, J Jed, and K Quillin made the figures.

Literature Cited

1. Alberch P, Gould SJ, Oster GF, Wake DB. 1979. Size and shape in ontogeny and phylogeny. *Paleobiology* 5:296–317
2. Alexander RM. 1971. *Size and Shape.* London: Edward Arnold
3. Alexander RM. 1983. *Animal Mechanics.* Oxford: Blackwell Sci. 2nd ed.
4. Alexander RM. 1990. Apparent adaptation and actual performance. *Evol. Biol.* 25:357–73
5. Alexander RM, Jayes AS. 1983. A dynamic similarity hypothesis for the gaits of quadrupedal mammals. *J. Zool.* 201:135–52
6. Arnold EN. 1995. Identifying the effects of history on adaptation: origins of different sand-diving techniques in lizards. *J. Zool.* 235:351–88
7. Arnold SJ. 1983. Morphology, performance and fitness. *Am. Zool.* 23:347–61
8. Arnold SJ. 1986. Laboratory and field approaches to the study of adaptation. See Ref. 54, pp. 157–79
9. Arnold SJ. 1988. Quantitative genetics and selection in natural populations: microevolution of vertebral numbers in the garter snake *Thamnophis elegans*. In *Proc. 2nd Int. Conf. on Quant. Genet.,*

ed. BS Weir, J Eisen, MJ Goodman, G Namkoong, pp. 619–36. Sunderland, MA: Sinaur
10. Arnold SJ, Alberch P, Csányi V, Dawkins RC, Emerson SB, et al. 1989. Group Report: How do complex organisms evolve? See Ref. 232a, pp. 403–33
11. Arnold SJ, Bennett AF. 1988. Behavioral variation in natural populations. V. Morphological correlations of locomotion in the garter snake, *Thamnophis radix*. *Biol. J. Linn. Soc.* 34:175–90
12. Balmer RT, Strobusch AD. 1977. Critical size of newborn homeotherms. *J. Appl. Physiol.* 42:571–77
13. Baron CJ. 1991. What functional morphology cannot explain: a morphogenetic model of sea urchins and a discussion of the role of morphogenetic explanations in evolutionary morphology. In *The Unity of Evolutionary Biology,* ed. EC Dudley, pp. 471–88. Portland, OR: Dioscorides
14. Batty RS. 1984. Development of swimming movements and musculature of larval herring *(Clupea harengus)*. *J. Exp. Biol.* 37:129–53
15. Benkman CW, Lindholm AK. 1991. The advantages and evolution of a morphological novelty. *Nature* 349:519–20
16. Bennett AF, Huey RB. 1990. Studying the evolution of physiological performance. *Oxford Surv. Evol. Biol.* 7:251–84
17. Berg HC, Purcell EM. 1977. Physics of chemoreception. *Biophys. J.* 20:193–219
18. Best BA. 1983. Mechanics of orientation in sea pens: a rotational one line joint. *J. Biomech.* 16:297
19. Black R. 1976. The effects of grazing by the limpet, *Acmaega insessa* on the kelp, *Egregia laevigata,* in the intertidal zone. *Ecology* 57:265–77
20. Bock WJ, won Wahlert G. 1965. Adaptation and the form-function complex. *Evolution* 19:269–99
21. Bonner JT. 1968. Size change in development and evolution. *The Paleontol. Soc. Mem.* 2:1–15
22. Bonner JT. 1981. Evolutionary strategies and developmental constraints in cellular slime molds. *Am. Nat.* 119:530–52
23. Bonner JT, ed. 1982. *Evolution and Development.* Berlin: Springer-Verlag
24. Bonner JT, Horn HS. 1982. Selection for size, shape, and developmental timing. See Ref. 23, pp. 259–76
25. Bushnell DM, Moore KJ. 1991. Drag reduction in nature. *Annu. Rev. Fluid Mech.* 23:65–79
26. Cadle JE, Greene HW. 1993. Phylogenetic patterns, biogeography, and the ecological structure of neotropical snake assemblages. In *Species Diversity in Ecological Communities: Historical and Geographical Perspectives,* ed. RE Ricklefs, D Schleter, pp. 281–93. Chicago: Univ. Chicago Press
27. Calder WA III. 1984. *Size, Function and Life History.* Cambridge, MA: Harvard Univ. Press
28. Carrington E. 1990. Drag and dislodgement of an intertidal macroalga: consequences of morphological variation in *Mastocarpus papillatus* Kutzing. *J. Exp. Mar. Biol. Ecol.* 139:185–200
29. Cheer AYL, Koehl MAR. 1988. Fluid flow through insect filters. *IMA J. Math. Appl. Med. Biol.* 4:185–99
30. Cheer AYL, Koehl MAR. 1988. Paddles and rakes: fluid flow through bristled appendages of small organisms. *J. Theor. Biol.* 129:17–39
31. Childress S, Koehl MAR, Miksis M. 1987. Scanning currents in Stokes flow and the efficient feeding of small organisms. *J. Fluid Mech.* 177:407–36
32. Cossins AR, Bowler K. 1987. *Temperature Biology of Animals.* London: Chapman & Hall
33. Cracraft J. 1990. The origin of evolutionary novelties: pattern and process at different hierarchical levels. See Ref. 167a, pp. 21–46
34. Craig DA. 1990. Behavioral hydrodynamics of *Cloeon dipterum* larvae (Ephemeropter: Baetidea). *J. North Am. Benthol. Soc.* 9:346–57
35. Currey J. 1984. *The Mechanical Adaptations of Bones.* Princeton, NJ: Princeton Univ. Press
36. Dadswell MJ, Weihs D. 1990. Size-related hydrodynamic characteristics of the giant scallop, *Placopecten magellanicus* (Bivalvia: Pectinidae). *Can. J. Zool.* 68:778–85
37. Daniel TL, Webb PW. 1987. Physical determinants of locomotion. In *Comparative Physiology: Life in Water and on Land,* ed. P Dejours, L Bolis, CR Taylor, ER Weibel. Padova: Liviana
38. Davidson LA, Koehl MAR, Keller R, Oster GF. 1995. How do sea urchins gastrulate? Distinguishing between mechanisms of primary invagination using biomechanics. *Development* 121:2005–18
39. Denny MW. 1987. Lift as a mechanism

of patch initiation in mussel beds. *J. Exp. Mar. Biol. Ecol.* 113:231–45

40. Denny MW. 1988. *Biology and the Mechanics of the Wave-Swept Environment.* Princeton, NJ: Princeton Univ. Press
41. Denny MW. 1993. *Air and Water.* Princeton, NJ: Princeton Univ. Press
42. Denny MW. 1994. Roles of hydrodynamics in the study of life on wave-swept shores. See Ref. 228a, pp. 169–204
43. Denny MW, Daniel T, Koehl MAR. 1985. Mechanical limits to the size of wave-swept organisms. *Ecol. Monogr.* 55:69–102
44. Emerson SB. 1991. The ecomorphology of Bornean tree frogs (family Rhacophoridae). *Zool. J. Linn. Soc.* 101:337–57
45. Emerson SB, Arnold SJ. 1989. Intra- and interspecific relationships between morphology, performance and fitness. See Ref. 232a, pp. 295–314
46. Emerson SB, Greene HW, Charnov EL. 1994. Allometric aspects of predator-prey interactions. See Ref. 228a, pp. 123–39
47. Emerson SB, Koehl MAR. 1990. The interaction of behavioral and morphological change in the evolution of a novel locomotor type: "Flying" frogs. *Evolution* 44:1931–46
48. Emerson SB, Travis J, Koehl MAR. 1990. Functional complexes and additivity in performance: a test case with "flying" frogs. *Evolution* 44:2153–57
49. Endler J. 1986. *Natural Selection in the Wild.* Princeton, NJ: Princeton Univ. Press
50. Endler JA. 1986. Defense against predators. See Ref. 54, pp. 109–34
51. Ennos AR. 1989. The effect of size on the optimal shapes of gliding insects and seeds. *J. Zool.* 219:61–69
52. Fauci L. 1996. Computational modeling of the fluid dynamics of locomotion. *Am. Zool.* In press
53. Feder ME, Lauder GV. 1986. Commentary and conclusion. See Ref. 54, pp. 180–89
54. Feder ME, Lauder GV, eds. 1986. *Predator-Prey Relationships: Perspectives and Approaches from the Study of Lower Vertebrates.* Chicago: Univ. Chicago Press
55. Feinsinger P, Colwell RK, Terbergh J, Chaplin SB. 1979. Elevation and the morphology, flight energetics, and foraging ecology of tropical hummingbirds. *Am. Nat.* 113:481–97
56. Fonseca MS, Fisher JS, Zieman JC, Thayer GW. 1982. Influence of the seagrass *Zostera marina* L., on current flow. *Estuar. Coast. Shelf Sci.* 15:351–54
57. Frazzetta TH. 1975. *Complex Adaptations in Evolving Populations,* Sunderland, MA: Sinauer. 267 pp.
58. Futuyma DJ. 1986. *Evolutionary Biology.* Sunderland, MA: Sinauer. 2nd ed.
59. Gans C. 1966. Some limitations and approaches to problems in functional anatomy. *Folia Biotheor.* 6:41–50
60. Gans C. 1986. Functional morphology of predator-prey relationships. See Ref. 54, pp. 6–23
61. Garland T Jr. 1983. The relation between maximum running speed and body mass in terrestrial mammals. *J. Zool.* 99:157–70
62. Garland T Jr. 1994. Why not to do two-species comparative studies: limitations on inferring adaptation. *Physiol. Zool.* 67:797–828
63. Garland T Jr, Losos JB. 1994. Ecological morphology of locomotor performance in squamate reptiles. See Ref. 228a, pp. 240–302
64. Gaylord B, Blanchette CA, Denny MW. 1994. Mechanical consequences of size in wave-swept algae. *Ecol. Monogr.* 64:287–313
65. Gilbert SF, Opitz JM, Raff RA. 1996. Resynthesizing evolutionary and developmental biology. *Dev. Biol.* 173:357–72
66. Glasheen JW. 1995. *Strange locomotion: from humans running on their hands to lizards running on water.* PhD diss. Dept. Org. Evol. Biol., Harvard
67. Glasheen JW, McMahon TA. 1996. Weight support on the water surface in Basilisk lizards. *Nature.* 380:340–42
68. Gould SJ, Lewontin RC. 1979. The spandrels of San Marco and the panglossian paradigm: a critique of the adaptationist programme. *Proc. R. Soc. London Ser. B* 205:581–98
69. Gould SJ, Vrba ES. 1982. Exaption: a missing term in the science of form. *Paleobiology* 8:4–15
70. Grant BR, Grant PR. 1989. Natural selection in a population of Darwin's finches. *Am. Nat.* 133:377–93
71. Grant BW, Porter WP. 1992. Modeling global macroclimatic contraints on ectotherm energy budgets. *Am. Zool.* 32:154–78
72. Greene HW. 1982. Dietary and phenotypic diversity in lizards: Why are some organisms specialized? In *Environmental Adaptation and Evolution,* ed. D

Mossakowski, G Roth, pp. 107–28. New York: Gustav Fisher
73. Greene HW. 1986. Diet and arboreality in the emerald monitor, *Varanus prasinus*, with comments on the study of adaptation. *Fieldiana Zool.* 31:1–12
74. Greene HW. 1986. Natural history and evolutionary biology. See Ref. 54, pp. 99–108
75. Greene HW. 1994. Homology and behavioral repertoires. In *The Hierarchical Basis of Comparative Biology*, ed. BK Hall, pp. 369–90. San Diego, CA: Academic
76. Hallam A. 1975. Evolutionary size increase and longevity in Jurasic bivalves and ammonites. *Nature* 258:493–96
77. Hanken J. 1985. Morphological novelty in the limb skeleton accompanies miniaturization in salamanders. *Science* 229:871–74
78. Hansen B, Tiselius P. 1992. Flow through the feeding structures of suspension feeding zooplankton: a physical model approach. *J. Plankton Res.* 14:821–34
79. Harrold C, Reed DC. 1985. Food availability, sea urchin grazing, and kelp forest community structure. *Ecology* 66:1160–69
80. Harvey PH, Pagel MD. 1991. *The Comparative Method in Evolutionary Biology*. Oxford: Oxford Univ. Press
81. Heinrich B. 1979. *Bumblebee Economics*. Cambridge, MA: Harvard Univ. Press
82. Heinrich B. 1993. *The Hot-blooded Insects: Strategies and Mechanisms of Thermoregulation*. Berlin: Springer-Verlag
83. Hickman CS. 1980. Gastropod radulae and the assessment of form in evolutionary paleontology. *Paleobiology* 6:276–94
84. Hickman CS. 1988. Analysis of form and function in fossils. *Am. Zool.* 28:775–93
85. Holbrook NM, Denny M, Koehl MAR. 1991. Intertidal "trees": consequences of aggregation on the mechanical and photosynthetic characteristics of sea palms. *J. Exp. Mar. Biol. Ecol.* 146:39–67
86. Holwill MEJ, Sleigh MA. 1967. Propulsion by hispid flagella. *J. Exp. Biol.* 47:267–76
87. Huey RB. 1982. Temperature, physiology, and the ecology of reptiles. In *Biology of Reptilia*, ed. C Gans, pp. 25–91. New York: Academic
88. Huey RB, Bennett AF. 1986. A comparative approach to field and laboratory studies in evolutionary biology. See Ref. 54, pp. 82–98
89. Huey RB, Bennett AF. 1987. Phylogenetic studies of coadaptation: preferred temperatures versus optimal performance temperatures of lizards. *Evolution* 41:1098–115
90. Huey RB, Kingsolver JG. 1993. Evolutionary responses to extreme temperatures in ectotherms. *Am. Nat.* 143:S21–S46
91. Huey RB, Stevenson RD. 1979. Integrating thermal physiology and ecology of ectotherms: a discussion of approaches. *Am. Zool.* 19:357–66
92. Hunter T. 1988. *Mechanical design of hydroids: flexibility, flow forces and feeding in* Obelia longissima. PhD dissertation Zool., Univ. Calif., Berkeley
93. Jablonski D. 1996. Body size and macroevolution. In *Evolutionary Paleobiology: Essays in Honor of James W. Valentine*, ed. D Jablonski, DH Erwin, JH Lipps, pp. 256–89. Chicago: Univ. Chicago Press. In press
94. Jablonski D, Bottjer DJ. 1990. The ecology of evolutionary innovation: the fossil record. See Ref. 167a, pp. 253–88
95. Jacobs DK. 1992. Shape, drag, and power in ammonoid swimming. *Paleobiology* 18:203–20
96. James FC, McCulloch CE. 1990. Multivariate analysis in ecology and systematics: Panacea or Pandora's Box? *Annu. Rev. Ecol. Syst.* 21:129–66
97. Jayne BC, Bennett AF. 1989. The effect of tail morphology on locomotor performance of snakes: a comparison of experimental and correlative methods. *J. Exp. Zool.* 252:126–33
98. Jayne BC, Bennett AF. 1990. Scaling of speed and edurance in garter snakes: a comparison of cross-sectional and longitudinal allometries. *J. Zool.* 220:257–77
99. Jensen JS. 1990. Plausibility and testability: assessing the consequences of evolutionary innovation. See Ref. 167a, pp. 171–90
100. Johnson AS. 1986. *Consequences of individual and group morphology: a hydrodynamic study of the benthic suspension-feeder* Phoronopsis viridis. PhD dissertation Univ. Calif., Berkeley
101. Johnson AS. 1990. Flow around Phoronids: consequences of a neighbor to suspension feeders. *Limnol. Oceanogr.* 35:1395–401
102. Johnson AS, Koehl MAR. 1994. Main-

tenance of dynamic strain similarity and environmental stress factor in different flow habitats: Thallus allometry and material properties of a giant kelp. *J. Exp. Biol.* 195:381–410

103. Kingsolver JG. 1983. Ecological significance of flight activity in Colias butterflies: implications for reproductive strategy and population structure. *Ecology* 64:546–51
104. Kingsolver JG. 1988. Thermoregulation, flight, and the evolution of wing pattern in Pierid butterflies: the topography of adaptive landscapes. *Am. Zool.* 28:899–912
105. Kingsolver JG, Koehl MAR. 1985. Aerodynamics, thermoregulation, and the evolution of insect wings: differential scaling and evolutionary change. *Evolution* 39:488–504
106. Kingsolver JG, Koehl MAR. 1989. Selective factors in the evolution of insect wings: response to Kukalova-Peck. *Can. J. Zool.* 67:785–87
107. Kingsolver JG, Koehl MAR. 1994. Selective factors in the evolution of insect wings. *Annu. Rev. Entomol.* 39:425–51
108. Kingsolver JG, Woods HA, Gilchrist G. 1995. Traits related to fitness. *Science* 267:396
109. Koehl MAR. 1977. Effects of sea anemones on the flow forces they encounter. *J. Exp. Biol.* 69:87–105
110. Koehl MAR. 1977. Mechanical diversity of the connective tissue of the body wall of sea anemones. *J. Exp. Biol.* 69:107–25
111. Koehl MAR. 1977. Mechanical organization of cantilever-like sessile organisms: sea anemones. *J. Exp. Biol.* 69:127–42
112. Koehl MAR. 1982. Mechanical design of spicule-reinforced connective tissues: stiffness. *J. Exp. Biol.* 98:239–68
113. Koehl MAR. 1984. How do benthic organisms withstand moving water? *Am. Zool.* 24:57–70
114. Koehl MAR. 1984. Mechanisms of particle capture by copepods at low Reynolds number. In *Trophic Interactions in Aquatic Ecosystems*, ed. DL Meyers, JR Strickler, pp. 135–160. Boulder, CO: Westview
115. Koehl MAR. 1986. Seaweeds in moving water: form and mechanical function. In *On the Economy of Plant Form and Function,* ed. TJ Givnish, pp. 603–34. Cambridge: Cambridge Univ. Press
116. Koehl MAR. 1989. From individuals to populations. In *Perspectives in Ecological Theory,* ed. J Roughgarden, RM May, SA Levin, pp. 39–53. Princeton, NJ: Princeton Univ. Press
117. Koehl MAR. 1990. Biomechanical approaches to morphogenesis. *Semin. Dev. Biol.* 1:367–78
118. Koehl MAR. 1992. Hairy little legs: feeding, smelling, and swimming at low Reynolds number. *Contemp. Math.* 141:33–64
119. Koehl MAR. 1995. Fluid flow through hair-bearing appendages: feeding, smelling, and swimming at low and intermediate Reynolds number. In *Biological Fluid Dynamics,* ed. CP Ellington, TJ Pedley. *Soc. Exp. Biol. Symp.* 49:157–82
120. Koehl MAR. 1996. Small-scale fluid dynamics of olfactory antennae. *Mar. Freshwater Behav. Physiol.* 27:127–41
121. Koehl MAR, Alberte RS. 1988. Flow, flapping, and photosynthesis of macroalgae: functional consequences of undulate blade morphology. *Mar. Biol.* 99:435–44
122. Koehl MAR, Hunter T, Jed J. 1991. How do body flexibility and length affect hydrodynamic forces on sessile organisms? *Am. Zool.* 31:A60
123. Koehl MAR, Powell TM. 1996. Effects of benthic organisms on mass transport at wave-swept rocky shores. *Eos, Trans. Am. Geophys. Union* 76:OS69
124. Koehl MAR, Wainwright SA. 1977. Mechanical adaptations of a giant kelp. *Limnol. Ocean.* 22:1067–71
125. LaBarbera M. 1986. The evolution and ecology of body size. In *Patterns and Processes in the History of Life,* ed. DM Raup, D Jablonski, pp. 69–98. Berlin: Springer-Verlag
126. LaBarbera M. 1989. Analyzing body size as a factor in ecology and evolution. *Annu. Rev. Ecol. Syst.* 20:97–117
127. Lande R, Arnold SJ. 1983. The measurement of selection on correlated characters. *Evolution* 37:1210–26
128. Lauder GV. 1981. Form and function: structural analysis in evolutionary morphology. *Paleobiology* 7:430–42
129. Lauder GV. 1982. Introduction. In *Form and Function: A Contribution to the History of Animal Morphology,* ed. ES Russell, pp. xi-xlv. Chicago: Univ. Chicago Press
130. Lauder GV. 1990. Functional morphology and systematics: studying functional patterns in an historical context. *Annu. Rev. Ecol. Syst.* 21:317–40
131. Lauder GV. 1991. Biomechanics and

evolution: integrating physical and historical biology in the study of complex systems. See Ref. 196, pp. 1–19
132. Lauder GV. 1991. An evolutionary perspective on the concept of efficiency: How does function evolve? In *Efficiency and Economy in Animal Physiology,* ed. RW Blake, pp. 169–84. Cambridge: Cambridge Univ. Press
133. Lauder GV. 1995. On the inference of function from structure. In *Functional Morphology in Vertebrate Paleontology,* ed. JJ Thomason, pp. 1–18. Cambridge: Cambridge Univ. Press
134. Lauder GV, Crompton AW, Gans C, Hanken J, Liem KF, et al. 1989. Group Report: How are feeding systems integrated and how have evolutionary innovations been introduced? See Ref. 232a, pp. 97–115
135. Lauder GV, Leroi AM, Rose MR. 1993. Adaptations and history. *TREE* 8:294–97
136. Lauder GV, Liem KF. 1989. The role of historical factors in the evolution of complex organismal functions. See Ref. 232a, pp. 63–78
137. Lauder GV, Wainwright PC. 1992. Function and history: the pharyngeal jaw apparatus in primitive ray-finned fishes. In *Systematics, Historical Ecology, and North American Freshwater Fishes,* ed. RL Mayden, pp. 455–71. Stanford: Stanford Univ. Press
138. Lehman JT. 1986. Grazing, nutrient release, and their importance on the structure of phytoplankton communitites. In *Trophic Interactions within Aquatic Communities,* ed. DG Meyers, JR Strickler, pp. 49–72. AAAS Sel. Symp. 85
139. Leversee GJ. 1976. Flow and feeding in fan-shaped colonies of the gorgonian coral, *Leptogorgia. Biol. Bull.* 151:344–56
140. Liem KF. 1973. Evolutionary strategies and morphological innovations: Cichlid pharyngeal jaws. *Syst. Zool.* 22:425–41
141. Liem KF. 1990. Key evolutionary innovations, differential diversity, and symecomorphosis. See Ref. 167a, pp. 147–70
142. Losos JB. 1990. Ecomorphology, performance capability, and scaling of West Indian *Anolis* lizards: an evolutionary analysis. *Ecol. Monogr.* 60:369–88
143. Losos JB. 1990. The evolution of form and function: morphology and locomotor performance in West Indian *Anolis* lizards. *Evolution* 44:1189–203
144. Losos JB, Miles DB. 1994. Adaptation, constraint, and the comparative method: phylogenetic issues and methods. See Ref. 228a, pp. 60–98
145. Losos JB, Sinervo B. 1989. The effects of morphology and perch diameter on sprint performance in West Indian *Anolis* lizards. *J. Exp. Biol.* 145:23–30
146. Loudon C, Best BA, Koehl MAR. 1994. When does motion relative to neighboring surfaces alter the flow through an array of hairs? *J. Exp. Biol.* 193:233–54
147. Manel JL, Dadswell MJ. 1993. Swimming of juvenile sea scallops, *Placopecten magellanicus* (Gmelin): a minimum size for effective swimming? *J. Exp. Mar. Biol. Ecol.* 174:137–75
148. Marden JH. 1995. How insects learned to fly. *The Sciences* 35:26–30
149. Marden JH, Kramer MG. 1994. Surface-skimming stoneflies: a possible intermediate stage in insect flight evolution. *Science* 266:427–30
150. Marden JH, Kramer MG. 1995. Locomotor performance of insects with rudimentary wings. *Nature* 377:332–34
151. Marrasé C, Costello JH, Granata T, Strickler JR. 1991. Grazing in a turbulent environment: energy dissipation, encounter rates, and efficacy of feeding currents in *Centropages hamatus. Proc. Natl. Acad. Sci. USA* 87:1653–57
152. Mayr E. 1983. How to carry out the adaptationist program? *Am. Nat.* 121:324–34
153. McMahon TA. 1984. *Muscles, Reflexes, and Locomotion.* Princeton, NJ: Princeton Univ. Press
154. McMahon TA, Bonner JT. 1983. *On Size and Life.* New York: Freeman
155. Menge BA, Berlow EL, Blanchette CA, Navarrete SA, Yamada SB. 1994. The keystone species concept: variation in interaction strength in a rocky intertidal habitat. *Ecol. Monogr.* 64:249–87
156. Mills LS, Soulé ME, Doak DF. 1993. The keystone-species concept in ecology and conservation. *BioScience* 43:219–24
157. Monteith JL, Unsworth MH. 1990. *Principles of Environmental Physics.* London: Edward Arnold. 2nd ed.
158. Moore JA. 1987. *Science as a Way of Knowing V—Form and Function.* Chicago: Am. Soc. Zool. 220 pp.
159. Morse DR, Lawton JH, Dodson MM, Williamson MH. 1985. Fractal dimension of vegetation and the distribution of arthropod body lengths. *Nature* 314:731–33
160. Müller GB. 1990. Developmental mech-

anisms at the origin morphological novelty: a side-effect hypothesis. See Ref. 167a, pp. 99–130
161. Müller GB, Wagner GP. 1991. Novelty in evolution: restructuring the concept. *Annu. Rev. Ecol. Syst.* 22:229–56
162. Muzik K, Wainwright SA. 1977. Morphology and habitat of five Fijian sea fans. *Bull. Mar. Sci.* 27:308–37
163. Nee S, Lawton JH. 1996. Body size and biodiversity. *Nature* 380:672–73
164. Neruda P. 1971. Enigmas. In *Neruda and Vaellejo: Selected Poems,* ed. R Bly, pp. 131–32. Boston: Beacon
165. Newell ND. 1949. Phyletic size increase, an important trend illustrated by fossil invertebrates. *Evolution* 3:103–24
166. Niklas KJ. 1992. *Plant Biomechanics: An Engineering Approach to Plant Form and Function.* Chicago: Univ. Chicago Press
167. Niklas KJ. 1994. *Plant Allometry: The Scaling of Form and Process.* Chicago: Univ. Chicago Press
167a. Nitecki MW, ed. 1990. *Evolutionary Innovations.* Chicago: Univ. Chicago Press
168. Nitecki MW. 1990. The plurality of evolutionary innovations. See Ref. 167a, pp. 3–18
169. Nobel PS. 1983. *Biophysical Plant Physiology and Ecology.* San Francisco: Freeman
170. Norberg UM. 1994. Wing design, flight performance, and habitat use in bats. See Ref. 228a, pp. 205–29
171. Nowell ARM, Jumars PA. 1984. Flow environments of aquatic benthos. *Annu. Rev. Ecol. Syst.* 15:303–28
172. Okamura B. 1984. The effects of ambient flow velocity, colony size, and upstream colonies on the feeding success of bryozoa. I. *Bugula stolonifera* (Ryland), an arborescent species. *J. Exp. Mar. Biol. Ecol.* 83:179–93
173. Okamura B. 1985. The effects of ambient flow velocity, colony size, and upstream colonies on the feeding success of bryzoa-II. *Conopeum reticalum* (Linnaeus), an encrusting species. *J. Exp. Mar. Biol. Ecol.* 89:69–80
174. Okamura B. 1988. The influence of neighbors on the feeding of an epifaunal bryozoan. *J. Exp. Mar. Biol. Ecol.* 120:105–23
175. Olson EC. 1965. Summary and comment. *Syst. Zool.* 14:337–42
176. Osse JWM, Drost MR. 1983. Hydrodynamics and mechanics of fish larvae. *Pol. Arch. Hydrobiol.* 36:455–66
177. Oster GF, Wilson EO. 1978. *Caste and Ecology in the Social Insects.* Princeton, NJ: Princeton Univ. Press
178. Paine RT. 1966. Food web complexity and species diversity. *Am. Nat.* 100:65–75
179. Paine RT. 1976. Size-limited predation: an observational and experimental approach with the *Mytilus-Pisaster* interaction. *Ecology* 57:858–73
180. Paine RT. 1994. *Marine Rocky Shores and Community Ecology: An Experimentalist's Perspective.* Oldendorf/Luhe, Germany: Ecology Inst. 152 pp.
181. Paine RT, Levin SA. 1981. Intertidal landscapes: disturbance and the dynamics of pattern. *Ecol. Monogr.* 51:145–78
182. Paine RT, Suchanek TH. 1983. Convergence of ecological processes between independently evolved competitive dominants: a tunicate-mussel comparison. *Evolution* 37:821–31
183. Patterson MR. 1991. The effects of flow on polyp-level prey capture in an octocoral, *Alcyonium Siderium. Biol. Bull.* 180:92–102
184. Pedley TJ, ed. 1977. *Scale Effects in Animal Locomotion.* London: Academic
185. Pennycuick CJ. 1992. *Newton Rules Biology.* New York: Oxford Univ. Press
186. Pentcheff ND. 1991. Resistance to crushing from wave-borne debris in the barnacle *Balanus glandula. Mar. Biol.* 110:399–408
187. Podolsky RD. 1993. Separating the effects of temperature and viscosity on swimming and water movement by sand dollar larvae (*Dendraster excentricus*). *J. Exp. Biol.* 176:207–21
188. Podolsky RD. 1994. Temperature and water viscosity: physiological versus mechanical effects on suspension feeding. *Science* 265:100–4
189. Porter WP, Grant BW. 1992. Modeling global macroclimatic constraints on ectotherm energy budgets. *Am. Zool.* 32:154–78
190. Porter WP, Mitchell JW, Beckman WA, DeWitt CB. 1973. Behavioral implications of mechanistic ecology: thermal and behavioral modelling of desert ectotherms and their microenvironment. *Oecologia* 13:1–54
191. Porter WP, Mitchell JW, Beckman WA, Tracy CR. 1975. Environmental constraints on some predator-prey interactions. In *Perspectives of Biophysical Ecology,* ed. DM Gates, RB Schmerl, pp. 347–64. New York: Springer-Verlag

192. Porter WP, Parkhurst DF, McClure PA. 1986. Critical radius of endotherms. *Am. J. Physiol.* 250:699–707
193. Power ME, Tilman D, Estes JA, Meyer BA, Bond WJ, et al. 1996. Challenges in the quest for keystones. *BioScience.* In press
194. Price MVe. 1986. Symposium: mechanistic approaches to the study of natural communities. *Am. Zool.* 26:3–106
195. Raff RA, Parr BA, Parks AL, Wray GA. 1990. Heterochrony and other mechanisms of radical evolutionary change in early development. See Ref. 167a, pp. 71–98
196. Rayner JMV, Wooton RJ, eds. 1991. *Biomechanics in Evolution.* Cambridge: Cambridge Univ. Press
197. Reilly SM, Lauder GV. 1992. Morphology, behavior, and evolution: comparative kinematics of aquatic feeding in salamanders. *Brain Behav. Evol.* 40:182–96
198. Reilly SM, Wainwright PC. 1994. Conclusion: ecological morphology and the power of integration. See Ref. 228a, pp. 339–54
199. Ricklefs RE, Miles DB. 1994. Ecological and evolutionary inferences from morphology: an ecological perspective. See Ref. 228a, pp. 13–41
200. Rigler FH. 1982. Recognition of the possible: an advantage of empiricism in ecology. *Can. J. Fish. Res. Aquat. Sci.* 39:1323–31
201. Rosenzweig ML, McCord RD. 1991. Incumbent replacement: evidence for long-term evolutionary progress. *Paleobiology* 17:202–13
202. Roth G, Wake DB. 1989. Conservation and innovation in the evolution of feeding in vertebrates. See Ref. 232a, pp. 7–21
203. Rothschild BJ, Osborn TR. 1988. Small-scale turbulence and plankton contact rates. *J. Plankton Res.* 10:465–74
204. Rubenstein DI, Koehl MAR. 1977. The mechanisms of filter feeding: some theoretical considerations. *Am. Nat.* 111:981–94
205. Schmidt-Nielsen K. 1984. *Scaling: Why is Animal Size So Important?* Cambridge: Cambridge Univ. Press
206. Schoener TW. 1986. Mechanistic approaches to community ecology: a new reductionism? *Am. Zool.* 26:81–106
207. Sebens KP, Done TJ. 1992. Water flow, growth form and distribution of scleraetinian corals: Davies Reef (GBR), Australia. *Proc. 7th Int. Coral Reef Symp.* 1:557–68
208. Seilacher A. 1973. Fabricational noise in adaptive morphology. *Syst. Zool.* 22:451–65
209. Shimeta J, Jumars PA. 1991. Physical mechanisms and rates of particle capture by suspension-feeders. *Oceanogr. Mar. Biol. Annu. Rev.* 29:191–257
210. Shimeta J, Jumars PA, Lessard EJ. 1995. Influences of turbulence on suspension feeding by planktonic protozoa; experiments in laminar shear fields. *Limnol. Ocean.* 40:845–59
211. Siemann E, Tilman D, Haarstad J. 1996. Insect species diversity, abundance and body size relationships. *Nature* 380:704–6
212. Slowinski JB, Guyer C. 1993. Testing whether certain traits have caused amplified diversification: an improved method based on a model of random speciation and extinction. *Am. Nat.* 142:1019–24
213. Sousa WP. 1984. The role of disturbance in natural communities. *Annu. Rev. Ecol. Syst.* 15:1353–91
214. Stanley SM. 1973. An explanation for Cope's rule. *Evolution* 27:1–26
215. Statzner B. 1988. Growth and Reynolds number of lotic macroinvertebrates: a problem for adaptation of shape to drag. *Oikos* 51:84–87
216. Statzner B, Holm TF. 1989. Morphological adaptation of shape to flow: microcurrents around lotic macroinvertebrates with known Reynolds numbers at quasi-natural flow conditions. *Oecologia* 78:145–57
217. Stearns SC. 1982. On fitness. In *Environmental Adaptation and Evolution,* ed. D Mossakowski, G Roth, pp. 3–17. New York: Gustav Fischer
218. Steneck RS, Dethier MN. 1994. A functional group approach to the structure of algal-dominated communities. *Oikos* 69:476–98
219. Stone R. 1996. Taking a new look at life through a functional lens. *Science* 269:316–17
220. Thomas FIM. 1994. Morphology and orientation of tube extensions on aggregations of the polychaete annelid *Phragmatopoma californica. Mar. Biol.* 119:525–34
221. Thomas FIM. 1994. Transport and mixing of gametes in three free-spawning polychaete annelids *Phragmatopoma californica* (Fewkes), *Sabellaria cementarium* (Moore), and *Schizobranchia insignis* (Bush). *J. Exp. Mar. Biol. Ecol.*

179:11–27

222. Van Valkenburgh B. 1994. Ecomorphological analysis of fossil vertebrates and their paleocommunities. See Ref. 228a, pp. 140–68
223. Videler JJ. 1995. Body surface adaptations to boundary-layer dynamics. In *Biological Fluid Dynamics, Soc. Exp. Biol. Symp.*, ed. CP Ellington, TJ Pedley, 49:1–20. London: Co. Biol.
224. Vincent JFV. 1990. *Structural Biomaterials*. Princeton, NJ: Princeton Univ. Press
225. Vogel S. 1993. When leaves save the tree. *Nat. Hist.* 102:58–62
226. Vogel S. 1994. *Life in Moving Fluids*. Princeton, NJ: Princeton Univ. Press. 2nd ed.
227. von Wahlert G. 1965. The role of ecological factors in the origin of higher levels of organization. *Syst. Zool.* 14:288–300
228. Wainwright PC. 1994. Functional morphology as a tool in ecological research. See Ref. 228a, pp. 42–59

228a. Wainwright PC, Reilly SM, eds. 1994. *Ecological Morphology: Integrative Organismal Biology*. Chicago: Univ. Chicago Press

229. Wainwright PC, Reilly SM. 1994. Introduction. See Ref. 228a, pp. 1–12
230. Wainwright SA, Biggs WD, Currey JD, Gosline JW. 1976. *Mechanical Design in Organisms*. Princeton, NJ: Princeton Univ. Press
231. Wainwright SA, Dillon JR. 1969. On the orientation of sea fans (genus *Gorgonia*). *Biol. Bull.* 136:130–39
232. Wake DB. 1982. Functional and evolutionary morphology. *Perspect. Biol. Med.* 25:603–20

232a. Wake DB, Roth G, eds. 1989. *Complex Organismal Functions: Integration and Evolution in Verbrates*. New York: Wiley & Sons

233. Wake DB, Roth G. 1989. The linkage between ontogeny and phylogeny in the evolution of complex systems. See Ref. 232a, pp. 361–77
234. Wake MH. 1992. Morphology, the study of form and function, in modern evolutionary biology. In *Oxford Surveys in Evolutionary Biology*, ed. D Futuyma, J Antonovics, pp. 289–346. New York: Oxford Univ. Press
235. Wasserthal LT. 1975. The role of butterfly wings in regulation of body temperature. *J. Insect Physiol.* 21:1921–30
236. Webb PW. 1986. Locomotion and predator-prey relationships. See Ref. 54, pp. 24–41
237. Weihs D. 1980. Energetic significance of changes in swimming modes during growth of larval anchovy *Engraulis mordax*. *Fish. Bull. Fish Wildl. Serv. US* 77:597–604
238. West-Eberhard MJ. 1989. Phenotypic plasticity and the origins of diversity. *Annu. Rev. Ecol. Syst.* 20:249–78
239. Westneat MW. 1995. Systematics and biomechanics in ecomorphology. *Environ. Biol. Fish.* 44:263–83
240. Will KL. 1995. Plecopteren surface-skimming and insect flight evolution. *Science* 270:1684–85
241. Williams TA. 1994. Locomotion in developing *Artemia* larvae: mechanical analysis of antennal propulsors based on large-scale physical models. *Biol. Bull.* 187:156–63
242. Williams TA. 1994. A model of rowing propulsion and the ontogeny of locomotion in *Artemia* larvae. *Biol. Bull.* 187:164–73
243. Williamson MH, Lawton JH. 1988. Fractal geometry of ecological habitats. In *Habitat Structure: The Physical Arrangement of Objects in Space*, ed. SS Bell, ED McCoy, HR Mushinsky, pp. 69–86. New York: Routledge, Chapman & Hall
244. Worcester SE. 1994. Adult rafting versus larval swimming: dispersal and recruitment of a botryllid ascidian on eelgrass. *Mar. Biol.* 121:309–17
245. Young CM. 1995. Behavior and locomotion during the dispersal phase of larval life. In *Ecology of Marine Invertebrate Larvae*, ed. LR McEdward, pp. 249–78. Boca Raton, FL: CRC

Annu. Rev. Ecol. Syst. 1996. 27:543–67

ADAPTIVE EVOLUTION OF PHOTORECEPTORS AND VISUAL PIGMENTS IN VERTEBRATES

Shozo Yokoyama and Ruth Yokoyama
Department of Biology, Syracuse University, 130 College Place, Syracuse, New York 13244

KEY WORDS: color vision, opsin, adaptive evolution, amino acid replacements, phylogeny

ABSTRACT

Animals may be camouflaged by a coloration that matches their surroundings or by a combination of color and shape. Some species make themselves conspicuous and rely upon bold and bright coloration as a means of warning off their potential predators. Population biologists have accumulated information on the adaptive significance of coloration for a large number of species. To elucidate the mechanisms underpinning such natural selection events, it is necessary to understand the visual systems of interacting organisms.

Molecular genetic analyses on the human opsin genes by Nathans and his colleagues made it possible to characterize the opsin genes of various vertebrates. A striking level of diversity in the opsin gene sequences reflects adaptive responses of various species to different environments. Comparative analyses of opsins reveal that gene duplications and accumulation of mutations have been important in achieving that diversity. The analyses also identify amino acid changes that are potentially important in controlling wavelength absorption by the photosensitive molecules, the visual pigments. These hypotheses can now be rigorously tested using tissue culture cells. Thanks to the molecular characterization of the opsin genes, it is now possible to study the types of opsins associated with certain environmental conditions. Such surveys will provide important first molecular clues to how animals adapt to their environments with respect to their coloration and behavior.

INTRODUCTION

Nature provides diverse photic environments, ranging from the darkness at the bottom of the ocean to bright light in the desert. Protective coloration, mimicry,

0066-4162/96/1120-0543$08.00

and sexual display have been well documented (27, 36, 45, 103, 106). Animal colorations are particularly important in territorial and courtship displays, which are often greatly enhanced by exaggerated and stereotyped movements and postures (39). Such a seemingly endless variety of colors and patterns become effective only when they are interpreted correctly by the members of the same species or by their predators. Consequently, vision has profound effects on the evolution of organisms by affecting fitness through such behaviors as mating, foraging, and predator avoidance.

Given the importance of vision, it is not surprising that animals have optimized their visual capabilities to suit their specific ecological niches. Thus, it is of considerable interest to understand how animals distinguish different colors and utilize both dim and bright light. The ability of vertebrates to see the visible spectrum of light is generally controlled by a set of photosensitive molecules, visual pigments, in two types of photoreceptor cells, rods and cones. Typically, rods function in dim light and cones are responsible for color vision. Each visual pigment consists of a chromophore and a transmembrane protein, called opsin. With the rapid accumulation of nucleotide sequences of the opsin genes in vertebrates and invertebrates, we can now start to better understand such basic biological phenomena. Comparative analyses of opsins reveal the importance of gene duplications and the accumulation of mutations in leading to the extant diverse opsins. The analyses also identify amino acid changes that are potentially important in controlling wavelength absorption by the visual pigments. Hypotheses derived from these analyses can now be manipulated and tested experimentally (24, 72, 73, 81, 94, 104, 109, 115). This provides a rare opportunity to study adaptive evolution of visual systems at both molecular and organismal levels.

In this chapter, we describe how animal body color and color patterns have interactively evolved with visual systems to maximize organism survival. Recent developments on the phylogenetic analyses of opsins in vertebrates are then reviewed.

MODIFICATION OF COLORS AND PATTERNS

During evolution, organisms have developed astonishingly diverse body colors and patterns uniquely suited to their ecological niches and lifestyles. The evolutionary importance of animal colorations relating to camouflage, mimicry, warning, and sexual display has been widely recognized among population biologists (e.g. see 27, 45, 82, 103, 106).

Industrial Melanism

Probably the most spectacular and convincing example of natural selection is industrial melanism. More than 70 species of moths in Europe have light and dark variants. Until the middle of the nineteenth century, the former were common, and the latter forms were recorded only as rare mutants. However, the industrial revolution profoundly altered the European countryside, polluting the environment, which led to the dark forms replacing the light ones. For the peppered moth, *Biston betularia*, there exist three phenotypes, the melanic *carbonaria,* the dark *insularia*, and the pale *typical.* These phenotypes are controlled by three alleles, with *carbonaria* dominant over both the others and *insularia* dominant over *typical.* In industrialized cities, the pale *typical* form became very rare within 100 years (57). The selective advantage of the darker forms over the lighter ones lies in their ability to remain on darkened tree trunks concealed from avian predators (57).

In 1956, clean air legislation was passed in England. Because of a reduction of industrial smoke and sulfur dioxide in many formerly polluted areas, dramatic declines have occurred in the frequency of melanic forms of *B. betularia* and other insects (17). This example of natural selection clearly demonstrates that adaptive coloration can occur very rapidly.

Mimicry

One of the best-studied examples of mimicry is the large orange and black North American monarch butterfly (*Danaus plexippus*). The monarch possesses cardiac glycosides, taken from milkweed, which disrupt the heartbeat and influence the area of the brain that controls vomiting in vertebrates. Consequently, when avian preditors eat an unpalatable monarch, they quickly become ill and regurgitate the butterfly (19). Morphologically similar butterflies, the palatable viceroy (*Limenitis archippus*) and unpalatable queen butterfly (*Danaus gilipous*), coexist in the same geographical region when the monarch migrates to the South. Because of the similarities in coloration and patterns of these butterfly species, predatory birds tend to avoid them altogether (18, 20).

Water Environments

Most of the light used in vision is generated by the sun. Visible light corresponds to photon wavelengths of 360 nm (ultraviolet in color) to 700 nm (deep red). Transmission of light by water is dependent on the wavelength of the light source and the presence of organic material that absorbs light. Fish living in shallow freshwater have access to a fairly broad spectrum of light. Inshore and inland waters are green, yellow-green, or even orange-brown because of phytoplankton and dissolved organic material (60). Many inshore and freshwater fishes possess brilliant red or orange body colors that have clear behavioral significance (64). However, in clear oceans and lakes the light becomes increasingly monochromatic so that at a depth of 200–1000 m, light is restricted to a narrow spectral band of blue or blue-green color. Accordingly, most deep-sea fish have black backs and silver underbodies to be less conspicuous against the background light (66, 67). Even so, they can have difficulty being camouflaged because they can be seen in silhouette due to the light from above. For these fishes, the use of bioluminescence turns out to be an excellent solution. Bioluminescence is an enzyme-catalyzed reaction, in which luciferases catalyze the oxidation of specific substrates, termed luciferins. The energy released is emitted as light. Generally, this bioluminescence matches the downwelling light, blue or blue-green color (44). In addition to blue-emitting light organs, three genera of deep-sea fishes (*Malacosteus, Pachystomias,* and *Aristostomias*) produce far-red light (29, 30, 84, 107). This far-red light could be used both for intraspecific communication and for covert illumination of prey at distances about ten times greater than the range of lateral line senses (53, 84). The deep-sea fish *Aristostomias tittamanni* has visual pigments that are sensitive to blue, green, red, and far-red lights (84), while many other deep-sea fishes are not equipped to see red and far-red lights. One fascinating aspect of this system is that far-red is too long a wavelength to be seen by most other fishes in the area (64).

Visual Effects of Color Patterns Within and Between Species

Examples considered so far demonstrate the adaptation of colorings and patterns to an animal's surroundings. Clearly, color patterns cannot be too bold because these might make the animal conspicuous prey. At the same time males cannot be too dull or they will be at a disadvantage in competition for females. How can a species maintain effective visual communication with mates while remaining inconspicuous to predators? It appears that animal color pattern is strongly affected by many factors, including background matching, predation intensity, predator vision, and sexual selection for effective mating (34).

A color pattern can be hidden, i.e. cryptic, if it resembles its background. One of the most thoroughly studied polymorphisms is that of the color and banding patterns in the shell of the land snail *Cepea nemoralis*, in which a Mendelian polymorphism appears to be maintained by environmental heterogeneity in different areas (for a review, see 51). The polymorphism is maintained by differential predation by thrushes among snails with different shell colors (e.g. 23, 26). When the predator is abundant, there is a high frequency of cryptic organisms, but where the predator is rare, there are fewer cryptic individuals (e.g. 51). Thus, intensity of camouflage also seems to depend on level of predation (for more examples, see 34). By collecting field data on ambient light spectra, water transition spectra, courtship and attack distance, and vision of a guppy, *Poecilia reticulata*, and several of its predator species, Endler (35) evaluated conspicuousness of the fish under various conditions. The data suggest that color patterns are relatively more conspicuous to guppies at times and places of courtship and relatively less conspicuous at the times and places of maximum predator risk. Thus, animal colorations are time- and space-dependent and can be adjusted relatively easily.

MECHANISM OF VISION

Human Visual System

When we look at the outside world, the primary event is that light is focused on an array of 125 million photoreceptor cells in the retina of each eye (46). All photoreceptors, rods and cones, have an outer segment, an inner segment, a perikaryal region containing the cell nucleus, and a terminal (e.g. see Figure 1). The retina contains about 100 million rods and 3 million cones. The rod outer segment contains about 1000 discrete layers of discs, each of which contains about 10,000 photosensitive molecules, visual pigments (101). Cone outer segments also have discs, but they are contiguous, formed by an infolding of the cone membrane. The visual pigments extend the entire thickness of the membrane, where the amino-terminal ends are always on the inside of the disk and the carboxy-terminal on the outside with seven transmembrane segments (95; Figure 2).

The visual pigments in the human rods, rhodopsins, absorb light maximally (λmax) at about 495 nm, and those in three types of cones have λmax values of 420 nm (blue- or short wavelength–sensitive; SWS), 530 nm (green- or middle wavelength–sensitive; MWS), and 560 nm (red- or long wavelength–sensitive; LWS) (10). Thus, using only the three cone pigments plus rhodopsin, we are able to discriminate the various colors and hues ranging from 400 to 700 nm. Each of our four types of visual pigments consists of a chromophore, 11-*cis* retinal (also known as vitamin A1 aldehyde), and a transmembrane protein, opsin (see Figure 2), which is encoded by one of the four distinct opsin genes (75, 77). The light sensitivity of the visual pigments is due to the chromophore, which is covalently bound to opsin (99, 100).

Human vision begins when a photon is absorbed by a visual pigment. Then, the 11-*cis* retinal physically changes its shape to all-*trans* retinal, and the photoexcited opsin activates transducin (a guanine nucleotide-binding protein, or G protein). The active subunit of transducin (α subunit) activates the effector cGMP phosphodiesterase, which then closes ion channels, hyperpolarizing the illuminated photoreceptors. The electrical signals emitted by the visual pigments reach the visual cortex of the brain, and we can visualize the outside world essentially instantaneously (71). Although a range of photic environments have led to the evolution of various visual systems among vertebrates, the basic mechanisms involved appear to be the same.

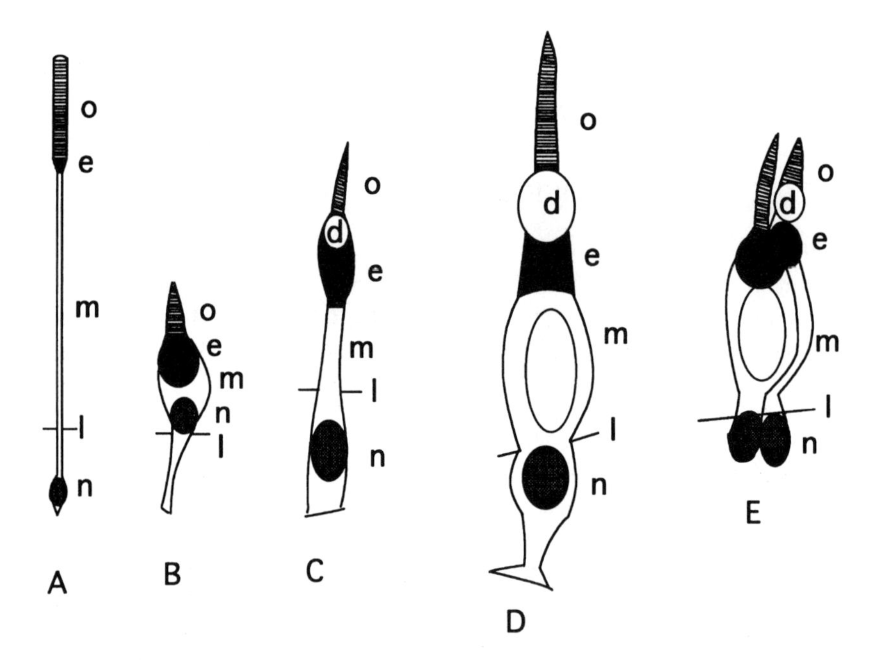

Figure 1 Rod and cone cells from a variety of animals: (A) rod of goldfish, *Carassius auratus*; (B) cone of goldfish; (C) cone of leopard frog, *Rana pipiens*; (D) cone of snapping turtle, *Chelydra serpentina*; (E) double cone of western painted turtle, *Chrysemys picta marginata.* (d) oil droplet; (e) ellipsoid; (l) external limiting membrane; (m) myoid; (n) nucleus; (o) outer segment (Redrawn from 103).

Rods and Cones

Most vertebrates have both rods and cones. The classical distinction between rods and cones is based upon the morphological differences of the outer segments. Typical rods have cylindrical inner and outer segments, whereas cones have a conically formed outer segment (Figure 1A, B). Physiologically, the rod photoresponse is 2–5 times slower than that of a cone but is 100 times more sensitive to light. In addition, rods detect a narrower range of background light intensities than do cones (for a review, see 108). The rods and cones have developed separate protein isoforms involved in the phototransduction process such as visual pigments, transducin, phosphodiesterase, and cGMP-gated ion channels.

Typically, a single type of rod and multiple types of cones provide the various color sensitivities, as exemplified by those in the human retina. However, some amphibians are known to have two types of rods, called the red rods and green rods based on their color (103). In lower vertebrates, there are often double photoreceptors (usually cones) that are closely connected, having contiguous inner segments. The two members may be similar in size (twins) or significantly different (doubles), with a principal member and an accessory, smaller member (Figure 1E). The variety of photoreceptor morphology present in various classes of vertebrates has been described (12, 28, 103).

Microspectrophotometry on a variety of fish has shown that the single cones usually contain blue-sensitive, and sometimes UV-sensitive, visual pigments. Double or twin cones usually consist of one member sensitive to red light and the other, to green light (65).

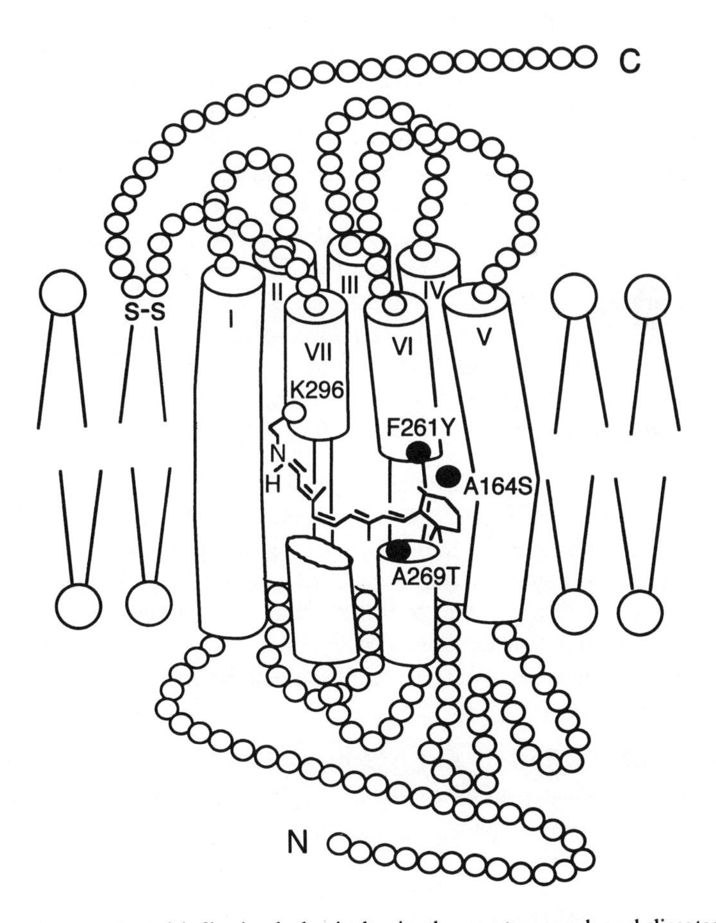

Figure 2 Schematic model of bovine rhodopsin showing the seven transmembrane helices together with the 11-*cis* retinal linked to K296 (modified from 1). Location of three amino acid changes involved in the differentiation of MWS and LWS opsins are shown in filled circles.

ADAPTIVE MECHANISMS OF LIGHT USE

Diverse visual systems have evolved to use all available light. In fact, the visual system is one of the best examples of obvious adaptation to specific ecological niches. Size and shape of eyes, photoreceptor organization, and color sensitivity provided by visual pigments can differ dramatically depending upon an animal's specific needs. For instance, the photic environment inhabited by fish is much more important in determining their visual capability than their species relatedness (60).

Photoreceptor Organization

PHOTORECEPTOR RATIOS The ratio of rods to cones is often indicative of an animal's life-style. Deep-water or nocturnal organisms need to maximize sensitivities to the available light rather than to color discrimination. Such organisms have done this in part by increasing the number of rods in the retina. For example, deep-sea fish have developed pure rod retinas, with the rod outer segments often elongated or composed of multiple rows (for a review, see 61). Some have even developed tubular eyes (66, 67). Likewise, nocturnal terrestrial animals have a preponderance of rods, whereas diurnal vertebrates with good color vision often have more cones. Two lizard species have taken this to extremes: The nocturnal gecko (*Gekko gekko*) has only rods (102), whereas the diurnal, highly visual American chameleon (*Anolis carolinensis*) has only cones (28).

A survey of the wavelength absorption by the visual pigments present in the cases of various fishes shows that freshwater species have more red-sensitive receptors than do blue-water species (62, 63, 65).

RETINOMOTOR ACTIVITY Most fish, frogs, and birds have the ability to adapt to light intensity by mechanistically repositioning the photoreceptor outer segments and melanin granules in the adjacent pigment epithelium. In light, the melanin granules migrate toward the photoreceptors, the rods elongate, and the cones contract. This places the cones first in line for light reception and shields the rods from bright light exposure. During darkness, the opposite movements occur. Retinomotor activity can be regulated both by circadian signals and by light (9).

Modifications Affecting Color Sensitivity

The range of wavelengths to which an animal is sensitive depends upon its visual pigments and the absorption and reflection characteristics of structures within the eye.

OIL DROPLETS/TAPETUM/LENS One fascinating characteristic of photoreceptor cells is that many amphibians, birds, and reptiles have colored oil droplets in the inner segment, just adjacent to the outer segment (see Figure 1C–E). For example, the retina of the chicken (*Gallus gallus*) consists of rods, and single and double cones that contain transparent, red, orange-yellow, blue, and green oil droplets. The light passes through the oil droplets, essentially providing colored filters that adjust the λmax of a photoreceptor cell according to the color of its oil droplet (14). The three types of cone visual pigments in the pigeon have λmax at 461, 514, and 567 nm. Using differently colored oil droplets, however, the birds have at least six types of cone with λmax of 440, 485, 567, 575, 589, and 619 nm (11). This fine discriminatory mechanism allows the pigeon to determine information on the ratio of chlorophyll to red pigments and the amount of chlorophyll or red pigments in a leaf (64). In the central field of the retina of the pigeon, the proportion of cones with a red oil droplet is about a quarter; another quarter are those with orange-yellow; and the rest contain green, blue-green, or colorless oil droplets. This proportion seems to be typical of many other herbivorous birds (64). Apparently, the pigeon uses the red oil-droplet region, known as the "red field" of the retina selectively (64). Colored oil droplets reduce sensitivity because only part of the spectrum is allowed to pass. Therefore, it is not surprising to find that the nocturnal birds have few colored oil droplets (103).

The deep-sea fish *Malacosteus*, already mentioned because of its red bioluminescence, uses two structural means for increasing its visual capability toward the red. Its eyes contain a red mirroring device, termed a tapetum, between the rods and pigment layer, and it reflects only long wavelengths. In addition, its lens is yellow, which will absorb light below ≈500 nm, thereby increasing the relative brightness of long-wave light (85). It is not uncommon that fish lenses preferentially absorb short wavelengths (31).

VISUAL PIGMENTS Visual pigment sensitivity is related to the ecological background of the animal. The best-studied species are fish because of their variety of clearly defined photic environments. Lake and ocean water selectively absorbs both short- and long-wave light, restricting the maximum transmission to a wavelength of ≈470 nm. It was gratifying to discover that most deep-sea fish have visual pigments absorbing at 470–480 nm (83).

In any particular place, inshore and inland water color changes frequently, depending upon factors such as rainfall and the time of day and year (64). Fish living in these types of environments have developed visual pigments tuned to their specific requirements; they generally have green and blue wavelength color visual pigments (62, 65). Freshwater is relatively transparent to red light. Significantly, freshwater fish also often have red visual pigments in addition to the blue and green visual pigments (62, 65). There are even fishes with different spectral sensitivity in the upper and lower retina. For instance, the

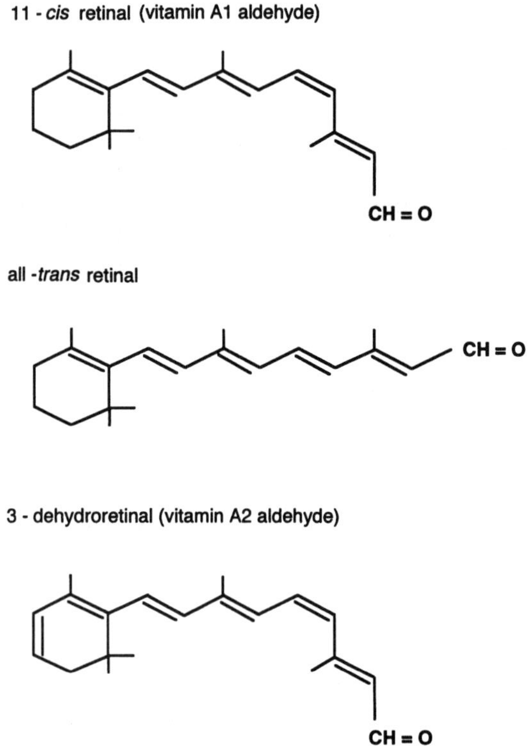

11 - *cis* retinal (vitamin A1 aldehyde)

CH = O

all -*trans* retinal

CH = O

3 - dehydroretinal (vitamin A2 aldehyde)

CH = O

Figure 3 The structures of 11-*cis* retinal, all-*trans* retinal, and 3-dehydroretinal. A bend or line indicates a carbon atom and associated hydrogen atoms (modified from 32).

guppy *Poecilia reticulata* has the upward-looking lower retina dominated by green-sensitive cones, good for detecting dark objects against downwelling light, while the upper retina contains red-, green-, and blue-sensitive cones, good for color vision. The male guppy makes sure that his color can be truly appreciated by performing his mating displays either in front of or below the female (60).

The visual pigment can be altered by opsin mutations (discussed in the following sections), or it can be red-shifted by using 3-dehydroretinal (or vitamin A2 aldehyde) as a chromophore instead of the usual 11-*cis* retinal (vitamin A1 aldehyde) (105; see also Figure 3). An additional double bond in A2 shifts the λmax to longer wavelength (lower energy) light. The relative proportions of these two chromophores have been demonstrated to change in certain lampreys, numerous fishes, and amphibians (e.g. see 8). These changes are apparently due not to dietary changes but to regulation of chromophore synthesis enzymes (89). Increasing the vitamin A2 usage is consistent with an anticipated change for a redder environment and can be brought about by environmental changes in light, season, migration, temperature, and hormone (58). Adult fish often have a higher ratio of vitamin A2 during the winter because of decreased light and/or temperature. However, chromophore switches also occur during metamorphosis of certain amphibians (28) and during migration from freshwater to saltwater (e.g. salmonids) or vice versa (e.g. the American eel). American (or European) eels *Anguilla rostrata* breed in the deep, blue waters of the Sargasso Sea, but they spend their immature adult life in freshwater rivers, a more reddish environment. These migratory fishes use vitamin A1–based and vitamin A2–based chromophores during their deep-ocean and freshwater phases, respectively. In this species, the A1 pigment with λmax of 501 nm transforms to the A2 pigment with λmax of 523 nm using the same opsin (7). Similar chromophore changes have been observed in different salmon species during their breeding migration between freshwater and saltwater (6).

With respect to their photoreceptors and visual pigments, the American chameleon is unique among terrestrial vertebrates, containing only cones (28, 98) and A2 pigments (88). The red visual pigments have λmax of 625 nm, which is some 50 nm further into the red than any other terrestrial vertebrates examined to date (88).

EVOLUTION OF RED AND GREEN COLOR VISION

When color perceptions of a diverse range of species are surveyed, only a limited number of species in the bony fishes, birds, reptiles, and primates are found to have full-fledged trichromatic color vision, while many extant vertebrates are color-blind (103). For example, most mammals have dichromatic color vision, having short wavelength (SWS) vision with either MWS or LWS vision (49). The full-fledged color vision in the various major vertebrate lineages has probably been developed independently (103, 111).

At the visual pigment level, two major factors influence the spectral absorption: (*a*) the type of chromophores used, i.e. either vitamin A1 or vitamin A2, and (*b*) the interaction between the chromophore and the opsin. For the latter, amino acid sequence differences among the opsins have important effects in determining the distinctive absorption spectra of the visual pigments.

Multiple Loci

In order to understand the molecular mechanisms underlying the difference between LWS and MWS vision, the LWS and MWS opsins from human and the Mexican fish *Astyanax fasciatus* were compared (111). The human LWS and MWS opsins are known to differ at 15 out of 364 residues (77), showing that some or all of these residues can make the difference between LWS and MWS vision. The evolutionary analyses of the available data suggested two things. First, the LWS and MWS color vision in the two species occurred by independent gene duplications followed by nucleotide substitutions (Figure 4). Second, the LWS opsins in both humans and fish evolved from the MWS opsins independently by three identical amino acid changes A180S (amino acid change A⟶S at residue 180), F277Y, and A285T, following the residue numbers of the human LWS/MWS opsins (Figure 4) (111). By characterizing 8 LWS/MWS sequences from primates, Neitz et al (78) also concluded that the spectral difference between MWS and LWS vision could be explained by the additive effects of A180S, F277Y, and A285T. All of these positions, which correspond to residues 164, 261, and 269 in bovine rhodopsin, are very near the chromophore (Figure 2).

This type of analysis has important implications because it identifies previously unsuspected important amino acid changes in wavelength absorption. In fact, Chan et al (24) tested the role of the above-mentioned three amino acid changes in the red-shift of λmax by modifying the corresponding three positions in bovine rhodopsin, which is 16 residues shorter than the human

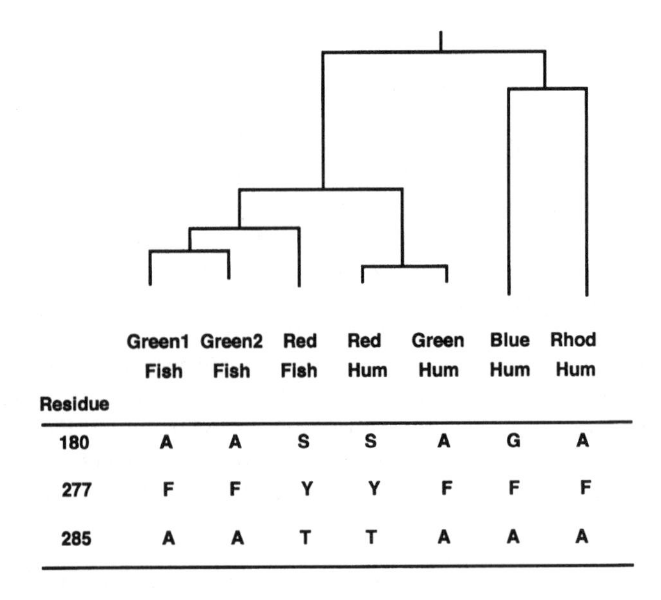

Residue	Green1 Fish	Green2 Fish	Red Fish	Red Hum	Green Hum	Blue Hum	Rhod Hum
180	A	A	S	S	A	G	A
277	F	F	Y	Y	F	F	F
285	A	A	T	T	A	A	A

Figure 4 Phylogenetic tree and amino acid compositions of LWS and MWS opsins of the Mexican cave fish, *Astyanax fasciatus*, and human, *Homo sapiens*.

LWS and MWS opsins. They observed that three changes, A164S, F261Y, and A269T, in this protein increased λmax values by 2, 10, and 14 nm, respectively, showing that these changes could explain the majority of the 30 nm difference between MWS and LWS vision. Qualitatively, the same conclusion has been reached by constructing human MWS and LWS opsin chimeras, except that the entire 30 nm of λmax-shift also requires Y116S, T230I, S233A, and F309Y (3). Furthermore, using the rhodopsin of the Mexican fish *A. fasciatus*, the reverse experiment with the Y261F mutation is shown to shift λmax toward blue by 8 nm (109). So far, virtually all known LWS opsins contain S180 (amino acid S at residue 180), Y277, and T285. However, an exception to this rule is a LWS/MWS opsin of goldfish, *Carassius auratus*, which has S180, Y277, and T285 but has an absorption spectrum of MWS rather than LWS (50). This suggests that there must be other amino acids that can determine red vs green vision. In addition, more recent analyses using additional opsin sequence data suggest that a LWS opsin, rather than a MWS opsin, is the ancestral form.

Multiple Alleles

In order to achieve LWS and MWS vision, animals usually require both LWS and MWS opsin genes. New World monkeys reveal an interesting exception, where they have multiple alleles at the one X-linked locus. For example, the squirrel monkey, *Saimiri sciureus*, is a dicromat, with two opsin loci, one encoding SWS opsin and another, MWS or LWS opsin (13). A fascinating feature about the latter locus is that it consists of three alleles. When these opsins are reconstituted with 11-*cis* retinal, the three types of resulting visual pigments have λmax of 532, 547, and 561 nm (78). The visual pigments with λmax of 532, 547 and 561 nm consist of amino acids (A180, F277 and A285), (A180, F277, and T285) and (S180, Y277, and T285), respectively (78). In this visual system, all males and homozygous females at this locus are dichromatic. However, if heterozygous females happen to have both pigments with λmax of 532 and 561 nm, they are trichromatic (12, 13).

EVOLUTION OF ULTRAVIOLET VISION

Many flowers have markings that guide the visiting insects toward nectar or pollen. For example, bees are attracted by the ultraviolet (UV) center of flowers (70). The colors of insect-pollinated flowers must have evolved to advertise their presence to insects. Certainly, "the overlap in spectral sensitivity of man and insects allows us to appreciate some of the visual displays not directed toward us" (64, p. 189).

There is evidence that some vertebrates also have UV vision (for a review, see 48). These vertebrates include hummingbirds (41), pigeons (33, 59), turtles (2), the tiger salamander (86); fish such as roach (5), Japanese dace (43), goldfish (16), and rainbow trout (15); and mammals such as rat, house mouse, gerbil, and gopher (47). UV sensitivity in some fish is known to decline during their lifespan. For example, UV pigments can be easily recorded in young brown trout, *Salmo trutta*, which live in shallow water and feed on plankton, but they are not detected in two-year-old fish, which live in deeper water where the UV light does not penetrate (15). Although the advantage of UV vision in vertebrates is not immediately clear, the petals of bird-pollinated flowers have substantial UV reflectance. This may provide attractive targets to species with UV vision. Similarly, the color patterns of fish and bird plumages are also highly reflective of UV, which may enhance the visibility of body coloration patterns to UV-sensitive individuals (21, 42).

In this context, the recent finding of UV vision in five closely related species of Puerto Rican anoline lizards is of interest (38). When the spectral reflectance of their throat dewlaps was measured, two species (*Anolis cristatellus* and *A. krugi)* showed a high degree of UV reflectance, two species (*A. evermanni* and *A. gundalachi*) exhibited low reflectance, and one species (*A. pulchellus*), an intermediate level of reflectance. Interestingly, the three species with UV-bright dewlaps live in microhabitats that are often exposed to direct sunlight, while the two species with no UV reflection inhabit the understory of closed-canopy forest, where little UV light is found. Thus, a clear relationship exists between the light conditions of the habitat and the extent of dewlap UV reflection. For these species, dewlaps are usually hidden and appear to be used almost exclusively in communication (38).

EVOLUTION OF OPSINS

The sequence information and the availability of the cDNA clones for the human opsin genes (77) have made it possible to clone and characterize the opsin genes from other species. So far, about 45 DNA and cDNA clones of opsin genes from diverse vertebrate species have been characterized. Currently known opsin genes are shown in Figure 5. Opsin genes classified as RH1, RH2, and SWS groups contain 5 exons, while those in MWS and LWS groups contain 6 exons. Among these, an opsin gene from American chameleon, *A. carolinensis,* spans 18.3 kb from start to stop codons, which is the longest opsin gene known in vertebrates (55). On the other hand, the corresponding length of the LWS opsin gene in the Mexican cavefish, *A. fasciatus,* is about 1600 bp, making it the shortest visual pigment gene known in vertebrates. A more recent survey, however, strongly suggests that the RH1 genes in some fish lineages seem to have lost all of the four introns (37).

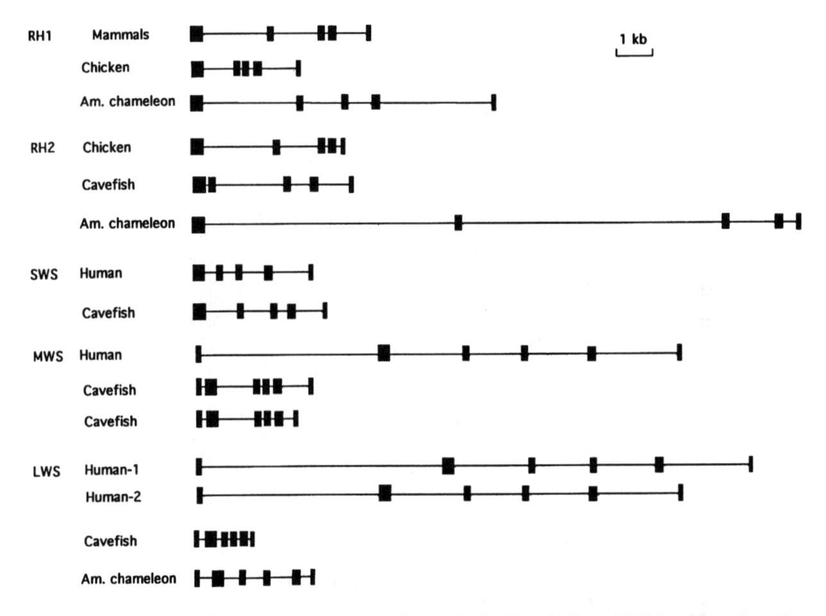

Figure 5 Structures of opsin genes from a variety of animals: chicken, *Gallus gallus*; American chameleon, *Anolis carolinensis*; cavefish, *Astyanax fasciatus*; human, *Homo sapiens*. Two human LWS opsin genes have intron-1 length polymorphism, one of them being 2 kb longer than the other (52). (For RH1, RH2, SWS, MWS, and LWS, see Figure 6.)

The 45 opsins deduced from DNA and cDNA sequences may be classified into four major groups (112, 113): RH1, RH2, SWS, and LWS/MWS clusters (Figure 6). From Figure 6, it becomes immediately clear that the divergence of the four major opsin groups predates that of different vertebrate lineages (e.g. see RH1 cluster). Thus, the ancestors of all vertebrates must have possessed all of these opsin genes. The SWS cluster may further be divided into two subgroups: (*a*) The blue opsins from mouse, human, and bovine, blue-1 opsin from American chameleon, and violet opsins from chicken and frog (SWS-I group); and (*b*) blue opsins from chicken, cavefish, and goldfish, and blue-2 opsin from American chameleon (SWS-II group). The RH1, RH2, LWS/MWS, SWS-I and SWS-II correspond to Rh, M2, L, S, and M1 groups in Okano et al's (79) classification. Note that in Okano et al (79), the SWS-II group is more closely related to the RH1 and RH2 groups than to SWS-I group, which differs from that in Figure 6. Since the bootstrap support for the SWS cluster is only 53%, the tree topology in Figure 6 is not as clear-cut as it may seem. However, when the LWS/MWS opsins are excluded from the analysis, the support of the SWS clustering becomes much higher (113).

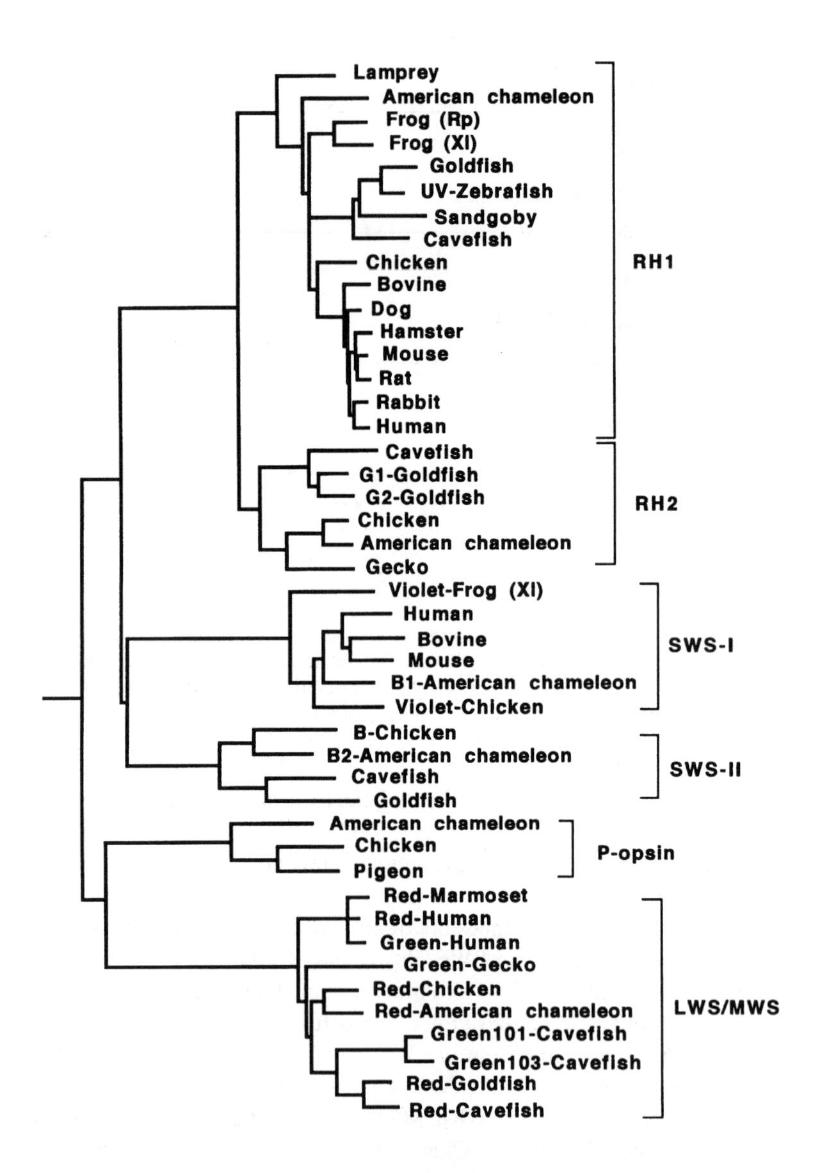

Figure 6 Phylogenetic tree for the vertebrate opsins, constructed by the NJ method (93), based on the numbers of amino acid substitutions per site. For the outgroup, Rh1, Rh2, Rh3, and Rh4 of *Drosophila melanogaster* were used. Lamprey, *Lamptera japonica*; American chameleon, *Anolis carolinensis*; frog (Rp), *Rana pipiens*; frog (Xl), *Xenopus laevis*; goldfish, *Carassius auratus*; zebrafish, *Brachydanio rerio*; sandgoby, *Pomatoschistus minutus*; cavefish, *Astyanax fasciatus*; chicken, *Gallus gallus*; bovine, *Bos taurus*; dog, *Canis familiaris*; hamster, *Cricetulus griseus*; mouse, *Mus musculus*; rat, *Rattus norvegicus*; rabbit, *Oryctolagus cuniculus*; human, *Homo sapiens*; gecko, *Gekko gekko*; pigeon, *Columba livia*. Virtually all DNA sequences are available from GenBank. Exceptions are B1- and B2-American chameleon (56).

RH1 Cluster

With the exception of RH1 opsin in American chameleon, the protein tree topology coincides with the organismal tree topology, where jawless lamprey, jawed fishes, amphibians, birds, and mammals emerged in that order. The bootstrap support for the RH1 opsin of American chameleon and the rest of vertebrates is 63%, and the tree topology is not reliable. The cause of this uncertain topology seems to be an accelerated rate of amino acid replacement in the RH1 opsin in the rod-less American chameleon. Immunocytochemical analysis suggests that this RH1 opsin is expressed only in the pineal gland, which controls circadian rhythms (40). The accelerated evolutionary rate might have been needed in establishing the pineal gland–specific function (54). Probably the most unexpected result in this cluster is that the UV-sensitive opsin (UV-Zebrafish) in zebrafish has λmax of 360 nm (91). All other opsins with known function are associated with λmax of about 500 nm (113). This UV opsin diverged from the RH1 opsin much after the land animal and bony fish lineages diverged by a single amino acid replacement (91, 113).

The rhodopsin cDNA from eyed freshwater fish, *A. fasciatus*, unlike any rhodopsin reported so far, encodes tyr at position 261, which corresponds to Y277 of the human LWS opsin and was implicated as important for the red-shift in wavelength absorption (as already noted). Because of this unusual characteristic, site-directed mutagenesis was used to produce the mutant Y261F. Expression of the Y261F mutant in cultured cells showed λmax of 496 nm, compared to 504 nm for the wild-type pigment (109). Microspectrophotometry of *Astyanax* rods regenerated with 11-*cis* retinal showed a λmax of 503 (FI Harosi, J Kleinschmidt, personal communication). Thus, the two analyses agree extremely well and indicate a mutation to produce a red-shift in the *Astyanax* wild-type rhodopsin gene.

RH2 Cluster

The RH2 cluster is a mixture of opsins with different functions. When the green opsin in chicken is reconstituted with 11-*cis* retinal, the resulting visual pigment has λmax of 508 nm (79). In nature, however, the chicken photoreceptor cell with these visual pigments has λmax value of 533 nm due to the presence of a green oil droplet in the cell (14). Similarly, when green-1 (G1-Goldfish) and green-2 (G2-Goldfish) opsins in goldfish are reconstituted with 11-*cis* retinal, the visual pigments have λmax of 506 and 511 nm, respectively, but they have MWS vision, probably because of vitamin A2 usage (50). The blue opsin in gecko simply has λmax of 467 nm.

Figure 6 shows that the American chameleon RH2 opsin is more closely related to the chicken green opsin than to the gecko blue opsin. The bootstrap support for the cluster of the two lizards is 1.0 and highly reliable. This tree topology differs from the organismal tree, where the two lizard species should be most closely related. This strongly suggests that the two lizard opsin genes have been derived from duplicate ancestral genes (55).

SWS Cluster

The SWS-I visual pigments in human and chicken have λmax values of 420 and 415 nm, respectively, while the SWS-II visual pigments in goldfish and chicken have λmax of 441 and 455 nm, respectively. So far, the λmax values associated with the other SWS opsins have not been determined. As already noted, many other fish, bird, amphibian, reptilian, and mammalian species also possess UV vision (48). The SWS-I visual pigments in mouse (25) and American chameleon (56) are also suspected to be responsible for UV vision. The duplication event of the rhodopsin genes in the ancestor of zebrafish seems to have occurred long after the divergence between fish and the other vertebrates (see RH1 cluster in Figure 6). This observation strongly suggests that UV vision in zebrafish and those in many other vertebrates have been achieved independently (113). Accordingly, the molecular mechanisms responsible for the development of UV vision in zebrafish may be entirely different from those in other vertebrates.

Figure 6 shows that the SWS-I opsins are evolving with a faster rate than the SWS-II opsins. The faster evolutionary rates of the SWS-I opsins might have been necessary to accomplish the lower wavelength absorption. The short wavelengths of absorption by the two SWS groups seem to have been achieved by different sets of amino acid replacements in the transmembrane regions (56).

LWS/MWS Cluster

In this cluster, two groups of opsins can be distinguished: (*a*) mammalian opsins (group I) and (*b*) nonmammalian opsins (group II). The bootstrap supports for the group I and group II are 99% and 90%, respectively. The difference between group I and II is unexpected because the reptiles and birds are more closely related to human than to fish, at the organismal level, and therefore we would expect the LWS/MWS opsins of the gecko, American chameleon, and chicken to be more closely related to those of mammals than to those in fishes. This unexpected tree topology of the LWS/MWS cluster may be explained by assuming that the group I and group II opsins descended from two different duplicated ancestral genes (114).

Figure 6 provides another interesting result. That is, gene duplication of the ancestral red and green opsin genes predates the divergence between the Mexican cavefish and goldfish. In the RH2 cluster, the opsin with unknown function in the cavefish is closely related to the green opsins of goldfish. These observations suggest that goldfish also has an additional gene that is orthologous to the green opsin genes of the cavefish. Southern hybridization analysis supports the hypothesis that the goldfish genome contains more than one copy of the LWS/MWS opsin gene (91), which remains to be isolated.

EVOLUTION OF PINEAL GLAND–SPECIFIC OPSINS

Virtually all eukaryotes and some prokaryotes are known to express daily rhythms in behavior, physiology, and biochemistry (4). The periodic nature of biological functions is under direct influence of both light/dark and an endogenous clock, which defines the rhythmicity of all physiological functions (4). The biological clock has been detected in the hypothalamic suprachiasmatic nucleus of mammals (69, 92), pineal gland of birds and reptiles (97), and retina of amphibians and birds (22). The dissociated pineal cell cultures reveal three major components: (*a*) an input pathway that is photosensitive; (*b*) a circadian oscillator that generates the rhythm; and (*c*) an output pathway that results in the synthesis of melatonin (97). These observations suggest that the pineal and retina are equipped with similar proteins that are involved in phototransduction.

Until recently, the functions of such opsin-related proteins in the pineal gland were not known. However, this picture changed drastically by the isolation of a pineal-specific photosensitive molecule from chicken (68, 80). Figure 6 shows that such P-opsins from chicken, pigeon (S Kawamura, S Yokoyama, unpublished data), and American chameleon (S Kawamura, S Yokoyama, unpublished data) are clearly related to opsins. The proteins deduced from the P-opsin genes are about 45% identical to the vertebrate retinal opsin amino acid sequence. When the full-length chicken cDNA was subcloned into an expression vector, expressed in cultured cells, and reconstituted with 11-*cis* retinal, the regenerated P-opsin demonstrated the wavelength of λmax of 470 nm (80), showing that it is SWS. Note that American chameleon contains two nonretinal opsins, RH1 and P-opsin. It will be of interest to evaluate how these two and possibly other opsins control the lizard's circadian rhythms.

CONCLUSION

Population biologists have described a large number of fascinating examples of animal colorations. These examples demonstrate that animals have developed a diverse level of body colors and patterns to blend them into backgrounds, to mimic and warn others, and to be effective in courtship. Field data on guppies and their predators suggest that animals' colorations are not necessarily permanent, but that they may be modified in time and space (35).

Vision researchers have also described a large number of equally fascinating vision systems, in which ecology and evolution have been recognized as important forces in generating such differences (64, 103). Variation can be detected at the eye, photoreceptor cell, visual pigment, and opsin structure levels. Many fish, frog, and bird species exhibit rearrangement of the positions of their photoreceptor cells depending upon light intensity (9, 103). Furthermore, physiologists have compiled extensive data on the absorption spectra of visual pigments from a diverse array of vertebrates (e.g. 63), which again reflect their light environments.

Molecular characterization of the human RH1, SWS, MWS, and LWS opsin genes (75, 77) opened the door for conducting genetic analyses of vision not only in human but also in other species. Molecular structures of the human LWS/MWS opsin genes and their association with red/green color blindness have revealed that unequal intragenic recombination between the two very similar genes is the major cause of color blindness (71, 74, 76, 77). Similarly, molecular analyses of the human RH1 opsin genes have identified more than 30 mutations associated with autosomal dominant retinitis pigmentosa, which causes a slow degeneration of the retina (74). With the rapid accumulation of nucleotide sequences of the opsin genes, we are now able to formulate experimental approaches to answer basic evolutionary questions: (*a*) How did different vertebrates attain their current dim and color vision? (*b*) How did paralogous genes achieve similar λmax values? (*c*) How did orthologous genes result in achieving different λmax values? Since these questions are closely related to the function of opsins, comparative sequence analyses also identify potentially important amino acid changes that may be responsible for shifts in λmax values of visual pigments (113). Thus, evolutionary approaches are central in understanding the ways in which visual pigments and photoreceptor cells achieve a wide range of wavelength absorption. Importantly, these hypotheses can be tested rigorously using site-directed mutagenesis, by expressing the mutagenized opsins in tissue culture cells, reconstituting with the chromophore 11-*cis* retinal, and measuring absorption spectra of the resulting visual pigments (24, 72, 73, 81, 94, 104, 109, 115). These in vitro results agree well with the data on absorption spectra of visual pigments obtained by many physiologists (49, 63), justifying the expression analyses of opsins in tissue culture cells.

So far, molecular genetic analyses of opsins, including site-directed mutagenesis studies, are concerned mostly with function. This is important because the function of opsins, including interactions between opsins and other proteins, plays a major role in phototransduction (for more details, see 110). However, to fully appreciate the evolution of animal coloration, we need to understand better both the genetic bases of body color and of color vision of interacting species. We are now at an exciting time when observations from many fields, providing very different techniques, can be used and synthesized to understand better the complexities involved in evolution. It is now feasible to evaluate the types of opsins and visual pigments associated with different environmental conditions, including their ecological backgrounds and interacting species. Such surveys will produce important organismal and molecular data to better understand how animals adapt to different environments.

To date, one of the best-established examples of adaptive change of proteins is that of crocodilian hemoglobin. This hemoglobin maintains high acidity when crocodiles stay under water for a prolonged period of time. This adaptation can be explained by only a few amino acid substitutions in key positions (87). The convergent evolution of stomach lysozyme of ruminants and foregut lysozyme of langur monkey can also be explained by five amino acid changes (96). However, it has not been easy to produce a clear-cut case of adaptive evolution in vertebrates at the molecular level. This is because it is difficult to find genetic systems in which the functional effects of mutations can be evaluated directly and unequivocally. Currently available data demonstrate clearly that the evolutionary changes of organisms' coloration and vision have been strongly controlled by natural selection. In this chapter, we have shown that such adaptation has occurred and can be studied at all levels, from molecules to morphology. Expression of different opsin mutants in cultured cells allows us to evaluate their effects on λmax rigorously, providing evolutionary biologists a rare opportunity to study adaptive evolution at the DNA, protein, cellular, and functional levels. In the future, the molecular information of a particular species needs to be related to the behaviors of its own and prey and predator species so that we can understand the evolution of each species and the effect of species' interactions.

ACKNOWLEDGMENTS

This work was supported by NIH grant GM42379.

Literature Cited

1. Applebury ML. 1990. Insight into blindness. *Nature* 343:316–17
2. Arnold K, Neumeyer C. 1987. Wavelength discrimination in the turtle, *Pseudemys scripta elegans. Vision Res.* 27:1501–11
3. Asenjo AB, Rim J, Oprian DD. 1994. Molecular determination of human red/green color discrimination. *Neuron* 12:1131–38
4. Aschoff J. 1981. *Handbook of Behavioral Neurobiology*. Vol. 4. *Biological*

Rhythms. New York: Plenum. 563 pp.

5. Avery JA, Bowmaker JK, Djamgoz MBA, Downing JEG. 1983. Ultraviolet sensitive receptors in a freshwater fish. *J. Physiol.* 334:23
6. Beatty DD. 1966. A study of the succession of visual pigments in Pacific salmon (*Oncorhyncus*). *Can J. Zool.* 44:429–55
7. Beatty DD. 1975. Visual pigments of the American eel, *Anguilla rostrata. Vision Res.* 15:771–76
8. Beatty DD. 1984. Visual pigments and the labile scotopic visual system of fish. *Vision Res.* 24:1563–73
9. Besharse JC, Iuvone PM, Pierce ME. 1988. Regulation of rhythmic photoreceptor metabolism: a role for post-receptoral neurons. *Prog. Retinal Res.* 7:21–61
10. Boynton RM. 1979. *Human Color Vision*. New York: Holt, Rinehart & Winston. 438 pp.
11. Bowmaker JK. 1977. The visual pigments, oil droplets and spectral sensitivity of the pigeon. *Vision Res.* 17:1129–38
12. Bowmaker JK. 1991. Evolution of photoreceptors and visual pigments. In *Evolution of the Eye and Visual Pigments,* ed. JR Cronly-Dillon, RL Gregory, pp. 63–81. Boca Raton, FL: CRC. 493 pp.
13. Bowmaker JK, Jacobs GH, Mollon JD. 1987. Polymorphism of photopigments in the squirrel monkey: a sixth phenotype. *Proc. R. Soc. Lond. B* 231:382–90
14. Bowmaker JK, Knowles A. 1977. The visual pigments and oil droplets of the chicken retina. *Vision Res.* 17:755–64
15. Bowmaker JK, Kunz YW. 1987. Ultraviolet receptors, tetrachromatic colour vision and retinal mosaics in the brown trout (*Salmo trutta*): age-dependent changes. *Vision Res.* 27:2102–8
16. Bowmaker JK, Thorpe A, Douglas RH. 1991. Ultraviolet-sensitive cones in the goldfish. *Vision Res.* 31:349–52
17. Brakefield PM. 1987. Industrial melanism: Do we have the answer? *Trends Ecol. Evol.* 2:117–22
18. Brower JVZ. 1958. Experimental studies of mimicry in some North American butterflies. Part 1. The monarch, *Danaus plexippus* and Viceroy, *Limenitis archippus archippus. Evolution* 12:32–47
19. Brower LP. 1969. Ecological chemistry. *Sci. Am.* 220:22–29
20. Brower LP, Cook LM, Croze HJ. 1967. Predator responses to artificial Batesian mimics released in a neotropical environment. *Evolution* 21:11–23
21. Burkhardt D. 1989. UV vision: a bird's eye view of feathers. *J. Comp. Physiol. A* 164:787–96
22. Cahill GM, Grace MS, Besharse JC. 1991. Rhythmic regulation of retinal melatonin: metabolic pathways, neurochemical mechanisms and the ocular circadian clock. *Cell Mol. Neurobiol.* 11:529–60
23. Cain AJ, Sheppard PM. 1954. Natural selection in *Cepea. Genetics* 39:89–116
24. Chan T, Lee M, Sakmar TP. 1992. Introduction of hydroxyl-bearing amino acids causes bathochromic spectral shifts in rhodopsin: amino acid substitutions responsible for red-green color pigment spectral tuning. *J. Biol. Chem.* 267:9478–80
25. Chiu MI, Zack DJ, Wang Y, Nathans J. 1994. Murine and bovine blue cone pigment genes: cloning and characterization of two new members of the S family of visual pigments. *Genomics* 21:440–43
26. Clarke B. 1960. Divergent effects of natural selection on two closely related polymorphic snails. *Heredity* 14:423–43
27. Cott HB. 1940. *Adaptive Coloration in Animals.* London: Methuen. 508 pp.
28. Crescitelli F. 1972. The visual cells and visual pigments of the vertebrate eye. In *Handbook of Sensory Physiology,* ed. HJA Dartnall, VII/1:245–363, Berlin: Springer-Verlag. 810 pp.
29. Denton EJ, Gilpin-Brown JB, Wright PG. 1970. On the "filters" in the photophores of mesopelagic fish and on a fish emitting red light and especially sensitive to red light. *J. Physiol.* 208:72P–73P
30. Denton EJ, Herring PJ, Widder EA, Latz MF, Case JF. 1985. The role of filters in the photophores of oceanic animals and their relation to vision in the oceanic environment. *Proc. R. Soc. Lond. B.* 225:63–97
31. Douglas RH, McGuigan CM. 1989. The spectral transmission of freshwater teleost ocular media—an interspecific comparison and a guide to potential ultraviolet sensitivity. *Vision Res.* 29:871–79
32. Dowling JE. 1987. *The Retina: An Approachable Part of the Brain.* Cambridge: Harvard Univ. Press. 282 pp.
33. Emmerton J, Delius JD. 1980. Wavelength discrimination in the "visible" and ultraviolet spectrum by pigeons. *J. Comp. Physiol.* 141:47–52
34. Endler JA. 1978. A predator's view of animal color patterns. *Evol. Biol.* 11:319–64
35. Endler JA. 1991. Variation in the appearance of guppy color patterns to guppies and their predators under different visual conditions. *Vision Res.* 31:587–608
36. Ferrari M. 1993. *Colors for Survival: Mimicry and Camouflage in Nature.*

Charlottesville, VA: Thomasson-Grant. 144 pp.
37. Fitzgibbon J, Hope A, Slobodyanyuk SJ, Bellingham J, Bowmaker JK, Hunt DM. 1995. The rhodopsin-encoding gene of bony fish lacks introns. *Gene* 164:273–77
38. Fleishman LJ, Loew ER, Leal M. 1993. Ultraviolet vision in lizards. *Nature* 365:397
39. Fogden M, Fogden P. 1974. *Animals and Their Colors*. New York: Crown. 172 pp.
40. Foster RG, Garcia-Fernandez JM, Provencio I, DeGrip WJ. 1993. Opsin localization and chromophore retinoids identified within the basal brain of the lizard *Anolis carolinensis*. *J. Comp. Physiol. A* 172:33–45
41. Goldsmith TH. 1980. Hummingbirds see near UV light. *Science* 207:786–88
42. Harosi FI. 1985. Ultraviolet-and violet-absorbing vertebrate visual pigments: dichromatic and bleaching properties. In *The Visual System,* ed. A Fein, JS Levine, pp. 41–55. New York: Liss. 198 pp.
43. Harosi Fl, Hashimoto Y. 1983. Ultraviolet visual pigment in a vertebrate: a tetrachromatic cone system in the dace. *Science* 222:1021–23
44. Herring PJ. 1983. The spectral characteristics of luminous marine organisms. *Proc. R. Soc. Lond. B* 220:183–217
45. Hinton HE. 1973. Natural deception. In *Illusion in Nature and Art,* ed. RL Gregory, EH Gombrich, pp. 6–159. London: Duckworth. 288 pp.
46. Hubel DH. 1988. *Eye, Brain, and Vision.* New York: Sci. Am. 240 pp.
47. Jacobs GH, Neitz J, Deegan HJF. 1991. Retinal receptors in rodents maximally sensitive to ultraviolet light. *Nature* 353:655–56
48. Jacobs GH. 1992. Ultraviolet vision in vertebrates. *Am. Zool.* 342:544–54
49. Jacobs GH. 1993. The distribution and nature of colour vision among the mammals. *Biol. Rev.* 68:413–71
50. Johnson RL, Grant KB, Zankel TC, Boehm MF, Merbs SL, Nathans J, Nakanishi K. 1993. Cloning and expression of goldfish iodopsin sequences. *Biochemistry* 32:208–14
51. Jones JS, Leith BH, Rawlings P. 1977. Polymorphism in *Cepea*: a problem with too many solutions. *Annu. Rev. Ecol. Syst.* 8:109–43
52. Jorgensen AL, Deeb SS, Motulsky AG. 1990. Molecular genetics of X chromosome-linked color vision among populations of African and Japanese ancestry: high frequency of a shortened red pigment gene among Afro-Americans. *Proc. Natl. Acad. Sci. USA* 87:6512–16
53. Kalmijn AJ. 1988. Hydrodynamic and acoustic field detection. In *Sensory Biology of Aquatic Animals,* ed. J Atema, RR Fay, AN Popper, WN Tavolga, pp. 83–130. New York: Springer-Verlag. 936 pp.
54. Kawamura S, Yokoyama S. 1994. Cloning of the rhodopsin-encoding gene from the rod-less lizard *Anolis carolinensis. Gene* 149:267–70
55. Kawamura S, Yokoyama S. 1995. Paralogous origin of the rhodopsinlike opsin genes in lizards. *J. Mol. Evol.* 40:594–600
56. Kawamura S, Yokoyama S. 1996. Phylogenetic relationships among the short wavelength-sensitive opsins of American chameleon (*Anolis carolinensis*) and other vertebrates. *Vision Res.* In press
57. Kettlewell HBD. 1973. *The Evolution of Melanism.* Oxford: Clarendon. 423 pp.
58. Knowles A, Dartnall HJA. 1977. Habitat, habit and visual pigments. In *The Eye,* ed. H. Davson, 2B:103–74. New York: Academic. 689 pp.
59. Kreithen ML, Eisner T. 1978. Ultraviolet light detection by the homing pigeon. *Nature* 272:347–48
60. Levine JS, MacNichol EF. 1982. Color vision in fishes. *Sci. Am.* 246:108–17
61. Locket NA. 1977. Adaptations to the deep-sea environment. In *Handbook of Sensory Physiology,* ed. F. Crescitelli, VII/5:67–192. Berlin:Springer-Verlag. 813 pp.
62. Loew ER, Lythgoe JN. 1978. The ecology of cone pigments in teleost fishes. *Vision Res.* 18:715–22
63. Lythgoe JN. 1972. The adaptation of visual pigments to their photic environment. In *Handbook of Sensory Physiology,* ed. HJA Dartnall, VII/1:566–603. Berlin: Springer-Verlag. 810 pp.
64. Lythgoe JN. 1979. *The Ecology of Vision.* Oxford: Clarendon. 244 pp.
65. Lythgoe JN. 1984. Visual pigments and environmental light. *Vision Res.* 24:1539–50
66. Marshall NB. 1954. *Aspects of Deep Sea Biology*. London: Hutchinson
67. Marshall NB. 1971. *Explorations in the Life of Fishes*. Cambridge, MA: Harvard Univ. Press. 204 pp.
68. Max M, McKinnon PJ, Seidenman KJ, Barrett RK, Applebury ML, et al. 1995. Pineal opsin: a nonvisual opsin expressed in chick pineal. *Science* 267:1502–6
69. Meijer JH, Rietveld WJ. 1989. Neurophysiology of the suprachiasmatic circadian pacemaker in rodents. *Physiol. Rev.*

69:671–707
70. Menzel R, Backhaus W. 1991. Colour vision in insects. In *The Perception of Colour,* ed. P Gouras, pp. 262–93. Boca Raton, FL: CRC. 314 pp.
71. Nathans J. 1987. Molecular biology of visual pigments. *Annu. Rev. Neurosci.* 10:163–94
72. Nathans J. 1990. Determinations of visual pigment absorbance: role of charged amino acids in the putative transmembrane segments. *Biochemistry* 29:937–42
73. Nathans J. 1990. Determinants of visual pigment absorbance: identification of the retinylidene Schiff's base counterion in bovine rhodopsin. *Biochemistry* 29:9746–52
74. Nathans J. 1992. Molecular genetics of human visual pigments. *Annu. Rev. Genet.* 26:403–24
75. Nathans J, Hogness DS. 1984. Isolation and nucleotide sequence of the gene encoding human rhodopsin. *Proc. Natl. Acad. Sci. USA* 81:4851–55
76. Nathans J, Piantanida TP, Eddy RL, Shows TB, Hogness DS. 1986. Molecular genetics of inherited variation in human color vision. *Science* 232:203–10
77. Nathans J, Thomas D, Hogness DS. 1986. Molecular genetics of human color vision: the genes encoding blue, green, and red pigments. *Science* 232:193–202
78. Neitz M, Neitz J, Jacobs GH. 1991. Spectral tuning of pigments underlying red-green color vision. *Science* 252:971–74
79. Okano T, Kojima D, Fukada Y, Shichida Y, Yoshizawa T. 1992. Primary structures of chicken cone visual pigments: vertebrate rhodopsins have evolved out of cone visual pigments. *Proc. Natl. Acad. Sci. USA* 89:5932–36
80. Okano T, Yoshizawa T, Fukada Y. 1994. Pinopsin is a chicken pineal photoreceptive molecule. *Nature* 372:94–97
81. Oprian DD, Asenjo AB, Lee N, Pelletier SL. 1991. Design, chemical synthesis, and expression of genes for the three human color vision pigments. *Biochemistry* 30:11367–72
82. Owen D. 1980. *Camouflage and Mimicry.* Chicago: Univ. Chicago Press. 158 pp.
83. Partridge JC, Archer SN, Lythgoe JN. 1988. Visual pigments in the individual rods of deep-sea fish. *J. Comp. Physiol. A* 162:543–50
84. Partridge JC, Douglas RH. 1995. Far-red sensitivity of dragon fish. *Nature* 375:21–22
85. Partridge JC, Shand J, Archer SN, Lythgoe JN, van Groningen-Luyben WAHM. 1989. Interspecific variation in the visual pigments of deep-sea fishes. *J. Comp. Physiol. A* 164:513–29
86. Perry RJ, McNaughton PA. 1991. Response properties of cones from the retina of the tiger salamander. *J. Physiol.* 433:561–87
87. Perutz MF. 1983. Species adaptation in a protein molecule. *Mol. Biol. Evol.* 1:1–28
88. Provencio I, Loew ER, Foster RG. 1992. Vitamin A2-based visual pigments in fully terrestrial vertebrates. *Vision Res.* 32:2201–8
89. Provencio I, Foster RG. 1993. Vitamin A2-based photopigments within the pineal gland of a fully terrestrial vertebrate. *Neurosci. Lett* 155:223–26
90. Register EA, Yokoyama R, Yokoyama S. 1994. Multiple origins of the green-sensitive opsin genes in fish. *J. Mol. Evol.* 39:268–73
91. Robinson J, Schmitt EA, Harosi FI, Reece RJ, Dowling JE. 1993. Zebrafish ultraviolet visual pigment: absorption spectrum, sequence, and localization. *Proc. Natl. Acad. Sci. USA* 90:6009–12
92. Rusak B, Bina KG. 1990. Neurotransmitters in the mammalian circadian system. *Annu. Rev. Neurosci.* 13:387–401
93. Saitou N, Nei M. 1987. The neighbor-joining method, a new method for reconstructing phylogenetic tree. *Mol. Biol. Evol.* 4:268–73
94. Sakmar TP, Franke RR, Khorana HG. 1989. Glutamic acid-113 serves as the retinylidene Schiff base counterion in bovine rhodopsin. *Proc. Natl. Acad. Sci. USA* 86:8309–13
95. Schertler GFX, Villa C, Henderson R. 1993. Projection structure of rhodopsin. *Nature* 362:770–72
96. Stewart C-B, Wilson AC. 1987. Sequence convergence and functional adaptation of stomach lysozymes from foregut fermenters. *Cold Spring Harbor Symp. Quant. Biol.* 52:891–99
97. Takahashi JS, Murakami N, Nikaido SS, Pratt BL, Robertson LM. 1989. The avian pineal, a vertebrate model system of the circadian oscillator: cellular regulation of circadian rhythms by light, second messengers and macromolecular synthesis. *Rec. Proc. Horm. Res.* 45:279–352
98. Underwood G. 1970. The eye. In *Biology of the Reptilia,* ed. C Gans, TS Parsons, pp. 1–97. New York: Academic. 374 pp.
99. Wald G. 1935. Carotenoids and the visual cycle. *J. Gen. Physiol.* 19:351–71
100. Wald G. 1955. The photoreceptor process in vision. *Am. J. Opthalmol.* 40:18–41

101. Wald G. 1968. Molecular basis of visual excitation. *Science* 162:230–39
102. Walls GL. 1934. The reptilian retina. *Am. J. Ophthalmol.* 17:892–915
103. Walls GL. 1942. *The Vertebrate Eye and Its Adaptive Radiation.* New York: Hafner. 785 pp.
104. Weitz CJ, Nathans J. 1993. Rhodopsin activation: effect of the metarhodopsin I-metarhodopsin II equilibrium of neutralization or introduction of charged amino acids within putative transmembrane segments. *Biochemistry* 32:14176–82
105. Whitmore AV, Bowmaker JK. 1989. Seasonal variation in cone sensitivity and short-wave absorbing visual pigments in the rudd *Scadinius erythrophythalmus. J. Comp. Physiol. A* 166:103–15
106. Wickler W. 1968. *Mimicry in Plants and Animals.* New York: McGraw-Hill. 253 pp.
107. Widder EA, Latz, MI, Herring PJ, Case JF. 1984. Far red bioluminescence from two deep-sea fishes. *Science* 225:512–13
108. Yau K-W. 1994. Phototransduction mechanism in retinal rods and cones. *Invest. Ophthalmol. Vis. Sci.* 35:9–32
109. Yokoyama R, Knox BE, Yokoyama S. 1995. Rhodopsin from fish, Astyanax: role of tyrosine 261 in the red shift. *Invest. Ophthalmol. Vis. Sci.* 36:939–45
110. Yokoyama S, Starmer WT. 1996. Evolution of G-protein coupled receptor superfamily. In *Human Genome Evolution,* ed. MS Jackson, G Dover, T Strachan. Oxford: Bios Sci. In press
111. Yokoyama R, Yokoyama S. 1990. Convergent evolution of the red-and green-like visual pigment genes in fish, *Astyanax fasciatus,* and human. *Proc. Natl. Acad. Sci. USA* 87:9315–18
112. Yokoyama S. 1994. Gene duplications and evolution of the short wavelength-sensitive visual pigments in vertebrates. *Mol. Biol. Evol.* 11:32–39
113. Yokoyama S. 1995. Amino acid replacements and wavelength absorption of visual pigments in vertebrates. *Mol. Biol. Evol.* 12:53–61
114. Yokoyama S, Starmer WT, Yokoyama R. 1993. Paralogous origin of the red-and green-sensitive visual pigment genes in vertebrates. *Mol. Biol. Evol.* 10:527–38
115. Zhukovsky EA, Oprian DD. 1989. Effect of carboxylic acid side chains on the absorption maximum of visual pigments. *Science* 246:928–30

Annu. Rev. Ecol. Syst. 1996. 27:569–95

MICROBIAL DIVERSITY: Domains and Kingdoms

David M. Williams and T. Martin Embley
The Natural History Museum, Cromwell Road, London SW7 5BD, United Kingdom

KEY WORDS: domains, Eucarya, Archaea, Bacteria, systematics,

ABSTRACT

With the discovery of the eukaryote nucleus, all living organisms were neatly divided into prokaryotes, which lacked a nucleus, and eukaryotes, which possessed it. As data derived directly from the genome became available, it was clear that prokaryotes were comprised of two groups, Eubacteria and Archaebacteria. These were subsequently renamed at the new taxonomic level of Domain as Bacteria and Archaea, with the eukaryotes named as the Eucarya Domain. The interrelationships of the three Domains are still subject to discussion and evaluation, as is their monophyly. Further data, drawn from various protein sequences, suggest conflicting schemes, and resolution may not be straightforward. Additionally, Bacteria and Archaea as well as Eucarya are largely based on organisms already in culture. Investigation of the potentially enormous quantity of uncultured organisms in nature is likely to have as broad-ranging implications as the exploration of new protein sequences.

> The jaguars, ants and orchids would still occupy distant forests in all their splendor, but now they would be joined by an even stranger and vastly more complex living world virtually without end.
>
> E. O. Wilson 1994. (109, p. 364)

INTRODUCTION

Ernst Mayr, in his *Growth of Biological Thought* (66), suggested that the Linnean classification of animals into vertebrates and invertebrates was a retrogressive step remedied by Cuvier who created instead six new classes of animals for the diverse invertebrate creatures: mollusks, crustaceans, insects, worms, echinoderms, and zoophytes. In addition, and perhaps of more revolutionary significance, Cuvier extended comparative zoology to the study of

0066-4162/96/1120-0569$08.00

the anatomy and classification of invertebrates, so that investigation of new features of invertebrate anatomy led to the discovery of new groups. According to Mayr (66), a second empirical revolution occurred with the introduction of the microscope, allowing not only a more detailed study of existing anatomical structures but the discovery of a diversity never before suspected.

As more of the microscopic living world was being discovered, efforts were extended to represent this diversity in a simple form relating all of life. Extending Darwin's wish to have life expressed as one phylogenetic tree, Haeckel (42) presented one of the first attempts at a genealogical classification of all organisms from this new evolutionary perspective—his now famous and often reproduced tree of life. Tree diagrams were used frequently before Haeckel, and they have remained the primary way to represent life's diversity, although they are now significantly less decorative than the Haeckelian "oak-trees."

Of significance to this review is not so much what Haeckel included in his tree of life but that which Haeckel probably never dreamt existed: the extensive microbial diversity hidden in the trunk of the "oak tree" under the humble name of Monera. The term "microbial" entered the literature to convey the idea of any group of taxa that happen to be microscopic (such as bacteria, unicellular protists, and unicellular fungi), and the term "protist" is usually used to convey the idea of a unicellular eukaryote. Hence, some overlap exists in the terminology, and phrases such as "microbial diversity" refer to all life that is neither plant, fungus, nor animal, while "protistan diversity" refers to all life that is neither "bacteria," plant, fungus, nor animal. The imprecision of these terms is indicative of the lack of knowledge in organizing the "lower" end of biotic diversity.

One of the useful aspects of Haeckel's (42) trees was that most of the branches were labeled and the relative closeness of the different branches offered a way of visualizing relationships. This notational aspect of trees has continued to the present when most if not all hierarchical taxonomic schemes can be rendered diagrammatically. In this review we retain the simple branching diagram as a way of depicting relationships with the understanding that taxon name and node on a tree are synonymous, in a fashion similar to the trees of Haeckel. What differs from the early diagrams of Haeckel and those of today is the direct connection between evidence (the data) and hypothesis (the tree), whereas sometimes Haeckel's trees were rather speculative.

Three properties of tree diagrams remain controversial:

1. What taxonomic rank should be assigned to the differing depth of bifurcating branches? That is, what is the empirical difference, if any, between a Kingdom, an Empire, and a Domain (10, 24, 67, 68, 108, 114, 115)?

2. What is the connection between the branching pattern and the possible process(es) that may have caused that pattern (11, 71, 72, 95, 118)?

3. What if the patterns of interrelationships are not truly branching and various organisms' histories are not accurately represented by a single branched diagram? As our focus is upon the relationships among taxa implied by particular trees we will not comment further on point (a) apart from noting that it appears largely to be a matter of convenience. Points (b) and (c) are inter-related and we comment on them below.

THE TREE OF LIFE

It was perhaps not until the mid 1900s that a significant division of the living world was made. The discovery of the unity of the eukaryote nucleus divided the world neatly into prokaryotes (those that lacked the nucleus) and eukaryotes (those that possessed it). This fundamental division of the living world has been significantly enhanced and elaborated with the arrival of DNA sequencing. Knowledge of the nucleotide base composition of DNA sequences allows direct access to data that can relate all of life, because all cellular organisms have a genome and share common molecules such as SSU (small subunit) rRNA (18). Access to these data has produced firmer ideas on the major divisions as well as on how those organisms are related. Like the revolutions before it, these data have presented as many surprises as solutions.

Life's Major Divisions

Woese and his colleagues (e.g. 112) were the first to investigate SSU rRNA data in a systematic way. They concluded that, as all organisms possessed SSU rRNA, this gene would be a perfect candidate for the universal chronometer of all of life. Their investigations led to the discovery that prokaryotes were composed of two separate groups, Eubacteria and Archaebacteria, quite distinct from Eukaryota. Eubacteria corresponded more or less with the traditional understanding of bacteria, while the new taxon Archaebacteria was composed of at least three quite different physiological groups: extreme halophiles, which grow in the presence of high salt concentrations, methanogens which metabolically reduce carbon compounds to methane, and extreme thermophiles that live at high temperatures and metabolize sulfur. Prior to Woese's (112) studies, Prokaryotes were simply (mis)understood as organisms that lacked the eukaryotic nucleus; it was clear what prokaryotes were not (they were not eukaryotes), but it was not clear exactly what they were.

Woese and colleagues (112) first recognized the three main divisions as Kingdoms, but they subsequently (114) renamed them at the new taxonomic level of Domain with Eucarya equivalent to Eukaryota, Bacteria equivalent to Eubacteria, and Archaea equivalent to Archaebacteria. Thus, currently recognized biotic diversity consists of three groups, two of which harbor microbial life exclusively (Archaea and Bacteria), while the third (Eucarya) contains a substantial quantity of microbial life (as unicellular protists). In the context of understanding biotic diversity, the relationships between and among "microbial" organisms present a significant challenge.

Although Woese (112) could delimit three major groups of cellular life, the data were silent as to which two of these three were more closely related to each other than to the third. What remained to be discovered was the order of appearance of these taxa.

Rooting the Tree

Haeckel's trees were notable in that they resembled a real tree complete with branches and trunk. Haeckel circumvented the problem of a universal root by naming the trunk after Monera, the taxon that contained all bacterial life known at that time; from within Monera the rest of life emerged. Trees from the analysis of DNA sequence data are primarily presented as unrooted. That is, with three taxa as terminals (such as Eucarya, Archaea, and Bacteria), there are three possible positions to place the root of the tree, yielding four possible interrelationships among the Domains: (*a*) Eucarya are more closely related to Archaea than they are to Bacteria, (*b*) Bacteria are more closely related to Archaea than they are to Eucarya, (*c*) Eucarya are more closely related to Bacteria then they are to Archaea, and finally (*d*) the interrelationships among Eucarya, Bacteria, and Archaea are unresolved or ambiguous, which simply restates the original problem.

If the three Domains circumscribe all of life, then only nonlife remains a "true" possibility for comparison. As a consequence trees from single genes such as SSU rRNA cannot be used to root trees but are used only to discover terminal groups. This seemingly intractable problem was first tackled by studies on ATPase genes (30, 48). The ATPases are composed of several different kinds of subunits, each related among themselves. The F-, V- and A-enzymes have catalytic and noncatalytic subunits thought to have arisen by an early gene duplication prior to the separation of the three Domains (30). Hence, the tree from the catalytic can be rooted with the noncatalytic (30, 48). The idea of rooting trees from gene duplications was first suggested by Schwartz & Dayhoff (94), subsequently developed by Fitch & Upper (22, 23) and generalized by Weston (107).

Using the subunits of ATPases, the root was placed on the Bacterial branch, making Eucarya and Archaea sister taxa (30, 48). However, accurately inferring gene duplications is possible only if all relevant taxa and their subunits are sampled. Originally, the ATPase subunit F- was thought to be present only in Bacteria, while the A/V subunits were thought to occur only in Archaea and Eucarya. However, subunit A/V has since been found in Bacteria (99a) and the F-subunit in Archaea (47, 90). Reanalysis of these data showed that the A/V ATPase indicate a close relationship between *Thermus* (formally in Bacteria) and some Archaea, rather than with Bacteria (99a); and the F-subunit indicated that *Methanosarcina* (formally in Archaea) was more closely related to Bacteria than to Archaea. Tsumtsumi et al (99a) inferred from the presence of archeal A/V-ATPase in Bacteria that all subunits would have been present in the common ancestor of both Bacteria and Archaea and that two gene duplications had occurred rather than one: one gene duplication responsible for the catalytic and noncatalytic forms and another duplication responsible for the A/V and F subunits in each of the catalytic and noncatalytic forms, thus effectively masking the "correct" root of the tree (25).

In response, Hilario & Gogarten (45) suggested horizontal transfer as a mechanism to account for the odd position of the anomalous archaeal and bacterial genes allowing the congruence between Eucarya and Archaea relative to Bacteria to remain, a possibility suggested by the SSU rRNA data. Difference of opinion rests upon interpretation of the resulting trees. For the ATPase subunits, either a number (one or two) of gene duplications have occurred, or the genes have been transferred horizontally across Domains (29, 31). While the ATPase story is complex and open to interpretation, further data were acquired that addressed to question of rooting.

One of the most convincing studies so far is that of Brown & Doolittle (5) who examined sequences from aminoacyl tRNA synthetases. These genes form a series of 20 enzyme families, each family the result of a gene duplication that supposedly predates the origin of the three Domains. With two separate gene families the resulting trees could be rooted in a fashion similar to that proposed for the ATPases. Analysis of the data and rooting the tree with the duplicated genes revealed convincing support for the sister group relationship of Archaea and Eucarya relative to Bacteria while it also recovered a monophlyletic Archaea.

Interrelationships Between Domains

Doolittle & Brown (18) summarized the data that addressed the question of interrelationships among the Domains from the perspective of sequence similarity. For the tree relating Eucarya and Archaea together relative to Bacteria, they listed in support Agininosuccinate synthetase (18), ATPase catalytic and noncatalytic subunits (30–32, 47, 48, 90), DNA polymerase B (18), elongation factor (EF) α/Tu and EF-G/2 (9, 13, 48), HMG-CoA reductase (18), ribosomal proteins L2, L3, L6, L10, L22, S9 and S10, and RNA polymerase II (50, 57, 84). For the tree relating Bacteria and Archaea together relative to Eucarya, they listed support from acetyl-coenzyme A synthetase, aspartate aminotransferase (18), citrate synthetase (18, 99), glutamate dehydrogenase II (2), glutamine synthetase (6, 51, 87), gyrase B, heat shock protein (HSP70) (39–41), and ribosomal proteins L11 and L17. For the tree relating together Eucarya and Bacteria relative to Archaea, they listed support from enolase, FeMn superoxidase dismutase, *his* C, glyceraldehyde-phosphate dehydrogenase (GAPDH) (64), malate dehydrogenase (MDH), phosphoglycerate kinase, and *trp* C. In the case of the glycolytic enzymes (enolase, GAPDH, and phosphoglycerate kinase), "horizontal transfer" has been invoked to explain the Eucarya/Bacteria relationship (70). Golding & Gupta (33) cite, in addition, deoxyribodipyrimidine photolyase, ferrodoxin, and pyrroline-5-carboxylate reductase as evidence for the Bacteria/Eucarya association. The data that do not distinguish between any of the competing hypotheses are carbamoyl-phosphate synthetase, prophobilinogen synthetase, ribosomal protein S15 (116), *trp* A, B, D and E. Although the preponderance of existing data support the tree relating Eucarya and Archaea relative to Bacteria, conflicting data are sufficient to suggest that this solution does not explain all of the data adequately. What of the conflicting data that support one of the alternative relationships such as the Bacteria/Archaea relationship and the Eucarya/Bacteria relationship? At least two explanations are possible: Either one or more of the Domains is not monophyletic, or one or more of the genes resulting in conflicting topologies are not orthologous.

MONOPHYLY OF THE DOMAINS

Of the 40 molecules that supply evidence of some kind for the interrelationships among the three Domains, Doolittle & Brown comment that for some "... at least one sequence did not support the monophyletic groups (B, A, and E) [B = Bacteria, A = Archaea and E = Eucarya] expected from rRNA data according to the neighbor-joining method." Of the 17 supporting the (B(E A)) tree, 2 apparently conflicted with the monophyly of the Domains; of the 9 supporting the (E (A B)) tree, 6 conflicted with the monophyly of the Domains; of the 7 supporting the (A (E B)) tree, 3 conflicted with the monophyly of the Domains; and of the 7 equivocal genes, 6 conflicted with the monophyly of the Domains. Thus, just under half of the molecules conflict with the accepted monophyly of Domains Bacteria, Archaea, and Eucarya.

Below we discuss some aspects of the various data that suggest a nonmonophyletic Archaea or Bacteria; Eucarya seems to be monophyletic with respect to all data sets.

The Monophyly of Archaea

Archaea are composed of extreme halophiles, methanogens, and extreme thermophiles (eocytes, 53, 56)—three quite different groups. The studies of Lake and his colleagues on the eocytes revealed characters of the ribosomes that prompted the idea that they were more closely related to Eucarya than to other

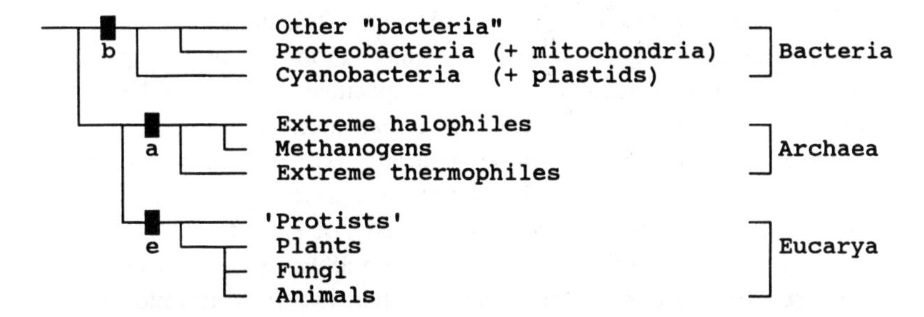

Figure 1 Archaebacterial tree. Relationships showing rooted tree with monophyletic Arachaea (node a). Abbreviations for other nodes are: b (for Bacteria) and e (for Eucarya).

Archaea or Bacteria (53, 56), contrary to published evidence from SSU rRNA sequence data which pointed to a monophyletic Archaea (77).

Subsequently, different analyses of SSU rRNA data led to two competing trees for the relationships among the Bacteria, Archaea, and Eucarya, of which one solution suggested the nonmonophyly of the Archaea (56).

The standard interpretation, called the "archaebacterial" tree, retains all the Domains as monophyletic. Within Archaea are two divisions: the Euryarcheota (114) composed of methanogens and halophiles, and Crenarcheota (114) composed of the eocytes which branch independently at the base of the archaeal group (Figure 1).

The alternative arrangement, obtained using a new method to analyze the data called evolutionary parsimony (52), is called the eocyte tree. Rather than three divisions, Lake et al (53, 56) suggested two. Parkaryotae (53) is composed of the extreme halophiles Bacteria and the methanogens; Karyotae (53) is composed of eocytes and Eucarya. Thus, while Bacteria and Eucarya remain monophyletic, Archaea do not (Figure 2).

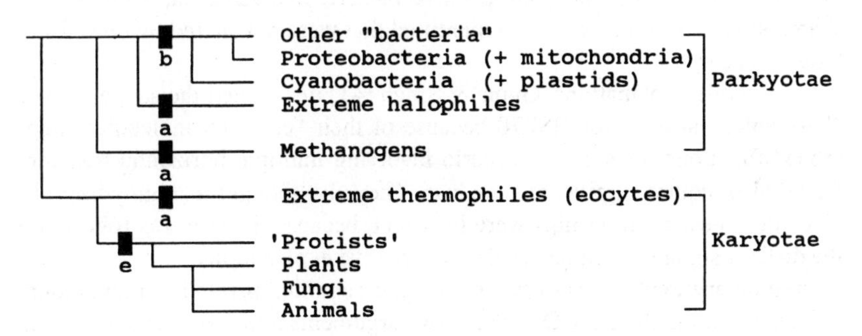

Figure 2 The "eocyte" tree. Relationships showing rooted tree with nonmonophyletic Archaea (nodes a); other abbreviations as in Figure 1.

In the most ambitious study undertaken so far (17a), analysis of about 3000 SSU rRNA sequences gave ambiguous results. In analyzing data, particular models of presumed character change can be specified. When the Jukes-Cantor (49) model was applied to the data, the Archaebacterial tree was recovered; when the Jin-Nei (48a) model was applied, the "eocyte" tree was recovered. de Rijk et al (17a) consider the Jin-Nei correction to be a more realistic approximation. In addition to the SSU rRNA study, de Rijk et al (17a) analyzed 225 LSU (large subunit) rRNA sequences (from 38 Eucarya, 16 Archaea, and 171 Bacteria; of the Bacteria, 31 were from plastids, 76 from mitochondria) which suggested the "Archaebacterial" tree regardless of whether the Jukes-Cantor or Jin-Nei correction was used. Yet, according to de Rijk et al (17a), none of the trees from the SSU or LSU RNA sequences could be considered robust, despite considerable effort on their part in collecting data and refining methodology.

While the RNA data appears ambiguous, new data from aminoacyl-tRNA sythetase genes (5) and TATA binding proteins (86) as well as data from the ATPases (30, 48) and elongation factor (EF) proteins (30) also suggest that Archaea are monophyletic.

The Monophyly of Bacteria

The most significant evidence for a nonmonophyletic Bacteria comes from the studies of the 70-kDa heat shock proteins (HSP70) (39–41). Heat shock proteins are understood to be among the most conserved proteins known, HSP70 being the most conserved among them (3). HSP70 is present in the cytoplasm, endoplasmic reticulum, mitochondria, and chloroplasts (in prokaryotic cells it is referred to as *Dna*K). Gupta & Singh's (41) analysis of 36 genes from 25 taxa suggested that, rather than a monophyletic Bacteria and Archaea, some gram-positive bacteria are more closely related to some Archaea, and some gram-negative bacteria are more closely related to Eucarya. With respect to the relationships of some gram-positive bacteria and Archaea, a number of shared signature sequences were identified that were not shared by other Bacteria or Eucarya.

Of possible explanations, Gupta & Singh (41) discounted the suggestion of "horizontal transfer" for HSP70 because of their "essential molecular chaperone" functions (cf 46a); a scenario involving multiple horizontal transfers would also require significant gene loss. They also thought it "highly improbable" that these relationships were incorrect, because the tree was robust and the distinct signature sequences defy alternative explanations.

In summary, evidence is sufficient to suggest conflicting relationships among, as well as within, the three Domains. Two arguments have been used to explain these conflicting data: (*a*) All the trees are correct and origin of the eukaryote cell is the result of a complex fusion of ancient genomes; or (*b*) the trees are incorrect because the various genes being compared are not homologous among themselves.

HOMOLOGY: EXPLANATION OR DISCOVERY?

Trees with Different Histories?

Initial explanations of conflicting topologies were prompted by the desire to reconcile the well-sampled RNA tree (still considered to be a universal chronometer for the origin of life—18) with the newer, less-well-sampled conflicting protein trees—all that would be required is to explain the differences. Not all genes necessarily reflect the history of organisms. To reconcile the various trees, different molecular histories need to be separated before the origin of the organisms can be reconstructed with reference to their actual history, presumed by many to be captured by the SSU rRNA data. However, according to Doolittle (17b), a shift of emphasis has occurred, so that efforts are now increasingly directed to "provide an explanation for why some archaebacterial sequences are like eukaryotic ones and other's eubacterial," as well as why some eukaryote genes are related to bacterial genes rather than to the expected archaeal genes.

In their original proposal of the interrelationships among the three Domains, Woese & Fox (112) suggested that the origin of the eukaryotic cell was probably via an amalgamation of an RNA-based progenote (a term proposed for the primitive hypothetical ancestor of prokaryotes and eukaryotes, 112) and various additional endosymbionts, the latter becoming organelles, the former donating the RNA translational machinery. The notion of an endosymbiotic origin of organelles, and consequently the origin of new taxa by these means, has had a long if meandering history (89). Revived in semi-modern form by Margulis (63), the idea established itself as credible after Schwartz & Dayhoff (94) reviewed the various molecular data available at that time. It now seems beyond doubt that both chloroplasts and mitochondria have their relationships among Bacteria, where chloroplast sequences are related to cyanobacteria and mitochondrial sequences are related to α-proteobacteria (37).

Extending the idea of genome fusion to other components of the eukaryotic genome was a simple matter (41, 43, 44, 56, 84, 96, 117). For instance, with respect to the implications of the eocyte tree, Lake et al (56) suggested that an engulfed organism, in his view an eocyte from an early endosymbiotic event, donated the ribosomal subunits to the eukaryotic cell, rather than the other way around.

With respect to the implications of the "archaebacterial" tree, to explain why Archaea appear more closely related to Bacteria than to Eucarya in the SSU rRNA tree, Sogin (96) suggested that the eukaryotic cell arose by fusion of an RNA progenote and an ancestral archaebacterium (not necessarily an eocyte) with a DNA genome. Pühler and colleagues (84) proposed a fusion hypothesis for the eukaryote cell, but between a eubacterium and an archaebacterium (again, not necessarily an eocyte); Gupta and his colleagues (41) offered a fusion scenario with a possible endosymbiont engulfment involving a gram-negative bacterium and an eocyte.

Each fusion scenario assumes all the trees to be correct for their recovered relationships. Forterre et al (25) noted that, if a particular protein phylogeny does not correspond with the prevalent view relating Archaea more closely to Eucarya than to Bacteria, then "either lateral gene transfer, altered rates of evolution or diverse fusion hypotheses have been proposed." Forterre et al (25), using data from DNA topoisomerases, EF genes, and the ATPases, concluded that there was only sufficient information to support the three Domains and that "proteins from archaebacteria, eubacteria and eukaryotes are more or less equidistant. . . ." Further, they suggested that all contradictions were caused by analyzing "a mixture of orthologous and paralogous proteins."

Fusion hypotheses rely upon the idea that horizontal transfer of genetic material between genomes of different lineages is possible. The relationships among genes that arose by horizontal transfer have been called xenologous (36). Sorting xenologous genes from paralogous and orthologous genes may be problematical, as in the case of the ATPase subunits. What exactly is the difference between orthology, paralogy, and xenology? Orthology and paralogy are considered "kinds" of homology (82) as is xenology (36). If these different categories of homology can be separated, it may be possible to substantiate the idea that there is a common, albeit complex, history for all genes.

Trees with Common Histories?

All data so far collected may indeed be telling the same story, and differences could be a methodological artifact. Two arguments have been offered in support of this view. First, as there is a diversity of tree-building methods to choose from, coupled with the likelihood that the processes behind DNA evolution will confound analysis even further, some tree-building methods may yield incorrect results. Second, there is the difficulty suggested above of discriminating between orthologous, paralogous, and xenologous sequences, confusion of which may also yield incorrect results with respect to species relationships.

Tree-Building Methods

A large number of tree-building methods are now available; some may differ more in detail than in substance. Whatever the differences in approach, tree-building methods are becoming increasingly sophisticated (e.g. 54, 55, 58, 60). Efforts to refine analysis may illuminate some areas of conflict. But, as Lake (54) recently noted, "... like all [tree-building] algorithms which have been tested, they [the resulting trees] were strongly affected by alignment biases." Aligning sequences allow base nucleotide identity to be proposed, and the accuracy with which sequences can be successfully aligned is a crucial first step, almost independent of the analytical method finally adopted. In short, poorly aligned sequences may be the cause of spurious results, rather than inappropriate analytical methods. For instance, Lake (53a) experimented with various alignment protocols and tree-building methods for elongation factor (EF-Tu) protein data, discovering that results were dependent on alignment. Both the archaebacterial and eocyte tree could be recovered from the same sequences if they were aligned differently.

Nevertheless, one expects a certain amount of signal to emerge from any dataset, and efforts to unravel in detail the mysteries of DNA evolution prior to analysis may serve only to invoke further "'hand-waving' arguments about accelerated rates of change in eukaryotic rRNAs against a backdrop of constant or even reduced rates of change in their protein coding regions" (96). Stubbornness to yield meaningful results may simply mean that the data are inappropriate for the questions addressed (for instance 25), in spite of efforts to "turn to a model that would help to wring truth from recalcitrant data" (83).

Orthology, Paralogy, and Xenology

Orthology, paralogy, and xenology have all been understood as kinds of homology, and their discrimination yields the most common explanations for conflict in competing trees.

Homology may appear more complex for molecular data than for morphological data as the trees may reflect the descent of species or the descent of genes. Orthologous sequences are the relationships among genes that reflect the descent of species and seem directly equivalent to the term homology as used in morphology (81). Paralogous sequences are said to reflect only the descent of genes rather than species. Xenologous sequences reflect in part the descent of species as they will be incongruent with species relationships recovered from orthologous genes. Thus, the protein sequences that are said to have arisen in a particular organism by "horizontal transfer" are xenologues.

The usual meaning of homology is common ancestry, and to a certain extent, common ancestry is a possible explanation for orthology, paralogy, and

Table 1 Comparison of molecular and morphological relations and their performance in Patterson's tests for homology (+) = pass; and (−) = fail, ... = no equivalent relation (28a, 81, 82).

Morphology	Congruence	Similarity	Conjunction	Molecules
Homology	+	+	+	Orthology
Homonomy	+	+	−	Paralogy
Complement	+	−	+	Complement
2 homologies	+	−	−	2 orthologies
Parallelism	−	+	+	Xenology
?	−	+	+	Synology
Multiparallel	−	+	−	Paraxenology
Homeosis	−	+	−	Pleurology
Convergence	−	−	+	convergence
Endoparasitism	−	−	−	......

xenology. However, common ancestry can serve only as explanation. What of the discovery of orthology and its partners? Patterson (81, 82) suggested three tests for their discrimination: similarity, conjunction, and congruence (Table 1).

Similarity allows a comparison between sequences to be made in the first place. Two or more sequences can be aligned and their similarity quantified; one may say that two sequences are 100% similar if all bases are identical, 50% similar if only half the bases are identical, and so on. Although this "measure" was once seen as synonymous with homology, it became clear that similarity on its own cannot distinguish between paralogy, orthology, or xenology (82).

Patterson's (82) conjunction test is used in morphology to distinguish between two structures deemed to be homologous because they are similar enough to merit further comparison. If both structures are present in the same organism at the same time, such as hairs or blood cells, they are generally referred to as serial homologues (82). In the case of sequence data, the presence of two genes on the same genome deemed to be similar enough to warrant further comparison may indicate gene duplication and help to identify paralogy, as in the case of the ATPases.

Congruence is the most decisive test in both morphology and molecules because it recognizes coincident features or genes respectively at coincident nodes on a particular tree as homologous (orthologous for molecules) (82). However, for molecular data this requires comparison of trees. For instance, Doolittle & Brown listed genes that support the different hypotheses of relationships among Bacteria, Archaea, and Eucarya. The hypothesis that Archaea and Eucarya are more closely related to each other than they are to Bacteria received support from seven different proteins. These seven correspond and are thus congruent for that particular hypothesis. The remaining genes conflict with this overall hypothesis. What can be said of the proteins whose trees conflict? In morphology, features that do not correspond are considered as instances of homoplasy; the features in question have a poor fit to the tree (or at least a poorer fit than the homologies). In molecular data, this suggests that paralogous and xenologous proteins also have a poor fit to the tree of all congruent proteins and both are instances of molecular homoplasy (81). In this respect paralogy is like orthology, differing by appearing at more than one node of a particular tree but not conflicting with it (75): Orthology and paralogy are both gene trees, with only orthologues tracking species relationships exactly. Invoking enough numbers of gene duplications will reconcile any gene tree with the congruent orthologous genes. Hence, conflict is eliminated or explained away. Xenology seems equivalent to parallelism in morphology (82), where a similar morphological structure appears on two different branches of a tree and conflicts with the branching order as recovered by the homologies. For instance, chloroplasts are an essential feature of photosynthetic organisms and are a property of those organisms, but the relationships of chloroplasts are among the Cyanobacteria, whereas the relationships of the organisms are among other eukaryotes (37). In this case, explanation by horizontal transfer seems reasonable, and the relationship is xenologous (37). A more controversial example is the origin of the ATPases (28a, 45).

With respect to xenology, Gogarten (28a) recently introduced the term synology for genes homologous due to possible endosymbiotic fusion events. Gogarten's inspiration was to separate homology due to simple lateral transfer (xenology) from homology due to "lineage-fission" (synology). Gogarten (28a) suggested that while xenology contributes nothing to our further understanding of taxon interrelationships, synology, which indicates a particular evolutionary event such as eukaryotic cell fusion, is useful. However, beyond explanation of a postulated event, it seems analytically impossible to separate synology from xenology, and fusion events remain explanations of conflict (Table 1).

Trees of taxon relationships from different proteins can be compared with each other, and orthology, paralogy, and xenology can be read from the nodes of a composite tree in the same way that homology, parallelisms, and convergences can be read from morphological trees. Homology (orthology and paralogy) and homoplasy (parallelisms and convergence, and xenology) are properties of trees and not data. Gene duplication explains away paralogous genes; horizontal transfer explains away xenologous genes.

It is beyond the scope of this article to detail all the possible ramifications of molecular homology. Suffice it to note that consideration of trees of relationships that may have differing histories is not a new problem and has been investigated with respect to biogeography (76, with geographic paralogy as a confounding factor) and parasites and their hosts (80) as well as to the more conventional questions of coevolution (4). Indeed, it is becoming such a broad topic in systematic enquiry that two of the reviews from the 1995 volume of the *Annual Review of Ecology and Systematics* dealt with the topic (17, 72a).

Data and their analysis are the most significant factors. From what source should those data come to address relevant problems? Some of these issues are explored below with reference to the effects on particular parts of the tree.

EXPLORATION OF UNCULTURED MICROBIAL DIVERSITY USING rRNA SEQUENCES

Microbiologists have always considered it possible that cultured microbial diversity would turn out to be only a small fraction of microbial diversity in nature. The belief stemmed from major discrepancies between what could be counted using microscopy and what could be isolated in laboratory culture (35). Selective isolation selects for those members of a prokaryote population that can grow under the particular conditions imposed. Methodologies whereby the hypothesized uncultured microbial diversity could be tackled came out of the same research program that had used SSU rRNA sequences to infer phylogenetic relationships between microorganisms that were in laboratory culture. Norman Pace and his colleagues (79) first suggested that it should be possible to isolate community RNA genes from any sample and to separate individual genes for subsequent phylogenetic analysis by gene cloning methods. The gene library produced would represent a sample of the SSU rRNA genes present in the environment at the time of sampling. Incorporation of these sequences into a phylogenetic analysis would permit the relationships of the uncultured taxa to the cultured taxa to be inferred within the framework of the SSU rRNA gene tree, and equally importantly, it would allow any new groups to be recognized.

It is worth briefly describing the methodology so that some of its strengths and limitations can be appreciated. The choice of SSU rRNA as a tool for making surveys of prokaryote diversity has been dictated by the large database of sequences for comparison, and by certain desirable features of its sequence (113). Some regions of SSU rRNA sequence are strongly conserved in all studied examples of cellular life, the inference being that they are ancestral; the mild expectation is that they will be present more or less unchanged in cellular life still to be discovered. They serve as universal polymerase chain reaction priming sites to allow a relatively unbiased sampling of extant diversity for taxonomic analysis, through the amplification and recovery of different SSU rRNAs with equal efficiency. In reality, the efficiency of PCR in sampling potentially complex mixtures is affected by, for example, gene concentration within individual cells (20) or within mixtures (103) and by primary sequence features internal to the priming sites (85).

Other features of SSU rRNA facilitate empirical tests for the presence of a sequence in a sample and allow estimates of abundance to be made. Highly variable regions of SSU rRNA sequences are good targets for specific probes and can be used to infer the relative abundance of a specific sequence in a sample by calibration with other more general probe responses (28, 97). A fluorescent dye attached to a probe, termed a phylogenetic stain (15), can be used to visualize intact cells in complex mixtures. This last technology provides the expectation that individual microorganisms that carry phylogenetically interesting SSU rRNA sequences can be identified and, in conjunction with micro manipulation and selective isolation, can be recovered into culture (e.g. 46). Thus, while the effects of environmental sampling have so far been made almost entirely via the topology of the SSU rRNA gene tree, it should in future be possible to broaden sampling for other genes from lineages (taxa) that appear to be especially interesting from their positions in the SSU rRNA tree.

Impact of Molecular Methods on Archaeal Diversity at Higher Levels

Within the SSU rRNA tree for Archaea, two kingdoms, the Crenarchaeota and Euryarchaeota, have been identified by Woese and coworkers (114). The two kingdoms are defined with reference to the root relative to Bacterial outgroup sequences, which split Archaea into two putative monophyletic groups, and by sequence idiosyncrasies that can be used to distinguish them. The main reason given for this rather bold step of naming two kingdoms (114) was the recognition that the few cultured and sequenced Archaea formed a relatively simple structure that could be divided into two and would serve as a focus for research, as "the kingdoms within the Archaea had never had appropriate names of any kind" (114). Thus the Crenarchaeota and the Euryarchaeota are defined exclusively using features of SSU rRNA sequences, and environmental sampling of Archaeal sequences has the potential to affect this interpretation.

The first two publications dealing with the investigation of uncultured Archaeal diversity using SSU rRNA sequences appeared almost simultaneously (14, 26) and reported much the same story. Deep and shallow water samples from the Pacific and Atlantic revealed the same two novel groups (termed marine Archaea group I and group II, Figure 3) of highly divergent Archaeal sequences (see also 27). Subsequent analysis has revealed the presence of Group I sequences in the stomachs of holothurians from the Atlantic abyss (69) and from coastal and deep waters off the west coast of Ireland (M Mularkey,

J McInerney, R Powell, personal communication). All of these sequences form a strongly supported monophyletic group with previously published group-I marine Archaea in maximum parsimony, maximum likelihood, and distance matrix analyses. A recent survey of sequence diversity within a single paddy field in Japan has reported Archaeal sequences that formed a sister group to the marine Archaea group I, suggesting their occurrence in nonmarine systems (100). Unfortunately, the sequence fragments for the paddy field clones are rather short (c. 300 bases), which has made assessment of their detailed position relative to other clones uncertain.

These data, from small samples taken over vast geographic distances, suggest that Group I and II marine Archaea are abundant components of marine picoplankton, a suggestion confirmed by recent data which suggest that at certain times of the year marine Archaea comprise about 34% of the prokaryote biomass in coastal Antarctic surface waters (16). The data also suggest that pelagic marine Archaea are broadly represented by these two groups. As suggested by Mullins and co-workers (73), this evident limited diversity, in the sense of the number of new monophyletic groups so far discovered, has the potential to greatly facilitate analysis of community dynamics using nucleic acid probes.

Analysis of the new environmental sequences can be expected to affect interpretation of Archaeal diversity based upon the SSU rRNA tree in at least two ways pertinent to our discussion: for example, by the discovery of higher order groupings and their potential to disrupt relationships between other such groups, and by the discovery of new combinations of sequence characters that affect the interpretation of what have been previously identified as group specific signatures.

→

Figure 3 Tree based upon SSU rRNA sequences from representative cultured Archaea including sequences from clone libraries prepared from different environmental samples. The tree is based upon analysis of 791 sequence positions using the Jukes & Cantor (49) correction and Neighbor Joining (88). Essentially the same topology was recovered in maximum parsimony analysis. Interestingly, analysis using the LogDet method (60) designed to deal with unequal base compositions between sequences, as occurs here, produced a similar topology but consistently placed environmental clones env.pJP27 and env.pJP78 on the Eucarya branch. [Key: env.pJP, clones from mud volcano area of Jim's Black Pool, hot spring , Yellowstone National Park (1); env.carna, Carna Bay bacterioplankton, west coast of Ireland (unpublished, M Mularkey, R Powell, University of Galway); env. pPM7, mid Atlantic sample (unpublished, J McInerney, R Powell); env.ANT, Antarctic bacterioplankton (16); env.WHAR, Woods Hole DNA clone (14); env.SBAR, Santa Barbara Channel bacterioplankton (14); env.C and env.P, coastal salt marsh clones (unpublished, M Munson, D Nedwell, TM. Embley). Published aligned sequences were obtained from the Ribosomal Data Base Project (62).]

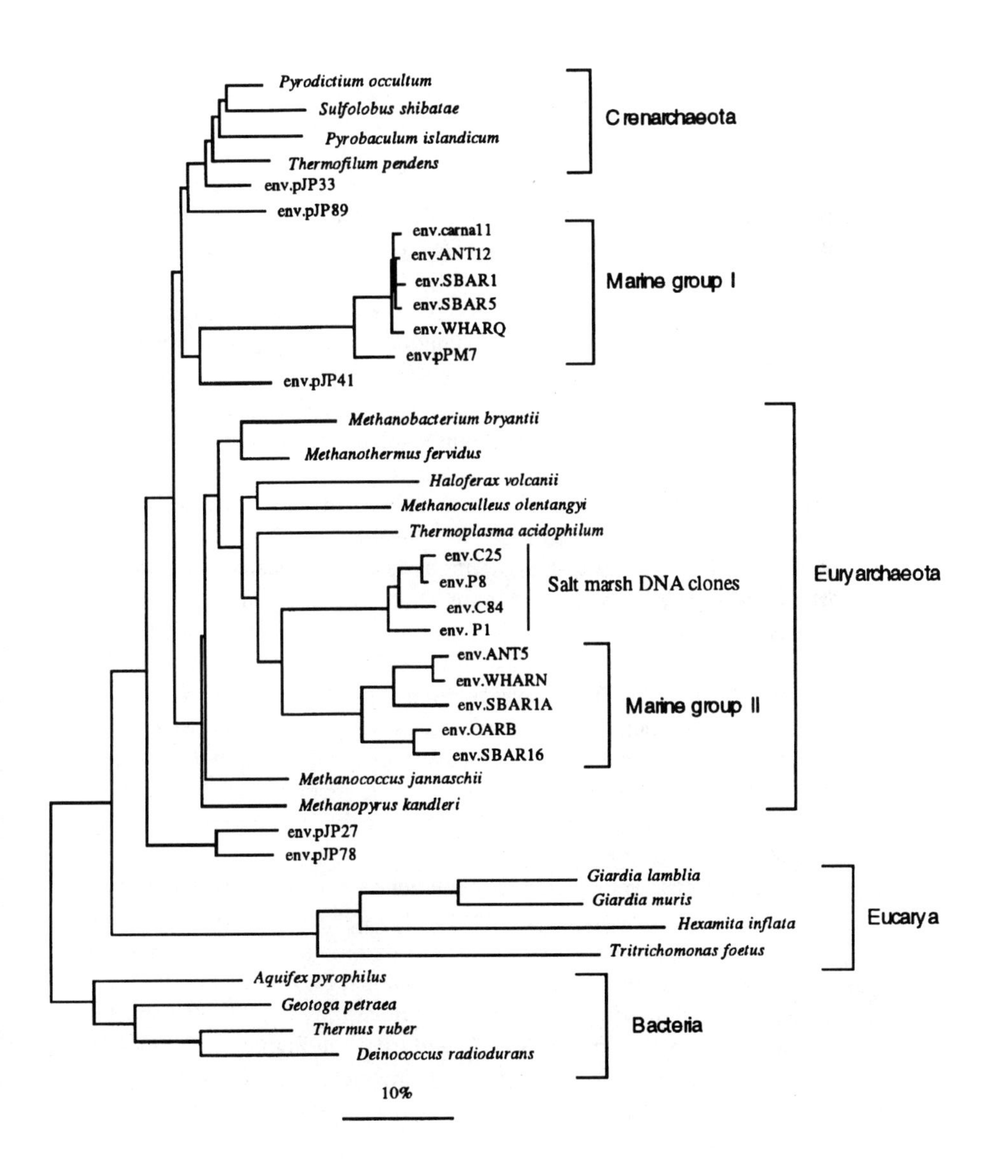
Pyrodictium occultum
Sulfolobus shibatae
Pyrobaculum islandicum
Thermofilum pendens
env.pJP33
env.pJP89
Crenarchaeota
env.carnal1
env.ANT12
env.SBAR1
env.SBAR5
env.WHARQ
env.pPM7
Marine group I
env.pJP41
Methanobacterium bryantii
Methanothermus fervidus
Haloferax volcanii
Methanoculleus olentangyi
Thermoplasma acidophilum
env.C25
env.P8
env.C84
env. P1
Salt marsh DNA clones
env.ANT5
env.WHARN
env.SBAR1A
env.OARB
env.SBAR16
Marine group II
Methanococcus jannaschii
Methanopyrus kandleri
Euryarchaeota
env.pJP27
env.pJP78
Giardia lamblia
Giardia muris
Hexamita inflata
Tritrichomonas foetus
Eucarya
Aquifex pyrophilus
Geotoga petraea
Thermus ruber
Deinococcus radiodurans
Bacteria
10%

Analysis of the marine Archaea group I sequences has so far failed to resolve their detailed higher order relationships (1, 14, 16, 26). McInerney et al (69) used an alignment of 819 nucleotide positions, including all planktonic and abyssal sequences for which data were then available (some are only short fragments). The taxonomic scope of the alignment was expanded to include Bacterial and Eucaryal sequences encompassing a range of G + C content values in an attempt to counter possible analytical bias due to thermophilic convergence among sequences. Parsimony and distance matrix analysis using all substitutions or transversions yielded only three conflicting hypotheses of a sister group relationship between marine Archaea group I sequences and (i) Archaea, (ii) Crenarchaeota, and (iii) Eucarya (with the Crenarchaeota closer to the Eucarya as in the eocyte tree). Bootstrap analyses indicated that none of these competing hypotheses was well supported. DeLong (14) searched for confirmation of the phylogenetic position of Group I marine Archaea from signature analysis, but with no conclusive resolution of the problem. The group-I marine Archaea possessed only 7 of 12 comparable Crenarchaeota signatures identified by Winker & Woese (110).

The position of the group II marine Archaea as a major lineage within the Euryarchaeota has been indicated in most published trees (14, 16, 27), and a specific relationship to the wall-less thermophile *Thermoplasma acidophilum* has commonly been inferred (as in Figure 3). Sequences related to group II marine Archaea have recently been detected in coastal salt marsh sediments, which are sometimes covered during high tides (M Munson, D Nedwell, TM Embley, unpublished data). The precise relationship of the salt marsh sequences as sister group to the group II marine Archaea (as in Figure 3), or as sister group to both the group II marine Archaea and *Thermoplasma acidophilum*, is not clearly resolved. Analysis of signatures for some members of the group II marine Archaea reveals that they share 8 of 12 compared positions previously identified as Euryarchaeota signatures (14).

A second major discovery of Archaeal relationships came from Barns, Pace and co-workers (1), who used primers designed to selectively retrieve a variety of Archaeal SSU rRNA sequences from 5 ml of sediment taken from a hot spring ("Jim's Black Pool") in Yellowstone National Park. Several of the recovered sequences showed no close phylogenetic affinity to cultivated species of Archaea (pJP33, pJP41, pJP89, pJP27, and pJP78 in Figure 3). These five sequences branched closer to the root of the Crenarchaeota stem than did sequences from cultured Crenarchaeota (1). Clone pJP89, the position of which with the Crenarchaea was strongly supported in 90% of maximum likelihood trees, contained 8 Crenarchaeota signatures and 7 Euryarchaeota signatures, of 16 positions sampled. It is interesting that, using bootstrapping, Barns et al (1) reported that pJP27 and pJP78 formed a weakly supported sister group to the marine Archaea group I sequences in maximum parsimony, distance matrix, and maximum likelihood analyses. In other analyses, pJP27 and pJP78 were the sister group to all other Archaea including the marine sequences (as in Figure 3 and as in the RDP maximum likelihood tree, 62). Signature analysis carried out by Barns et al (1) revealed that pJP27 and pJP78 shared only 6 of 16 recognized Crenarchaeota signatures, but 8 of 16 Euryarchaeota signatures. Barns et al (1) interpreted this mixture as consistent with the hypothesis that they constitute a distinct lineage from the marine Archaea group-I that, an expanded analysis demonstrated, contained 10 of 16 Crenarchaeota signatures.

It is already apparent that the few environmental sequences available so far have the capacity to change the interpretation of Archaeal relationships based upon 16S rRNA sequences. The new sequences have brought new combinations of bases at sites previously used to distinguish between the two established kingdoms of Archaea. It is clear from their papers that Winker & Woese (110) regarded the signatures defining the Crenarchaeota and Euryarchaeota not as autapomorphies for these groups but as characteristic compositions that could distinguish them within the framework of the Archaea. Whether the claim that they are truly characteristic will survive examination of more environmental data remains to be seen, especially if the pace of discovery of new Archaea lineages continues. The group I marine Archaea and the hot spring clones JP27 and 78 form sufficiently distinct groups in the 16S rRNA tree that it is possible, if the process of naming continues, each will need to be considered as a new grouping akin to the Crenarchaeota and Euryarchaeota (78). The status of pJP89, with its interesting combination of signatures and its potential position as sister group to other Creanarchaea, is also of interest. One can argue from the perspective of traditional microbiological practice (74) that it would be desirable to have viable cultures representing each of these, before they are formally named as new taxonomic groups. This seems to us a sensible position because living cultures have advantages in allowing other genes and physiology to be studied and thus providing an important resource for future study. However, within the logic of defining Archaea taxonomic groups on the sole basis of 16S rRNA sequence features (110), these potential new taxa can be described and recognized just as readily as can be cultured taxa. Interestingly, cloned SSU rRNA genes from marine group I Archaea were recently used to identify DNA fragments from these organisms in a genomic library prepared from environmental DNA (98). Further sequencing identified an elongation factor EF-2 sequence, which when analyzed with a small sample of other elongation factor sequences, supported a sister group relationship between Marine Archaea group I and three Crenarchaeota sequences.

It can be expected that further changes in the organization and appearance of the 16S rRNA tree for Archaea will occur as sequences accumulate from new environmental analyses and from the analysis of cultured taxa with new combinations of phenotypic features such as *Methanopyrus kandleri*, an extremely thermophilic methanogen that branches deep within the Euryarchaeota (8).

BACTERIAL HIGHER ORDER DIVERSITY

In summarizing his own work and that of his coworkers over a period of approximately 20 years, Carl Woese (111) recognized 12 major bacterial groups (which he termed phyla) on the basis of the analysis of over 500 partial (i.e. ribonuclease T1 catalogues) and approximately 50 complete 16S rRNA sequences. These 12 phyla (Figure 4), with over 1500 full-length prokaryote 16S rRNA sequences (62), have now been supplemented by several new groups of roughly equivalent phylogenetic depth. Woese and coworkers (114) have suggested that most of these could eventually be classified as new kingdoms.

The phylum containing *Aquifex* and *Hydrogenobacter* is of particular interest since it is the closest to the root of the bacterial tree in recently published SSU rRNA trees rooted on Archaeal outgroups (7, 62) and, more tentatively, because outgroup sequences were more difficult to infer in a recently published tree based upon analysis of RecA sequences (19). In analyses of small (62, 101) and large subunit rRNA sequences (17a, 91, 101), RecA sequences (19) and EF-Tu/EF-α sequences (91) lineages comprising *Thermotoga* and relatives, and *Thermus* and *Deinococcus*, are also deep branching.

The order of divergence of the remaining phyla in the rRNA trees (SSU and LSU) appears to be uncertain, changing with the species sampled and the method used (17a, 101). The use of genes other than ribosomal RNA, including ATP synthase, elongation factor Tu (61), HSP-60 (102), and RecA (19), has not resolved branching order between the remaining bacterial phyla. Species sampling has not been representative of all phyla for all genes, although these four genes are probably the best sampled after the rRNAs.

The integrity of most of phyla suggested by Woese (111) has survived further sampling of SSU rRNA sequences, but the monophyly of the gram-positive bacteria (111), which comprises two groups based upon either high or low $G + C$ contents in their DNA, has been questioned from the perspective of some analyses of large and small subunit rRNA sequences (17a, 61, 101) and from analysis of RecA (19), ATP-synthase, and EF-Tu sequences (61). The tree for HSP-60 recovers all sampled gram-positive bacteria along with cyanobacteria as a strongly supported monophyletic group, but with only moderate (59%) bootstrap support for the gram-positives forming a sister group (102). In a less detailed tree based upon HSP-70 (41), the gram-positive bacteria were

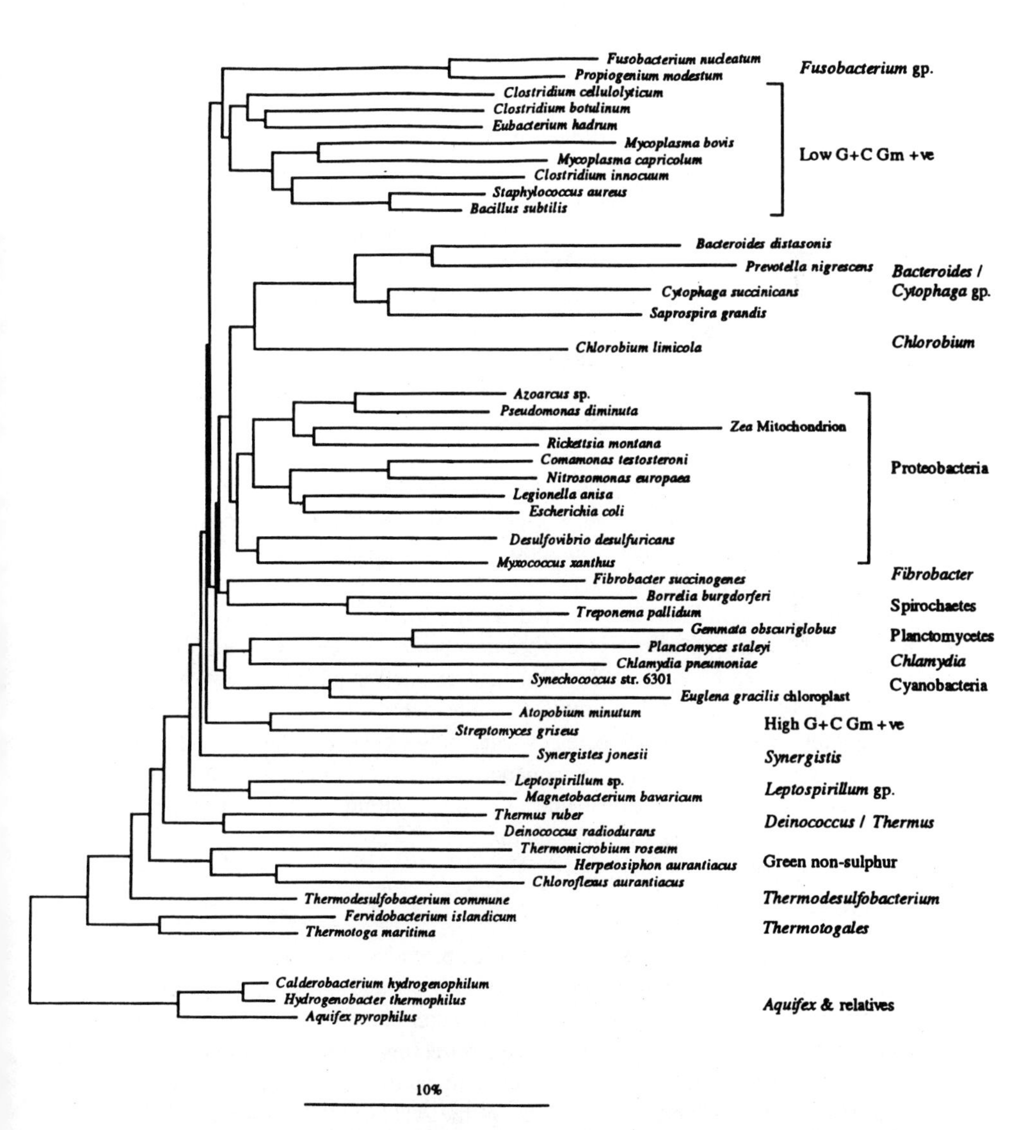

Figure 4 Tree based upon SSU rRNA sequences and showing the relationships between major bacterial phyla. The tree is based upon analysis of 1100 sequence positions using the Jukes & Cantor (49) correction and Neighbor Joining (88). Published aligned sequences were obtained from the Ribosomal Data Base Project (62).

recovered as a clade which also contained some sequences from methanogenic and halophilic Archaea.

Exploration of the diversity of uncultured bacterial diversity has revealed some new lineages which, by their depth of branching, might be considered as forming the basis of new phyla, as well as revealing new sequences that extend considerably the diversity of existing phyla (28, 59, 104). The impact of sequences from environmental gene libraries on the topology of the SSU rRNA tree for Bacteria has not been so spectacular as for the Archaea, possibly because the Bacterial tree was already much more developed. Some monophyletic groups of predominantly uncultured bacteria related to the cyanobacteria or to the alpha or gamma proteobacteria appear to be very abundant in the oceans and are almost certain to be ecologically important in these systems (73).

Analysis of the phylogenetic relationships of bacteria that have been cultured is continuing to yield important results. For example, sequencing of 16S rRNA genes from species of the genus *Clostridium*, which is largely defined on the basis of anaerobiosis and the ability to form endospores, has revealed that it is highly divergent throughout the gram-positive phylum and is intermixed with other spore-forming and non-spore-forming genera (12). Determined and imaginative attempts to isolate members of the morphologically conspicuous and highly divergent *Planctomyces* and *Verrucomicrobium* (92, 93) have yielded a large number of new cultures for laboratory study. Sequencing the 16S rRNA genes from these isolates (105, 106) has revealed significant new diversity among the Planctomycetales, previously considered a rather depauperate phylum comprising only a few species. These studies have also demonstrated that a group of environmental clone sequences from soil, which were previously considered to represent a novel line of descent within the Bacteria (59), are specifically related to *Verrucomicrobium.*

CONCLUSIONS

The general analytical basis of systematics provides a vocabulary for the interpretation of results, with homology (orthology) and taxa being products of discovery. Resolution of interrelationships is complicated by incongruence between trees from different genes, perhaps reflecting a distinct history of those genes or perhaps reflecting an inadequacy in sampling with incomplete gene histories being identified, as in the case of the ATPases. Either way, judicious sampling seems to be the general solution.

Sampling of environments using molecular biology provides a means of collecting new sequence information in a relatively unbiased way; it is not hindered by expectations of potential relationships, and in this way acts blind. Almost all current interpretations are made with reference to the SSU rRNA gene tree, which can provide clues from the position of uncultured sequences guiding their isolation (46) or the sampling of additional gene fragments (98). Results so far, particularly on the Archaeal tree, are encouraging and indicate that diversity in this domain far outstrips early expectations.

In the editorial for the first issue of *Molecular Phylogenetics and Evolution*, Goodman noted that "we are at the threshold of a new age of exploration that promises to greatly increase our knowledge of the history and ongoing evolution of the ramifying lines of life" (34). This new age of exploration and the data it will provide can be made comprehensible only in the light of biology's oldest and most significant discipline: systematics.

Literature Cited

1. Barns SM, Fundyga RE, Jeffries MW, Pace NR. 1994. Remarkable archaeal diversity detected in a Yellowstone National Park hot spring environment. *Proc. Natl. Acad. Sci. USA* 91:1609–13
2. Benachenhou-Lahfa N, Forterre P, Labedan B. 1993. Evolution of glutamate dehydrogenase genes: evidence for two paralogous protein families and unusual branching patterns of the archaebacteria in the universal tree of life. *J. Mol. Evol.* 36:335–46
3. Boorstein WR, Ziegelhoffer T, Craig EA. 1994. Molecular evolution of the HSP70 multigene family. *J. Mol. Evol.* 38:1–7
4. Brooks DR, McLennan DA. 1991. *Phylogeny, Ecology and Behavior.* Chicago: Univ. Chicago Press
5. Brown JR, Doolittle WF. 1995. Root of the universal tree of life based upon ancient aminoacyl-tRNA synthetase gene duplications. *Proc. Natl. Acad. Sci. USA* 92:2441–45
6. Brown JR, Masuchi Y, Robb FT, Doolittle WF. 1994. Evolutionary relationships of bacterial and archaeal glutamine synthetase genes. *J. Mol. Evol.* 38:566–76
7. Burggraf S, Olsen GJ, Stetter KO, Woese CR. 1992. A phylogenetic analysis of *Aquifex pyrophilus*. *Syst. Appl. Microbiol.* 15:352–56
8. Burggraf S, Stetter KO, Rouviere P, Woese CR. 1991. *Methanopyrus kandleri*: an archaeal methanogen unrelated to all other known methanogens. *Syst. Appl. Microbiol.* 14:346–51
9. Cammarano P, Palm P, Creti R, Ceccarelli E, Sanangelantoni AM, et al. 1992. Early evolutionary relationships among known life forms inferred from elongation factor EF-2/EF-G sequences: phylogenetic coherence and structure of the archaeal domain. *J. Mol. Evol.* 34:396–405
10. Cavalier-Smith T. 1992. Bacteria and eukaryotes. *Nature* 356:570
11. Cavanaugh CM. 1994. Microbial symbiosis: patterns of diversity in the marine environment. *Am. Zool.* 34:79–89
12. Collins MD, Lawson PA, Willems A, Cordoba JJ, Fernandez-Garayzabel J, et al. 1994. The phylogeny of the genus *Clostridium*: proposal of five new genera and eleven new species combinations. *Int. J. Syst. Bacteriol.* 44:812–26
13. Creti R, Ceccarelli E, Bocchetta M, Sanangelantoni AM, Tiboni O, et al. 1994. Evolution of translational elongation factor (EF) sequences: reliability of global phylogenies inferred from EF-1-α(Tu) and EF-2(G) proteins. *Proc. Natl. Acad. Sci. USA* 91:3255–59
14. DeLong EF. 1992. Archaea in coastal marine environments. *Proc. Natl. Acad. Sci. USA* 89:5685–89

15. DeLong EF, Wickham GS, Pace NR. 1989. Phylogenetic stains: ribosomal RNA-based probes for the identification of single cells. *Science* 243:1360–63
16. DeLong EF, Wu KY, Prezelin BB, Jovine RVM. 1994. High abundance of Archaea in Antarctic marine picoplankton. *Nature* 371:695–97
17. De Queiroz A, Donoghue MJ, Kim J. 1995. Separate versus combined analysis of phylogenetic evidence. *Annu. Rev. Ecol. Syst.* 26:657–81
17a. de Rijk P, van de Peer P, van den Broeck I, de Wachter R. 1995. Evolution according to large ribosomal subunit RNA. *J. Mol. Evol.* 41:366–75
17b. Doolittle RF. 1995. Of Archae and Eo: What's in a name? *Proc. Natl. Acad. Sci. USA* 92:2421–23
18. Doolittle WF, Brown JR. 1994. Tempo, mode, the progenote and the universal root. *Proc. Natl. Acad. Sci. USA* 91:6721–28
19. Eisen JA. 1995. The RecA protein as a model molecule for molecular systematic studies of bacteria: comparison of trees of RecAs and 16S rRNAs from the same species. *J. Mol. Evol.* 41:1105–23
20. Farelly V, Rainey FA, Stackebrandt E. 1995. Effect of genome size and rrn copy number on PCR amplification of 16S rRNA genes from a mixture of bacterial species. *Appl. Environ. Microbiol.* 61:2798–801
21. Deleted in proof
22. Fitch WM, Upper K. 1987. The phylogeny of tRNA sequences provides evidence for ambiguity reduction in the origin of the genetic code. *Cold Spring Harbor Symp. Quant. Biol.* 52:759–67
23. Fitch WM, Upper K. 1988. The evolution of life—an overview of general problems and a specific study of the origin of the genetic code. In *Evolutionary Processes and Metaphors,* ed. MW Ho, S Fox, pp. 33–47. New York: Wiley
24. Forterre P. 1992. Neutral terms. *Nature* 355:305
25. Forterre P, Benachenhou-Lahfa N, Confalonieri F, Duguet M, Elie C, et al. 1993. The nature of the last universal ancestor and the root of the tree of life, still open questions. *BioSystems* 28:15–32
26. Fuhrman JA, McCallum K, Davis AA. 1992. Novel major archaebacterial group from marine plankton. *Nature* 356:148–49
27. Fuhrman JA, McCallum K, Davis AA. 1993. Phylogenetic diversity of subsurface marine microbial communities from the Atlantic and Pacific Oceans. *Appl. Environ. Microbiol.* 59:1294–302
28. Giovannoni SJ, Britschgi TB, Moyer CL, Field KG. 1990. Genetic diversity in Sargasso Sea bacterioplankton. *Nature* 345:60–63
28a. Gogarten JP. 1994. Which is the most conserved group of proteins? Homology-orthology, paralogy, xenology, and the fusion of lineages. *J. Mol. Evol.* 39:541–43
29. Gogarten JP. 1995. The early evolution of cellular life. *TREE* 10:147–51
30. Gogarten JP, Kibak H, Dittrich P, Taiz L, Bowman EJ, et al. 1989. Evolution of the vacuolar H+-ATPase: implications for the origin of eukaryotes. *Proc. Natl. Acad. Sci. USA* 86:6661–65
31. Gogarten JP, Starke T, Kibak H, Fichman J, Taiz L. 1992. Evolution and isoforms of V-ATPase subunits. *J. Exp. Biol.* 172:137–47
32. Gogarten JP, Taiz L. 1992. Evolution of protein pumping ATPases: rooting the tree of life. *Photosyn. Res.* 33:137–46
33. Golding GB, Gupta RS. 1995. Protein-based phylogenies support a chimeric origin for the eukaryotic genome. *Mol. Biol. Evol.* 12:1–6
34. Goodman M. 1992. Editorial. *Mol. Phylogenet. Evol.* 1:1–2
35. Gottschal JC, Harder W, Prins RA. 1991. Principles of enrichment, isolation, cultivation and preservation of bacteria. In *The Prokaryotes,* ed. A Balows, HG Trüper, M Dworkin, W Harder, K-H Schleifer, pp. 149–96. New York: Springer-Verlag
36. Gray GS, Fitch WM. 1983. Evolution of antibiotic resistance genes: the DNA sequence of a kanamycin resistance gene from *Staphylococcus aureus*. *Mol. Biol. Evol.* 1:57–66
37. Gray MW. 1992. The endosymbiont hypothesis revisited. *Int. Rev. Cytol.* 141:233–357
38. Deleted in proof
39. Gupta RS, Aitken K, Falah M, Singh B. 1994. Cloning of *Giardia lamblia* heat shock protein HSP70 homologs: implications regarding origin of eukaryotic cells and of endoplasmic reticulum. *Proc. Natl. Acad. Sci. USA* 91:2895–99
40. Gupta RS, Golding GB. 1993. Evolution of HSP70 gene and its implications regarding relationships between archaebacteria, eubacteria and eukaryotes. *J. Mol. Evol.* 37:573–82
41. Gupta RS, Singh B. 1994. Phylogenetic analysis of 70 kD heat shock protein sequences suggests a chimeric origin for the eukaryotic cell nucleus. *Curr. Biol.*

4:1104–14
42. Haeckel E. 1866. *Generelle Morphologie der Organismen.* Berlin: Georg Reimer
43. Hartman H. 1984. The origin of the eukaryotic cell. *Specul. Sci. Technol.* 7:77–81
44. Hartman H. 1992. The eukaryotic cell: evolution and development. In *The Origin and Evolution of the Cell,* ed. H Hartman, K Matsuno, pp. 3–12. Singapore: World Sci.
45. Hilario E, Gogarten JP. 1993. Horizontal transfer of ATPase genes—the tree of life becomes a net of life. *BioSystems* 31:111–19
46. Huber R, Burggraf S, Mayer T, Barns SM, Rossnagel P, et al. 1995. Isolation of a hyperthermophilic archaeum predicted by in situ RNA analysis. *Nature* 376:57–58
46a. Hughes AL. 1993. Nonlinear relationships among evolutionary rates identify regions of functional divergence in heat-shock proteins 70 genes. *Mol. Biol. Evol.* 10:243–55
47. Ihara K, Mukohata Y. 1992. The ATP synthetase of *Halobacterium salinarium* is an archaebacteria type as revealed from the amino acid sequences of its two major subunits. *Arch. Biochem. Biophys.* 286:111–16
48. Iwabe N, Kuma K, Hasegawa M, Osawa S, Miyata T. 1989. Evolutionary relationship of archaebacteria, eubacteria, and eukaryotes inferred from phylogenetic trees of duplicated genes. *Proc. Natl. Acad. Sci. USA* 86:9355–59
48a. Jin L, Nei M. 1990. Limitations of the evolutionary parsimony method of phylogenetic analysis. *Mol. Biol. Evol.* 7:82–102
49. Jukes TH, Cantor CR. 1969. Evolution of protein molecules. In *Mammalian Protein Metabolism,* ed. HN Munro, pp. 21–132. New York: Academic
50. Klenk H-P, Zillig W, Lanzendorfer M, Grampp B, Palm P. 1995. Location of protist lineages in a phylogenetic tree inferred from sequences of DNA-dependent RNA polymerases. *Arch. Protistenkd.* 145:221–30
51. Kumada Y, Benson DR, Hillemann D, Hosted TJ, Rochefort DA, et al. 1993. Evolution of the glutamine synthetase gene, one of the oldest existing and functioning genes. *Proc. Natl. Acad. Sci. USA* 90:3009–13
52. Lake JA. 1987. A rate-independent technique for analysis of nucleic acid sequences: evolutionary parsimony. *Mol. Biol. Evol.* 4:167–91
53. Lake JA. 1988. Origin of the eukaryotic nucleus determined by rate-invariant analysis of rRNA sequences. *Nature* 331:184–86
53a. Lake JA. 1991. The order of sequence alignment can bias the selection of tree topology. *Mol. Biol. Evol.* 8:378–85
54. Lake JA. 1994. Reconstructing evolutionary trees from DNA and protein sequences: paralinear distances. *Proc. Natl. Acad. Sci. USA* 91:1455–59
55. Lake JA. 1995. Calculating the probability of multitaxon evolutionary trees: Bootstrappers Gambit. *Proc. Natl. Acad. Sci. USA* 92:9662–66
56. Lake JA, Henderson E, Clark HW, Matheson AT. 1982. Mapping evolution with ribosome structure: intralineage constancy and interlineage variation. *Proc. Natl. Acad. Sci. USA* 79:5948–52
57. Langer D, Hain J, Thuriaux P, Zillig W. 1995. Transcription in Archaea: similarity to that in Eucarya. *Proc. Natl. Acad. Sci. USA* 92:5768–72
58. Lento GM, Hickson RE, Chambers GK, Penny D. 1995. Use of spectral analysis to test hypotheses on the origins of pinnipeds. *Mol. Biol. Evol.* 12:28–52
59. Liesack W, Stackebrandt E. 1992. Occurrence of novel groups of the domain *Bacteria* as revealed by analysis of genetic material isolated from an Australian terrestrial environment. *J. Bacteriol.* 174:5072–78
60. Lockhart PJ, Steel MA, Hendy MD, Penny D. 1994. Recovering evolutionary trees under a more realistic model of sequence evolution. *Mol. Biol. Evol.* 11:605–12
61. Ludwig W, Neumaier J, Klugbauer N, Brockmann E, Roller C, et al. 1993. Phylogenetic relationships of Bacteria based on comparative sequence analysis of elongation factor Tu and ATP-synthase beta-subunit genes. *Antonie van Leeuwenhoek J. Microbiol. Serol.* 64:285–305
62. Maidak BL, Larsen N, McCaughey J, Overbeek R, Olsen GJ, et al. 1994. The ribosomal database project. *Nucleic Acids Res.* 22:3485–87
63. Margulis L. 1970. *Origin of Eukaryotic Cells.* New Haven: Yale Univ. Press
64. Markos A, Miretsky A, Muller M. 1993. A glyceraldehyde-3-phosphate dehydrogenase with eubacterial features in the amitochondriate eukaryote, *Trichomonas vaginalis. J. Mol. Evol.* 37:631–43
65. Deleted in proof

66. Mayr E. 1982. *The Growth of Biological Thought.* Cambridge, MA: Harvard Univ. Press
67. Mayr E. 1990. A natural system of organisms. *Nature* 348:491
68. Mayr E. 1991. More natural classification. *Nature* 353:122
69. McInerney JO, Wilkinson M, Patching JW, Embley TM, Powell R. 1995. Recovery and phylogenetic analysis of novel archaeal rRNA sequences from a deep-sea deposit feeder. *Appl. Environ. Microbiol.* 61:1646–48
70. Michels PAM, Opperdoes FR, Hannaert V, Wiemer EAC, Allert S, et al. 1992. Phylogenetic analysis based on glycolytic enzymes. *Belgian J. Bot.* 125:164–73
71. Mindell DP. 1992. Phylogenetic consequences of symbiosis: eukarya and eubacteria are not monophyletic taxa. *BioSystems* 31:53–62
72. Mindell DP. 1993. Merger of taxa and the definition of monophyly (reply to J. Zrzavy and Z. Skala). *BioSystems* 31:130–33
72a. Morrone JJ, Crisci JV. 1995. Historical biogeography: introduction to methods. *Annu. Rev. Ecol. Syst.* 26:373–401
73. Mullins TD, Britschgi TB, Krest RL, Giovannoni SJ. 1995. Genetic comparisons reveal the same unknown bacterial lineages in Atlantic and Pacific bacterioplankton communities. *Limnol. Oceanogr.* 40:148–58
74. Murray RGE, Stackebrandt E. 1995. Taxonomic note: implementation of the provisional status Candidatus for incompletely described prokaryotes. *Int. J. Syst. Bacteriol.* 45:186–87
75. Nelson G. 1994. Homology and systematics. In *Homology: The Hierarchical Basis of Comparative Biology,* ed. BK Hall, pp. 101–49. San Diego: Academic
76. Nelson G, Ladiges PY. 1996. Paralogy in cladistic biogeography and analysis of paralogy-free subtrees. *Am. Mus. Novit.* 3167:1–58
77. Olsen GJ. 1987. Earliest phylogenetic branching comparing rRNA-based evolutionary trees inferred with various techniques. *Cold Spring Harbor Symp. Quant. Biol.* 52:825–37
78. Olsen GJ, Woese CR. 1993. Ribosomal RNA: a key to phylogeny. *FASEB* J. 7:113–23
79. Pace NR, Stahl DA, Lane DJ, Olsen GJ. 1985. Analysing natural microbial populations by rRNA sequences. *Am. Soc. Microbiol. News* 51:4–12
80. Page RDM. 1994. Maps between trees and cladistic analysis of historical associations among genes, organisms and areas. *Syst. Biol.* 43:58–77
81. Patterson C. 1982. Morphological characters and homology. In *Problems of Phylogenetic Reconstruction,* ed. KA Josey, AE Friday, pp. 21–74. London: Academic
82. Patterson C. 1988. Homology in classical and molecular biology. *Mol. Biol. Evol.* 5:603–25
83. Patterson C. 1994. Null or minimal models. In *Models in Phylogeny Reconstruction,* ed. RW Scotland, DJ Seibert, DM Williams, pp. 125–55. London: Academic
84. Pühler G, Leffersy H, Gropp F, Palm K, Klenk K-H, et al. 1989. Archaebacterial DNA-dependent RNA polymerases testify to the evolution of the eukaryotic nuclear genome. *Proc. Natl. Acad. Sci. USA* 86:4569–73
85. Reysenbach A-L, Giver LJ, Wickham GS, Pace NR. 1992. Differential amplification of rRNA genes by polymerase chain reaction. *Appl. Environ. Microbiol.* 58:3417–18
86. Rowlands T, Baumann P, Jackson SP. 1994. The TATA-binding protein: a general transcription factor in Eukaryotes and Archaebacteria. *Science* 264:1326–29
87. Saccone C, Gissi C, Lanave C, Pesole G. 1995. Molecular classification of living organisms. *J. Mol. Evol.* 40:273–79
88. Saitou N, Nei M. 1987. The neighbor-joining method: a new method for reconstructing phylogenetic trees. *Mol. Biol. Evol.* 4:406–25
89. Sapp J. 1994. *Evolution by Association. A History of Symbiosis.* Oxford: Oxford Univ. Press
90. Schafer G, Meyering-Vos M. 1992. F-type or V-type? The chimeric nature of archaebacteria ATP synthetase. *Biochim. Biophys. Acta* 1101:232–35
91. Schleifer K-H, Ludwig W. 1989. Phylogenetic relationships among bacteria. In *The Hierarchy of Life,* ed. B Fernholm, K Bremer, H Jornvall, pp. 103–17. Amsterdam: Excerpta Medica
92. Schlesner H. 1987. *Verrucomicrobium spinosum* gen. nov., a budding, peptidoglycan-less bacterium from brackish water. *Syst. Appl. Microbiol.* 10:177–80
93. Schlesner H. 1994. The development of media suitable for microorganisms morphologically resembling *Planctomyces* spp. *Pirellula* sp., and other Planctomycetales from various aquatic habitats using dilute media. *Syst. Appl. Microbiol.* 17:135–45

94. Schwartz RM, Dayhoff MD. 1978. Origins of prokaryotes eukaryotes, mitochondria and chloroplasts. *Science* 199:395–403
95. Skala Z, Zrzavy J. 1994. Phylogenetic reticulations and cladistics: discussion of methodological concepts. *Cladistics* 10:305–13
96. Sogin ML. 1991. Early evolution and the origin of eukaryotes. *Curr. Opin. Genet. Dev.* 1:457–63
97. Stahl DA, Flesher B, Mansfield HR, Montgomery L. 1988. Use of phylogenetically based hybridization probes for studies of ruminal microbial ecology. *Appl. Environ. Microbiol.* 54:1079–84
98. Stein JL, Marsh TL, Wu KY, Shizuya H, DeLong EF. 1996. Characterisation of uncultivated prokaryotes: isolation and analysis of a 40-kilobase-pair genome fragment from a planktonic marine Archaeon. *J. Bacteriol.* 178:591–99
99. Sutherland KJ, Henneke CM, Towner P, Hough DW, Danson MJ. 1990. Citrate synthetase from the thermophilic archaebacterium *Thermoplasma acidophilum*: cloning and sequencing of the gene. *Eur. J. Biochem.* 194:839–44
99a. Tsutsumi S, Denda K, Yokoyama K, Oshima T, Date T, Yoshida, M. 1991. Molecular cloning of genes encoding major two subunits of a eubacterial V-type ATPase from *Thermus thermophilus*. *Biochim. Biophys. Acta* 1098:13–20
100. Ueda T, Suga Y, Matsuguchi T. 1995. Molecular phylogenetic analysis of a soil microbial community in a soybean field. *Eur. J. Soil Sci.* 46:415–21
101. van de Peer Y, Neefs J-M, de Rijk P, de Vos P, de Wachter R. 1994. About the order of divergence of the major bacterial taxa during evolution. *Syst. Appl. Bacteriol.* 17:32–38
102. Viale AM, Arakaki AK, Soncini FC, Ferreyra RG. 1994. Evolutionary relationships among eubacterial groups as inferred from GroEL (chaperonin) sequence comparisons. *Int. J. Syst. Bacteriol.* 44:527–33
103. Wagner A, Blackstone N, Cartwright P, Dick M, Misof B, et al. 1994. Surveys of gene families using polymerase chain reaction: PCR selection and PCR drift. *Syst. Biol.* 43:250–61
104. Ward DM, Bateson MM, Weller R, Ruff-Roberts AL. 1992. Ribosomal RNA analysis of microorganisms as they occur in nature. *Adv. Microb. Ecol.* 12:219–86
105. Ward N, Rainey FA, Stackebrandt E, Schlesner H. 1995. Unravelling the extent of diversity within the order Planctomyceteales. *Appl. Environ. Microbiol.* 61:2270–75
106. Ward-Rainey N, Rainey FA, Schlesner H, Stackebrandt E. 1995. Assignment of hitherto unidentified 16S rDNA species to a main line of descent within the domain Bacteria. *Microbiology* 141:3247–50
107. Weston P. 1994. Methods for rooting cladistic trees. In *Models in Phylogeny Reconstruction,* ed. RW Scotland, DJ Siebert, DM Williams, pp. 25–155. London: Academic
108. Wheelis ML, Kandler O, Woese CR. 1992. On the nature of global classification. *Proc. Natl. Acad. Sci. USA* 89:2930–34
109. Wilson EO. 1994. *Naturalist.* Washington, DC: Island Press
110. Winker S, Woese CR. 1991. A definition of the domains Archaea, Bacteria and Eucarya in terms of small subunit ribosomal RNA characteristics. *Syst. Appl. Microbiol.* 14:305–10
111. Woese CR. 1987. Bacterial evolution. *Microbiol. Rev.* 51:221–71
112. Woese CR, Fox GE. 1977. Phylogenetic structure of the prokaryotic domain: the primary kingdoms. *Proc. Natl. Acad. Sci. USA* 74:5088–90
113. Woese CR, Gutell R, Gupta R, Noller HF. 1983. Detailed analysis of the higher-order structure of 16S-like ribosomal ribonucleic acids. *Microbiol. Rev.* 47:621–69
114. Woese CR, Kandler O, Wheelis ML. 1990. Towards a natural system of organisms: proposal for the domains Archaea, Bacteria, and Eucarya. *Proc. Natl. Acad. Sci. USA* 87:4576–79
115. Woese CR, Kandler O, Wheelis ML. 1991. A natural classification. *Nature* 351:528–29
116. Yang D, Gunther I, Matheson AT, Auer J, Spiker G, et al. 1991. The structure of the gene for ribosomal protein L5 in the archaebacterium *Sulpholobus acidocaldarius*. *Biochimie* 73:679–82
117. Zillig W, Schnabel R, Stetter KO. 1985. Archaebacteria and the origin of the eukaryotic cytoplasm. *Curr. Top. Microbiol. Immunol.* 114:1–18
118. Zrzavy J, Skala Z. 1993. Holobionts, hybrids and cladistic classification (reply to D. P. Mindell). *BioSystems* 31:127–30

Annu. Rev. Ecol. Syst. 1996. 27:597–623

THE GEOGRAPHIC RANGE: Size, Shape, Boundaries, and Internal Structure

James H. Brown, George C. Stevens, and Dawn M. Kaufman
Department of Biology, University of New Mexico, Albuquerque, New Mexico 87131

KEY WORDS: biogeography, distribution, ecological biogeography, geographic range, macroecology, range limits

ABSTRACT

Comparative, quantitative biogeographic studies are revealing empirical patterns of interspecific variation in the sizes, shapes, boundaries, and internal structures of geographic ranges; these patterns promise to contribute to understanding the historical and ecological processes that influence the distributions of species. This review focuses on characteristics of ranges that appear to reflect the influences of environmental limiting factors and dispersal. Among organisms as a whole, range size varies by more than 12 orders of magnitude. Within genera, families, orders, and classes of plants and animals, range size often varies by several orders of magnitude, and this variation is associated with variation in body size, population density, dispersal mode, latitude, elevation, and depth (in marine systems). The shapes of ranges and the dynamic changes in range boundaries reflect the interacting influences of limiting environmental conditions (niche variables) and dispersal/extinction dynamics. These processes also presumably account for most of the internal structure of ranges: the spatial patterns and orders-of-magnitude of variation in the abundance of species among sites within their ranges. The results of this kind of "ecological biogeography" need to be integrated with the results of phylogenetic and paleoenvironmental approaches to "historical biogeography" so we can better understand the processes that have determined the geographic distributions of organisms.

INTRODUCTION

If there is any basic unit of biogeography, it is the geographic range of a species. Most biogeographic research is the study of the structure and dynamics of

0066-4162/96/1120-0597$08.00

geographic ranges: their sizes, shapes, boundaries, overlaps, and locations. Biogeographers study the spatial patterns of dispersion of ranges, the temporal patterns of changes in ranges, the relationships between ranges and phylogenies, and the processes that produce these patterns. Some biogeographers are concerned with a particular region, so they may consider only a portion of the range of one or more species. Others are interested in the distributions of multiple species: of clades, taxonomic groups, functional groups, or overall species diversity. Nevertheless, nearly all biogeographic research is attempting to answer questions about the processes that determine the location in space and the shifts in time of the ranges of species.

In the present chapter we review what is known about the patterns and processes that characterize the ranges of species. We are concerned primarily with variation among species in a clade, taxon, or functional group in the size, shape, and internal structure of ranges. Because these features of ranges are all related to the environmental factors and ecological processes that limit distribution and abundance, we are also concerned with range boundaries: their location and configuration in space and their changes over time. Each major topic is divided into two subsections. In the first, "Patterns," we review the empirical information on the quantitative characteristics of ranges and their relationships to other variables. While space does not permit us to mention all of the relevant data and published studies, we try to survey what is known and to provide many citations. In the second subsection, "Processes," we discuss the mechanisms that have been or might be invoked to account for the patterns. Here we are necessarily more speculative, because many of the patterns have only recently been discovered, and hypotheses about mechanisms are still being developed and evaluated. We hope that our chapter will stimulate the effort to better characterize the patterns and to better understand the mechanistic processes that produce them.

A Digression: Defining Species and Ranges

If the geographic range of a species is a basic unit of biogeography, then biogeographic research will depend on how species and their ranges are characterized. The definition of species has been complicated in recent years by two important advances in phylogenetic systematics. One is the use of molecular genetic information for phylogenetic reconstructions and taxonomic revisions. Studies of variation at the molecular level have often revealed genetic discontinuities within taxa that had formerly been considered to be single species (e.g. 60). The second complication has come from the introduction of new evolutionary (92) and phylogenetic (22) species concepts. The former would define as a species any population that is sufficiently isolated from other populations so as to be an independent evolutionary unit. The latter would consider as a species any population in which a unique derived character (apomorphy) is fixed. Application of these new species concepts has resulted in the splitting of species into multiple new species based on their distinctive molecular genetic characteristics. These changes in systematics have the merits of making the definition of species consistent with the theory and practice of reconstructing phylogenetic relationships using molecular data, but they create practical problems for practicing taxonomists and all other individuals who are faced with the task of identifying living organisms in the field or their preserved remains in fossil deposits or museum collections. There are also problems at higher levels of classification, because phylogenetic systematists are using the concept of clade, defined as all of the descendants of common ancestor, to redefine traditional taxonomic groups.

We do not mean to be critical of these developments. Indeed, the recent advances in phylogenetic reconstruction—together with recent studies of earth history and fossil organisms—are leading to greatly increased understanding of the history of plant and animal distributions. We would, however, make two pleas. First, we emphasize the need for practical, operational, and reasonably standardized species definitions, so that biogeographers and other scientists can identify their organisms and can apply the advances in phylogenetic reconstruction to their own studies. In the meantime, quantitative studies of geographic ranges will have to be based on existing taxonomic classifications and published studies that provide standardized range maps or other data on the distributions of many species. Second, we ask that the zeal for applying phylogenetic reconstructions to comparative ecological and biogeographic studies be tempered by the realization that other factors affect abundance, distribution, and diversity. The constraints of phylogeny certainly influence contemporary ecological relationships and geographic distributions, but ecology and geography also influence phylogeny. It is no more reasonable to demand that phylogenetically explicit analyses be included in comparative ecological and biogeographic studies than it is to demand that earth history and ecology be incorporated explicitly in phylogenetic reconstructions. Phylogenetic analyses can make important contributions to comparative biogeography (e.g. 87), but they are neither necessary nor sufficient to address all of the interesting questions.

Efforts to do comparative, and especially quantitative, research in biogeography are also complicated by problems of defining and mapping geographic ranges (11, 31, 32, 88). Like attempts to define and classify species, efforts to characterize geographic ranges of species necessarily involve reducing a complex phenomenology to a greatly simplified abstraction. The real units of geographic ranges are the complex spatial and temporal patterns in which individual organisms are dispersed over the earth. Any maps or other characterizations of the geographic ranges of species necessarily simplify such complex distributions.

Most comparative biogeographic research relies on data compiled from the literature or from other sources, such as museum specimens and biological surveys. The original data on distributions of species, and the range maps that are constructed from them, have problems of precision, accuracy, and interpretation. The range is most often mapped as an irregular area. Such "outline maps" are often so simplified that they do not depict either holes within the range boundaries where a species does not occur or islands around the perimeter where isolated populations are found. Somewhat more precision is afforded by "dot maps" that plot each location where a species has been recorded. Most published range maps attempt to define the historical range of a species. This means that the range encompasses all localities where a species is known to have regularly occurred in the past, including areas where it formerly was present but is now extinct and areas that it has recently colonized. Unless the map is up to date, however, it may not include all locations where a species has recently been recorded. The mapped range usually does not incorporate records of occurrence that are judged to represent individual organisms that have dispersed or been transported by humans beyond the normal distribution of a species.

Given the problems in defining species and their ranges, a naive reader may wonder whether there is any point in trying to do research in comparative biogeography—of trying to quantify patterns and to understand the processes that produce them. While it is important to be aware of these problems, they are hardly crippling. Indeed, problems of precision, accuracy, and completeness of information are common to most ecological and systematic research. When, as is usually the case with geographic ranges, there are orders of magnitude of variation among the entities being compared, small differences owing to human factors are not likely to be important.

An Historical Perspective

Considering the long history of biogeography and the central place of the geographic range in biogeographic research, it is surprising that most comparative studies of the characteristics of ranges have been done within the last 15 years. The earliest biogeographers, including de Candole, Wallace, Hooker, and Darwin, were concerned with factors that limited distribution and influenced species composition and diversity, but they rarely focused explicitly on geographic ranges. Perhaps the first person to do so was Willis (94), whose treatise *Age and Area* quantified the areas of geographic ranges of species in several taxonomic groups, pointed out the wide variance and distinctive shape of the frequency distributions, and advanced the hypothesis that the areas reflected the age of the taxa and thus the time since they differentiated from an ancestor. While Willis's ideas seem quaint today, he must be regarded, along with Arrhenius (4) who worked on species-area curves, as one of the pioneers of quantitative biogeography.

For most of the twentieth century, research on geographic ranges was directed primarily toward trying to identify the environmental factors responsible for range boundaries of particular species. This work was often motivated by practical concerns about what limited the distribution of commercially valuable plants (e.g. 56), invasive weeds and insect pests (e.g. 1, 91), or potential agents of biological control (e.g. 25, 50). Connell's (20) classic experimental investigation of the factors limiting the distribution of the barnacle *Chthamalus stellatus* remains one of the most thorough and rigorous studies. Connell's work was typical of much of the research on range boundaries, however, in that it was on a small spatial scale, focused on a few limiting factors, and had a more ecological than biogeographic flavor.

The discipline of biogeography gained new vigor in the second half of the twentieth century, stimulated in large part by the contributions of Darlington, Croizat, MacArthur & Wilson, Nelson, Platnick, and Rosen. In 1977, Sydney Anderson published the first of several papers based on measuring the areas of the mapped geographic ranges of vertebrates in North America and Australia (2, 3). Primary credit for stimulating interest in quantitative studies of geographic ranges, however, must go largely to Rapoport, whose creative and insightful monograph *Aerography* was published in English in 1982 (73). Rapoport not only anticipated nearly all of the ideas in the present review article, he also investigated many of them by collecting data and performing elegantly simple analyses.

The last decade has seen a gratifying increase in comparative and quantitative studies of geographic ranges. Such studies have been greatly facilitated by advances in computer technology. Large computerized data bases compiled from museum collections, standardized biological surveys (e.g. the North American Breeding Bird Survey and Butterfly Survey), and other records provide detailed accounts of occurrence for many taxa. Published range maps, often prepared from these data bases, provide relatively standardized representations of the distributions of many species. Other data bases make available information on geography, geology, climate, soils, vegetation, and other environmental variables from earth-based and remotely sensed sources. Development of computer hardware (e.g. scanners and digitizers) and software (e.g. statistical and graphics packages, and Geographic Information Systems) permit the quantification, representation, and analysis of distributional patterns. Advances in mathematical and simulation modeling (e.g. nonlinear dynamics, cellular automata, and agent-based models) facilitate understanding of complex, spatially explicit processes. While the recent availability of data bases and of new analytical and modeling tools has already contributed greatly to biogeographic research, they promise even greater contributions in the future.

SIZE OF RANGE

Patterns

There is enormous variation in the sizes of geographic ranges of individual species. Among the smallest are the natural distributions of the Soccoro isopod (*Thermosphaeroma thermophilum*) and the Devil's Hole pupfish (*Cyprinodon diabolis*), each of which occurs in a single freshwater spring with a surface area of less than 100 m^2 (M Molles, personal communication; 68). Among the largest ranges are those of several marine organisms, such as the blue whale (*Balaenoptera musculus*), which include most of the world's unfrozen oceans, areas on the order of 300,000,000 km^2. Among terrestrial organisms, species with very large native ranges include the peregrine falcon, barn owl, and osprey, which are widely distributed over all of the continents except Antarctica. Of course, modern *Homo sapiens* is now one of the most widely distributed species, and humans have carried several species of symbionts and exotics with them as they have spread over the entire earth.

Two features of the variation in range size are especially interesting. One is its sheer magnitude. Just for comparison, the 12 orders of magnitude variation in area of geographic range is much greater than the variation in genome size, but much less than the variation in body size among all living organisms (about 6 and 21 orders of magnitude, respectively; 11). The other is that this variation is of two types. For some organisms, the geographic range is approximately the same size as the home range of an individual organism, so that the species is composed of a single, freely interbreeding population. This is true not only for species with tiny ranges, such as Soccoro isopod and Devil's Hole pupfish mentioned above, but also for some of the animals with the largest ranges, such as some pelagic marine fishes, seabirds, and whales. For the majority of organisms, however, the geographic range is many orders of magnitude larger than the ambit of an individual, and the species is comprised of numerous populations, isolated by distance and often by geographic barriers to dispersal.

The frequency distribution of range sizes among the species in a large clade or taxonomic group has a distinctive shape. Many species have small- to moderate-sized ranges, and a few have very large ones. This pattern was first documented by Willis (94), but it has been confirmed by many subsequent investigators working on many different kinds of organisms (e.g. 2, 31, 34, 55, 70, 73). The shape of the frequency distribution is highly right-skewed when plotted with area on a linear axis, but more normal-shaped or perhaps even left-skewed

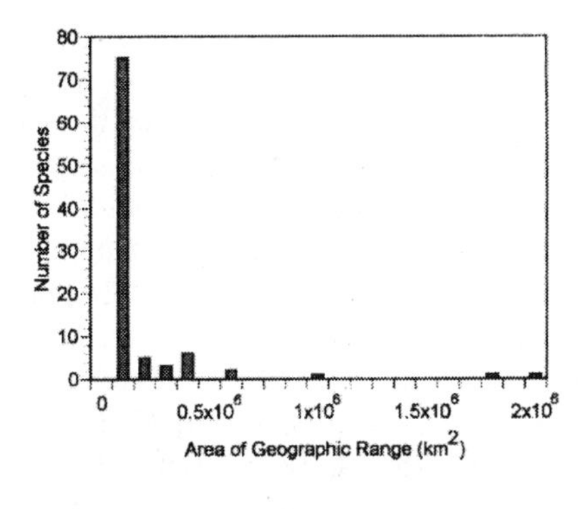

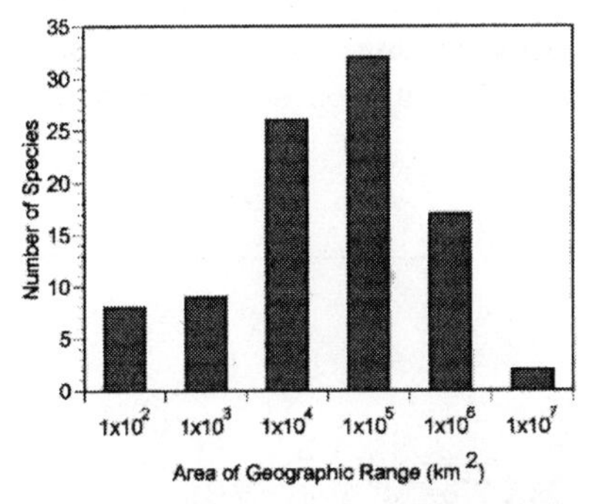

Figure 1 Frequency distributions of the areas of the geographic ranges of pines of the world (95 species of *Pinus*). The same data are plotted on a linear axis (*above*); and on a logarithmic axis (*below*). These distributions are similar to those observed for birds, mammals, and other organisms (11, 30).

when plotted on a logarithmic axis (see example for pines in Figure 1). Within the limits of the accuracy of the range maps and taxonomy, this pattern appears to be very general. While we hesitate to call it universal, we know of no clear exceptions.

Superimposed upon or embedded within this general distribution of range sizes are additional patterns. The range of variation is specific to particular taxonomic or functional groups of organisms. This is apparent at several levels. Closely related species, such as congeners, tend to have range sizes more similar than those of more distantly related species (11, 57). This suggests that intrinsic characteristics of the organisms inherited from their common ancestors influence the ecological interactions that limit geographic distribution. The influence of taxonomic and functional constraints on range size are even more apparent when very distantly related and dissimilar organisms are compared.

For example, among terrestrial and freshwater organisms, the smallest ranges of vascular plants and fishes (< 1 km^2) are several orders of magnitude less than the smallest ranges of birds and mammals (on the order of 10,000 km^2). Similar variation occurs among marine taxa, where studies have related some of it to dispersal capabilities. For example, marine mollusk species that do not have a planktonic larval phase in their life cycle tend to have smaller ranges than do species with more readily dispersed planktotrophic larvae (45, 58).

There are additional patterns of variation in range size with characteristics of the organisms. Several investigators have explored the relationship between range size and body size (9, 11, 13, 14, 30, 34, 35, 36, 37, 69) and range size and abundance (5, 6, 7, 10, 13, 14, 34, 37, 69). These relationships can be characterized by correlations: i.e. in most cases there are highly significant positive correlations between range size and both body mass (Figure 2) and some measure of average population density (Figure 3). There is usually,

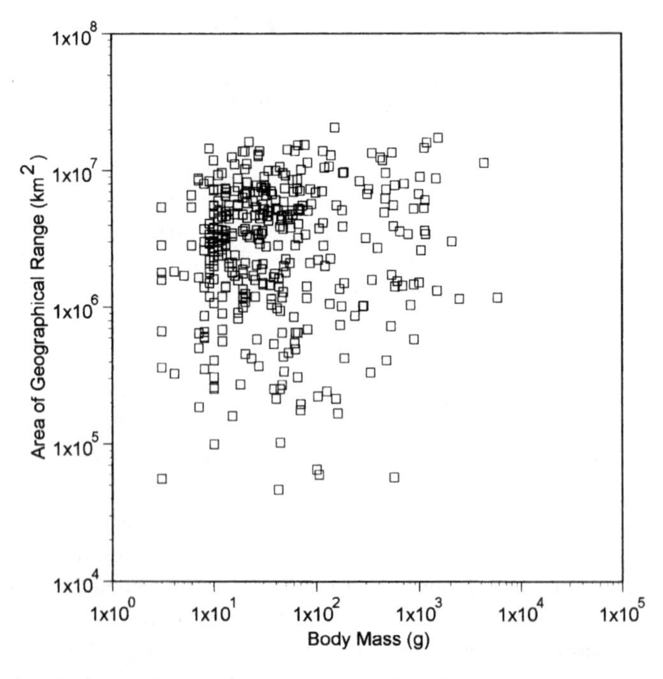

Figure 2 Relationship between area of geographic range and body size on logarithmic axes for 391 species of North American land birds (from Brown & Maurer 1987). Note that while there is a marginally significant positive correlation ($r = 0.08$, $0.05 < P < 0.10$), the data points tend to fall within a triangular space so that most species of large body size have large geographic ranges.

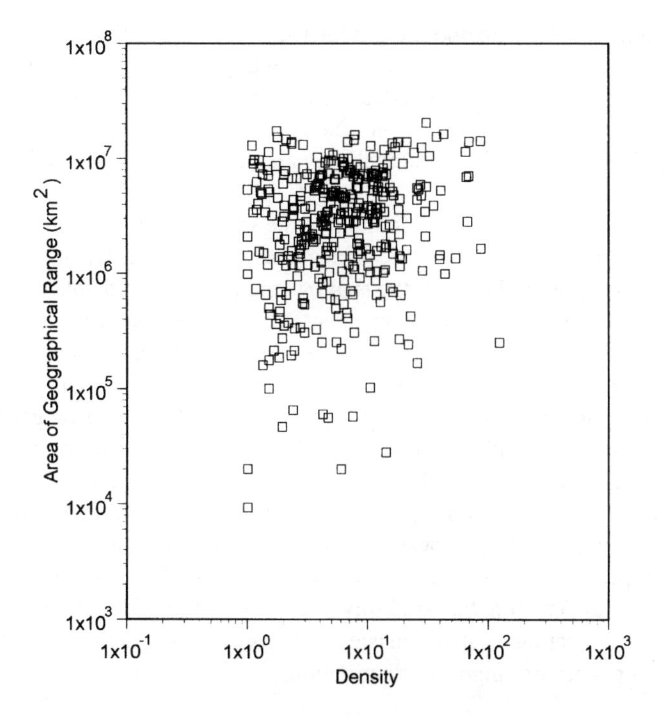

Figure 3 Relationship between area of geographic range and average population density on logarithmic axes for the same species of North American land birds as in Figure 2 (from Brown & Maurer 1987). There is a significant positive correlation ($r = 0.18$; $P < 0.01$); most of the data fall within a triangular space so that most species with high abundances have large ranges.

however, a great deal of residual variation. Brown & Maurer (9, 11, 13, 14) have pointed out that it may be more informative to consider the overall pattern of variation in bivariate plots on logarithmic axes (Figures 2 and 3). While there is considerable variation, it is constrained to certain combinations of the variables. Some of these constraints may be absolute: For example, the maximum size of geographic range may be constrained by the area of the continent over which the organisms are distributed. Other constraints may be relative or probabilistic: For example, it is not impossible for an organism of large body size to have a very small geographic range, but it is highly unlikely.

One of the most interesting patterns of variation in range sizes occurs in ecogeographic gradients: Range size tends to decrease with decreasing latitude and decreasing elevation in terrestrial environments and to increase with increasing depth in marine environments. The latitudinal pattern was documented for subspecies within species by Rapoport (73) and subsequently shown to hold for species within higher taxonomic groups (82). The generality of the relationship and its occurrence in elevational and depth gradients have been explored by Stevens (82, 83, 84), who called the empirical pattern "Rapoport's Rule." The pattern is apparent when range size is measured either as the area of the geographic range or as the latitudinal, elevational, or depth range of the species distribution.

The majority of studies have found that the species with the smallest ranges are consistently confined to the tropical end of latitudinal gradients, the lower end of elevational gradients, or the shallow end of depth gradients (28, 61, 64, 70, 84, 87). It is probably not coincidental that the regions with the smallest ranges are also the regions of highest species diversity for the taxon. Thus in the genus *Pinus*, which is an exception to the typical latitudinal gradient of species diversity, both the highest species richness and the smallest geographic ranges occur at mid-latitudes (85). The few studies that question the empirical generality of Rapoport's Rule are either of marine organisms (77, 80) or for the continent of Australia (81). For marine taxa, depth range rather than latitudinal range most influences the range of environmental conditions that a species experiences (84). The fact that Australia has low species diversity in its arid center and that climatic variability peaks at mid-latitudes (M Westoby, personal communication) is also consistent with its exception to the more common latitudinal pattern of range size distributions.

Processes

Empirical patterns, such as the seemingly general relationships between range size and other variables presented above, call for mechanistic explanations. Before postulating specific mechanisms, however, it is important to show that there is a pattern to be explained. The apparent pattern should be tested against the null hypothesis that the observed distribution is simply the result of sampling or some other random process (41). Unfortunately, there is usually not just one applicable null hypothesis. There are multiple possible null hypotheses that incorporate different amounts of information about the system, and consequently these null models differ in the mechanisms and degree of "biological realism" that are implicitly assumed. At a first level, it is important to test a frequency distribution against a normal or lognormal distribution, and a bivariate relationship against a bivariate normal (or bivariate lognormal) distribution, because random sampling and other kinds of simple stochastic processes tend to produce normal distributions. Most of the patterns discussed above are readily distinguished from univariate normal or bivariate normal distributions. A possible exception is the univariate frequency distribution of range sizes on a logarithmic axis for some taxa; further testing and exploration of this pattern is warranted.

More complicated null models that incorporate more of the structure of the data and/or more deterministic mechanisms are usually easy to construct but harder to reject (41). For example, Colwell & Hurtt (19) have developed several alternative null hypotheses for the latitudinal version of Rapoport's Rule. These make different assumptions about how geographic ranges are distributed on and constrained by the spherical geometry and basic geography of the earth. Some of these null models produce patterns similar to Rapoport's Rule. It is important to note, however, that failure to reject even a simple null hypothesis does not necessarily mean that an empirical pattern is due simply to uninteresting random processes. Perhaps the best example comes from quantitative population genetics, where a normal-shaped frequency distribution of a trait can usually be assumed to reflect not small random errors of sampling or measurement but rather the additive influence of many genes with small effects.

While it will be worthwhile to continue to test some of the empirical patterns of range size further against null models, it is also appropriate to develop and test hypotheses about deterministic processes that may have produced the patterns. Two classes of mechanisms have typically been invoked to account for the patterns described above. One class might be termed dynamic processes of colonization and extinction (and sometimes speciation). The other class might be called niche processes or mechanisms of limitation by environmental variables. These two classes of processes are not mutually exclusive. Indeed, they are both operating simultaneously in many cases. But often one or the other may be sufficient to explain a particular pattern. Thus, for example, to account for the fact that most organisms of large body size have large geographic ranges, it seems reasonable to invoke a high probability of extinction due to small total population size. Similarly, ecological limiting factors can probably account for most of the variation in distribution and abundance of any given species, and hence for the size and boundaries of its geographic range. However, the critical factors and parameter values can be expected to be different for each species, and to be difficult and costly to measure. Many studies have identified one or a few of the limiting niche dimensions (see above), but we are not aware of any study that has quantified the entire niche of a species and then tested the ability of this characterization to predict the size and limits of the geographic range.

On the other hand, there are cases where an explanation based on just one process is unsatisfying. An example is an explanation for the relationship between range size and abundance or latitude that is based solely on colonization/extinction dynamics. It has been suggested that locally dense populations are more likely to export emigrants, which in turn are more likely to colonize other areas, both "source" habitats capable of sustaining populations and "sink" habitats requiring a continual influx of immigrants in order to persist

(46–49). But even though such a mechanism might be plausible to account for the maintenance of the positive correlation between distribution and abundance, many investigators would not be satisfied with it. They would want to know why some species are more abundant and more widely distributed in the first place. Such a question would seemingly have to be answered by information about the environmental factors that affect population growth and dispersal. Similarly, one might invoke lower extinction rates in the tropics to account for the smaller geographic ranges there. While this might provide a partial explanation, most investigators would want to know what it is about tropical environments that enables species with small ranges—and perhaps also low population densities—to persist. Is it constancy of current climatic conditions, absence of pandemic diseases, lack of large-scale historical disturbances, or some other factor or combination of factors?

SHAPE OF RANGE

Patterns

Maps of geographic ranges show enormous variation in their shapes. In fact, despite the emphasis of many historical biogeographers on similar distributions and congruent area cladograms, the differences in the shapes and locations of ranges are perhaps more striking than the similarities—and this is as true of the ranges of clades or higher taxonomic groups as it is of individual species. Nevertheless, there appear to be some general patterns.

Rapoport noticed that despite the orders of magnitude variation in the areas of ranges of North American mammals, the periphery-to-area ratio remained relatively constant. That is, when he measured the perimeter of the range boundary and the area encompassed within that boundary, he found that the ratio of the two variables was approximately 10 and did not vary with range size.

However, Rapoport's observation about perimeter-to-area ratios should not divert attention from the great variation in the shapes of ranges: some are compact and globular whereas others are long and attenuated. A simple way to convey much information about shape is to plot two distances across the range as a function of each other. It would be interesting to do this for the longest and shortest linear dimensions. Brown & Maurer (11, 14) did something slightly different but perhaps equally informative. They plotted maximum north-south distance as a function of maximum east-west distance, thus referencing variation in shape with respect to geography. The result is the kind of graph shown in Figure 4, in which the line of equal distances has been plotted for reference. In such a graph, ranges with equal dimensions, such as circles or squares, would

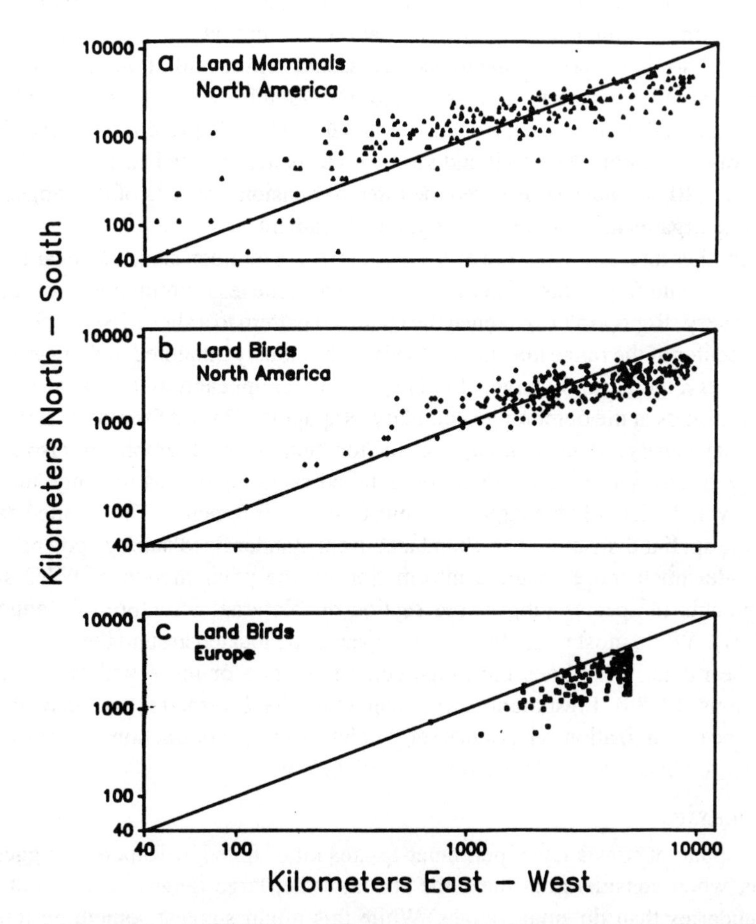

Figure 4 Shapes of geographic ranges of North American and European land birds and North American terrestrial mammals. Each data point represents a species. The maximum north-south dimension of the range is plotted against the maximum east-west dimension (both on logarithmic axes), so that the smallest ranges are in the lower, left-hand part of the graphs. The diagonal lines show equal dimensions, so that points above the lines correspond to ranges that are attunuated in a north-south direction, and points below the lines correspond to ranges that are elongate in an east-west direction. (From Brown & Maurer 1989)

fall along this line, with small ranges to the left and large ones to the right; ranges attenuated in a north-south direction will fall above the line, whereas those attenuated east-west will fall below the line. The few such graphs that have been compiled show interesting patterns. Although there is considerable scatter, North American mammals, birds, and reptiles all show a consistent trend: Small ranges tend to be oriented north-south, whereas large ones tend to be aligned east-west. European mammals and birds present an interesting contrast, with both the small and large ranges being oriented east-west. Other studies (40, 84) have considered the three-dimensional shapes of the ranges of marine organisms in relation to ecological gradients.

Another interesting feature of range shape is the number, size, and location of the holes and fragments. The range tends to become less continuous toward the periphery. Rapoport (73) likened the idealized pattern to a slice of Swiss cheese. The center of the range may be relatively continuously inhabited, but toward the periphery, increasingly large, closely spaced holes appear until they coalesce to form islands at the outer range boundary. Rapoport (73) and Stevens & Enquist (85) have analyzed the number, size, and location of the geographically isolated range fragments that are plotted on detailed range maps. Most range maps, however, depict only the largest and most isolated fragments. The detailed data on the spatial distribution of abundance from standardized surveys potentially provide much more accurate information on the phenomenon of holes and fragments in species ranges (see section on "Internal Structure of Ranges" below). While most range fragments appear to be located around the periphery of the range, sometimes the range consists of two or more widely isolated portions (12, 88). Presumably these disjunctions have formed as a result of long-distance colonization, vicariance, or wholesale range contraction as discussed under the heading of "Range Boundaries" below.

Processes

The apparent constancy of perimeter-to-area ratios noted by Rapoport suggests that, when considered at the same spatial scale, large ranges have smoother boundaries than do small ranges. While this might suggest something interesting about colonization-extinction dynamics, ecological limiting factors, or some combination of these, it is first important to evaluate an alternative hypothesis: that it simply reflects the unintentional bias of the map makers. There is an inherent tendency to draw maps with a fractal structure, including more detail about boundaries (and other features) as the spatial scale decreases (65, 75). In many cases, including Hall's (44) treatise on North American mammals which Rapoport used for many of his analyses, small ranges are typically mapped at greater magnification than large ones. Thus the apparent constancy of periphery-to-area ratios over a wide range sizes should be reevaluated. If the fractal inclinations of map makers can be circumvented, however, the periphery-to-area ratio is one simple measure of range shape that warrants further study.

Brown & Maurer (14) suggested that the patterns of north-south and east-west orientation described above reflect the physical geography of the continents. The east-west orientation of large ranges in both North America and Europe was suggested to reflect the ultimate influence of the major east-west oriented belts of climate and vegetation on species with large ranges. The difference between the continents in the orientation of small ranges (north-south in North America and east-west in Europe) was suggested to reflect the influence of environmental variables associated with smaller-scale geographic features such as the orientation of major mountain ranges, river valleys, and coastlines, in determining the boundaries of smaller ranges. These ideas could be pursued further with analyses of range shapes in relation to abiotic and biotic environmental variables in other kinds of organisms in other geographic and ecological settings such as on other continents or in the oceans.

RANGE BOUNDARIES

Patterns

We cannot review here the enormous literature on the factors that have been implicated to set boundaries on the geographic ranges of species (but see 30 and references therein). Most of the studies focus on specific environmental conditions that appear to limit local distribution along one edge of the range. These environmental factors typically appear to be specific to the species or higher taxon being studied. There has been little attempt to review, reanalyze, and synthesize the results of all the relevant studies, so there may be more general patterns than those considered below.

One pattern that may have considerable generality concerns the relative importance of abiotic and biotic limiting factors along different margins of species ranges (59). Dobzhansky (24) and MacArthur (63) suggested that biotic interactions tended to limit distribution and abundance at lower latitudes, whereas abiotic factors were more likely to be limiting at higher latitudes. Intertidal ecologists have developed a similar paradigm: Abiotic factors related to exposure to physical stress between tides set the upper limits of distribution, whereas predation and interspecific competition set the lower limits (21). Similar patterns appear to occur in other ecological gradients such as elevational and aridity gradients in terrestrial environments and depth gradients in aquatic environments; in one direction along the gradient the distributions of species are limited by increasing physical stress, while in the other direction they are limited by increasing numbers and impacts of biological enemies. These patterns of range limitation in geographic and ecological gradients appear to be related not only to each other, but also to Rapoport's Rule, because the direction of gradient in which biotic factors appear to limit distributions is the direction in which sizes of the ranges of species decrease and the numbers of other species increase (85).

A second general feature of many range boundaries is that they are extremely dynamic. While some boundaries such as those corresponding to coastlines and other major, relatively permanent geographic features may appear to remain relatively constant, other boundaries are constantly shifting (see examples in 12, 38, 53). All kinds of patterns can be observed. Some species ranges have expanded along one or more boundaries, while others have contracted, and still others have shifted back and forth. Probably the best documentation of range boundary shifts is available at two contrasting spatial scales. On the one hand, the recent fossil record documents many range shifts that accompanied the global changes in glacial geology, climate, and vegetation during the Pleistocene, and especially within the last 10,000 years following the retreat of the last continental ice sheets and the development of an interglacial climatic regime (18, 23, 27, 42). On the other hand, museum collections and ecological surveys document many range shifts within the last two or three centuries, most undoubtedly caused in part by human activities (1, 17, 26, 29, 43, 52, 53, 62, 67, 86, 88, 89, 91).

Both the fossil and the written record document several kinds of range shifts. One is the relatively gradual, incremental expansion or contraction of the distribution along an existing range boundary. Another involves the long-distance dispersal of one or a few individuals across a "biogeographic barrier" to found a new and isolated population. Alternatively, the formation of a barrier may break up a once continuous distribution and create isolated, disjunct populations. Biogeographers refer to these latter two kinds of range changes, which involve either the crossing or formation of barriers, as dispersal and vicariant events, respectively. Finally, there are collapses of ranges: rapid contractions of once widespread species to one or a small number of isolated sites. At least at the extremes, the large changes that result in populations isolated by barriers after long-distance dispersal or range collapse are distinct from the incremental expansions and contractions that occur around the edges of the range.

Processes

The edges of geographic ranges are set primarily by ecological factors that limit local distribution and abundance. The numerous case studies of range boundaries document the many kinds of abiotic and biotic factors that can limit individual species. Most of these studies have inferred that a particular factor is limiting, because the range boundary is closely correlated with a particular value of the parameter (e.g. 74, 78, 79), but some are based on more direct evidence such as experimental manipulations (e.g. 20) or observations of range shifts in response to environmental changes (e.g. 67). The environmental factors limiting ranges are so varied that it is hard to generalize about them. Each species has a unique ecological niche: a set of environmental variables that limit abundance and distribution because survival and reproduction can occur only within a certain range of parameter values. Any niche variable, either independently or in interaction with other variables, can determine a local range boundary. The boundary of the entire geographic range, especially if the range is large, is set by multiple niche variables limiting local or regional distribution at different locations around the periphery (12). While the role of multiple limiting factors is trivially obvious—e.g. many species have part of a range boundary at a coastline and part of one inland—the total number of niche variables responsible for the entire range boundary warrants theoretical and empirical study. We have done some preliminary computer simulations that suggest that the number of independent environmental variables that have an important influence on distribution and abundance of a species may be modest, on the order of five to ten. If there are too many parameters, each of which is distributed independently in space and can assume values preventing the occurrence of a species, then the places where the species can live will be few and widely separated.

One general pattern of range limitation mentioned above is the relative importance of abiotic and biotic limiting factors in ecological gradients. In most ecological and geographic gradients the majority of species appear to find one direction to be physically stressful and the other to be biologically stressful, and as physical stress diminishes, there is a corresponding decrease in the average size of the range and an increase in overall species diversity. While the empirical patterns are becoming increasingly clear, it is difficult to sort through the correlated phenomena to develop and test hypotheses about causal mechanisms (but see 11, 21, 59, 63, 78, 79, 82). Efforts to do so soon encounter some of the big unresolved questions about the ecological and biogeographic processes that generate and maintain the spatial patterns of biological diversity: questions such as "What does it mean to say that an environment is physically harsh or stressful?", and "What is the relationship between species diversity and the number and strength of interspecific interactions?" This should be a fruitful area for research, but with the caveat that satisfying answers probably will not come easily.

The view that multiple niche variables largely set the boundaries of geographic distributions accords well with what is known about one class of range shifts, the local and incremental expansion or contraction of distribution. van den Bosch et al (89) have modeled one kind of such shifts, the rapid range expansion of a colonizing species such as an introduced exotic. Their model makes standard exponential population growth spatially explicit by incorporating a dispersal parameter. When this model is parameterized for well-studied invading populations, it seems to fit the observed pattern of range expansion quite well. Since this model and the data on invading species suggest that the rate of range expansion is exponential or nearly so in the absence of environmental limits, and since most species are not rapidly expanding their ranges, we can infer that most existing range boundaries are set by limiting environmental variables.

When cases of incremental range expansion and contraction have been studied, changes in critical environmental conditions that have made previously unfavorable areas habitable or vice versa have often been identified (e.g. 67, 86). When effects of environmental changes have been investigated, usually the range shifts of species are highly individualistic. This is true in the case of the range shifts that have occurred in response to the large changes in climate and other abiotic conditions at the Pleistocene-Holocene transition (18, 23, 42) and of recent range shifts that have occurred in response to activities such as predation and habitat alteration (62). Less frequently, there are coincident shifts in the ranges of multiple species, which appear to be caused by changes that have occurred in one or more environmental variables (often in a suite of correlated variables) that are important niche variables for these species (e.g. 29, 67).

The exact position of the range boundary is determined by the interaction of the population processes of birth, death, and dispersal with the spatial and temporal variation in the environment. It is possible to imagine a variety of circumstances, modeled by Pulliam (71), such that the peripheral populations represent some combination of sources, where births exceed deaths and emigration exceeds immigration, and of sinks, where the opposite conditions obtain. Whether a population is a source or sink will depend on the local environmental conditions and on the proximity of and rate of exchange of dispersing individuals with other populations. It is possible to imagine a variety of situations, ranging from highly vagile organisms, such as some birds, in which most of the populations near the range boundary are sinks (43), to more sedentary organisms (or good habitat selectors), such as some plants, in which most of the peripheral populations occur in local patches of favorable environment. Unfortunately, the necessity for detailed data on demography and dispersal make it difficult to distinguish sources from sinks empirically.

Different kinds of processes must be invoked to account for the other class of range shifts: long-distance dispersal across a barrier to found an isolated population. The success of introduced species in so many parts of the world indicates that many, probably most, species do not live everywhere they can; barriers to dispersal prevent their occurring in distant but otherwise habitable areas. There is a large literature on range expansion by means of such barrier-crossing dispersal, which is exemplified by, but not limited to, colonizing islands (16, 93) and invading exotics (26, 52). At this point it must be emphasized that the distinction made above between the two classes of range shifts emphasizes the extremes of a continuum. The entire subject of metapopulation dynamics focuses on the processes and consequences of small-scale colonization and extinction (39), and these processes undoubtedly occur around the boundaries of many geographic ranges.

INTERNAL STRUCTURE OF THE RANGE

Patterns

Most maps of geographic ranges and most quantitative studies based on range maps ignore the area inside the range boundary. Exceptions are contour maps and other kinds of maps that show variation in abundance within the range. Most such maps have been published fairly recently, following the compilation of large computerized data sets and the development of Geographic Information Systems (GIS) to reference, map, and analyze spatially explicit data. Such maps, and the extensive data on spatial variation in abundance from which they are derived, provide invaluable information on the "internal structure" of the geographic ranges of species. Investigation of this internal structure—searching for patterns and erecting and testing null and mechanistic hypotheses—promises to contribute greatly to understanding most of the phenomena related to sizes, shapes, and boundaries of ranges that are discussed above. While studies of the internal structure are just in their infancy, some interesting results have ben obtained.

First of all, there appears to be wide variation in abundance within the range. As mentioned above, many areas within the boundaries of published range maps are uninhabited, i.e. local abundance is zero. But this is still a great oversimplification, because there is typically enormous variation in abundance at those localities where the species occurs (11, 15, 67). While the abundance of some rare species may vary from zero to a few individuals, that of some classes of rare species (see 33, 72) and most common ones may vary by several orders of magnitude. The frequency distribution of abundance among multiple sites throughout the range appears to have a characteristic shape: Zero or a very few individuals occur at most locations, but tens or hundreds are found at a few sites (Figure 5; 15). For example, for most common passerine birds censused in the North American Breeding Bird Survey (BBS), the modal number of

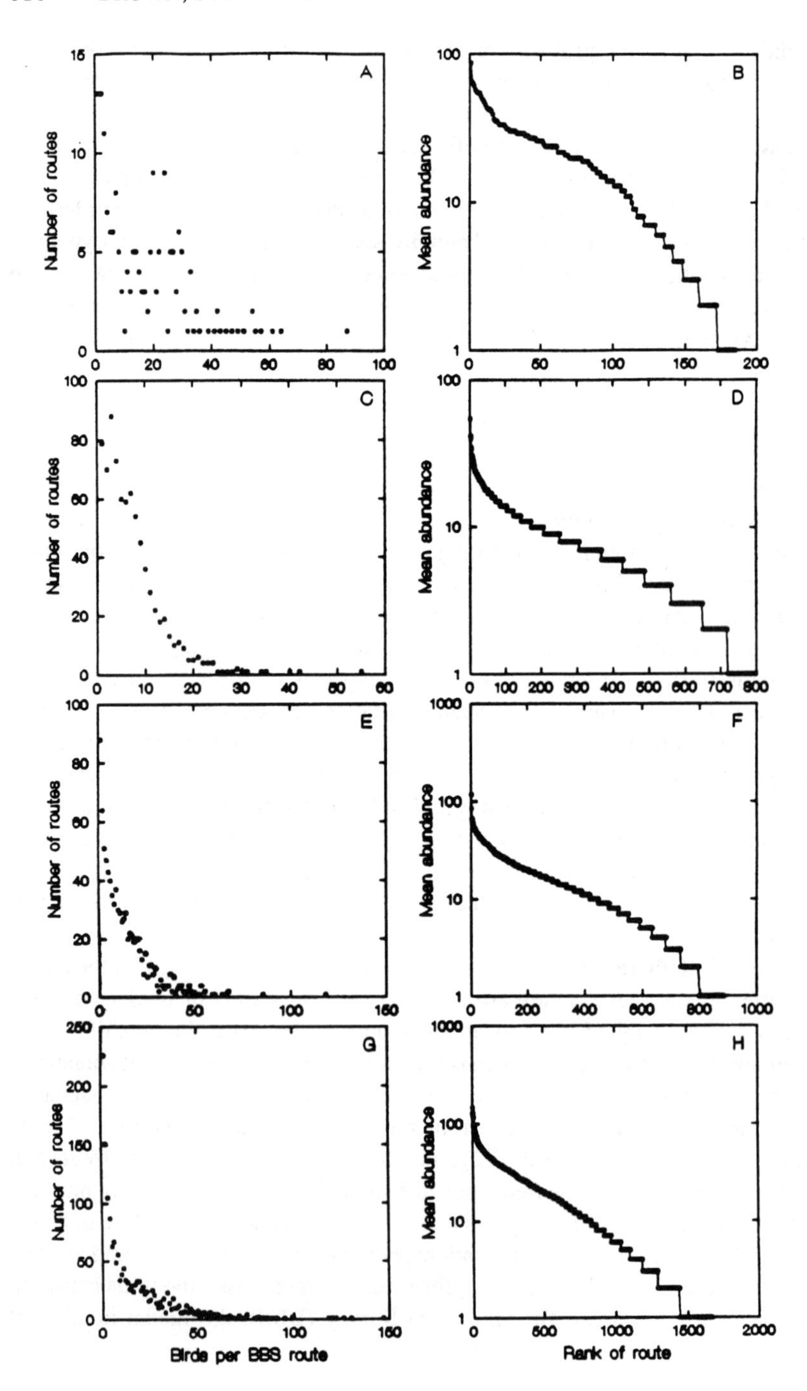
A
B
C
D
E
F
G
H
Number of routes
Mean abundance
Birds per BBS route
Rank of route

individuals counted at a site is typically one (and almost always less than five), although the maximum number can be in the hundreds or even thousands.

Our studies of avian distributions based on the BBS have revealed at least five patterns in the spatial distribution of abundance within the range (10, 11, 15, 67; see also 17, 27, 51, 53, 54). First, there is spatial autocorrelation: Abundances tend to be more similar among nearby localities than distant ones. Second, while there are some changes in abundance at local sites over time, at many sites abundances of particular bird species have remained quite similar over the last 20 years. Third, spatial variation in abundance tends to be greatest near the center of the range, where the sites of highest abundance but also sites with zero and low abundance occur. Abundance tends to be uniformly low near the boundaries of the range. Fourth, an exception to the previous pattern is that when the range boundary coincides with a coastline, abundance often tends to be relatively high right up to the coast rather than decreasing as the boundary is approached as it does toward the other edges of the range (67). Fifth, comparisons among species, even closely related, ecologically similar ones, show that the both the spatial patterns of abundance and the temporal changes in abundance at sites are highly species specific (8).

Maurer (66) has taken a somewhat different approach to characterize the internal structure of the range. He used spatially explicit BBS data on abundances of bird species to calculate a fractal dimension of distribution, which provides information on the degree and size of fragmentation. This approach has other applications, such as analyzing the fractal nature of range boundaries to distinguish between artefacts of mapping and real features of the scale and pattern of distribution.

Processes

Much research needs to be done to explore the generality of the above patterns with respect to different kinds of organisms and geographic regions and to develop and test hypotheses about causal mechanisms. Nevertheless, it appears that most of the spatial variation in abundance reflects the influence of spatial and temporal variation in environmental variables on population dynamics, both local population regulation and metapopulation colonization-extinction dynamics. Clearly most of the variation is not just random noise. There is too

Figure 5 Frequency distribution of abundance of four species of North American land birds among census sites (Breeding Bird Survey routes) distributed across their geographic ranges. The data are plotted as both arithmetic frequency distributions (*left*) and logarithmically scaled ranked abundances (*right*). Species are: A,B, scissor-tailed flycatcher, *Tyrannus forficatus*; C,D, Carolina chickadee, *Parus carolinensis*; E.F, Carolina wren, *Thryothurus ludovicianus*; G,H, red-eyed vireo, *Vireo olivaceous*. (From Brown et al 1995)

much regular pattern, and we have in some cases tested and rejected simple null models (15). We suspect that ultimately it will be possible to explain most of the variation in any particular species with a model of how spatial and temporal variations in limiting niche parameters affect local and regional population dynamics. But the species-specificity of the patterns of variation suggests there will be little generality at this level of analysis. Each species will have a unique niche, defined by particular environmental variables and ranges of parameter values, and each species will have unique population dynamics, a consequence of its life history and dispersal characteristics.

There may, however, be some generality in the ways that important environmental factors, especially abiotic ones such as climate, geology, and soils in terrestrial systems or oceanographic conditions in marine systems, vary and covary in space and time. There may also be, as mentioned in the section on range boundaries, some generality in the number of important niche dimensions that limit the distribution and abundance of a species, and in the way that these interact with each other and with population dynamic processes to produce the orders-of-magnitude variation and the spatial and temporal patterns of abundance. It should be possible to explore these questions through computer simulation modeling and empirical studies of a few selected species.

THE ROLE OF HISTORY

Except for some introductory comments on phylogeny, no mention of history has appeared in this review. Brown (11) has pointed out that the word "history" is used in two different and sometimes confusing ways by biogeographers, ecologists, and evolutionary biologists. There is the history of place: the changes in geology, climate, and other environmental factors that are extrinsic to particular kinds of organisms. Then there is the history of lineage: the changes in the intrinsic characteristics of organisms that have been inherited from their ancestors. The history of place is not usually affected by the history of lineage (although some kinds of organisms do substantially alter the environment for themselves and for other organisms). The history of lineage, however, is profoundly influenced by the history of place, because the intrinsic characteristics of organisms were molded in part by interactions with past environments.

Many characteristics of species and multispecies clades, including the size, shape, boundaries, and internal structures of their geographic ranges, reflect the influences of both the history of place and the history of lineage. The characteristics of past environments have acted as selective agents to influence the environmental requirements and tolerances, and the demographic, life history, and dispersal characteristics of contemporary organisms, and these characteristics in turn affect the geographic range. The history of place has also influenced past colonization, speciation, and extinction events in ways that may affect present geographic distributions, for example through changes in the kind, location, and severity of barriers to dispersal. The complex, interacting influences of the histories of place and lineage on characteristics of geographic ranges is a fertile area for research. Unfortunately, limited space and expertise preclude a more thorough treatment here.

We believe that the apparent division between "ecological" and "historical" biogeography inhibits a thorough, synthetic understanding of the patterns of distribution and the contemporary and historical processes that have produced them. Too often ecological biogeographers have ignored the influences of past environments and phylogenetic constraints on current distributions. We are guilty of this to some extent in this review. Too often historical biogeographers have focused so exclusively on phylogenetic history that they have ignored the influence of past and present environments. Too often both ecological and phylogenetic biogeographers have ignored the insights into past distributions and environments that can only come from studies of the fossil record. There are, however, encouraging signs of the emergence of a synthetic perspective that incorporates information from phylogenetic reconstructions, the fossil record, and ecological studies to provide a more complete understanding of the processes that have shaped geographic distributions (76, 90).

CONCLUSION

Although the geographic range of a species has always been a basic unit of biogeographic research, only recently has a synthetic view of the range begun to emerge. This view is largely owing to the revitalization of biogeography as a modern quantitative science, with a rigorous empirical and theoretical basis. It has been facilitated by the accumulation of large quantitative data bases, the development of computer software for statistical and spatially explicit analyses, and advances in mathematical and computer simulation modeling. These advances have not only made available many data on the spatial distributions of organisms, they have also contributed to the discovery of empirical patterns in the characteristics of species ranges and their relationships with other variables.

The discovery of quantitative patterns in the characteristics of ranges has led inevitably to the search for the causal processes and the development and testing of hypotheses about the mechanisms. While there is much more to be learned about the patterns and especially about the processes, a synthesis is emerging. The geographic range is the manifestation of complex interactions between the intrinsic characteristics of organisms—especially their environmental tolerances, resource requirements, and life history, demographic, and dispersal attributes—and the characteristics of their extrinsic environment—in particular those features whose variation in space and time limit distribution and abundance. The consequences of these interactions influence all characteristics of geographic ranges: their sizes, shapes, boundaries, and internal structures.

We conclude by emphasizing two major areas of research where quantitative studies of geographic ranges have the potential to make a major contribution that will extend beyond biogeography and result in wide interdisciplinary influence. One contribution will be to a synthesis between the earth sciences and the biological sciences. The geographic ranges of organisms provide a wealth of information on the complex relationships between the physical environment of the earth and the biological characteristics of the organisms that live on the earth—and between earth history and the history of life. The other contribution will be to a synthesis between biogeography and the other basic and applied biological sciences concerned with biodiversity. The present and past relationships of the geographic ranges of species to the environment, both abiotic conditions and other organisms, provide insights into the fundamental processes that determine distribution, abundance, and ultimately diversity. There are intriguing patterns in the sizes, shapes, and boundaries of ranges in relation to the latitudinal and other geographic gradients of species diversity. Further study of these relationships should contribute both to increased understanding of the processes that generate and maintain diversity and to practical efforts to conserve biodiversity.

ACKNOWLEDGMENTS

Our biogeographic research has been supported by grants BSR-8807792 and DEB-9318096 from the National Science Foundation. We thank many colleagues and collaborators, especially B Enquist, K Matthew, B Maurer, D Mehlman, and D Richardson, for generously sharing their ideas, data, and analyses.

Literature Cited

1. Anderson S. 1977. Geographic ranges of North American terrestrial mammals. *Am. Mus. Novit.* 2629:1–15
2. Anderson S, Marcus LF. 1992. Aerography of Australian tetrapods. *Aust. J. Zool.* 40:627–51
3. Andrewartha HG, Birch LC. 1954. *The Distribution and Abundance of Animals.* Chicago: Univ. Chicago Press
4. Arrhenius O. 1921. Species and area. *J. Ecol.* 9:95–99
5. Bock CE. 1984. Geographical correlates

of rarity vs. abundance in some North American winter landbirds. *Auk* 101:266–73
6. Bock CE. 1987. Distribution-abundance relationships of some North American landbirds: a matter of scale? *Ecology* 68: 124–29
7. Bock CE, Ricklefs RE. 1983. Range size and local abundance of some North American songbirds: a positive correlation. *Am. Nat.* 122:295–99
8. Bohning-Gaese K, Taper ML, Brown JH. 1995. Individualistic species responses to spatial and temporal environmental variation. *Oecologia* 101:478–86
9. Brown JH. 1981. Two decades of homage to Santa Rosalia: toward a general theory of diversity. *Am. Zool.* 21:877–88
10. Brown JH. 1984. On the relationship between abundance and distribution of species. *Am. Nat.* 124:255–79
11. Brown JH. 1995. *Macroecology*. Chicago: Univ. Chicago Press
12. Brown JH, Gibson AC. 1983. *Biogeography*. St. Louis, MO: Mosby
13. Brown JH, Maurer BA. 1987. Evolution of species assemblages: effects of energetic constraints and species dynamics on the diversification of North American avifauna. *Am. Nat.* 130:1–17
14. Brown JH, Maurer BA. 1989. Macroecology: the division of food and space among species on continents. *Science* 243:1145–50
15. Brown JH, Mehlman D, Stevens GC. 1995. Spatial variation in abundance. *Ecology* 76:2028–43
16. Carlquist S. 1965. *Island Life*. Garden City, NY: Nat. Hist. Press
17. Caughley G, Grice D, Barker R, Brown B. 1988. The edge of range. *J. Anim. Ecol.* 57:771–85
18. Cole KL. 1982. Lake Quaternary zonation of vegetation in the eastern Grand Canyon. *Science* 217:1142–45
19. Colwell RK, Hurtt GC. 1994. Nonbiological gradients in species richness and a spurious Rapoport effect. *Am. Nat.* 144:570–95
20. Connell JH. 1961. The influence of interspecific competition and other factors on the distribution of the barnacle *Chthamalus stellatus*. *Ecology* 42:410–23
21. Connell JH. 1975. Some mechanisms producing structure in natural communities: a model and evidence from field experiments. In *Ecology and Evolution of Communities*, ed. ML Cody, JM Diamond, pp. 460–90. Cambridge: Harvard Univ. Press
22. Cracraft J. 1989. Speciation and its ontology: the empirical consequences of alternative species concepts for understanding patterns and processes of differentiation. In *Speciation and Its Consequences*, ed. D Otte, JA Endler, pp. 27–59. Sunderland, MA: Sinauer
23. Davis MB. 1986. Climatic instability, time lags, and community disequilibrium. In *Community Ecology*, ed. J Diamond, TJ Case, pp. 269–84. New York: Harper & Row
24. Dobzhansky T. 1950. Evolution in the tropics. *Am. Sci.* 38:209–11
25. Dodd AP. 1959. The biological control of prickly pear in Australia. In *Biogeography and Ecology in Australia*, ed. A Keast. *Monogr. Biol.* 8. The Hague: Dr W Junk
26. Drake JA, Mooney HA, di Castri F, Groves RH, Kruger FJ, et al. 1989. *Biological Invasions*. New York: Wiley & Sons
27. Enquist BJ, Jordan MA, Brown JH. 1995. Connections between ecology, biogeography, and paleobiology: relationships between local abundance and geographic distribution in fossil and recent mollusks. *Evol. Ecol.* 9:586–604
28. France R. 1992. The North American latitudinal gradient in species richness and geographical range of freshwater crayfish and amphipods. *Am. Nat.* 139:342–54
29. Frey JK. 1992. Response of a mammalian faunal element to climatic changes. *J. Mamm.* 73:43–50
30. Gaston KJ. 1990. Patterns in the geographical ranges of species. *Biol. Rev.* 65:105–29
31. Gaston KJ. 1991. How large is a species' geographic range? *Oikos* 61:434–38
32. Gaston KJ. 1994. Measuring geographic range sizes. *Ecography* 17:198–205
33. Gaston KJ. 1994. *Rarity*. London: Chapman & Hall
34. Gaston KJ. 1996. Species-range-size distributions: patterns, mechanisms and implications. *Trends Ecol. Evol.* 11:197–201
35. Gaston KJ, Lawton JH. 1988. Patterns in body size, population dynamics and regional distributions of bracken herbivores. *Am. Nat.* 132:622–80
36. Gaston KJ, Lawton JH. 1988. Patterns in the distribution and abundance of insect populations. *Nature* 331:709–12
37. Gaston KJ, Lawton JH. 1990. Effects of scale and habitat on the relationship between regional distribution and local abundance. *Oikos* 58:329–35
38. Gibbons DW, Reid JB, Chapman RA. 1993. *The New Atlas of Breeding Birds*

in Britain and Ireland: 1988–1991. London: T & AD Poyser
39. Gilpin M, Hanski I, eds. 1991. *Metapopulation Dynamics*. London: Academic
40. Glover RS. 1961. Biogeographical boundaries: the shapes of distributions. In *Oceanography*, ed. M Sears. *Publ. Am. Assoc. Adv. Sci.* 67:201–28
41. Gotelli NJ, Graves GR. 1996. *Null Models in Ecology*. Washington, DC: Smithson. Inst. Press
42. Graham RW. 1986. Responses of mammalian communities to environmental changes during the late Quaternary. In *Community Ecology*, ed. J Diamond, TJ Case, pp. 300–13. New York: Harper & Row
43. Grinnell J. 1922. The role of the "accidental." *Auk* 39:373–80
44. Hall ER. 1981. *The Mammals of North America*, Vols. I & II. New York: Wiley & Sons. 2nd ed.
45. Hansen TA. 1980. Influence of larval dispersal and geographic distribution on species longevity in neo-gastropods. *Paleobiology* 6:193–207
46. Hanski I. 1982. Dynamics of regional distribution: the core and satellite species hypothesis. *Oikos* 38:210–21
47. Hanski I. 1991. Single-species metapopulation dynamics: concepts, models and observations. See Ref. 39, pp. 17–38
48. Hanski I, Gyllenberg M. 1991. *Two General Metapopulation Models and the Core-Satellite Species Hypothesis*. Lulea Univ. Technol., Dep. Appl. Math., Res. Rep. 4
49. Hanksi I, Kouki J, Halkka A. 1993. Three explanations of the positive relationship between distribution and abundance of species. In *Species Diversity in Ecological Communities*, ed. RE Ricklefs, D Schluter, pp. 108–16. Chicago: Univ. Chicago Press
50. Harris P, Peschken D, Milroy J. 1969. The status of biological control of the weed *Hypericum perforatum* in British Columbia. *Can. Entomol.* 101:1–15
51. Hedderson TA. 1992. Rarity at range limits; dispersal capacity and habitat relationships of extraneous moss species in a boreal Canadian National Park. *Biol. Conserv.* 59:113–20
52. Hengeveld R. 1989. *Dynamics of Biological Invasions*. London: Chapman & Hall
53. Hengeveld R. 1990. *Dynamic Biography*. Cambridge: Cambridge Univ. Press
54. Hengeveld R, Haeck J. 1982. The distribution of abundance. I. Measurements. *J. Biogeogr.* 9:303–16
55. Hesse R, Allee WC, Schmidt KP. 1951. *Ecological Animal Geography*. New York: Wiley & Sons. 2nd ed.
56. Hocker HW Jr. 1956. Certain aspects of climate as related to the distribution of loblolly pine. *Ecology* 37:824–34
57. Jablonski D. 1987. Heritability at the species level: analysis of geographic ranges of cretaceous mollusks. *Science* 238:360–63
58. Jablonski D, Lutz RA. 1983. Larval ecology of marine benthic invertebrates: paleobiological implications. *Biol. Rev.* 58:21–89
59. Kaufman DM. 1995. Diversity of New World mammals: universality of the latitudinal gradients of species and bauplans. *J. Mammal.* 76:322–34
60. Knowlton N. 1993. Sibling species in the sea. *Annu. Rev. Ecol. Syst.* 24:189–216
61. Letcher AJ, Harvey PH. 1994. Variations in geographical range size among mammals of the Palearctic. *Am. Nat.* 144:30–42
62. Lomolino MV, Channell R. 1995. Splendid isolation: patterns of geographic range collapse in endangered mammals. *J. Mammal.* 76:335–47
63. MacArthur RH. 1972. *Geographical Ecology: Patterns in the Distribution of Species*. New York: Harper & Row
64. Macpherson E, Duarte MC. 1994. Patterns in species richness, size, and latitudinal range of East Atlantic fishes. *Ecography* 17:242–48
65. Mandelbrot BB. 1977. *Fractals: Form, Chance and Dimension*. San Francisco: Freeman
66. Maurer BA. 1994. *Geographical Population Analysis: Tools for the Analysis of Biodiversity*. Oxford: Blackwell Sci.
67. Mehlman D. 1995. *The spatial distribution of abundance: analysis of the geographic range*. PhD thesis. Univ. New Mex., Albuquerque
68. Miller RR. 1948. The cyprinodont fishes of the Death Valley system of eastern California and southwestern Nevada. *Misc. Publ. Mus. Zool. Univ. Mich.* 68:1–155
69. Morse DR, Stork NE, Lawton JH. 1988. Species numbers, species abundance and body length relationships of arboreal beetles in Bornean lowland rain forest trees. *Ecol. Entomol.* 13:25–37
70. Pagel MD, May RM, Collie AR. 1991. Ecological aspects of the geographical distribution and diversity of mammalian species. *Am. Nat.* 137:791–815
71. Pulliam HR. 1988. Sources, sinks, and

population regulation. *Am. Nat.* 132:652–61
72. Rabinowitz D. 1981. Seven forms of rarity. In *The Biological Aspects of Rare Plant Conservation,* ed. J Synge, pp. 205–17. Chichester: Wiley
73. Rapoport EH. 1982. *Aerography: Geographical Strategies of Species.* Oxford: Pergamon
74. Repasky RR. 1991. Temperature and the northern distributions of wintering birds. *Ecology* 72:2274–85
75. Richardson LF. 1961. The problems of contiguity: an appendix of statistics of deadly quarrels. *Gen. Syst. Year* 6:139–87
76. Riddle BR. 1996. The molecular phylogeographic bridge between deep and shallow history in continental biotas. *TREE* 11:207–11
77. Rohde K, Heap M, Heap D. 1993. Rapoport's rule does not apply to marine teleosts and cannot explain latitudinal gradients in species richness. *Am. Nat.* 142:1–16
78. Root T. 1988. Environmental factors associated with avian distributional boundaries. *J. Biogeogr.* 15:489–505
79. Root T. 1988. Energy constraints on avian distributions and abundances. *Ecology* 69:330–39
80. Roy K, Jablonski D, Valentine JW. 1994. Eastern Pacific molluscan provinces and latitudinal diversity gradient: no evidence for "Rapoport's Rule". *Proc. Natl. Acad. Sci. USA* 91:8871–74
81. Smith FDM, May RM, Harvey PH. 1994. Geographical ranges of Australian mammals. *J. Anim. Ecol.* 63:441–50
82. Stevens GC. 1989. The latitudinal gradient in geographic range: How so many species coexist in the tropics. *Am. Nat.* 133:240–56
83. Stevens GC. 1992. The elevational gradient in altitudinal range: an extension of Rappoport's latitudinal rule to altitude. *Am. Nat.* 140:893–911
84. Stevens GC. 1996. Extending Rapoport's rule to Pacific marine fishes. *J. Biogeog.* In press
85. Stevens GC, Enquist BJ. 1996. Macroecological limits to the abundance and distribution of *Pinus.* In *Ecology and Biogeography of* Pinus, ed. DM Richardson. Cambridge: Cambridge Univ. Press
86. Taulman JF, Robbins LW. 1996. Biogeography of the nine-banded armadillo (*Dasypus novemcinctus*) in the United States: what caused the current range expansion and where will it end. *J. Biogeogr.* In press
87. Taylor CM, Gotelli NJ. 1994. The macroecology of Cyprinella: correlates of phylogeny, body size, and geographic range. *Am. Nat.* 144:549–69
88. Udvardy MDF. 1969. *Dynamic Zoogeography: With Special Reference to Land Animals.* New York: Van Nostrand-Reinhold
89. van den Bosch F, Hengeveld R, Metz JAJ. 1992. Analyzing the velocity of range expansion. *J. Biogeogr.* 19:135–50
90. Wagner WL, Funk VA. 1995. *Biogeographic Patterns in the Hawaiian Islands.* Washington, DC: Smithson. Inst. Press
91. White TCR. 1976. Weather, food, and plagues of locusts. *Oecologia* 22:119–34
92. Wiley EO. 1981. *Phylogenetics: The Theory and Practice of Phylogenetic Systematics.* New York: Wiley
93. Williamson M. 1981. *Island Populations.* Oxford: Oxford Univ. Press
94. Willis JC. 1922. *Age and Area.* Cambridge: Cambridge Univ. Press

SUBJECT INDEX

A

B

C

G

H

I

M

N

O

Q

R

T

Z

CUMULATIVE INDEXES

CONTRIBUTING AUTHORS, VOLUMES 23–27

CHAPTER TITLES, VOLUMES 23–27